G8T812 35

PLANTS
AN INTRODUCTION TO MODERN BOTANY

THIRD EDITION

PLANTS
AN INTRODUCTION TO MODERN BOTANY

Victor A. Greulach & J. Edison Adams
Professors of Botany, University of North Carolina

JOHN WILEY & SONS, INC.
NEW YORK LONDON SYDNEY TORONTO

This book was set in Baskerville. It was set, printed and bound by Vail-Ballou Press, Inc. The designer was Jerry Wilke. The drawings were designed and executed by John Balbalis with the assistance of the Wiley Illustration Department. Picture research was done by Stella Kupferberg. Regina R. Malone supervised production.

Copyright © 1962, 1967, and 1976, by John Wiley & Sons, Inc.

All rights reserved. Published simultaneously in Canada.

No part of this book may be reproduced by any means, nor transmitted, nor translated into a machine language without the written permission of the publisher.

Library of Congress Cataloging in Publication Data:

Greulach, Victor A
Plants : an introduction to modern botany.

Includes bibliographical references and index.
1. Botany. I. Adams, Joseph Edison, 1903– joint author. II. Title.
QK47.G83 1975 581 75-16134
ISBN 0-471-32769-7
Printed in the United States of America

10 9 8 7 6 5 4 3 2 1

PREFACE

This book has been written for one-semester or one-quarter courses in general botany. We believe that it will be suitable both for students who will specialize in botany or related sciences and for the larger number of students who will take general botany as a required or elective part of a program of liberal education and will probably take no other courses in botany. We hope, however, that the book will assist the instructor in stimulating an interest in the plant sciences at least among the more capable students of the latter group, and that as a result, some of them will decide to pursue careers in the plant sciences.

Although the book is designed for a one-term course, it can be effectively used for the first term of a year course in botany by devoting the second term to a more extensive treatment of the Plant Kingdom, or for a year course in general biology in conjunction with a one-term zoology text. We have tailored the book for one-term courses principally by treating the Plant Kingdom only in sufficient depth to give the student a good understanding of the scope and diversity of the plant life on Earth as illustrated by common representatives of the larger plant groups and by omitting inadequate descriptions of a few plant families. The chapter on sexual reproduction contains detailed life cycles of representatives of various groups of plants, selected to express the general theme rather than to overwhelm the student with the infinite variety of detail.

The organization of this book differs from that of most other general botany texts in several respects. (1) A section on the structural organization of plants with only limited references to physiological processes is followed by a section devoted to physiology and physiological ecology. We believe this organization is less confusing to the student than the customary one, and that the student will profit by having some understanding of the different levels of plant structure before becoming involved with physiological processes. Perhaps this organization will help students avoid conclusions that photosynthesis and transpiration occur only in leaves, translocation only in stems, and absorption only in roots. (2) The related topics of reproduction, heredity, and evolution have been grouped into one section. (3) More than the usual emphasis has been placed on physiology, ecology, and genetics, with the inclusion of many of the important recent advances in these areas. Although some instructors may feel that certain of these topics are treated in too great detail, we believe that the material presented is not too comprehensive for an introductory college course and that it will provide for a clearer understanding than would a more skeletonized treatment, particularly in a one-term course with limited lecture and discussion time available.

In writing the book we have assumed that the student will have had an introductory course in chemistry, either in high school or college. For those students who have never taken chemistry we have included an appendix outlining elementary basic chemical concepts essential for a reasonable understanding of the chemistry used in the text. This is, of course, no substitute for a course in chemistry but we believe it will enable intelligent students without a chemical background to use the text with a reasonable degree of understanding. This appendix may also be useful to students who wish to review their chemistry.

In general, our aim has been to present an interesting and up-to-date account of the more important concepts and principles of modern botany with emphasis on the dynamic aspects that are currently receiving so much attention in research laboratories. We have tried to limit extensive and detailed factual information to those items that contribute toward an understanding of important concepts and principles. The historical introductions to photosynthesis, mineral nutrition, and genetics are included to give the student some idea of the ways botanical concepts have developed through continuing research. We regret that space limitations have not permitted a similar historical treatment of more topics.

The bibliographical references at the end of each chapter are largely comprised of recent popular books and magazine articles, and these have been selected with a view to providing a ready means of access to a more extensive treatment of certain topics by the interested student.

In the preparation of this third edition, we have retained the general organization of the previous editions and have added two more sections. "*The Plant Kingdom*" has been expanded from a chapter to a section. Nevertheless the detail provided is still substantially less than in the larger general botany textbooks, so our book continues to be more suited to one-semester or one-quarter courses. The chapter on plants of the past has been shifted to this new section, since its emphasis is on types of ancient plants rather than on evolutionary processes. The treatment of plant classification has been expanded and placed in a chapter of its own. It now provides students with a better understanding of the problems involved in classifying organisms as well as offering a survey of the various schemes of broad classification of plants that have been proposed. This edition contains a greatly expanded and updated treatment of ecology, which has attained increasing importance and attention in recent years and which is more essential than any other biological discipline for those whose biological education will span only one college year. In addition the new Chapter 1 is devoted mainly to the ecological importance of plants to man.

Many of the other chapters have been extensively revised and brought up to date. Most of the illustrations used in previous editions have been retained, including almost all of the drawings by Peggy-Ann Kessler Duke and Marion Seiler. However, a substantial number of new illustrations has been added. These include new electronmicrographs, photomicrographs, photographs, graphs, diagrams, and a few new drawings. We believe that these new illustrations, together with the newly added tables, will substantially increase the effectiveness of the textbook. At the request of many of the users of our previous editions we have added a glossary.

An entirely new set of reviewers examined the manuscript for the third edition carefully and critically and made many valuable suggestions. Four botanists reviewed the entire manuscript. These include L. A. Larson and Gene A. Pratt and two others who prefer to remain anonymous. William F. Little of the University of North Carolina Department of Chemistry reviewed Chapter 7 and the Chemistry Appendix. We thank the members of the editorial staff of John Wiley for their continued help and cooperation and for putting up with our failures to meet various deadlines. In particular, we express appreciation to Stella Kupferberg, Photo Researcher, and to Regina R. Malone, Production Supervisor, of the Wiley staff for their very high order of expertise in their respective fields, and for their infinite patience.

And, finally, special thanks are due Frances E. Silliman, who critically reviewed many chapters and performed many of the onerous tasks associated with authorship.

Victor A. Greulach
J. Edison Adams
Chapel Hill, North Carolina, 1975

TO THE STUDENT

As you begin your study of botany, you should be aware that not everything you read about plants in books and magazine articles is true. For some reason, more pseudoscience has appeared in print as regards plants than any other aspect of nature. This pseudobotany includes such things as expositions on the feelings of plants, the influence of prayer on plants, improved plant growth in response to people who treat them kindly, the ESP of plants, and completely unscientific discourses on such subjects as the sex life of plants. Also, much of the theory about organic gardening is simply not true, even though the cultural procedures may be good. Much of the theory about health foods is also unacceptable. Finally, many of the popular books and articles about plants are filled with teleological explanations of plant behavior.

More information on this can be found in two papers by outstanding botanists, Frank Salisbury (*BioScience, 24,* 201, 1974) and Arthur W. Galston (*BioScience, 24,* 415–416, 1974).

This book, like other botany textbooks and many excellent articles by botanists (such as those listed as related readings at the end of each chapter in this book), make every effort to present the true theories and facts about plants as revealed by scientific research. Of course, they do contain occasional errors and perhaps some information and theories that have been made obsolete by continuing scientific research, but every effort is made to avoid such things and present a truly scientific and accurate information. We hope you will learn to distinguish between botany (the scientific study of plants) and the fantasy, misinformation, and superstitions of pseudobotany.

CONTENTS

PLANTS
AN INTRODUCTION TO MODERN BOTANY

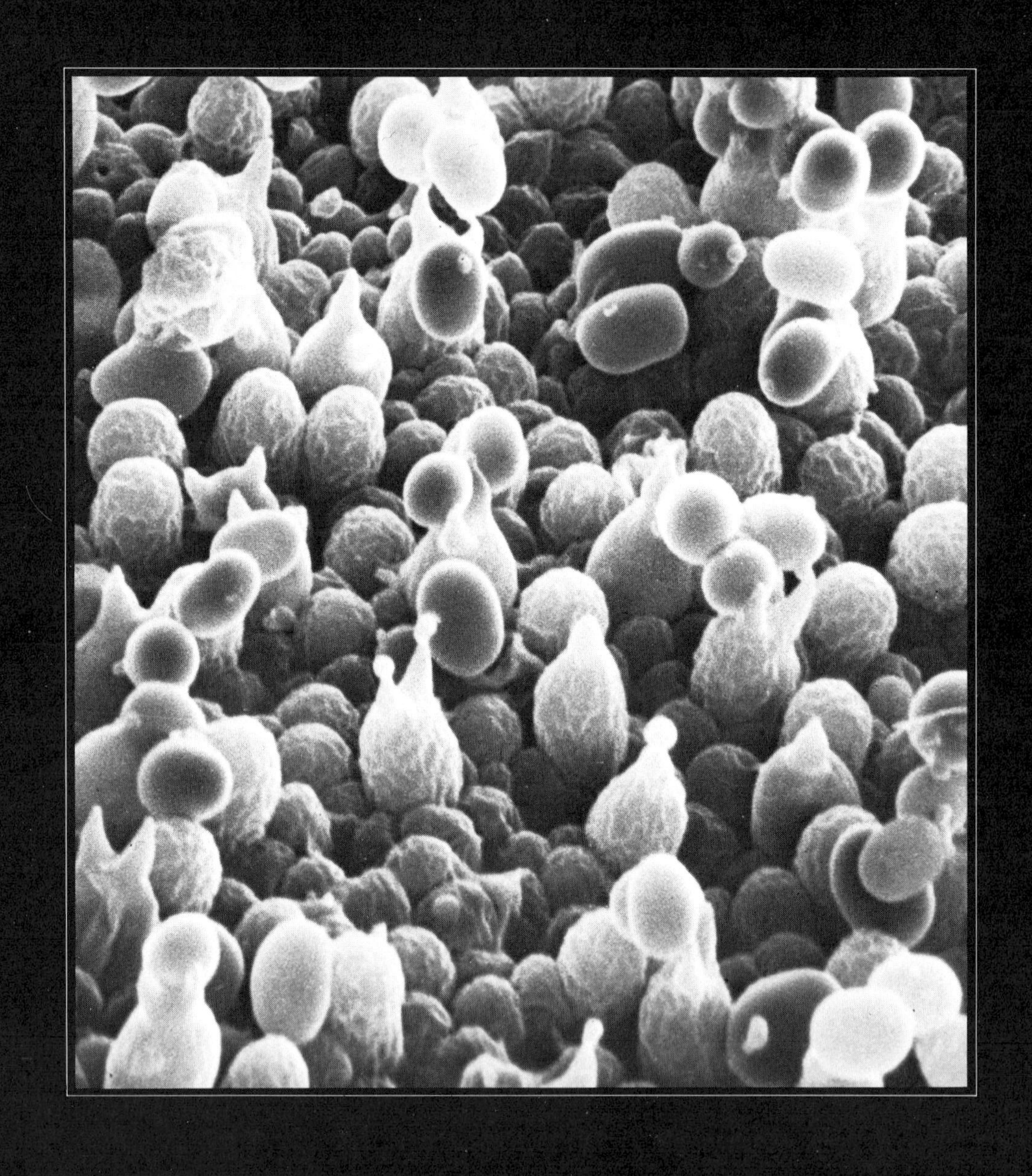

MAN AND THE WORLD OF PLANTS

1

PHOTO OPPOSITE: SPORES AND BASIDIA OF CULTIVATED MUSHROOM, *AGARICUS BISPORUS*. (MAGNIFICATION 1450 ×)

1
OUR COMPLETE DEPENDENCE UPON PLANTS

The stresses and complications of modern living have brought us to an increasing awareness of our dependence on the environment for our very existence as well as for the comforts to which we have grown accustomed. Thus, we have come to realize that we are not the independent, free-living beings we have often supposed ourselves to be. Through the process of becoming civilized we have acquired a wide variety of skills that enable us to modify our environment, in a limited way, and to utilize it to meet our needs and desires. However, rather than reducing our dependence on the many elements of the environment, these new skills may be said to have increased it. The very technologies that appear to give us greater freedom actually tie us more and more closely to the environment, for they consist essentially of progressively more intensive and extensive processing of natural materials.

Although no one of the elements of the environment can be regarded as more important than another, all being equally important in man's existence, it is true that plants, comprising

Earth's vegetation, have been more amenable to man's exploitation and have contributed more to the attainment of our modern mode of life.

It is through their role as primary producers of food for man and other animals that plants assume a special place among the elements of our environment, for from them directly or indirectly come all of our foods. The proteins, carbohydrates, and fats, as well as important accessory nutrients such as certain vitamins and minerals, are all made available to us and all other animals through green plants. In a forest, grassland, desert, lake, ocean, or any other natural community, photosynthetic plants provide the basic food supply and constitute the broad base of a "food pyramid" (Fig. 1.1). At the second level are organisms that secure their food directly from the plants, including herbivorous animals and many different kinds of organisms living on the plants as parasites, saprophytes, or symbionts. At the third level are carnivorous animals that use the herbivorous animals and other second-level organisms as food, including the parasites of these second-level organisms. At the fourth and subsequent levels are carnivorous animals that live on other carnivores, and again, parasites. Finally, the saprophytic bacteria and fungi and the scavenger animals live on the remains of dead plants and animals.

As food is transferred through the food chain and up through the pyramid, it is reduced in quantity at each successive level, since all organisms use a considerable portion of their food supply in respiration. The energy freed from foods by respiration and expended by each organism is also lost as far as the next level of consumer organisms are concerned, so that at each step up the pyramid a smaller and smaller portion of the energy originally derived from sunlight during photosynthesis is still available. Consequently, the total bulk of organisms at any level in the food pyramid is less than in the next lower level. In a plant-herbivore-carnivore food chain there is also an increase in size of the animals and a decrease in their numbers as the food pyramid is ascended from one level to the next. Food chains leading from hosts to parasites, however, go in the reverse direction, from a few larger organisms to many smaller ones.

Each biological community has only a limited level of productivity, dependent in the final analysis on the amount of photosynthesis possible in the community, and there is a definite limit to the total bulk of living organisms that can be supported at each level in the food pyramid. In a mature and balanced community the actual bulk of organisms at each level is at the possible maximum.

Although saprophytic bacteria and fungi and perhaps scavenger animals may be considered to operate somewhat outside the main food pyramid, they play a very important role in the general biological system by consuming dead organisms and converting their constituent substances into simple compounds such as carbon dioxide, ammonia, water, and mineral salts that can be used by green plants and so may again flow through a food chain. Wood-rotting fungi and termites may, for example, be regarded as destructive organisms by man, but in nature they play very useful roles.

Although the flow of matter through organisms is cyclic, any particular atom moving from the environment into organisms, from one organism to another, and back into the environment many times, the flow of energy through a biological community is a one-way process.

The food pyramid is applicable, in a modified form, to man and his agricultural plants and animals as well as to natural biological communities. Since man is herbivorous as well as carnivorous he may be considered to belong to both the second and third levels of the agricultural food pyramid. None of our domestic animals is strictly carnivorous (although some of our seafood is), so the agricultural food pyr-

Fig. 1.1 This simplified food pyramid of Antarctic life begins with microscopic floating algae (phytoplankton) and ends with man. The energy loss at each step is about 90%. One thousand pounds of phytoplankton support 100 pounds of zooplankton, 10 pounds of whale, and 1 pound of man. (Redrawn from Scientific American, *January 1958, with permission of the publishers.)*

amid does not extend beyond the third level, unless we wish to consider human parasites. In accordance with the general principles of food chains and pyramids, any region can support a larger human population if the bulk of the food supply comes directly from the crop plants rather than indirectly through meat, milk, eggs, and other animal products. In many parts of Asia and other regions where there is a pressure of the population on the food supply, the great bulk of the human food consists of rice and other plant products. There humans are living almost entirely at the second food level.

One of man's major problems at the present time is that of providing adequate food for an ever-increasing human population. Thomas Malthus, in his essay on population published in 1789, predicted that by the first half of the present century there would be worldwide starvation because the increase in food production could not possibly keep pace with the increase in human population. Such a situation has developed in certain regions, but not on a worldwide basis. Malthus had failed to anticipate the great increase in agricultural productivity made possible by the application of scientific discoveries. Many students of the problem decided that continued agricultural advances and more effective use of the potentially large supply of food from the oceans would be able to support the increasing human population indefinitely.

However, as a result of the greatly increased rate of human population growth during the present century (Fig. 1.2), brought about in part as a result of the increase in life expectancy through applications of the medical sciences, many biologists and social scientists have now reached the conclusion that it is just a matter of time until the predictions of Malthus come true, despite the most intensive agricultural production possible or such procedures as massive culturing of algae for animal and human consumption. One point worthy of note is that in any country where there is a pressure of population on the available food supply an increase in available food has the potential to bring with it a further increase in population. Another point to be considered is that as a population increases, more and more land is withdrawn from agricultural production by the growth of large urban complexes.

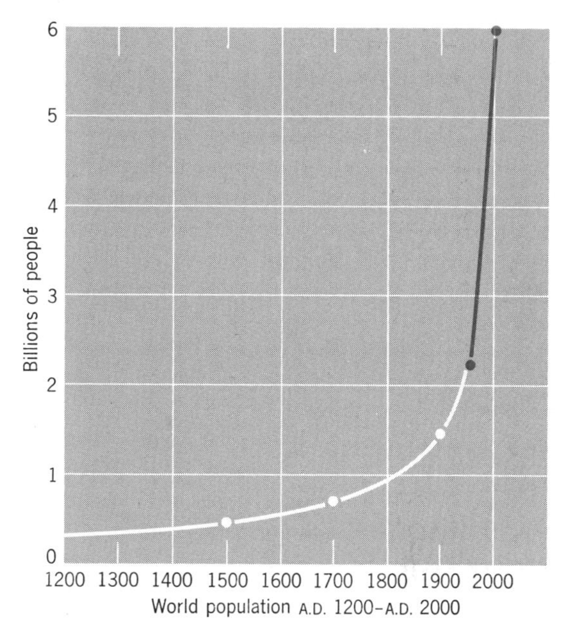

Fig. 1.2 Increase in the human population of Earth between A.D. *1200 and 1950, with projection to* A.D. 2000.

Even if a world population of 6 billion, predicted by the year 2000, can be fed, there is a definite possibility of a shortage of water or even of living space. There is a limit to the human population that Earth can support, and if the increase is not stopped by reduction of the birth rate, it will be stopped by lack of food or water or perhaps complete disruption of the ecosystem. Faced with these threatening possibilities, it is obvious that man must strive to understand the nature of the ecosystem of which he is a part.

However, it is not only in the production of food that plants are indispensable to man. Photosynthesis, the food-producing process carried on by green plants, has a profound influence upon Earth's atmosphere. Oxygen, a by-product of photosynthesis, is used by both plants and animals in the utilization of food for energy in aerobic respiration. Although large amounts of oxygen are removed from the atmosphere by such nonbiological processes as the rusting of metals and the combustion of organic materials such as fuels for domestic and industrial purposes, as well as by respiration, the concentration of that element in the lower layers of the atmosphere remains fairly constant at about 20%. Were it not for photosynthesis the supply of atmospheric oxygen would be depleted and aerobic respiration would cease.

The carbon dioxide in the atmosphere and the waters of the Earth are the sources of the elements that enter into the construction of organic materials in photosynthesis. Respiration by living plants and animals and the combustion of organic fuels return carbon dioxide to the environment. Although the concentration of carbon dioxide in the air is only .03%, the reserve supply is so great that there is no likelihood of its exhaustion by photosynthesis. Modern research has shown that the amount of atmospheric carbon dioxide is increasing at the rate of .7% per year, as a result of man's prodigious use of the fossil fuels, coal, gas, and oil in industrial operations. Since the composition of the atmosphere has a profound effect on the quantity and quality of solar radiation reaching the surface of the Earth, long continuation of man's present activities could have a significant effect on our environment with respect to temperature levels and all light-dependent processes. Such problems involving environmental changes and the degree of adaptability of animals and plants to such changes are under intensive study by students of the ecosystem.

Consideration of our modern civilization, so strongly based on energy-consuming technologies, invites brief mention here of our dependence on plants for the things that we consider essential to existence (Fig. 1.3). This and other aspects of our plant-dependence will be dealt with in a subsequent chapter.

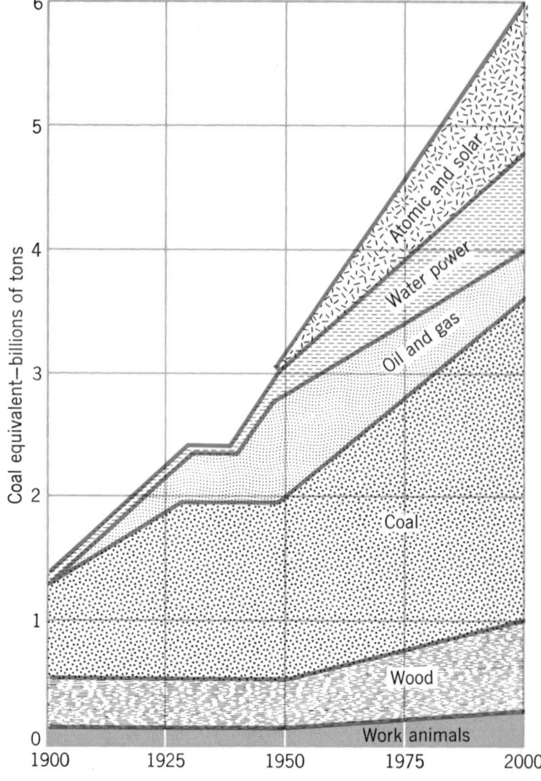

Fig. 1.3 Sources of energy used by man in home, industry, and transportation, projected to A.D. *2000. Note that at the present time almost three-quarters of our available energy comes from plants, directly or indirectly.*

With these broad and important considerations in mind, it is clear that we must have some basic knowledge of ourselves and our relationship to the elements of our environment that influence our daily living so strongly, so that we may intelligently make our choices and voice our convictions with respect to man's exploitation of this environment. In the following pages, we will inquire into the nature of one of these elements, plant life. But it will be remembered that it is only one of the factors in the total environmental complex.

RELATED READING

Anonymous. "The world food crisis," *Time, 104*(20), 66–83, November 11, 1974.

Borlog, N. E. "Millions could die," *Readers' Digest, 105*(632), 118–120, December 1974.

Caudill, H. M. "Farming and mining: there is no land to spare," *Atlantic Monthly, 232*(3), 85–95, September 1973.

Commoner, B. "Motherhood in Stockholm." *Harper's, 244*(6), 49–54, June 1972.

Freedman, R. and B. Berleson. "The human population." *Scientific American, 231*(3), 31–39. 1974.

Galston, A. W. "The specter behind the statistics." *Natural History, 83*(4), 6–15, 1974.

Hardin, G. "Living on a lifeboat," *BioScience, 24,* 561–568, 1974.

Lappé, F. M. "Fantasies of famine, *Harper's, 250* (1497), 51–90, February 1975.

Reed, N. P. "Spacecraft Earth is overloaded," *Readers' Digest, 105*(632), 115–117, December 1974.

Revelle, R. "Food and population," *Scientific American, 231*(3), 161–170, 1974.

Wade, N. "Green revolution: creators still quite hopeful on world food." *Science, 185,* 844–845, 1974.

2
USES OF PLANTS BY MAN

In the previous chapter, man's fundamental dependence on plants was discussed primarily from the standpoint of food supply. Although food supply, obviously, is of transcendent importance to man's existence, other aspects of the plants' importance in human economy should be mentioned. Many of these have their origin in modern man's advances in technology and his further exploitation of Earth's plant cover. Some of them have come about through his developing aesthetic sense and through increasing understanding of the nature of plants.

FOODS

Fields and orchards yield such important staples as apples, grapes, cereal grains, potatoes, tomatoes, beans, and sugar. The chief food staple of over half the people on Earth is rice (Fig. 2.1), with wheat and maize serving as staples for almost all the rest. These plants are grasses. Add to them other grasses, such as rye, oats, sugar cane (Fig. 2.2) and the pasture grasses for animal feeding, and the importance of just one family of plants can be visualized. Valuable food oils are obtained from olives, peanuts, maize, cottonseed, and soybeans. Large quantities of cotton and

Fig. 2.1 Rice terraces in western Java.

Fig. 2.2 Cutting sugarcane in Florida. Crude sugar is obtained by evaporating the juice expressed from the cane.

soybean oils are converted to solid fats by hydrogenation and used in the manufacture of oleomargarine. Modern methods of distribution have made several tropical fruits, such as pineapple, banana, avocado and date, common articles of diet throughout the temperate regions (Figs. 2.3 and 2.4). Although not important for their nutritive values, we prize highly tea, coffee, and chocolate (Figs. 2.5 and 2.6) and such flavorings and condiments as those prepared from the fruits of pepper and vanilla, the flower buds of cloves, the seeds of mustard, the bark of cinnamon, the leaves of peppermint and sage, and the roots of licorice and horseradish. The food fish of the sea and fresh waters are nourished by microscopic marine plants. Even the strictly carnivorous animals are dependent on green plants, for ultimately, as shown in the "food pyramid," in a series of "eater and eaten," an herbivorous member will be found.

Fig. 2.3 (*Top left*) *Bananas ready for harvesting in Guatemala.*

Fig. 2.4 (*Top right*) *Harvesting pineapple in Jamaica.*

Fig. 2.5 (*Right*) *Picking coffee berries in Costa Rica. The seed, freed of fruit pulp, is dried in the sun and shipped as "green" coffee. After roasting, it is the coffee of commerce.*

Fig. 2.6 Gathering cacao pods in Panama. The roasted seeds when finely ground are chocolate.

RAW MATERIALS FOR INDUSTRY

Our utilization of plant materials for industrial purposes stands next in importance to their use as food. Wood is still our chief structural material for buildings. The Douglas fir trees of the great western forests furnish about a quarter of our sawtimber and a like amount is derived from various species of pines from the southern and western forests. Various hardwoods, such as birch, maple, gum, and walnut from our eastern forests as well as mahogany, rosewood, and others with beautiful color and markings from tropical regions furnish the solid stocks and veneers for the furniture industry. In recent years, the amount of timber converted to plywood has very greatly increased owing to the development of better gluing and bonding methods. The greater size range, superior strength, and light weight are features of plywood that make it more suitable than ordinary lumber for certain types of construction.

The mechanical and chemical conversion of wood to pulp for the manufacture of paper and certain synthetic textiles consumes enormous quantities of timber. Sawmill and pulping wastes may be chemically treated to yield

Fig. 2.7 Collecting turpentine from tapped longleaf pine. Distillation of turpentine, an oleoresin, (shown in glass cylinder) yields oil ("spirit") of turpentine and rosin.

Fig. 2.8 Collection of liquid latex from tapped rubber tree in Thailand. Coagulated latex is crude rubber.

plastics, glues, and special bonding agents. The distillation of wood yields a great variety of valuable industrial chemicals such as methanol and acetic acid. Turpentine and rosin (Fig. 2.7), linseed oil, tanning materials, rubber (Fig. 2.8), and cork are just a few of the many other industrial raw materials we secure from plants.

TEXTILES AND CORDAGE

Many species of plants yield natural fibers for the manufacture of fabrics and cordage. Such fibers as cotton (Fig. 2.9) and flax have a very long history of such uses. Despite the inroads of the modern synthetic fibers, cotton is still the most important fiber in the world. Besides its preeminence in the textile industry, cotton is used to make cording for automobile tires and light duty rope. The manufacture of disposable absorbent tissues and even some of the competing rayon consume significant portions of the cotton crop. It may be mentioned, incidentally, that oil and protein extracted from the cotton seed (after removal of the fibers) are important articles of commerce, the latter especially in the compounding of animal feeds. Flax, now of minor commercial importance in textiles, is used in the production of certain high-quality special fabrics and some special items such as firehose, where its pliability, fineness, durability, and tensile strength

Fig. 2.9 Cotton bolls ready for harvesting. The cotton fibers are hairs growing from the surface of the cotton seed.

are especially valuable.

Other fibers such as hemp and jute are used for coarse fabrics like burlap and in rug manufacture. Manila hemp, derived from a species closely related to the banana, is superior for heavy duty ropes and twines and, along with sisal, constitutes the most important raw material of the cordage industry.

MEDICINES

The early history of botanical science is to a very large extent concerned with man's interest in the use of plant materials for the treatment of human ills. The records of the pre-Christian Egyptians, Greeks, and Romans discuss several thousand species of plants commonly used as medicinal agents. Modern botanical science is said to have started during the fifteenth to seventeenth centuries with the studies and writings of the herbalists, who devoted themselves to the description and illustration of thousands of plant species. Although their interest in plants was chiefly medical, the herbalists were careful observers, and a surprisingly large proportion of the botanical conclusions drawn from their studies is still valid today. Modern chemistry has produced a vast array of synthetic drugs that have replaced many of the old and well-established botanicals, but others are still important items in our *materia medica.* Belladonna, opium, nux vomica, and cinchona are the sources of the important alkaloids, atropine, morphine, strychnine, and quinine, respectively. Tea and coffee owe their stimulant properties to their content of the alkaloid, caffeine. Digitalis, containing several active principles known as glycosides, is used extensively in the treatment of certain

Fig. 2.10 A colony of green mold, Penicillium chrysogenum, *a source of penicillin. The beads of liquid seen on the surface of the colony contain measurable amounts of the antibiotic.*

heart disorders. Botanical and medical researchers are ever on the lookout for new plant drugs, and in recent years, under the stimulation of new botanical and chemical knowledge, have become increasingly interested in reinvestigating many old-fashioned plant drugs that have fallen into disuse. One of the most exciting chapters of medical discovery began in 1928 with the observation by Sir Alexander Fleming that a green mold, *Penicillium* (Fig. 2.10), destroyed certain disease-causing bacteria. From this observation grew a whole new field of medical and botanical research which has produced a growing list of important antibiotics. Streptomycin, penicillin, terramycin, and aureomycin are well-known antibiotics important in the control of specific bacterial and rickettsial diseases.

Indiscriminate use of some medicinal plant materials has given rise to sociological and health problems that concern us greatly at the present time. Some of these have long been useful in regular medical practice for the treatment of human ills, but may become habit-forming in prolonged and uncontrolled use. Among these is heroin, a chemical modification of the alkaloid morphine derived from opium. Opium is the dried latex obtained from the capsules of a species of poppy grown chiefly in the Orient (Fig. 2.11). Cocaine is an alkaloid extracted from the leaves of a shrub native to the Andean regions of South America. Hashish is a resinous extract obtained from the flowering tops of Indian hemp, native to southern Asia, and the small leaves and tops of a related species usually

Fig. 2.11 Collection of opium from the capsules of the opium poppy, Papaver somniferum. *A shallow, circumcisal cut around a fully grown but unripe capsule opens the latex tubes, permitting the latex to exude. The dried latex on the surface of the capsule is removed by scraping.*

growing wild in temperate regions of North America and Europe yields marijuana or "pot" (Fig. 2.12). It may be noted, however, that widespread use of these narcotics, in crude or refined form, is a long-established practice in certain cultures, generally in the regions where the drug plants are grown. Our alarm stems, of course, from the extremely rapid increase in the abuse of narcotics in our society.

Fig. 2.12 The marijuana plant, Cannabis sativa.

INSECTICIDES

Man's struggle against destructive and annoying insects seems unending. Many species of plants, used in the form of extraces or powders, have been shown to possess some insecticidal value. Two of the most potent insecticides of plant origin are rotenone and pyrethrins. The former is extracted from the roots of a member of the bean family native to South America and the Far East; the latter are prepared from the unopened flower heads of certain daisy-like plants native to southeastern Europe. Both of these plants are extensively cultivated as commercial sources of insecticides. Unlike the modern synthetic insecticides such as DDT and Chlordane, and the compounds of copper and arsenic, rotenone and pyrethrins are relatively nontoxic to man and the higher animals and are thus safer for general use on certain food crops as dust or spray applications to combat the Mexican bean beetle, certain aphids, European corn borer, and other crop-destroying insects. The housefly, mosquito, cockroach, flea, tick, and other obnoxious insects also succumb to these insecticides. Tobacco dust and crude extracts of tobacco, by-products of the tobacco industry, are effectively used as contact poisons for certain soft-bodied insect pests because of the nicotine content. Indiscriminate use of several highly toxic insecticides has given rise to serious ecological problems in the destruction of bees and other pollinating insects. Some of these insecticides persist in the environment for long periods of time and when used in great quantities may be washed by rain into streams and lakes where they poison valuable food fishes.

FUELS

The great stores of energy represented by the deposits of coal, petroleum, and natural gas were laid down mostly during the Carboniferous period, about 300 million years ago. These fossil fuels are our principal energy source (Fig. 1.3). The remains of the luxuriant vegetation of those times became buried by sedimentary deposits and were subjected to tremendous pressure and high temperature. The loss of soluble minerals by leaching and the compression of the remaining carbon compounds yielded the seams of coal interlayered with sheets of slate. The organic remains that constitute the coal are the bodies of now extinct giant clubmosses, horsetails, tree ferns, and primitive seed plants, often so compressed and distorted as to be practically unrecognizable. The clues to the nature of the ancient coal-forming plants come from the study of well-preserved plant remains that became embedded in the associated layers of sedimentary rock.

Petroleum is commonly believed to be largely of marine origin and to represent principally the accumulation of oil that occurs as tiny droplets in the bodies of microscopic one-celled plants called diatoms. During the long period of underground storage, the oil has undergone considerable chemical change. The study of microscopic fossils in the rocks of petroleum-yielding regions often gives clues to where oil wells may profitably be drilled.

Natural gas occurs in coal beds, but the most important commercial supplies are found in petroleum wells. It is a by-product of the peculiar chemical and physical reactions involved in the production of the coal and petroleum. The formation of these important fuels is believed to have been effected under conditions of limited availability of oxygen and hence incomplete decay.

In many forested regions of the world, wood is still the principal fuel because of its cheapness.

VALUABLE ACTIVITIES OF NONGREEN PLANTS

Many nongreen plant species have been found useful for their very special properties. Many species of fungi and bacteria are responsible for the familiar process of decay whereby organic wastes and the dead bodies of plants and animals are reconverted to simple substances. These organisms, incapable of making their own primary food as do green plants, consume organic matter as food. The simple products of decay are thus returned to the environment from whence they may enter the metabolic cycles of still other organisms, principally green plants.

Some of our common foodstuffs, such as cheeses and sauerkraut, owe their distinctive flavors to the presence of metabolic end products of certain species of bacteria and molds. The alcoholic beverage industry is dependent on yeast, a fungus that produces ethyl alcohol as it uses simple sugars in anaerobic respiration. Yeast is used in the leavening of bread dough, by virtue of its production of carbon dioxide gas in respiration. Other fungi and bacteria similarly produce a wide variety of organic acids, alcohols, and other important substances including the antibiotics already mentioned. Some free-living soil bacteria and others in the roots of certain species of the bean family are the agents for the fixation of free atmospheric nitrogen in such form that this important element becomes available to growing green plants. This is the basis for the common practice of enriching the soil by growing crops of legumes, such as clover or lespedeza, in whose roots the nitrogen-fixing bacteria live. Other bacterial species take part in converting unusable nitrogen compounds resulting from decay to usable nitrogen compounds. The natural or artificial addition of partially decayed plant remains, that is, humus, to productive soils improves or helps maintain fertility. Gardeners commonly improve the structure of soil by adding humus in the form of peat moss or leaf compost. In nature, the same effect is produced by fallen leaves and twigs.

AESTHETIC VALUES

Plants also affect our lives in ways less material than those cited. The recreational value of the great forested areas of our country is well recognized. Each year, millions of vacationers visit the national parks such as Great Smoky Mountains, Yellowstone, and Yosemite. These parks and the national forests offer facilities for relaxation and education and superb scenery to satisfy our aesthetic sense. The aesthetic and recreational values of the home flower garden to the suburban dweller, and the shady park or even the window box to the city dweller, are not to be underrated. Our environment is made more pleasing through the professional efforts of florists, landscape gardeners, and nurserymen.

HARMFUL EFFECTS

It must be added, finally, that although most plants are beneficial to man, some are harmful. The rank growth of certain vines and weeds chokes pastures, overruns cultivated fields and gardens, and thus constitutes an important economic problem. The cost of weed control measures plus the reduction in productivity of crop plants due to weed competition is reliably estimated to be at least 5 billion dollars a year.

Some bacteria and fungi cause plant diseases and so are responsible for further major economic losses. Tomato and cucumber wilt and cotton root rot are examples of bacterial diseases. Apple scab, wheat rust, and potato blight are a few of the fungal diseases of economic plants. The U. S. Department of Agriculture has estimated that in a normal year, the loss in yield of barley, wheat, maize, and oats due to common diseases aggregates over 2.3 billion bushels (Fig. 2.13). Similarly, disease accounts for the annual loss of 65 million bushels of potatoes, 12 million bushels of apples, 81 million pounds of tobacco, and nearly 16 thousand tons of grapes. In decreased production alone, without accounting for the cost of control measures, the annual bill for crop diseases comes to 2.8 billion dollars. Add an annual 3.6

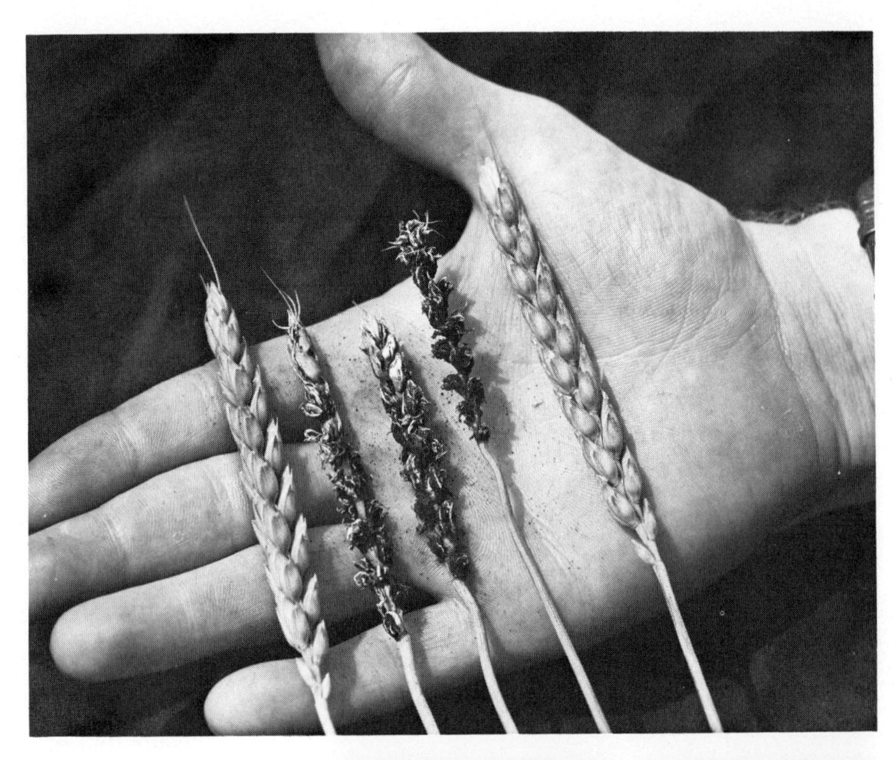

Fig. 2.13 Brown loose smut in wheat.

Fig. 2.14 Typical damage caused by corn leaf blight.

billion dollar loss attributable to insect damage to the losses caused by weeds and diseases, and the total is truly staggering. To express this food loss in another way, it has been estimated that an amount of produce equivalent to the normal yield from 120 million acres never reaches the consumer (Figs. 2.14, 2.15, 2.16, 2.17, and 2.18).

Fungi also cause serious diseases of many of our desirable trees. The Dutch elm disease and the chestnut blight are notorious examples, the latter being responsible for the virtual extinction of the American chestnut. The decay of stored lumber is similarly attributable to fungi (Fig. 2.19).

Some species of plants affect man adversely in a direct and personal way. Contact with the familiar poison ivy, poison oak, or poison sumac causes serious skin irritation. Many people are allergic to the pollen of flowering plants such as oak and ragweed, and suffer acutely from hay fever and other reac-

Fig. 2.15 (Top left) Complete destruction of an ear of maize by corn smut.

Fig. 2.16 (Left) Apples infected with scab. The smaller specimen shows the deforming effect of the disease.

Fig. 2.17 (Top right) Brown rot on peach.

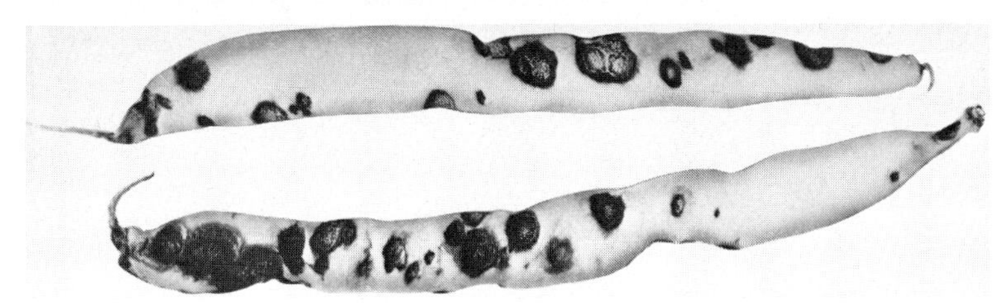

Fig. 2.18 Anthracnose of bean, a disease caused by fungus Colletotrichum.

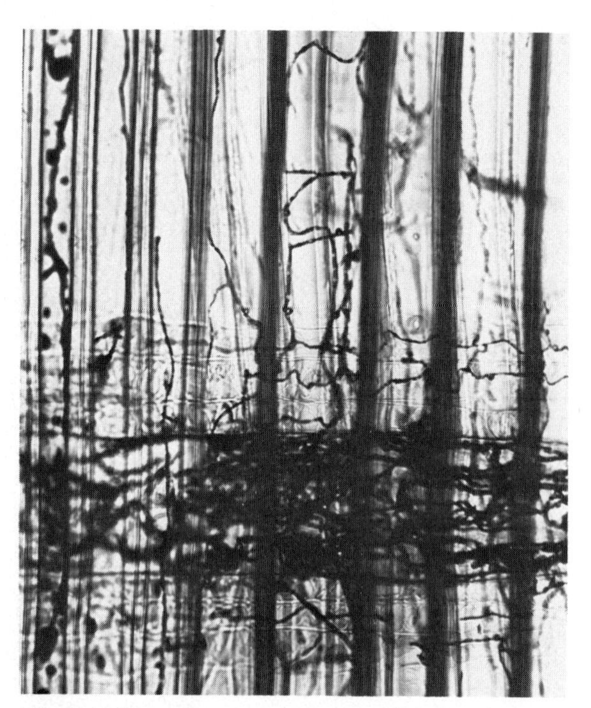

Fig. 2.19 Decay fungus, Lenzites, *in cells of Southern yellow pine lumber.*

tions during the season when such plants are in flower. Many of the common vegetable foods cause allergic reactions in some individuals. Even certain species of algae are reported to cause allergic reactions and gastrointestinal disorders in man. Mushroom fanciers who collect their own supply must learn to avoid certain extremely poisonous species. Some medicinally valuable plants may be fatally toxic when used in immoderate doses. Serious diseases of man, such as typhoid, tuberculosis, cholera, streptococcus infection, and tetanus are caused by bacterial species. Domesticated animals are similarly afflicted by various bacterial diseases. Species of fungi are responsible for the common skin diseases, athlete's foot and ringworm in man, and for a variety of serious pulmonary infections. Avoidance and medical counter-measures are our only defence against these widespread, harmful species.

TO CONSERVE OUR RESOURCES

The foregoing section, by means of a few examples, suggests the magnitude of our dependence on plants and the products of plants. As our study progresses, other important relationships between plants and man will suggest themselves. Our tremendous consumption of and dependence on the products of forest, field, and grassland make these assemblages of plants our greatest natural resource. The dire predictions of some authors that we shall be faced with critical shortages of food and raw materials as the world's population continues to increase could come true, unless we learn to use our resources intelligently and practice conservation in its broadest sense.

Many people erroneously think of conservation as a kind of large-scale hoarding. Conservation is an attempt at a wise, effective use of a resource for the greatest ultimate good. It aims to provide the largest amount of return from a resource for the longest possible time through avoiding waste and depletion and through restoration and improvement.

The problems of conservation are often very complex, involving many different kinds of activity and the use of principles and methods from several fields of science. Genetic research may develop new disease-resistant or more productive varieties of crop plants. Chemical and physiological research may create more effective fertilizers, weed killers, insecticides, and fungicides. These, along with the more popularly known conservation activities, such as soil erosion control and forest fire prevention, are all parts of the total conservation program.

The Soil Resource

Most of our plant resources flourish on the thin mantle of fertile soil on Earth's crust, from which they obtain the necessary water

and minerals and in which they are stably situated. The relationship of plant to soil is an intimate one, and the quality and quantity of plant product reflects the degree of soil fertility. It follows, therefore, that maintenance and improvement of productive soil make up a significant part of our conservation efforts.

The loss of fertile soil by the physical action of wind (Fig. 2.20) and water (Fig. 2.21) is one of the great problems arising from man's exploitation of his environment. The production of field crops, the timbering of forests, uncontrolled forest fires (Fig. 2.22), careless strip-mining (Fig. 2.23), and even the grazing of the grasslands may entail serious disturbance or destruction of the protective vegetational cover and thus expose the soil to possible erosion by wind or water (Figs. 2.24 and 2.25). Well-forested areas and undisturbed grasslands are subject to practically no erosion by wind, and erosion by water is significantly reduced.

The vegetation acts as a protective cover in

Fig. 2.20 *The exposed roots of sagebrush show that six inches of topsoil have been removed by wind action following a burn. Picture taken one year after the burning.*

Fig. 2.21 *Erosion in central Tunisia caused by flooding after destruction of natural vegetation.*

Fig. 2.22 *Destruction of young forest and exposure of soil to further damage by erosion.*

Fig. 2.23 *Destruction of natural vegetation by strip mining.*

Fig. 2.24 *Fence line separates overgrazed and depleted range (right and foreground) from lightly grazed range.*

Fig. 2.25 *Erosion of pasture land in Turkey by overgrazing and wind.*

Fig. 2.26 Erosion following clear cutting on very steep slope. Note breakdown of stream bank and severe silting of stream.

several ways. The spreading leafy branches of the plants reduce the force of rain falling upon the soil. The water falls gently upon the surface, with a minimum of dislodgement of soil particles. The mat of accumulated forest litter derived from fallen leaves and branches acts similarly and, in addition, retards the lateral flow of water, thus reducing the movement of soil by violent runoff. The widely spreading roots of plants increase the porosity and absorptive capacity of the soil and thus also reduce the surface flow. Destructive floods often result from the choking of the natural drainage channels with water-borne soil. Washed-down soil from eroding watersheds may fill reservoirs, interfere with the generation of electric power, and in some instances render rivers unnavigable. The fish population suffers also from the silting of streams and lakes. The eroded land, of course, is unfit for agricultural use since the fertile topsoil has been lost. Under severe conditions the land may, depending on the structure of the subsoil, become deeply gullied (Fig. 2.26).

The Water Resource

The vegetational cover is of equal importance in its influence on water resources. By reducing the rate of runoff and increasing absorption by the soil during periods of high rainfall or rapid melting of snow, the vegetational cover equalizes the flow of streams through the seasons. Absorbed water may ultimately find its way to greater depths, thus replenishing or maintaining the underground reserves. The rapid depletion of ground water reserves has become a critical problem in some sections of the country where deep well irrigation is practiced. Such depletion imposes a limitation on future agricultural and industrial development.

The Forest Resource

The Forest Service of the U. S. Department of Agriculture estimates that by the year 2000 we shall require almost twice our present wood production to meet the needs of a population of 275 million people. To assure the continued benefits of our forest resource and provide for the future demand, we must maintain, improve, and increase our productive forests by intelligent management. Various federal and state governmental agencies and foresighted commercial producers, through the best forestry practices in the national and state forests and on private lands, have clearly demonstrated the value of enlightened management of the forests. However, the more than 4 million small-tract owners who control and exploit one-half of our forest lands do not generally engage in the best practices. Since the public-owned forests and those of the larger commercial concerns cannot meet the future demand, it is the smaller owner who must be educated to the importance of proper forest management, that he may make up the future deficiency.

Various methods are employed in timber management to insure the greatest productivity. In some types of producing forests, only the mature trees are selected for harvesting. Young growing timber is left for further growth and future cutting. Natural reseeding from the remaining trees makes such stands of timber self-perpetuating (Fig. 2.27). Other types of forest yield greater returns under a

Fig. 2.27 Natural reseeding from mature trees left standing after selective cutting.

Fig. 2.28 Block cutting in Montana forest. When young trees are well established in cleared blocks, other blocks will be cut.

method known as strip- or block-cutting. In this method all trees are removed from relatively small sections of the forest and the bare ground is either naturally reseeded from the older surrounding trees, or young nursery-grown trees are planted (Figs. 2.28, 2.29, 2.30, 2.31). The strip- or block-cutting method is necessary where seedlings of the more valuable timber species are intolerant of shade. Proper thinning of timber stands to reduce competition and stimulate growth, and the removal of diseased trees and undesirable competing species are also important in profitable forest management (Fig. 2.32).

Extensive reforestation projects must also be undertaken in an effort to meet the future demand for timber. It is estimated by the Forest Service that about 52 million acres of potentially productive forest land need to be restocked. This addition to the present forest reserve would, under proper management, meet the projected need for the year 2000. An additional 5 million acres could profitably be planted for erosion control, watershed protection, and improvement of wildlife habitat.

In many parts of the world, concerted efforts at reforestation of areas long devoid of tree growth are under way. The resulting forests will reduce the ravages of soil erosion and provide fuel, timber, and useful employment (Fig. 2.33).

Through the destructive effects of disease, insects, and fire, the annual loss in productivity of the forests amounts to nearly 36 billion board feet. This amount closely approaches the net timber growth. Thus, almost half of the

Fig. 2.29 *(Top) A clear-cut block in a forest of Douglas fir in 1940.*

Fig. 2.30 (Bottom left) *Same scene as in Fig. 2.29 in 1955. Natural reseeding from mature trees in surrounding uncut areas.*

Fig. 2.31 *(Bottom right) Same scene as in Fig. 2.30 in 1970. Forest has become almost mature in 30 years. In modern forest management, aerial seeding or planting of nursery-grown young trees would accelerate the process of forest regeneration.*

Fig. 2.32 Proper thinning for improved growth of western yellow pine.

total timber producing potential of the forests is lost through the action of destructive agents. As a result of intensive effort in prevention and control, fire has been reduced, in recent years, to the third in importance among the three destructive agents, and still accounts for a loss of 7.5 billion board feet. Disease leads the list, causing a loss of nearly 20 billion board feet, and insect damage accounts for an additional loss of over 8.5 billion board feet.

The Field Resource

Evaluating land for agricultural use is an important first step in a comprehensive conservation program. General fertility, slope, and drainage features are factors to be considered. Steeply sloping land or land of low fertility are better left with the natural vegetational cover. Eroded land on steep slopes should be planted to grass or trees. Erosion may thus be checked

Fig. 2.33 Forestry trainees working with pine seedlings in tree nursery for afforestation project in the highlands of Ecuador.

and further damage prevented. Special planting practices such as contour planting, strip cropping, and terracing permit the production of crops on gentler slopes with minimal danger of erosion. In contour planting of row crops, the furrows are laid off on a level across the face of the slope. Water is caught by the furrows at various levels and seeps slowly into the soil where it becomes available to the growing crop and contributes to the underground water reserve. In strip-cropping, broad bands or strips of a ground-covering crop such as grass, clover, or alfalfa alternate with strips of clean-cultivated row crops. The close cover of alfalfa or clover prevents water from coursing down the slope and eroding gullies. The strips are laid out on contours, and a regular rotation of crops is followed. Terracing, effective on rolling land, consists of throwing up on contours a series of broad ridges with shallow channels above to trap the water and prevent its rapid runoff down slope.

Serious reduction of fertility is a certain result of long continued removal of the annual crops without restorative treatment. Every pound of crop removed from the land, whether it be cabbages or cattle, reduces the fertility of the soil. The successful producer realizes this fact and follows a definite plan of soil restoration designed to meet the needs of his main crop. Different crops make different demands on the soil. Sometimes a crop rotation plan involving the growing of legumes such as clover or alfalfa at certain intervals, is employed to replenish the nitrogen supply in the soil. The addition of other elements, notably phosphorus and potassium, are also periodically made to maintain adequate stocks of these nutrients. The best practice is to have frequent recourse to soil analyses and to use fertilizers of the kind and amount required for the intended crop.

The Grassland Resource

About 40% of the land area of the continental United States is natural grassland devoted to grazing. Carefully planned and controlled grazing assures a continuous supply of forage for cattle and sheep. Placing too many animal units upon a grazing area, or grazing too early in the season, weakens or kills the grass, bares the soil, and exposes it to erosion by wind or water. The destructive dust storms of the western plains during years of severe drought are traceable to mismanagement of grazing lands or the cultivation of land better left in natural grass.

Restoration and improvement of the grasslands are a major part of the conservation programs of the federal and state departments of agriculture. Experiments to develop methods of reseeding with adapted grasses are under way, improved grazing practices are being devised, and new methods of reducing wind erosion on cultivated lands are being studied.

Finally, a moment's reflection on the matters set forth in this chapter will make clear that there are social as well as economic implications. Malpractice in the exploitation of our resources may lead to lower productivity or exhaustion, and consequently to lower standards of living for the people. Economic disaster, starvation, and death often attend the failure of the sole food crop produced under a "one-crop" economy. Interruption or reduction of output or interference with the normal distribution of raw materials in world trade may mean ruin for a people. It is no wonder, then, that modern nations aspire to self-sufficiency in these things.

Broader education of the consuming public in these vital matters and intensive popular support for broad, coordinated programs of resource conservation are an urgent need.

RELATED READING

Baker, H. G. *Plants and Civilization*. Belmont, Calif.: Wadsworth 1965.

Dovring, F. "Soybeans," *Scientific American, 230*(2), 14–20, February 1974.

Fisher, H. L. "Rubber," *Scientific American, 195*(5), 75–88, November 1965.

Hall, F. K. "Wood pulp," *Scientific American, 230*(4), 52–62, April 1974.
Harpstead, D. G. "High-lysine corn," *Scientific American, 225*(2), 34–43, August 1971.
Lessing, L. P. "Coal," *Scientific American, 193*(1), 58–67, July 1955.
Putnam, J. J. and B. Dale. "Timber: how much is enough?" *National Geographic, 145,* 485–510, 1974.
Reed, T. B. and R. M. Lerner. "Methanol: a versatile fuel for immediate use," *Science, 182,* 1299–1304, 1973.
Rose, A. H. "Beer," *Scientific American, 200*(6), 90–100, June 1959.
Rose, A. H. "Yeasts." *Scientific American, 202*(2), 136–142, February 1960.
Salamon, R. N. "The influence of the potato," *Scientific American, 187*(6), 50–56, December 1952.
Shapley, D. "Sorghum: miracle grain for the world food shortage?" *Science, 182,* 147–148, 1973.
Solheim, W. G. II. "An early agricultural revolution," *Scientific American, 226*(4), 34–41, April 1972.
Wilkes, H. G. "Maize and its wild relatives," *Science, 177,* 1071–1077, 1972.
Wittwer, S. H. "Maximum production capacity of food crops," *BioScience, 24,* 216–224, 1974.
Wood, A. L. "Cinnamon—spice that changed history," *Natural History, 68,* 578–591, 1959.
Youngken, H. W. "Botany and medicine," *American Journal of Botany, 43,* 862–869, 1956.

THE PLANT KINGDOM

2

PHOTO OPPOSITE: MALE CONE OF MONTEREY PINE, *PINUS RADIATA*. (MAGNIFICATION 35 ×)

3
PLANT CLASSIFICATION

In Chapters 1 and 2 we gained some idea of the importance of plants to man. These plants range in size and form from very simple, microscopic organisms to huge trees. Modern estimates place the number of kinds or **species** comprising **the Plant Kingdom** at about 350,000. These have come to be recognized by many people over a very long period of time as a result of careful observation and intensive study. When this diverse assemblage of plants is closely studied, it becomes evident that these species can be grouped or classified according to similarities in appearance, behavior, or composition. By thus classifying, we facilitate study of the Plant Kingdom, for a group embracing many similar plants may be thought of as a unit. To a certain extent, what is characteristic of one member would be characteristic of other members of the group with respect to the criteria chosen as the basis for classification. Of course, classifications are man-made and may be designed to suit man's convenience or reflect his primary interest in plants. Thus, a classification of plants could be based on such features as size, food value, harmfulness, medical properties, or habitat. Certainly, early man's concept of plant groups was based on such utilitarian considerations. However useful such groupings may be, modern botanists believe that the most comprehensive understanding of the Plant Kingdom may be obtained by devising a classification that expresses their evaluation of *relationships* among plants and thus reflects the course of evolutionary development. Several such classifications have been proposed in the past and others will be in the future, each succeeding one based on new knowledge of plants and on new *interpretations* of data. The perfect system of classification has not yet been devised, for our knowledge of plants is still far from complete.

Detailed studies of the structure, physiology, genetic behavior, and fossil records of plants have yielded the data upon which a modern classification may be based. One such scheme for classification is outlined in Table 3.1. The Plant Kingdom is conceived to consist of 15 rather distinct groups called **divisions.** Each plant species may be assigned, on the basis of its characteristics, to a division. A division is an assemblage of plant species ordinarily believed to have had a common origin in some remote ancestral form and to show characteristics interpreted as a distinctive, *major* evolutionary trend, or track of evolutionary development. Within most of the divisions, smaller groups called **classes** may be recognized. These represent smaller circles of relationship or lines of development within the limits of the division. Table 3.1 indicates only a few of the classes. The possible interrelationships among the divisions may, at present, be unknown or very difficult to evaluate; at any rate, their origins are obscured in the remoteness of time. However, it is sometimes desirable to distribute divisions among three larger groups to reflect their attained general level of evolution as measured by the kind and

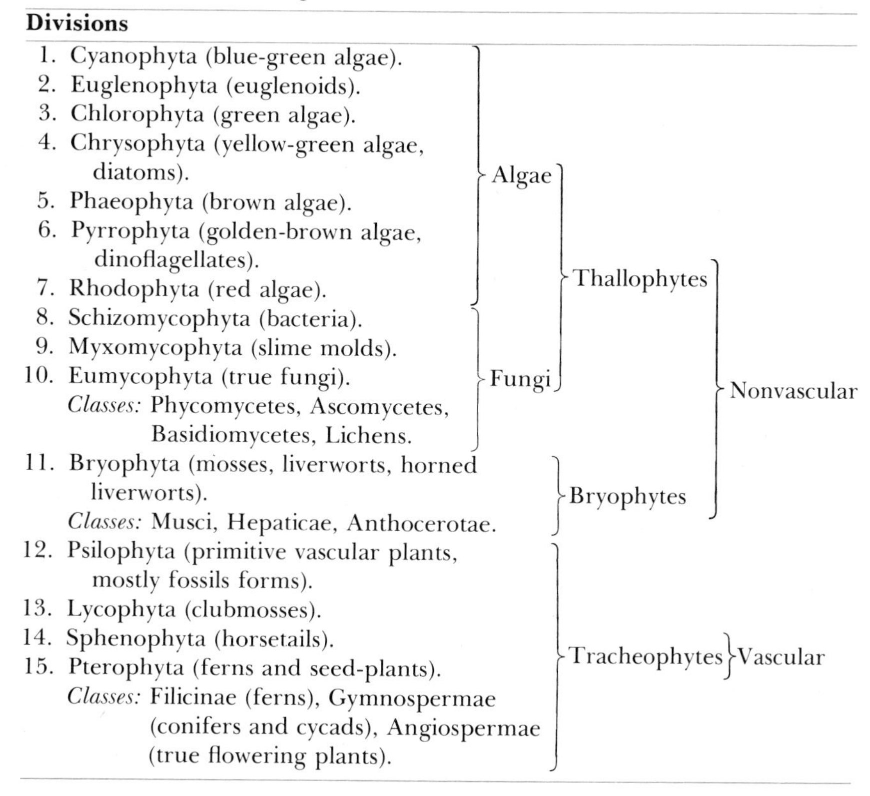

Table 3.1 **The Plant Kingdom**

Divisions				
1. Cyanophyta (blue-green algae).	Algae	Thallophytes	Nonvascular	
2. Euglenophyta (euglenoids).	Algae	Thallophytes	Nonvascular	
3. Chlorophyta (green algae).	Algae	Thallophytes	Nonvascular	
4. Chrysophyta (yellow-green algae, diatoms).	Algae	Thallophytes	Nonvascular	
5. Phaeophyta (brown algae).	Algae	Thallophytes	Nonvascular	
6. Pyrrophyta (golden-brown algae, dinoflagellates).	Algae	Thallophytes	Nonvascular	
7. Rhodophyta (red algae).	Algae	Thallophytes	Nonvascular	
8. Schizomycophyta (bacteria).	Fungi	Thallophytes	Nonvascular	
9. Myxomycophyta (slime molds).	Fungi	Thallophytes	Nonvascular	
10. Eumycophyta (true fungi). *Classes:* Phycomycetes, Ascomycetes, Basidiomycetes, Lichens.	Fungi	Thallophytes	Nonvascular	
11. Bryophyta (mosses, liverworts, horned liverworts). *Classes:* Musci, Hepaticae, Anthocerotae.		Bryophytes	Nonvascular	
12. Psilophyta (primitive vascular plants, mostly fossils forms).		Tracheophytes	Vascular	
13. Lycophyta (clubmosses).		Tracheophytes	Vascular	
14. Sphenophyta (horsetails).		Tracheophytes	Vascular	
15. Pterophyta (ferns and seed-plants). *Classes:* Filicinae (ferns), Gymnospermae (conifers and cycads), Angiospermae (true flowering plants).		Tracheophytes	Vascular	

degree of differentiation of the plant body.

In Chapters 4 and 5, the chief features of the major plant groups are briefly summarized and illustrations of typical representatives provided.

LESSER CATEGORIES OF CLASSIFICATION

Although it is important to view the Plant Kingdom in its largest aspects and to summarize it in some all-embracing scheme of classification, it is true that the common plants growing around us are thought of in terms somewhat less inclusive than the division or class. Just as we have seen that the plant kingdom may be subdivided into divisions, and the divisions into classes, so botanists recognize the need for progressively lesser and lesser categories. This need becomes obvious when, on an afternoon's stroll, we encounter, for example, six different kinds of oaks, three kinds of maples, and three kinds of pines.

Quite naturally, we give names to the things we know and use; so we apply names to the plants around us. First, these names are likely to be chosen with reference to some quality or attribute of the plant. A descriptive name, an allusion to a use, or a fanciful reference may, and often does, suffice as a **common name.** Thus "Blood Root," to describe the red sap exuding from the plant when cut, and "Mayflower," to describe the time of flowering, are well-chosen common names for the respective plants. But the same plants may be known by different and equally good names in other localities, or indeed, the same names may be applied to altogether different plants. There are said to be eighteen different species or kinds of plants growing in the United States that pass under the name "Snakeroot." It is clear that confusion would be unending if there were not some standard system of naming. The scientific study of plants requires a precise identification of the plants. To this end, in botanical practice, one valid name is given to each plant species, and it bears that name throughout the world, wherever scientific study of plants is pursued. The names given are in Latin or are occasionally derived from Greek. Botanists, regardless of nationality, can understand precisely what plant is indicated by its **scientific name.**

The scientific name of a plant consists of two parts, the name of the **genus** followed by the **specific epithet.** Thus the sugar maple is named *Acer saccharum.* To insure greater accuracy in the designation of plant species, the identity of the author of the plant's valid scientific name is appended, usually as an abbreviation: for example, *Acer saccharum* Marsh. (for Humphrey Marshall, the botanist who first described and named this species of maple). A species is a single *kind* of plant in which all the individual members are similar in most of the details of structure and behavior, and whose distinguishing characteristics are passed on through a succession of generations. For example, besides the sugar maple, there are many other maples such as the scarlet maple, *Acer rubrum* L. (for Carolus Linnaeus), and the Sierra maple, *Acer glabrum* Torr. (for John Torrey). All of these and many more species of maples constitute the genus *Acer* (maples).

Some genera are related to other genera just as some species are related to other species. Therefore the grouping of related genera gives a higher category of classification called the **family.** Related families constitute an **order.** Related orders constitute a **subclass** or **class** and related classes, a **division.** In botanical practice the names of the various categories of classification from division down through family are given distinctive endings, as shown below.

The complete classification of the sugar maple, therefore, would be:

Division Ptero*phyta*
Class Angiosperm*ae*
Subclass Dicotyledon*eae*
Order Sapind*ales*
Family Acer*aceae*
Genus *Acer*
Species *saccharum.*

By inspection of the complete classification, we can relate this plant to all others.

In view of the large number of plant species already known and the estimated 5000 new ones described each year, an orderly system for classifying them is a necessity. Indeed, it helps to make such a diverse group of organisms comprehensible. Although no one person can know all the species of plants, he can know a great deal about them if he knows how they fit into a scheme of classification based on relationships. As new knowledge of plants is gained from researchers in the several fields of botany, it becomes the basis for the possible devising of new and possibly better schemes of classification. Classification is not an end in itself but an attempt to present a current summary of our knowledge of plants.

WHICH ORGANISMS ARE PLANTS?

Biologists hold a diversity of opinions as to just which organisms should be included in the plant kingdom. Traditionally, all organisms were classified into only two kingdoms: plant and animal. Table 3.1 is based on this broad concept of plants, and in this book we are considering all the organisms listed in this table as plants.

However, as more information about the lower organisms was accumulated and as it became evident that sharp distinctions between the plant and animal kingdoms were difficult to make, some biologists decided that there should be a third kingdom (Protista) that included the lower organisms. But again there was difficulty in drawing sharp lines between kingdoms. Some biologists thought that Protista should include only the protozoa, bacteria, blue-green algae, and certain unicellular or filamentous algae, and the fungi. Others wished to include in Protista all the algae, thus limiting the Plant Kingdom to bryophytes and vascular plants.

Further consideration of the lower plants led some biologists to the conclusion that the Kingdom Protista was far too diverse a group, and that one or two additional kingdoms should be established. The proposal that the bacteria and blue-green algae be placed in a separate kingdom (Monera) received broad acceptance because these organisms differ from all others in having procaryotic rather than eucaryotic cells. Then R. H. Whittaker and others proposed that the fungi differ so much from other organisms that they should be placed in a kingdom of their own. Acceptance of this proposal would result in the recognition of five kingdoms of organisms: Monera, Protista, Fungi, Plantae, and Animalia. However, there is still considerable difference of opinion as to what Protista should include, ranging from just the protozoa, slime molds, and some of the unicellular algae (other than the blue-green algae) to all these groups plus all the other algae. Table 3.2 summarizes the various attempts to divide organisms into kingdoms.

All this sounds confusing, but it should be realized that all schemes of classification are man-made and are not necessarily natural entities. The schemes of classification proposed by biologists vary from one another, not only at the kingdom level but also at the lesser levels of classification. As stated previously, biologists strive to attain a system of classification that groups together the organisms that are most closely related to one another, and there are bound to be differences of opinion as to which system does this best. Nevertheless, some system of classification is essential when dealing

Table 3.2 **Outline of Three Schemes of Classifying Organisms into Kingdoms**

<table>
<tr><td rowspan="4">Plantae
Bacteria
Fungi
Algae
Bryophytes
Vascular plants</td><td rowspan="3">Protista
Protozoa
Bacteria
Slime molds
Fungi
Some algae</td><td>Monera
Bacteria
Blue-green algae</td></tr>
<tr><td>Protista
Protozoa
Slime molds
Some algae</td></tr>
<tr><td>Fungi</td></tr>
<tr><td>Plantae
Green algae [a]
Red algae [a]
Brown algae [a]
Bryophytes
Vascular plants</td><td>Plantae
Green algae [a]
Red algae [a]
Brown algae [a]
Bryophytes
Vascular plants</td></tr>
<tr><td>Animalia
Protozoa
Multicellular animals</td><td>Animalia
Multicellular animals</td><td>Animalia
Multicellular animals</td></tr>
</table>

[a] Placed in the Protista by some.

with the vast and diverse array of organisms, just as it is in dealing with the numerous books in a large library. A library may use either the Dewey decimal system or the Library of Congress system. There are differences of opinion as to which is better, but either one works, and both are natural systems based on subject matter rather than artificial systems based on such things as the size or color of the books. Similarly, biological classifications must be natural systems based on relationships between organisms. Any such system is usable from a practical standpoint so long as the systems are not mixed.

RELATED READING

Bell, C. R. *Plant Variation and Classification.* Belmont, Calif.: Wadsworth, 1967.

Heywood, V. H. *Plant Taxonomy.* New York: St. Martin's Press, 1967.

Whittaker, R. H. "On the broad classification of organisms," *Quarterly Review of Biology, 34,* 210–226, 1959.

4
NONVASCULAR PLANTS

THALLOPHYTES

The thallophytes, embracing the first ten divisions, are structurally relatively simple. They lack the differentiation into roots, stems, and leaves that we recognize in the familiar flowering plants. In size, however, they range from microscopic bacteria to seaweeds many meters long. They possess no specialized absorbing organs and generally no conductive tissues. Water and other substances pass directly into the plants through all surfaces. In reproduction, these simple plants display a variety of methods. In some, whose bodies are a single cell, reproduction may be accomplished by a simple **fission,** or division into two cells. In some multicellular and some unicellular thallophytes, one or more of the cells may, by division, produce a number of **spores,** each of which may develop into a new plant. Many kinds reproduce sexually, that is, through the formation of fusing sex cells, or **gametes.** Botanists regard the thallophytes as primitive plants because of the simplicity of their structure, or lack of great differentiation. Presumably, these plants have been derived with little change in appearance and structure from their ancestral types.

The first seven divisions of thallophytes are usually collectively called **algae** (singular, **alga**). The divisions of algae are quite diverse and are thought to have had independent origins, or, rather, their interrelationships are obscure. Algae have in common the green pigment **chlorophyll,** which enables them to manufacture their food by a process called **photosynthesis.** Essentially all plants of the remaining three divisions of thallophytes lack chlorophyll or other photosynthetic pigments and, with a very few exceptions, are unable to produce their own food. Plants that can make their food are said to be **autotrophic;** those that cannot, **heterotrophic.** The nongreen thallophytes, with the exception of the bacteria, are commonly classed together under the name **fungi** (singular, **fungus**). The relationships among the divisions and classes of fungi are uncertain and perhaps remote. About 110,000 species of thallophytes are known at the present time and doubtless thousands are yet to be discovered, mostly among the microscopic forms.

Algae

The algae are chiefly aquatic, occurring in freshwater lakes and streams, in stagnant swamps, and in the sea. Some algae occur in moist situations on land such as wet soil and moist surfaces of trees and rocks. Some live in symbiotic association with other organisms, as in the **lichens,** where the alga provides the food and the fungus assists in the absorption of water.

As a group the algae are of great economic value. Of chief importance is the fact that they are the basic source of food for fish and other aquatic animals. The world's most productive fishing grounds are in cool seas where algae flourish the most abundantly. Free-swimming and free-floating algae constitute, along with small animal species, the **plankton** of seas and lakes; the algal species growing attached to the bottom constitute the **benthon.** Algae are also important to fish because they remove carbon dioxide from the water and restore the oxygen supply in the process of photosynthesis. Somewhat over 18,000 species of algae are known.

In some parts of the world, particularly the Orient, the larger seaweeds are consumed directly as food by man. Some marine species are rich sources of iodine, and from various brown and red algae a number of commercially important substances such as **algin** are extracted and used in the manufacture of adhesives and as stabilizers and emulsifiers in the baking, pharmaceutical, and cosmetic industries. **Agar,** a jellylike extract from species of red algae, serves as a base for culture media in bacteriological and mycological laboratories, and like algin, as a stabilizer and emulsifying agent.

Cyanophyta. The **blue-green algae** commonly appear as a bluish green smear on moist surfaces of rocks and bark of trees, on and below the surface of damp soil. Some species occur in fresh or salt water, and some live on the rocks in hot springs. About 1500 species are known.

The body of a cyanophyte is a small cell occurring singly or in variously shaped clusters or in filaments. Groups of cells or filaments are commonly embedded in a gelatinous sheath (Fig. 4.1). The predominant bluish green color is due to the presence of the pigments chlorophyll *a* and **phycocyanin.** Other pigments, such as **carotenoids** (yellow, orange, or browns) and **phycoerythrin** may also be present. When the latter is present in large amounts, the plants appear reddish. The pigments occur not in organized plastids, as in the higher plants, but in a lamellated peripheral zone of the cell, the **chromoplasm.** No highly organized nucleus is present.

Some species of blue-green algae are able to fix atmospheric nitrogen, and thus contribute to soil fertility. Excessive growth of some species in reservoirs imparts a disagreeable odor and an unpleasant, "fishy" taste to the water.

Euglenophyta. The **euglenoids** can be

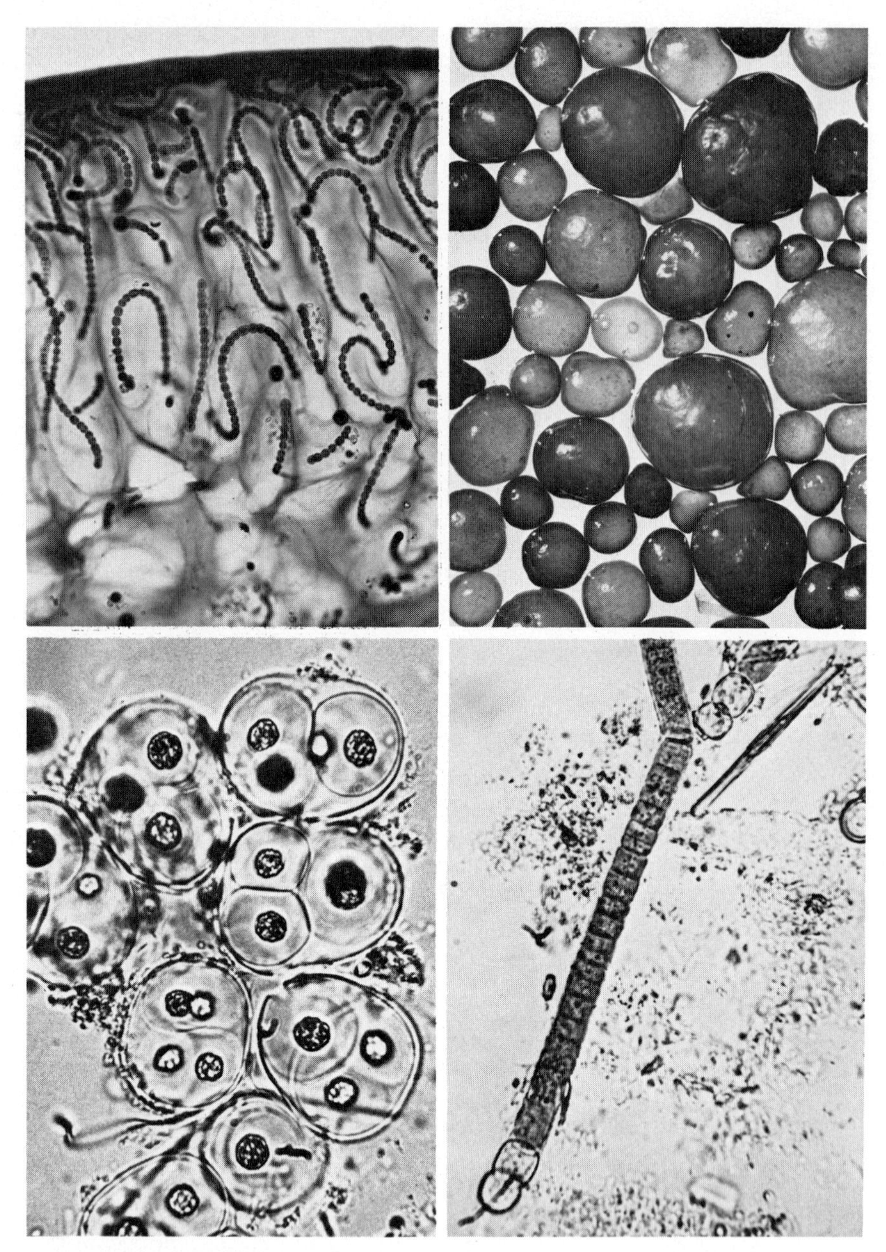

Fig. 4.1 Blue-green algae: (Top left) microscopic section through part of a colony of Nostoc, *showing each filament embedded in a gelatinous matrix. (Top right) Colonies of* Nostoc *as they occur in quiet, shallow pools. (Bottom left) Colonies of* Gloeocystis *with their enveloping gelatinous sheath. (Bottom right) Filaments of* Tolypothrix. *(Top, copyright, General Biological Supply House, Inc., Chicago; bottom, courtesy, Carolina Biological Supply Co.)*

found in fresh water in which decaying organic matter is present.

Most species are single, swimming cells with one to three whip-like **flagella** (Fig. 4.2). The cell is usually without a rigid cell wall. The pigments, chlorophyll *a* and *b* and some carotenoids, give the plants a grassy green color, and are contained in organized **plastids.** Unlike the blue-green algae, an organized nucleus is present in the cell. In the genus *Euglena* the flagella arise from a "gullet" at the anterior end of the cell. A light-sensitive red spot is present

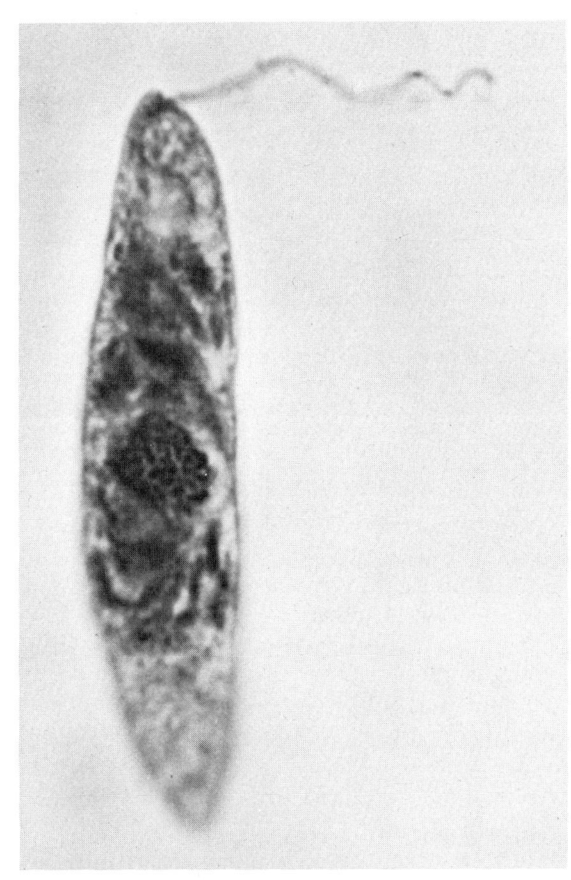

Fig. 4.2 Euglena, *a euglenophyte. Movement of the anterior flagellum propels the organism. Dark bodies surrounding the central nucleus are chlorophyll-bearing plastids. (Copyright, General Biological Supply House, Inc., Chicago.)*

near the "gullet." Some euglenoids actually take in solid food through the gullet, and thus resemble some simple animals. It has been suggested by some biologists that the euglenoids may be plantlike animals that have evolved chlorophyll and so become autotrophic. There are about 300 species in Euglenophyta.

The euglenoids are important members of the plankton in lakes and streams and thus constitute a part of the food of fish and other aquatic animals.

Chlorophyta. The **green algae** are found abundantly in fresh water and on moist soil, rock surfaces, and tree trunks. Many are found in the salt water of the seas.

Many species are unicellular, the cells occurring singly, or variously associated in colonies (Figs. 4.3 and 4.4), whereas in others the cells are arranged in filaments which may be attached to the bottom of shallow pools, or may be free floating (Fig. 4.5). In some species, the plant body is a leaflike sheet of cells from one to several cell layers in thickness (Fig. 21.3). The plants are mostly grassy green, the pigments in the chloroplasts being chlorophyll *a* and *b*, along with carotenoids. The chloroplasts are of a variety of shapes, bearing, in many species, protein bodies, the **pyrenoids** that are related to starch formation. The cellulose cell wall and the nucleus of the cell are similar to those of the higher plants. The green algae are generally believed to be the most likely source of higher green plants in the long course of plant evolution. Nearly 6000 species of green

Fig. 4.3 Four species of desmids. Note that each cell has a median constriction dividing it into two "semicells," the nucleus situated in the connecting isthmus. (Copyright, General Biological Supply House, Inc., Chicago.)

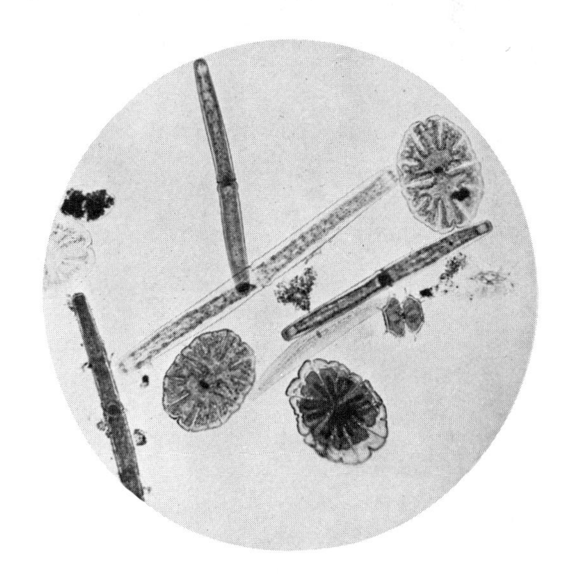

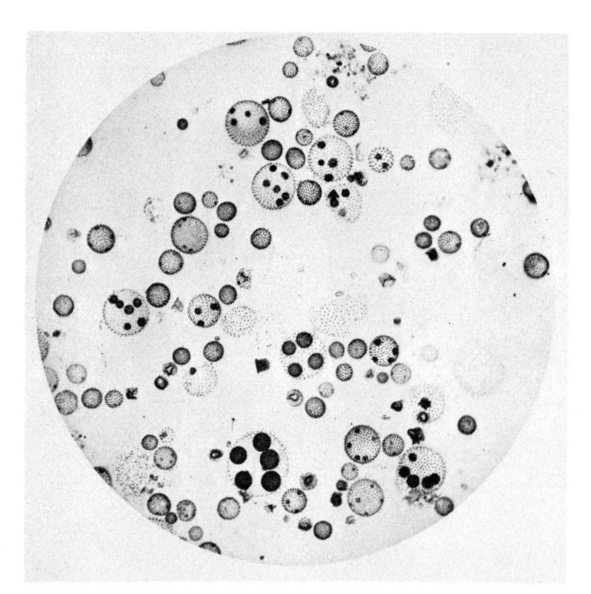

Fig. 4.4 A field of spherical colonies of Volvox, *a motile green alga. The larger colonies may consist of thousands of biflagellate cells held together in a one-layered hollow sphere by a gelatinous secretion. Young colonies are present inside several of the older ones. (Copyright, General Biological Supply House, Inc., Chicago.)*

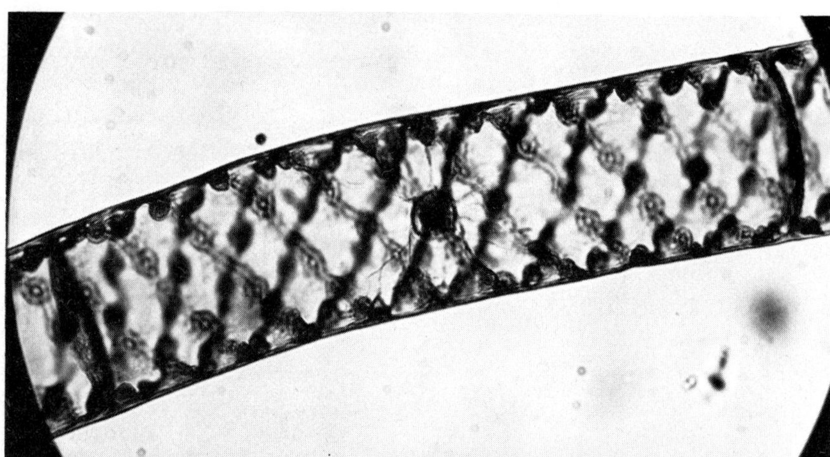

Fig. 4.5 A single cell from a long filament of a species of Spirogyra, *a free-floating filamentous chlorophyte. Note the helical, ribbonlike chloroplasts "beaded" with pyrenoids, and the central nucleus.*

algae are known.

The economic importance of chlorophytes relates to their abundance in the plankton. They are a primary source of food for fish and contribute to the purification of the water through the absorption of carbon dioxide and production of oxygen during the photosynthetic process. Extensive experimentation is under way to test the feasibility of large-scale artificial culture of certain species as a means of producing food for direct human consumption. In some agricultural areas devoted to the production of field crops, the extensive use of chemical fertilizers has resulted in the leaching of large quantities of soluble phosphates and nitrates into adjacent streams, lakes, and ponds. In response to the higher than normal concentration of these plant nutrients, algae have proliferated to the point of clogging the streams and interfering with the normal circulation of water in lakes and ponds, to the detriment of fish life.

Chrysophyta. The **yellow-green algae** and diatoms are predominantly freshwater

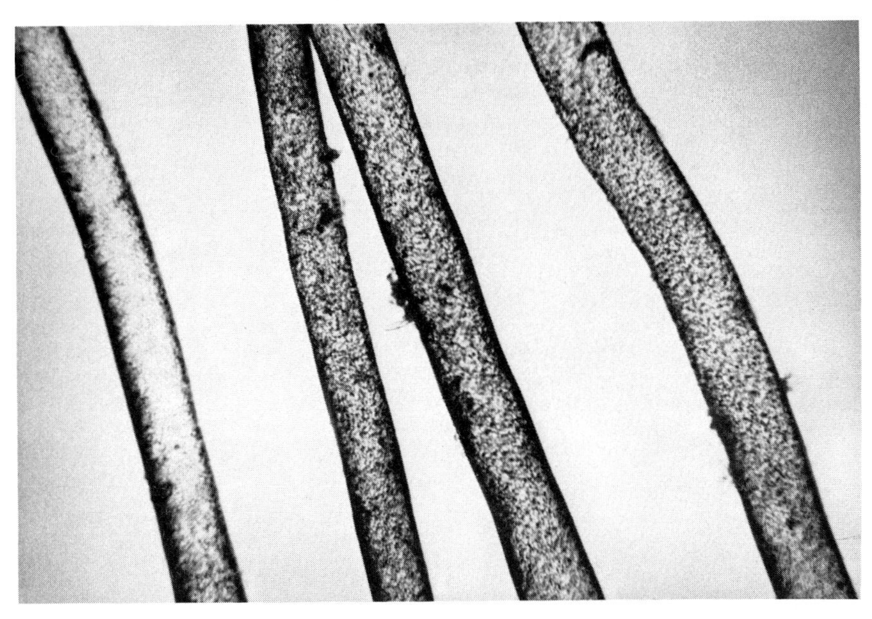

Fig. 4.6 Vaucheria, *a coenocytic chrysophyte. Note absence of cross walls and the many free nuclei. (Courtesy, Carolina Biological Supply Co.)*

plants although some occur in marine waters and on moist soil.

Most species of chrysophytes are unicellular, sometimes motile, or the cells may form variously shaped colonies. Filamentous species are either multicellular or the plant body is a continuous tubular structure with many nuclei and without cross walls **(coenocytic)** (Fig. 4.6). The cell walls are mostly pectinlike with little cellulose and in many species are impregnated with silica. The plants are yellowish green or golden brown in color. The pigments are contained in plastids and consist of chlorophyll *a* and, in some species, chlorophyll *c* or chlorophyll *e,* along with a high proportion of yellow or brown carotenoids. Starch is rarely present as a food reserve. There are more than 6000 species of chrysophytes, most of them diatoms (Figs. 4.7 and 4.8, see p. 47).

The siliceous-walled diatoms are important members of the marine plankton and thus are a food source for marine organisms. In several parts of the world, extensive deposits of **diatomaceous earth** have resulted from the accumulation of the siliceous walls of marine diatoms. Such areas were at one time covered by the sea and later raised by geologic action. Diatomaceous earth is quarried for commercial use in the manufacture of abrasives and polishes, insulation for high temperature apparatus, as filtering agents in oil and sugar refining, and as an absorbent in industrial explosives (Fig. 4.9, see p. 47).

Phaeophyta. The **brown algae** are almost entirely marine, growing attached to rocks along the shores of the cooler seas. Most of them grow in relatively shallow water, but some species occur in depths of almost 90 m. A few are free floating and occur in the open ocean. The famous Sargasso Sea of the South Atlantic Ocean is an extensive area inhabited by a free-floating species of *Sargassum.* Along the northern coasts, the leathery brown rockweeds are common, growing on rocks exposed at low tide (Fig. 4.10, see p. 47).

The plants are multicellular and range in size from microscopic branched filaments to very large, leathery, frondlike or bladelike forms often attaining 60 m in length. Attachment to rocks or to the ocean floor is accomplished by means of a branched **holdfast,** the broad blades being kept afloat by air-filled

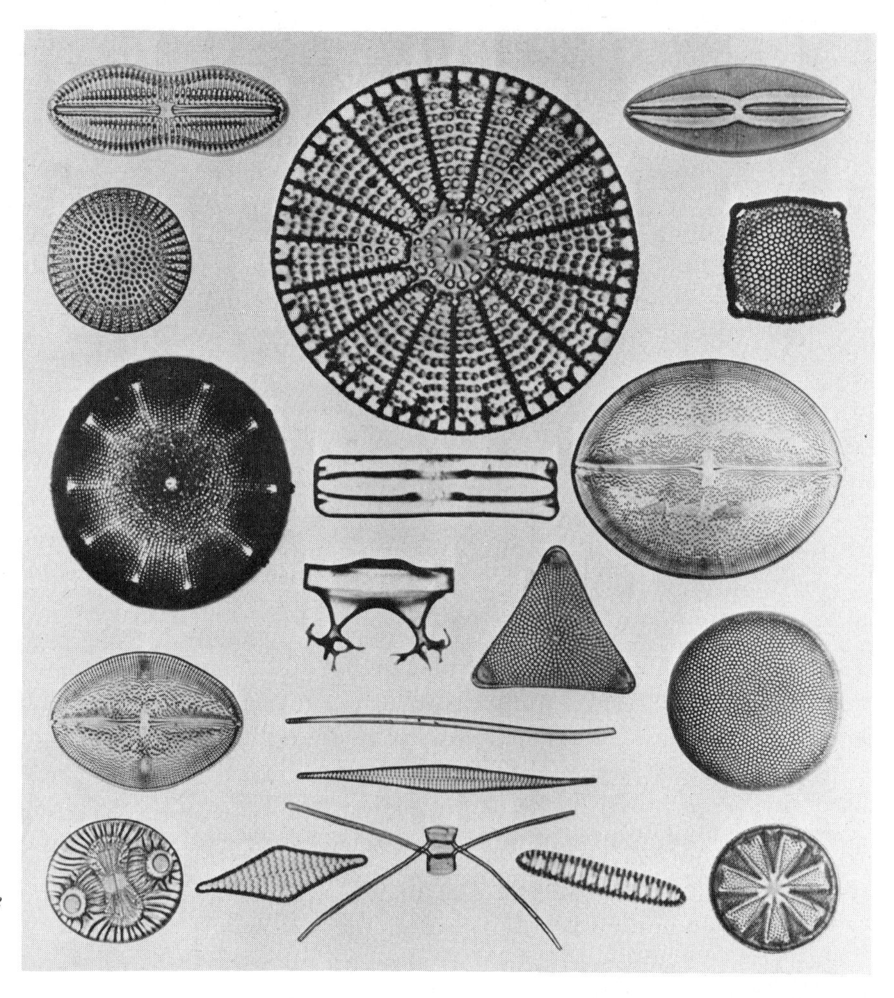

Fig. 4.7 A selection of species of diatoms (chrysophytes) showing the intricate patterns of their siliceous walls.

bladders (Fig. 4.11, see p. 48).

Some cellular differentiation occurs in the larger species. The cellulose cell walls are overlaid by a mucilaginous layer of **algin.** The characteristic brown color of the plants is produced by the presence of plastids containing the brown pigment **fucoxanthin** and other carotenoids along with chlorophylls *a* and *c*. Somewhat more than 1000 species of phaeophytes are known.

The larger species of brown algae serve as important articles of diet in Japan. Algin, extracted from the plants, is commercially valuable as a stabilizer and emulsifier in the drug and food industries.

Pyrrophyta. The **pyrrophytes** are chiefly flagellate, unicellular marine organisms, although a few species inhabit fresh water. Some species are filamentous and nonmotile.

The cells have yellowish green to golden brown plastids containing chlorophylls *a* and *c* and a special series of yellow pigments (xanthophylls). Cell walls may be absent, or present and cellulosic.

The dinoflagellates, represented by about 1000 species, constitute the largest class of pyrrophytes (Fig. 4.12, see p. 49). In the warmer seas the dinoflagellates are said to be the sec-

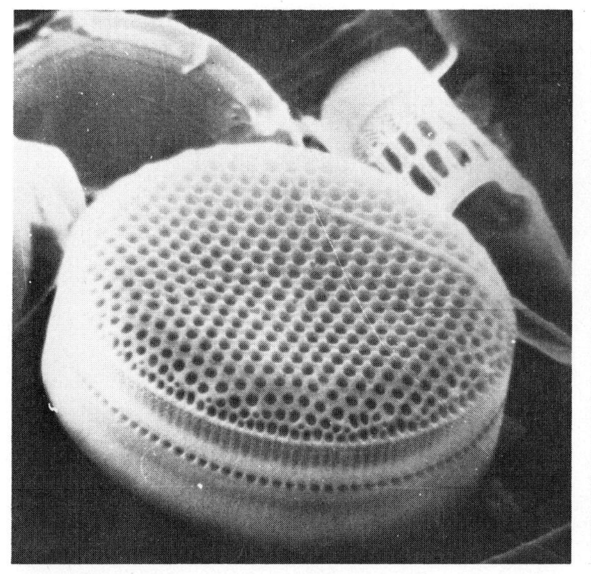

Fig. 4.8 (Top left) A scanning electron micrograph of a diatom species, showing sculpturing of siliceous wall. 2600X.

Fig. 4.9 (Bottom) Quarrying diatomaceous earth in the Lompoc Valley, California.

Fig. 4.10 (Top right) A brown alga, Ascophyllum, *growing on rocks in the intertidal zone. Note the bladders that keep the fronds afloat at high tide.*

ond (after diatoms) most important elements in the plankton, and are thus valuable as a food source for marine animals. Certain species of pyrrophytes are responsible for the occasional "red tides" experienced on the Florida coast and elsewhere.

Rhodophyta. The **red algae,** represented by about 3000 species, are predominantly marine plants of tropical and subtropical regions, usually growing attached to rocks in

Fig. 4.11 Brown algae. (Top) Nereocystis, *floating in a laboratory tank. Note the large floating bladder and the holdfast at the base of the stipe, by which it is attached to rocks along the shore. (Bottom)* Fucus vesiculosus. *Note small floating bladders on the fronds. (Top, courtesy, Carolina Biological Supply Co.)*

the intertidal zone or to the ocean floor at depths as great as 180 m. Certain species whose bodies become calcified contribute to the formation of the reefs of tropic seas.

Most red algae are multicellular. The plants are either flat sheets, ribbons, delicate cylinders, or they are finely subdivided and featherlike (Figs. 4.13 and 4.14). The plants are relatively small, rarely exceeding a few centimeters in length. The cells are commonly multinucleate and contain plastids bearing chlorophylls *a* and *d* and carotenoids along

Fig. 4.12 (*Top left*) Peridinium leonis, *an "armored" dinoflagellate. The cellulosic wall consists of a definite number of plates arranged in a specific pattern. Scanning electron micrograph, 1000X.*

Fig. 4.14 (*Top right*) Grinnellia, *a red alga. (Courtesy Carolina Biological Supply Co.)*

Fig. 4.13 (*Bottom*) Polysiphonia, *a red alga. (Copyright, General Biological Supply House, Inc., Chicago.)*

with a red pigment **phycoerythrin,** and sometimes a blue one, **phycocyanin.** In most species the red pigment is present in an amount sufficient to mask the other pigments, thus imparting to the plants the characteristic red color. **Floridean starch** is a unique form of reserve carbohydrate found in the cells of the red algae and, along with chlorophyll *d,* sets the red algae well apart from the other algae. The walls of the cells are composed of cellulose and mucilaginous substances.

Agar, extensively used as a culture medium in bacteriological and mycological techniques and as a stabilizer in food and drug manufacture, is extracted from species of *Gelidium* and other genera. Chocolate milk drinks are commonly stabilized with carrageenin, an emulsifying agent extracted from *Chondrus,* the "Irish Moss" of the North Atlantic coasts.

Fungi

The fungi, with a few exceptions, are heterotrophic plants, and thus, unlike the algae, are incapable of manufacturing their own foods. They are limited to substrates yielding organic foods. Some of them, called **parasites,** secure their food from the living bodies of other plants or animals, whereas others, called **saprophytes,** consume as food the dead tissues of other organisms or the organic products of other living plants or animals. Many species are found in waters and in moist soils rich in organic matter. Autotrophic members of this group, chiefly among the bacteria, for the most part obtain energy for the manufacture of organic compounds from the oxidation of inorganic substances, such as ferrous compounds, hydrogen sulfide, and ammonia. Such organisms are termed **chemosynthetic.** A few bacterial species utilize light as the energy source for organic construction, and are thus photosynthetic. Estimates of fungal species number run as high as 90,000.

The traditional grouping of Schizomycophyta, Myxomycophyta and Eumycophyta as "fungi" should, perhaps, not be construed as indicative of a close relationship among them. On evidence derived from studies of cell wall composition and organization of "nuclear" material within the cells, some workers believe the bacteria to be more closely allied with the blue-green algae and possibly ancestral to them. To reflect this, in some schemes of classification, the blue-green algae and bacteria are assigned the status of classes under a common division. Many organisms interpreted by other workers as links between the bacteria and true fungi are known.

The fungi exert a profound influence upon man, directly and indirectly. The special activities of many of these organisms are put to good use in the processing of many food items, such as tea, coffee, cheese, and vinegar, and in the production of important industrial chemicals, such as acetone, citric acid, and alcohols. Vitamin B_2 and many antibiotics are derived from fungi. The decomposition of organic matter through putrefaction and decay and the maintenance of soil fertility through nitrogen transformation are other beneficial actions. The details of some of these activities are discussed in later chapters. A few species of the so-called higher fungi, the mushrooms, are esteemed as foods, although they are not highly nutritious.

Schizomycophyta. Most species of **bacteria** are unicellular, occurring in rodlike **(bacillus),** spiral **(spirillum),** or spherical **(coccus)** forms, singly or in characteristic aggregations. The rodlike and spiral-shaped cells may possess numerous slender protoplasmic projections called flagella, whose wavelike or beating motion propels them through liquid medium (Fig. 4.15). Motility is attained by some bacilli that lack flagella by a twisting motion of the cell. Some species occur as simple or branched filaments resembling some of the simpler true fungi.

The cells of cocci rarely exceed 3μ in di-

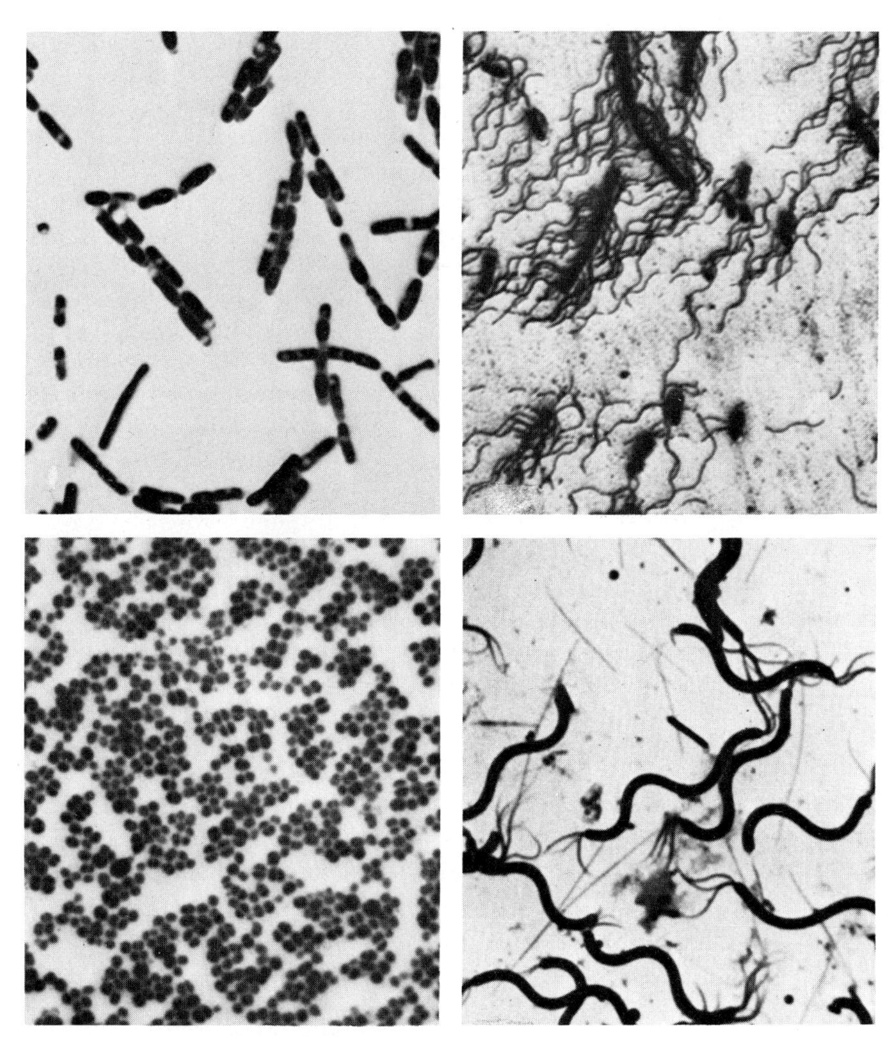

Fig. 4.15 Forms of bacteria. (*Top left*) Bacillus subtilis, *the hay bacillus.* (*Top right*) Salmonella typhosa, *a rodlike bacillus with flagella, which causes typhoid fever in man.* (*Bottom left*) Staphylococcus aureus, *a coccus form which causes boils.* (*Bottom right*) Spirillum volutans, *a spiral form found in marsh water.* (*Copyright, General Biological Supply House, Inc., Chicago.*)

ameter, while bacilli and spirilla, of similar width, are usually much longer. The cell wall of apparently the majority of bacterial species is composed of proteins, related polypeptides and lipids, along with certain amine-substituted hexose sugars related to chitin. The cells are often encapsulated, singly or in groups by a slimy gelatinous substance usually of a polysaccharide character. In a few species the chief component of the cell wall is cellulose. Organized plastids and nucleus similar to that of the higher plants are absent, although the nuclear material appears to be present in a dispersed condition. In some species the nuclear material is clumped in the center of the cell, so as to resemble a simple, and possibly primitive, kind of nucleus.

The extremely small size of bacteria coupled with a narrow range of form differences makes it difficult or impossible to distinguish among many species on the basis of morphological characters alone. Physiological characteristics often differ widely among morphologically similar species, and thus the bacteriologist makes extensive use of such characteristics in identification. Some species, in general the common saprophytic ones responsible for decay, are quite catholic in their tastes, whereas others often of a parasitic and pathogenic nature are highly specific, utilizing only certain organic compounds, parasitizing only certain host species or even certain specific tissues of the host. Whereas most bacterial species are dependent upon organic substrates, a few species are autotrophic. The green sulfur bacteria and the purple sulfur bacteria are *photosynthetic,* deriving their energy for the construction of organic food from light absorbed through the agency of special bacterial pigments. Others are *chemosynthetic,* deriving their energy from the oxidation of inorganic materials such as ammonia, nitrates, ferrous iron, and hydrogen sulfide.

The more than one thousand species of bacteria are very widely distributed, occurring wherever suitable substrate and physical environment permit. Some of them possess a remarkable tolerance to extremes of environmental factors well beyond the tolerance of most other organisms. In the general cycle of nature the bacteria play a significant role. Their participation in the **nitrogen cycle,** by which this essential element is made available to green plants through the processes of decay, ammonification, nitrification, and nitrogen fixation probably far overshadows their importance as the agents of destructive diseases of man and his agricultural crops and domesticated animals. Our use of bacteria as the means for certain kinds of important industrial processing has been alluded to in Chapter 2.

Myxomycophyta. The **slime molds** occur on fallen decaying leaves, old tree stumps, and rotting wood, in moist and shady situations, and on moist soil containing organic matter.

The vegetative body, called a **plasmodium,** consists of a multinucleate mass of protoplasm bounded by a very thin and flexible membrane (Fig. 4.16). The plasmodium may be colorless, yellow, red, or purplish and may attain an area of several square centimeters. By a creeping and flowing amoeboid movement, the plasmodium moves about on the surface of the substrate, engulfing bacteria and other solid food particles which become enclosed within vacuoles formed by the invagination of the exterior membrane. After a period of vegetative activity, the plasmodium mounds up to form one or several commonly stalked **sporangia** of distinctive form and frequently brightly colored (Figs. 4.16 and 4.17). Spores released from the sporangium give rise ultimately to the plasmodium (Fig. 4.16).

A few parasitic species of slime molds are of economic importance, causing such diseases of crop plants as clubroot of cabbage and powdery scab of potato. About 500 species of slime molds are known.

Eumycophyta. The **true fungi** are parasitic or saprophytic plants growing on moist substrates yielding organic food. Like bacteria, some species have a high specificity toward food sources.

The plant body of most species consists of single or variously massed filaments. The fungal filaments, called **hyphae,** are usually much branched, often forming extensive interwoven or felted systems that cover or pervade the substrate. A mass of hyphal branches constitute the **mycelium.** Specialized absorbing hyphae called **haustoria** and anchoring hyphae called **rhizoids** that penetrate the substrate are characteristic of many species. A hypha may be a single, elongated, multinucleate cell or a multicellular filament. Some species are unicellular, with the cell of more usual proportions. Cell walls of the eumycophytes are sometimes of cellulose but more commonly of a **chitin** resembling that found in the bodies of insects.

The true fungi obtain their food supply from the organic substrate through the digestive action of enzymes that are produced within the fungus but diffuse outward to act on the food. The digested and dissolved food is

Fig. 4.16 *(Top) Plasmodium of the slime mold* Stemonitis fusca *photographed one hour before fruiting. (Bottom) Fruiting bodies (sporangia) of* Stemonitis. *(Bottom, copyright, General Biological Supply House, Inc., Chicago.)*

Fig. 4.17 Sporangia of slime molds. (*Top*) Diachaea leucopoda. (*Courtesy, Carolina Biological Supply Co.*) (*Bottom*) *A species of* Arcyria. (*Copyright, General Biological Supply House, Inc., Chicago.*)

then absorbed by the fungal cells.

One class of eumycophytes called **Phycomycetes,** or algalike fungi, includes about 1200 species, ranging from small unicellular forms to ones whose bodies consist of branched, multinucleate hyphae without cross walls **(non-septate hyphae).** A number of these are parasitic on crop plants, fish, and insects. *Phytophthera infestans,* a phycomycete, is the cause of the late blight of potatoes, and was responsible for the disastrous crop failure in Europe in 1845 and a subsequent population shift of significant proportions. Some parasitic forms such as the **downy mildews** are the cause of the downy mildew of grapes. Without soil sterilization or the use of fungicides, seedlings are often killed by "damping off" because of species of *Pythium* in the soil. The **water molds** are commonly seen, appearing as cottony tufts, growing parasitically on fish and insects, or as saprophytes on lifeless animal and plant debris. The well-known blackmold, *Rhizopus,* is common on moist bread and overripe fruit (Fig. 4.18).

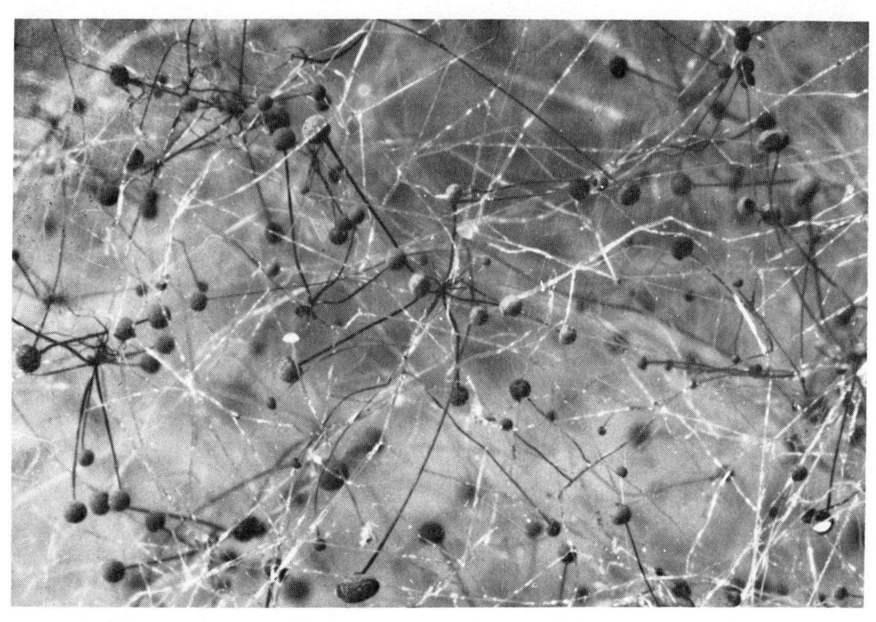

Fig. 4.18 Hyphae and sporangia of the black mold, Rhizopus, *from a culture grown on moist bread. (Courtesy, Carolina Biological Supply Co.)*

Fig. 4.19 (Bottom) A group of yeast cells, Saccharomyces, *an ascomycete. Two of the cells have produced young yeast cells by "budding."*

A second class, the **Ascomycetes,** or sac fungi, containing about 15,000 species, are structurally quite diverse. The plant body may consist of a single cell, as in yeast (Fig. 4.19), extensive loose masses of separate **hyphae,** the segments containing from few to many nuclei, as in *Penicillium* (Fig. 4.20), *Aspergillus,* and other molds. The cell walls are chiefly chitinous. In many of the larger forms, the fruiting body may be a conspicuous and elaborate structure as in the cup fungus, *Peziza* (Fig. 4.21), and the morel, *Morchella* (Fig. 4.22), common on forest litter. The name sac fungus is derived from the fact that spores which form a part of the sexual reproductive cycle are produced in small saclike sporangia **(asci),** four or eight per ascus. Some of them reproduce asexually by vegetative **budding** (Fig. 4.19), or by the formation of clusters of terminal spores called **conidia** at the tips of the hyphae (Fig. 4.20).

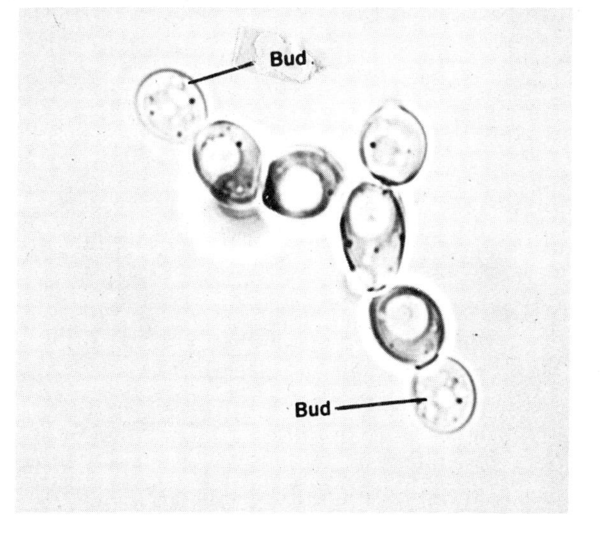

Some species of yeast are of great economic importance, as in the baking and brewing industries and in the commercial production of alcohol, whereas others are responsible

Fig. 4.20 Penicillium, *an ascomycete. (Top) Photograph of a glass model of* Penicillium notatum *as it appears in a laboratory culture. The plant body of mycelium is a loose mat of septate hyphae. Conidia, borne in chains at the tips of certain hyphae become detached and germinate under favorable conditions, giving rise to new mycelia. (Bottom) photomicrograph of a section through host tissue parasitized by a species of* Penicillium. *(Bottom, courtesy, Carolina Biological Supply Co.)*

for certain diseases in man and various crop plants. The parasitic organisms responsible for chestnut blight and Dutch elm disease, referred to in Chapter 2, are ascomycetes, as are the saprophytic blue and green molds found growing on citrus fruits, bread, other foodstuffs, and leather. Species of *Penicillium* are important in the culture of certain types of blue cheese and in the production of the antibiotic, penicillin (Figs. 4.20 and 2.10).

Many higher plants are subject to attack by **powdery mildews,** and cereal crops may be damaged by several species of *Claviceps* which invade the young developing grain, replacing it with **ergot,** a hard mycelial mass called a **sclerotium.** The sclerotium, an overwintering device in the life history of the fungus, contains several toxic principles. Flour milled from heavily ergotized grain has been the cause of many deaths in times past in years of

Fig. 4.21 Cup fungus, Peziza, *an ascomycete. The asci with ascospores arise on the inner surface of the cup. (Courtesy, Carolina Biological Supply Co.)*

Fig. 4.22 The common, edible and highly prized morel, Morchella, *an ascomycete, photographed in its natural habitat on moist hardwood forest soil. (Courtesy, Carolina Biological Supply Co.)*

heavy infestation. Some species, such as the morel and the subterranean truffle, are highly regarded as gastronomic delicacies in some circles.

The **Basidiomycetes,** or club fungi are a third important class of the true fungi. About 15,000 species are included in the class. The mycelium usually consists of an irregular mass of septate hyphae that penetrate the substrate, either living host tissue or lifeless organic matter. The **basidiospores,** produced as part of the sexual reproductive cycle, are typically borne in groups of four on a club-shaped hyphal tip, the **basidium.** In the parasitic smuts (Fig. 2.15) and rusts, which are responsible for enormous losses in grains and other crops such as pine, cherries, and beans, masses of spores burst out of the host tissues. The saprophytic mushrooms and toadstools produce a conspicuous **sporophore** or **sporocarp** consisting of compact hyphae. The spore-bearing basidia are borne on delicate platelike gills or on the walls of tiny pores on the underside of the

Fig. 4.23 A poisonous mushroom, Amanita phalloides, *called the "Death Angel." Basidia with basidiospores are borne on the delicate gills of the cap. The mushroom arises from a subterranean mycelium.*

Fig. 4.24 Agaricus campestris, *the commercial edible mushroom. (Top) Young "buttons" growing on commercial substrate. The web-like mycelium is shown at right. (Bottom) Mushrooms ready for harvest.*

fleshy cap (Figs. 4.23, 4.24, 4.25*a, b,* and 4.26). In the familiar puffballs, the basidia are produced inside a closed, globe-shaped sporocarp, the spores escaping through a rupture or pore (Fig. 4.27). Some species also produce conidia. In the rusts, particularly, these may be highly specialized and of more than one type. Black stem rust of wheat, white pine blister rust, and cedar-apple rust are examples of the phenomenon of alternate host parasites among the basidiomycetes. These organisms complete certain stages of their life cycles in one host and other stages of their life cycles in another host (Chapter 17).

The cell walls of the septate hyphae are chitinous. The hyphal segments of the fruiting body are generally binucleate.

Many species of mushrooms are edible and considered to be delicacies, each with its own distinctive flavor, but others are virulently poisonous. *Agaricus campestris,* the common field mushroom, is the species most generally eaten. It is extensively cultivated and appears in the market canned and in the fresh condition seasonally. No simple test or rule of thumb is available to the nonexpert for distinguishing the poisonous from the edible mushrooms. Only the expert, therefore, should be en-

Fig. 4.25 Common species of basidiomycetes. (a) Cantharellus cinnabarinus, *edible.* (b) Clitocybe illudens, *the Jack-o-lantern mushroom, a phosphorescent species growing in characteristic dense colonies at the base of oak trees.* (c) Clavaria pyxidata, *the cup-tip coral mushroom.*

Fig. 4.26 A pore fungus growing on tree bark. The tough and woody sporophores arise from a mycelium growing in the bark. Basidia with basidiospores are borne on the lining of tiny tubes on the underside of the cap.

trusted with the collection of wild mushrooms for human consumption.

A fourth group of fungi, called the **Imperfect Fungi,** consists of a large number of species whose life histories are incompletely known. The discovery and recognition of the missing sexual reproductive stages of the life cycles of many of these forms will permit their ultimate assignment to the proper class. Thus the group, Imperfect Fungi, is a kind of temporary assignment. Through further study many of these forms will be removed from the group, but many others have doubtless completely lost their sexual stages through evolu-

Fig. 4.27 Mature sporocarps of the puffball, Lycoperdon, *growing on rotting wood. The wall of the sporocarp is very thin; contact with passing animals or even raindrops causes the puffball to act like a bellows, forcing the basidiospores out through the apical pore.*

tionary specialization. Although assignment to one of the three main classes of fungi is uncertain, most of the Imperfect Fungi are believed to be ascomycetes. Many forms are destructive parasites on certain crop plants, whereas others are responsible for serious diseases of man, such as ringworm, athlete's foot, and some pulmonary infections.

Lichens. Lichens constitute a special group of thallophytes, for they are composite plants consisting of a symbiotic association of a fungus and an alga. The algal member is usually a blue-green or green alga and the associated fungus is most commonly an ascomycete, although a few lichens have a basidiomycete component. The plants are common on the trunks of trees, on rocks, and occasionally on moist soil. Lichens occur as dry crusty patches **(crustose),** leaflike scales **(foliose),** or as erect, much branched tufts **(fruticose)** (Fig. 4.28 and 4.29). They may be gray-green, yellow-orange, white, black, or brown.

Although nominally considered as a class of eumycophytes, the species of lichens may more properly be assigned to other classes of fungi. The fungal component generally dominates the vegetative and reproductive characteristics of the lichens. The hyphae of the fungus surround the algal cells and receive nourishment from them (Chapter 17). For most of the lichen fungi this relationship is an obligatory one, for they do not complete their life cycle when grown apart from the alga. The algal member, however, is capable of independent existence. In most lichens the fungus completes a characteristic sexual reproductive cycle, the resulting spores yielding mycelia that entrap new algal cells. Many lichens, however, reproduce asexually by the formation of small clusters of algal and fungal cells, called **soredia** (Chapter 20). Detached from the parent plant, the soredia are distributed by air currents, fall upon suitable terrain and grow directly into a new lichen plant.

Lichens are of limited direct economic value. Certain species yield blue, yellow, or brown dyestuffs. Litmus, an indicator dye employed in chemical work to determine relative acidity and alkalinity, is extracted from certain species. On the arctic tundra, lichens serve as forage for caribou and reindeer. In some regions, lichens are used in limited amounts as food for man. About 15,000 species of lichens are known.

BRYOPHYTES

Whereas the thallophytes, as we have seen, comprise a large assemblage of species that fall into several rather diverse and probably remotely related groups, the bryophytes constitute a single, more homogeneous division. The

Fig. 4.28 A foliose or leafy lichen growing on bark of tree. (Right) A close-up. (Copyright, General Biological Supply House, Inc., Chicago.)

Fig. 4.29 Cladonia rangiferina, *the reindeer moss, a fruticose or shrubby lichen, growing on forest soil. (Courtesy, Carolina Biological Supply Co.)*

most familiar representatives of the division are the **mosses** (Class **Musci**) that grow in green cushionlike patches on shady stream banks and in moist woods. Mosses are usually rather small, seldom exceeding a few centimeters in height. They seem more like familiar plants than most of the thallophytes, for the plants have an elongate stemlike axis bearing many small, delicate, green, cushiony structures and are anchored in the soil by a branched rootlike system of **rhizoids** (Fig. 4.30). Mosses are generally found only in moist places, or where they may be subject to frequent wetting. They have no true absorbing roots and no special conducting tissue in their small "stems" and "leaves." Absorption of water can take place through all the plant's surfaces. Thus, although the mosses externally tend to resemble the more familiar higher plants, they are much simpler in construction. The vegetative body of the **liverworts** (Class **Hepaticae**) is commonly a somewhat fleshy, leaflike **thallus** growing prostrate on the moist soil (Figs. 4.31, 4.32, and 4.33). Some of these with slender elongated axes and delicate cush-

Fig. 4.30 (Left) A colony of the moss Funaria. *(Right) Drawing of a single fruiting moss plant from a cushionlike colony.*

Fig. 4.31 Marchantia, *a liverwort. (Below) Male plants at the center, female at left. (Bottom) Thalli of* Marchantia *bearing gemmae cups containing budlike gemmae.*

iony appendages resemble prostrate mosses (Fig. 4.32). As in the mosses, no specialized conductive tissue or absorbing structures are present. A very small third class, the **Anthocerotae** or **horned liverworts** (Fig. 4.33), should be mentioned here chiefly because the spore-producing phase of the plant possesses a persistent **meristem** (zone of new cell formation) and specialized aerating tissues similar to those found in the vascular plants.

Most bryophytes reproduce sexually. The two phases that characterize all sexual cycles are particularly clearly marked. The reproductive organs producing the sex cells (gametes) are more complex than those of most of the thallophytes, being multicellular and invested by a jacket of sterile cells. Following fusion of the gametes, a multicellular **embryo** is formed which is nourished for a time by the parent plant. A detailed discussion of the sexual process will be provided in Chapter 21. Some species reproduce asexually by fragmentation or by the production of budlike offshoots called

gemmae (Fig. 4.31), but not by means of asexual spores, such as the conidia of some fungi.

The bryophytes, represented by a little over 20,000 species, are a small group compared with the thallophytes and tracheophytes. They are of little direct economic importance. As early colonizers on bare rocky surfaces, by mechanical action and the deposit of organic matter they slowly build up a fertile soil layer that may support other plants. Peat deposits are to a considerable extent the accumulated, partially decomposed remains of mosses.

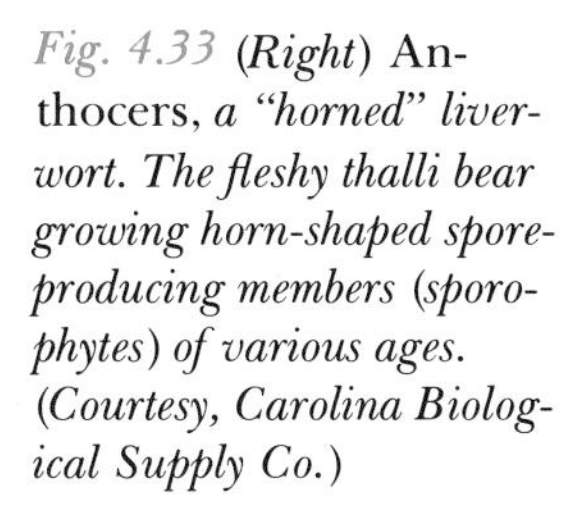

Fig. 4.32 Calopogeia, *a "leafy" liverwort. (Courtesy, Carolina Biological Supply Co.)*

Fig. 4.33 (Right) Anthocers, *a "horned" liverwort. The fleshy thalli bear growing horn-shaped spore-producing members (sporophytes) of various ages. (Courtesy, Carolina Biological Supply Co.)*

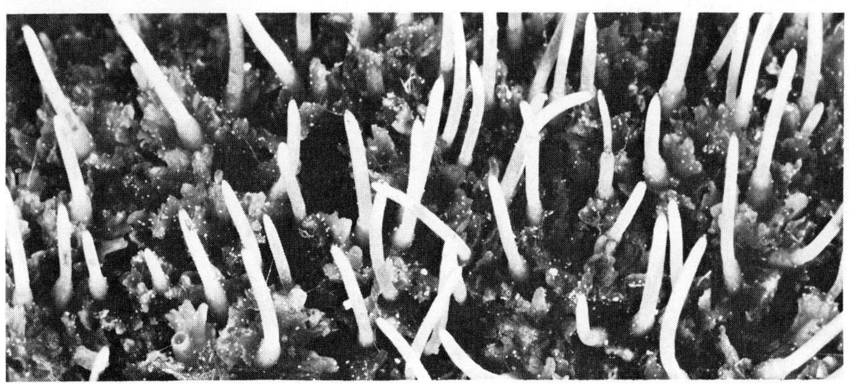

RELATED READING

Echlin, P. "The blue-green algae," *Scientific American, 214*(6), 74–81, June 1966.

Emerson, R. "Molds and men," *Scientific American, 186*(1), 28–32, January 1952.

Harris, J. E. "Diatoms serve modern man," *Natural History, 65,* 64–71, 1956.

Hunter, S. H. and J. J. A. McLaughlin. "The red tide," *Scientific American, 199*(2), 92–98, August 1958.

Lamb, I. M. "Lichens," *Scientific American, 201*(4), 144–156, October 1959.

North, W. J. and B. Littlehales. "Giant kelp: sequoias of the sea," *National Geographic, 142,* 251–268, 1972.

Sokolov, R. "A moral tale," *Natural History, 83*(5), 38–45, 1974.

Zahl, P. A. "The bizzare world of fungi," *National Geographic, 128,* 502–527, 1965.

Zahl, P. A. "Algae: the life givers." *National Geographic, 145,* 361–377, 1974.

5
VASCULAR PLANTS

Tracheophytes are the true land plants, for they are well differentiated into specialized organs which enable most of them to occupy dry land effectively. The **roots** not only fix the plant firmly in the soil but also serve as effective *absorbing* structures capable of deep penetration into the soil whence the supply of water and minerals comes. The **stems** and **leaves,** as well as the roots, have well-developed **vascular tissues** through which substances are transported from one part of the plant to another. The specialized nature of the cells of the vascular tissues and their arrangement within the stem enable the stem to stand erect (particularly in woody plants) and, as in trees, to support tremendous weight and attain large size. The leaves are the principal site of food manufacture. The plant is protected against excessive loss of water, and in some instances against mechanical injury, by the presence of specialized protective tissues on outer surfaces. Tracheophytes reproduce sexually and also by vegetative means (Chapters 20 and 21). The reproductive structures are generally borne upon the ends of stems and branches or upon leaves in a position favorable to the dispersal of spores or seeds. The spore-producing phase of the life cycle is the dominant one, consisting of an axis with a meristem at each end. Through the action of the meristems in producing new cells the axis is lengthened and roots, stems, and leaves are formed. Long continued activity of these terminal meristems accounts for the large size and long life of many species. For the most part, the gamete-producing phase is small, inconspicuous, and short lived. In most of the species, the fusion of gametes is not dependent upon physical wetness.

The tracheophytes, numbering somewhat more than 220,000 species, are the dominant plant type in our modern flora. Botanists are generally agreed that the vascular plants had a common origin and that the divisions of tracheophytes represent major divergent lines of development from a remote ancestral vascular form. A most remarkable consistency prevails in details of form and organization of the unique vascular tissues of tracheophytes. The ultimate origin of vascular plants from green algae is generally accepted. The fossil record of the tracheophytes, although only fragmentary, is extensive, and from it we have gained considerable insight into the course of evolution of the group. We owe this useful record of ancient forms largely to the qualities of the cells that compose the vascular tissue, as well as to the size of the organisms.

Psilophyta

The Psilophyta are represented in the modern flora by only two genera and a few species, mainly tropical in distribution. Several fossil genera are known from Devonian or Silurian sedimentary rocks over 300 million years old (Chapter 6). The plant body consists of a forked-branched, erect vascular stem arising from a horizontal underground stem, or **rhizome.** Typical roots are absent and the functions of absorption and anchoring are served by the branching system of rhizomes. All the plants are small, seldom exceeding several centimeters in length, and some of them are epiphytic, that is, they grow attached to the surfaces of larger plants, such as trees, but are not parasitic upon them. The photosynthetic requirements are met by the green stem; the leaves are usually very small and without vascular tissue or only scantily provided with it. In the extinct species, spores were produced in sporangia borne at the tips of the major branches of the stem; in the modern genus *Psilotum* sporangia are borne at the tips of very short lateral branches. Although of no economic value, the extant psilophytes are of interest for they are widely regarded as the modern representatives of the earliest and most primitive tracheophytes known. At least they are very simple vascular plants and, as might be expected, they are the subject of much discussion concerning their relationship to the earliest known fossil representatives and to the other divisions of tracheophytes (Fig. 5.1).

Lycophyta

The lycophytes, known generally as the **club mosses,** are represented in our modern flora by about 1000 species, mainly of the genera *Lycopodium* (Figs. 5.2 and 5.3) and *Selaginella* (Fig. 5.4). Most species are inhabitants of moderately moist situations in forest shade. The main stems of many species trail along the ground surface, often attaining a length of several meters, while others are shallowly subterranean and shorter. Lateral ascending branches usually rise to a height of two or three decimeters. The stems are usually densely clothed with small scalelike or awl-shaped leaves containing a single unbranched central vein. The roots arise from the underside of the horizontal stem in a random fashion, throughout its length, the plants often forming, thus, a dense, well-anchored mat. Some species are adapted to extreme seasonal dryness, as the "resurrection plant," a species of *Selaginella* that rolls up into a tight dry ball in times of drought. Sporangia containing spores are usually borne in terminal cones or **strobiles** at the tips of branches, each

Fig. 5.1 (*Top*) Psilotum, *a leafless and rootless psilophyte growing on dead tree stump.* (*Bottom*) *Close-up of terminal branches bearing sporangia.* (*Courtesy, Carolina Biological Supply Co.*)

Fig. 5.2 (*Top*) Lycopodium obscurum *growing in an open woodland habitat. Note the terminal strobili consisting of scalelike spore-bearing leaves.*

Fig. 5.3 (*Bottom*) *Aerial stems of three common species of* Lycopodium, *club mosses.* (*Left to right*) L. lucidulum, L. alopecuroides, L. glabelliforme. *In* L. lucidulum *the sporangia are borne in fertile zones along the stem; in the others they are aggregated into terminal strobili or cones.* (*Courtesy, Carolina Biological Supply Co.*)

sporangium upon or in close association with a leaf. A few species of *Lycopodium* also reproduce vegetatively by the formation of detachable budlike structures.

The lycophytes are an ancient group with a fairly continuous fossil record running back to Devonian and Silurian times. Whereas the modern species are all small plants, the stems of many extinct species attained truly arborescent proportions through the action of a persistent lateral meristem which continued through years to add additional wood, in the manner of our modern trees. Indeed, judging by the fossil record, these arborescent lycophytes were the dominant members of the late Paleozoic forests (Table 6.1). Another feature of special interest was the development of a true seed by many extinct members of the Carboniferous period. Since the lycophytes constitute a distinct division separate from that containing the modern seed plants, it follows that seed formation arose independently in the two evolutionary lines. Prerequisite to seed formation is the production of two kinds of spores, sexually differentiated. This condition is called **heterospory.** About half of the living species of lycophytes, of the genera *Selaginella* and *Isoetes,* are heterosporous. However, no seeds are produced by living lycophytes.

Fig. 5.4 Selaginella apus, *a small club moss. (Courtesy, Carolina Biological Supply Co.)*

Modern club mosses are of little or no economic importance. The extinct members, however, as dominant elements of the Carboniferous forests, were major contributors to the accumulated plant residues which became coal.

Sphenophyta

The sphenophytes are represented in the modern flora only by about 25 species of the genus *Equisetum,* the **horsetails.** They occur chiefly in wet soils or in shallow water at the margins of ponds, although a few are adapted to dry habitats. The plants are of distinctive aspect, the greenish, conspicuously jointed stems rising up to six decimeters from a branched subterranean rhizome. The stems, usually fairly firm, are longitudinally ridged, rarely exceed a centimeter in diameter and bear whorls of small, scalelike, nongreen leaves at the joints or nodes. The walls of the epidermal cells of the stems of most species have deposits of silica and are usually coarse and rough to the touch. Compressed into pads, the rough, silica-laden stems are sometimes used for scouring kitchen utensils, and this use has given rise to the common name, "scouring rushes," often applied to these plants. The

Fig. 5.5 Equisetum telmatei, *a sphenophyte, showing both vegetative (in background) and reproductive shoots. The reproductive shoots are about 25 centimeters tall. Note the whorled arrangement of small leaves and the jointed stems. (Courtesy, Carolina Biological Supply Co.)*

spores are produced in sporangia borne in whorls on highly modified lateral branches arranged in compact terminal strobiles or cones (Fig. 5.5).

The fossil record of the sphenophytes may extend back to the Devonian period (Table 6.1) when conspicuously jointed-stemmed forms, tentatively assigned to this division, were contemporaneous with the early psilophytes and lycophytes. The most extensive development of the sphenophytes occurred in the coal-forming period when some species attained the stature of trees. The modern *Equisetum* is interpreted as a much reduced remnant of an earlier, more robust form.

Pterophyta

The last division, Pterophyta, is the largest, most diversified, and economically the most directly important, yielding most of the useful products of field, forest, and grassland. The pterophytes, making up more than two-thirds of the Plant Kingdom, are the dominant members of the flora over most of the land areas of the world. The plant body, in general, is well differentiated into roots, stems, and leaves, all containing well-developed vascular tissues. Through the action of persistent terminal meristems and special lateral ones, many species continue to grow for many years and may attain very large size. The spore-producing phase of the life cycle is always dominant, the gamete-producing phase small and ephemeral.

The division comprises three classes: (1) Filicinae or ferns, (2) Gymnospermae, represented by the pines, hemlocks, spruces and other cone-bearing trees, the cycads, and *Ginkgo,* and (3) Angiospermae or flowering plants.

Filicinae. There are somewhat more than 6000 species of modern ferns. Most of the familiar ones have short and usually horizontal subterranean stems. Typically the leaves (the familiar fronds) are quite large and extensively veined and arise seasonally in clusters at the ends of the perennial branched stem (Figs. 5.6 and 5.7). The leaves of former seasons wither and decay. A few species, the "tree ferns," have

Fig. 5.6 Ferns growing on moist floor of spruce forest.

erect stems that may attain the stature of small trees. The stems bear the scars of fallen leaves of former seasons (Fig. 5.8). Spores are borne in sporangia arranged in clusters on the underside of the leaves (Fig. 5.9). Except in a few species that are aquatic, the spores are all alike. The ferns are the modern representatives of an ancient race of vascular plants. Certain fossil species contemporary with the early psilophytes of the Devonian period are the probable ancestral forms from which the ferns were derived. During the great coal-forming period, ferns were a conspicuous element of the flora. Certain fernlike plants of that time produced seeds, but these properly are placed in the class Gymnospermae. The modern ferns, however, do not produce seeds. Their inclusion in the division Pterophyta along with the seed plants (gymnosperms and angiosperms) is based primarily upon anatomical features relating to the nature of the complex multiveined leaf. Although predominantly terrestrial plants, it should be pointed out that they have retained certain ancient aquatic habits of reproduction that explain, in large measure, their general restriction to moist habitats.

Gymnospermae. The classes Gymnospermae and Angiospermae together constitute the **Seed Plants.** Excluded from this grouping, of course, are the fossil seed-bearing lycophytes previously mentioned, whose seeds differ in certain developmental details from those of the gymnosperms and angiosperms, and

Fig. 5.7 Young expanding leaves of the Cinnamon Fern, Osmunda. *"Fiddleheads" are characteristic of most ferns.*

Fig. 5.8 Tree ferns in Australian eucalyptus forest. Specimen in foreground is about 7.5 meters tall. Most modern ferns have short and often subterranean stems.

Fig. 5.9 Polypodium, *showing underside of leaflets bearing clusters (sori) of sporangia. (Copyright, General Biological Supply House, Inc., Chicago.)*

whose stem and leaf structure differs from that of the pteriophytes.

In the gymnosperms the seeds are typically produced in an exposed manner (*gymnos,* naked; *sperma,* seed) on leaves that are commonly highly modified and often aggregated into a cone or strobile. The view is almost generally held among botanists that two separately evolving lines of development may be recognized within the class. One of these, sometimes called **cycadophytes,** includes the modern cycads (Fig. 5.10) and their supposed ancestors, the seed-ferns **(pteridosperms)** of late Paleozoic age. They have retained several features of fernlike aspect, conspicuously, the large pinnately divided leaves and a stem with a large pith and a relatively thin layer of wood. Modern cycads, numbering about 60 species are distributed in tropical and subtropical regions. They are of little economic importance, although their ancestors, the pteridosperms, were major contributors to the coal deposits throughout the world.

The second line, often referred to as **coniferophytes,** includes such familiar modern cone-bearers as the pines, hemlocks, firs, redwoods, and many others, as well as *Ginkgo.* Included also is a large group of fossil forms, typified by the genus *Cordaites* of Carboniferous age. These and the modern coni-

Fig. 5.11 (Top) A gymnosperm, Douglas fir, Pseudotsuga taxifolia, *in Colorado. Note the tall spirelike crown, typical of most conifers.*

Fig. 5.10 (Bottom) A cycad, Cycas circinalis, *a primitive gymnosperm with male cone, growing in cultivation. Male and female cones are borne on separate plants.*

ferophytes differ strikingly from the cycadophytes in having undivided, usually small leaves, a thick, woody, much branched stem capable of massive secondary thickening and a small pith. These features, so different from those of the ferns and cycadophytes, are commonly given as grounds for excluding the ferns and seed-ferns from the ancestral line of the coniferophytes. Their ultimate origins possibly lie among the early Devonian psilophytes, but the manner of transition is uncertain. The coniferophytes, represented in our modern flora by about 450 species are of considerable economic importance. Such genera as *Pinus* (pines), *Abies* (firs), *Pseudotsuga* (Douglas firs) (Fig. 5.11), and *Sequoia* (redwoods) yield the great bulk of structural timber in the northern

hemisphere, while several species of *Araucaria* serve similarly in the southern hemisphere. The woods of certain genera have special properties that make them valuable for special uses, as in packing boxes, paper pulp, and some furniture. *Ginkgo biloba,* the maidenhair tree (Fig. 5.12), is the sole survivor of an ancient line with a fossil record going back to the Permian period. A native of China, *Ginkgo* is widely planted as an ornamental tree in parks and along city streets where it seems quite tolerant of the noxious dirt and fumes of modern city traffic. Only the male trees are usually planted, because the ripe pulpy seeds of the female trees emit a disagreeable odor.

Angiospermae. The angiosperms or flowering plants are the most numerous, numbering more than 200,000 species. The herbs, grasses, shrubs, and broad-leaf trees from which most of our foodstuffs and many industrial materials come are flowering plants. A distinctive structural feature is the **flower** (Fig. 5.13), which consists essentially of a condensed or shortened stem bearing whorls or spirals of modified "leaves," and in which the seeds are produced within a closed caselike structure, the **ovulary** (Chapters 9 and 21). The name angiosperm (*angeion,* enclosing vessel; *sperma,* seed) in fact, points up one of the important contrasts with the gymnosperms. The ovulary develops into a true fruit, a structure characteristic of the angiosperms. There are other distinctive features that are better discussed in another context. These will be shown in later chapters. Compared with the other groups of plants, the angiosperms are considered from fossil evidence to be of relatively recent origin, the oldest authentic representatives occurring in the late Mesozoic era (Table 6.1). No fossil evidence is available that clearly suggests the pathways by which they may have originated from older groups of vascular plants.

Fig. 5.12 Ginkgo biloba, *a primitive gymnospermous tree widely planted as an ornamental in parks and along streets. The "fruits," really naked seeds, and most of the leaves are borne on short spur shoots. Male and female reproductive structures are borne on separate trees. The ripe seeds are foul smelling, so usually only male trees are cultivated.*

For the angiosperms in particular, it is desirable to carry the classification one step further at this point by mentioning the two recognized **subclasses, Monocotyledoneae** and **Dicotyledoneae.** Many other classes throughout the Plant Kingdom are similarly divided into subclasses by the taxonomists, but these are of special prominence in our modern angiospermous flora. The Monocotyledoneae (Figs. 5.14, and 5.15), often referred to cryptically as "monocots" include such familiar plants as orchids, irises, lilies, grasses, sedges, and palms. Typically, the embryo within the seed of these plants possesses *one* well-developed **cotyledon** or seed leaf (Chapters 9 and 10). With notable exceptions, the leaves

Fig. 5.13 *A selection of flower types.* (*Top left*) Prunus, *cherry;* (*Top right*) Nymphaea, *water lily;* (*Bottom left*) Chrysanthemum, *daisy;* (*Bottom right*) Cypripedium, *Lady's Slipper.*

are narrow, the veins running parallel from base to apex. In most of them the vascular tissue of the stem is relatively scanty and the stem is incapable of massive diametric growth through the formation of additional vascular tissues (Chapter 9). In contrast, Dicotyledoneae (Fig. 5.16) or "dicots" possess *two* cotyledons (Chapters 9 and 10). The leaves, except in many small herbs, are typically broad and the leaf veins are usually arranged in riblike or netlike patterns. The vascular tissues of the stem are typically arranged in a circular pattern, and the perennial trees and shrubs among them are capable of producing considerable additional vascular tissue through the action of lateral meristems, thus increasing the

Fig. 5.14 Agave, *a monocotyledon of the lily family.*

Fig. 5.15 Giant bamboo, a monocotyledon of the grass family.

Fig. 5.16 Spanish moss, Tilandsia, *a dicotyledon, growing on cypress trees, a gymnosperm, in a Louisiana swamp. It is not a moss, but a member of the pineapple family. After the Spanish moss is cured, it is used for upholstery stuffing.*

diameter of the stem (Chapter 9). With the exception of species of pine, fir, and other coniferous gymnosperms, which furnish the major portion of commercial structural timber, the woody dicots supply practically all of our requirements for wood. Ash, oak, maple, walnut, hickory, mahogany, tulip poplar are the mainstay of the furniture industry, used either as structural material or as a source of veneers. Dicot trees, shrubs and vines, such as apple, pear, avocado, blueberry, and grape yield important fruits. Most garden crops, such as beans, squash, spinach, cabbage are dicots.

RELATED READING

Arditti, J. "Orchids," *Scientific American, 214*(1), 70–78, January 1966.

Ayensu, E. S. "Beautiful gamblers of the biosphere (orchids)," *Natural History, 83*(8), 37–45, 1974.

Bastin, E. W. "The ginkgo: strange survivor," *Nature Magazine, 43,* 410, 1950.

Gerald, J. H. "Horsetails," *Natural History, 62,* 352–354, 1953.

Hodge, H. "Tree ferns," *Natural History, 65,* 88–91, 1956.

Meeuse, B. J. D. "The voodoo lily," *Scientific American, 215*(1), 80–88, July 1966.

Wilhelm, S. "The garden strawberry: a study of its origin," *American Scientist, 62,* 264–271, 1974.

6
PLANTS OF THE PAST

At one time most biologists subscribed to the belief that the species of plants and animals now living had existed without change since the beginning of life on Earth, and that each species was independently created. However, during the past two centuries, discoveries resulting from careful observation of and experimentation with living organisms have made it increasingly clear that most of the species now living have existed for a relatively short time, that many species formerly living are now extinct, and that through change one species has given rise to others. The history of the Plant Kingdom, therefore, is the history of change. That is, our modern flora and fauna are the products of organic evolution from ancestral species. Presumably all plants and perhaps all living things can be traced back to a single common ancestor, the first relatively simple unicellular organism. Although this view is supported by the fundamental biochemical, physiological, and cellular similarities among all living organisms, some biologists believe that life may have originated at several different times and places and that thus not all organisms would actually be related in the true sense.

Our present-day flora is certainly not the end of change. If we could look 2 billion years into the future, we might find that further evolution had resulted in changes as great as those accomplished in the past 2 billion years, assuming that life would still be present on Earth. However, despite the changes that have occurred in the evolution of plants, throughout the Plant Kingdom there runs a profound core of conservatism, both structural and physiological.

FOSSILS

The record of plant evolution is the accumulation of ancient plant remains **(fossils)** in the sediments and sedimentary rocks of Earth's crust. Most plant remains were deposited in swamp or bog areas where they grew; some were carried by streams to the site of deposition in the sediments of river deltas or lakes. Plant remains may be fossilized only if local conditions prevent their decay. Quick coverage by water, sediments, or ash excludes oxygen and thus preserves the plant tissues for subsequent formation of fossils. Plants that grew in swamps, bogs, or on pond margins and near the sea, were the most likely to be fossilized, whereas plants of inland arid regions were less likely to be preserved.

In general, soft-bodied plants would have had less chance of preservation than those with much woody tissue. Algae, fungi, mosses, and delicate herbs are scarce in the fossil record, even though their ancient habitats were of the proper sort. The more delicate structures were liable to breakage or extreme distortion under the conditions of fossilization, and thus they may escape recognition. As a result of conditions at the time of and following the deposition of plant remains, fossils may occur in a number of forms. One of the most common is an **impression** formed when a plant part, such as a leaf, falls upon wet sand or clay. Sediments press the leaf into the soft substratum and an impression is made. The impression remains after the leaf disintegrates and the sediments, under pressure, consolidate to hard rock (Fig. 6.1). In some circumstances, the embedded plant may become chemically altered to leave only its carbon residue. Such fossils are known as **compressions** (Fig. 6.2). In these, the cellular structure of the part is not preserved. Compressions are common in strata associated with coal beds. Another type of fossil is produced by the infiltration of the tissues with minerals from the water and sediments surrounding the buried plant part. The process may be so complete that the cell walls are literally changed to stone. Fossils of this type are **petrifactions** (Fig. 6.3). When they are cut into thin sections and examined under a microscope, much of the fine structural detail may be seen. In the Petrified Forest in Arizona and other places in the west, there are large petrified trunks of ancient trees now exposed on the surface.

Fossils and Earth History

To reconstruct a wholly satisfying picture of the progress of the Plant Kingdom from only the fossil record is impossible. It can hardly be doubted that the earliest representatives of the Plant Kingdom were simple and delicate in structure. Yet tangible proof of this in the form of fossil remains is scanty, for these are the very forms that largely would have escaped fossilization. Much of the evidence upon which we build the early part of our story of plant evolution, therefore, is derived from other kinds of plant study.

Although the fossil record is obviously incomplete, the sequence of deposition and the relative ages of the fossils are fairly accurately known. Sedimentation occurred in great cycles of time leaving strata of rock. The fossil plant species dominating the various strata are usually quite different from the plants of today, and from each other. Thus the fossil record does not present a series of gradually changing forms. Instead, the record is punctuated by many gaps.

Fig. 6.1 (*Left*) Neuropteris, *a Carboniferous seed-fern. Impression of a leaflet exposed on the surfaces of a split nodule.*

Fig. 6.2 (*Left*) Annularia, *a Carboniferous horsetail. Compression of a jointed, leafy stem.*

Fig. 6.3 (*Opposite*) *A petrofaction of a giant lycopod stem from the Carboniferous. (Copyright, General Biological Supply House, Inc., Chicago.)*

Earth is very old. Geologists have estimated its age to be at least 4.5 billion years and probably somewhat more. The earliest rocks were formed of molten materials and contain no evidence of living organisms. Presumptive evidence of the earliest occurrence of living organisms is found in sedimentary rocks more than 2 billion years old as deposits of certain biologically significant organic compounds and as deposits of limestone believed to have been formed by organisms comparable to modern lime-secreting algae. Recent studies of **Precambrian** (Table 6.1) sediments in southern Ontario have revealed the presence of a varied assemblage of structurally well preserved fossil microorganisms that are unques-

Table 6.1 **Geologic Time Scale**

Eras	Periods	Millions of Years from Present	Major Developments in Plant Life as Shown by Fossil Record
Cenozoic	Quarternary	Present to 2	Extinction of many trees through climatic changes. Increase in herbaceous flora.
	Tertiary late	2 to 65	Dwindling of forests; climatic segregation of floras. Rise of herbaceous plants.
	Tertiary early		Development of many modern angiospermous families. Rise and worldwide extension of modern forests.
Mesozoic	Cretaceous	65 to 136	Angiosperms gradually become dominant, some modern angiospermous types represented. Gymnosperms decline.
	Jurassic	136 to 225	Earliest known angiosperms. Cycads and conifers dominant; primitive gymnosperms disappear.
	Triassic		Increase of cycads, ginkgo, and conifers. Disappearance of seed ferns.
Paleozoic	Permian	225 to 345	Waning of arborescent clubmosses and horsetails. Early cycads and conifers.
	Carboniferous		Extensive coal-forming forests of giant clubmosses, horsetails, and seed ferns. Primitive gymnosperms.
	Devonian	345 to 430	Early vascular plants: psilophytes, primitive clubmosses, horsetails, and ferns. Early forests of arborescent clubmosses.
	Silurian		Algae dominant. First direct evidence of land plants.
	Ordovician	430 to 570	Marine algae dominant. Fungi evolve?
	Cambrian		Some modern algal groups established.
Precambrian		570 to 4600	900—Green algae? 3000—Bacteria and blue-green algae? 4000—Origin if life?

tionably bacteria and blue-green algae which strikingly resemble some of the modern genera of those groups. Confidently estimated to be 2 billion years old, these constitute the earliest known occurrence of identifiable cellular organisms in the fossil record (Fig. 6.4). Late Precambrian sediments in central Australia, about 900 millions years old, have recently been found to contain well-preserved fossils of unicellular and multicellular filamentous forms that are believed to be green algae,

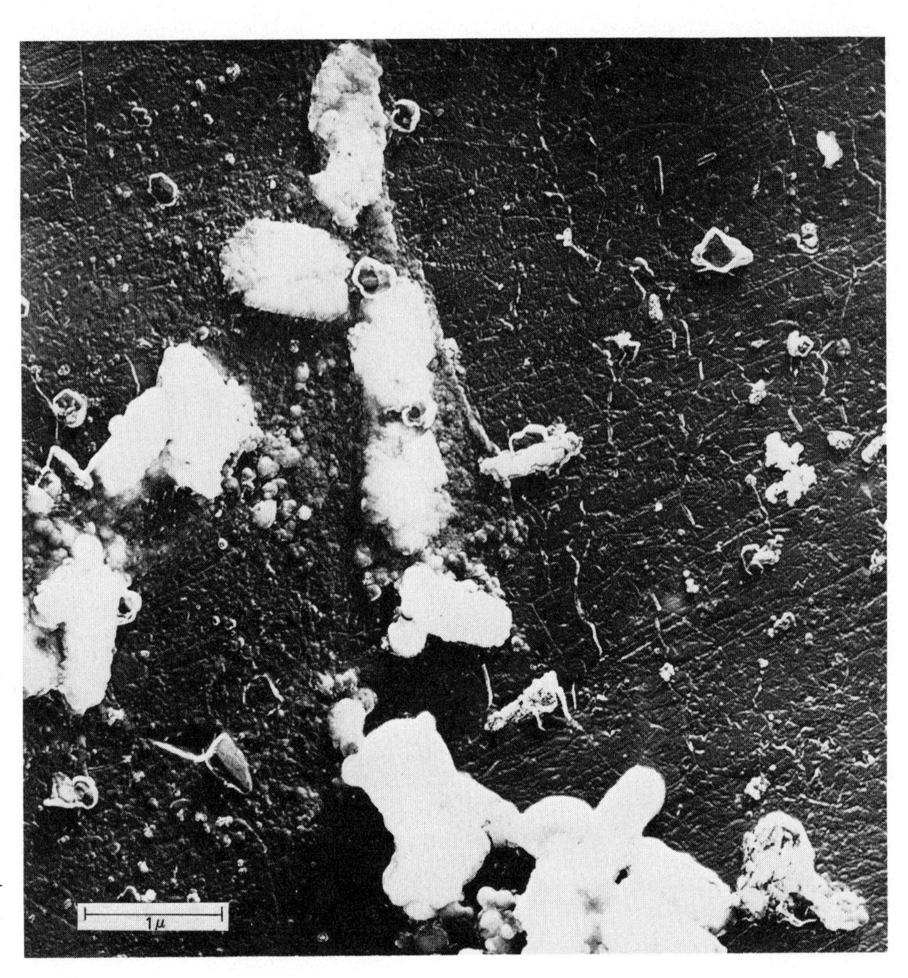

Fig. 6.4 Rod-shaped bacteria in Gunflint chert in southern Ontario, aged about 2 billion years. The photograph is a negative print of an electron micrograph of platinum-carbon surface replicas.

along with blue-greens (Fig. 6.5). If the identifications are correct, the Australian deposits contain the earliest known record of green algae.

The variety of forms among these early plants clearly indicates that evolutionary diversification had long been under way at that distant time and that they represent, in the history of the evolution of plants from a presumed unicellular ancestral organism, forms that were already old. During late Precambrian times, animals as advanced as simple worms and crustaceans were present.

It has been suggested that anaerobic bacteria, possibly similar in form and physiology to certain modern species, were among the earliest cellular organisms. Speculation on the probable character of the early environment is consistent with this suggestion. This and the further suggestion that bacteriochlorophyll, the photosynthetic pigment of certain bacterial species, may have been the evolutionary precursor of chlorophyll *a*, the dominant photosynthetic pigment of algae and higher plants, brings the bacteria into prominence in considerations of the origin of life and the evolution of green plants.

The Cambrian and Ordovician periods of the Early **Paleozoic** era (from 570 million to 430 million years ago) furnish scanty plant remains, but enough is known about them to make it clear that the algae were already well established. Forms representing some of the modern groups of algae were in existence at that time. The Animal Kingdom was represented by higher invertebrates, corals, starfish, and early vertebrates.

The origin of land plants is doubtless to be sought among the algae. No fossils are known that are clearly transitional. It may be supposed that green algae with features that prevented loss of water and death by desiccation became established on land and then developed anchorage and an erect habit. By Silurian and early Devonian periods of the Middle Paleozoic (430 million to 345 million years ago), there appeared a group of small leafless and rootless forms which in their simplicity are

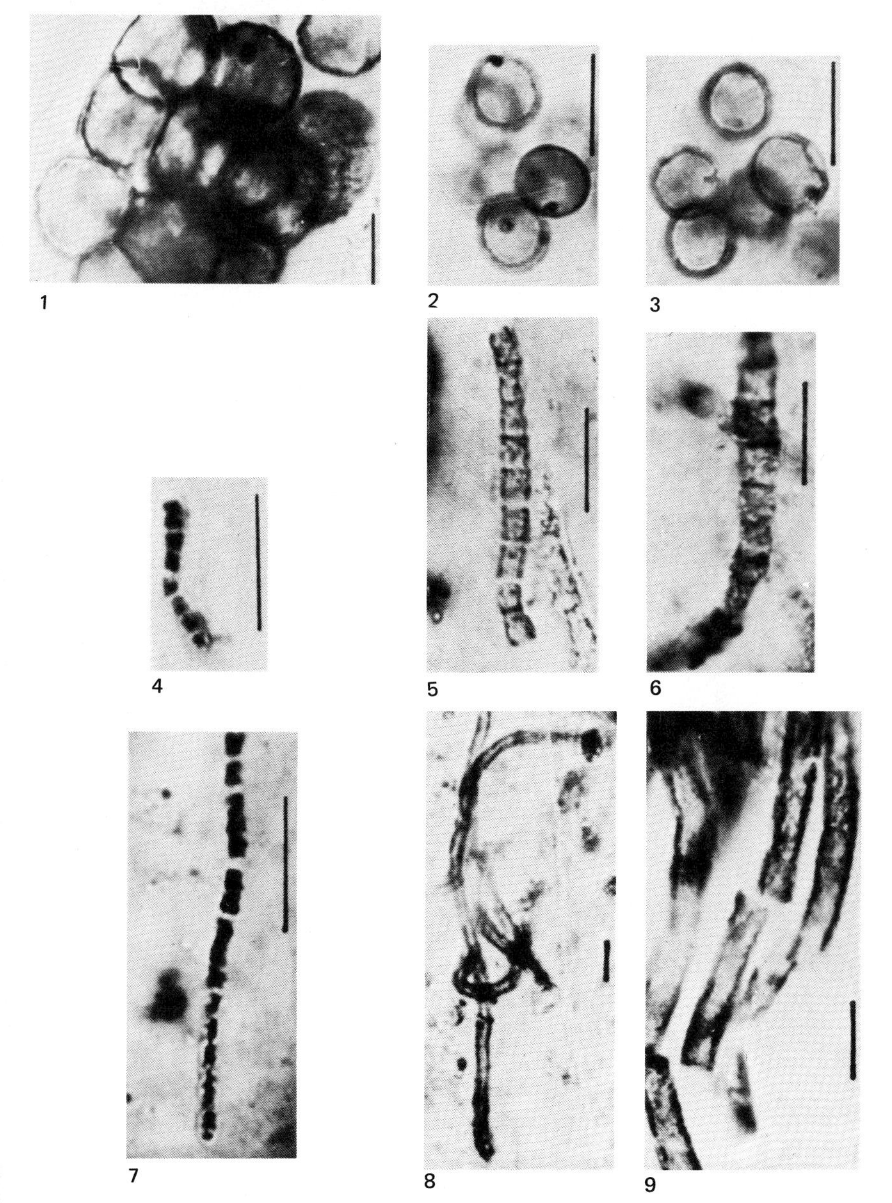

Fig. 6.5 Fossil microorganisms in Bitter Springs chert of central Australia, aged about 900 million years. Interpreted as blue-green and green algae, these are among the oldest known fossils showing good cellular preservation, and the earliest known occurrence of green algae. Thin sections of rock were photographed with transmitted light. (1, 2, 3) *Spheroidal cells resembling resting zygotes of a green alga like* Chlamydomonas. (4, 7) *Unbranched, septate filaments, possibly enclosed in a gelatinous sheath, resembling the blue-green algae* Oscillatoria *and* Nostoc. (5, 6) *Large, unbranched, septate filaments resembling green algae such as* Ulothrix. (8, 9) *Nonseptate filaments bearing a resemblance to certain coenocytic yellow-green algae.*

Fig. 6.6 Glass model of Rhynia, *a psilophyte, one of the earliest vascular plants. Devonian.*

algalike, but whose creeping stems and slender erect branches contain the simplest of vascular tissue. The erect leafless branches were photosynthetic and bore typical stomata. These plants, called the psilophytes (Psilophyta), are the first true land plants in the fossil record (Fig. 6.6). It is probable that these early land plants represented the sporophytic phase of the life cycle. The transition from the gametophyte-dominant algae to the sporophyte-dominant vascular plant is a vast one to cover and probably involved the evolution of an extensive series of intermediate forms unknown in the fossil record. The early psilophytes evidently did not remain on earth very long, for they disappear from the fossil record before the end of the Devonian period. However, two modern genera of plants, *Psilotum* and *Tmesipteris,* are believed to be descendants of the ancient psilophytes. A recently discovered fossil, *Baragwanathia,* of the Silurian period, poses a problem in that it antedates the psilophytes but strongly resembles the lycopods, a group generally considered more advanced than the psilophytes.

The Devonian was a period of mild and equable climate, a time for rapid plant development. The first representatives of the modern club mosses (Lycophyta) and horsetails (Sphenophyta) appeared. The lycopods were especially vigorous, for they attained heights of eight meters or more and formed extensive forests. In late Devonian, early ferns (Pterophyta) occurred that were probably the forerunners of our modern ferns. Lungfishes, scorpions, sharks, and early amphibians represented the Animal Kingdom.

Great coal-forming forests characterized the Carboniferous period of the Late Paleozoic (345 million to 225 million years ago). Strata of rock in association with coal seams yield abundant fossils of ferns, tree-size club mosses and horsetails, pteridosperms (seed-bearing but fernlike), and primitive gymnosperms (Fig. 6.7). Increasing and widespread aridity and in

Fig. 6.7 Reconstruction of a Carboniferous landscape, showing arborescent club mosses (1, 2, 3, 4), *horsetails* (5, 6), *ferns* (7, 8, 9), *seed-ferns* (10, 11), *gymnosperms* (12).

some places extensive glaciation characterized the succeeding Permian period. During the Permian period many Carboniferous forms dwindled and became extinct. Fossils of primitive reptiles, insects, and early terrestrial vertebrates represent the animals of the time.

The **Mesozoic** era (225 million to 65 million years ago) witnessed significant changes in the vegetation. During the early portion, under somewhat arid and warm conditions, most of the giant lycopods and horsetails disappeared, along with the seed ferns. The remaining club mosses, horsetails, and ferns were small species similar to those of the present. Higher types of gymnosperms, such as conifers, cycads, and *Ginkgo* arose and rapidly became the dominant plants. The first angiosperms appeared in Middle Mesozoic under fluctuating climatic conditions, developed rapidly, and by the close of the era had become very abundant as the gymnosperms gradually lost ground. Many modern types of flowering plants are represented in the Late Mesozoic flora. Fig, oak, holly, magnolia, sassafras, willows, maples, and many others occurred in the forest. During the Mesozoic era, the giant reptiles appeared and disappeared, specialized insects were numerous, and primitive mammals appeared for the first time.

The Early Tertiary period of the **Cenozoic** era (65 million years ago) was a time of humid, warm climate which supported a rich and widespread flora, with woody angiosperms dominant. During the Late Tertiary the flora became restricted in distribution and segregated into climatic types, as general cooling occurred and climatic zones became established. The redwoods and some associated species disappeared from the inland areas made dry by mountain building. At the close of the Tertiary several species disappeared as a result of the increasing aridity in western North America. *Metasequoia* (Fig. 6.8), a relative of the redwoods and bald cypress, was widespread in the

Fig. 6.8 Fossil branchlet of Metasequoia *from Tertiary deposits in Oregon. (Courtesy, Carolina Biological Supply Co.)*

northern hemisphere during the Tertiary period. It was believed to be extinct, until a few years ago when living representatives were discovered in southwestern China.

Extensive glaciation occurred in Pleistocene (2 million years ago) in the northern hemisphere, profoundly affecting the distribution of many plant species. New herbaceous angiosperms arose and evolved rapidly during this time. Their short life cycle had considerable survival value under conditions of progressive cooling. Developments in the Animal Kingdom during Early Tertiary include the rise of modern birds and higher mammals, and in Late Tertiary the appearance of early man.

THE THEME OF EVOLUTION

Obviously the fossil record fails to provide the complete story of plant evolution. Gaps of considerable magnitude, unbridged by transitional forms between the old and the new, occur at many levels in the fossil record. However, studies in the fields of morphology, physiology, and genetics have provided the unifying theme of the story and given some insight into the mechanisms involved.

Morphological study of plants supports the concept of relationship among plant species. Similarities in form and structure indicate relationship. In general, the greater the resemblances, the closer the relationship, and close relationship suggests common ancestry. Thus, two closely related species such as the white oak and the post oak had a common ancestor in the rather recent past, geologically speaking, whereas more distantly related species such as the white oak and the sugar maple would have common ancestors only in the more remote past. Structural and behavioral patterns persisting in a long series may be assumed to identify a line of evolutionary development from an ancestral type. Thus the vascular tissue is remarkably uniform in composition, basic arrangement, and function throughout the tracheophytes, despite the differences in the various types of steles. The basic uniform structural pattern in flowers and the individual role in reproduction played by the several parts persists through a host of modifications. The invariable repetition of the same theme in the sexual reproductive cycles of plants, involving gametes, gametic fusion, meiospore formation, the alternation of phases, and all the rest, suggests that differences in some of the details of execution are only secondary modifications. Thus, sexually reproducing plants have a deep-lying relationship, and these organisms may have evolved one from another in a sequence revealed by their secondary modifications. That is, in spite of the evident versatility or changeability of living things, they are essentially conservative and resistant to fundamental change.

Physiological studies also aid in the understanding of plant relationships. One of the most conspicuous is the similarity of pigmentation in the members of various groups. This matter has received much attention in systematic studies among the algae. Although the algae are all chlorophyll bearing, there are at least ten different kinds of chlorophyll whose presence or absence individually or in combination characterize certain of the phyla. Auxiliary pigments, such as phycophaein and fucoxanthin, are also employed in the evaluation of relationship. Similarly, special aspects of chemical composition may be used in studies of this kind. Similarities or differences in the nature of such constituents as proteins, volatile oils, resins, and alkaloids, may provide valuable data on relationships.

The study of the **genetic behavior** of plants in both natural and artificial circumstances offers evidence in support of the concept of plant relationships and evolution by orderly cumulative change in living organisms. The fact of interfertility of certain plant species indicates a basic compatibility of the hereditary material of the species, just as infertility generally suggests a fundamental incompatibility, and thus a remoteness of relationship. Whether such compatibilities are expressed in terms of viable offspring produced or in terms of chromosome pairing at meiosis makes little difference, since the fact of crossing creates the new genetic complexes upon which the selective action of environment may play. From such selections come the new populations with still newer genetic complexes.

Geneticists frequently note the spontaneous appearance of new types (mutants) of plants among old familiar parent stock. Such changes may be the result of gene mutations, chromosome alterations such as fragmentation and deletion, or aberrant chromosome behavior. Mutations have been artificially induced by various experimental techniques, involving radiation and chemical treatment. Such mutations, representing relatively permanent changes in the hereditary material, may be passed on to subsequent generations and thus enter the evolutionary stream. Mutations may entail conspicuous changes in form or they may be expressed as relatively inconspicuous physiological changes which alter some aspect of a plant's metabolism. Man has created many new varieties of crop plants by hybridization and selection of mutant stocks. The difficult problem of the taxonomist in recognition and delimitation of poorly defined species may reflect a natural plasticity, that is, proclivity for change, or a compatibility with other species.

Origin of Species

The concept of organic evolution has been in the minds of men at least since the time of the early Greek philosophers, and since then many biologists have sought an explanation of the mechanisms by which organisms change. Our modern understanding of evolution (still incomplete) stems largely from the work and writings of Charles Darwin (1809–1882). In 1859 he proposed a theory of evolution based on three main points.

1. A population of organisms is capable of producing many more offspring than its site can possibly support.

2. All offspring cannot survive under the conditions of limited space and shortage of materials necessary to support them, and thus they compete with each other. Only a small proportion of the total offspring survive the competition, and these are adapted to their particular environment; the unfit perish.

3. Not all individuals of a species are alike. The successful competitors possess highly favorable variations that give them an advantage in the same environment over the unsuccessful ones with less favorable variations. The favorable variations are transmitted to succeeding generations, further variations arise, followed by competition and selection of the best adapted, and so on, generation after generation. In this way, according to Darwin, many new species of organisms come into being, each more favorably adapted to its environment than its predecessor. In the process, many ill-adapted individuals disappear.

A population of a species frequently shows many variant forms. Some of these are direct **environmental variations** induced solely by environmental factors acting on the body tissues of the plant. For example, if a plant is grown in the shade, its height and the size of its leaves might be different from those of a plant grown in full sunlight. Such variations are not heritable, for they are not controlled by the hereditary substance of the reproductive cells. Selection, acting on such variations, is not, of course, significant in evolution. On the other hand, some variants in the population will transmit their special feature to a succeeding generation irrespective of environmental factors. Such inherited variations are **genetic variations** and they are important in the evolution of the species.

Biologists are familiar with the fact that species are often distinguishable one from another on the basis of features that have no imaginable survival value, that is, they are **non-adaptive characters.** Such characters are, of course, genetically controlled and heritable. However, they are not accounted for in natural selection as Darwin conceived it. It is possible, of course, that such species possess **adaptive characters** of other kinds that are not obvious to man.

It was Darwin's view that evolutionary changes were accomplished as a result of an infinite number of small, more or less continuous and cumulative variations. Some critics of the Darwinian theory have held that small variations commonly within the range of normal variability of the species are insignificant in bringing about major changes, and that only large variations could be significant in natural selection. Both large and small variations do occur in species populations. Some of the latter are of such magnitude as to constitute new types, or new species.

The emphasis on the importance of large variations as a means to species formation came chiefly from the work of the Dutch botanist Hugo de Vries (1848–1935). Among plants of certain species of evening primrose he observed the spontaneous appearance of individuals sufficiently different to be regarded as distinct species. Changes of such magnitude de Vries called **mutations,** and he regarded them as the basic units of evolution on which natural selection may act.

With the theories of Darwin and de Vries as bases, modern workers in the fields of cytology and genetics have striven to relate the observed facts of evolution to the physical basis of reproduction in an effort to establish the causes and mechanisms of evolution. All of the facts of evolution are not yet explained and are unlikely to be until much more is known about the nature of the hereditary material. Meanwhile, certain general statements may be made to summarize the current views on evolutionary mechanisms.

The chief hereditary materials are the genes and these largely control the structure and metabolism of the organism. The genes are mutable and the chromosomes in which they reside are subject to physical alteration and aberrant behavior. Such mutations bring about changes of various magnitudes in structure and metabolism. Very large gene mutations or excessive derangement of the chromosome complement may give rise to a fatal imbalance between organism and environment, or the variants may be better adapted to an environment than the parent form, and in a stable environment may supersede the parent form. Changes in environment change the adaptive value of variations. Plants with new gene combinations resulting from hybridization and polyploid forms would, of course, be similarly selected on the basis of their adaptive characters. Many additional secondary factors, such as geographic and ecologic isolation, time of flowering, adaptation to pollinating agents, and so forth, have important influences on the origin of new species.

Origin of Life

Darwin and other students of evolution presented much evidence for the evolutionary origin of species from other species, but they were unable to take the final step by providing any reasonable scientific hypotheses regarding the origin of life on earth. For many years this seemed to be an insoluble scientific problem. The first breech in this impasse was made by the Russian biochemist A. I. Oparin in his book *Origin of Life,* published in 1936. Oparin pointed out that, although life cannot arise by spontaneous generation under present environmental conditions, the environmental conditions of the lifeless world approximately four billion years ago were probably such that nonbiological synthesis of organic compounds and thus the origin of life could occur. Oparin and other scientists subsequently modified and extended his hypotheses, and a number of scientists, including S. L. Miller and S. W. Fox, provided convincing experimental evidence that the hypotheses were valid.

It is believed that before life appeared on earth the atmosphere was very different from our present atmosphere. It was composed primarily of hydrogen, ammonia (NH_3), methane (CH_4), cyanide (HCN), and water vapor. Other gases were probably present in smaller amounts, but there was little or no carbon dioxide (CO_2), or nitrogen (N_2) and no oxygen (O_2). The gases of the atmosphere contained all the chemical elements essential for the synthesis of organic compounds except for essential mineral elements such as phosphorus, potassium, and iron, and these were present in the rocks and oceans of the earth. Before life could exist these chemical elements would have to be combined into the various organic compounds essential for life, such as proteins, carbohydrates, fats, and nucleic acids. At the present time, and throughout most of the history of life on earth, all naturally occurring organic compounds have been products of organisms but, of course, before the origin of life there were no organisms. Thus there must have been nonbiological synthesis of the essential organic compounds.

In addition to the essential chemical elements, the synthesis of organic compounds requires energy, and it appears that the energy was supplied by electrical discharges and radiation, particularly ultraviolet radiation. The prebiological atmosphere permitted much more ultraviolet from the sun to reach the earth than does our present atmosphere. The synthesis of organic compounds continued for long periods of time, and it is believed that they accumulated and became more concentrated in quiet pools of water isolated from the open oceans. This nonbiological synthesis was possible only because there was no oxygen in the atmosphere. If oxygen had been present it would have broken down (oxidized) the compounds about as fast as they were produced.

With the essential compounds present the next step in the origin of life would have to be their organization into simple cells. This appears to be a highly improbable event, but experiments by S. W. Fox and others have shown that when solutions of proteins are treated in certain ways the proteins aggregate into small spheres (microspheres) about the size of cells. The microspheres have an amazing number of structural similarities to cells, including double membranes and vacuoles. They may divide and have limited enzyme activity. However, they are not cells and cannot carry on the many life processes of cells, but still they do indicate that the first cells could have arisen in a similar manner. Other experiments have shown that when mixtures of gases like those in the primitive atmosphere are enclosed in a container, and subjected to an electrical discharge, amino acids and other organic compounds are formed. Thus the theory of the origin of life has considerable experimental support.

Once living cells were formed they could use the surrounding organic compounds as food. As in present-day organisms, this food could be used in building cell structures and as a source of energy made available by oxidation of some of the food in the process of respiration. Since there was no oxygen available the respiration had to be anaerobic, as in yeasts and some bacteria. Anaerobic respiration can release only a small amount of energy from food, in contrast with aerobic respiration that is carried on by the vast majority of plants and animals and requires oxygen. Also, as the number of living cells increased, there was an increased use of the surrounding organic compounds and the possibility that they would be used in respiration faster than they could be synthesized by the nonbiological processes. Thus the first life on earth was in a rather precarious situation and in no position to evolve into any very large or very active organisms.

What saved the day was the evolution in some of the cells of the capacity for producing chlorophyll, enabling them to use light energy from the sun in the production of an abundant and continuing supply of food. The beginning of photosynthesis had another important effect—the production of oxygen. As the oxygen content of the atmosphere increased further nonbiological synthesis of organic compounds became impossible, and from that time on all organisms became dependent on green plants for their food. The oxygen in the atmosphere also permitted the evolution of aerobic respiration and so, in time, the evolution of large and complex organisms, and particularly the active animals that could not possibly secure enough energy by anaerobic respiration.

RELATED READING

Andrews, H. N. "Early seed plants," *Science, 142,* 925–931, 1963.

Baghoorn, E. S. "The oldest fossils," *Scientific American, 224*(5), 30–42, May 1971.

Calvin, Melvin and G. J. Calvin. "Atom to Adam," *American Scientist, 52,* 163–186, 1964.

Keosian, John. *The Origin of Life.* New York: Reinhold, 1964.

Schopf, J. W., T. D. Ford, and W. J. Breed. "Microorganisms from the Late Precambrian of the Grand Canyon, Arizona," *Science, 179,* 1319–1321, 1973.

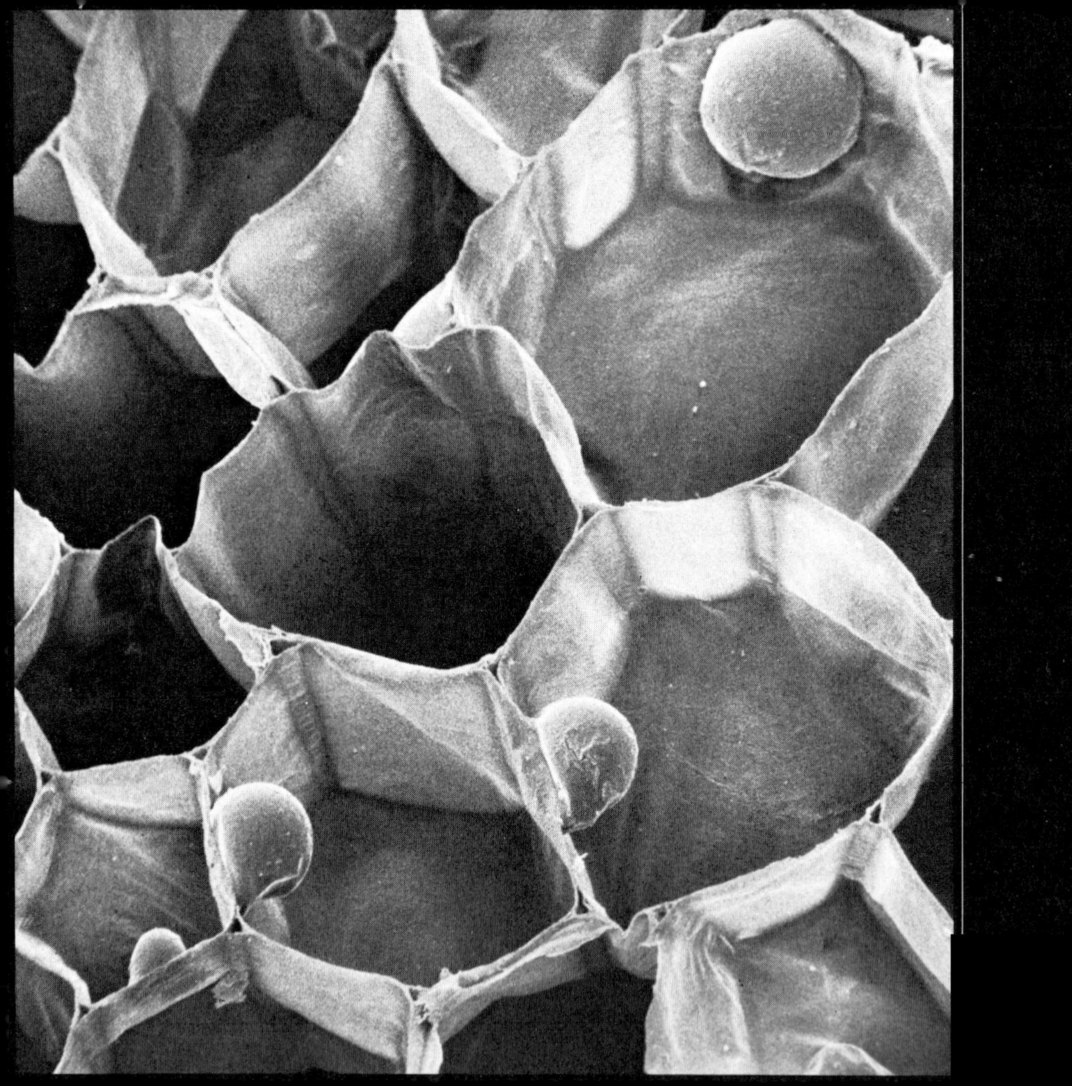

LEVELS OF PLANT ORGANIZATION

3

PHOTO OPPOSITE: PARENCHYMA CELLS IN THE PETIOLE OF CELERY, *APIUM GRAVEOLENS L.* (MAGNIFICATION 3400 ×)

7
THE ATOMIC AND MOLECULAR LEVELS

The term **organism** gives a clue to a most important characteristic of living things: their high degree of structural organization. The life processes of an organism occur within the framework of this structural organization, and the normal growth, development, and behavior of an organism are just as dependent on a certain specific assembly of structures as is the normal operation of a machine such as an automobile or a computer. However, the structures of an organism are not only the site of its life processes but also products of the life processes. There is, then, a very close interrelation between the structure and processes of organisms.

Observation of the leaves, flowers, and other organs of a plant reveals something of its structural organization, and microscopic examination of its tissues and cells further reveals a high degree of fine internal organization. A single living cell is composed of many precisely organized structures, many of the details being so minute that only an electron microscope will reveal them. It is becoming increasingly clear that many, if not all, of the biochemical processes of cells are localized in specific cell structures. In turn, these minute units of structure are composed of organized aggregates of molecules, which themselves are organized from atoms bonded together in a specific fashion. In a plant, then, we have a hierarchy of organizational levels, each basic to the ones above it: subatomic, atomic, molecular, molecular aggregate, cell structure, cell, tissue, organ, organ system, and organism. Finally, organisms are in turn organized into biological communities. Although unicellular organisms attain only the cellular level of organization, even they are remarkably highly organized entities.

In this chapter we begin our discussion of the various levels of plant organization with a consideration of the atoms and the more important kinds of molecules and molecular aggregates that constitute the basic building units of plants. Consideration of some of the biochemical reactions involving these molecules will come in later chapters. The remainder of this section of the book will be devoted to a discussion of the higher levels of plant organization.

Since this chapter, and several of the succeeding chapters, assume some knowledge of elementary chemical concepts, you may wish to study the chapter on chemistry in the appendix to this book before proceeding with your reading.

THE CHEMICAL ELEMENTS OF PLANTS

Chemical analysis of plant tissues generally shows the presence of about 40 different chemical **elements,** but only 15 elements are definitely known to be essential constituents of vascular plants. Three elements—carbon, hydrogen, and oxygen—make up about 99% of the total fresh weight of plants (Fig. 7.1). Nitrogen is the next most abundant element in plants, followed by potassium, phosphorus, calcium, sulfur, magnesium, and iron, respectively. The other essential elements are required by plants only in minute quantities and generally occur in plants only as traces. Of these **trace elements** boron, copper, manganese, molybdenum, and zinc are definitely known to be essential for vascular plants and sodium, chlorine, and cobalt may be essential. Several other elements such as iodine are essential for higher animals, whereas a variety of rare elements have been reported to be essential for at least some fungi. The roles of the various elements in plants are discussed in Chapter 13.

None of the elements essential for vascular plants is rare, oxygen and hydrogen being among the most abundant of all elements and nitrogen constituting the bulk of the earth's atmosphere. The most striking difference in the proportion of elements in organisms in contrast with the nonliving world is the high percentage of carbon in organisms (Fig. 7.1). Carbon constitutes only a small fraction of 1% of the nonliving world. There is nothing unique, however, about the atoms making plants and animals—indeed, they have been and will again be constituents of the air, waters, soil, or rocks of the Earth.

THE CHEMICAL COMPOUNDS OF PLANTS

To find substances that are really characteristic of organisms, we must turn from the atomic to the molecular level of organization. Although the **inorganic compounds** of plants—water, gases, salts, acids, and bases—are also constituents of nonliving things and are not characteristically biological, the **organic compounds** found in nature are all products of living organisms (or have been derived from these products). All organic compounds contain carbon, and we may further limit them to compounds that also contain hydrogen. Oxygen, nitrogen, and other elements are also constituents of some organic compounds. About 95% of the dry weight of plants

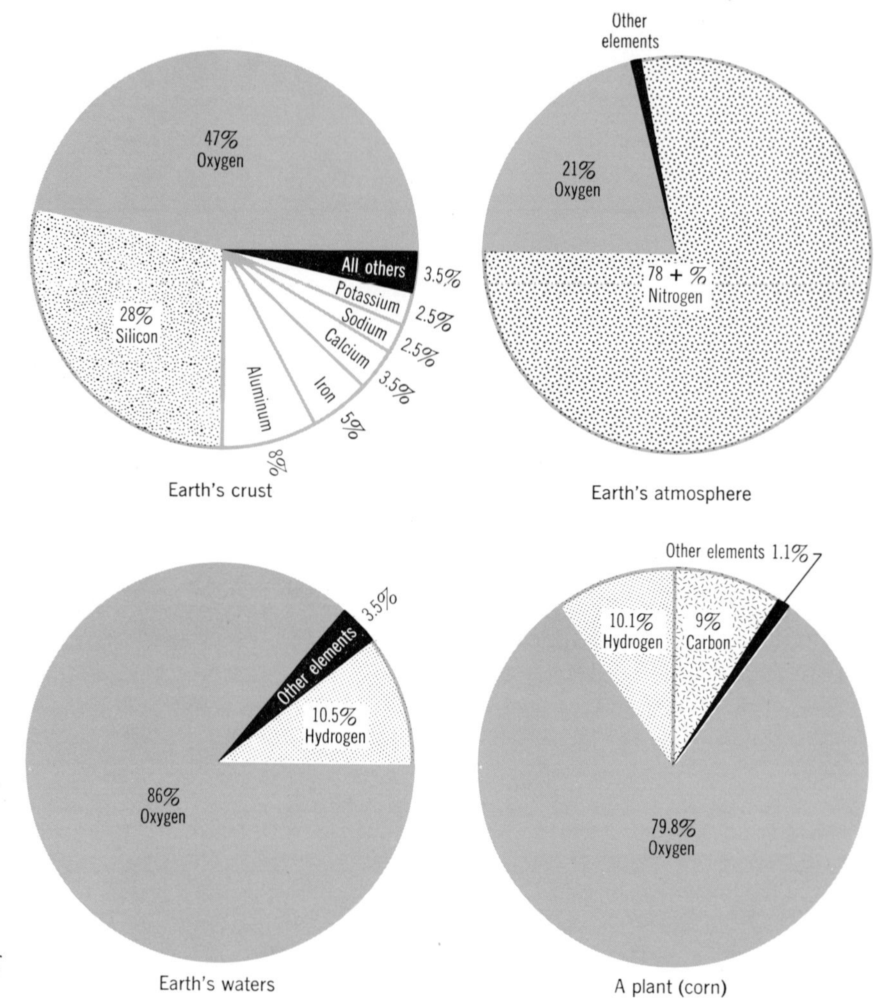

Fig. 7.1 Graphic representation of the percentage by weight of the more abundant elements in a corn plant compared with the distribution of elements in three parts of the nonliving world.

(what is left after the water is removed) consists of organic compounds, principally proteins, carbohydrates, fats and other lipids, and nucleic acids. Plants produce hundreds of other classes of organic compounds, many of them extremely important in the structure and processes of plants, but all of these together generally constitute only a small fraction of the total dry weight. Without organic compounds, life would not exist, but certain inorganic compounds are also essential to life. It may be noted that organisms are not only the source of all organic compounds found in nature at the present time but also of at least one inorganic substance—the oxygen of the atmosphere, a product of photosynthesis. Before there were photosynthetic plants on earth, the atmosphere probably contained no oxygen gas. As life was originating in the distant past, organic compounds apparently were being synthesized nonbiologically. The first organisms were assembled from some of these organic compounds. Then the organisms used some of the remaining organic compounds as food.

INORGANIC CONSTITUENTS OF PLANTS

Water

More than 90% of the living structures of cells **(protoplasm)** consists of water, the walls of most living plant cells are impregnated with water, and a large part of the volume of many plant cells is occupied by vacuoles filled with cell sap (water with various substances dissolved in it). In vascular plants there is also considerable water in the conducting tubes of the wood on its way from the roots to the upper parts of the plant, and the spaces between cells are commonly almost, if not completely, saturated with water vapor. Water constitutes 80% or more of the total fresh weight of plants, except perhaps of large trees.

Water is just as essential to life as are the characteristic organic compounds. It not only plays an essential structural role but also is indispensable as a solvent and a participant in many important biochemical reactions. No organism can carry on its life processes actively without an adequate quantity of water.

Water has several unusual properties in relation to heat. Its specific heat, heat of vaporization, and heat of fusion are all much higher than those of most other substances of similar molecular weight (Table 7.1). The **specific heat** of a substance is the number of gram calories of heat energy required to raise the temperature of one gram 1°C. The specific heat of water is 1 cal, about twice the specific heat of the common liquids in Table 7.1 and many others including oils. The specific heat of water is four times that of air or aluminum and ten times that of iron. The **heat of vaporization** of a substance is the number of calories required to convert one gram from a liquid to a gas without an increase in temperature. Note the high heat of vaporization of water compared with that of the other liquids in Table 7.1. (The heat of vaporization varies with temperature—as do the other two factors. The heat of vaporization of water is 585 cal at 20°C.) The unusually high heat of vaporization of water results largely because much energy is required to break its hydrogen bonds. The **heat of fusion** of a substance is the number of calories required to convert one gram of the substance from solid to liquid at its melting point, without an increase in temperature. Again, note the high heat of fusion of water compared with the other liquids in Table 7.1.

The unusually great amount of heat energy required to heat and evaporate water and to melt ice is of great importance in making life on Earth possible. These properties of water make large bodies of water heat and cool slowly, and this in turn greatly reduces the temperature fluctuations of land areas even

Table 7.1 **Specific Heats, Heats of Vaporization, and Heats of Fusion of Selected Liquids** [a]

Substance	Specific Heat at 15°C	Heat of Vaporization at the Boiling Point	Heat of Fusion at the Melting Point
Acetone	0.52	125	20
Benzene	0.40	94	30
Ethyl alcohol	0.58	204	25
Glycerol	0.54	—	48
Methyl alcohol	0.56	263	16
Water	1.00	539	80

[a] The values given are gram calories per gram of substance and are rounded off.

some distance away. Plants and moist soils exposed to direct sunlight heat up much less than they would if water did not have such a high specific heat and heat of vaporization. Indeed, if life were based on some liquid other than water it could probably not survive on Earth, even if the liquid were otherwise suitable.

One other unusual property of water that favors life in streams and lakes is that although it contracts as the temperature decreases (and so becomes somewhat heavier), it begins to expand at 4°C. Thus, ice is lighter than the liquid water and floats to the surface. Aquatic plants and animals can continue to live under the ice through the winter. If water were like other ordinary liquids which continue to contract and get heavier as the temperature decreases, ice would form at the bottom of a lake and extend up from there, possibly freezing all the water in it. It is possible that the ice toward the bottom would remain frozen throughout the summer.

Gases

Although gases are not really structural components of plants, various gases are always present within and between the cells of plants. The most abundant and important gases in plants are water vapor (H_2O), oxygen (O_2), carbon dioxide (CO_2), and nitrogen (N_2). Oxygen and carbon dioxide are both used and produced by plants, but the nitrogen of the atmosphere can be used directly only by nitrogen-fixing bacteria and blue-green algae (and the few species of nitrogen-fixing fungi).

Salts

All of the essential elements except carbon, hydrogen, and oxygen generally enter plants as the **ions** of salts. Salts are primarily important in plants as the source of elements used in the synthesis of a variety of essential organic compounds, although their ions also play a number of other important roles (Chapter 13). Proteins contain nitrogen, each molecule of chlorophyll contains an atom of magnesium and four of nitrogen, many very important organic compounds contain phosphorus, and some enzymes contain iron, to give just a few examples.

Salts generally occur as ions, not only in solution but also in their crystals, and it is as ions that salts are absorbed, transported, and used by plants. There are at least ten different **cations** and seven different **anions** in plants (Appendix, Table A.1) that can combine with one another and form 70 different salts. However, it is both more convenient and more accurate to consider the ions in both the soil solution and in plants individually rather than as the salts they can form.

Although most salts in plants are dissolved in water, plant cells may contain crystals of rather insoluble salts such as calcium oxalate and calcium sulfate. Calcium oxalate crystals (Fig. 7.2) are particularly abundant in the cells of *Caladium*, elephant ear, jack-in-the-pulpit, and other members of the same family (Araceae) and also in the small aquatic vascular plants known as duckweeds.

When dried plant tissue is burned com-

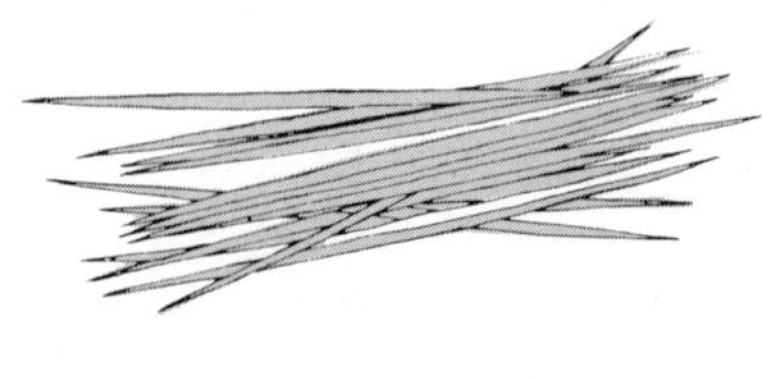

Fig. 7.2 Two of the several shapes of calcium oxalate crystals formed in plant cells.

pletely, only the mineral elements remain in the white ash. The burning oxidizes all organic compounds to carbon dioxide and water and the nitrogen present is also oxidized to a gas. The remaining ash is composed of oxidized mineral elements that the plants absorbed from the soil as ions.

Acids and Bases

Plants contain all the ions essential for the formation of most of the common inorganic acids such as hydrochloric (HCl), sulfuric (H_2SO_4), nitric (HNO_3), and phosphoric (H_3PO_4), and most of the common inorganic bases such as sodium hydroxide (NaOH), potassium hydroxide (KOH), calcium hydroxide ($Ca(OH)_2$), and ammonium hydroxide (NH_4OH). However, there is no appreciable concentration of the strong acids or bases because their ions are not present in large quantities and because plant cells are highly buffered by the presence of many organic acids and other weak acids and their salts. As a result, most of the hydrogen ions (H^+, or more properly hydronium ions, H_3O^+) characteristic of acids are incorporated in the molecules of the poorly dissociated weak acids, whereas excess hydroxyl ions (OH^-) from strong bases are incorporated into water (HOH) molecules through neutralization by acids. Protoplasm is generally just on the acid side of neutrality, being buffered at a pH of about 6.8, and any very marked deviation from this pH results in a disruption of protoplasmic structure and life processes. Plant juices may, however, be quite acid, orange juice having a pH of about 3 and lemon juice a pH of about 2. Although many of the weak acids in the buffer systems of plants are organic acids synthesized by the plants, at least two of the weak acids of plant buffer systems are inorganic: phosphoric acid (H_3PO_4) and carbonic acid (H_2CO_3). Buffer systems play a very important role in plant cells by preventing marked fluctuations in pH.

ORGANIC CONSTITUENTS OF PLANTS

The organic compounds of plants not only constitute the great bulk of their dry weight but are also much more numerous and varied than the inorganic constituents and are synthesized by the plants themselves. Most organic compounds have larger and more complex molecules than do inorganic compounds. Some organic compounds contain hundreds or even thousands of atoms. For many years chemists believed that organic compounds could be synthesized only by living cells, but in 1824 the German chemist, Wöhler, synthesized urea from inorganic substances and since then chemists have synthesized many other organic compounds. However, it is still true that all naturally occurring organic compounds found on Earth today were either synthesized by living organisms or derived from compounds so synthesized.

In the following pages, we consider representatives of the classes of organic compounds that are the most abundant in plants: carbohydrates, proteins, fatty substances (lipids), and organic acids. Later on in the book we shall from time to time have occasion to refer to other organic compounds that play important roles in plants.

Carbohydrates

The carbohydrates include the sugars and the high molecular weight polysaccharides derived from sugars. Sugars play many important basic roles in the life of plants. Sugars are produced by photosynthesis and used in respiration. They are used in making polysaccharides such as starch and cellulose, and are converted into the thousands of other organic compounds of plants including organic acids, fats, proteins, and nucleic acids. There are hundreds of different sugars, but all of them are sweet, white powders or crystals that are soluble in water.

Monosaccharides. The most fundamental and simplest sugars are the monosaccharides, or simple sugars. The simple sugars cannot be digested (hydrolyzed) into other sugars with smaller molecules, and they are the basic building units from which all other carbohydrates are assembled. Monosaccharides usually have two hydrogen atoms and one oxygen atom for each carbon atom in their molecules ($C_nH_{2n}O_n$). Although there are monosaccharides with various numbers of carbon atoms (mostly 3 to 9), the 6-carbon sugars **(hexoses)** are the most abundant in both plants and animals.

Of the many hexoses, two deserve particular mention: **glucose** and **fructose** (Fig. 7.3). Glucose, also called dextrose and grape sugar, is the principal sugar in animal blood as well as one of the more abundant sugars in plants. Glucose assumes considerable importance in plants as the substance from which both starch and cellulose are synthesized. Fructose, also called levulose and fruit sugar, is found in plants in relatively large amounts. It plays an important role in various metabolic processes and, like glucose, is used in the synthesis of more complex carbohydrates.

Since glucose, fructose, and the other hexoses all have the same chemical formula $C_6H_{12}O_6$, the question may arise as to how they can be different chemical compounds. The answer is that, although all hexoses are made of the same number and kinds of atoms, these atoms are put together in different patterns, just as many different kinds of houses can be built from a certain number of bricks, boards, and nails (Appendix). Figure 7.3 shows how glucose and fructose differ in structure.

The 5-carbon sugars **(pentoses)** are much less abundant in plants than the hexoses, but no less important. They have the general formula $C_5H_{10}0_5$. Pentoses are used in important

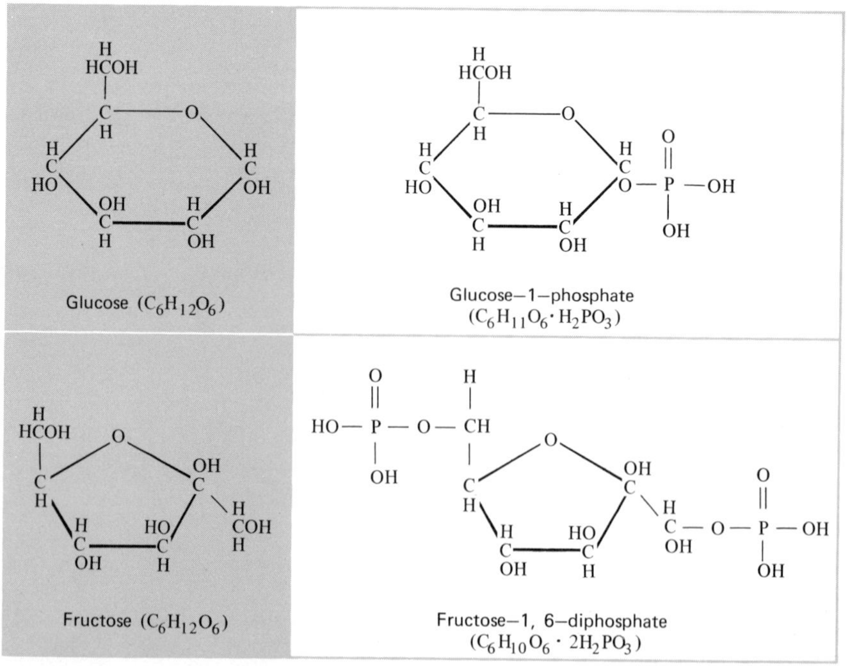

Fig. 7.3 Structural formulae of two hexoses and a phosphate ester of each.

Glucose ($C_6H_{12}O_6$)

Glucose–1–phosphate ($C_6H_{11}O_6 \cdot H_2PO_3$)

Fructose ($C_6H_{12}O_6$)

Fructose–1, 6–diphosphate ($C_6H_{10}O_6 \cdot 2H_2PO_3$)

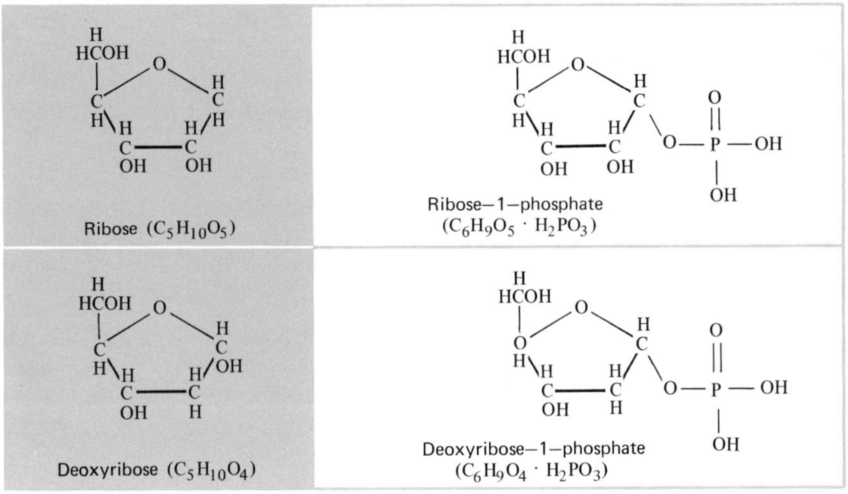

Ribose ($C_5H_{10}O_5$)

Ribose–1–phosphate ($C_6H_9O_5 \cdot H_2PO_3$)

Deoxyribose ($C_5H_{10}O_4$)

Deoxyribose–1–phosphate ($C_6H_9O_4 \cdot H_2PO_3$)

Fig. 7.4 Structural formulae of ribose, deoxyribose, and their 1-phosphate esters. Note that deoxyribose has one less oxygen atom than ribose.

synthetic reactions almost as rapidly as they are formed. Several different pentoses are used in making complex carbohydrates found in cell walls and in plant gums and mucilages. **Ribose** and **deoxyribose** (Fig. 7.4) are important pentoses, since they are constituents of the nucleic acids which carry the genetic codes. Deoxyribose is a constituent of DNA (deoxyribonucleic acid) and ribose is a constituent of RNA (ribonucleic acid). Another pentose, **ribulose,** is involved in photosynthesis.

Although the 3-carbon sugars (**trioses,** $C_3H_6O_3$) are not found in plants in large quantities, they play central roles in both respiration and photosynthesis as well as in various related processes (Chapter 14). Phosphorylated triose is the first carbohydrate produced by photosynthesis.

The various sugars can participate in many biochemical processes only when phosphorylated. In phosphorylated sugars, phosphate groups ($-H_2PO_3$) have replaced one or two of the hydrogens of the sugar molecule. The chemical formulae of several of the phosphorylated sugars are given in Fig. 7.3. For simplicity the phosphate group is sometimes written in formulae as –Ⓟ.

Disaccharides. Two molecules of simple sugars may become joined together, forming a somewhat more complex type of sugar known as a **disaccharide.** Although disaccharides can be made from various simple sugars, the most abundant ones are all made from hexoses. **Sucrose** (ordinary table sugar) is made from one glucose and one fructose molecule. It is one of the most plentiful sugars in plants. **Maltose,** composed of two molecules of glucose, is abundant in germinating grains and other tissues in which starch is being digested. Sucrose and maltose, as well as all other hexose disaccharides, have the formula $C_{12}H_{22}O_{11}$. Disaccharides contain two less hydrogen atoms and one less oxygen atom than the two molecules of monosaccharides used in their synthesis, a net loss of H_2O.

Figure 7.5 shows a molecule of maltose that is formed from two molecules of glucose. Another disaccharide present in plants **(cellobiose)** is also produced from glucose, but this is a different glucose isomer from that used in making maltose. It is α-D glucose that combines to form maltose, while β-D glucose forms cellobiose. These two isomers of D glucose are possible only when the glucose molecule is in the ring form. Note in Fig. 7.5 that the bond between the α-D glucose molecules is different from that linking the β-D glucose molecules. Thus, cellobiose is a disaccharide distinct from maltose. The enzyme that catalyzes the synthesis of maltose is not effective in the synthesis of cellobiose, and different enzymes are also necessary to hydrolyze the two disaccharides. Plants also make sugars from three **(trisaccharides)** to six **(hexasaccharides)** or more molecules of simple sugars.

Polysaccharides. Dozens, hundreds, or even thousands of monosaccharide molecules may be linked together into long chains, which are sometimes branched. These giant molecule carbohydrates are called polysaccharides. Plants synthesize a considerable number of different polysaccharides, mostly from either hexoses or pentoses. The most abundant polysaccharides in plants are starch and cellulose, both made from glucose.

Like maltose, starch is made from α-D glucose. Because the oxygen bridges linking the glucose molecules are all on one side (Fig. 7.6), the starch molecules have a helical configuration (Fig. 7.7). One kind of starch **(amylose)** has unbranched molecules while another kind **(amylopectin)** has branched molecules. Both kinds of starch are generally produced by a plant. Like cellobiose, cellulose is made from β-D glucose.

These differences in the molecular structure of starch and cellulose are responsible for some of the distinct differences in their physical properties. The long, straight cellulose molecules can be bound together into strong fibers (Fig. 7.8) that are ideal cell-wall structural materials (Fig. 8.12). The helical and often branched starch molecules cannot be bound together into fibers, but do cluster into grains (Fig. 14.23) of microscopic size that accumulate in plant cells. Many other polysaccharides be-

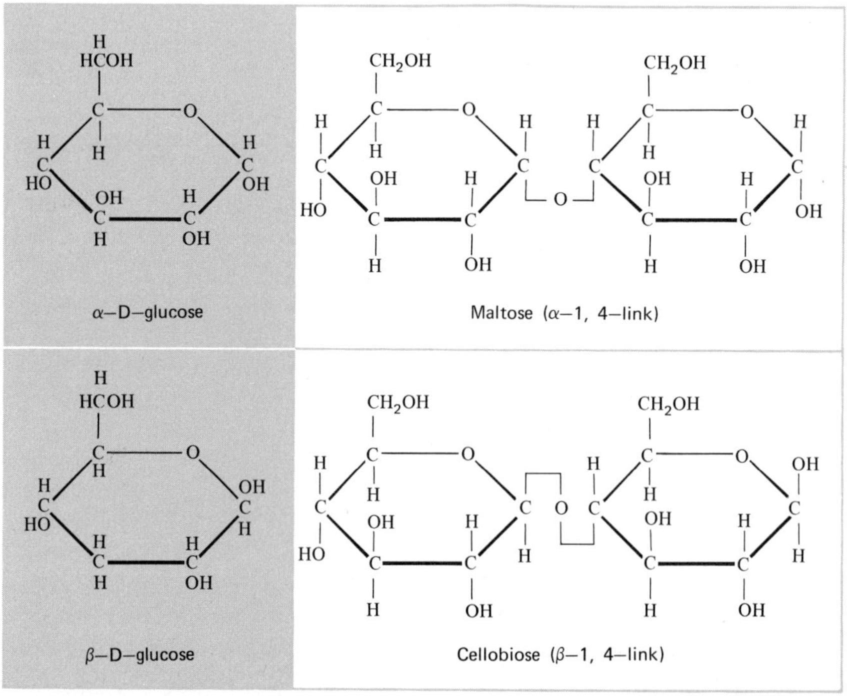

Fig. 7.5 Structural formulae of α-D-glucose and β-D-glucose and the disaccharides derived from each one (maltose and cellobiose respectively).

α–D–glucose

Maltose (α–1, 4–link)

β–D–glucose

Cellobiose (β–1, 4–link)

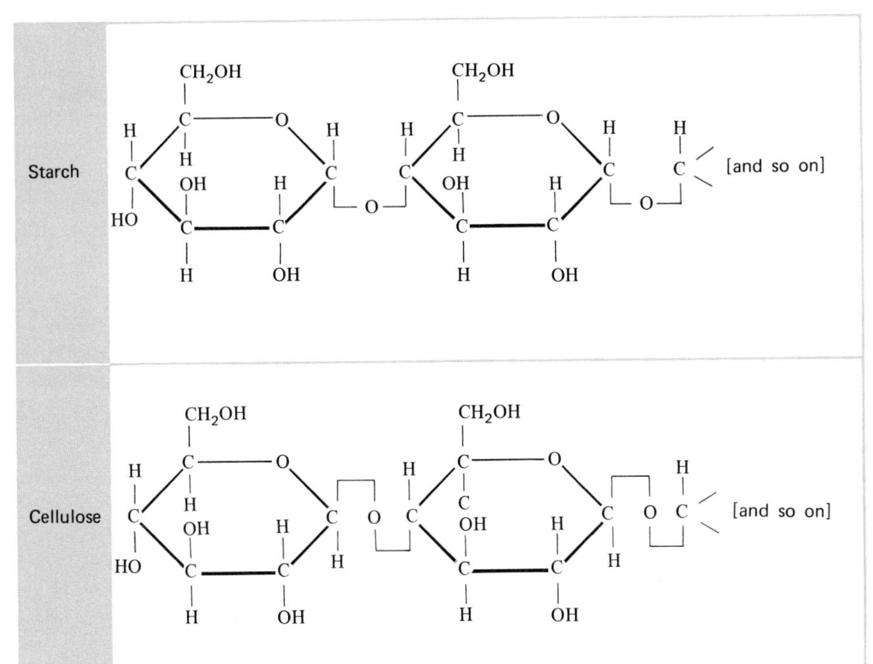

Fig. 7.6 *Structure of very small parts of a starch molecule and a cellulose molecule. Each compound contains on the order of 1000 glucose residues. Starch is synthesized from α-D-glucose and cellulose from β-D-glucose.*

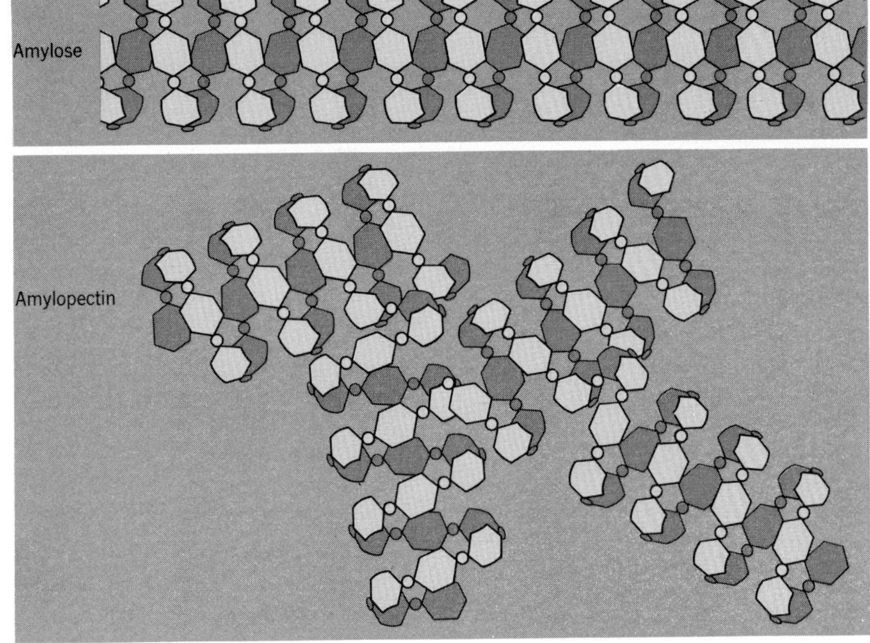

Fig. 7.7 *Diagram showing the helical arrangement of glucose residues in small portions of two types of starch, amylose and amylopectin. Amylose is an unbranched helix, whereas amylopectin is highly branched. Each turn of the helix is composed of six glucose residues (hexagons). (Drawing by Evan L. Gillespie, reproduced from* Principles of Plant Physiology *by James Bonner and A. W. Galston, 1952, through the courtesy of the authors and publisher, W. H. Freeman & Co.)*

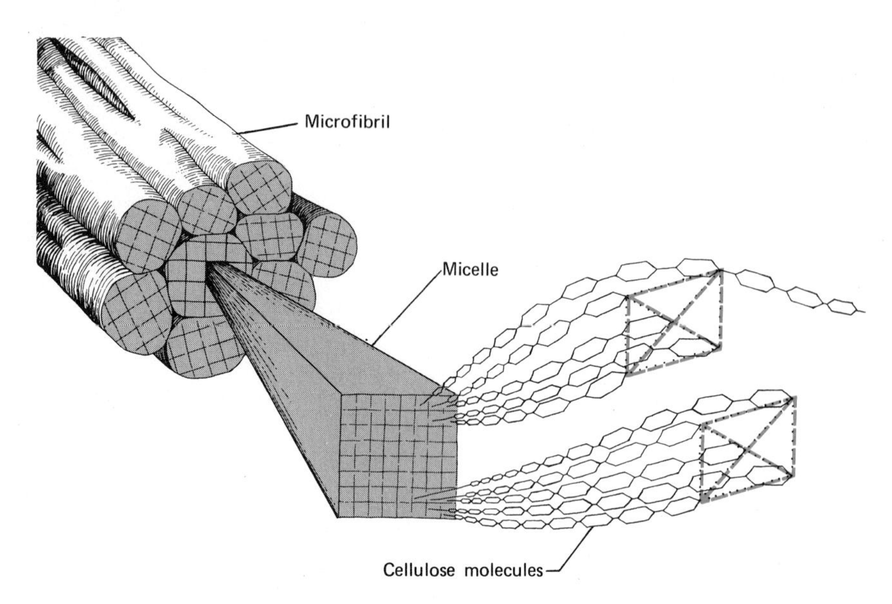

Fig. 7.8 Diagram of structure of a cellulose fibril. The basic units of five cellulose molecules each are bound into crystalline micelles, which in turn make up the branched microfibrils. (Modified from a drawing by Evan L. Gillespie in Principles of Plant Physiology *by James Bonner and A. W. Galston, through the courtesy of the authors and publisher, W. H. Freeman & Co.)*

sides starch and cellulose are synthisized by plants, a good many of them being cell-wall constituents. **Inulin** is a crystalline polysaccharide composed of fructose residues and is found most abundantly in certain members of the sunflower family. Most polysaccharides, unlike sugars, are neither sweet nor freely soluble in water.

Organic Acids

Many different organic acids occur in plants, and many of them play vital roles in the life processes of plants. Like inorganic acids, organic acids produce hydronium ions when in solution, the hydrogen being provided by the carboxyl (−COOH) group characteristic of organic acids. Organic acids are generally weak, since they are not highly ionized. Here we shall consider representatives of three classes of organic acids: the fatty acids, the "plant" acids, and the amino acids.

Fatty Acids. The **fatty acids** are composed of chains of carbon atoms linked together with a single −COOH group at the end of the chain. The fatty acids of plants are built up from acetic acid ($CH_3 \cdot COOH$) derivatives linked together and so all naturally occurring fatty acids have an even number of carbon atoms in them: 4, 6, 8, and on up to 20 or so. The structure of several fatty acids is given in Fig. 7.16. Fatty acids differ from one another in the number of carbons in the chain and also in the number of hydrogen atoms attached to the various carbon atoms. If all the carbon atoms of a fatty acid hold as many hydrogen atoms as possible, the fatty acid is **saturated,** as, for example, caproic acid: $CH_3 \cdot CH_2 \cdot CH_2 \cdot CH_2 \cdot CH_2 \cdot COOH$. If, however, some of the carbons in the chain hold only one hydrogen atom instead of two, the acid is unsaturated: $CH_3 \cdot CH = CH \cdot CH_2 \cdot CH_2 \cdot COOH$. Unsaturated fatty acids are more common in plants than in animals. Fatty acids are generally oily liquids. They are important as constituents of fats, waxes, and other lipdis.

Plant Acids. The so-called "plant acids" are a rather diverse group chemically, but all have relatively small molecules with usually six or fewer carbon atoms in them (Fig. 7.9). They generally have from one to three −COOH groups per molecule. Among the plant acids are citric, malic, succinic, and fumaric. Acetic acid may be considered either as a fatty acid or a plant acid. Plant acids have long been known to be particularly abundant in fruits and in the leaves of succulent plants like *Sedum,* but in recent decades they have come to be recognized as important substances in respiration and other life processes of both plants and animals.

Some of the plant acids contain a carbonyl or ketone group (>C=O) in addition to –COOH groups and are known as **keto acids.** In addition to playing a part in respiration, the keto acids may be converted into amino acids by the substitution of an $-NH_2$ (amine) group from ammonia and an –H for the oxygen of the >C=O group.

Amino Acids. The **amino acids** are important as the building units of proteins. There are many different amino acids, but only 20 or so are known to be constituents of proteins. All amino acids used in making proteins have the $-NH_2$ group that characterizes amino acids attached to the carbon that is adjacent to the carboxyl (–COOH) carbon. Thus the general structure of amino acids may be represented as $R\cdot\underset{NH_2}{\underset{\cdot}{CH}}\cdot COOH$, the R representing the remaining atoms of the molecule. R is different for each kind of amino acid. The simplest amino acid is glycine, and in it R is simply a hydrogen atom: $\underset{NH_2}{\underset{\cdot}{CH_2}}\cdot COOH$. The structures of several other amino acids are shown in Fig. 7.10. Several amino acids contain sulfur and at least a few of these sulfur-containing amino acids are found in protein molecules.

Fig. 7.9 Structural formulae of a few of the important polycarboxylic ("plant") acids.

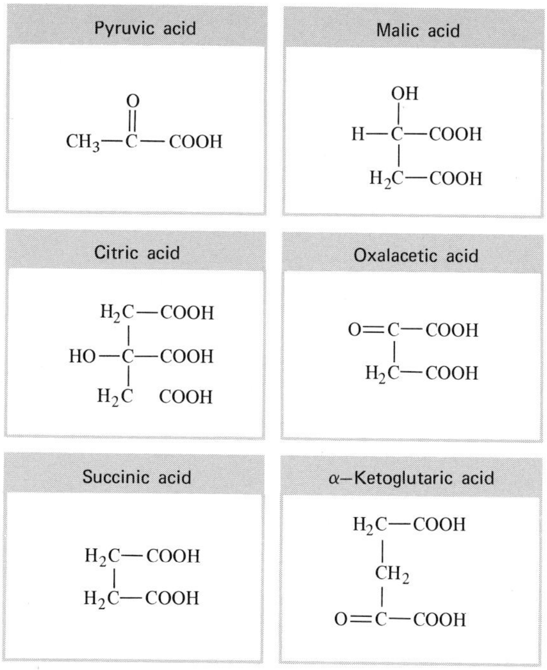

Pyruvic acid

$$CH_3-\overset{\overset{O}{\|}}{C}-COOH$$

Malic acid

$$\begin{array}{l} \quad\;\; OH \\ \quad\;\; | \\ H-C-COOH \\ \quad\;\; | \\ \; H_2C-COOH \end{array}$$

Citric acid

$$\begin{array}{l} \quad H_2C-COOH \\ \qquad | \\ HO-C-COOH \\ \qquad | \\ \quad H_2C \quad COOH \end{array}$$

Oxalacetic acid

$$\begin{array}{l} O=C-COOH \\ \quad\;\; | \\ \; H_2C-COOH \end{array}$$

Succinic acid

$$\begin{array}{l} H_2C-COOH \\ \quad | \\ H_2C-COOH \end{array}$$

α–Ketoglutaric acid

$$\begin{array}{l} \; H_2C-COOH \\ \quad\;\; | \\ \quad\; CH_2 \\ \quad\;\; | \\ O=C-COOH \end{array}$$

Proteins

Proteins are the most abundant organic compounds of protoplasm and, along with the nucleic acids, are probably more responsible for its characteristic properties than any other class of substances. Like starch and cellulose, proteins are giant molecules **(macromolecules).** A protein molecule is composed of hundreds to thousands of amino acid residues linked together, the –COOH of one amino acid being attached to the $-NH_2$ of the next one. This is known as a **peptide linkage.** The –OH and –H lost prior to the linkage process by the amino acids (Chapter 14) may be considered to represent the net production of a molecule of water. Since 20 different amino acids may be used in protein synthesis, millions of different proteins are possible, just as many words can be constructed from the letters of our alphabet. In contrast, starch and cellulose (composed of long chains of glucose residues) would be comparable to a single letter repeated hundreds of times, for example *ggggggggg*· · ·.

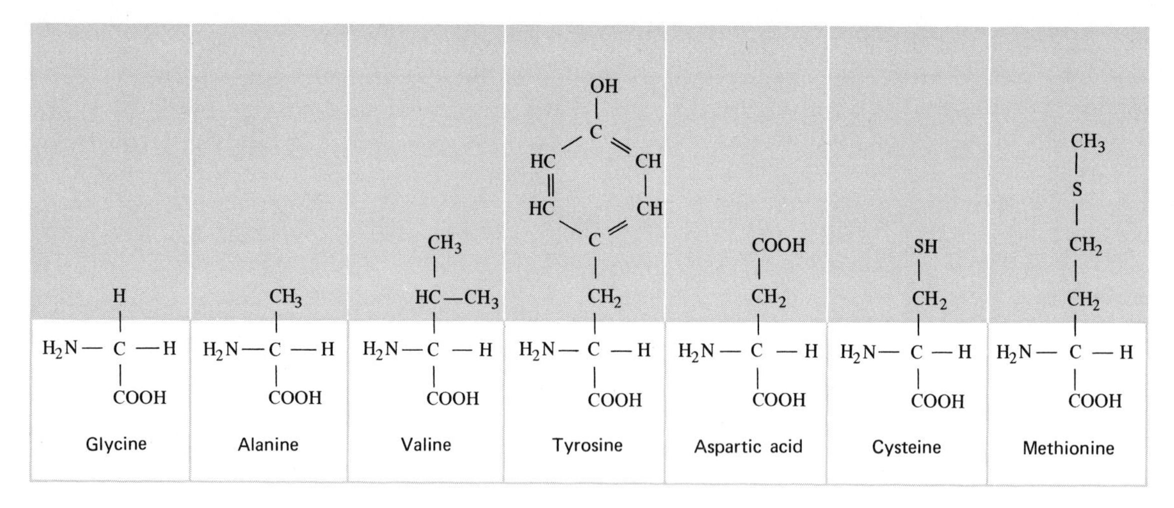

Fig. 7.10 Structural formulae of several of the amino acids used in protein synthesis. All have the basic structural features of α-amino acids, and differ from one another in the portions of the molecules in the shaded area.

Since each species of plant and animal has many kinds of proteins that are different from those of all other species it is evident that there are many millions of distinct kinds of protein in existence. Molecules of all kinds of protein are large, but the range of molecular weights is wide—from a few thousands in some kinds to several million in others. Protein molecules are not only large but also complex in structure. The primary structure of proteins is determined by the sequence of the different kinds of amino acid residues in the chain, each kind of protein having its own specific sequence of a characteristic number of amino acids. The substitution of a different amino acid for even a single one in the chain will produce a different protein, and may cause a drastic change in the catalytic capability of those proteins that are enzymes. The secondary structure of most proteins is the twisting of the chain into a helix known as an alpha helix (Fig. 7.11). The adjacent turns of the helix are bound together by hydrogen bonds between amino acids in one turn and those above or below them in the adjacent turn. In addition, cysteine components of the chain may bond with cysteines above or below them through their sulfur atoms, thus forming disulfide bonds (−S−S−). The tertiary structure of a protein molecule results from the complex folding of the alpha helix so that a structure resembling a badly tangled piece of string results (Fig. 7.11). However, despite the tangled appearance, each kind of protein has a very characteristic tertiary (as well as a primary and secondary) structure. Hydrogen and disulfide bonds connect adjacent portions of the twisted helices together, just as they do the adjacent turns within a helix. These bonds may be broken by heat or various chemicals, resulting in loss of the tertiary structure. Such proteins are said to be denatured, and they have changed physical properties and loss of enzyme activity.

Enzymes, the organic catalysts (Appendix) essential for the activation of practically all the biochemical reactions of organisms, are proteins. Some enzymes have nonprotein **coenzymes** associated with them. These coenzymes play essential roles in the catalytic action of the enzymes. Like other catalysts, enzymes affect the rate of chemical reactions, generally greatly increasing them. Although an enzyme is not consumed in a reaction and appears unchanged at the end, it does enter into temporary combination with the reacting substances. Enzymes are very specific, each one reacting only with substances with a certain molecular configuration. Thus, an enzyme that can hydrolyze starch is unable to hydrolyze cellulose. A living cell contains numerous different enzymes, probably well over a thousand. The kinds of enzymes present in any particular organism determine the reactions that can take place in it. Although some processes that are activated by enzymes in plants, such as the breakdown of starch into glucose, can be carried on outside of organisms without enzymes, high temperatures and acids or other strong reagents are usually required.

Most enzymes act within cells, but some enzymes may be secreted by cells and act outside of them, for example, the digestive enzymes of animals and fungi. Enzymes can be extracted from tissues, and some of these extracts can catalyze their characteristic reactions independently of the organism that produced them. **Diastase,** a crude extract of starch-digesting enzymes from germinating grains or molds, is sold in many drugstores as well as by chemical supply houses. **Papain,** extracted from fruits of the tropical papaya, is a protein-digesting enzyme sold by grocery stores as a meat tenderizer. Enzymes can be isolated from the crude extracts in pure crystalline form (Fig. 7.12) and over a hundred different pure enzymes have been prepared.

The **nucleoproteins** constitute another extremely important class of proteins, consisting of proteins linked with nucleic acids. Chromosomes are basically nucleoprotein structures. Enzymes and nucleoproteins constitute a substantial portion of the protoplasmic proteins. In contrast with these biologically active proteins are the apparently inert proteins that accumulate as crystals in seeds and sometimes in other parts of plants.

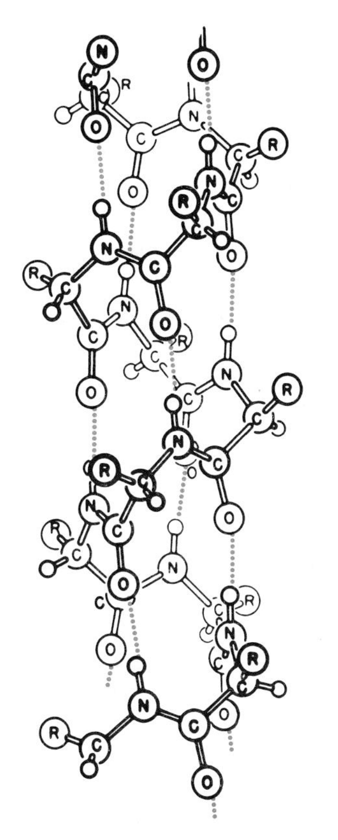

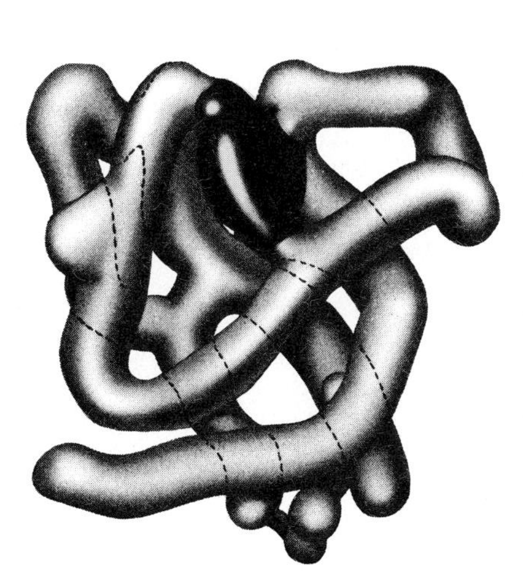

Fig. 7.11 *(Left) Diagram of a portion of a protein alpha helix, showing the orientation of several amino acid residues. The dotted lines represent hydrogen bonds. (Right) Model of a protein molecule, showing the highly folded tertiary structure. The convoluted tubule represents the alpha helix without showing the atomic details as in the left hand diagram. (Left: after Pauling. Right: after Kendrew.)*

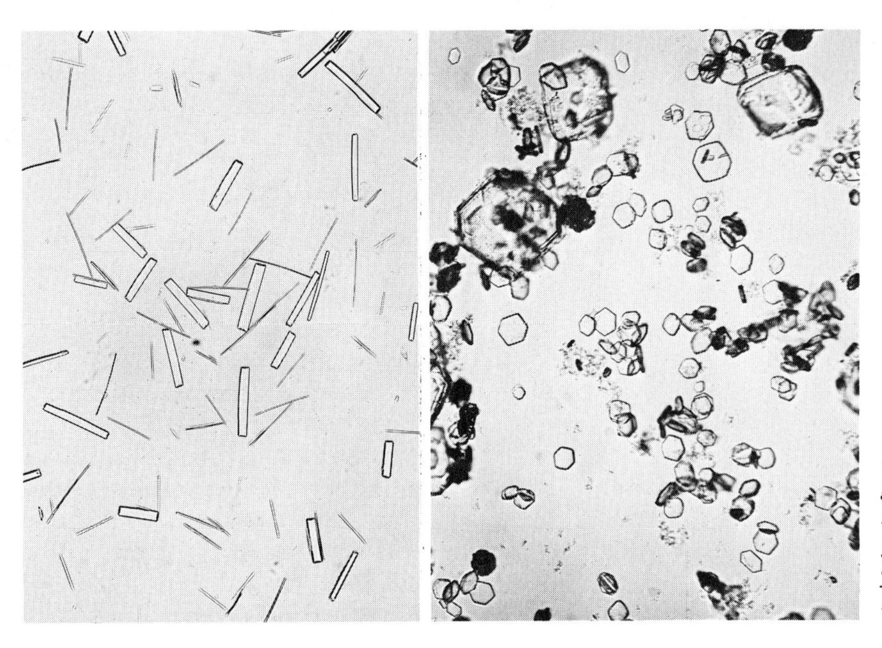

Fig. 7.12 Photomicrographs of crystals of two enzyme proteins. (Left) Inorganic pyrophostase from yeast. (Right) Trypsin inhibitor from soybean.

Lipids

The **lipids** are another important class of plant constituents and, like others we have discussed, are essential components of plant cells. Lipids are fatty or oily substances and include the **fats, waxes,** and **phospholipids.** These are not soluble in water, but are soluble in fat solvents such as gasoline, ether, and carbon tetrachloride. Although fats are insoluble in water, they can form **emulsions** consisting of microscopic fat globules uniformly dispersed in water, as in homogenized milk. **Emulsifying agents** such as soap, detergents, egg yolk, and gums prevent the fat globules from merging and floating to the surface as an oily layer. Fats in cells are usually present as emulsions.

Fats. The fats are compounds formed from **glycerol** (glycerine) and **fatty acids,** with three molecules of fatty acids reacting with each molecule of glycerol. The molecular structure of one fat is shown in Fig. 7.13. The many specific kinds of fats found in plants differ from each other in their constituent fatty acids. Commonly, at least one of the three fatty acids is different from the others. If a fat is liquid at room temperature, it is referred to as an **oil.** Most plant fats are oils, for example, corn, peanut, cottonseed, and olive oils. Oils are generally unsaturated, whereas solid fats are saturated. Drying oils used in making paints, such as linseed and tung oils, have an extra high degree of unsaturation. All such oils should not be confused with petroleum oils, an entirely different class of compounds (hydrocarbons).

Soaps are salts of fatty acids and are made by allowing fats to react with potassium hydroxide (KOH) or sodium hydroxide (NaOH, commonly called lye). Glycerol is also produced by the reaction.

Fat molecules contain proportionately less oxygen and more hydrogen than the molecules of carbohydrates, and as a result contain much more energy. A pound of any fat contains about twice as much energy (calories) as a pound of any carbohydrate. Consequently, fats are an excellent storage form of food.

Phospholipids. Like fats, **phospholipids** are formed from glycerol and fatty acids, but a phosphoric acid molecule replaces one of the three fatty acids. Although cells contain much less phospholipid than fat, the phospholipids are probably more essential cell components. They are involved in cell membrane, chloroplast, and general protoplasmic structure and in important biochemical reactions. Much of their structural importance is probably related to the fact that whereas the bulk of the molecule is fat soluble, the phosphoric acid group is water soluble, which enables phospholipids to stabilize fat-water junctions as an emulsifier. The more complex phospholipids such as lecithin have additional compounds attached to the phosphoric acid unit.

Fig. 7.13 Molecular structures of three fatty acids (upper left), two alcohols (upper right), a triglyceride or fat (center), and a wax (bottom).

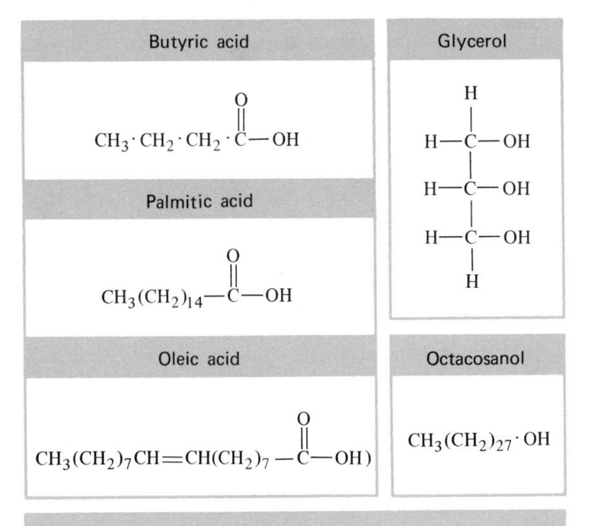

Waxes. The **waxes** are also made from fatty acids and alcohols, but unlike the other lipids they do not contain the alcohol, glycerol. Instead, a fatty acid molecule unites with a long straight chain alcohol having 24 to 36 carbon atoms and only one −OH group at the end (Fig. 7.13). Plant waxes are primarily important as constituents of **cutin,** which covers the epidermis of leaves, stems, and fruits and of **suberin,** the waterproofing material of the walls of cork cells. Plant waxes are used in making the more expensive and better grades of polishes for shoes, floors, furniture, and automobiles.

Nucleic Acids

Among the most complex and most important compounds in plants and animals are the **nucleic acids** (Figs. 7.14 and 7.15). The genes, or hereditary potentialities transmitted from one generation to another, are now known to be nucleic acids. Nucleic acids also constitute the genetic material of viruses. The chromosomes in the nuclei of cells contain **deoxyribonucleic acid** (DNA) and it is this nucleic acid that makes up the essential part of the genes. DNA controls the production of another kind of nucleic acid, **ribonucleic acid** or RNA, which is also present in cells and plays an essential role in the synthesis of proteins.

DNA molecules are long double helices built up from hundreds of smaller units known as **nucleotides.** Each nucleotide is made from a molecule of deoxyribose (a 5-carbon sugar with a −H substitued for one of the −OH groups of ribose, hence the *deoxy-*), a molecule of phosphoric acid, and a molecule of one of

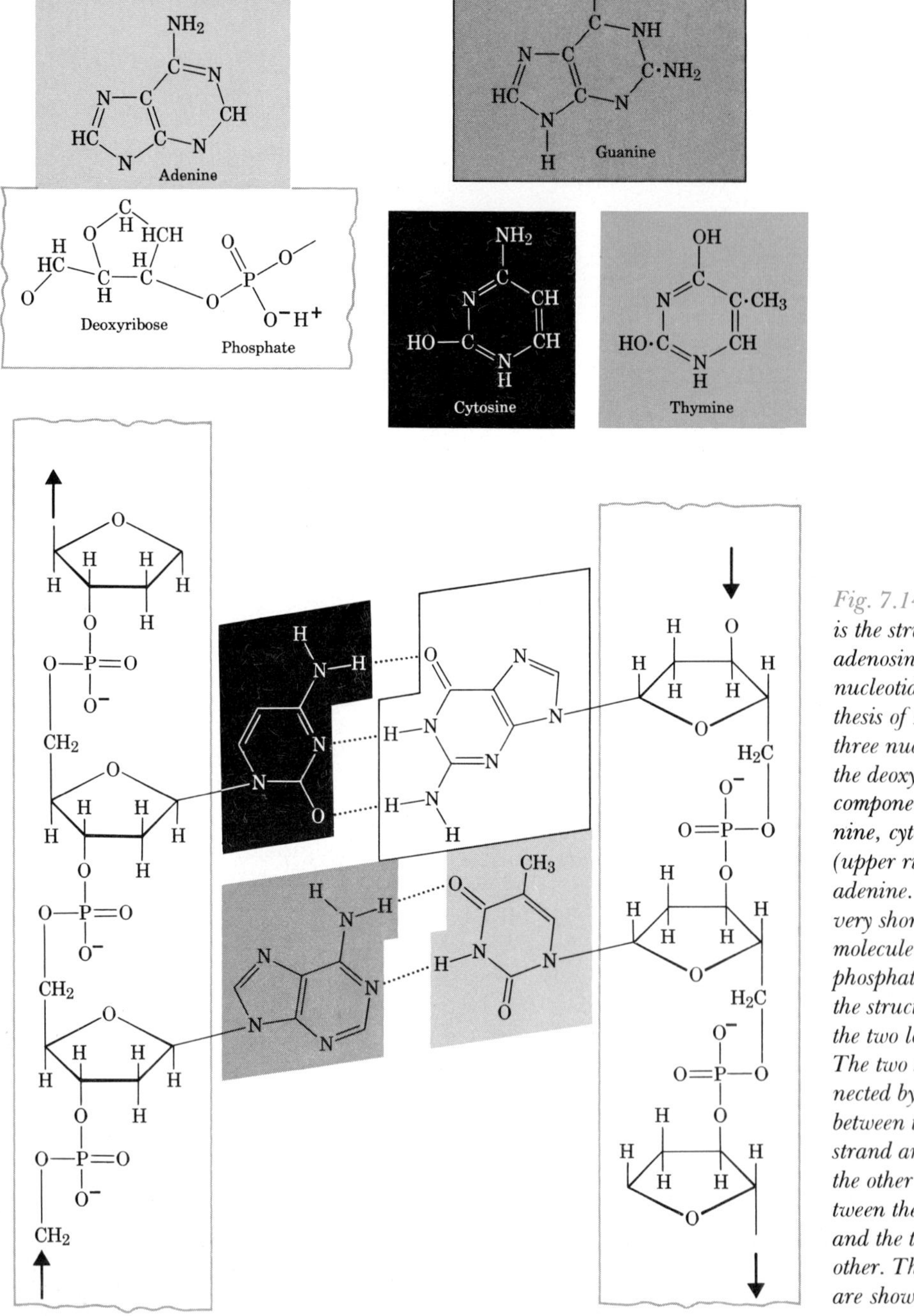

Fig. 7.14 At the upper left is the structural formula of adenosine, one of the four nucleotides used in the synthesis of DNA. The other three nucleotides also have the deoxyribose phosphate component, but have guanine, cytosine, or thymine (upper right) in place of the adenine. At the bottom is a very short portion of a DNA molecule. The deoxyribose phosphate chains make up the structural continuity of the two long helical strands. The two strands are connected by hydrogen bonds between the cytosine of one strand and the guanine of the other strand, and between the adenine of one and the thymine of the other. The hydrogen bonds are shown as broken lines.

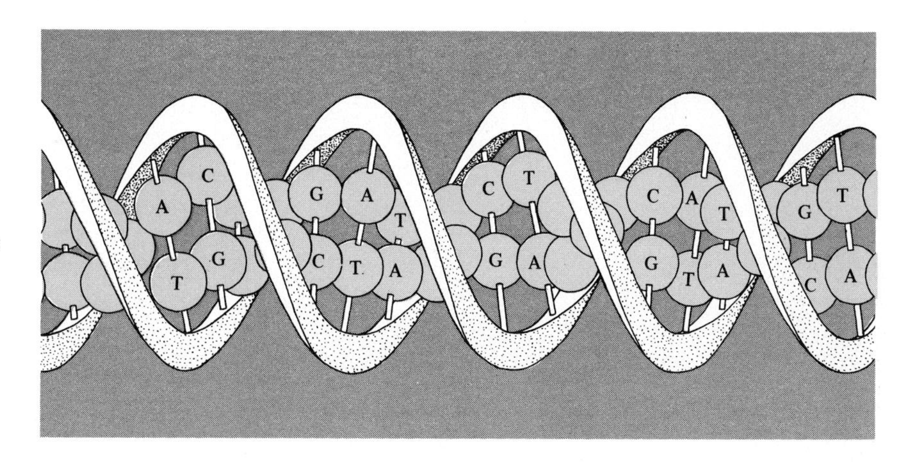

Fig. 7.15 Diagram of a small portion of a DNA molecule. The deoxyribose phosphate units making up the two helical strands are not shown individually, but the spheres represent the purine and pyrimidine bases that link the two helical strands of the molecule. A = adenine, C = cytosine, G = guanine, T = thymine. (Courtesy of The Fisher Scientific Company.)

four organic bases: adenine, guanine, cytosine, or thymine (Fig. 7.14). The four kinds of nucleotides differ only in their organic bases. In the DNA molecule, the phosphoric acid of one nucleotide is linked to the deoxyribose of the next, resulting in a long chain of alternating deoxyribose and phosphoric acid residues. The organic bases are attached to the deoxyribose molecules, and extend as side chains to the inside of the DNA molecule. Each DNA molecule is composed of two such sugar-phosphate chains with the organic bases of one chain joined to those of the other chain by hydrogen bonds, the whole molecule being twisted into a double helical structure. Thymine can be bonded only to adenine, and guanine only to cytosine. Since the base pairs thus formed can be arranged along the molecule in any sequence, they can provide a coding system that carries genetic information.

The molecular composition of RNA differs from that of DNA in two basic respects: it contains the sugar ribose instead of deoxyribose and the base uracil in place of thymine. Whereas DNA occurs primarily in the nuclei of cells (some also being present in chloroplasts and mitochondria), RNA is found both inside and outside of the nuclei. There are at least three different kinds of RNA: **messenger RNA** (mRNA), **transfer RNA** (tRNA), and **ribosomal RNA** (rRNA). Like DNA, the mRNA molecules are composed of long chains of nucleotides, but there is only a single strand instead of the double helix of DNA. The mRNA is produced in the nucleus on a DNA template and moves to the cytoplasm where, when in contact with ribosomes, it determines the sequence of amino acids incorporated into the protein molecule for which it carries the code (message) obtained from the DNA of the nucleus. The ribosomes are composed of rRNA and protein. The tRNA molecules are smaller than the mRNA molecules and are folded several times (Fig. 7.16). Each kind of tRNA can bond with only one kind of amino acid and also carries a nucleotide code complementary to one kind of mRNA code. Thus the amino acid-tRNA complex insures that the amino acid is added to the growing protein molecule at the precise point designated by the mRNA code (and, in turn, the DNA code). All this will be considered in greater detail in Chapters 14 and 23.

Other Important Compounds

Although the carbohydrates, proteins, lipids, and organic acids constitute the great bulk of the organic compounds synthesized by plants, many of the less abundant constituents

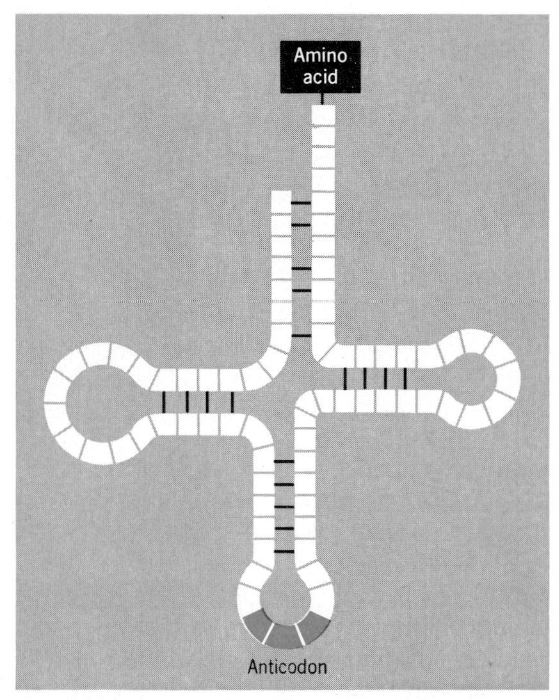

Fig. 7.16 The cloverleaf model of transfer RNA structure. The single RNA strand folds back on itself, forming four paired base regions (C-G, U-A), three unpaired loops, and a free, unpaired end to which only one kind of amino acid can attach. The bottom loop includes a triplet anticodon that can bond only with a mRNA codon for the specific amino acid attached to the tRNA molecule.

are of great importance in the biochemical processes of plants. Among these are the hormones, vitamins, antibiotics, chlorophylls, carotenoids, hydrogen carriers such as NAD (nicotinamide adenine dinucleotide), and energy carriers such as ATP (adenosine triphosphate). These substances will be considered in later chapters. Lignin, a complex ring compound, is abundant in the walls of cells in wood.

Some plants also synthesize a variety of other compounds including the anthocyanins (the red, blue, violet, and pink pigments of flowers and other plant structures), alkaloids (e.g., caffeine, nicotine), tannins, essential oils, (volatile oils such as peppermint and wintergreen), turpentine, resins, and rubber latex. Such substances, some of them produced by a limited number of plant species, are often of considerable commercial value but have generally been considered to be of little or no importance in the plants producing them. Further research may, however, reveal that they play roles of some importance.

MOLECULAR AGGREGATES

The molecules of gases and of many liquids exist individually in a more or less random and unorganized fashion. The molecules of most solids and some liquids, however, cluster together in definitely organized patterns thus forming structures varying in size from those visible to the unaided eye to those too small to be detected by electron microscopes. These **molecular aggregates,** although by no means restricted to organisms, constitute an important level of organization in plants and animals which is intermediate between individual molecules and cell structures such as plastids, nuclei, and walls. Starch grains and cellulose fibrils are both molecular aggregates, as are crystals and colloidal particles.

Colloids. Many of the structural components of protoplasm are present as **colloids.** Colloidal particles are small clusters of molecules (or sometimes, as in starch or proteins, single giant molecules) dispersed through a solid, gas, or liquid. Fog, mucilage, gelatine desserts, and most clay soils are all different types of colloidal systems. In cells the medium through which colloidal particles are dispersed is usually water. The dispersed particles of a colloid range in size between 0.001 μ and 0.1 μ (μ = **micron,** 1/1000 of a millimeter or 1/25,400 in.). The dispersed molecules and ions of a true solution, as of a salt, are less than 0.001 μ in diameter, whereas the droplets of liquid dispersed through an emulsion and the solid particles dispersed in a suspension such as muddy water are more than 0.1 μ in diameter. The particles of a suspension and the droplets of emulsions are visible under an ordinary microscope, but both colloid particles and the molecules and ions of solutions are too small to be seen through ordinary microscopes. Colloid particles may, however, be visible by means of an electron microscope and show up as bright dots of light diffracted from a beam at right angles to the tube of a light microscope.

Dispersed colloid particles are electrically charged, each particle having numerous charges rather than a few such as carried by ions. Since all the particles of any particular colloid have the same charge—either positive or negative—they repel each other and do not clump together and settle out. Some colloidal particles are covered with a jacket of compactly arranged water molecules. This also prevents clumping of the particles. Some colloids are no more viscous than solutions whereas others, like glue, are very viscous.

Certain colloidal systems, particularly gelatin and other protein colloids, can exist in either a liquid (**sol**) or a jellylike (**gel**) state and change from one to the other with changes in temperature and other factors. In gelatin, and probably other gels, the semisolid consistency is a result of the aggregation of the colloid particles into a network of fibers. The protein colloids of protoplasm generally are in a state of flux between the sol and the gel forms.

The colloidal state facilitates chemical reactions, since the small colloid particles provide relatively immense total surface areas where the reacting substances can make contact. As a substance is divided into smaller and smaller particles, the total surface area increases and the volume remains constant (Fig. 7.17, Table 7.2). A cube of protein 1 cm on a side has a surface area of 6 cm^2 whereas the same volume of protein in the form of colloidial particles 10^{-7} cm (0.001 μ) on a side has a total surface area of 6×10^7 (60 million) cm^2 (Table 7.2). Other colloid properties of biological importance are the charges on the particles, the solgel transformations, and the ability of colloid particles to attract and hold (**adsorb**) molecules and ions on their surfaces.

Crystals. That substances such as salts and sugars form **crystals** is a matter of general knowledge, but it is not so generally recognized that many other substances including enzymes and other proteins and even viruses may

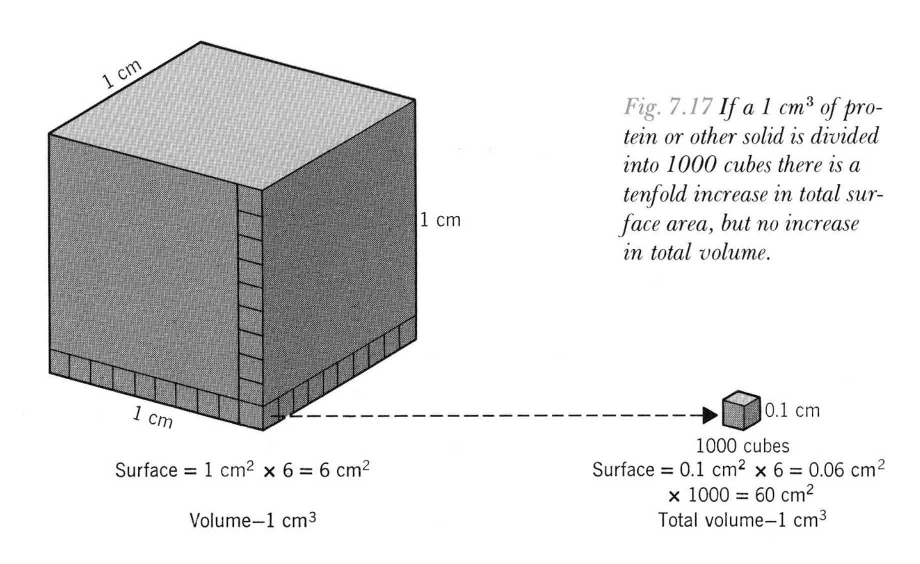

Fig. 7.17 If a 1 cm^3 of protein or other solid is divided into 1000 cubes there is a tenfold increase in total surface area, but no increase in total volume.

Table 7.2 **Increase in Total Surface of Cubes When a Cube 1 cm on a Side is Subdivided into the Indicated Number of Cubes**

Number of Cubes	Cube Diameter, cm	Surface Area per Cube, cm^2	Total Surface Area of Cubes, cm^2	Total Volume, cm^3
1	1.0	6.0	6	1
1,000	0.1	0.06	60	1
1,000,000	0.01	0.006	600	1
10^9	10^{-3}	6×10^{-6}	6×10^3	1
10^{12}	10^{-4}	6×10^{-8}	6×10^4	1
10^{15}	10^{-5}	6×10^{-10}	6×10^5	1
10^{18}	10^{-6}	6×10^{-12}	6×10^6	1
10^{21}	10^{-7}	6×10^{-14}	6×10^7	1

also be crystalline (Figs. 7.15, 7.18, and 7.19). The molecules making up a crystal are arranged in precise geometrical patterns characteristic of the substance, and as a result crystals of a compound have typical shapes that can be used in identifying the substance. Although most substances in cells are either in solution or in colloidal form, salt crystals are not uncommon in cells. Most of these are crystals of calcium salts. Seeds frequently contain proteins in crystalline form.

VIRUSES

A level of organization higher than that of molecular aggregates is represented by viruses and the various structures within cells such as chloroplasts, chromosomes, and mitochondria. Such cell structures will be considered in the next chapter. It is, however, difficult to decide where a consideration of viruses should be placed if, as in this book, a separate chapter cannot be devoted to them. Although they are much more complex and highly organized than any ordinary molecular aggregate and have certain properties of life, they have not attained a cellular level of organization and lack many of the characteristics that are common to all true organisms. Whether they should be regarded as the most complex natural nonliving things or the most primitive living things is a matter of opinion.

Although viruses vary considerably in size, they are all so small that they can readily pass through fine filters that hold bacteria and cannot be seen through the most powerful light microscopes. Electron microscopes, however, reveal the fact that any particular kind of virus has a characteristic structure that may be quite different from that of other kinds (Fig. 7.20). Although rods and spheres are the most common shapes, some viruses have more complicated structures such as heads and stalks. All viruses contain a nucleic acid core and a protein coat, and some also contain other substances such as lipids. However, they do not contain the wide variety of substances characteristic of even simple organisms such as bacteria, nor do they have structures such as differentially permeable membranes that are characteristic of all cells.

The nucleic acid of a virus provides it with a genetic code for the production of certain enzymes, and for making copies of itself. However, no virus is able to produce enzymes or replicate itself alone. This occurs only when a virus enters a living host cell and directs it to produce new virus nucleic acid copies, protein coats, and thus new virus particles. A virus it-

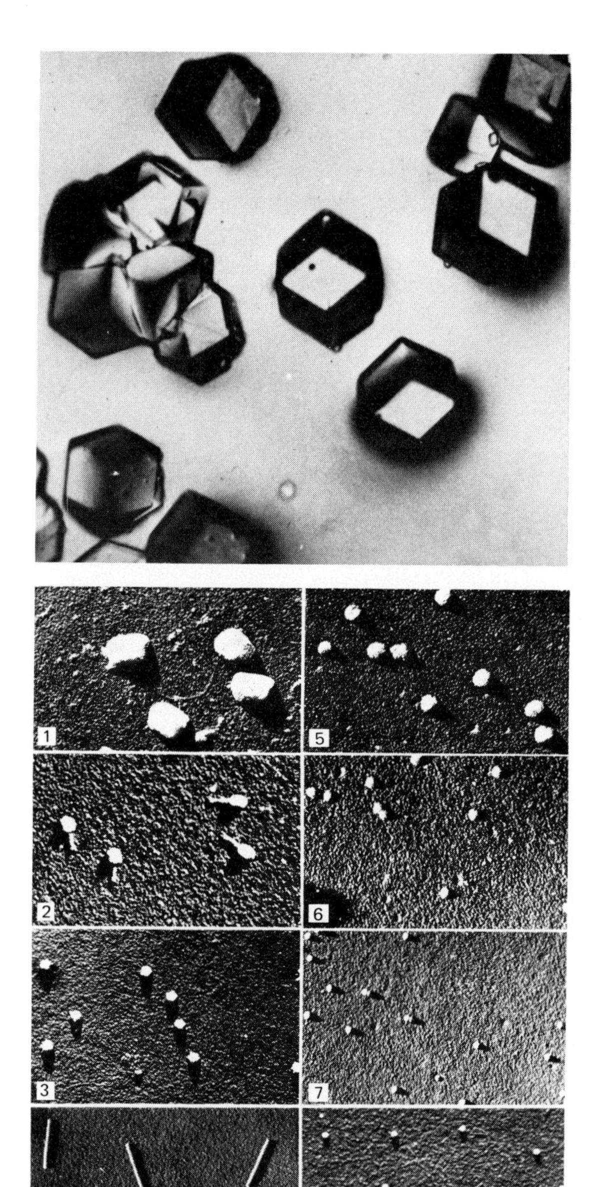

Fig. 7.18 (Left) Crystals of tomato bushy stunt virus (100X).

Fig. 7.20 Electron micrographs of several different kinds of virus. (1) vaccina, (2) T-2 bacteriophage, (3) T-3 bacteriophage, (4) tobacco mosaic, (5) influenza, (6) Shope papilloma, (7) tomato bushy stunt, (8) polio. Note the 1 micron scale at the lower right.

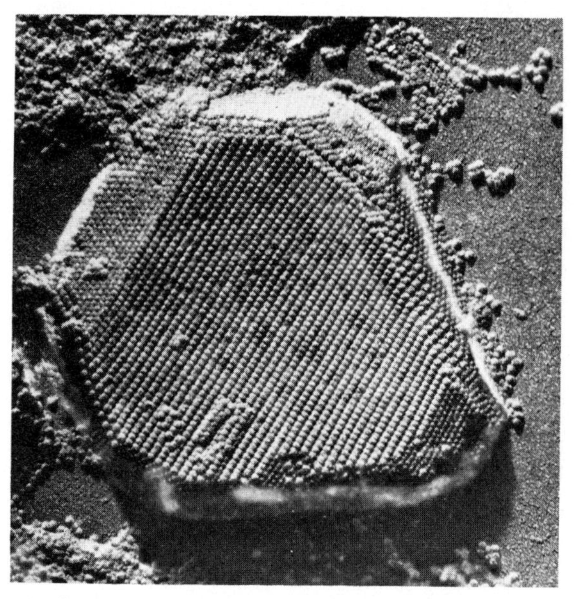

Fig. 7.19 Electron micrograph of a crystal of tobacco necrosis virus, showing the regular arrangement of the individual viruses composing the crystal (23,500X).

self can carry on none of the metabolic processes, such as respiration, that are characteristic of organisms. Viruses also differ from true organisms in that they can form crystals. When the crystals dissolve, the virus retains its full activity.

The presence of a virus in the cells of a host organism alters the metabolism of the cells by coding for enzymes (and so metabolic processes) not normally produced by the cells. The altered metabolism may result in serious disease or even disruption of the cells or death. Among the virus diseases of human beings are polio, influenza, and small pox, and there is evidence that viruses cause at least some kinds of cancer. Viruses also cause many serious plant diseases including tobacco mosaic (Fig. 7.21), tobacco necrosis, and tomato bushy stunt. Bacteriophages are viruses that attack bacteria and

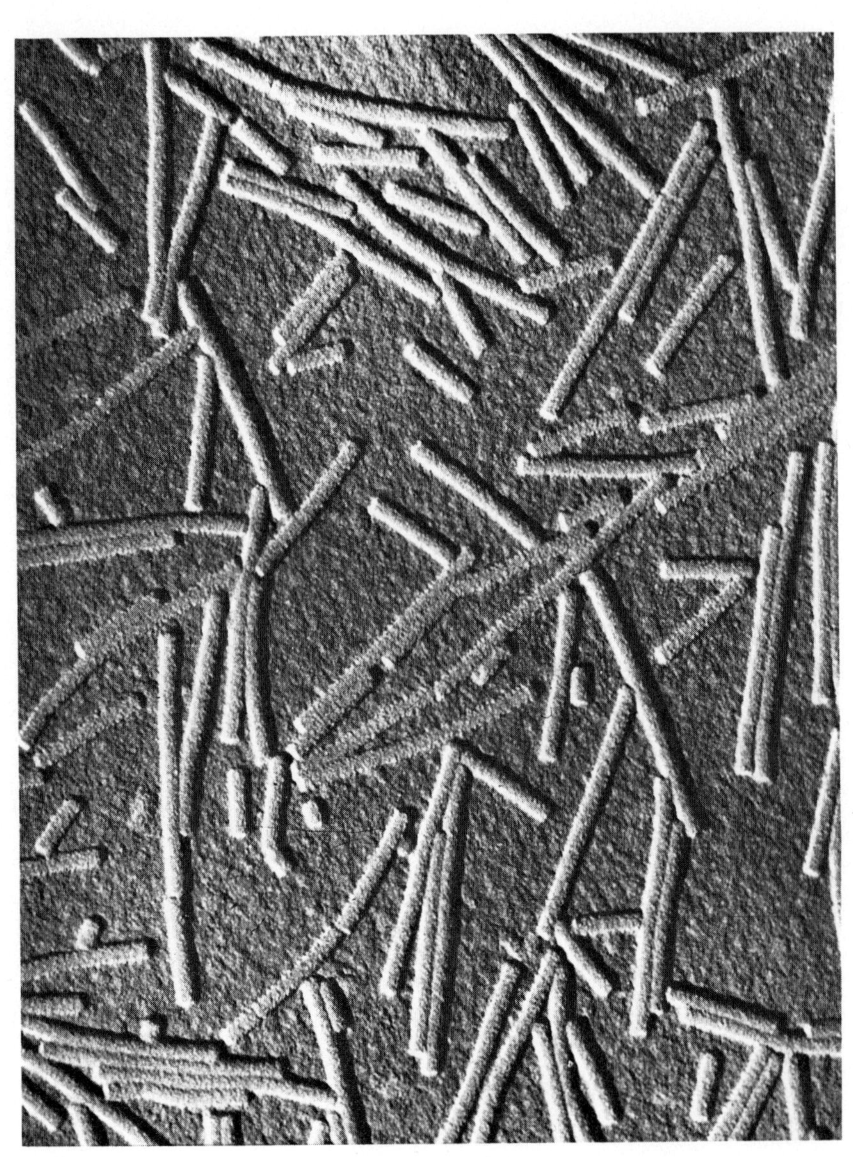

Fig. 7.21 High resolution electron micrograph of tobacco mosaic virus.

cause disruption of their cells (Fig. 7.22). However, some viruses have little or no serious effect on their host organisms and may bring about such conditions as variegated flowers or leaves in some species of plants. For example, the variegated floral patterns of some tulips are caused by a virus.

IN PERSPECTIVE

Has this chapter dealt with botany or chemistry? The answer is both. Although we have been considering chemical substances, they are, after all, the basic structural units of

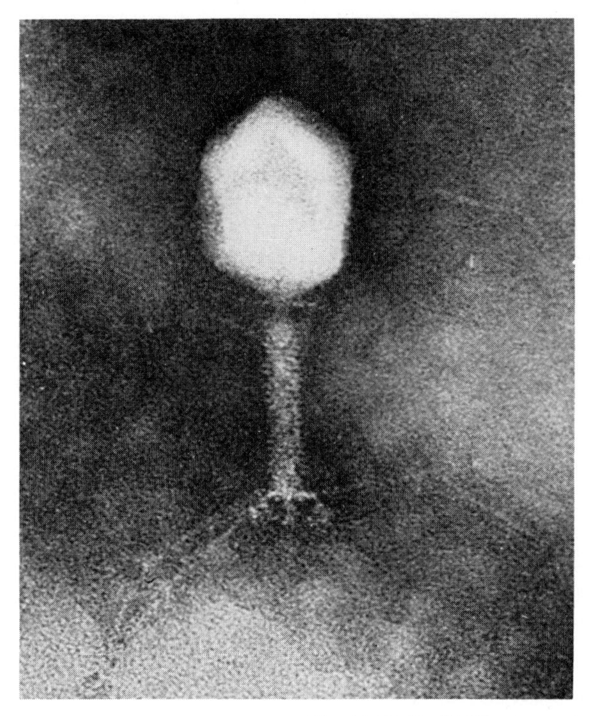

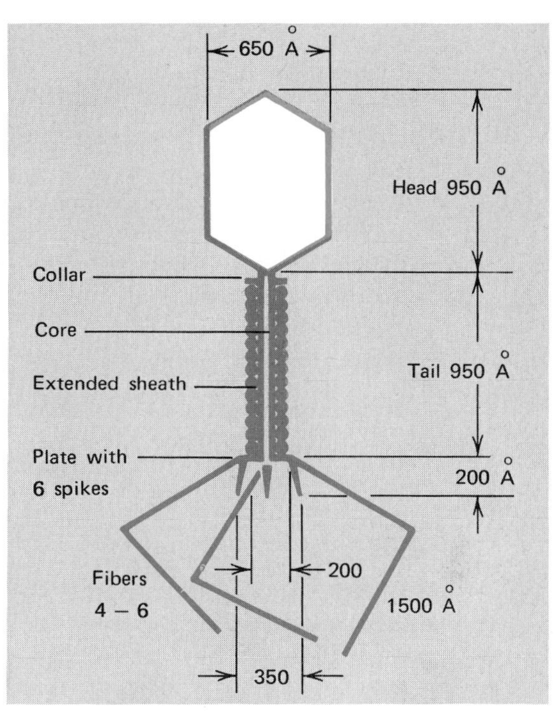

Fig. 7.22 (Left) Electron micrograph of a T-4 bacteriophage virus. (Right) Drawing of the virus with the structural details labeled and the dimensions given in Ångström units (1 Å = 0.1 nm = 0.0001 μ). The DNA is in the head, enclosed in a protein sheath, and the tail is protein. The virus attaches to a bacterium by its tail and the DNA is injected into the bacterium through the tubular tail.

plants, and some knowledge of them is just as essential for an understanding of plants as is some knowledge of the larger structural units fashioned from them. The molecules of cellulose, protein, lipids, and the other organic constituents of plants are, of course, just as characteristic products of plant activities as are protoplasm, cell walls, tissues, and organs. And, in the final analysis, plant activities occur at the molecular level whether these activities relate to biochemical reactions or physical energy transformations.

There is no single chemical level of plant organization, but rather a series of organizational levels of increasing complexity and particle size. From the electrons, protons, and neutrons of atoms, we go up to the atomic level of organization. Atoms, in turn, are organized into molecules and molecules may be linked together to form giant molecules. Next we come to molecular aggregates such as colloidal particles, crystals, the droplets of emulsions, and starch grains. A still more complex level of structural organization is found in viruses and a variety of cell structures such as chromosomes, mitochondria, and plastids. Each succeeding level of chemical organization is less characteristic of matter in general and more characteristic of living matter than the previous one, but it is difficult to determine the point at which organization ceases to be strictly chemical and assumes the properties of life. Some biologists feel that no unit smaller than a cell can really be considered to be alive, whereas others feel that viruses should be regarded as living entities.

As basic as the chemical level of organization is to an understanding of the structure and activities of plants, we must, of course, consider the higher levels of plant organization to have even a reasonably good picture of the nature of plants. A mixture of all the compounds making up a bean plant in just the right proportions would be a nourishing soup, but it would no more be a bean plant than a pile of the right amounts of steel, glass, gold, and jewels would be a watch. If, however, we provide a bean plant with water, oxygen, carbon dioxide, and the essential mineral elements, it will fashion these simple substances into its characteristic molecules, molecular aggregates, cells, tissues, and organs, a self-assembly job so characteristically limited to living things that it is ridiculous to consider a comparable accomplishment by any mechanism such as a watch. Nor could we imagine a watch reproducing itself, as a bean plant characteristically does. Like a watch, however, a bean plant must have a continuing supply of energy if it is to be a functioning entity, and both the watch and the plant can function only within a suitably organized structural system.

RELATED READING

Clevenger, S. "Flower pigments," *Scientific American, 210*(6), 85–92, June 1964.

Fowden, L. "The non-protein amino acids of plants," *Endeavour, 21,* 35–42, 1962.

Frank, S. "Carotenoids," *Scientific American, 194*(1), 80–86, January 1956.

Haagen-Smit, A. J. "Essential oils," *Scientific American, 189*(2), 70–75, August 1953.

Horne, R. W. "The structure of viruses," *Scientific American, 208*(1), 48–56, January 1963.

Lambert, J. B. "The shapes of organic molecules," *Scientific American, 222*(1), 58–70, January 1970.

Nord, F. F. and W. J. Schubert. "Lignin," *Scientific American, 199*(4), 104–113, October 1958.

Phillips, D. C. "The three-dimensional structure of an enzyme molecule," *Scientific American, 215*(5), 78–90, November 1966.

Preston, R. D. "Cellulose," *Scientific American, 197*(3), 156–168, September 1957.

Robinson, T. "Metabolism and function of alkaloids in plants," *Science, 184,* 430–435, 1974.

Thomas, M. "Vegetable acids in higher plants," *Endeavour, 10,* 160–165, 1951.

8
THE CELLULAR LEVEL

As we have seen in Chapter 3, the Plant Kingdom is made up of living organisms displaying such a diversity of form and range in size that by outward appearances, they seem to have little in common. But in spite of this diversity they do share important structural and functional attributes. As living organisms, large and small, they demonstrate those functional properties that are the mark of living things. That is, they maintain themselves, grow and differentiate in a controlled and orderly fashion, are responsive to varying conditions of environment, and give rise to new generations like themselves through characteristic reproductive processes. As the chemist seeks to explain the reactions of his materials through identification and understanding of the nature of some unique unit, so the biologist comes to recognize that the seat of the multiplicity of chemical and physical interactions that manifest themselves as attributes of the living organism is a unique physical unit, the **cell.**

The cell as a structural unit was first observed by Robert Hooke in 1665 in thin slices of cork and other plant tissues. In Hooke's words: ". . . I could exceedingly plainly perceive it to be all perforated and porous, much like a Honeycomb . . . these pores, or cells, were not very deep, but consisted of a great many little boxes, . . . Nor is this kind of texture peculiar to Cork only; for upon examination with my microscope, I have found that the pith of an Elder, or almost any other tree, the inner pulp or pith of the Cany hollow stalks of several other Vegetables: as Fennel, Carrets, Daucus, Bur-docks, Teasels, Fearn . . . &c have much such a kind of Schematisme. . . ." As so frequently happens in the realm of scientific inquiry, the fuller significance of Hooke's discoveries came many years later. In 1839, Schleiden and Schwann formulated the cell theory in which, on the basis of their own studies and those of older and contemporary observers, they postulated that the bodies of all plants and animals are composed of cells and the products of cell activity. In this great generalization they focused attention on the cell as a fundamental unit of life. With the later recognition by other observers that cells come only from preëxisting cells, came the realization that the story of life in all of its manifestations is basically the story of the cell.

STRUCTURE OF PLANT CELLS

Plant cells, with few exceptions, are too small to be seen by the unaided eye. Yet even though microscopic, they vary widely in size. The smallest known cell is a bacteria-like organism, *Mycoplasma,* which measures 0.1μ ($^1/_{250000}$ in.) in diameter. Far beyond the range of visibility with the light microscope, this organism was discovered by use of the electron microscope. The smallest cells visible with the light microscope ($0.2–5.0\ \mu = {}^1/_{125000}–{}^1/_{5000}$ in.) are found among the bacteria. In the more familiar higher plants the general range in diameter is 10–100 μ. Exceptional in size are some fiber cells of flax and cotton hairs that may attain a length of 5 cm. Cells also vary greatly in shape, as will be shown later in this chapter. However, with all of these variations, all living plant cells are fundamentally alike in that each consists of a minute bit of living substance called **protoplasm,** encased, with certain exceptions, in a nonliving wall of *cellulose.* A variety of differentiations of the protoplasm may occur, and the cellulose wall may be variously modified by the presence of other substances (Fig. 8.1). Such specializations in internal organization and wall constitution, along with specialization of shape (from the simple spherical), are associated with the performance of special functions. In view of this, it is not possible to identify any one cell as "typical" of all cells, but we can visualize a "composite" cell that would embrace many cellular features.

Progress in our knowledge of the cell is largely dependent upon the development and perfection of newer and more sophisticated instruments and methods of examination. This is true in all fields of scientific inquiry, but it is especially conspicuous in the development of modern biology. Through the application of new knowledge and techniques from the fields of chemistry and physics to the study of the cell we have expanded and improved our understanding of living organisms. The perfection of the electron microscope is a good example of this. The electron microscope is capable of resolving objects with a diameter of 0.001 μ. By this means, many details of *fine structure* of the cell have become known, and the future holds great promise of further revelations.

THE PROTOPLAST

The term **protoplast** is useful in designating the individual portion of the protoplasm of a plant contained within the limits of a single cell. Chemically, protoplasm is an organization of organic and inorganic compounds. The chief organic compounds are proteins, fats, carbohydrates, and organic acids. Proteins are the most abundant, sometimes constituting a

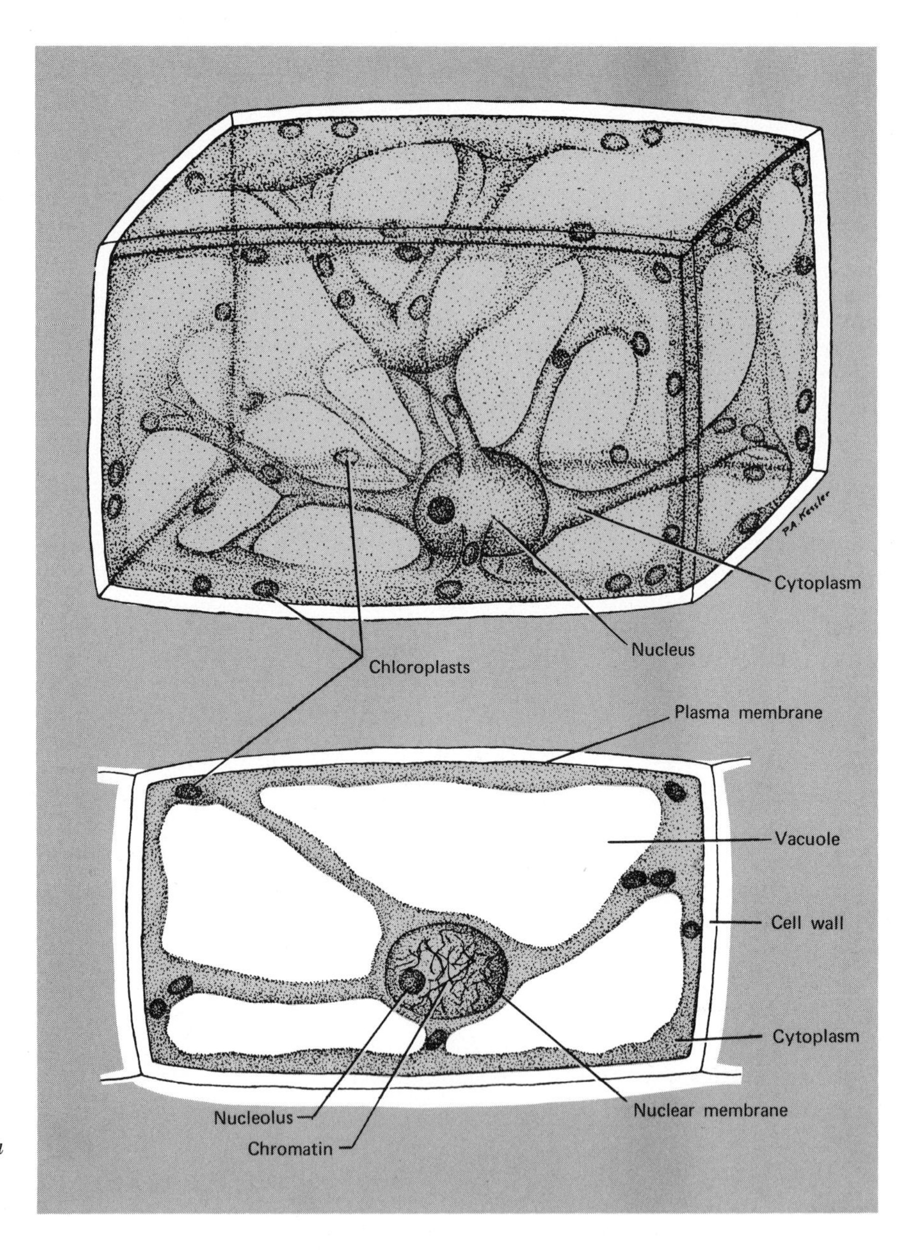

Fig. 8.1 Semidiagram of a green plant cell. Sectional view below.

third of the dry weight of protoplasm. The inorganic compounds are water and various salts. Water may constitute as much as 85–90% of the fresh weight of active protoplasm, inorganic salts commonly less than 1%. The high proportion of water in protoplasm is significant. In general, the higher the water content, the higher the rate of metabolic activity. This is shown in the phenomenon of seed germination. So long as seeds are kept dry they remain quiescent, but when supplied with water the protoplasm becomes more active and the seed germinates. This would certainly suggest that protoplasmic activity is in part, at least, related to chemical reactions, for it is well known that many chemicals fail to react in the anhydrous condition, but do so readily when brought together in the presence of water. It must not be thought, however, that a haphazard mixture of water, inorganic salts, and organic compounds will yield living protoplasm. On the contrary, it is a complex and highly organized chemical and physical system. The inorganic salts and most of the carbohydrates are water soluble and present in **true solution,** whereas the proteins and fats are present as **colloid** and **emulsoid dispersions.**

Detailed study of the protoplast, sometimes by the use of special methods, reveals a number of well-defined structural differentiations. Most of them are of universal occurrence in living cells, performing basic cell functions. A few, when present, endow the cell with special capabilities.

The Cytoplasm and Plasma Membranes. The preponderant bulk of the protoplast consists of the **cytoplasm,** a translucent, colorless, viscous mass. When seen with the light microscope it may appear hyaline or somewhat granular. Recent studies by electron microscopy have shown the cytoplasm to have a fine structure involving an extensive and complex system of membranes (Figs. 8.2*e, i* and 8.3*f*). This internal membrane complex, called the **endoplasmic reticulum,** is thought to be an essentially continuous, intricately convoluted membrane system. Small, dense, globular particles **(ribosomes)** rich in ribonucleic acid (RNA) are often found associated with the membrane, imparting a granular appearance. The ribosomes, on which enzymes and other proteins are synthesized in a pattern determined by the RNA, may also occur free in the cytoplasm. It is believed that the endoplasmic reticulum, by virtue of its tremendous surface, provides the means for the localization of the chemical reactions involving syntheses, secretions, and conversions characteristic of the cell.

The outer limits of the protoplast are defined by the presence of a delicate, two-layered **plasma membrane.** The membrane is believed to be intimately connected with the endoplasmic reticulum and an integral part of the cell's membrane system (Fig. 8.3*e*). The living membrane is capable of growing as the cell enlarges and can, within limits, repair itself when mechanically damaged. Chemical studies have shown it to be composed of proteins and lipids. A distinctive property of the membrane, called **differential permeability,** by which some substances may readily traverse the membrane and others cannot, or do so only very slowly, is probably, in part, related to the molecular arrangement and spacing of the proteins and lipids. Such differential permeability is not fixed, but is subject to change. Discussion of the significance of this important property appears in later chapters.

The Nucleus. The most conspicuous of the differentiated protoplasmic structures of the protoplast is the **nucleus,** a spherical or ellipsoid body averaging about 15 μ in diameter (Fig. 8.1). It was discovered by Robert Brown in 1831 and observed to be of general occurrence in living cells. Brown named it "nucleus" (Latin, kernel) to suggest its importance.

The nucleus is delimited by a two-layered **nuclear membrane** which appears, from electron micrographs, to represent an enveloping inward extension or fold of the cell's integrated membrane system (Figs. 8.2*c* and 8.3*g*). An interesting feature of the nuclear membrane is the presence of numerous pores that may be channels of communication between the nucleus and the surrounding cytoplasm. The preponderant bulk of the nucleus is a gel-like mass, the **nuclear sap,** or karyolymph, through which extends a diffuse tangle of separate, fine threads of **chromatin** (Fig. 8.1). The chromatin threads are colorless and not generally visible in living cells but may be rendered so by the use of appropriate dyes. In properly stained

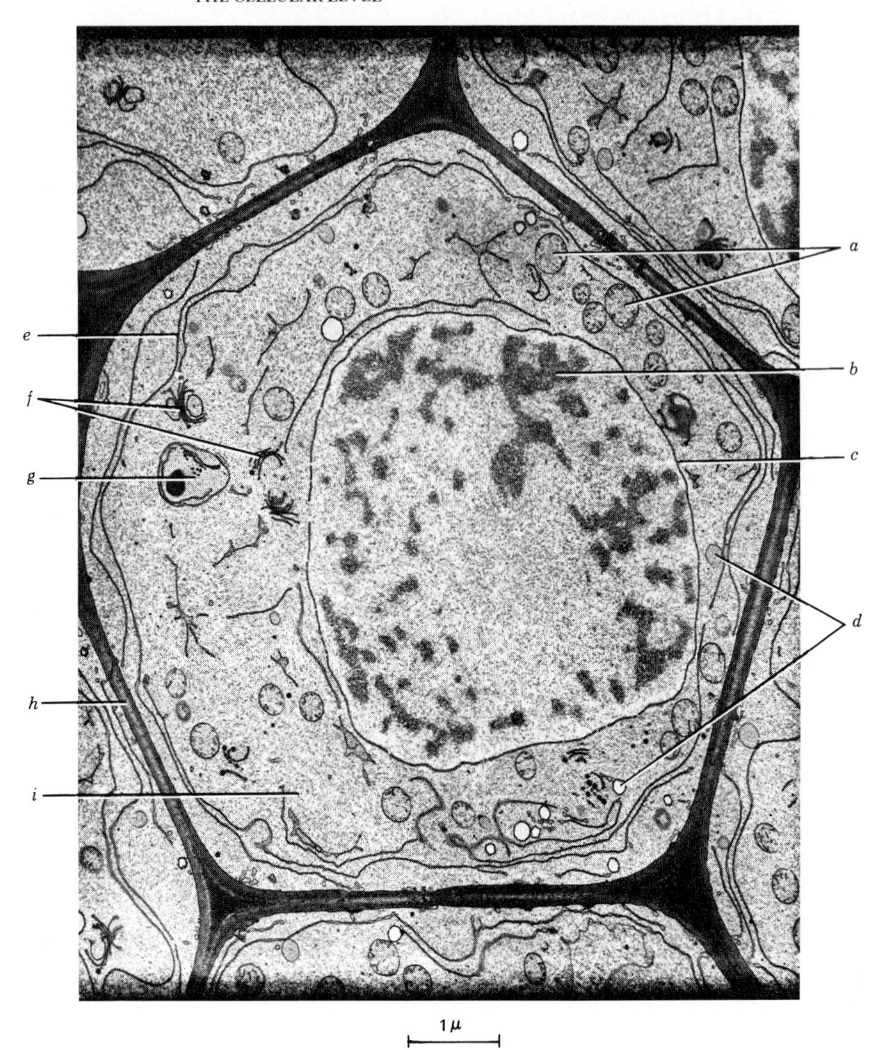

Fig. 8.2 Electron micrograph of section through a meristematic rootcap cell of maize showing (a) *mitochondria,* (b) *chromatin,* (c) *nuclear membrane,* (d) *cytoplasmic inclusions,* (e) *endoplasmic reticulum,* (f) *Golgi apparatus,* (g) *starch-storing leucoplast,* (h) cell walls, (i) *cytoplasm (5000X).*

material, certain coarsely granular portions of the chromatin stain darker than the threads. These are called **chromocenters** and are thought to consist of a special type of chromatin. The chromatin is the most important constituent of the nucleus, for it is the hereditary material. In the division of the nucleus the chromatin is resolved into a characteristic number of **chromosomes** that in their precise manner of distribution carry to the daughter nuclei characteristic complements of genes, the hereditary units borne within them. Chromatin consists of a complex of proteins, ribonucleic acid (RNA), and deoxyribonucleic acid (DNA).

Finally, each nucleus also contains one or more globular bodies apparently without membrane. These are the **nucleoli** (Fig. 8.1). They originate at certain positions (nucleolar-organizing regions) on specific chromosomes of the complement and are rich in proteins and RNA, along with some DNA. Nucleoli appear relatively homogeneous when viewed by light microscopy. Electron micrographs reveal two structural components, a complex of chromatin filaments, probably related to the chromatin of the chromosome to which the nucleolus is attached, and an amorphous matrix. During division of the nucleus the nucleoli become disorganized, the amorphous material apparently assuming a diffuse form and becoming intimately associated with the chromosomes. New nucleoli are organized at the close of nuclear division. Little of a precise nature is known concerning the function of the nucleolus. The RNA of the nucleolus, derived from nuclear synthesis, migrates to the cytoplasm and takes part in protein syntheses. Thus the nucleolus may serve as an intermediary in RNA transfer. Recent studies involving destruction of the nucleolus by radiation techniques indicate that it is essential for nuclear division.

The nucleus is the center of hereditary

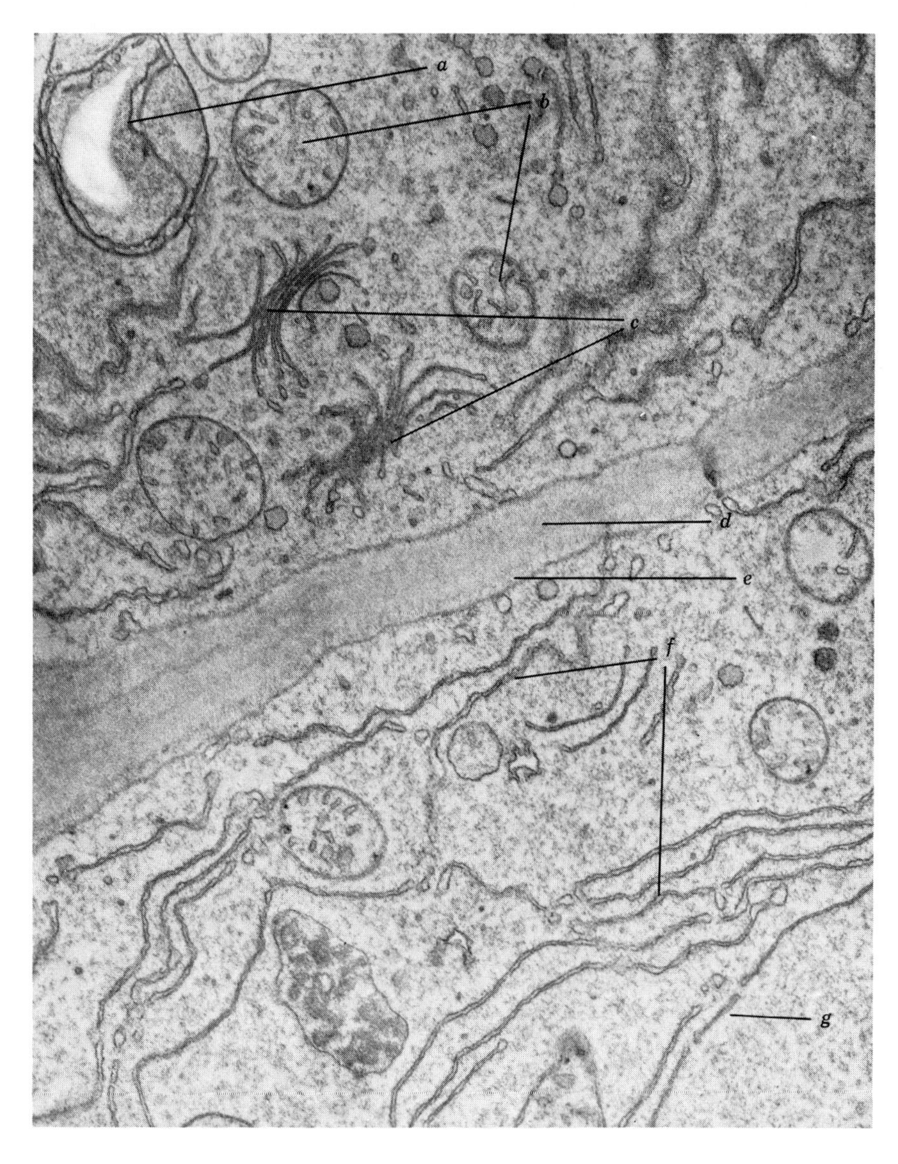

Fig. 8.3 Electron micrograph of section through cells of maize root showing (a) *leucoplast,* (b) *mitochondria,* (c) *Golgi apparatus,* (d) *young walls of contiguous cells,* (e) *plasma membrane,* (f) *endoplasmic reticulum,* (g) *nuclear membrane* (22,000X).

control over the cell. This control is exerted by the DNA of the chromatin, which determines the specific kinds of RNA produced according to its coded structure. The RNA is extruded from the nucleus into the cytoplasm where, attached to ribosomes, enzymes and other proteins are synthesized in a pattern dictated by the RNA. Thus the nucleus controls the processes of a cell by determining the kinds of enzymes and other proteins in the cell (Chapter 14).

The nucleus described above is characteristic of the cells of most organisms. It should be pointed out, however, that in the blue-green algae and bacteria a definitely defined nucleus, with nuclear membrane, nucleoli, and chromosomes, is lacking, but there are strands of DNA that carry the genetic code.

The cells of some species of algae and fungi, at least in certain phases of their life cycles, are multinucleate. These special occurrences will be discussed in appropriate context.

The Plastids. Also embedded in the cytoplasm are a variety of differentiated bodies that are functionally specialized and highly important in the cell economy. These conspicuous bodies, called plastids, are of various forms and some of them are distinctive for their special pigmentation. Plastids are believed to develop from smaller, less well-defined precursors, **proplastids,** which are able to increase their number by division. In division of the cell, however, the distribution of proplastids to the daughter cells is random. Whether the specific types of plastids arise from special types of proplastids or from a generalized proplastid that matures in a variety of specific patterns is unknown. Division of mature plastids has been observed in some plants.

In most of the cells of leaves and other green parts, the dominant type of plastid is the **chloroplast.** The chloroplast is green because of its content of pigments called **chlorophylls.** Several chemically different chlorophylls are known, but the chief ones, occurring in the green algae, mosses, and vascular plants, are chlorophyll *a* and chlorophyll *b* (Fig. 14.5). Associated with chlorophylls in the chloroplasts are **carotenes** and **xanthophylls,** pigments of yellow, orange, or red color. About two-thirds of the pigments of the chloroplast are chlorophylls, hence the dominant green color. In the higher plants the chloroplasts are generally discoid or saucer-shaped and numerous, whereas in some plants like the algae they may have elaborate shapes, such as the helical bands in *Spirogyra* (Fig. 21.5), a cylindrical net in *Oedogonium* (Fig. 21.5), and others. Under the light microscope the chloroplast presents a homogeneous or granular appearance and sometimes contains small starch grains. With the electron microscope the structure is seen to be highly organized. Each chloroplast consists of a large number of disc-shaped **grana,** each composed of a stack of thin platelets or lamellae bearing the chlorophyll on their surfaces as a very thin layer. The grana are embedded in a vacuolated, granular **stroma** or matrix with few or no lamellae (Figs. 8.4 and 8.5). We depend on the chloroplast with its enormous chlorophyll-bearing surface for our energy

Fig. 8.4 Electron micrograph of chloroplasts of spinach, showing 40 to 60 grana per chloroplast.

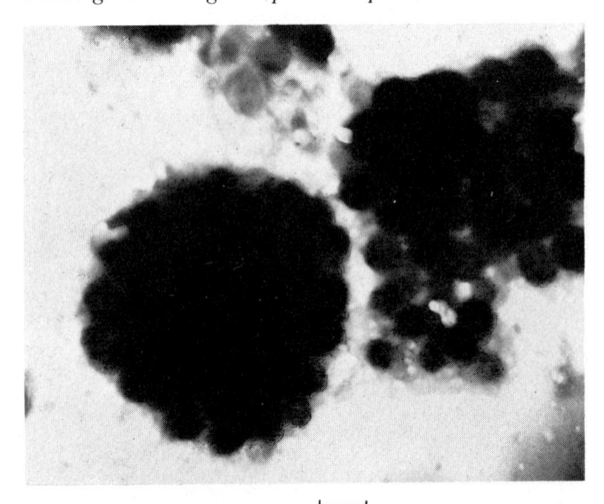

supply, for it is the chlorophylls that absorb radiant energy from the sun. This energy is converted to chemical bond energy and eventually incorporated in the sugar produced by photosynthesis, which may later be oxidized with release of the energy.

A second type of plastid, yellow, orange, or red in color, occurs in the cells of such material as the yellow petals of sunflower, the root of carrot, or the ripe fruit of sweet pepper or tomato. These are **chromoplasts** (Fig. 8.6). They are angular, spherical, lobed, or rodlike. The coloration is due to xanthophylls, carotenes, and related compounds. The color change during the ripening of certain fruits such as tomato or sweet pepper is brought about by the rapid destruction and nonreplacement of chlorophylls in the chloroplasts and the simultaneous development of red xanthophyll. The change in leaf color from

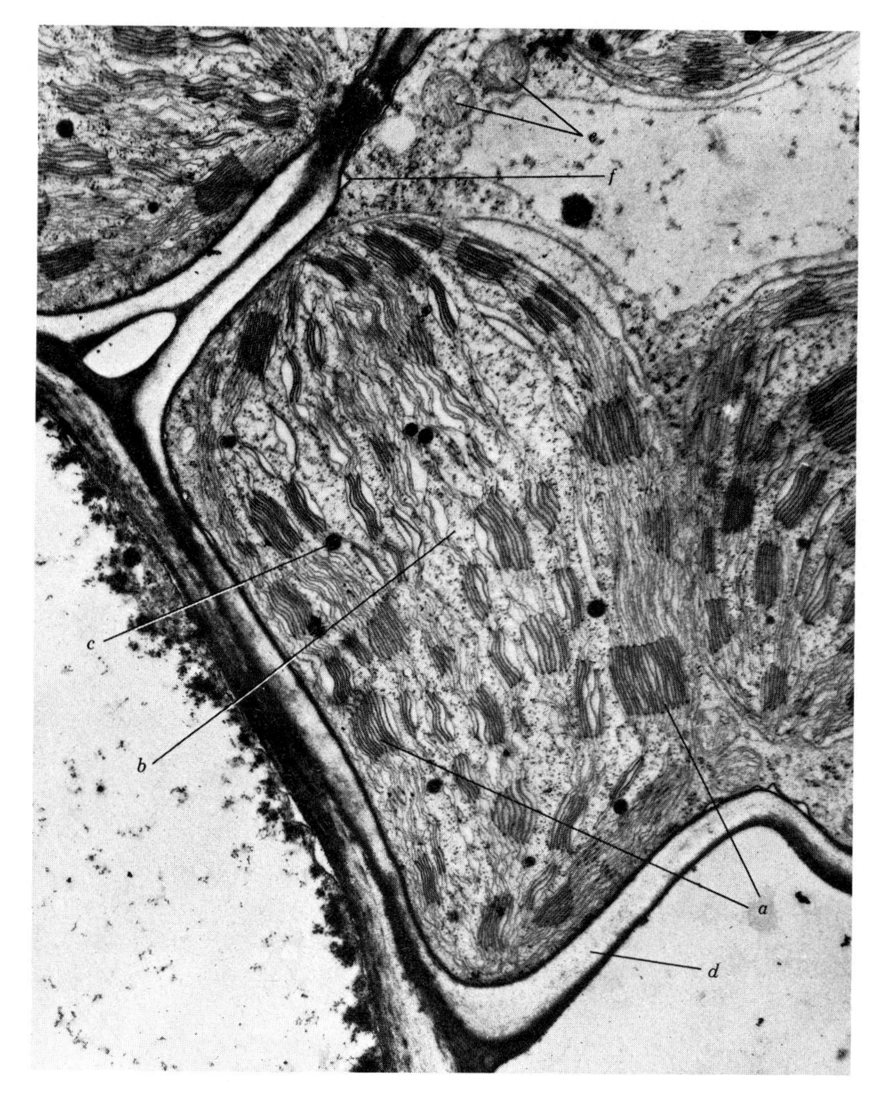

Fig. 8.5 Electron micrograph of chloroplasts in cells of grass leaf showing (a) *grana,* (b) *stroma,* (c) *lipoid body. Also visible are* (d) *cell wall,* (e) *mitochondria,* (f) *plasma membrane.*

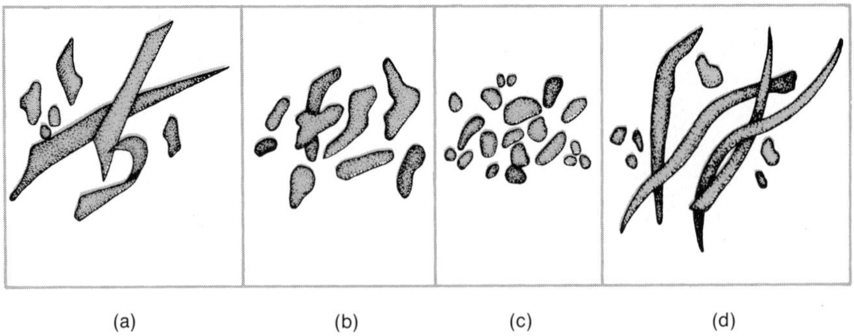

Fig. 8.6 Chromoplasts. (a) *From root of carrot* (b) *from fruit of tomato* (c) *from petal of dandelion,* (d) *from fruit of rose.*

(a) (b) (c) (d)

green to yellow in autumn results from the breakdown of chlorophyll, thus unmasking the color of the carotene and xanthophyll of the chloroplasts. In effect, therefore, the chromoplasts in such cases are degraded or altered chloroplasts. Most chromoplasts are not of this origin but apparently develop directly from proplastids. Little is known concerning the special function of chromoplasts in the plant. Animals, however, can synthesize vitamin A from the carotene they obtain in the plants they eat.

A third type of plastid is the colorless **leucoplast** (Fig. 8.3*a*). Leucoplasts are usually difficult to detect without special staining methods. They are the most common in the cells of underground parts and in tissues not exposed to light. They occur in various irregular shapes and, indeed, may change their shape from time to time. They generally present a granular appearance and often contain large starch grains. The accumulation of starch in the roots and tubers of plants is a common function of leucoplasts. Some leucoplasts are related to oil accumulation, as in the seeds of peanut, corn, and castor bean.

Mitochondria. Virtually all living cells contain in the cytoplasm large numbers of very small (0.2–3.0 μ), filamentous, rod-shaped, or globular bodies called **mitochondria** (Fig. 8.7 and 8.2*a*). These structures are rendered visible by the light microscope only after special staining or by the use of special optical equipment. Electron micrographs reveal that each

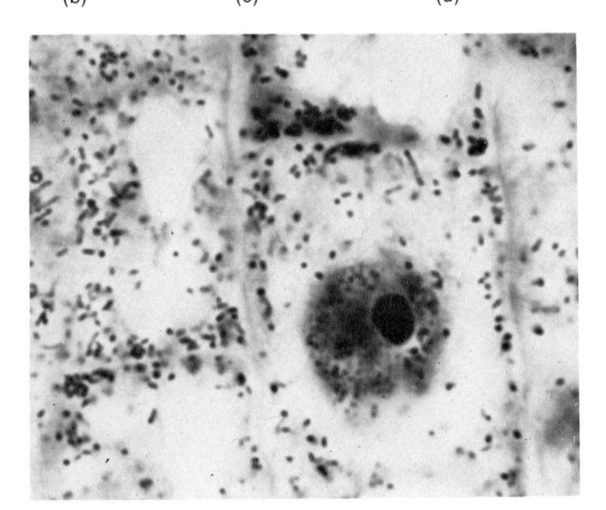

Fig. 8.7 Mitochondria in cytoplasm of cells.

mitochondrion is bounded by a two-layered membrane. The outer layer covers the surface smoothly, whereas the inner invades the fluid body of the mitochondrion in extensive saclike folds and convolutions, called **cristae,** thus creating an extensive inner surface (Figs. 8.3*b* and 8.8). The mitochondrial membrane is composed of lipoproteins, and appears to be similar to the membrane complex of the cell. Within the fluid matrix of the mitochondrion, organic acids derived from the breakdown of complex food molecules are progressively oxidized into carbon dioxide and water through many successive steps (respiration), each controlled by a specific enzyme. The energy released in the oxidations eventually comes to

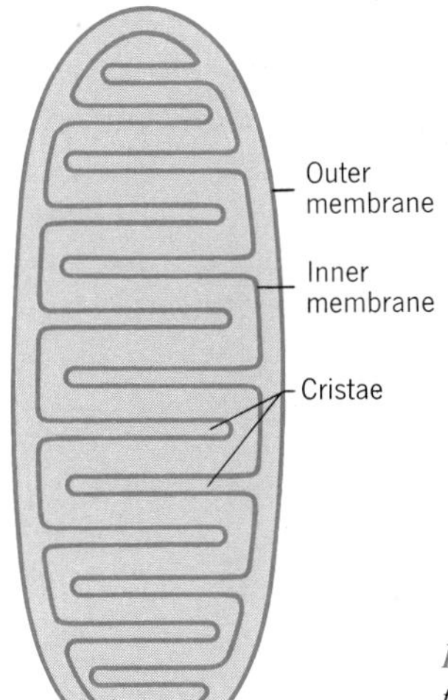

Fig. 8.8 Diagram of a mitochondrion as shown by electron microscopy.

reside in molecules of adenosine triphosphate (ATP) as high-energy bonds, through the action of phosphorylating enzymes associated with the inner layer of the membrane. ATP then passes outward to other parts of the cell where it becomes available for doing work. The mitochondria, as the respiratory centers of the cell, have been quite aptly called by one author, "the powerhouses of the cell."

The Golgi Apparatus. Each plant cell contains one to several of the organelles known as the Golgi apparatus. A Golgi apparatus is composed of a stack of membranous, hollow, discs (cisternae). There are usually three to eight cisternae per apparatus, but there may be more in some species. Pinched off from the sides of the cisternae are spherical, membranous vesicles. In plants, the main role of the Golgi apparatus appears to be the synthesis of the cellulose and other cell wall substances. After synthesis in the cisternae, these are incorporated in the Golgi vesicles and transported to the plasma membrane, where the vesicles merge with the membrane and release the cell wall substances that are then incorporated in the wall. Other roles of the organelles have been suggested for plants, and apparently they are quite different from the roles played by the Golgi of animal cells.

Ribosomes. The ribosomes are nonmembranous organelles generally attached to endoplasmic reticulum, although some are also found free in the cytoplasm. They are roughly spherical, each one being composed of two subunits of unequal size. Ribosomes are composed of protein and ribonucleic acid, are formed in the nucleus, and enter the cytoplasm through the pores of the nuclear membrane. They play an important role as the site of protein synthesis, which will be discussed in later chapters. Ribosomes are also found in chloroplasts and mitochondria. These are somewhat smaller than the more numerous ribosomes associated with the endoplasmic reticulum.

Microtubules. The microtubules are very long, thin nonmembranous cylinders of protein that are generally abundant in cells. Microtubules are generally straight and rather brittle. In the cytoplasm, microtubules are principally found near the plasma membrane and are thought to play a role in cytoplasmic streaming and perhaps also in the orientation of cellulose fibers in the walls. The spindle fibers that form during cell division are composed of microtubules, which also participate in plant cell formation. Flagella are also partly composed of microtubules. The contraction of spindle fibers and flagella are believed to result from the sliding of one microtubule along another.

PROCARYOTIC CELLS

The cell organelles that have just been described are characteristic of **eucaryotic cells** (prefix *eu-*, true; *karyon,* nut or nucleus) that compose the bodies of most organisms. Many of these organelles are present in both plant and animal cells.

However, the cells of the Monera (bacteria and blue-green algae) lack most of these organelles: a definitely organized nucleus, as previously noted, and also plastids, mitochondria, and Golgi bodies. In such cells, designated **procaryotic** (prefix *pro-*, before, referring to the probable *primitive* organization of the cell components), the DNA is present as thin strands not associated with histone proteins as in eucaryotic cells. In photosynthetic bacteria and the blue-green algae the pigments are in membranes that lack grana and that often lie parallel with the plasma membrane. The ribosomes of procaryotic cells are smaller than those of eucaryotic cells, except those in chloroplasts and mitochondria.

Because the ribosomes and DNA of chloroplasts and mitochondria are similar to those of the Monera, and for other reasons, some biologists believe that eucaryotic cells originated when cells of larger size were invaded by early bacteria and blue-green algae, the symbiotic relationship giving rise to mitochondria (from bacteria) and chloroplasts (from blue-green algae).

Description and discussion of plant cells in the remainder of this chapter apply to eucaryotic plant cells.

CELL INCLUSIONS

The active metabolism of the cell entails the construction and consumption of substances that may at various times and in varying amounts be present within the cell. Some of these substances assume characteristic forms within the cell and are readily visible. They are the products of cell action and are called **cell inclusions** (Fig. 8.2*d*).

Most conspicuous, especially in mature plant cells, is the **vacuole** (Fig. 8.1). The vacuole consists of water and dissolved substances including sugars and salts. The vacuolar limits are defined by a one-layered **vacuolar membrane.** Vacuoles arise in young cells as small droplets, probably in the loops or pockets of the convoluted endoplasmic reticulum. As they enlarge during growth of the cell, they coalesce to form a larger vacuole. Commonly assuming a central position in the cell, the enlarging vacuole displaces the cytoplasm, so that in mature living cells the cytoplasm often appears as a thin peripheral layer. The important role played by the vacuole in the interchange of materials between the cell and its environment will be discussed in later chapters.

Other important inclusions are such substances as **starch** (Figs. 8.3*a* and 8.9). Starch grains may develop in leucoplasts and chloroplasts in cells. As a starch grain enlarges, the body of the leucoplast becomes greatly distended and envelops the grain as a very thin inconspicuous film. The tuber of potato and the roots of many plants accumulate starch that may be used in the next season's early growth. Many kinds of seeds are rich in starch that, upon germination, serves as nourishment for the young, rapidly growing plant. Enough of such foods is usually present to last until the new season's growth is well started and the plant can begin to make more.

Crystals are another kind of inclusion. These occur in various forms (Fig. 8.10). They are composed chiefly of calcium oxalate. The crystals form within the cytoplasm and become surrounded by a cytoplasmic membrane. Silica and calcium carbonate are also occasionally present as crystalline inclusions.

Oil droplets and **fat globules** commonly occur in the cytoplasm, as in the cells of the endosperm and cotyledons of certain seeds.

THE CELL WALL

The protoplasts of plants are, with few exceptions, encased by a **cell wall** composed principally of cellulose along with varying

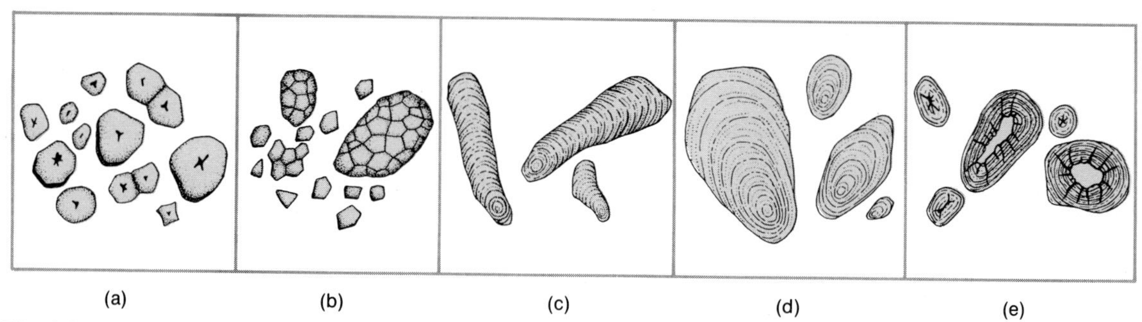

Fig. 8.9 Starch grains. (a) *from maize,* (b) *from rice,* (c) *from banana,* (d) *from potato,* (e) *from bean.*

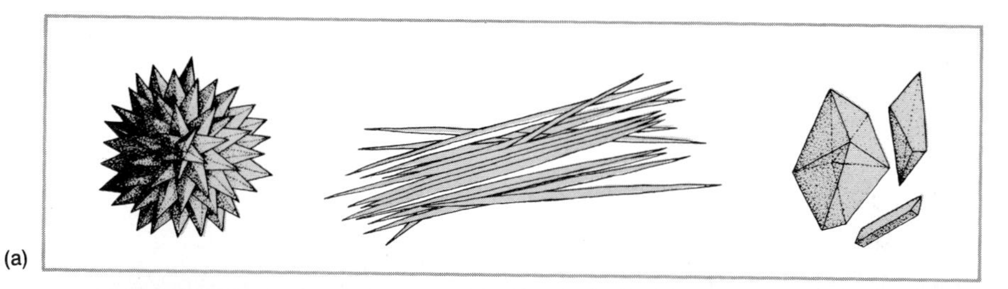

Fig. 8.10 (a) *Drawing of calcium oxalate crystals,* (b) *Scanning electron micrograph of a complex calcium oxalate crystal within a leaf cell of* Griselinia littoralis. *(6000X).*

Fig. 8.11 Plasmodesmata traversing the thick cell walls of the endosperm of a persimmon seed. (Copyright, General Biological Supply House, Inc., Chicago.)

amounts of pectin. The cell wall is secreted by the living protoplast and deposited upon its surface, that is, on the outside of the plasma membrane (Fig. 8.1). The cell wall is ordinarily continuous. However, in multicellular plants, abutting walls of adjacent protoplasts are traversed by extremely fine protoplasmic strands which thus interconnect the protoplasts. These protoplasmic connections are the **plasmodesmata** (singular, **plasmodesma**) (Fig. 8.11). The cellulose wall is a rather tough layer, and its presence imparts a certain degree of rigidity and stability of form to the cell. The importance of the cell wall in the manner and habit of growth of a plant and its general lack of motility have already been mentioned.

Growth of the plant cell is typically marked by three processes: construction of additional protoplasm, development and enlargement of vacuoles, and increase in area of the cell wall to encompass the increasing volume of the cell. In the process of cell growth the original cell wall becomes stretched and new layers of cellulose are deposited on its inner surface by the living protoplast at a rate such that the net thickness of the cell wall is not decreased; in some instances the thickness may actually increase as the wall extends its area. The deposition of wall material in successive layers, that is, by **apposition,** is believed to be the principal method of wall thickening, although it is probable that some new wall material is also deposited *within* the growing wall, that is, by **intussusception.** Cell walls that become quite thick often present a laminated appearance because of differences in density of successive layers.

The cell wall initially present and during growth of the cell is called the **primary wall.** Many types of cells, especially those that are metabolically active at maturity, may have only a primary wall. However, other types, after attaining their characteristic size and shape, continue through a secondary development consisting in part of the deposition of a **secondary wall.** This may take the form of uniform and sometimes great thickening of the wall, or the deposition of wall material in a distinctive pattern, with more or less restricted areas remaining unthickened. Such unthickened areas, if

Table 8.1

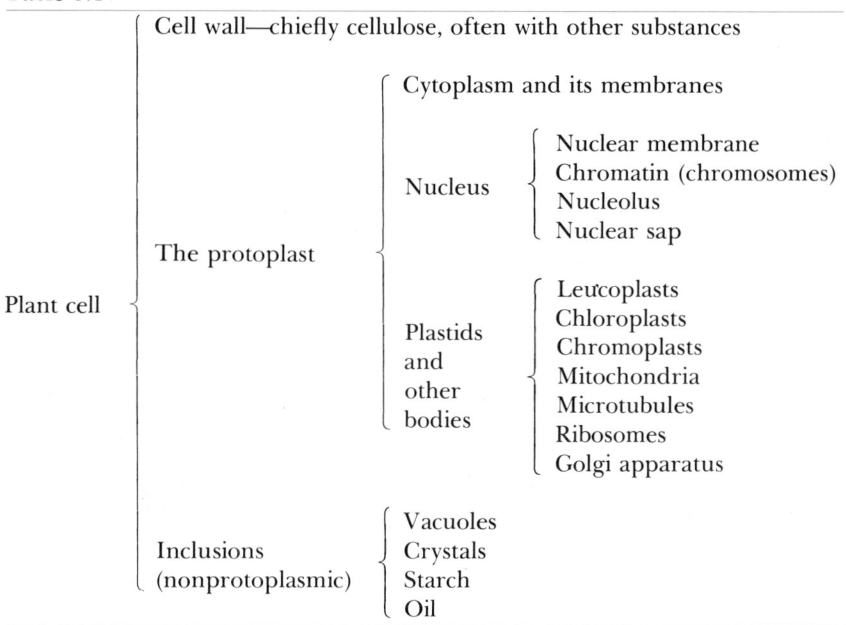

Plant cell	Cell wall—chiefly cellulose, often with other substances		
	The protoplast	Cytoplasm and its membranes	
		Nucleus	Nuclear membrane
			Chromatin (chromosomes)
			Nucleolus
			Nuclear sap
		Plastids and other bodies	Leucoplasts
			Chloroplasts
			Chromoplasts
			Mitochondria
			Microtubules
			Ribosomes
			Golgi apparatus
	Inclusions (nonprotoplasmic)	Vacuoles	
		Crystals	
		Starch	
		Oil	

they are of relatively small extent, are called **pits** (Fig. 8.15 and 8.17*d*). The chiefly cellulosic cell walls, both primary and secondary, may in secondary development become modified chemically by impregnation with other substances. **Lignin,** a complex substance that contributes strength and hardness, is found chiefly in the walls of cells that have a supportive function. **Suberin** and **cutin,** each composed of a mixture of waxy substances, commonly occur in certain exterior plant cells where they impart an impervious quality to the walls. Many other kinds of wall modifiers are found in specialized cells and in certain species. Waxes, fats, volatile oils, resins, tannins, and even inorganic mineral substances such as silica and calcium carbonate are common.

Studies of primary and secondary cell walls with the electron microscope show interesting differences in their microstructure. The cellulose molecule consists of a very large number of glucose units united in long latticelike chains, the **crystal lattices.** Groups or fascicles of crystal lattices constitute the cellulose fibrils (Chapter 7). In the primary wall the fibrils are loosely associated, giving the appearance of an open fabric (Fig. 8.12*a*). The fibrils of the secondary wall are more compact and parallel and generally follow a helical pattern about the cell (Fig. 8.12*b*). In successively deposited layers of fibrils in the secondary walls of cotton hairs, for example, the direction of the helix differs. Modifying substances deposited during secondary development would occupy the space between the fibrils.

Table 8.1 summarizes the relationship of the parts of a generalized plant cell.

CELL TYPES

In maturation, plant cells attain a variety of shapes and wall characteristics that are related to their functions in the plant body. This will be shown in Chapter 9 where the organiza-

(a) 

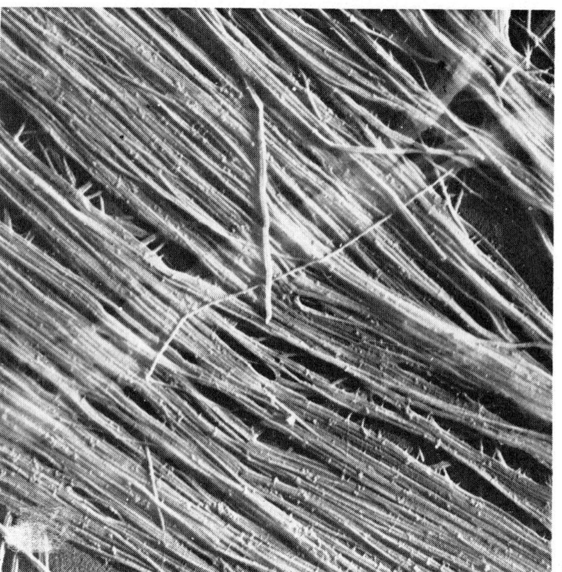 (b)

Fig. 8.12 Electron micrographs of cellulose fibrils in primary wall (a) *and secondary wall* (b) *of plant cell.*

tion of cells into functioning tissues is discussed. It is proper at this point, however, to enumerate and describe briefly the more common types of cells that may be found in a vascular plant.

Structurally, the simplest type of cell is the **parenchyma cell** (Fig. 8.13*a*). In general, parenchyma cells are essentially isodiametric and possess a thin wall and a long-active protoplast. However, some departure from the usual form and thin wall is encountered in parenchyma cells of specialized tissues. Most of the basic metabolic and reproductive functions of the plant are performed by cells of this type. Parenchyma cells containing chloroplasts are sometimes called **chlorenchyma cells** (Fig. 8.13*b*).

Collenchyma cells are generally rectangular as seen in longitudinal section. They retain a protoplast at maturity, and the cellulose wall is unevenly thickened by the deposition of additional cellulose and pectic compounds in certain areas (Fig. 8.14). Collenchyma cells are the first-formed specialized supporting cells to develop in the young differentiating stem tip, contributing a temporary, nonrigid support.

Sclerenchyma cells are of two general shape classes, that is, isodiametric, called **sclereids** (Fig. 8.15*a*), and elongated, called **sclerenchyma fibers** (Fig. 8.15*b, c*). These classes are not always sharply distinguishable, for sclereids are frequently moderately elongated and sometimes highly branched. Fibers, however, are generally slender, many times longer than wide and have tapering ends. Sclerenchyma cells have thick, strongly lignified secondary walls which may be pitted. At maturity these cells usually lose their protoplasts and thus play no active part in the plant metabolism. Sclerenchyma cells are commonly disposed in masses or in association with other cell types. Their chief function is support, and they provide resistance to compression and flexing forces. Sclereids are frequently associated in thick layers as in nut shells where they have great protective value.

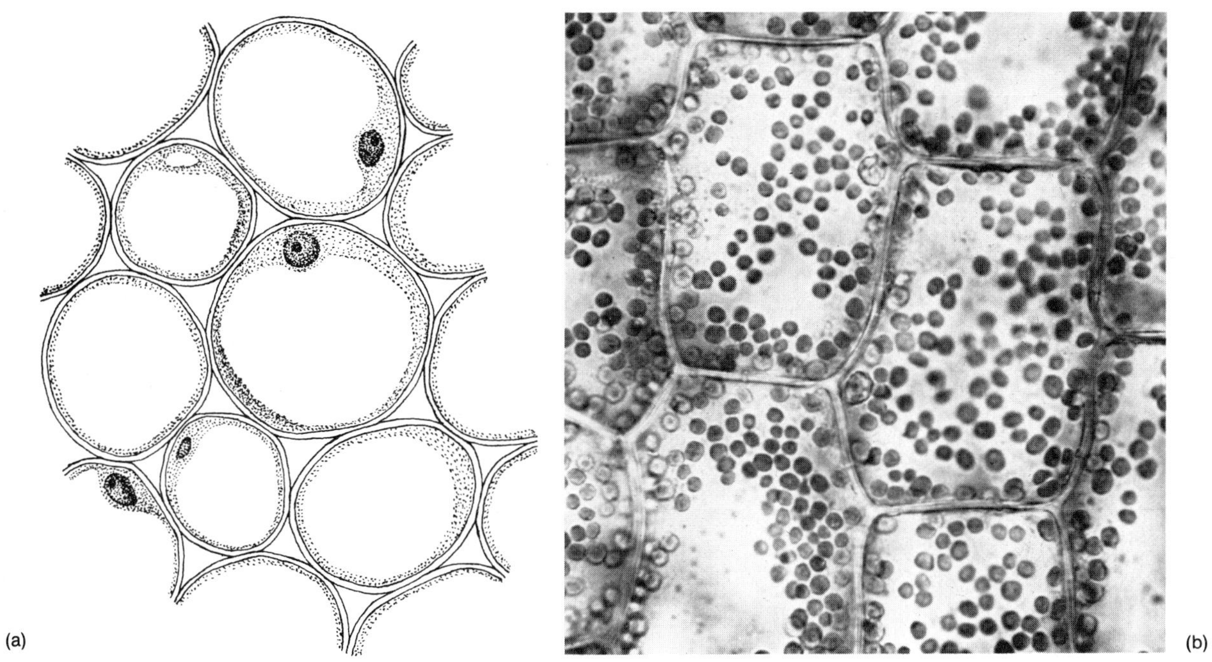

Fig. 8.13 (a) *Drawing of parenchyma cells from pith of sunflower stem, in sectional view, showing thin primary walls and active protoplasts. Note the large central vacuole.* (b) *Chlorenchyma of* Elodea *leaf, photographed at 430X.*

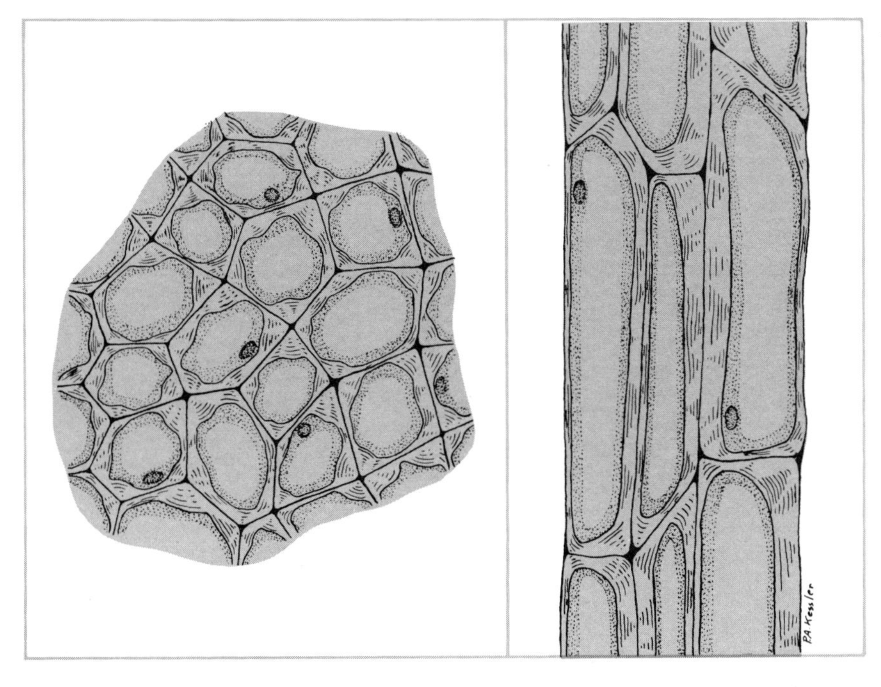

Fig. 8.14 Drawings of collenchyma cells from stem of mint. (a) *In cross section,* (b) *in longitudinal section.*

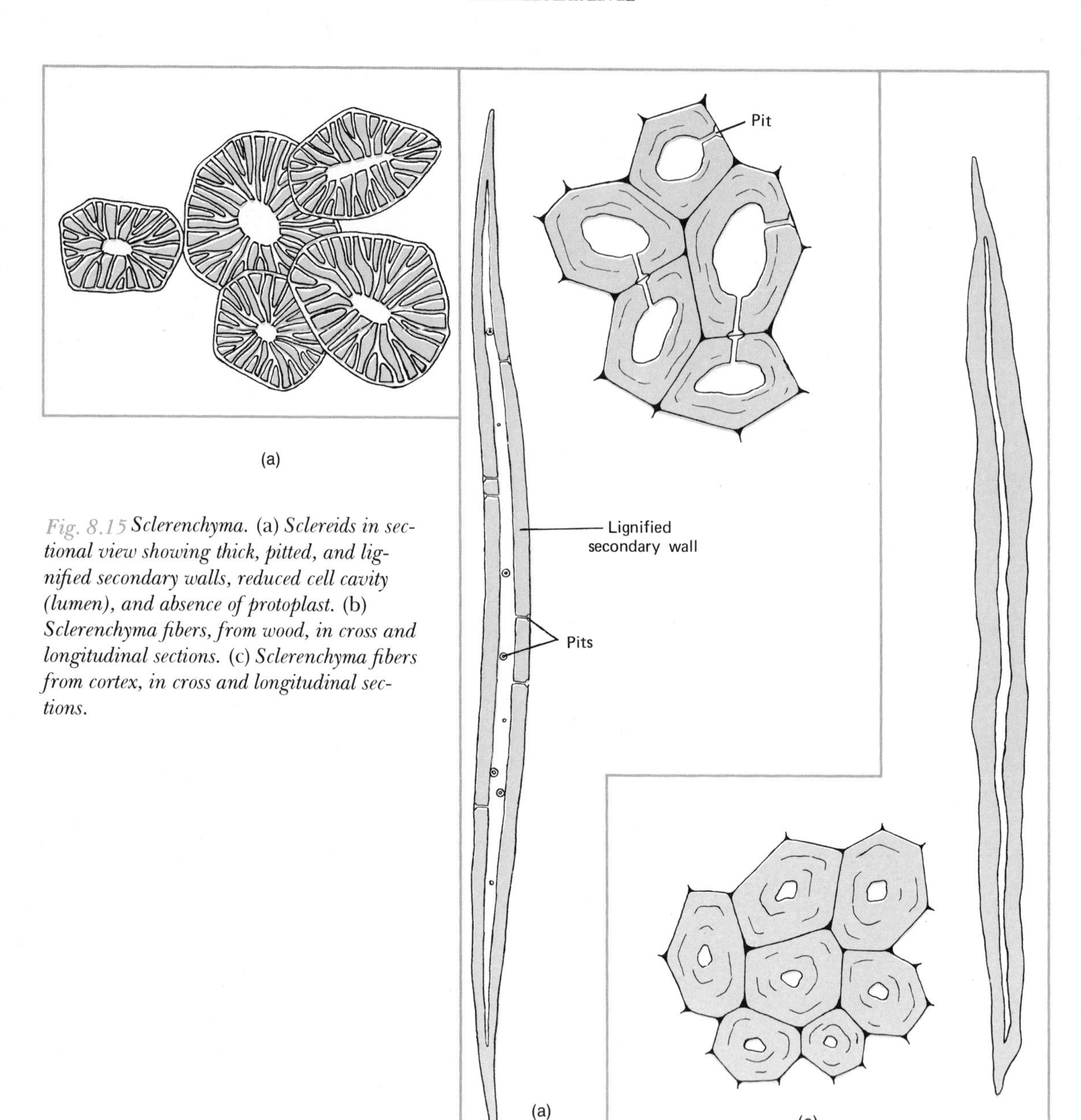

Fig. 8.15 Sclerenchyma. (a) *Sclereids in sectional view showing thick, pitted, and lignified secondary walls, reduced cell cavity (lumen), and absence of protoplast.* (b) *Sclerenchyma fibers, from wood, in cross and longitudinal sections.* (c) *Sclerenchyma fibers from cortex, in cross and longitudinal sections.*

Familiar sclerenchyma fibers are those of flax and hemp, used for textiles and cordage.

Epidermal cells are typically tabular (Fig. 8.16*a*). They are associated in a continuous, usually one-layered "sheet" on the surface of leaves, young roots, and stems. The outer walls of the epidermal cells of stem and leaves are modified by the deposition of cutin. This wall modification reduces the excessive water evaporation. The epidermal cells of young roots are not cutinized, and absorption of water from the soil takes place directly through the sur-

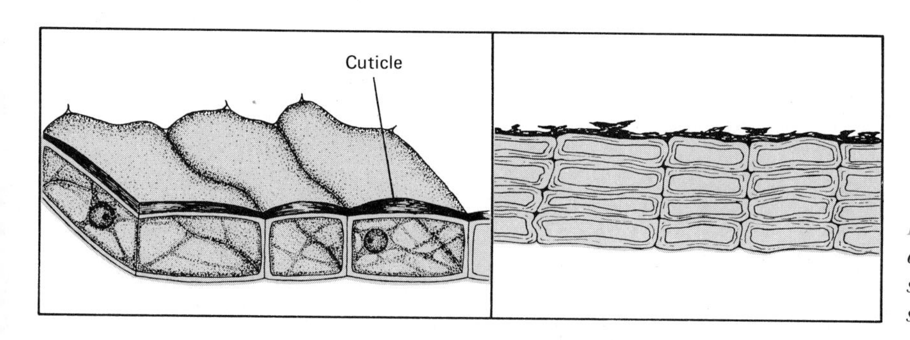

Fig. 8.16 (a) *Epidermal cells seen in three-dimensional view.* (b) *Cork cells seen in cross section.*

face. In most seed plants, some of the epidermal cells of young roots are outwardly extended as the **root hairs** (Fig. 12.11). The absorbing surface is thus greatly increased. At maturity epidermal cells retain their active protoplasts.

Cork cells are tabular, with all walls suberized (Fig. 8.16*b*). They occur in thick layers on the outer surfaces of older stems, branches, and roots of woody plants where their spongy texture and waterproof character provide some protection against mechanical injury and drying. At maturity the protoplasts die and the cells become air filled, which accounts for the lightness and compressibility of cork. The uses of cork for bottle stoppers and as a buoyant material in life preservers, as well as for floor coverings and insulation, are related to these properties.

The cell types so far described are chiefly supporitve, protective, and metabolic in function. We now consider a series of highly specialized cells comprising the distinctive cell types of the vascular tissues of the tracheophytes. Among these types is the **tracheid,** a long tapering cell with lignified wall thickenings (secondary wall) laid down in a variety of patterns (Fig. 8.17). The protoplast disappears at maturity.

The tracheid is a dual purpose cell, serving in both conduction and support. The cells occur in groups with much overlapping and in such a way that the pits in walls of adjacent cells are opposite each other. Water passes from cell to cell through the pits. The tracheid's supporting value is related to the wall thickness and the length and arrangement of the cells. The common gymnosperms, such as pine, depend on this cell type for conduction and support. The wood of such trees is composed almost wholly of tracheids. The development of the tracheid was very important in the evolution of terrestrial plants, for it provided the supportive and conductive facility that made possible the evolution of erect, tall stems.

During the long course of evolution, the tracheid gave rise to two other cell types that individually have assumed the tracheid's functions and have largely replaced the tracheid in the most highly evolved vascular plants. One of these, the **vessel element,** was derived through modifications involving slight shortening and widening of the cell and the loss or perforation of the end walls, which gives it an open tubular form (Fig. 8.18). The vessel element wall may be secondarily thickened in any of the patterns shown by the tracheids, and the protoplast is absent at maturity. The vessel elements are arranged end-to-end with little or no overlapping, and thus form a continuous, open pipelike structure, called the **vessel.** A second evolutionary derivative of the tracheid is the **wood fiber** in which all of the supportive features of the tracheid have been intensified (Fig. 8.15*b*). The wood fiber is longer than the tracheid and more slender and tapering. It has greater overlapping, more extensive and heavier wall thickening and lignification. Pits are

Fig. 8.17 Tracheids. A variety of patterns of wall thickening: (a) *annular,* (b) *spiral,* (c) *scalariform,* (d) *pitted. Note that the pits, as shown in* (d) *are in register, that is, they occur in pairs in adjacent cells.*

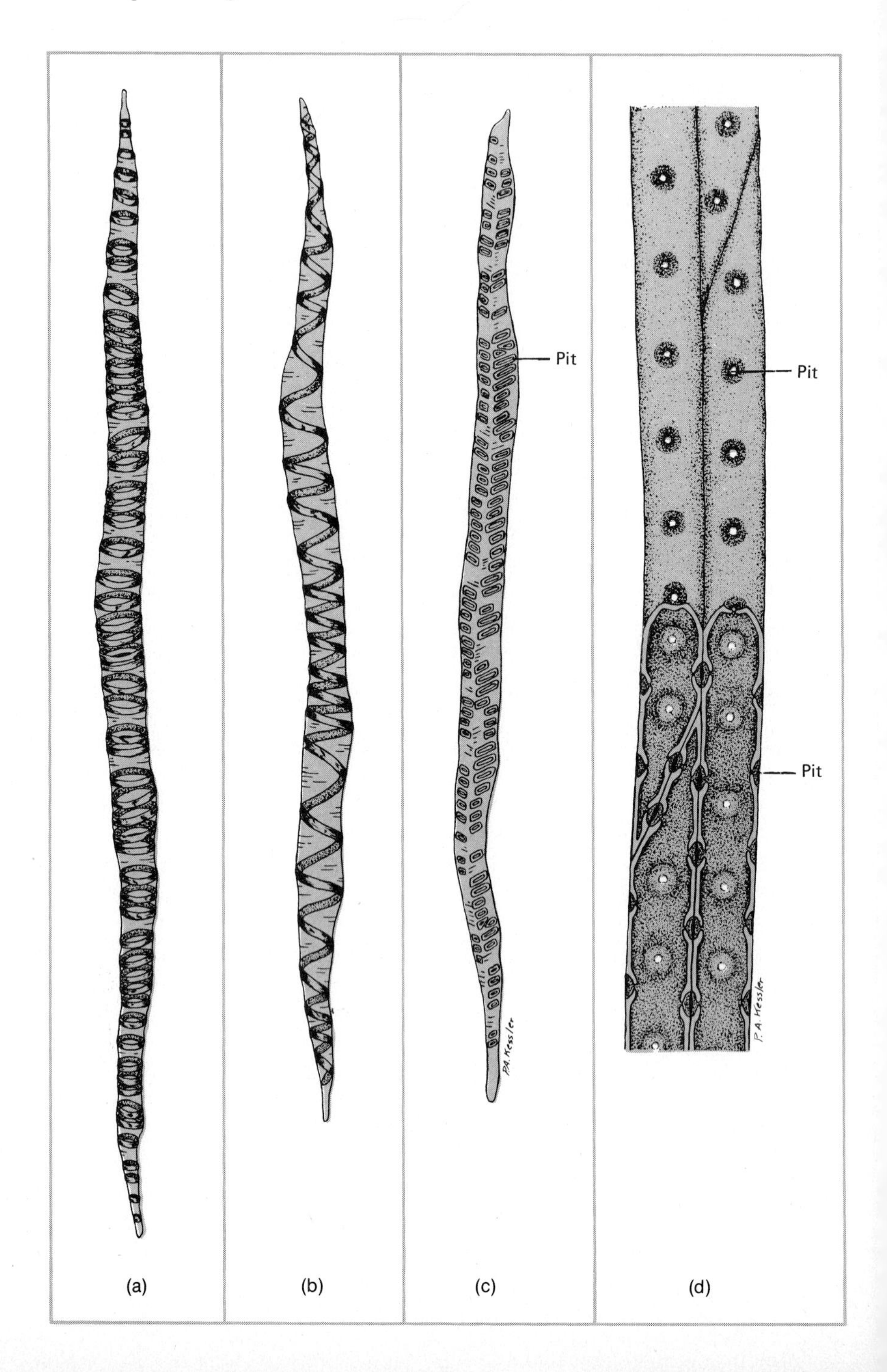

reduced in number and size, or are completely lost. Thus, these two derivatives of the tracheid, the vessel element, and the wood fiber, are complementary and together perform the dual functions of the original tracheid. Vessels and wood fibers together occupy, in function and position, the same place in the most highly evolved vascular plants that the tracheids do in the lower vascular plants.

Whereas the tracheid and the vessel con-

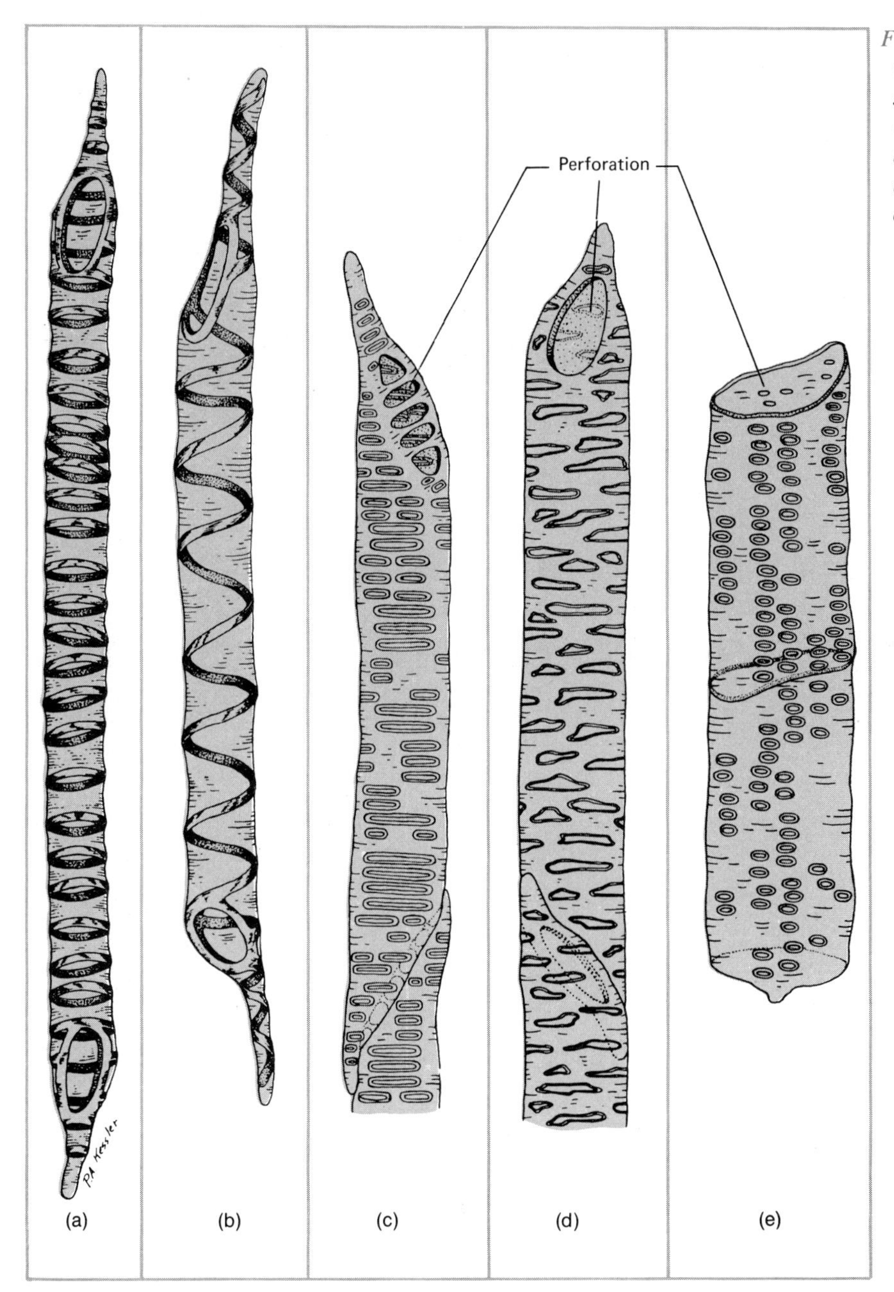

Fig. 8.18 Vessel elements. (a) *Annular,* (b) *spiral,* (c) *scalariform,* (d) *reticulate,* (e) *pitted. Note that vessel elements are perforated at the ends, whereas tracheids are imperforate.*

duct water and dissolved substances from roots to leaves, a structure of different design carries foodstuffs from leaves to roots. In angiosperms the translocation of organic substances takes place through **sieve tubes** (Fig. 8.19). A sieve tube consists of a vertical row of elongated, specialized cells, called **sieve-tube elements.** A distinctive feature is the multiperforate end walls, the **sieve plates.** The cytoplasm of adjacent sieve-tube elements appears to be connected through the pores of the sieve plates. The interconnecting cytoplasmic strands are each surrounded by a thin sleeve of a substance called **callose** where they pass through the sieve-plate pores. The nucleus of the developing sieve-tube element disintegrates as the cell becomes mature although the cytoplasm remains. The **companion cell,** an associate and sister cell of the sieve-tube element, has a dense protoplast and conspicuous nucleus. In gymnosperms, sieve-tube elements and companion cells do not occur; their places are taken by the less specialized **sieve cell** with tapering, overlapping ends and sieve areas mostly on the lateral walls.

It may be understood from the foregoing that the body of the vascular plant is composed of a considerable variety of cell types. These types are characterized by special form, wall construction, and function. All of these types are the products of cell evolution, having been derived directly or indirectly from unspecialized, parenchymalike cells. Simple plants are composed entirely or mostly of relatively unspecialized cells, whereas highly complex plants are composed of specialized cells of many kinds.

THE FORMATION OF NEW CELLS

At maturity, cells composing a stem, root, or leaf of a plant have fairly well-defined maximal sizes. Continued growth of a plant depends, therefore, on a continuous supply of new cells that in their turn become mature.

Fig. 8.19 Sieve-tube elements and companion cells in cross and longitudinal sections.

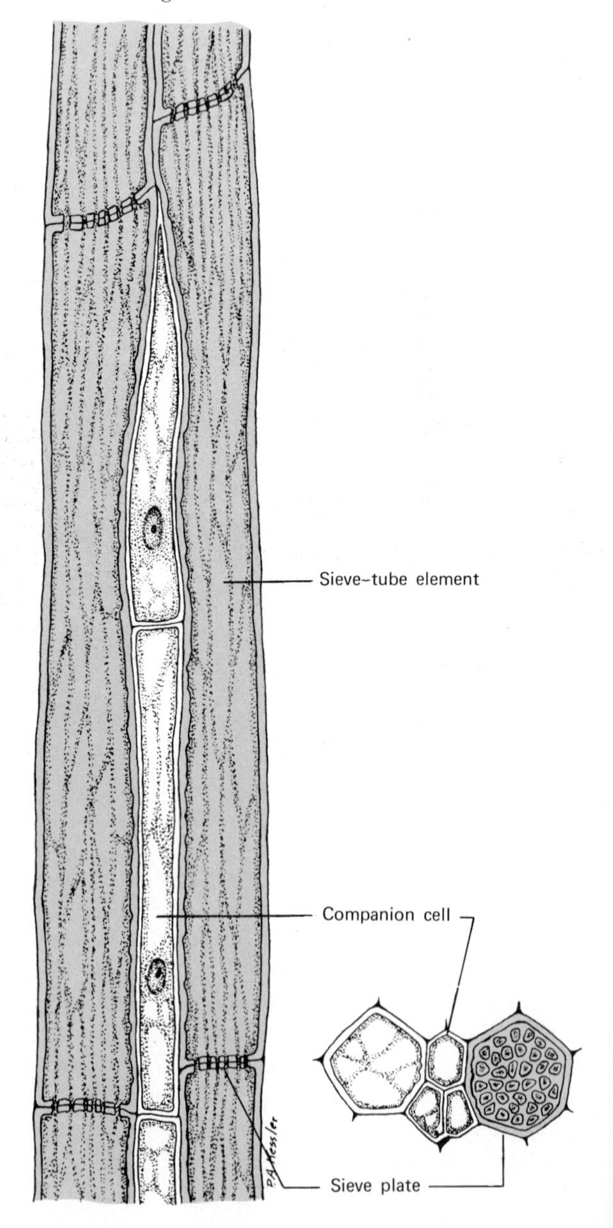

In an actively growing plant part, such as the tip of a stem or root, the supply of new cells is obtained by the *division* and duplication of existing cells. With respect to the individual cells involved, division is obviously a reproductive operation, but with respect to the plant, it is a part of the growth process. The division of a cell is a complicated two-step process. The first step entails the derivation of two nuclei from the original one by a process of replication. This apparent division of the nucleus is called **mitosis** (Fig. 8.20*a–g*). The second step, properly called **cytokinesis** (Fig. 8.20*h–j*), consists of dividing the cytoplasm into two parts, each containing one of the new nuclei, and the formation of a cell wall on the new surface of each new cell. The two new cells thus formed are each about half the size of the parent cell. By the production of new protoplasm and cell wall material and the uptake of water during an ensuing period, the cells grow to the original size. Both of these new cells may repeat the division process, or one or both of them may become specialized. The specialized cells so formed constitute an addition to the basic, permanent tissues of the plant whereas those that remain unspecialized continue to produce new cells. Continued growth of a stem would be possible, of course, only if some unspecialized cells remain. Specialized cells, for the most part, do not divide.

Mitosis

Mitosis proceeds with great precision and is a very striking phenomenon. At the beginning of mitosis, the apparently tangled and interwoven mass of chromatin threads undergoes a gradual reorganization to give a number of slender threads which promptly shorten and thicken and become discernible as discrete bodies. These chromatin bodies, which quickly become enveloped with a matrix, are the **chromosomes** (Fig. 8.20*b*). The number formed in the nucleus of each root, stem, or leaf cell is characteristic for a given species. For example, the number of chromosomes in cells of the tobacco plant, *Nicotiana tabacum,* is 48. In onion, *Allium cepa,* it is 16. Each chromosome consists of a pair of spirally coiled or closely appressed threads of chromatin, the **chromonemata,** embedded in the chromosomal matrix. Further thickening and shortening of the chromosomes often obscure the distinctness of the chromonemata. It is believed that at this time the chromonemata straighten and lie closely side by side. At a point where the chromosome is somewhat constricted, the chromonemata appear to be joined. This specialized point on the chromosome is marked by the presence of a small hyaline or clear zone, called the **centromere** or **kinetochore** (Fig. 8.20*d*).

In many species of plants, the individual chromosomes of a complement may be identified by their characteristic size and form. Length of the chromosome and position of the centromere are factors in determining chromosome identity. A species may have 16 chromosomes in the cells of its root or stem, but only 8 types may be distinguished. That is, each type is represented by 2 chromosomes (a pair).

The organization of the chromosomes, as described, is accompanied by other significant changes. The first of these is the gradual disappearance of the nuclear membrane. Second, there appear in the cytoplasm, at opposite poles of the cell, groups of fine radiating fibers. The fibers gradually extend toward the equator of the cell, and as they do so, the nucleoli gradually disappear. Eventually the fibers form a spindle-shaped figure appropriately called the **spindle** (Fig. 8.20*c, d*). The spindle fibers are believed to be composed of long-chain protein molecules. The chromosomes, now seen clearly to consist of two associated halves, **chromatids,** move to the center of the spindle in such manner that their centromeres come to lie in a horizontal plane

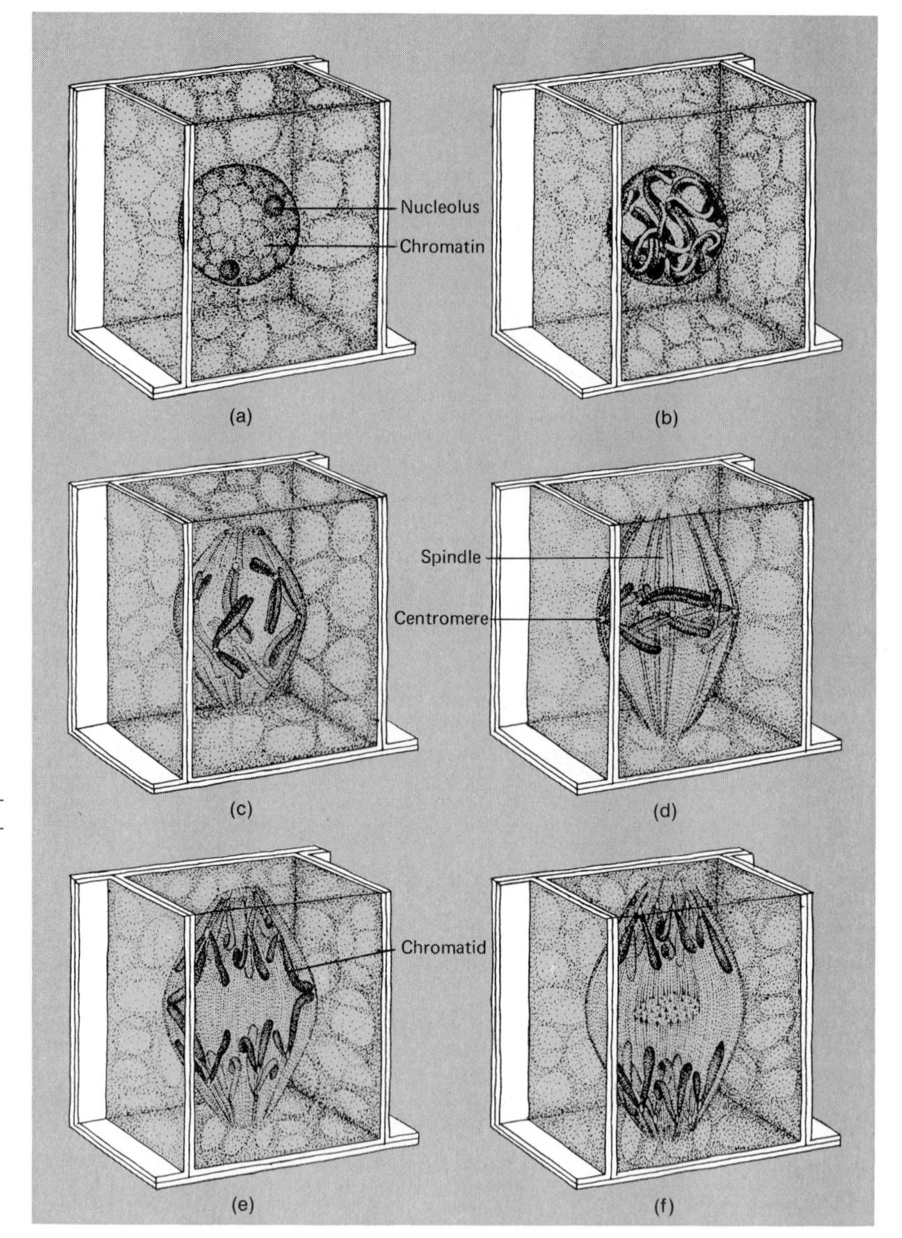

Fig. 8.20 Mitosis and cytokinesis shown diagrammatically. Front and top wall of cell removed. Explanation in text.

at the equator (Fig. 8.20*d*). Each chromatid contains the chromatin of a single chromonema.

The fibers of the spindle are apparently of two sorts, for some of them extend from pole to pole whereas others extend only from the poles to the equator where they attach to the chromatids at the centromere. By the apparent contraction of the attached fibers, the sister chromatids derived from each chromosome

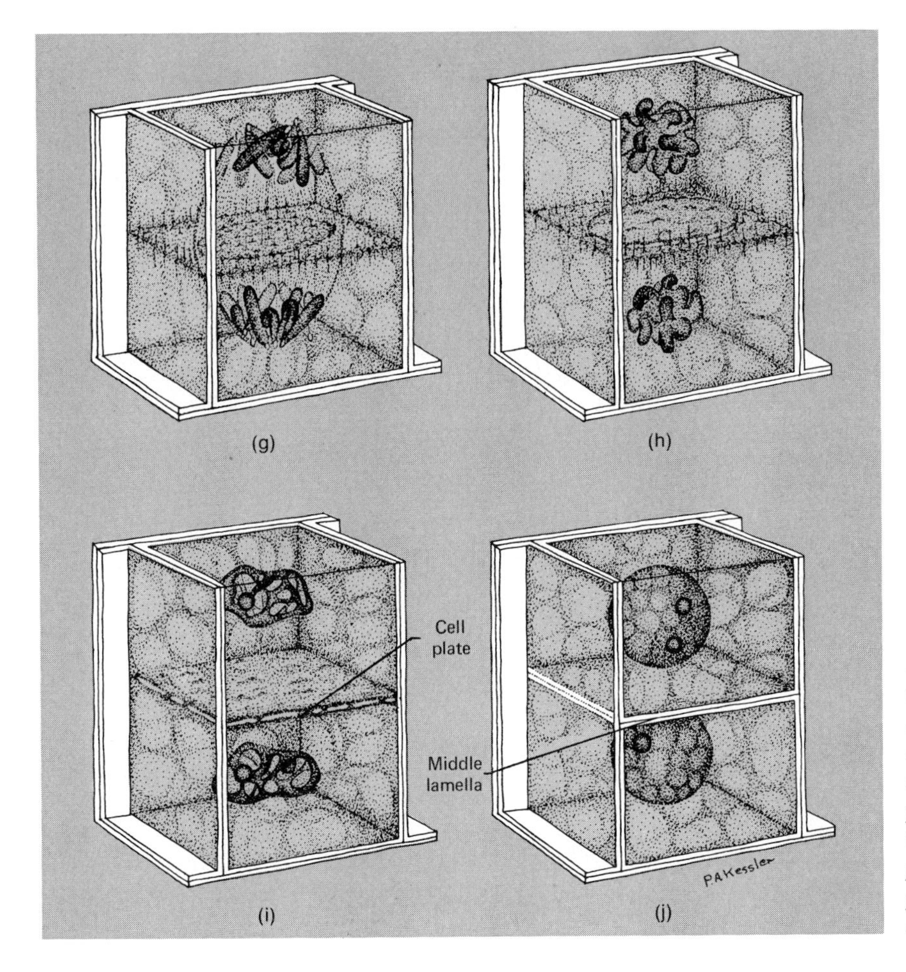

Fig. 8.21 (*Below*) *Mitosis in the root tip of onion as seen in selected stained longitudinal sections.* (a) *Late prophase,* (b) *metaphase,* (c) *anaphase,* (d) *late anaphase,* (e) *telophase,* (f) *cell division complete.* (*Courtesy, Carolina Biological Supply Co.*)

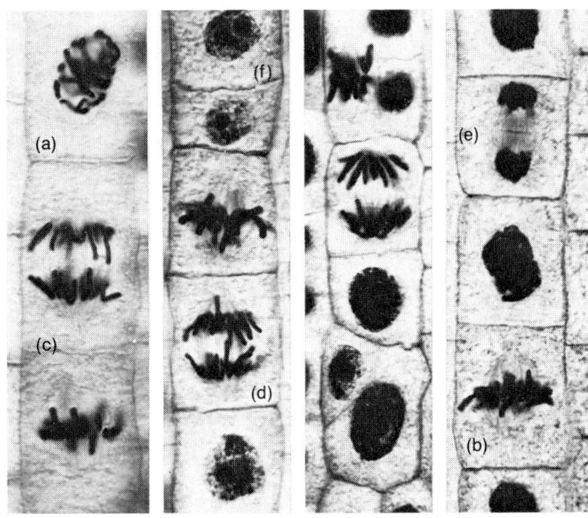

become separated and move to opposite poles (Fig. 8.20*e–g*). Arrival at the poles is quickly followed by a reconstitution of two new nuclei by reappearance of nuclear membranes and nucleoli and the gradual return of the chromatids (daughter chromosomes) to the thin, threadlike form and tangled disposition (Fig. 8.20*h–j*). Mitosis is completed when the two new nuclei have been formed.

It should be emphasized that the process just described has brought about an exact duplication of the chromatin of the parent nucleus and its distribution to the new nuclei in equal portions. Each chromatid becomes a fully organized chromosome, and thus each of

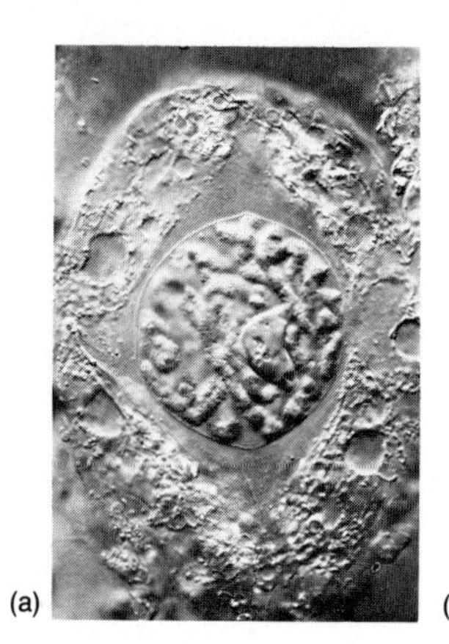
(a)

(b)

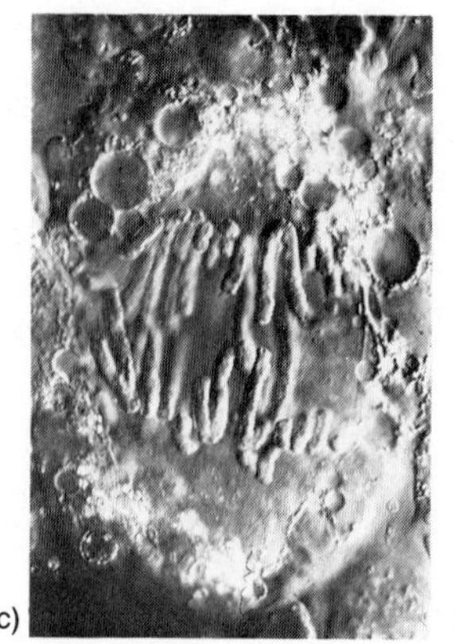
(c)

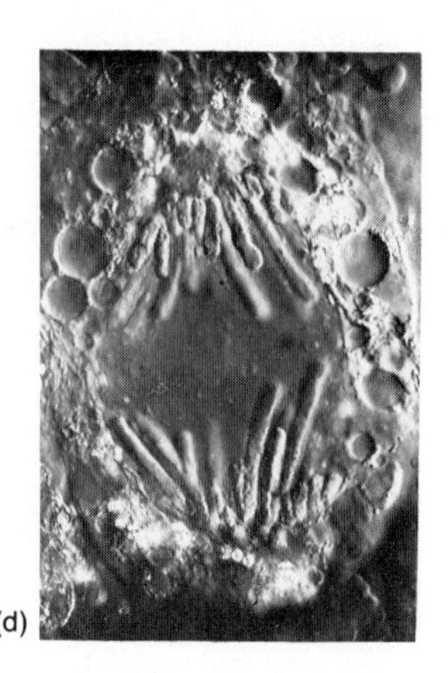
(d)

the two nuclei formed has the same number of chromosomes as the parent nucleus. For this reason mitosis may accurately be called an *equational* nuclear division. Special importance attaches to the precise distribution of the chromatin, for the chromosomes bear, in linear arrangement, the hereditary units or **genes.** Mitosis, therefore, endows each new cell with equal hereditary potentialities. The chromosomes evidently maintain their identities through the period between mitotic divisions, for the same chromosome set reappears in subsequent mitoses. The duplication of the chromatin thread and formation of chromonemata probably occurs during the period just prior to the onset of mitosis, and consists of the synthesis of new deoxyribonucleic acid (DNA) and proteins. Mitotic division is thus equational not only in a quantitative sense but in a qualitative sense as well.

It is possible to recognize, during the course of mitosis, certain standard configurations. For convenience these phases have been given names. Thus the early preparatory phase during which the chromosomes are becoming organized as discrete structures is termed the **prophase** (Figs. 8.20*a–c* and 8.21*a*). This phase is usually rather prolonged, and presents a series of stages that may be designated **early, mid-,** or **late** prophase. The equatorial position of the chromosomes is the distinctive mark of the **metaphase** (Figs. 8.20*d* and 8.21*b*). This phase is generally of relatively short duration, for the chromatids begin to separate almost at once. The period of movement of chromatids from equator to the poles is the **anaphase** (Figs. 8.20*e, f* and 8.21*c, d*). **Early, mid-,** and **late** anaphases may also be recognized. The final phase of mitosis is signaled by the arrival of the chromatids at the poles of the spindle and their subsequent change into the diffuse, tangled form of the nondividing nucleus. The changes culminating in the formation of the new nuclei constitute the **telophase** (Figs. 8.20*h, i* and 8.21*e*). Once begun, the mitotic process runs through to completion without pause, although the total time required may vary greatly from cell to cell and from species to species (Fig. 8.22*a–g*).

Cytokinesis

Cytokinesis, or the actual division of the protoplast, usually begins immediately after the completion of mitosis. The fibers of the remaining spindle appear to thicken and coalesce in the equatorial region while becoming fainter at their ends (Fig. 8.20*f–h*). Minute vesicles form at the equator and unite to form a thin **cell plate** within the limits of the disappearing spindle (Fig. 8.20*g, h*). As additional fibers are formed in the peripheral zone beyond the spindle, the cell plate is extended to the lateral walls (Figs. 8.20*g* and 8.22). The

(e) 

(f) 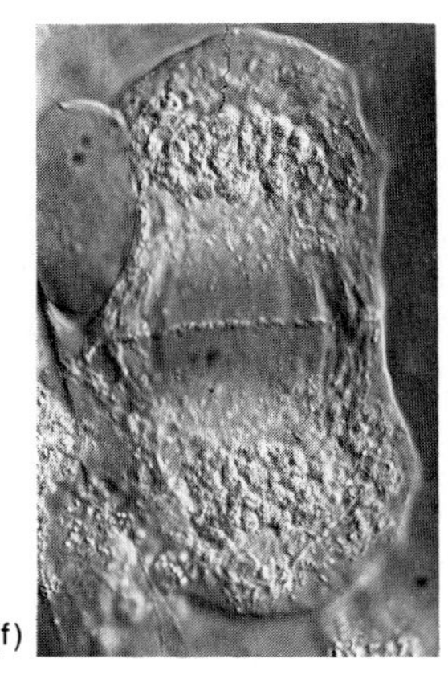

(g) 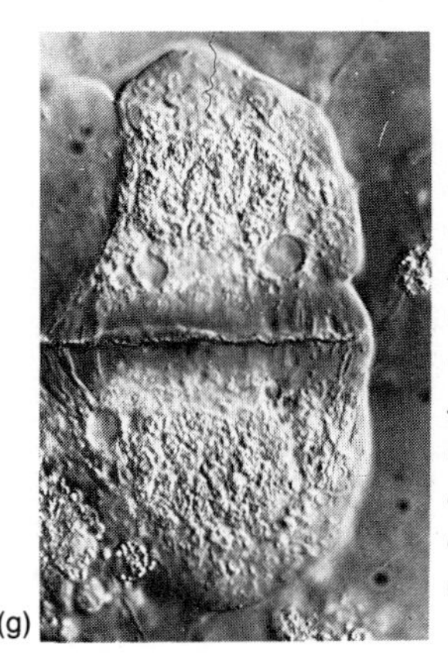

Fig. 8.22 Division of a cell of endosperm of Haemanthus, *the blood lily, as revealed by scanning electron microscopy.* (a) *Prophase.* (b) *Metaphase.* (c) *Anaphase.* (d) *Early telophase.* (e) *Late telophase, showing faint beginnings of cell plate formation.* (f, g) *Stages in development of cell plate and reconstitution of nuclei. Time lapse from* (a) *to* (g) *158 minutes.*

cell plate, composed of pectic substances secreted by the dividing protoplast, thus divides the protoplast in two. To what extent or in what manner the spindle fibers may be involved in formation of the cell plate is unknown. Plasma membranes form on the new surfaces thus created and become continuous with the membrane of the original cell at the lateral walls. New cellulose walls are quickly secreted on the outer surface of each new plasma membrane, flanking the pectic layer. The pectic layer, soon modified chemically, becomes an intercellular **middle lamella.** The new walls join the older lateral walls, and as the daughter cells enlarge, the lateral walls break at the edges of the middle lamella, and the daughter cells become complete and physically distinct (Figs. 8.20*j* and 8.21*f*).

The duration of cell division has been found, from studies of living cells with phase microscopy, to vary from several minutes to a few hours, depending on the organism and the kind of cell. The speed of the process is, however, markedly influenced by environmental conditions, notably temperature. At ordinary room temperature, most dividing cells of the higher plants complete the process in a period of one to one and a half hours duration.

RELATED READING

Bryan, J. "Microtubules," *BioScience, 24,* 701–711, 1974.

Buvat, R. *Plant Cells.* New York: McGraw-Hill, 1969.

Cohen, S. S. "Mitochondria and chloroplasts revisited," *American Scientist, 61* 437–445, 1973.

Fox, C. F. "The structure of cell membranes," *Scientific American, 226*(2), 30–38, February 1972.

Garrett, R. A. and H. G. Wittmann. "Structure and function of the ribosome," *Endeavour,* 32, 8–14, 1973.

Margulis, L. "Symbiosis and evolution," *Scientific American, 225*(2), 48–57, August 1971.

Marx, J. L. "Microtubules: versatile organelles," *Science, 181,* 1236–1237, 1973.

Mazia, D. "The cell cycle," *Scientific American, 230*(1), 54–64, 1974.

Rich, A. "Polyribosomes," *Scientific American, 209*(6), 44–53, December 1963.

9
THE TISSUE AND ORGAN LEVELS

In the discussion of the types of plant cells in Chapter 8, it was pointed out that such structural features as size, shape, wall thickness, and plastid types are related to the function of a cell. It will be found, however, upon detailed study of a seed plant body, that these specialized cells rarely serve alone, but are associated and act in conjunction with other cells of the same type. Thus, vessel elements are arranged end to end, forming a conducting unit of some length. Similarly, fibers in groups lend support quite beyond that possible for a single fiber. Food manufacturing in the plant is performed by masses of chlorophyll-containing cells. Frequently, two or more cell types may be associated in the performance of a general function. Such working aggregations of cells, related in structure or in function, constitute the **tissues** of the plant. Tissues are said to be **simple** when they are composed of one type of cell and **complex** when composed of two or more types.

The tissues occur in certain combinations and in certain spatial relationships in the several major parts or **organs** of the plant: the stems, leaves, roots, and reproductive structures. Each organ has a characteristic form and carries on a major function of the plant. The form and function of the organ are, of course, related to the type and arrangement of its component tissues and cells. It will be seen that the final disposition of all tissues and organs is the result of an orderly and progressive development that begins at the tip of the stem and root.

The continued growth and development of a stem or root depends on a continuous supply of new cells that may ultimately mature and take their places as functioning specialized cells. This supply is provided by the action of a group of perpetually youthful cells at the tips of the axis. Such a group of cells constitutes a formative tissue known as a **meristem;** and the cells composing a meristem are **meristematic cells.** A meristem located at the tip of a root or stem is called an **apical** meristem to distinguish it from others that occur elsewhere in the plant. The cells of the apical meristem are typically small, densely packed, thin walled, and not conspicuously vacuolate.

THE STEM

The production of new cells is accomplished by the division of each meristematic cell into two cells, as described in Chapter 8. That the identical new cells may, however, have different fates can be seen by the examination of a median longitudinal section of the growing stem tip (Fig. 9.1). The hemispheric mass of cells at the summit is the apical meristem below which **leaf primordia** occur, the youngest and smallest primordia being at the top. The older and longer primordia typically overlap and envelop the apical meristem. The cells of the leaf primordia are meristematic for a limited time, in contrast to those of the apical meristem. Thus, although the stem tip may grow indefinitely, the leaves of the plant soon reach a maximal size. The place of origin of a leaf primordium upon the stem is called a **node,** and the portion of the stem between two successive nodes is an **internode.** The angle formed by a leaf primordium and the internode above is the **axil** of the leaf, and in this position the primordium of a **lateral bud** normally develops. The lateral, or **axillary,** bud may develop into a branch that duplicates the structure of the main stem.

Elongation of the stem occurs chiefly as a result of the enlargement of the cells of the internodes, beginning just below the first node and extending downward some distance. The "bursting" of the dormant winter buds of woody plants results from the rapid elongation of the internodes as the higher temperatures and increased water availability return in spring.

Primary Tissues of the Stem

The development of the mature stem from the embryonic cells just below the apical meristem proceeds in an orderly manner. One of the first indications of maturation is the differentiation of a one-cell-thick surface layer. This is accomplished by tangential divisions of the surface cells. The outer cells thus produced quickly mature as epidermal cells. Thus, early in stem development a protective surface layer is provided. Often a few cell layers interior to the epidermis mature as collenchyma, which helps support the delicate stem tip. In a cross section cut a short distance behind the tip, several isolated groups of small cells may be seen arranged in a circle (Fig. 9.2). These groups have been formed by the continued longitudinal division of cells in those areas, and are conspicuous because of the dense cytoplasm and small cross-sectional area of the cells so formed. In a longitudinal section of the stem tip, these groups are seen to be strands of elongated cells—the **provascular strands.**

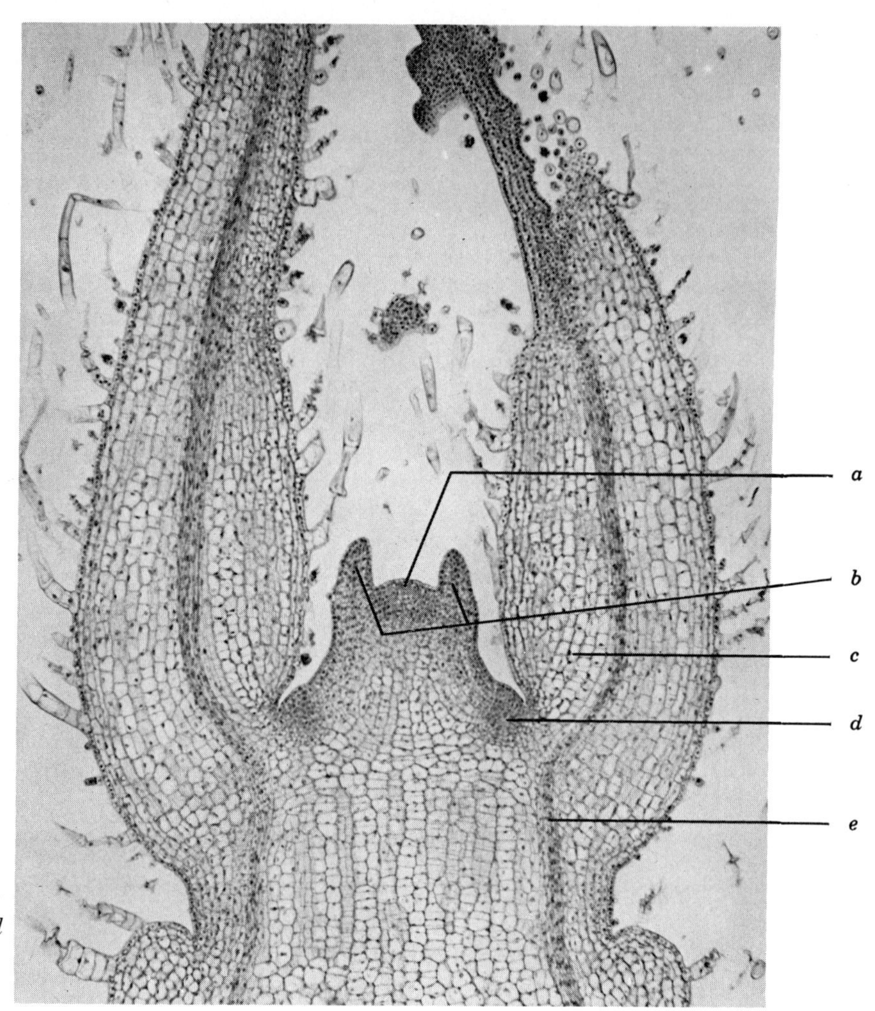

Fig. 9.1 Photomicrograph of a median longitudinal section of the stem tip of Coleus. (a) *Apical meristem,* (b) *leaf primordia at first node,* (c) *young leaf at third node,* (d) *lateral bud primordium in axil of young leaf,* (e) *provascular strand. (Copyright, General Biological Supply House, Inc., Chicago.)*

These become the first vascular tissues of the stem by maturation of the cells into several specialized types.

The maturation begins by differentiation occurring first at the outer margins of the strand and somewhat later at the inner margins. Proceeding inward from the point of initiation, the cells of the outer half of the strand mature into sieve-tube elements (with companion cells) or sieve cells and specialized, small parenchyma cells. The provascular cells near the inner margin mature into annular or spiral vessel elements or tracheids, and these are followed successively outward by a sequence of scalariform, reticulate, and pitted members often associated with specialized parenchyma and a few wood fibers.

The differentiated tracheids or vessel elements and associated cells occupying the inner half of the strand constitute the **primary xylem** or **primary wood.** The sieve-tube elements and associated cells make up the **primary phloem.** Primary xylem and primary phloem together constitute the **primary vascular tissue.** With the establishment of the vascular tissue the stem section becomes clearly

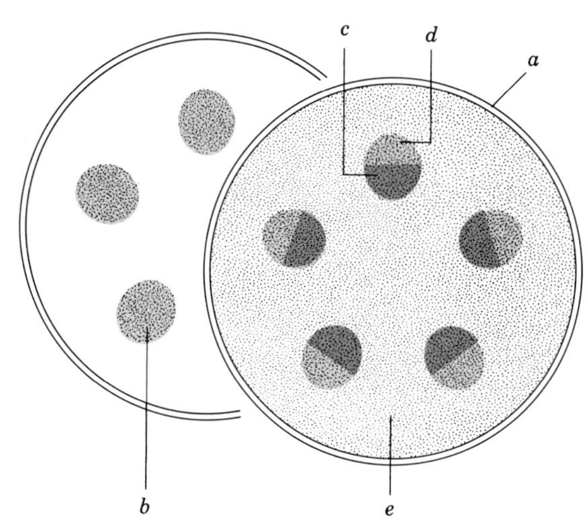

Fig. 9.2 Diagram of a cross section of a young stem tip, showing position of (a) *epidermis,* (b) *provascular strands,* (c) *first-formed xylem,* (d) *first-formed phloem.*

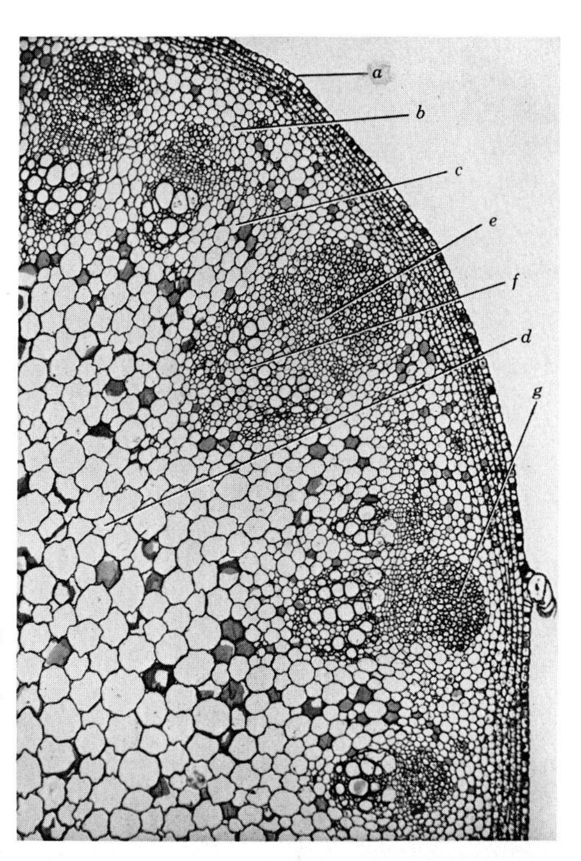

Fig. 9.3 Photomicrograph of cross section of stem of sunflower. (a) *Epidermis,* (b) *cortex,* (c) *pith ray,* (d) *pith,* (e) *primary phloem,* (f) *primary xylem,* (g) *cortical fibers. (Copyright, General Biological Supply House, Inc., Chicago.)*

zoned. The epidermis is underlaid by a broad zone of parenchyma cells extending inward to the vascular tissue. This is the **cortex.** Interior to the vascular tissue lies a core of parenchyma, the **pith.** The sheets of radially extended parenchyma are the **pith rays.** Each strand of vascular tissue, consisting of xylem and phloem, is a **vascular bundle** (Fig. 9.3). Sometimes the originally isolated provascular strands unite laterally, producing a hollow cylinder that matures as a complete cylinder of primary vascular tissue.

At the nodal regions of the stem, where leaves and branch buds arise, the vascular tissues of the stem connect with those of the leaf and branch bud. The vascular strands leading to leaves and branches are called **leaf traces** and **branch traces.** Where a trace departs from the primary vascular tissue of the stem, it leaves a gap or opening in the vascular cylinder just above its point of departure. The gap is appropriately referred to as a **leaf gap** or **branch gap.** When the gaps are rather tall, extending over more than one internode, or when the internodes are very short, the gaps overlapping at successive nodes dissect the vascular cylinder into many separate bundles, as viewed in a cross section. In secondary growth of stems (to be described in the next section) leaf and branch gaps are closed and buried deeply by the overlay of secondary tissues.

Some species, such as the grasses, possess in addition to the usual apical meristem, a persistent meristematic zone located at the base of each internode. These **intercalary meristems** may remain active for some time, long after the other parts of the internode have fully matured. Growth of the cells produced by such meristems contributes greatly to the rapid elongation of the stem. Eventually intercalary meristems cease to function and their cells mature into permanent tissues.

The tissues described are called **primary tissues** because they are formed first and directly from cells produced in the growing point of the stem. Many short-lived and delicate annual plants have only these primary tissues.

Secondary Tissues of the Stem

The stems of trees and shrubs and also those of the coarser annual plants have a capacity for continued production of tissues beyond the primary ones. This **secondary growth** may be of long duration and account for the continuing increase in diameter of the trunks of trees and stems of shrubs year after year, or it may be short-lived and contribute the "woodiness" to the stem bases of coarse annuals. In perennial stems the stages in the development of the primary tissues are the same as described previously, except that the outward maturation of the primary xylem and the inward maturation of the primary phloem stop short of meeting at the center of the provascular strand. There is thus left between primary xylem and primary phloem a short arc of provascular cells. Upon completion of primary growth, usually during the first few weeks of a growing season, the residual provascular cells divide actively, and thus become a **secondary meristem** called the **vascular cambium** (Fig. 9.4*e*). It is a secondary meristem because it arises, after the primary growth is complete, from cells originally formed by the apical

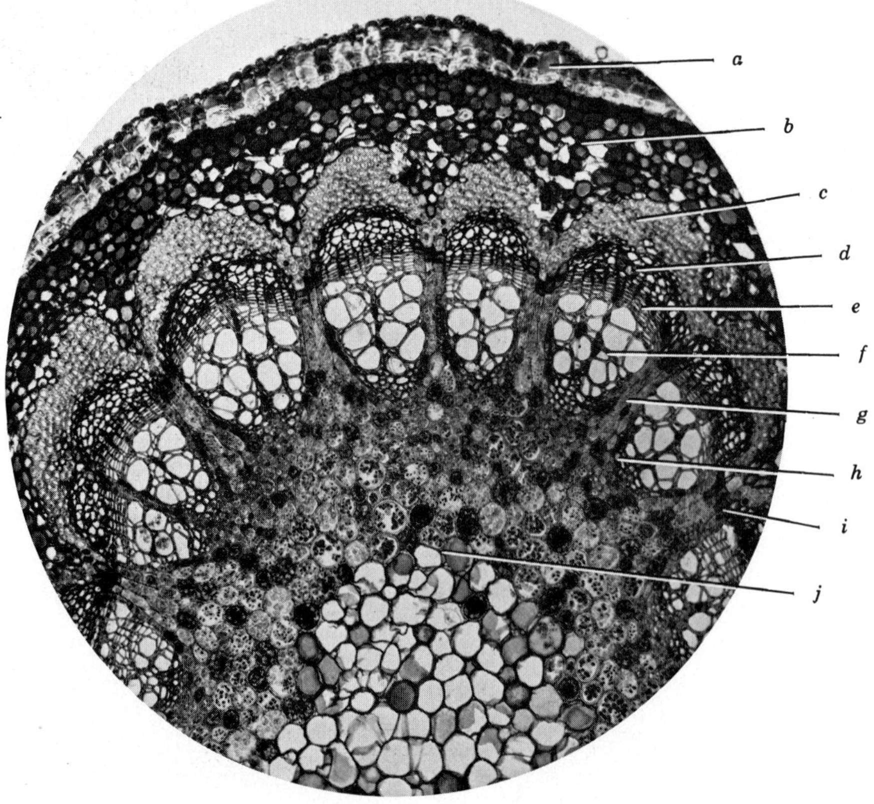

Fig. 9.4 Photomicrograph of cross section of sycamore stem in first year of growth. (a) *Cork,* (b) *cortex,* (c) *cortical fibers,* (d) *secondary phloem and remnants of primary phloem,* (e) *vascular cambium,* (f) *secondary xylem,* (g) *pith ray,* (h) *primary xylem,* (i) *interfascicular cambium,* (j) *pith. (Copyright, General Biological Supply House, Inc., Chicago.)*

meristem. The new cells produced by the division of cambial cells mature as **secondary xylem** elements and as **secondary phloem** elements (Fig. 9.4*d, f*). Secondary xylem is laid down on the outer side of the primary xylem, and secondary phloem on the inner side of the primary phloem (Fig. 9.4*d, h*). Of the two cells resulting from the division of a cambial cell, only one matures; its sister cell remains meristematic.

The vessel elements or tracheids of secondary xylem are of the pitted type. Occasional wood fibers and wood parenchyma cells are also produced by cambial action. Sieve-tube elements or sieve cells of the secondary phloem do not differ significantly from those of the primary and are associated with phloem fibers and phloem parenchyma. Secondary xylem elements are typically heavily lignified. They are hard and comparatively incompressible, in contrast to the thin-walled and delicate phloem elements. Spatial adjustments required as a result of the accumulation of secondary vascular tissues would, therefore, be made at the expense of primary phloem which is progressively crushed as new secondary tissues press outward. Primary phloem is thus replaced by secondary phloem (Fig. 9.4*d*).

As the secondary xylem continues to accumulate, the complete cambium cylinder increases circumferentially by an occasional radial division of some of its cells. Thus, in effect, the cambium moves outward as the secondary xylem increases. The radial extent of the phloem remains about constant since older secondary phloem is crushed by new secondary phloem, just as the primary phloem was crushed by the first-formed secondary phloem.

In stems where primary vascular bundles are separated by pith rays, the sections of vascular cambium within the bundles sometimes become connected by a meristematic strip developing across the ray. These new secondary meristems are **interfascicular,** that is, they lie between the vascular bundles (Fig. 9.4*i*). In some stems the interfascicular cambium produces more ray tissue, thus lengthening the rays as the vascular bundles increase radially. Such stems would preserve the individuality of the vascular bundles. In other stems, the interfascicular cambium produces secondary xylem and phloem. Thus, separate primary vascular bundles may become united by secondary tissues into a continuous cylinder.

Secondary rays, called **vascular rays,** which supplement the pith rays, are also formed by the cambium (Fig. 9.5*f*). These may be initiated at various points in the secondary xylem and extend outward to or through the phloem. They have limited vertical extent and are like small sheets of parenchyma inserted

Fig. 9.5 *Photomicrograph of cross section of sycamore stem in fourth year of growth.* (a) *Cork,* (b) *secondary phloem,* (c) *annual growth ring of third year,* (d) *summer wood of second annual growth ring,* (e) *spring wood of second annual growth ring,* (f) *vascular ray,* (g) *primary xylem,* (h) *pith. (Copyright, General Biological Supply House, Inc., Chicago.)*

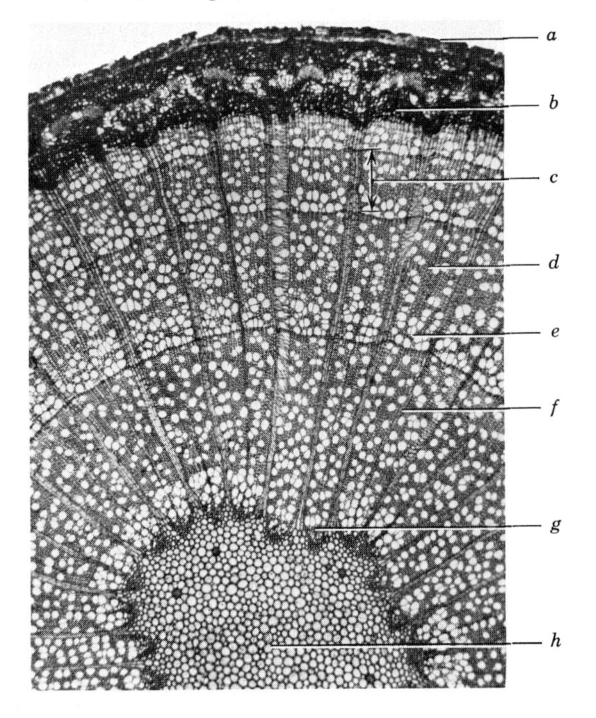

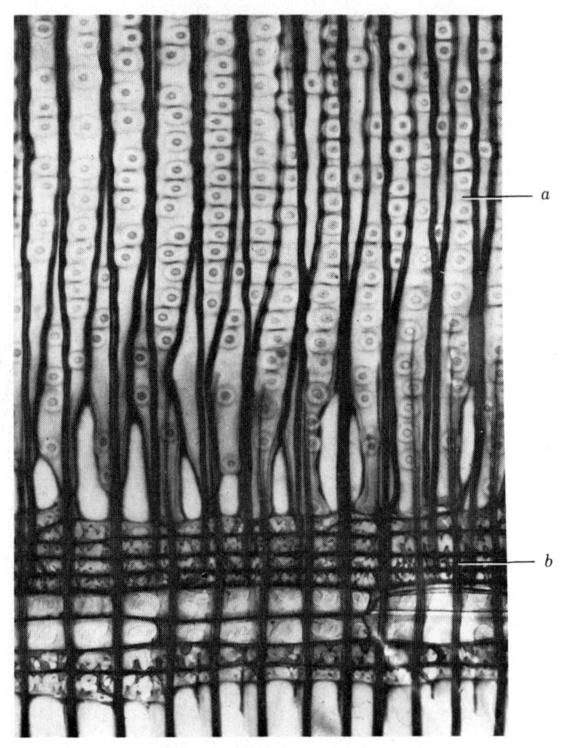

Fig. 9.6 Secondary wood of pine seen in radial section. (a) *Tracheids, with bordered pits,* (b) *vascular ray.*

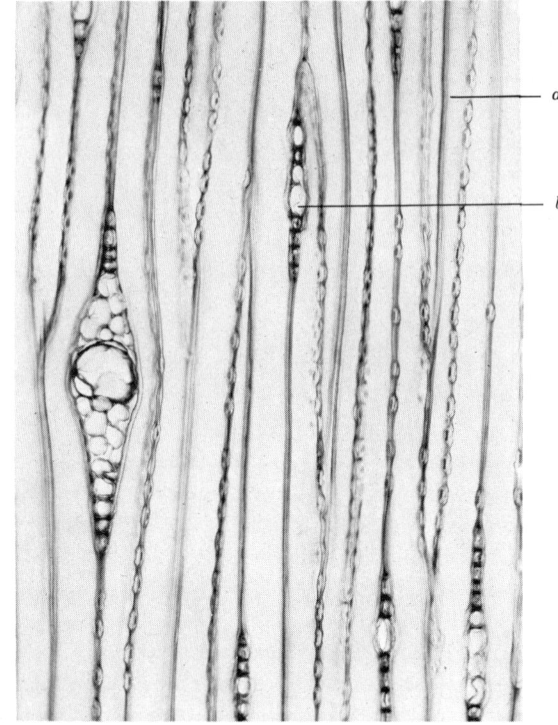

Fig. 9.7 Secondary wood of pine seen in tangential section. (a) *Tracheids,* (b) *vascular ray.*

among the vascular elements (Figs. 9.6 and 9.7). Like the pith rays, they are pathways of lateral movement of materials. Starch and crystals frequently accumulate in ray cells.

Only those tissues interior to the cambium are truly permanent, whereas those exterior to it are replaced as growth in diameter of the stem proceeds. Increase in diameter of the perennial stem, therefore, is related directly to the production of secondary xylem. In temperate regions where the seasonal cycle presents periods alternately favorable and unfavorable for growth, the secondary xylem is usually laid down in clearly marked concentric bands. Each band normally represents one year's production of wood and is called an **annual growth ring** (Fig. 9.5*c*). The demarcation of one annual growth ring from another reflects the fact that cambial activity is vigorous in spring, and the prevailing more favorable conditions for cell growth result in xylem elements of larger size than in summer (Fig. 9.9). In gymnosperms, such as pine, the tracheids formed in spring are large, whereas those formed in summer are smaller (Fig. 9.8). In angiosperms, such as oak, the **spring wood** may have a very high proportion of vessels, whereas the **summer wood** has a predominance of wood fibers. Thus the abrupt change in size or type of the cells marks the boundary between summer wood of one year's growth and the spring wood of the next year's. In any one annual growth ring, of course, the spring wood lies interior to the summer wood (Fig. 9.5*d, e*). Annual growth rings are generally absent or indistinct in wood formed in regions where favorable growing conditions prevail throughout the year. The radial widths of annual growth rings and the proportions of spring to summer wood are usually various, reflecting differences in environmental conditions from year to year as the rings were formed by the cambium. Conditions such as drought, low temperature, poor soil aeration, and excessive competition, which adversely affect plant growth, are generally reflected in the formation of narrow rings.

Although we call each year's production of secondary xylem a "ring," it is, in fact, an open-end cone of xylem, enclosing and extending beyond the cone of the previous year. Thus, the secondary xylem of a tree trunk may be properly visualized as a series of successively longer open-end cones stacked one over another, the longest and outermost being the last formed (Fig. 9.10). Since usually only one growth ring is formed each year, a count of the rings at the base of the trunk, where all rings are present, indicates the age of the tree. The age of a branch may be similarly determined, since, in structure, it duplicates that of the main stem.

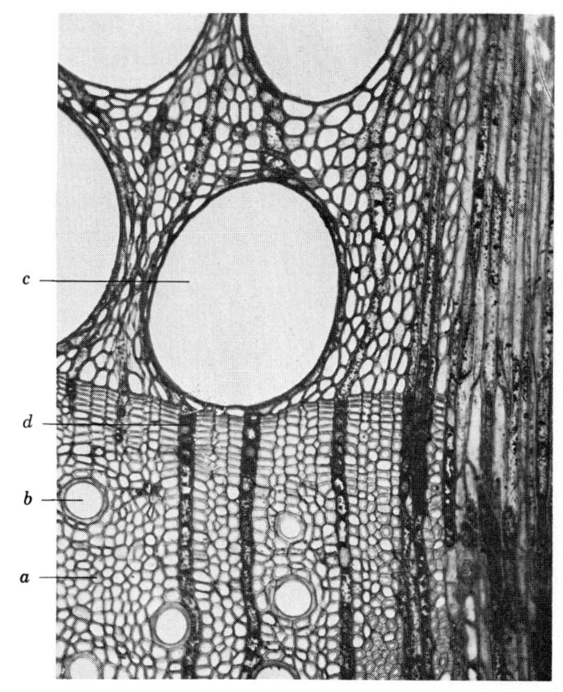

Fig. 9.9 Secondary wood of oak seen in cross section. (a) *Fibers of summer wood,* (b) *vessel of summer wood,* (c) *vessel of spring wood,* (d) *vascular ray.*

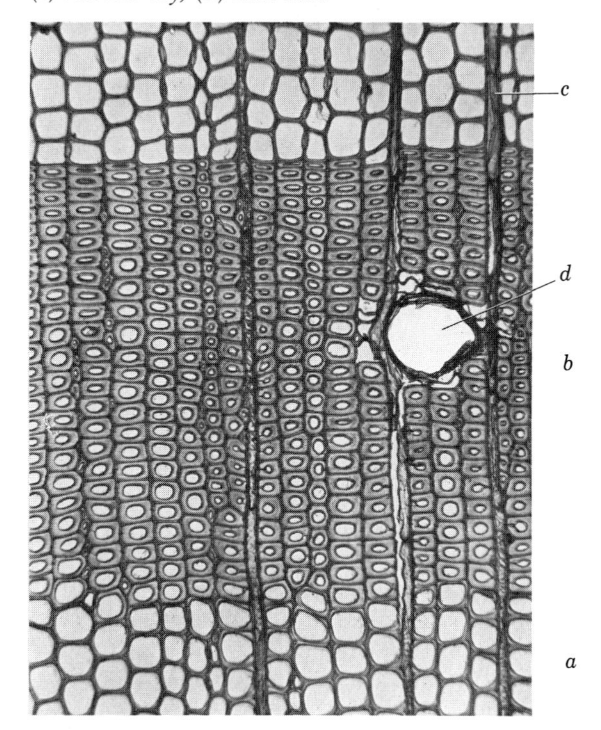

Fig. 9.8 Secondary wood of pine seen in cross section. (a) *Tracheids of spring wood,* (b) *tracheids of summer wood,* (c) *vascular ray,* (d) *resin duct.*

Branches originating in the primary stem sometimes die for one reason or another after a few years' growth and break off at the surface of the trunk. The internal section of such branches, extending inward to the primary tissues, will be buried ever more deeply by the continued secondary growth of the trunk. The buried, dead branch, hardened by the accumulation of resins, gums, and tannins, will be recognized as a knot in lumber sawed from the trunk.

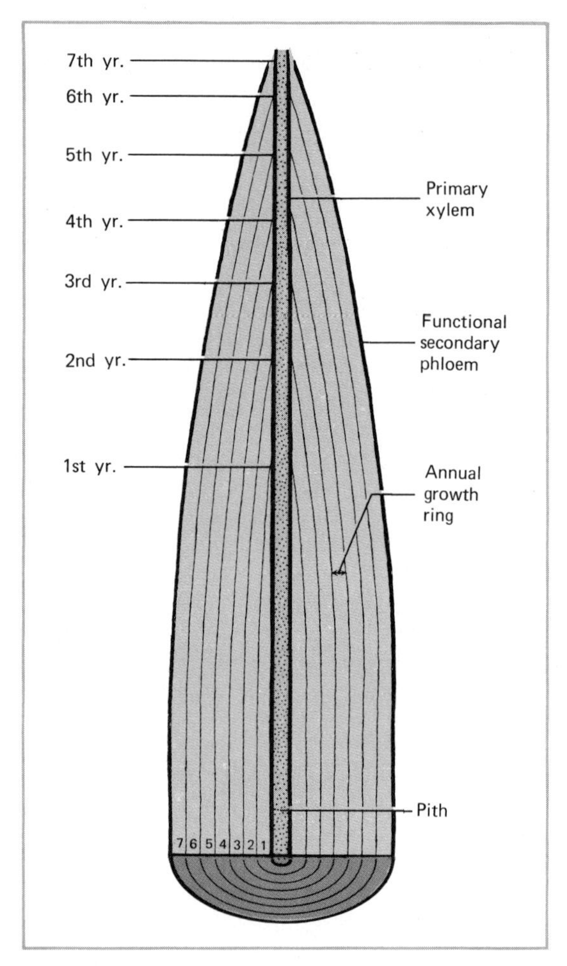

Fig. 9.10 Diagram showing position of successive annual growth rings and their relation to other vascular tissues in a perennial woody stem.

In most species of trees, conduction of water and dissolved substances by the secondary xylem becomes limited to the outer or younger portion of that tissue as the tree grows older. Usually only a few or several annual growth rings are active in conduction at any one time. This active portion of the xylem is called the **sapwood.** In the older rings, nearer the center of the trunk, the vessels commonly become occluded by the intrusive growth of surrounding parenchyma cells through the pits in the walls of the vessels, and by deposits of gums and resins. The inactive, nonconducting wood is called the **heartwood.** As new sapwood is added each year by the cambium, the older sapwood is converted to heartwood at about the same rate. Thus the radial thickness of the cylinder of sapwood remains fairly constant through the years while the heartwood increases (Fig. 9.11).

In most species of trees, the heartwood undergoes chemical change through the accumulation of a variety of substances, such as resins, oils, gums, and tannins. The accumulation of such materials commonly causes the heartwood to become deeply colored and harder than the sapwood. In Red Cedar, and other species, the deposit of oils and tannins also confers a resistance to decay and insect attack. Red Cedar posts are highly prized for fencing because of this. Planks and beams cut from the heartwood of timber species are stronger and more durable than those cut from the sapwood.

The woods from different species of trees differ greatly in their suitability for specific uses. Density, hardness, flexibility, shock resistance, compression strength, texture, resistance to warping, and checking, are some of the important attributes of wood that may determine its usefulness (Fig. 9.12). These qualities are related to such anatomical features as size, kinds, proportions, arrangement, and wall thickness of the cells composing the secondary xylem. The distinctive grain of woods used for cabinetwork and paneling is related mostly to the vertical and radial extent of the vascular rays, the length and straightness of the wood fibers, and the plane in which the log is sawed.

In most woody stems when the secondary vascular development begins, another secondary meristem arises in the outer part of the cortex. As seen in a transverse section, a continuous or sometimes broken circle of cells of the cortex close to the epidermis or just be-

Fig. 9.11 Cross section of the trunk of a red oak showing more than 90 growth rings. (p) *Position of pith,* (h) *heartwood,* (s) *sapwood,* (c) *position of vascular cambium,* (b) *bark, consisting of a thin layer of active phloem, inactive phloem, and cork. (Courtesy, Carolina Biological Supply Co.)*

Fig. 9.12 (Bottom) Relative density of two kinds of wood. Balsa, right; quebracho, left.

neath it become meristematic. Divisions of these cells occur in the tangential plane, the cells cut off on the outside usually maturing as cork cells, in regular radial rows, whereas those on the inside remain meristematic. In some divisions the inner daughter cells mature as parenchymalike cells, the **phelloderm;** the outer cells remain meristematic. This secondary meristem is the **cork cambium** or **phellogen** (Fig. 9.13*a, b*). After several divisions meristematic action stops and a new phellogen arises deeper in the cortex. Cork cells are dead and have suberized walls. Because of the suberized cell walls, the cork layers prevent the passage of water and thus the living epidermis and cortical cells exterior to it dry up and eventually fall away. Ultimately, all of the original cortex is lost and the successively deeper layers

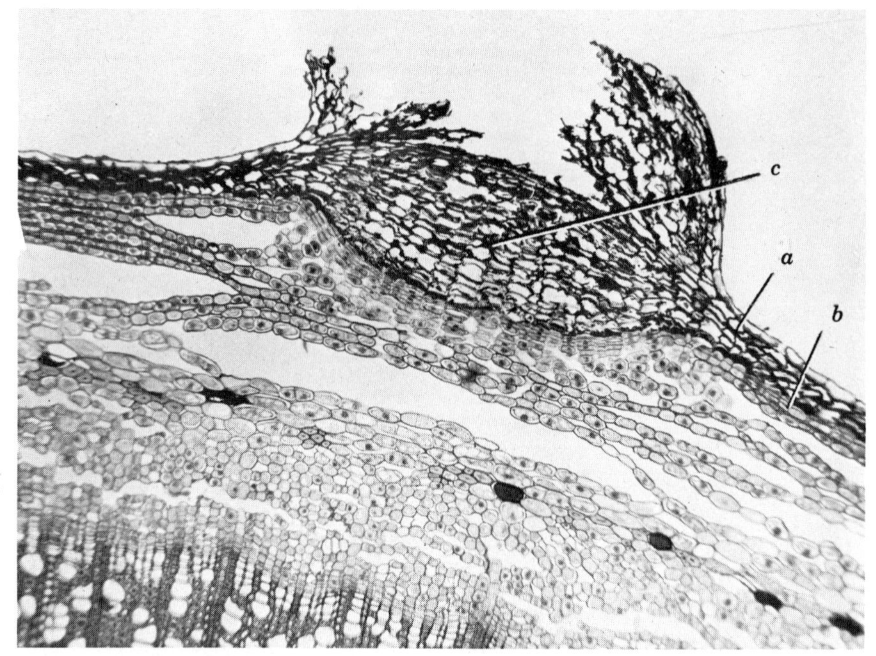

Fig. 9.13 Cork and lenticel formation seen in cross section of stem of alder. (a) *Cork,* (b) *phellogen,* (c) *lenticel tissue formed by active phellogen below. (Copyright, General Biological Supply House, Inc., Chicago.)*

of cork cambium arise in the older secondary phloem. The smooth surface of compact cork cells on young stems is interrupted in small, commonly lens-shaped areas by masses of soft tissue composed of thin-walled, loosely organized cells. These patches of contrasting tissue, the **lenticels,** are produced by more intensive activity of the phellogen in those areas (Fig. 9.13*c*). The cells formed toward the outside do not mature as normal cork cells. The loose organization of the lenticel tissue is believed to permit aeration of the deeper tissues.

The surface of the cork usually becomes fissured and rough as the growth of the inner part of the stem presses outward, since it is composed of dead cells and thus cannot accommodate by growth. Its rates of production of new cork and sloughing off of old are about equal in mature trees, thus the thickness of the layer, which is fairly characteristic of mature specimens of a given species, remains about constant, ranging from a few millimeters in the beech to several decimeters in the redwoods. As a protective covering tissue, the cork is effective against mechanical impact because of the spongy and elastic qualities of its dead, air-filled cells, and against excessive loss of water from the stem because of its suberized walls. It also has insulating properties and affords some protection against the high temperatures of forest fire. The resilient, fine-textured commercial cork is obtained from the cork oak, *Quercus suber.*

All tissues exterior to the vascular cambium are collectively called the **bark.** Bark is, then, not a tissue in the usual sense, but a composite made up of the functioning phloem, cork cambium, cork, and the inactive remains of older phloem and cortex. The manner in which cork is formed differs in some detail from species to species. This results in the development of distinctive patterns of ridges, splits, and cracks on the outer surface of the bark. The experienced woodman can identify many trees by the appearance of the bark surface. Even a casual observer can distinguish furrowed barks of trees such as oaks and ashes from the smooth barks of birch or cherry.

Herbaceous and Woody Stems

If vascular cambium is absent or cambial activity is limited to the first year, stems are **herbaceous** rather than woody. The herbaceous stem pattern is thought, in general, to have been derived in evolution from the woody, through reduction of the amount of vascular tissue produced. The tissues of an herbaceous stem are similar in composition and arrangement to the primary tissues of a woody stem. All intergradations between extreme woodiness and extreme herbaceousness may be encountered among the dicotyledonous angiosperms. The gymnosperms are generally woody. Among flowering plants, the ultimate in reduction of vascular tissue is found in stems of some monocotyledons. One representative of this group, the stem of corn, shown in cross sectional view in Fig. 9.14, has scattered vascular bundles lacking cambium (Fig. 9.15). All tissues in such stems are primary. The proportion of vascular tissue to soft parenchyma is very small. Cork is absent, of course, in stems consisting wholly of primary tissues.

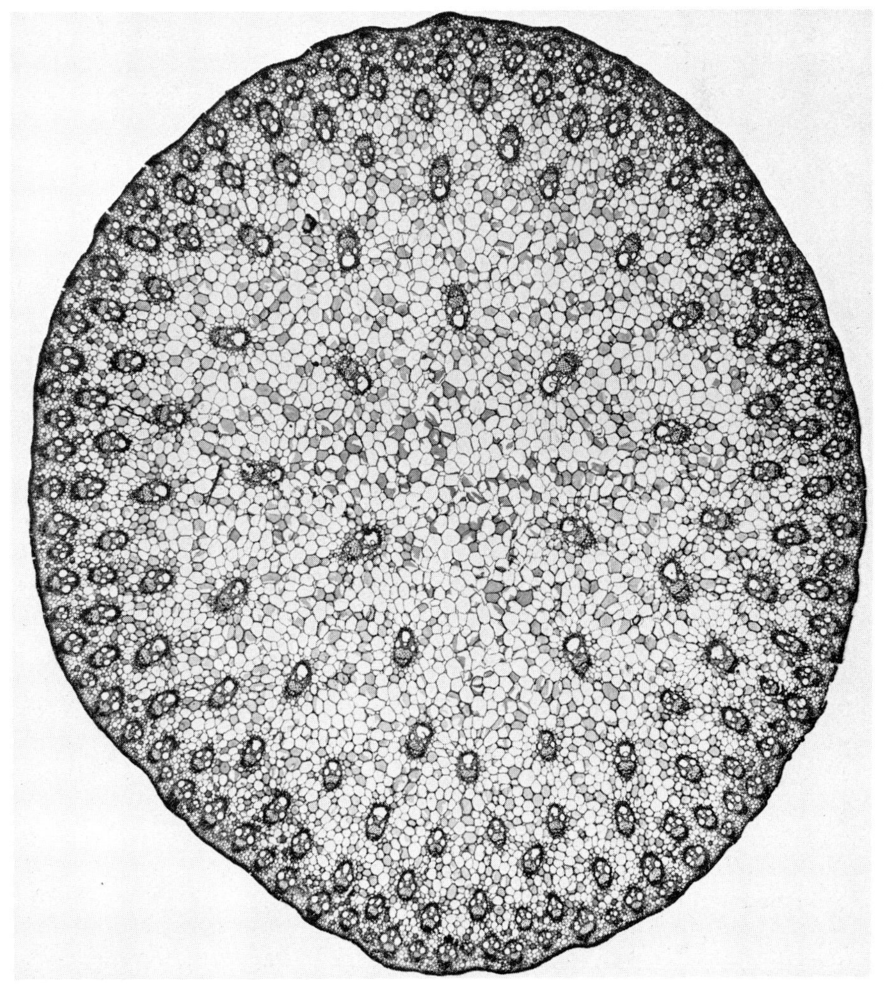

Fig. 9.14 Photomicrograph of maize stem, in cross section, showing scattered vascular bundles. (Courtesy, Carolina Biological Supply Co.)

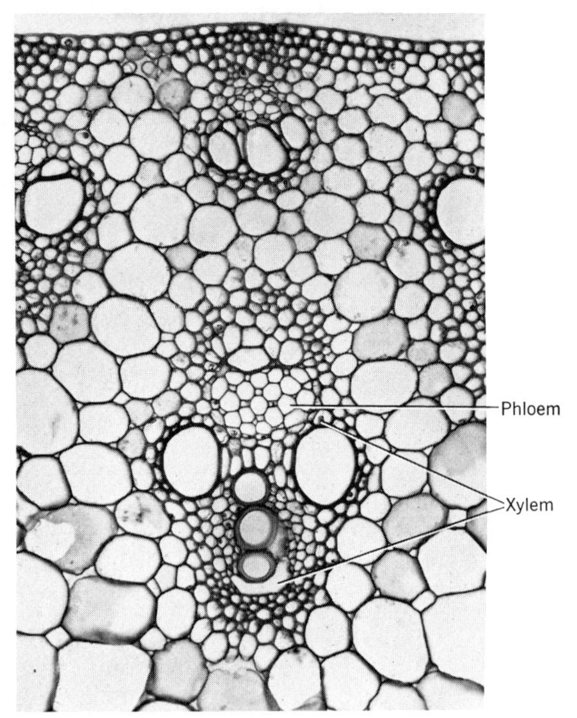

Fig. 9.15 Photomicrograph of maize stem in cross section, showing detail of vascular bundle. (Copyright, General Biological Supply House, Inc., Chicago.)

External Features of the Stem

Externally, the mature stem reflects the internal development we have just discussed. The **terminal bud** at the apex of the woody stem is a condensed stem closely enveloped by scalelike leaves, the **bud scales** (Figs. 9.16 and 9.17). The bud scales are often hairy, tough, or covered with gummy or resinous secretions. The overlapping bud scales effectively protect the delicate enclosed stem tip from desiccation and mechanical injury. We have already noted that in the growing stem tip, the alternation of nodes and internodes gives the stem a jointed appearance. At the nodes two lateral structures occur, namely, a **leaf** and a **lateral bud** (Fig. 9.18) in the leaf axil. The organization of the young stem within the bud was completed toward the end of the previous growing season. The opening of the terminal bud and the early growth of embryonic stem is brought about by enlargement of the cells in the internodal regions, and reflexing of the bud scales. The separation of the bud scales is the result of a rapid rate of growth on the inner faces of the scales, near the bases. Bud scales usually fall promptly after the bud opens, leaving a circle of small thin scars at the points of attachment on the stem. These circles of **bud scale scars** may remain visible on the surface for several years, or until the secondary growth of the stem obliterates them (Fig. 9.16*f*). Since a ter-

Fig. 9.16 Stem of horsechestnut in winter condition. (a) *Terminal bud,* (b) *internode,* (c) *node,* (d) *leaf scar,* (e) *lateral bud,* (f) *terminal bud scale scars. (Copyright, General Biological Supply House, Inc., Chicago.)*

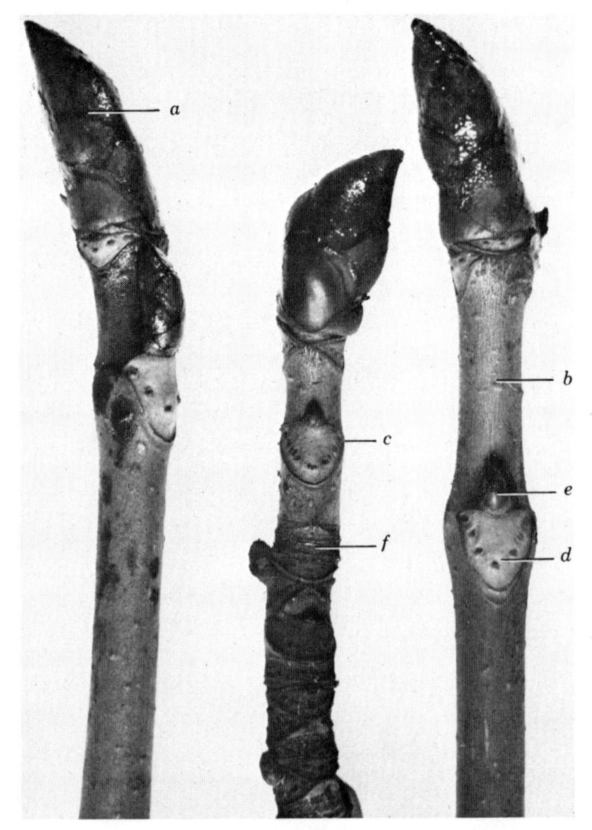

Fig. 9.17 Median longitudinal section of terminal mixed bud of horsechestnut, showing terminal flower cluster and later foliage branches.

minal bud is formed each year and leaves its scale scars the following spring, the age and rate of elongation of young stems and minor branches may be determined by inspection.

The lateral bud is a rudimentary stem and may develop into a full size branch which matches the main stem in structure and growth potential. Thus as the main stem grows taller, the branches elongate. The oldest, and usually therefore the longest, occur at the bottom of the crown, the youngest at the top. This accounts for the conical or spirelike shape so commonly observed in trees such as the pines, firs, and poplars. In most broad-leaved species, as the tree grows older, the rate of growth of the main stem diminishes and the branches grow more vigorously. In time the branches equal or overtake the main stem and thus produce a rounded or irregular crown (Fig. 9.19).

Specialized Stems

The stems of many species of angiosperms are highly specialized and sometimes bear little resemblance to ordinary stems. Stems of most

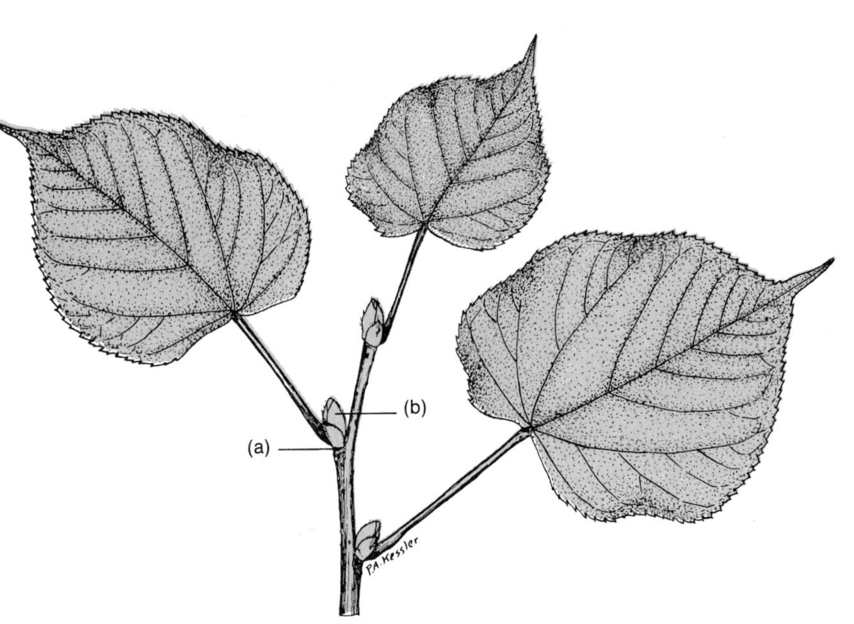

Fig. 9.18 Stem of linden. (a) *Node,* (b) *lateral bud in axil of leaf.*

Fig. 9.19 Black walnut, Juglans nigra, *showing the rounded crown typical of most broad-leaf species. Compare with Fig. 5.11.*

species of cactus, bearing reduced spinelike leaves, are expanded and green and function as the photosynthetic organs of the plant. Cactus stems store large quantities of water. The creeping stems of strawberry, **stolons,** develop new plants where the nodes rest upon the soil. The strawberry may be vegetatively propagated by setting out the small plantlets. The leaves of iris are borne at the ends of a fleshy underground stem called a **rhizome.** The **tuber** of the potato is a shortened subterranean stem whose nodes are marked by the "eyes." Each "eye" is a branch bud standing in the axil of a reduced scalelike leaf. Considerable food, mostly starch, accumulates in the tuber and is consumed by the growing sprout (branch bud) when the "seed pieces" are used for vegetative propagation. The use of stems and other plant parts in vegetative propagation is discussed in Chapter 20. The **thorns** of hawthorn are normally leafless branches situated in a leaf axil. Stems may be greatly elongated and twining and thus serve as a supporting device, as in morning glory, or support may be provided by modification into **tendrils** as in grapevine.

THE LEAF

The growth of leaf primordia and small leaves at the stem tip into mature organs takes place rapidly after the opening of the terminal bud. Division, enlargement, and maturation of the leaf cells occur fairly uniformly throughout the structure, so that by the time the leaf is fully grown, all of its tissues are mature. A conspicuous vascular strand commonly occupies the central part of the leaf and constitutes the main or **midvein.** From the midvein, smaller **secondary veins** depart. As a leaf is seen in cross section, the kinds of tissues present and their arrangement suggest a strongly flattened primary stem. The midvein is continuous with the vascular cylinder of the stem. The vascular tissue of the leaf consists of primary xylem and primary phloem, with the phloem typically on the lower side. The vein is commonly surrounded by a mass of parenchyma, sometimes with small amounts of collenchyma or fibrous tissue above and below. Parenchyma cells that contain many chloroplasts make up the major part of the leaf's bulk. It is called the **mesophyll** and is the chief food manufacturing tissue of the plant. Two layers of the mesophyll may commonly be distinguished. The upper layer, composed of one or more rows of vertically elongated cells, is the **palisade layer.** The lower layer of spherical or irregular, loosely associated cells is the **spongy layer** (Fig. 9.20).

Surrounding the whole leaf is the **epidermis,** which is continuous over all surfaces of the leaf and with the epidermis of the young stem. Epidermal cells are typically without chloroplasts. Those on the lower side of the leaf are usually smaller than those on the upper side. Dispersed among the epidermal

Table 9.1 **Number of Stomata per Square Centimeter of Leaf Surface**

Leaf	Upper Surface	Lower Surface
Scarlet oak	0	104,000
Black walnut	0	46,000
Apple	0	30,400
Oats	2,500	2,300
Kidney bean	4,000	17,600
Corn	7,840	9,440
Bluegrass	16,000	10,400
Alfalfa	17,440	14,240

cells are small pores called **stomata** (singular, **stoma**). Table 9.1 shows the distribution and density of stomata in some common species. Stomata are intercellular channels through which gases diffuse between the atmosphere and the interior of the leaf. The stomata communicate with large intercellular spaces among the cells of the mesophyll. If a strip of leaf epidermis is removed and examined under the microscope, each stoma is seen to be flanked by a pair of crescent shaped cells containing chloroplasts. These are the **guard cells** whose slight alteration in shape, through changes in

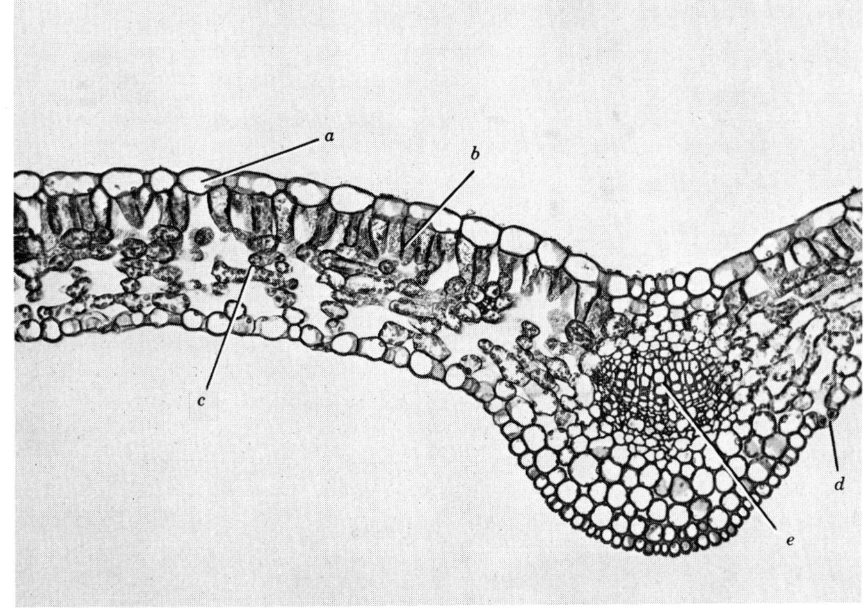

Fig. 9.20 *Leaf cross section showing usual mesomorphic features.* (a) *Epidermis with thin cuticle,* (b) *palisade layer of the mesophyll,* (c) *spongy layer of the mesophyll,* (d) *stoma, flush with the surface,* (e) *midvein. (Copyright, General Biological Supply House, Inc., Chicago.)*

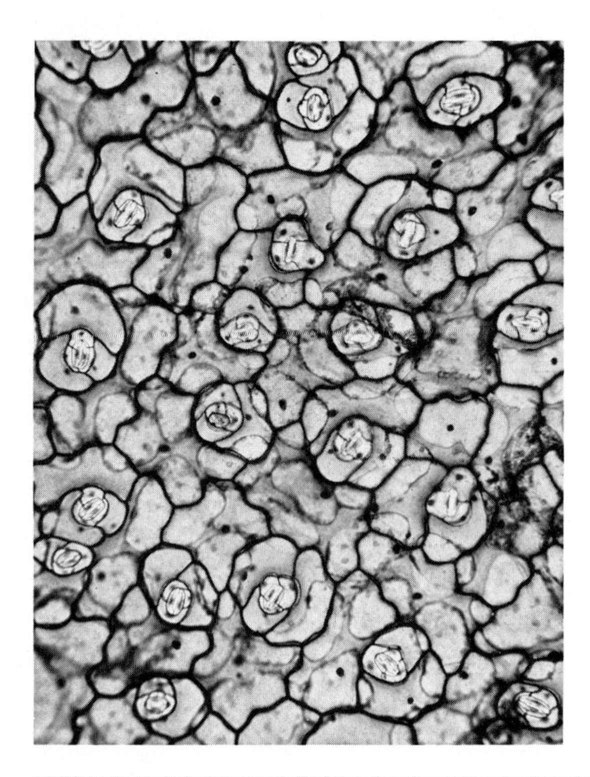

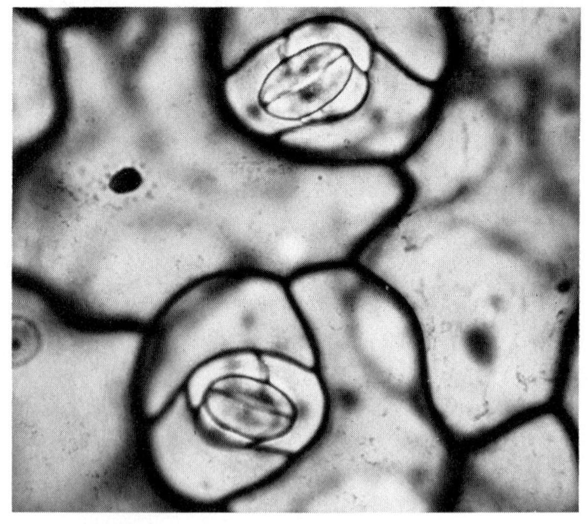

Fig. 9.21 *Epidermis of* Sedum *leaf, surface view, showing stomata partially closed. Note small subsidiary cells surrounding each pair of guard cells. (Left) photographed at 120X, (Right) at 350X.*

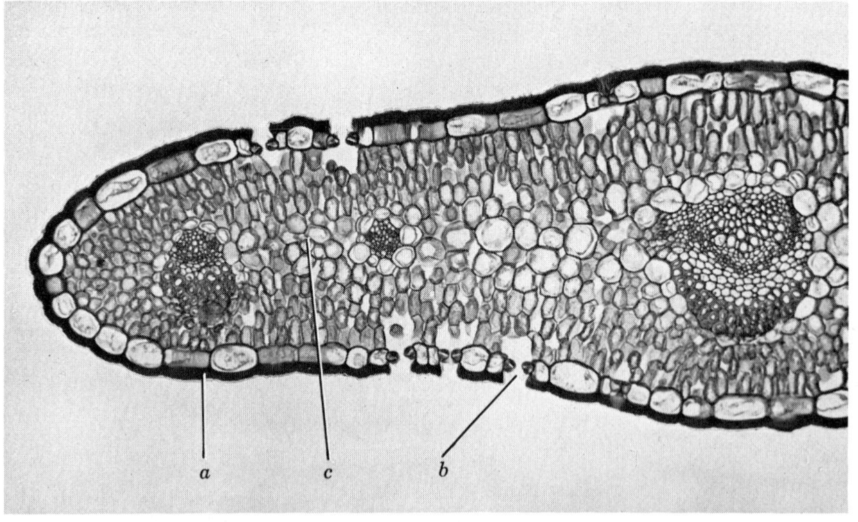

Fig. 9.22 *Leaf of* Dianthus *in cross section, showing xeromorphic features.* (a) *Heavy cuticle,* (b) *sunken stomata,* (c) *compact, poorly differentiated mesophyll. (Copyright, General Biological Supply House, Inc., Chicago.)*

their water content, regulate the size of the stomatal opening (Fig. 9.21). Stomata may occur on either or both surfaces of leaves, although in a given species the distribution is fairly characteristic.

The leaves of some species adapted to dry habitats commonly show a variety of **xeromorphic modifications** that tend to conserve water. In such forms the cuticle is thick, the stomata are depressed well below the general

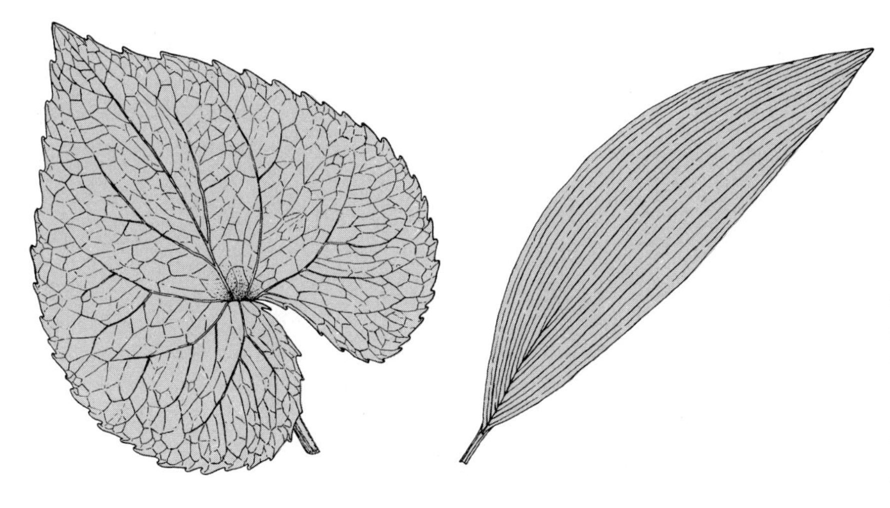

*Fig. 9.23 Common patterns of leaf venation. Left: palmate-netted; right: parallel. See also Fig. 9.25*a, *pinnate.*

Fig. 9.24 Stem of quince, showing stipules at base of petioles.

leaf surface, and the leaf is often thick and fleshy (Fig. 9.22).

In gross structure the leaf typically consists of a flat, expanded **blade** supported on a slender **petiole,** or sometimes **sessile** and attached directly to the stem (Fig. 9.25*b, c*). A system of veins, the vascular supply of the leaf, traverses the blade in a variety of characteristic patterns (Fig. 9.23 and 9.25*a*). Two small, leaflike appendages, the **stipules,** may be present at the base of the leaf in some species (Fig. 9.24). One or more leaves may be borne at a node, giving arrangements called alternate, opposite, or whorled (Fig. 9.25). When the blade is a single piece, the leaf is simple; when it is subdivided into free leaflets, it is compound (Fig. 9.26). Some common leaf shapes and types of leaf margins are shown in the figures cited. The gross features of the leaf are generally characteristic of a species and are helpful in plant identification.

The majority of broad-leaved trees and shrubs produce a new crop of leaves each spring and lose them in the autumn. Such plants are described as **deciduous.** The profound effect of annual leaf fall upon the landscape in forested areas is one of the most striking aspects of plant behavior. Important factors in initiating leaf fall are thought to be shortening of the days and reduced water availability, resulting in a decrease in growth hormone production. Under these conditions a special layer of weak, thin-walled cells, called the **abscission layer,** develops across the petiole at its base (Fig. 9.27). This layer soon disin-

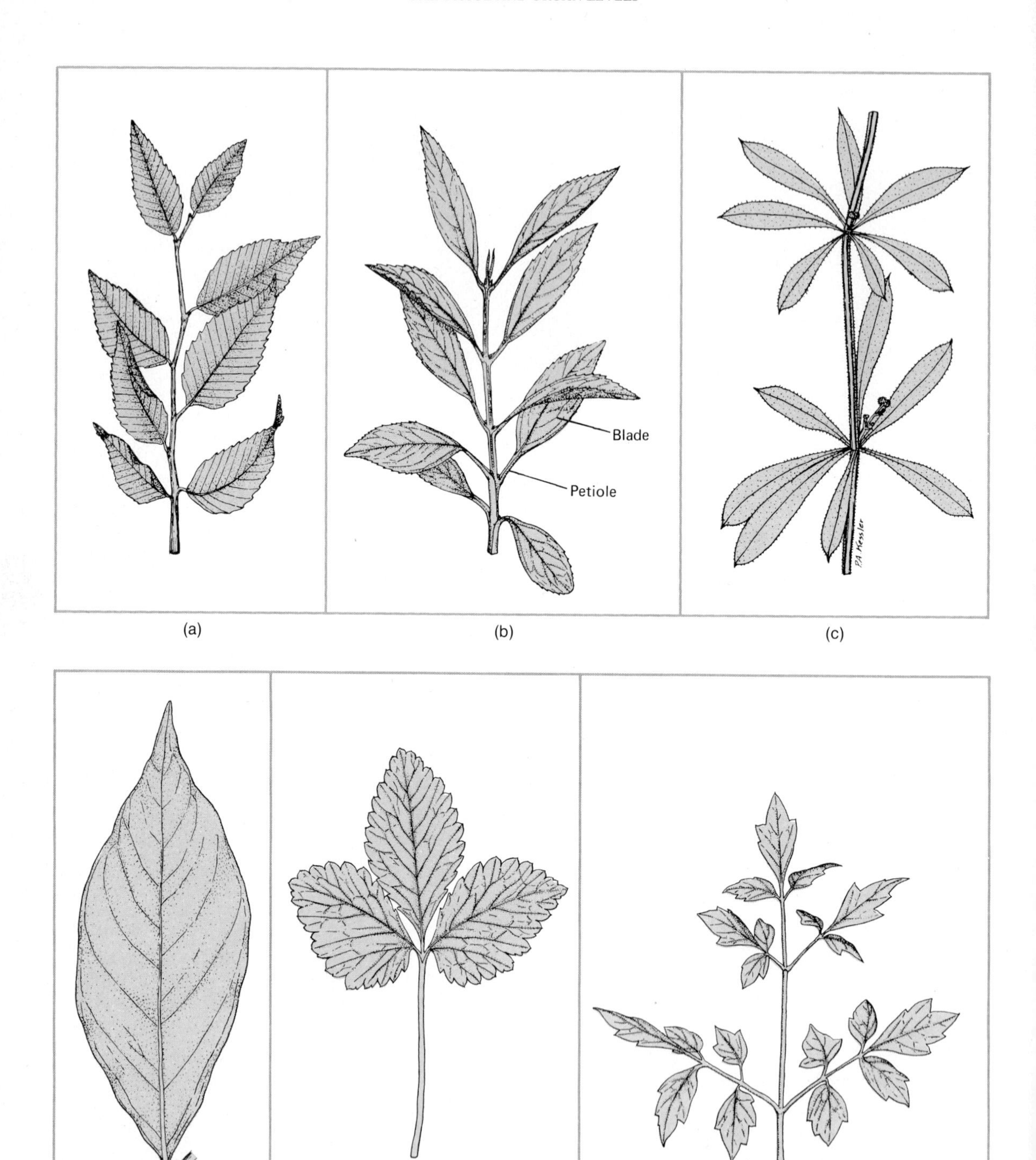
Blade
Petiole
P.A. Kessler
(a)
(b)
(c)
(a)
(b)
(c)

Fig. 9.25 (*Opposite top*) *Leaf arrangements.* (a) *Alternate, one leaf at a node,* (b) *opposite, two leaves at a node,* (c) *whorled, three or more leaves at a node.*

Fig. 9.26 (*Opposite bottom*) *Common leaf forms:* (a) *simple,* (b) *palmately compound,* (c) *bipinnately compound.*

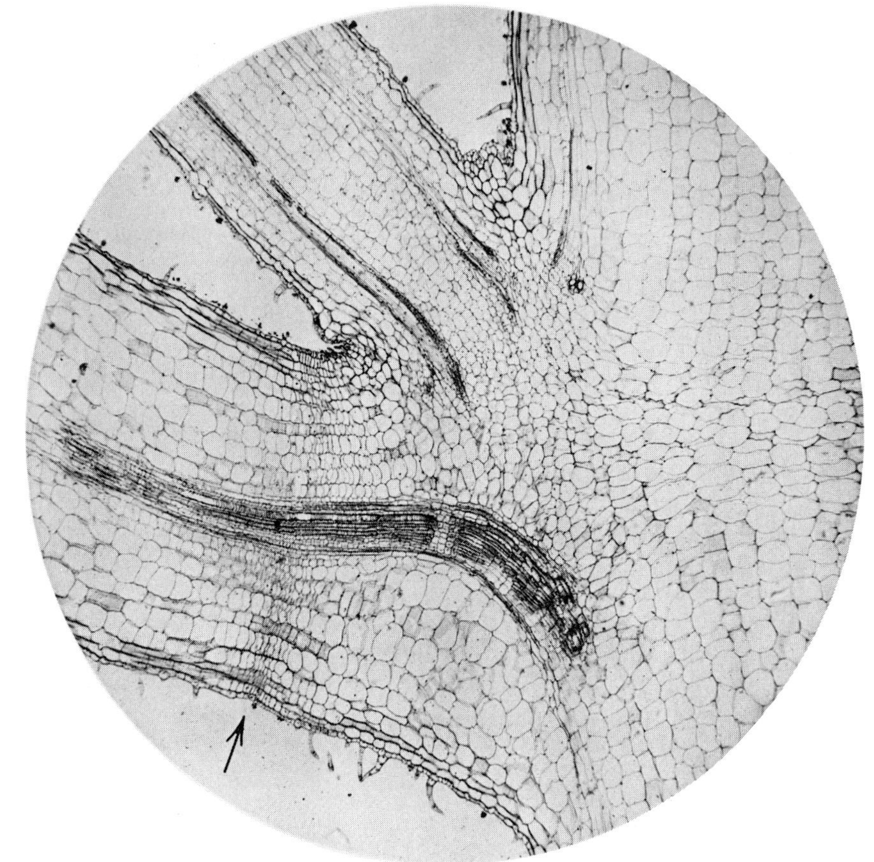

Fig. 9.27 Coleus. *Abscission layer developing in base of petiole. Note small branch in leaf axil. (Copyright, General Biological Supply House, Inc., Chicago.)*

tegrates, leaving only the vascular tissues of the petiole supporting the leaf. Air currents and the action of frost may cause the vascular tissue to break, and the leaf falls. Prior to leaf fall, however, a layer of cork develops below the abscission layer and seals the leaf scar. A few species, such as the white oak, are unusual in this respect for, although its leaves die in autumn, they remain attached to the stems, often throughout the winter. This behavior results from the absence of an abscission layer and failure of the vascular tissues to break.

The leaves of **evergreens** may persist on the plant for a few years but are eventually shed, although not all at one time. New leaves are produced each year.

Striking changes in leaf color often precede leaf abscission. It will be recalled that chloroplasts contain, in addition to the green chlorophylls, yellow and orange pigments. Chloroplasts are green because the chlorophyll predominates. In autumn new chlorophyll production ceases and as the chlorophyll decomposes, the yellow and orange pigments are revealed. In this manner the brilliant golden hues of the hickories, tulip trees, and some maples appear. The hard maples, sweetgums, and dogwoods turn vivid red in autumn. The red coloration is attributable to the development of **anthocyanins** in the vacuoles of the leaf cells. High sugar content of the leaves appears to be conducive to anthocyanin production. Abrupt lowering of temperature reduces the rate of sugar removal from the leaves, and thus may serve as a factor in anthocyanin formation.

Specialized Leaves

Many peculiar and interesting structural and functional specializations of leaves may be found among the angiosperms. These specialized leaves are often quite unlike ordinary leaves, and may perform functions additional to or in place of photosynthesis. Probably the most interesting specializations are those of the leaves of the insectivorous plants. The pitcher plant, Venus' flytrap, and the sundew are examples. Although the leaves of these plants are green and photosynthetic, they supplement the usual food supply by catching, digesting, and absorbing portions of the bodies of insects. Insects crawl into the pitchers of the pitcher plant and fall to the bottom where they are drowned in water containing protein-digesting enzymes (Fig. 9.28). The leaf of the Venus' flytrap folds up quickly when senstitive hairs on the surface are touched by crawling insects (Fig. 9.29). The insect, unable to escape because of the interlocking spines on the leaf margin, is digested by enzymes secreted by the leaf cells. The sundew leaf bears a large number of gland-tipped tentacles that curl over the insect and smear it with mucilaginous digestive secretion (Fig. 9.30).

Fig. 9.28 Pitcher plant, Sarracenia. *(Courtesy, Carolina Biological Supply Co.)*

Specialized leaves frequently serve as storage organs, protective and supporting devices, and reproductive structures. Thick, fleshy leaves, as in the popular "hen-and-chickens" and *Portulacca,* store large volumes of water in their cells. Many plants of this type are well adapted to dry habitats. The fleshy bulb scales of the onion store food and water which are consumed in the development of the flowering stem in the second season (Fig. 9.31*b*). The enclosing scales of dormant buds are much reduced, tough-textured leaves providing a protective cover for the growing point of the stem. The spines of the common barberry are modified leaves whose axils bear short, leaf-bearing branches (Fig. 9.31*a*). The terminal leaflets of the compound leaf of vetch and garden pea are reduced to tendrils, useful for support (Fig. 9.31*c*). Many members of the fleshy-leaved stonecrop family may be vegetatively propagated by merely placing a leaf upon the soil. Roots quickly grow from the broken leaf base and a new plant is thus started. The well-known *Kalanchoë* of the same family produces small plantlets, complete with roots, stems, and leaves, in the notches of the leaf margin (Fig. 20.10). These small plants fall to the ground and become established. Many popular house-plants such as African Violet and Gloxinia may be propagated by leaf-cuttings.

Fig. 9.29 Venus's-flytrap, Dionea. *The trap is sprung by touching the trigger hairs on inner surface of the specialized leaf. (Courtesy, Carolina Biological Supply Co.)*

Fig. 9.30 Sundew, Drosera. *Note the gland-tipped hairs on upper surface of leaves. (Copyright, General Biological Supply House, Inc., Chicago.)*

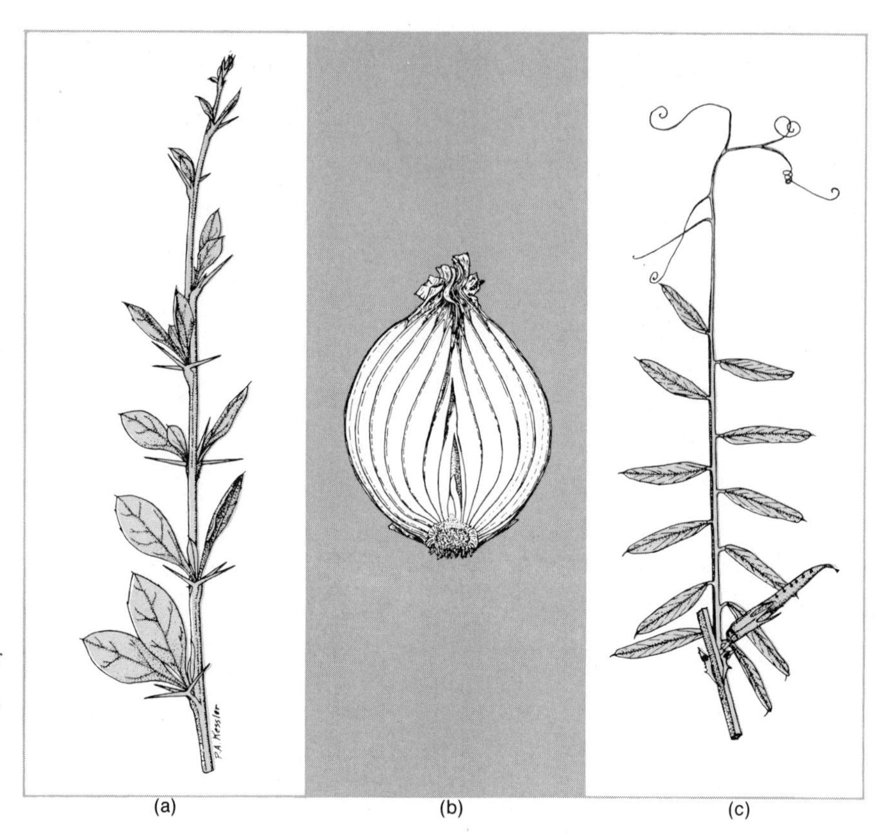

Fig. 9.31 Modified leaves: (a) *thorns of barberry, note young foliage shoot growing in their axils;* (b) *fleshy scales (leaves) of the bulb of onion, note their origin on shortened stem;* (c) *tendrils (terminal leaflets) of vetch.*

THE ROOT

The third major vegetative organ of the seed plant is the **root.** Situated typically below ground, it commonly escapes attention, yet it is an integral part of the total plant, performing a number of important functions. Indeed, the root is the first part of the young growing plant to establish intimate contact with the environment as the seed germinates and is thus part of a critical relationship in the establishment of the new plant. Throughout the life of the plant, the root system anchors the plant in the soil and absorbs water and soluble minerals.

The apical meristem of the root, unlike that of the stem, is covered by a conical **root cap** of parenchyma cells, which provides mechanical protection to the apical meristem as the root grows forward through the soil (Figs. 9.32*a* and 9.33*a*). The outer cells of the root cap are worn away as the root grows, but are replaced by new cells formed by the apical meristem (Fig. 9.33).

New cells destined to become part of the primary tissues of the root are also produced by the apical meristem. These cells undergo a period of enlargement, chiefly elongation, and finally mature into specialized cells. Although the sequence of events is essentially the same as that described for the growing stem, it is somewhat more easily seen in the root. Roots have no nodes and internodes, and the structural complications arising in the stem from the presence of young leaves and buds and the departure of vascular traces to them are, of course, absent. Thus the growing root tip,

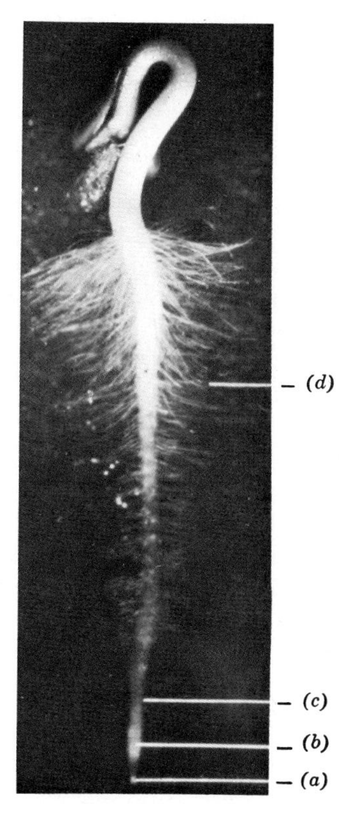

Fig. 9.32 (*Left*) *Growing root tip of radish,* Raphanus sativus. (a) *Root cap,* (b) *region of new cell formation,* (c) *region of cell elongation,* (d) *region of root hairs.*

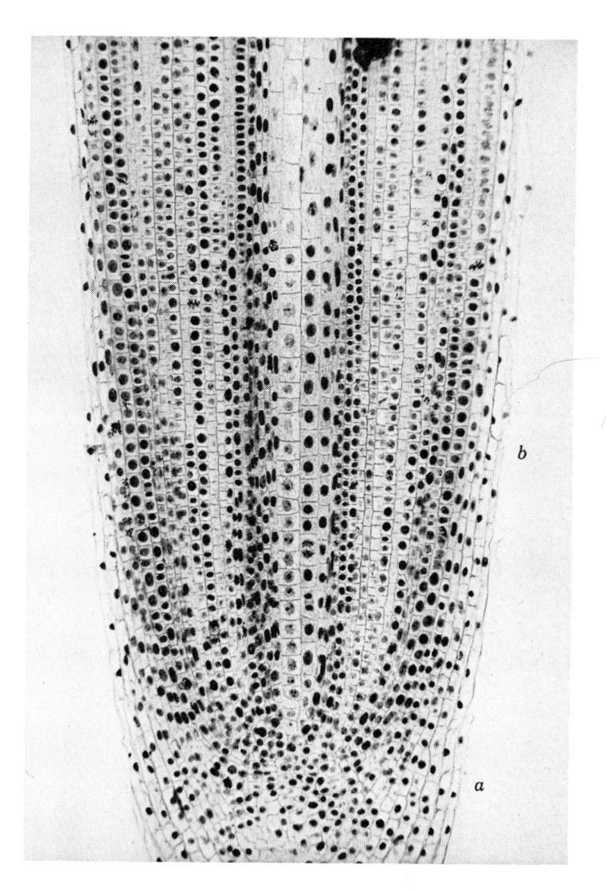

Fig. 9.33 (*Right*) *Root tip of onion, median longitudinal section.* (a) *Root cap,* (b) *region of cell division. (Copyright, General Biological Supply House, Inc., Chicago.)*

when viewed in median longitudinal section, shows clearly the terminal region of **new cell formation** covered by the root cap, a second region of **cell enlargement,** and a third region of **cell maturation** (Figs. 9.33 and 9.34).

The outermost cell layer of the young root is the epidermis. It must be noted, however, that the epidermal cells of the root are not cutinized as are those of the stem. The epidermis, therefore, may serve as an absorbing surface. Just behind the region of enlargement (Fig. 9.32*c*), most of the epidermal cells develop a hairlike protuberance on the outer surface, thus effecting a tremendous increase in the absorbing surface (Fig. 9.32*d*). These outgrowths, called **root hairs,** are actually extensions of the epidermal cells, and as they grow outward the cells become highly vacuolate, the nucleus commonly occupying a terminal position in the growing hair. As the root tip grows forward, new root hairs are produced. Older root hairs collapse and die at about the same rate that new ones are produced, their period of activity usually being limited to a few days. In their outward growth the hairs penetrate the soil to a maximal distance of about 15 millimeters and become closely applied to the surface of soil particles. The extreme delicacy of the root hairs makes it practically impossible to remove a plant from the soil without destroying most of them, unless a large ball of soil is taken up with the plant. In the transplanting of trees and shrubs, it may not be practical to try to preserve this intimate root-soil relationship. Therefore, reliance must be placed on the ability of the plant to promptly develop absorbing surfaces by producing new root tips. In general, it is true that the less the roots are

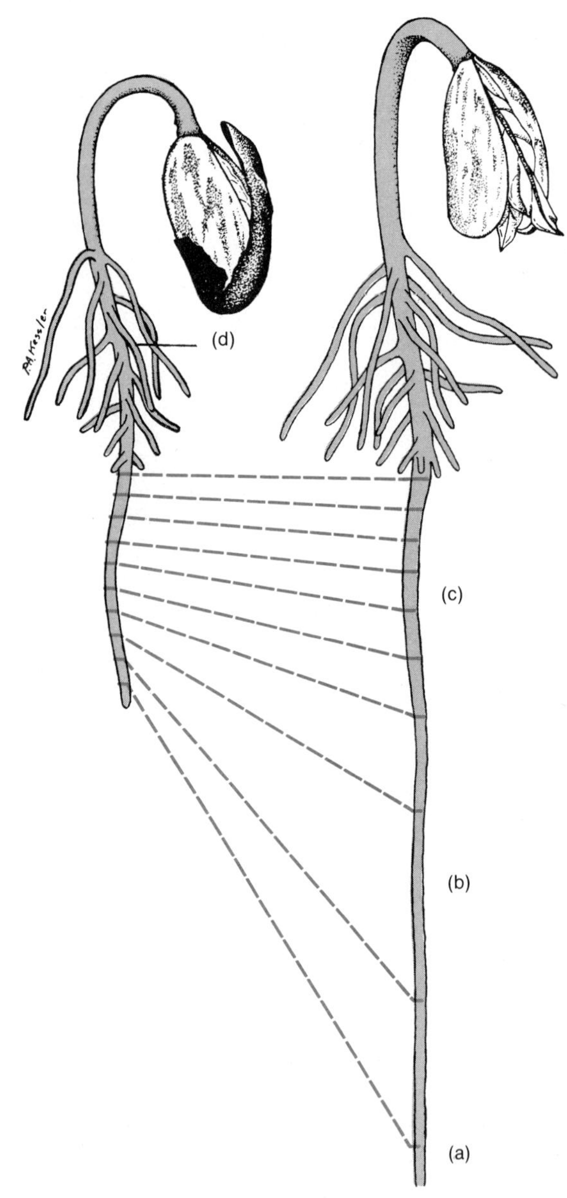

Fig. 9.34 Distribution of growth in roots. (a) *Region of new cell formation, including root cap cells;* (b) *region of cell enlargement;* (c) *region of cell maturation;* (d) *secondary roots.*

disturbed, the better the chances of survival after transplantation.

The pattern of maturation of the internal tissues of the root offers some interesting points of contrast with that of the stem. Although the same cell types and tissues are involved, certain differences of arrangement must be noted. In early maturation, a circle of separate provascular strands arises. In gymnosperms and the dicotyledonous angiosperms, alternate strands become extended inwardly and laterally until they meet at the center and form a rayed or star-shaped figure, as seen in cross-sectional view. Between the points of the star are situated the other alternate set of provascular strands. The star-shaped figure represents, of course, a fluted column. Final maturation begins at the points of the star with the conversion of the provascular cells to tracheids or vessel elements, and continues inwardly, ultimately forming a solid column of primary xylem (Fig. 9.35*c*). At the same time, maturation of the remaining provascular strands yields primary phloem (Figs. 9.35*b* and 9.36*e*). In the monocotyledonous angiosperms, the primary xylem is limited to separate radiating lines of vessels (Fig. 9.36*d*). It will be noted that the alternating or radial arrangement of primary xylem and primary phloem is in contrast to the so-called collateral arrangement seen in the stem (Fig. 9.3). Note also that the primary xylem begins its maturation on the other margin of the provascular strands in the root, whereas it begins on the inner margin in the stem. Where the primary xylem development extends to the center, there is no pith.

Immediately exterior to the primary vascular tissue is a narrow zone of parenchyma cells, called the **pericycle** (Fig. 9.35*d*). From this layer, at positions immediately exterior to the first-formed xylem, branch roots arise by localized meristematic action. The newly developing branch roots thus arise internally and push outward through the cortex and epider-

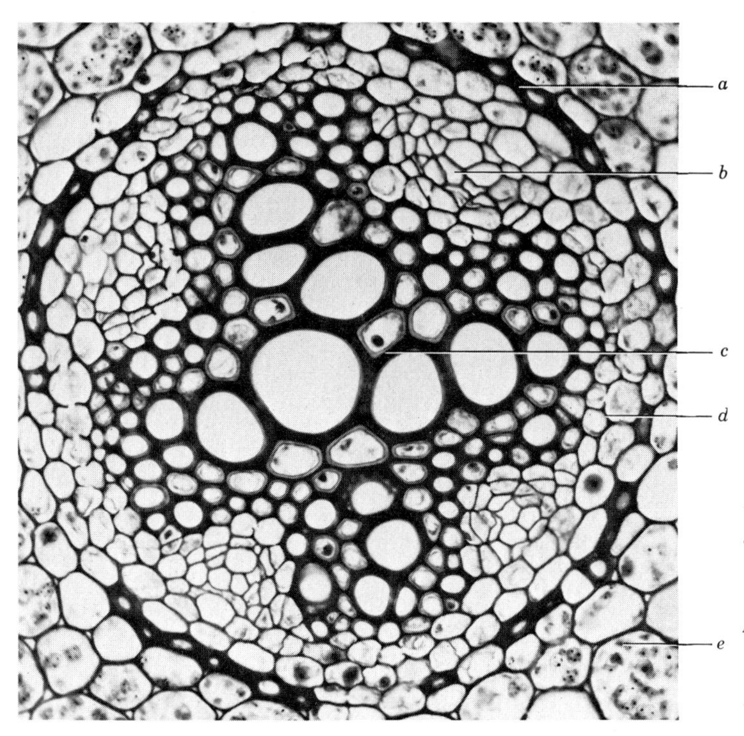

Fig. 9.35 (*Left*) *Photomicrograph of vascular tissue of root of buttercup.* (a) *Endodermis,* (b) *primary phloem,* (c) *primary xylem,* (d) *pericycle,* (e) *cortex.* (*Courtesy, Carolina Biological Supply Co.*)

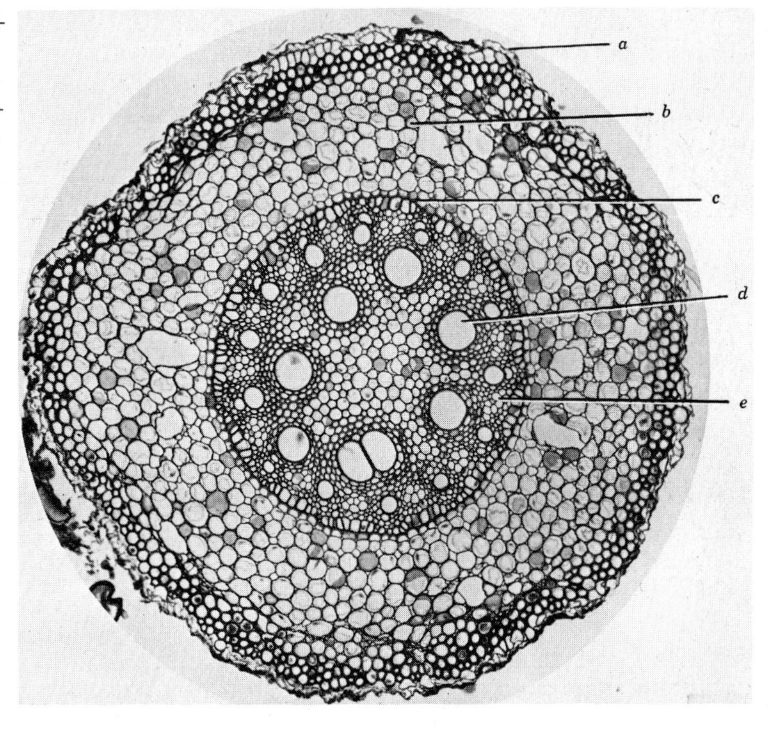

Fig. 9.36 (*Right*) *Photomicrograph of maize root in cross section.* (a) *Epidermis,* (b) *cortex,* (c) *endodermis,* (d) *primary xylem,* (e) *primary phloem.* (*Copyright, General Biological Supply House, Inc., Chicago.*)

mis. The origin of branch roots should be compared with that of stem branches. The latter, it will be recalled, arise superficially and the departure of branch and leaf traces leaves gaps or openings in the vascular cylinder.

The pericycle of the root is separated from the broad cortex by a single, almost complete layer of small cells, the **endodermis** (Figs. 9.35*a* and 9.36*c*). The endodermal cells have, most commonly, thickened and suberized radial walls. The endodermis is generally interpreted as the innermost layer of the cortex and is believed to function in the passage of absorbed substances from the cortex to the conducting elements of the xylem, but the manner of its action is not understood. The cortex of the root is similar to that of the stem, except that it usually has no chloroplasts (Fig. 9.36*b*). The epidermis, as we have noted, is the outermost tissue, serving most importantly as the absorption surface (Fig. 9.36*a*).

Secondary growth in roots is effected by the action of a vascular cambium arising partly from residual provascular cells in the bays between the primary xylem arms and partly from cells of the pericycle at the ends of the arms. Production of secondary xylem begins earlier in the bays, and this action soon rounds out the cambium circle. Thereafter, secondary phloem and xylem are produced as in woody stems. Continued secondary growth results in rupture and eventual loss of endodermis, cortex, and epidermis.

Cork tissue is developed from a phellogen originating initially in the pericycle, and later from the older phloem. The cork fissures and sloughs off in the same manner as in the stem. Thus, as a result of secondary growth, older roots come to resemble woody stems as regards both kinds and arrangement of tissues, although certain details of early development are different.

The root emerges from the germinating seed as a cylindrical structure. This is the first, or **primary root.** After a period of growth, small branch roots, originating in the pericycle, may be produced from it at points above (behind) the zone of root hairs. These branch roots are called **secondary roots** (Fig. 9.34*d*), the word secondary here referring to the time of their appearance. Secondary roots may soon produce branches of their own, and so on until an elaborate root system is developed. If, in this process, the original or primary root remains dominant, the root system is described as a **tap root system.** In some species the primary root quickly loses dominance by exaggerated growth of several secondaries, all of which are about the same size. The root system thus takes on a spreading form and is known as a **diffuse root system.** The individual roots of either the tap or diffuse systems may remain slender and become tough and woody through secondary growth, or they may become very thick and fleshy. The roots of carrot and *Dahlia* are examples of fleshy tap root and fleshy diffuse root, respectively.

Roots frequently arise from plant parts other than primary roots or their branches. Such roots are called **adventitious** (Fig. 9.37). Adventitious roots may arise at higher points on the stem and, as in the English ivy, provide the means of clinging to vertical surfaces. Propagation of household plants, such as geraniums, *Coleus,* and many others, is made possible by the production of adventitious roots at the base of stem cuttings. African violets are commonly propagated by leaf cuttings from which adventitious roots and buds grow. The roots of potato plants grown from pieces of "seed" potato are also of adventitious origin, since the potato tuber is a rhizome, or modified stem. Many species of grass spread widely by the growth of slender rhizomes just beneath the soil surface, the roots arising adventitiously at the nodes of the rhizome. The Irish potato is a rhizome, whereas the sweet potato is a true root. The adventitious "prop-roots" of maize at the base of the stem are important in holding the plant erect against wind action.

Fig. 9.37 Adventitious roots of Pandanus.

THE FLOWER

So far in this chapter we have dealt with the vegetative tissues and organs of vascular plants. In most species the orderly vegetative development described culminates in the formation of characteristic reproductive organs. Reproduction by means of seeds is a complicated process involving complex and highly specialized structures. In angiosperms, seeds are produced on specialized stems or branches called **flowers.** Flowers differ in size, shape, color, arrangement, number of parts, and in many other features and yet have much in common in their basic structural plan (Fig. 9.38). The many kinds of flowers owe their variety to specialization of their parts.

The typical floral parts are attached in characteristic fashion to a short axis, the **torus** (sometimes called the **receptacle**). Typically, four kinds of floral parts are borne upon the torus, always in the same order. Beginning at the base of the flower, we may find a circle or whorl of **sepals.** These are most commonly green and quite leaflike and sometimes very small. However, they are large and white or brightly colored in some well-known plants, such as the lily. All of the sepals together constitute the **calyx.**

Above the calyx is a whorl of conspicuous **petals,** often white or colored. These make up the **corolla** and are usually the most noticeable part of the flower. The beauty of a rose or a morning glory is due to the color and form of the corolla. In some species the corolla may be absent and the calyx may be the conspicuous member, as in the anemone. Or, the calyx, as well as the corolla, may be colored and distinguishable from the corolla only by its position on the torus. The popular Christmas plant, *Poinsettia,* the flowering dogwood, and the calla lily are deceptive to the casual observer, for the conspicuous petallike members are not parts of the flower at all, but specially modified leaves or **bracts** that surround a whole cluster of very small and inconspicuous flowers.

The filamentous or sometimes club-shaped structures occurring above the corolla are the **stamens.** Each consists of an **anther** at the tip of a slender stalk or **filament. Pollen** is produced in the anthers. The complement of stamens is often arranged in two whorls when the number is twice that of the petals, or in one whorl when the number is the same or less. Figure 9.39 shows an interesting condition in *Nymphaea,* a water lily, where intergradations between typical petals and stamens exist. The separate members shown here occur along a radius of the flower. This series of intergrades is not a mere developmental sequence, but an expression of the basic homology of stamens and petals. Many species of flowering plants

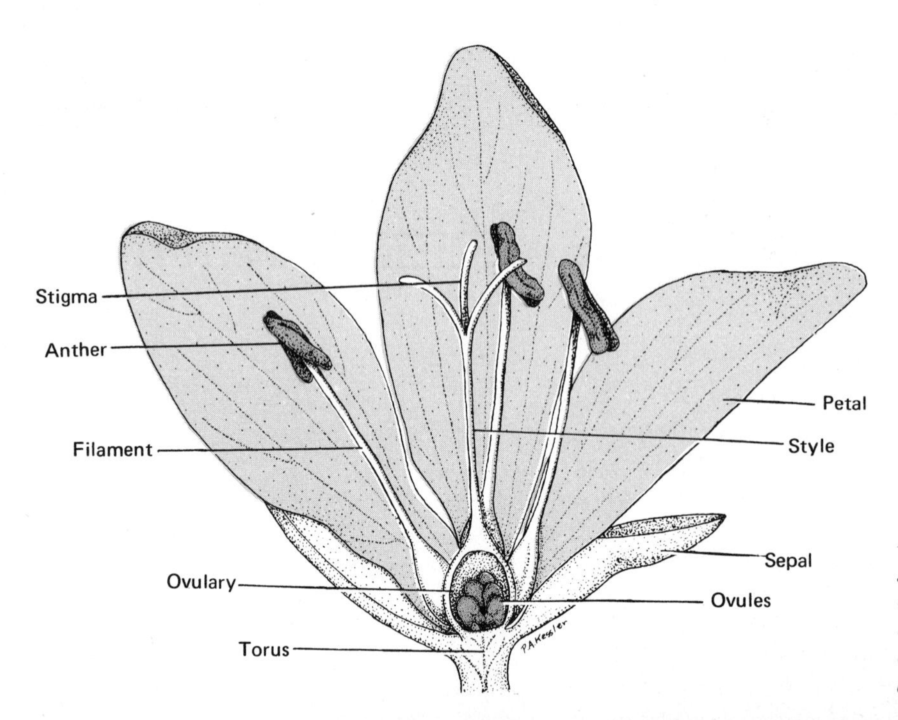

Fig. 9.38 (*Left*) *Semidiagram of a generalized dicotyledonous flower in longitudinal section. Only three of the five petals, stamens, and stigma lobes are shown.*

Fig. 9.39 (*Right*) *Flower parts of the water lily,* Nymphaea, *showing intergradations between petals and stamens.* (*Copyright, General Biological Supply House, Inc., Chicago.*)

reveal similar intergradations between and among floral whorls.

The topmost or central position of the flower is occupied by the **pistil,** a flask-shaped structure with an enlarged basal part called the **ovulary,** above which extends a slender neck, the **style,** surmounted by a somewhat enlarged tip, the **stigma.** The flowers of some species have more than one pistil (Fig. 9.43). The pistil, and more especially the ovulary, often

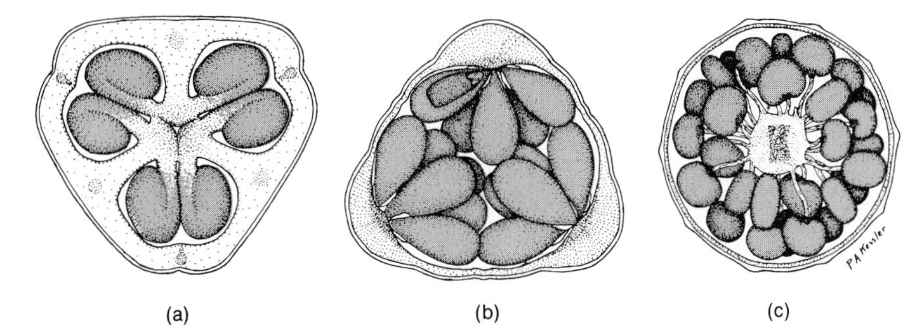

Fig. 9.40 Patterns of ovule arrangement in the ovularies of lily, violet, and chickweed. (a) *axile,* (b) *parietal,* (c) *free central.*

shows external ridges or grooves which suggest that it is composed of united parts. When these parts are separated, they commonly have a form suggesting small, modified leaves that have united by their edges. The leaflike components of the pistil are **carpels.** The pistil may consist of one carpel as in the bean or pea, when the pistil is said to be **simple,** or it may consist of more than one as in lily or mallow, when it is called **compound.** The cavity of the ovulary contains one or more **ovules** (future seeds) attached to a **placenta** (Fig. 9.40).

All flowers are essentially alike in their basic architecture. However, a wide variety of structural modification can be seen. For example, the petals are often fused, producing a saucerlike or bell shaped corolla, as in the African violet or morning glory, whereas in the pinks the sepals are fused and the petals are distinct. In many species, certain flower parts may be characteristically lacking, as in willow (Figs. 9.41 and 9.42), which has neither calyx

Fig. 9.41 Pistillate flowers of willow. Each cluster (catkin) bears many flowers, each consisting only of a pistil.

Fig. 9.42 Staminate flowers of willow. Each catkin bears many flowers, each consisting of two stamens.

nor corolla, or *Hepatica,* which has a conspicuous, brightly colored calyx but no corolla (Fig. 9.43). In hibiscus and in legumes, the stamens are more or less fused into a tube surrounding the pistil. Sometimes the stamens are attached to the upper surface of the petals, so they appear to arise from them, as in primrose. In apple and pear, the pistil is partly embedded in the torus and the fused stamens, petals, and sepals appear to originate from the top of the ovulary. In corn and squash and many other species, the flowers are unisexual, having either stamens or pistils, but not both. The staminate flowers of corn are in the tassels, the pistillate flowers in the ears. In many species having unisexual flowers, such as willow, the staminate and pistillate flowers occur on separate plants (Figs. 9.41 and 9.42). Such details, along with others, are used in the identification of flowering plants, and when considered in the light of certain established principles, make possible the arrangement of species in systems of classification expressing a possible evolutionary history of flowering plants.

The pistils and stamens of the flower are often called the **essential organs,** for they are the ones directly involved in seed production, although the corolla and the calyx may be useful in indirect ways.

Flowers are borne upon a plant in a characteristic arrangement known as the **inflorescence.** The inflorescence may consist of a solitary flower borne at the summit of an unbranched stalk, as in tulip, or of many flowers borne singly in the axils of reduced leaves along an elongated stem, as in snapdragons (Fig. 22.11) and lily of the valley. The stem may be profusely branched forming a large open cluster, as in lilac and many grasses, or extremely shortened with the flowers crowded into compact clusters of various shapes, as in clover, milkweed, and willow (Figs. 9.41 and 9.42). The degrees and patterns of branching of the inflorescence axis and the relative lengths of the individual flower stalks **(pedicels)** account for the great variety of inflorescence types. The tassels of corn plants are inflorescences of staminate flowers and the ears are inflorescences of pistillate flowers, each grain with its silk constituting an individual pistillate flower (Fig. 22.19). In the very compact flower heads (Fig. 22.16) of members of the sunflower family, the inflorescence often consists of two kinds of flowers. In some, like sunflowers and daisies, the center flowers have very small tubular corollas, and the marginal ones have enlarged strap-shaped corollas (rays). In other species, such as chrysanthemum and zinnia, most or all of the flowers have conspicuous rays. In still other species the heads are made up entirely of tubular flowers.

Fig. 9.43 Flower of Hepatica. *Note the many monocarpellate pistils at the center, and many stamens.*

THE FRUIT

Although seeds are ripened ovules, **fruits** are basically the ripened ovularies that contain the seeds. However, as we shall see, certain other parts of a flower may also contribute to the structure of fruits in some species. In most species pollination is essential, not only for fertilization of the egg and the subsequent development of the embryo and seeds but also for the development of the fruit. True fruits are limited to the angiosperms. The fruits of different species are of many different sizes, shapes, structures, and textures, and fruits are classified into a number of types on the basis of these differences. Practically all fruits in the common sense of the term, such as apples, oranges, grapes, cherries, raspberries, and watermelons are fruits in the botanical sense, but

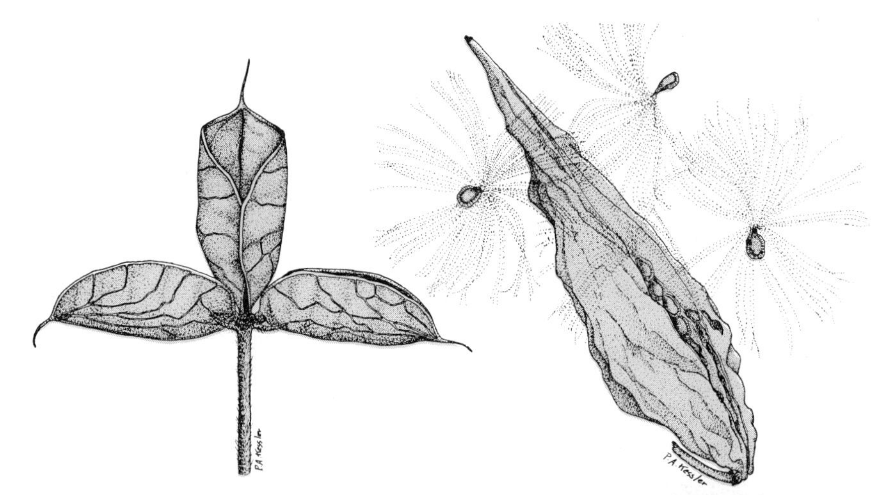

Fig. 9.44 Left: fruits of larkspur; right: milkweed.

many things not commonly called fruits, such as bean or pea pods, okra, tomatoes, milkweed pods, sunflower "seeds," and even the outer layers of corn or wheat grains are true fruits botanically. We shall now consider some of the more important types of fruits.

In one group of fruits, the ovulary wall is essentially dry when ripe. These may be called **dry fruits** and certain specific types may be recognized in this category. Descriptions of some of the more common dry fruits follow.

The **follicle** develops from a pistil composed of a single carpel. When the fruit is ripe, it spontaneously splits open (dehisces) along one side. The ripened seeds are usually shaken out by movement of the plant stem resulting from wind action or mechanical contact. The fruits of milkweed, magnolia, and larkspur are follicles. The seeds of milkweed are equipped with a feathery tuft of hairs that facilitates their dispersal by wind (Fig. 9.44).

The **legume** is characteristic of the bean family. It is monocarpellate and typically dehisces along two sides when ripe. Because of tensions created in the ripening fruit as it dries, the dehiscence sometimes occurs with explosive force, thus throwing the seeds some distance from the plant (Fig. 9.45). Peas, beans, clover, alfalfa, and lespedeza are sometimes

Fig. 9.45 (Bottom) Dehiscing seed pod (legume) of vetch. Multiple exposure shows separation of the two valves. The movements occurring explosively scatter the seeds over a considerable area. (Courtesy, Carolina Biological Supply Co.)

collectively called "legumes" for their fruits are, indeed, legumes. Other examples are wisteria, locust, and peanut. The fruit of the peanut deserves special mention because it is indehiscent and ripens below ground. The developing fruit of the peanut is forced into the soil by the rapid growth of the pistil stalk.

The **capsule** develops from a multicarpellate pistil which at maturity frees the seeds by splitting along several lines, or by the opening of pores. Examples are the fruits of poppy, lily, and cotton. The seeds are usually shaken out by wind or mechanical contact.

The **achene** is a small indehiscent monocarpellate fruit derived wholly from the ovulary, and with a single seed attached to the fruit wall at one point. The so-called "seeds" of strawberry and fig are really achenes (Fig. 9.46). The "seeds" of sunflower, beggartick, dandelion, and others of the sunflower family are also fruits. They have been called achenes, but they are bicarpellate and their walls are composed of other tissues in addition to the ovulary wall (Fig. 9.47). Achenes may be widely dispersed by adhering to the fur of animals or to clothing by means of small hooks or spines. Achenes that are associated with edible structures, as in strawberry or fig, may be swallowed by animals or birds, passed through their digestive tracts without damage to the achene, and ultimately voided at some distance away from the parent plant. The ovulary wall of such fruits matures into a hard, resistant fruit coat.

The **samara** is indehiscent, with a prominent wing developed from the ovulary wall. The samaras of maple (Fig. 9.48) and ash, although fairly heavy, may be carried a considerable distance by strong wind currents.

The **nut** is one seeded and the fruit wall or part of it becomes stony or very woody at maturity. The fruit wall is composed of the ovulary wall fused with other flower parts. Examples are the fruits of oak, hickory, and walnut. The seeds of some of these species are attractive food for rodents which bury many more nuts than they eat. These large round fruits are buoyant in water and may be dis-

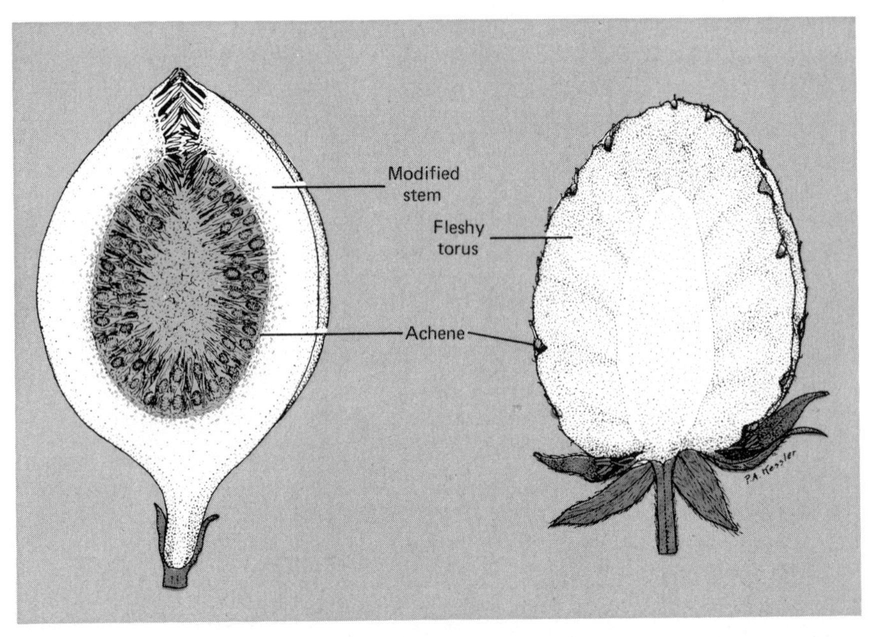

Fig. 9.46 Fruits of fig, left, and strawberry, right, in longitudinal section. The true fruits are the achenes or so-called "seeds"; the fleshy, edible parts are modified stem or torus. Each achene of the fig is derived from a single, separate flower, whereas many achenes are produced by a single strawberry flower.

persed by streams, or they may be moved along by the flow of surface water.

The **grain** is tricarpellate, one-seeded fruit with the seed coat completely fused to the wall of the fruit. Examples are the grains of rice, wheat, corn (Fig. 9.50*b*), and other grasses.

In the second group of fruit types, one or more layers of the fruit coat become soft and **fleshy** when ripe. Many fruits of this type are attractively flavored and sweet and serve as food. This aids in seed dispersal by animals, because the seeds may be eaten along with the fruit and passed through the digestive tract unharmed.

The **berry** is the most common of the fleshy types of fruit. One or many seeds are surrounded by a fleshy fruit wall derived from the wall of a one- to several-carpellate ovulary. Grapes and tomatoes are familiar examples. In the tomato (Fig. 9.49), the placental tissues are also fleshy. Sometimes the outer layers are leathery and contain many oil glands, as in the citrus fruits. The juicy pulp of the orange consists of multicellular juice-filled sacs developed upon the inner surface of the ovulary wall. The fruits of members of Cucurbitaceae, the squash family, such as watermelon, cucumber, and pumpkin are berrylike. The leathery fruit walls consist of ovulary wall plus tissue of the torus which has grown to it. The placental tissue and inner layers of the ovulary wall remain fleshy in watermelon, whereas in squash and

Fig. 9.47 (Top) Fruit of Tragopogon, *like that of dandelion, has a parachute formed from an adherent calyx. (Copyright, General Biological Supply House, Inc., Chicago.)*

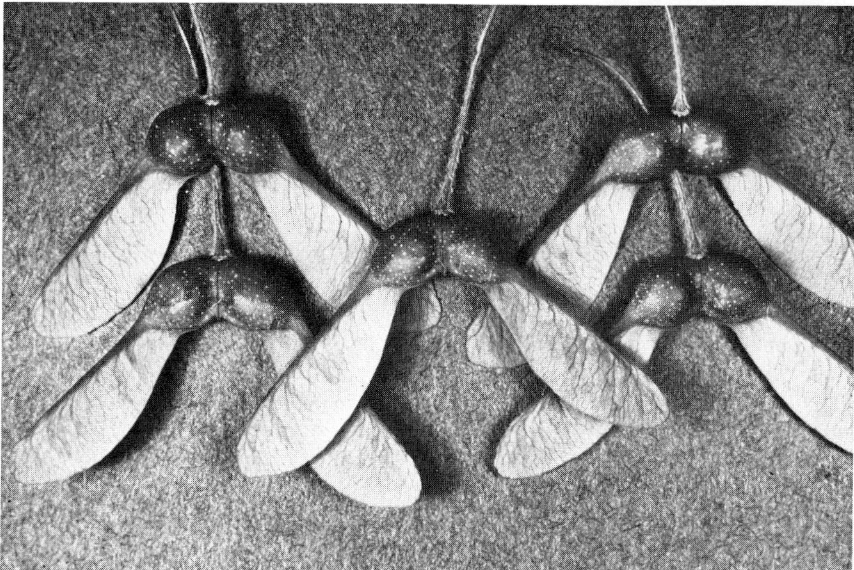

Fig. 9.48 (Left) Fruits of the sugar maple. The ovulary wall is modified into a flat wing. When ripe the fruit splits along the vertical center line into two winged, one-seeded sections.

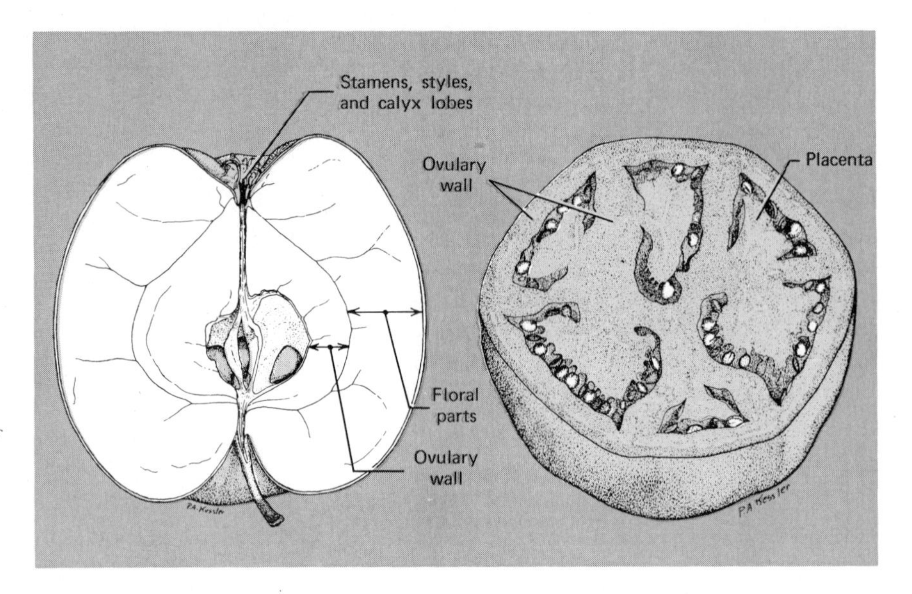

Fig. 9.49 Left: the fruits of apple (a pome) in longitudinal section. Right: tomato (a berry) in cross section.

pumpkin they break down into a fibrous mass.

Cherry, peach, and olive are examples of the so-called stone fruits, or **drupes.** The drupe, derived from a single carpel, is like a one-seeded berry except that the inner layer of the fruit coat is modified into a hard, bony layer (the "stone" or "pit"). The "kernel" of the peach is the complete seed, with a brown, membranous seed coat.

The **pome,** represented by the apple (Fig. 9.49) and pear, is more complex than most of the foregoing types, for it consists not only of ovulary wall, but also of torus and floral tissues that have grown up around it. Pomes are composed of five or more carpels. In development, the inner layers of the ovulary wall become cartilaginous, forming the core of the fruit.

Many other specialized forms of fruit are known and the more familiar ones selected for this discussion are subject to almost infinite variation in detail of structure and in arrangement. The blackberry, for example, is a cluster of small drupes produced from numerous pistils on the enlarged torus of a single flower. The many achenes of the strawberry are also the product of a single flower, but each achene of the fig is derived from a separate flower. The pineapple and mulberry, on the other hand, are the fused fruits of many separate flowers.

THE SEED

The seed is not customarily treated as an *organ* of the plant, being usually considered as a special category under the general title of "reproduction." Such treatment, however, fails to emphasize the structural relationship between the seed with its contained embryo and the plant that bears it, and with the plant to which it is to give rise. It is, therefore, advantageous to discuss the seed and embryo in this context, so that the relationship will be clearly understood.

We have already noted that seeds are ripened ovules. Since the ovules of angiosperms are borne within ovularies that ripen into fruits, the seeds of angiosperms are borne within fruits. The seeds of gymnosperms, however, are borne on the upper surface of the scales of the cones rather than within fruits. Since seeds contain the embryo plants of the new generation and generally considerable accumulated food, they are immensely important structures, both biologically and economically. Seeds vary greatly in size, from giant seeds such as those of the coconut to almost microscopic seeds such as those of orchids. For our study of seed structure we shall select three seeds of sufficient size for easy observation: bean, corn, and castor bean.

The exterior covering of the bean seed (Fig. 9.50*a*) is a tough glossy layer of high protective value. This is underlaid by a thin membranous layer. When the seed is soaked in water these two layers, the **seed coats,** may be easily removed, revealing the **embryo.** Separation of the two fleshy **cotyledons** (hence, **dicotyledonous**) reveals that they are oppositely attached near the summit of a tapering axis. The part of the axis above the attachment of the cotyledons is the **epicotyl,** consisting chiefly of a pair of folded miniature leaves enclosing a growing point. This is the first bud of the embryo and is called the **plumule.** Below the cotyledons extends the **hypocotyl,** the lower tip of which is the **radicle.** The fleshy cotyledons contains reserve food, chiefly starch.

In the seeds of some other dicotyledonous species, such as the castor bean (Fig. 9.50*c*), the reserve food is present in a fleshy tissue, the **endosperm,** which envelops the embryo and lies beneath the seed coats. The cotyledons of such forms are usually membranous.

The embryo of corn (Fig. 9.50*b*) displays interesting contrasts with those described previously, although the basic plan of organization is the same. The embryonic axis consists of an epi- and a hypocotyledonary region, with the cotyledon attached laterally. There is only one cotyledon (hence, **monocotyledonous**), which is rather like a heavy shield placed between the embryonic axis and the massive endosperm. The plumule is very small and is encased by a conical sheath called the **coleoptile.** The embryonic root is similarly encased in a sheath, the **coleorhiza.** The radicle emerges first, followed shortly thereafter by emergence of the

Fig. 9.50 Dissection of common seeds to show detail of embryo. (a) *bean with one cotyledon removed,* (b) *grain of maize (really a fruit whose wall is fused with the seed coats), in longitudinal section,* (c) *castor bean, with one cotyledon removed.*

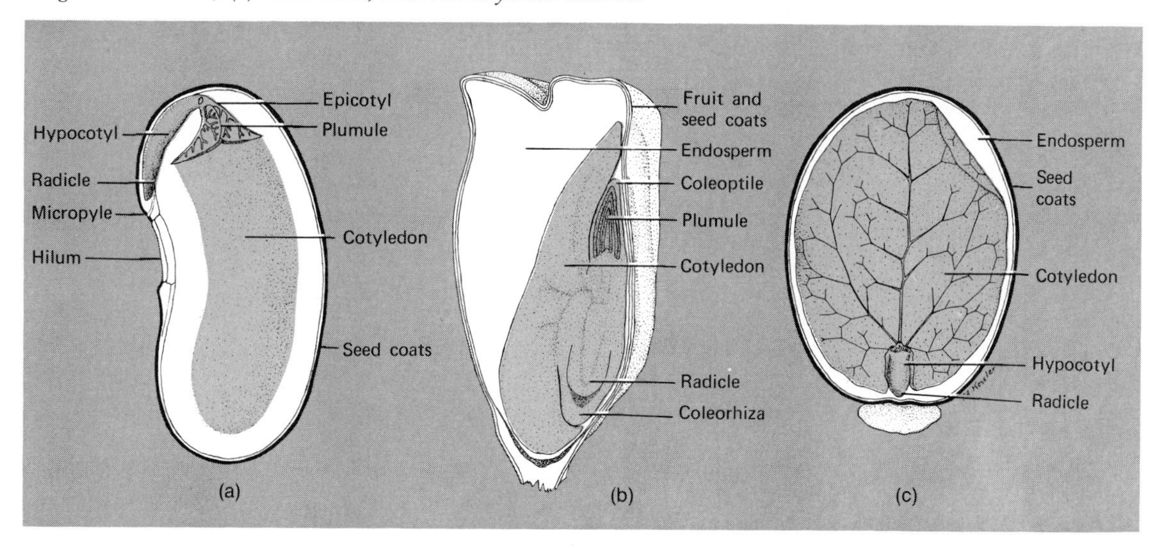

coleoptile. The cotyledon, although somewhat fleshy, does not contain the chief food reserve, but acts as a disgesting and absorbing structure that transfers food from endosperm to growing embryo.

We may wonder what the cotyledons really are. It is difficult to determine this in the bean or pea because the cotyledons are specialized as food reservoirs. But in some seeds such as castor bean where the cotyledons are thin, careful dissection of the embryo and observation at germination give a clue. The cotyledons of the castor bean embryo are not food reservoirs, for the food is present in the thick enveloping endosperm. Dissection of the castor bean seed to show the surface view of the cotyledons reveals a pattern of veins very like that found in a leaf (Fig. 9.50*c*). During germination, the cotyledons at first absorb food from the endosperm. As the seedling emerges from the ground, the cotyledons enlarge considerably and develop chlorophyll, thus acquiring the obvious characteristics of leaves. The cotyledons remain active on the plant for several weeks, serving, along with the leaves later produced by the epicotyl, as part of the general photosynthetic equipment. Thus, cotyledons are to be regarded morphologically as leaves, even though their exact nature may be obscured by specialization, as in the bean and pea. It is in recognition of this that cotyledons are sometimes called "seed leaves." Squash and cucumber seeds are, like the bean, without endosperm. In behavior at seedling stage their fleshy cotyledons are, in a way, intermediate between the bean and castor bean, for they become quite leaflike and green and persist for some time as photosynthetic organs.

Seed coats are frequently specialized and may facilitate dispersal by wind or animals. The winged seeds of some species of pine (Fig. 21.19) and *Paulownia* and the plumed seeds of milkweed (Fig. 9.44) are familiar examples of wind-dispersed types. In some the seed coat is adhesive (as in mistletoe) or becomes mucilaginous when wetted (as in mustard) and adheres to the feet of birds. Hairy seeds may cling to the bodies of animals and be transported for some distance.

The primary importance of seed coats is, however, that they protect the embryo plant against such hazards as excessive desiccation, mechanical injury, and the digestive juices of animals. The seed coats of some species are impermeable to water or oxygen or are hard and mechanically resistant, and thus bring about types of seed dormancy (Chapter 15) that may have considerable survival value. The capacity of seeds for remaining viable over considerable periods of time and through environmental conditions unfavorable for growth is one of the principal factors responsible for the dominance of seed plants in the land vegetation. This capacity results to a great extent from the presence of seed coats.

RELATED READING

Ashby, E. "Leaf shape," *Scientific American, 181* (4), 22–29, October 1949.
Epstein, E. "Roots," *Scientific American, 228* (5), 48–58, May 1973.
Fritts, H. C. "Tree rings and climate," *Scientific American, 226* (5), 92–100, May 1972
Hall, F. K. "Wood pulp," *Scientific American, 230* (4), 52–62, 1974.
Juniper, B. E. "The surfaces of plants," *Endeavour, 18,* 20–25, 1959.
Williams, S. "Wood structure," *Scientific American, 188* (1), 64–68, January 1953.

10 THE ORGANISMAL LEVEL

In previous chapters we have considered plants from the levels of their constituent molecules, cells, tissues, and organs. Although the information gained there is important in the understanding of plants, it must be realized that molecules, cells, tissues, and organs are only parts of a complete, functioning **organism.** Although the organs, tissues, or even individual cells of an organism may be separated from it and kept alive in artificial culture for some time, these do not constitute an organism any more than an automobile engine removed from an old car and used to operate a rotary saw is an automobile. An organism is the free-living being that maintains itself in nature, and develops and functions according to a specific plan. It has some means of obtaining food, which it can utilize as a source of energy or as material for growth and maintenance. Each species of organism also reproduces itself. An organism is something more than the simple sum of its constituent cells, tissues, and organs. It is a coordinated functioning unit—a living thing.

An organism may exist in quite different forms as it passes through the various stages of its life. Perhaps the best known examples of this are the insects that exist successively as fertilized egg, larva, pupa, and adult winged insect, or the frog that is first

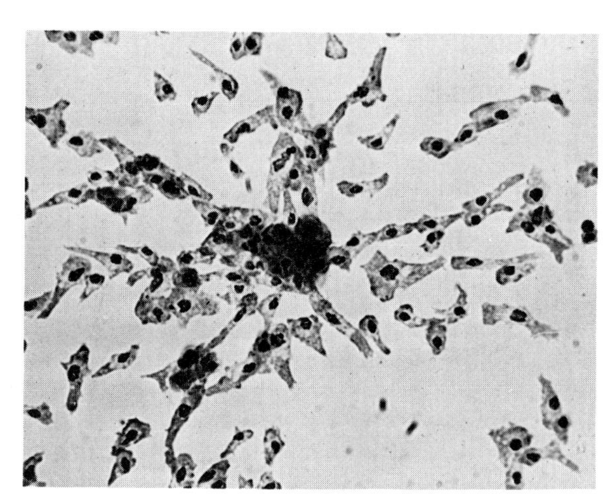

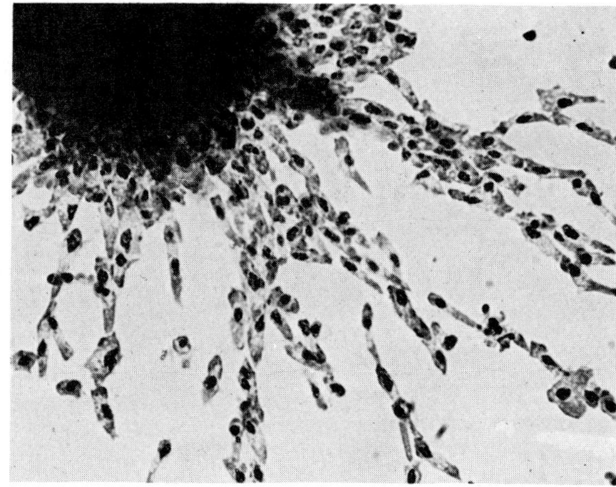

Fig. 10.1 The individual amoebalike cells of the cellular slime mold Dictyostelium discoideum *migrating toward a central collecting point where they aggregate into a multicellular mass. In photograph at the right the process is further advanced than in the one at the left.*

a fertilized egg, then a tadpole, and finally an adult frog. Similar alterations in form occur in the plants. Ferns, for example, are successively unicellular spores, tiny heart-shaped prothalli, fertilized eggs, and finally the familiar fern plants, as they pass through the various stages of their life cycle (Fig. 21.13). Slime molds are successively amoebalike single cells, multinucleate mobile masses of protoplasm, sporangia, and spores (Figs. 10.1 and 10.2).

UNICELLULAR AND COLONIAL ORGANISMS

In Chapters 4 and 5 we considered examples of the kinds of plant organisms that inhabit the earth. From this it should be evident that organisms may be of varying degrees of size and complexity. The simplest organisms—unless one wishes to consider the viruses as organisms—consist of a single cell, as in the bacteria and some algae and fungi. Such **unicellular** organism have, of course, no tissues or organs or even any specialization of cells; all essential life processes occur within the single cell. In these unicellular organisms the tissue and organ levels of organization are absent, and the cell *is* the organism.

Among the algae and fungi are some organisms that may be considered as **colonial,** for example, the spherical green alga, *Volvox,* and some of the filamentous algae and fungi. Although colonial organisms may consist of a few to several hundred cells, each cell in the colony can carry on all essential life processes and will generally continue to live in nature if separated from the other cells of the colony. Streptococcus and staphylococcus bacteria are colonial forms (Fig. 4.15).

TRUE MULTICELLULAR ORGANISMS

The final stage in the evolution of structural organization is the true **multicellular** type, in which some degree of cell specialization has occurred. Although cells or tissues may be removed from a multicellular organism and kept alive in culture, they lack some of the potentialities of the complete organism. Some-

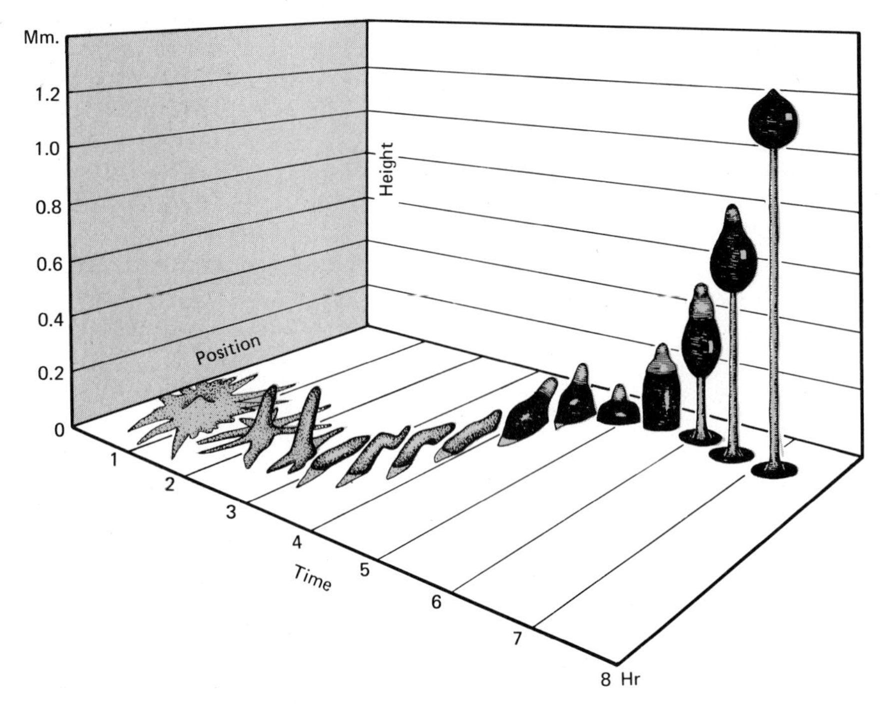

Fig. 10.2 Development of the aggregated plasmodium of Dictyostelium *(left) into a crawling "slug" and then into a stalked sporangium. The time scale is approximate. The spores released from the sporangium develop into the unicellular amoebae shown in Fig. 10.1. (Redrawn from John T. Bonner.)*

times such isolated parts of an organism can regenerate the missing parts and so become a complete new organism, as in a stem cutting that forms roots (Chapter 20). Unless such regeneration occurs, however, the detached portion of the organism cannot survive long as a living entity in nature.

Multicellular plants are of varying degrees of structural complexity, from relatively simple algae and fungi through the mosses and liverworts to the highly organized vascular plants with their roots, stem, and leaves. Even more complex organization is attained in the higher animals. One essential feature of multicellular organisms is some method of coördination of the various cells, tissues, and organs into a smoothly operating entity as the organism carries on its life processes and develops from a fertilized egg into a mature individual. In animals such coördination is accomplished by both the nervous system and hormones, whereas in plants coördination is brought about largely by hormones (Chapter 15) and other chemical agents.

SOME PROBLEMS RELATING TO THE ORGANISM CONCEPT

It is not difficult to recognize that a man or an oak tree, an elephant or a bean plant are living organisms. However, sometimes it is more difficult to determine whether we are dealing with an organism or not. Is a virus a living organism, or is it just an immensely complex chemical entity? A lichen meets all of the usual requirements for classification as an organism, including the ability to reproduce itself, yet the lichen is a complex of a fungus and an alga (Chapter 17). Although the unicellular algal component of the lichen can reproduce itself by cell division, the fungus has not yet been observed to reproduce itself when isolated. Is the fungus a separate organism? Is the lichen an organism?

Cells of several higher plants, including bean and carrot, have been isolated and kept alive in culture for some time. These cells, under certain conditions, may divide into other isolated cells, thus reproducing themselves in the same say that a bacterial cell does. Are they unicellular organisms? Certainly they are not bean or carrot plants, even though they contain all the hereditary potentialities of the plants from which they came. We can say that they are not organisms because they can continue to exist only under certain cultural conditions provided by man. But if we accept this, the question then arises as to whether some of our cultivated plants deserve to be called organisms since they can survive and flourish only under man's cultural care. In the course of selecting for traits desirable to man, other traits that permitted their ancestors to survive in nature have been lost.

Some organisms can exist in two quite different forms. For example, some of the fungi that cause human infections may occur in a filamentous form while living as parasites on their host, but when cultured they become unicellular organisms.

Plants propagated by cuttings or other vegetative methods pose another problem regarding the organism concept. We may get dozens of individual coleus plants from a single plant by cutting off branches, placing them in water or moist sand until they develop roots, and then potting them. We ordinarily regard each plant produced in this way as a separate individual organism, distinct from the parent plant and each other. Yet, in another sense, they are merely portions of a single organism with identical hereditary potentialities. In contrast, coleus plants raised from seeds are products of sexual reproduction, begin life as a single cell, and commonly have assortments of hereditary potentialities different from those of either parent.

A somewhat different problem relating to the organism concept arises in connection with cultivated apple trees and other plants propagated by grafting (Chapter 20). Here a branch from the tree being propagated is generally grafted to a wild apple seedling. Thus the roots and lower trunk of the resulting tree are from one organism and the upper trunk, leaves, and branches from another. Is the tree a single organism, or is it two organisms? Certainly, each portion of the grafted tree has quite different hereditary potentialities, yet we would ordinarily consider it as a single organism.

The foregoing examples could be added to, but they may make it clear that the concept of an organism is not always as easy to define as would at first appear. In a practical sense, however, we have no difficulty in identifying the myriads of individual plants and animals as organisms, each able to carry on the processes that permit it to live for a while, and generally having the capacity for reproducing its kind before it meets the eventual fate of every organism—death.

If any single level of plant organization is more important than another it is the organism level, for only when this level of organization is attained is there a living entity that can survive and reproduce in nature and constitute a functioning unit in the biological communities of the Earth. As we study the molecules, cells, tissues, and organs of plants or the life processes that occur within a cell, we may tend to lose sight of the fact that these are just isolated aspects of life and that it is the organism as a whole that is the natural living entity. This we should avoid if possible.

Another point worth considering is that only individual plant and animal organisms exist in nature—species, genera, families and other taxonomic categories are human concepts designed to provide a logical classification on the basic of apparent closeness of relationship, and to facilitate our dealing with the numerous organisms that inhabit the Earth. Species may, however, merge into one another without sharp dividing lines, and whether a certain group of plants belongs to one species or two or to one genus or two may be a matter of difference of opinion among the experts. However we may choose to classify them, the concrete entities are individual organisms, living, reproducing, and interacting with other organisms.

ORGANISMS IN TIME AND SPACE

Organisms have been described, facetiously but quite appropriately, as "four-dimensional worms," existing not only in the three dimensions of space but also in the fourth dimension, time (Figs. 10.1 to 10.5). A satisfactory general concept of any plant as an organism can be obtained only when we consider what it has been and what it will become, rather than just what it is at a particular point of time. We devote the remainder of this chapter to a consideration of various stages in the life cycle of a few selected plants, from their beginning as individuals through youth, maturity, and reproduction and on to eventual death, in an effort to clarify somewhat the concept of plants as four-dimensional organisms. We will frequently refer to material previously discussed, and anticipate the more detailed treatments of plants as living, growing, developing, and reproducing entities in subsequent chapters.

From among the many diverse kinds of plants, we select two quite different representatives for consideration: *Dictyostelium discoideum,* one of the primitive and relatively simple cellular slime molds, and *Phaseolus vulgaris,* the common garden bean. Comparative references to other species of the complex and highly evolved flowering plants will be made.

The spores of *Dictyostelium* germinate into unicellular "amoebae" very similar to ordinary amoebae. They move and engulf food by pseudopodia and even reproduce by cell division as amoebae do. However, when their population becomes rather great and the food supply becomes exhausted the amoebae suddenly begin to move toward a central collecting point, aggregating into a multicellular mass (the **plasmodium**) (Fig. 10.1). At first the plasmodium is rather diffuse, but it soon becomes transformed into an elongated cylindrical "slug" with a pointed front end that glides along for a time before it settles down and develops into a stalked sporangium (Fig. 10.2). The spores produced in the sporangium then give rise to the next generation of amoebae. *Dictyostelium* is a good example of the diverse forms that an organism may assume in the course of its life. John T. Bonner of Princeton University and other biologists have devoted much time to the investigation of the development of this slime mold, with particular reference to the factors causing aggregation and the differentiation of specific parts of the slug into the sporangium and its stalk.

In a seed plant such as the bean, a new individual arises when a sperm from a pollen grain unites with an egg inside an ovule, the resulting fertilized egg developing into the embryo plant of the seed. However, the beginnings of the plant as an *independent* individual may be considered to start with the detachment of the seed from the parent plant, or perhaps with the initiation of photosynthetic food production by the young seedling that has developed from the embryo during germination.

A viable (live) bean seed will germinate after it has imbibed sufficient water, provided that the temperature is suitable and there is an adequate supply of oxygen. The radicle or embryonic root rapidly elongates and breaks through the softened seed coats of the swollen seed, growing downward into the ground and

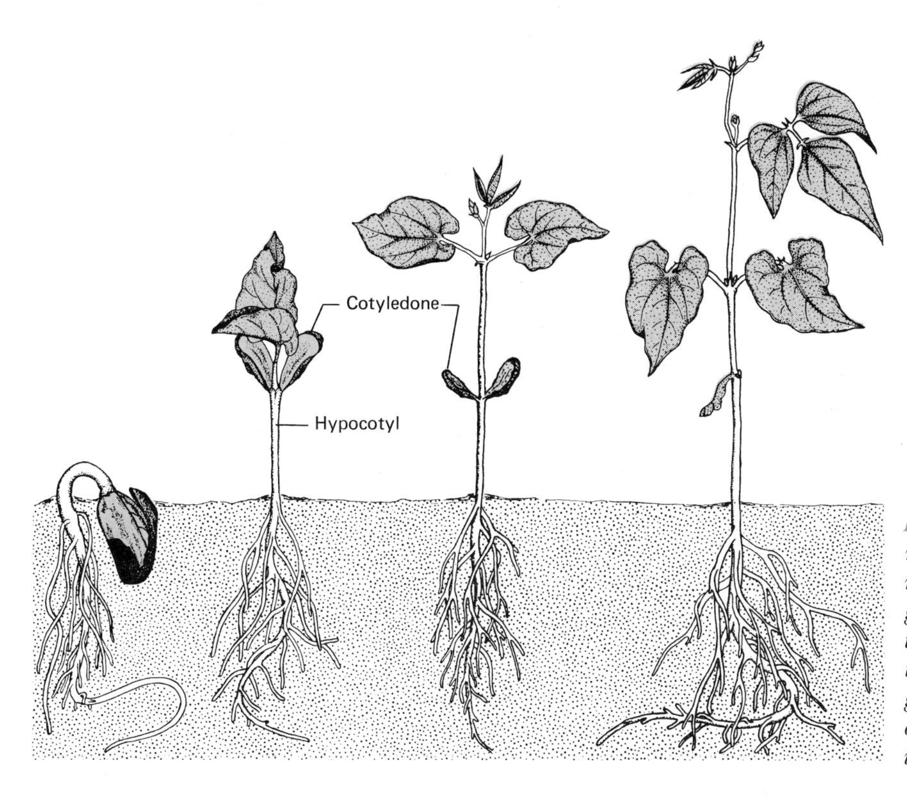

Fig. 10.3 Stages in the germination and early development of the bean. The elongation and straightening of the hypocotyl brings the cotyledons into position above ground. Note the withering of the cotyledons as the food in them is consumed.

producing root hairs. Soon secondary branch roots develop from this primary root. Firm anchorage in the soil and an extensive absorbing surface are thus established early. The upper portion of the hypocotyl arches and elongates, raising the cotyledons above the ground (Fig. 10.3). Once the hypocotyl is exposed to light its hook opens or straightens out and the small plumule between the cotyledons rapidly expands into the stem and first true leaves of the young plant. Elongation of the internodes results in the lengthening of the stem. As the food in the cotyledons is consumed in the rapid early growth, the cotyledons shrivel and fall to the ground, the food supply of the seedling now being provided by photosynthesis in the leaves and stems.

In the garden pea, a plant in the same family as the bean, the cotyledons are also fleshy and filled with food and, as in the bean, there is no endosperm in the seed. The pea cotyledons are, however, not raised above the ground during germination (Fig. 10.4), as they are in the bean, because the hypocotyl does not elongate. Otherwise, germination follows a course similar to that of the bean. The castor-bean, a plant that has endosperm in its seeds, illustrates a somewhat different modification of germination (Fig. 10.5). The thin, leafy cotyledons at first absorb food from the endosperm of the germinating seed, but after they emerge above ground they expand considerably, develop chlorophyll, and for several weeks remain attached to the plant and carry on photosynthesis like the true foliage leaves. The cotyledons of the castor bean reveal more clearly than do those of the bean or pea the true nature of cotyledons as modified leaves.

In the germinating grains of maize and other grasses, both the cotyledon and endosperm remain below ground. When the first internode above the cotyledon has elongated

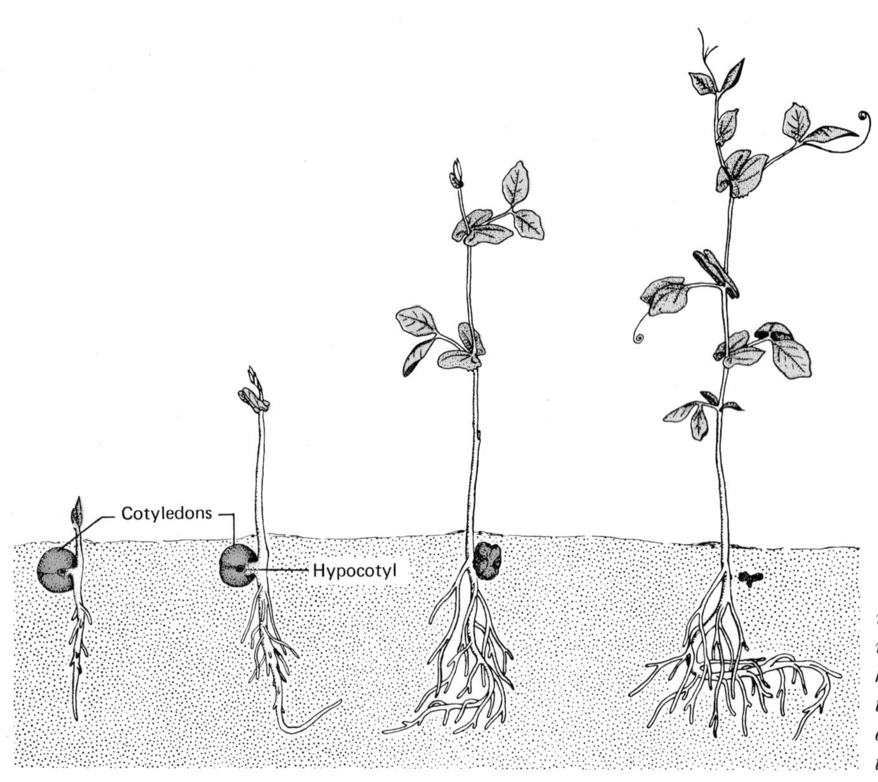

Fig. 10.4 Stages in the germination and early development of the pea. Since the hypocotyl does not elongate, the cotyledons remain underground, in contrast with the bean and castor bean.

enough to expose the tip of the coleoptile (Fig. 9.50*b*) above ground, further elongation ceases, and the rolled-up leaves of the plumule rapidly grow through the coleoptile tip, expanding into the first true foliage leaves. In corn, adventitious roots have meanwhile developed at the first node and these roots, rather than the short-lived primary root developed from the radicle, contribute toward the root system of the mature plant. Except for minor differences in detail, embryo structure and germination in other angiosperms are similar to the examples described. It should be clear that an embryo is really a miniature plant with all its major vegetative structures already defined, and that germination is not a "coming to life" but rather an acceleration of metabolic processes of the embryo resulting in a resumption of growth.

With the completion of germination, the seedling plant continues to grow, rather slowly at first, then much more rapidly for a month or more, and then at a somewhat reduced rate. Cell divisions in the apical meristems provide the new cells that enlarge and mature into the various types of cells that make up the tissues and organs of the plant. As growth continues the plant is synthesizing food by photosynthesis. Much of this food is used in growth as a source of energy or in building the new cells, tissues, and organs. Food production, however, exceeds food consumption and the excess food accumulates in the plant.

After a period of vegetative growth and development, which in the bean plant lasts for a month or so, flower buds appear and develop into flowers. Botanists have long been interested in the change from vegetative to reproductive development, and have done much research in an effort to elucidate the factors involved. In some species, initiation of flower primordia has been found to be controlled by environmental factors such as day length or temperature (Chapter 15), whereas in others such as bean, the only prerequisite seems to be the completion of a certain minimal period of vegetative development. Most trees do not bloom until they are several years old, and the well-known century plant is noted for its years of development before it blooms.

Once the flowers are open, the processes of pollination, fertilization, and seed and fruit development take place. Seed production is a complicated process involving many important structural details and specific processes that will be outlined in some detail in Chapter 21. Here we present only the gross features.

The mature pollen, freed by the opening of the pollen sacs of the anthers, is transferred to the stigma of the same or another flower of the species (pollination), commonly through the agency of wind or insects. Some flowers, such as those of the bean and pea, are usually self-pollinated. The pollen on the stigma develops a pollen tube that grows through the tissues of the stigma, style, and sometimes the ovulary and into the ovularian cavity (Fig. 9.38). When the pollen tube reaches an ovule, the tip of the tube ruptures, and one of the sperm in the tube is delivered to the egg within the ovule. The egg and sperm fuse (fertilization), and the fertilized egg (the zygote) develops into the embryo. As a result of other cell fusions and divisions (Chapter 21), the endosperm is produced, and meanwhile the integuments of the ovule undergo modification into seed coats. Thus, a seed consisting of embryo, endosperm, and seed coats is produced. As has been noted, however, the endosperm of some plants such as bean and pea does not persist into the mature seed.

One of the first conspicuous changes in

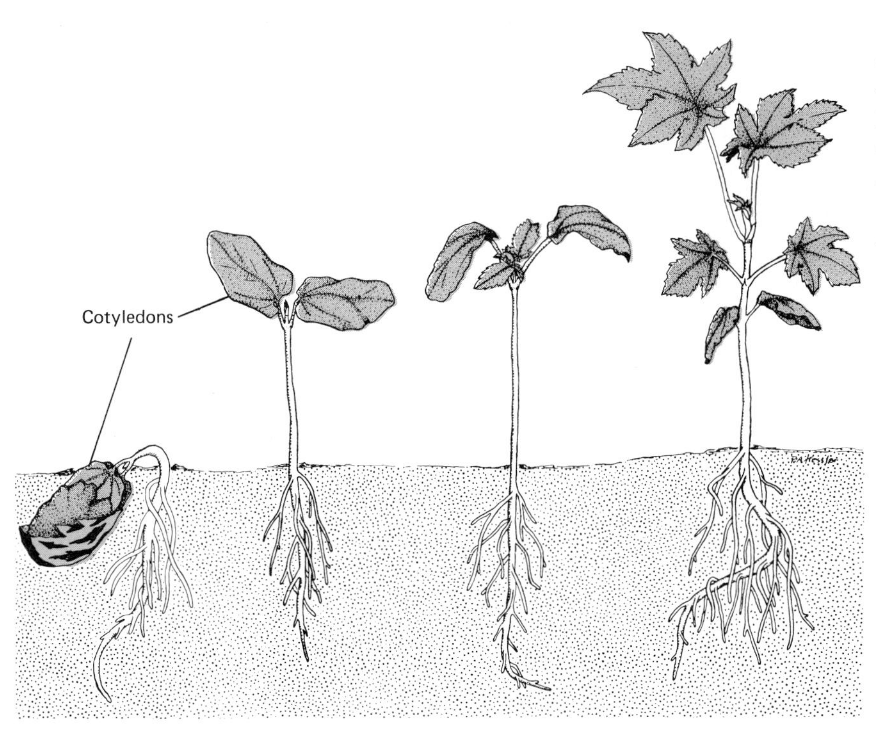

Fig. 10.5 Stages in the germination and early development of the castor bean. The thin cotyledons expand, develop chlorophyll and carry on photosynthesis for some time before they finally abscise.

the flower following pollination is the withering and fall of the petals and stamens. This is soon followed by the rapid enlargement and modification of the ovulary (and sometimes adjacent parts) into a fruit. If pollination and fertilization do not occur, the flower usually withers and abscises. The dependence of fruit development on pollination and fertilization was for many years a mystery, but it is now known that auxin and perhaps some other plant growth hormones (Chapter 15) provided by the pollen tube and the growing embryo are essential for fruit development and retention. However, in seedless fruits sufficient auxin is apparently available from other sources.

With the production of fruit containing viable seed, the life cycle of the individual plant may be said to have been completed. Species that have a short life span (annuals and biennials) generally produce only a single crop of seeds, but those with a longer life span (perennials) may produce seeds for many years before they finally die.

Despite the great differences between the slime molds and seed plants, they both illustrate characteristics common to the life cycles of most organisms: a succession in time of diverse structural organization as the individual progresses from its initiation, through youth and maturity, to reproduction and finally death. The fourth dimension—time—may be only a few hours, as in the slime molds, or thousands of years, as in the giant sequoias. Each stage in development, individually and successively, is a reflection of the coordination and control inherent in the dynamic and changing organism. No one stage of the life cycle, but rather the summation of all stages, defines the living organism.

RELATED READING

Bonner, J. T. "Volvox: a colony of cells," *Scientific American, 182*(5), 52, May 1950.

Koller, D. "Germination," *Scientific American, 200*(4), 75–84, April 1959.

Norman, A. G. "The uniqueness of plants," *American Scientist, 50*, 436–449, 1962

Spencer, P. W. "Season of the apple," *Natural History, 83*(5), 38–45, 1974.

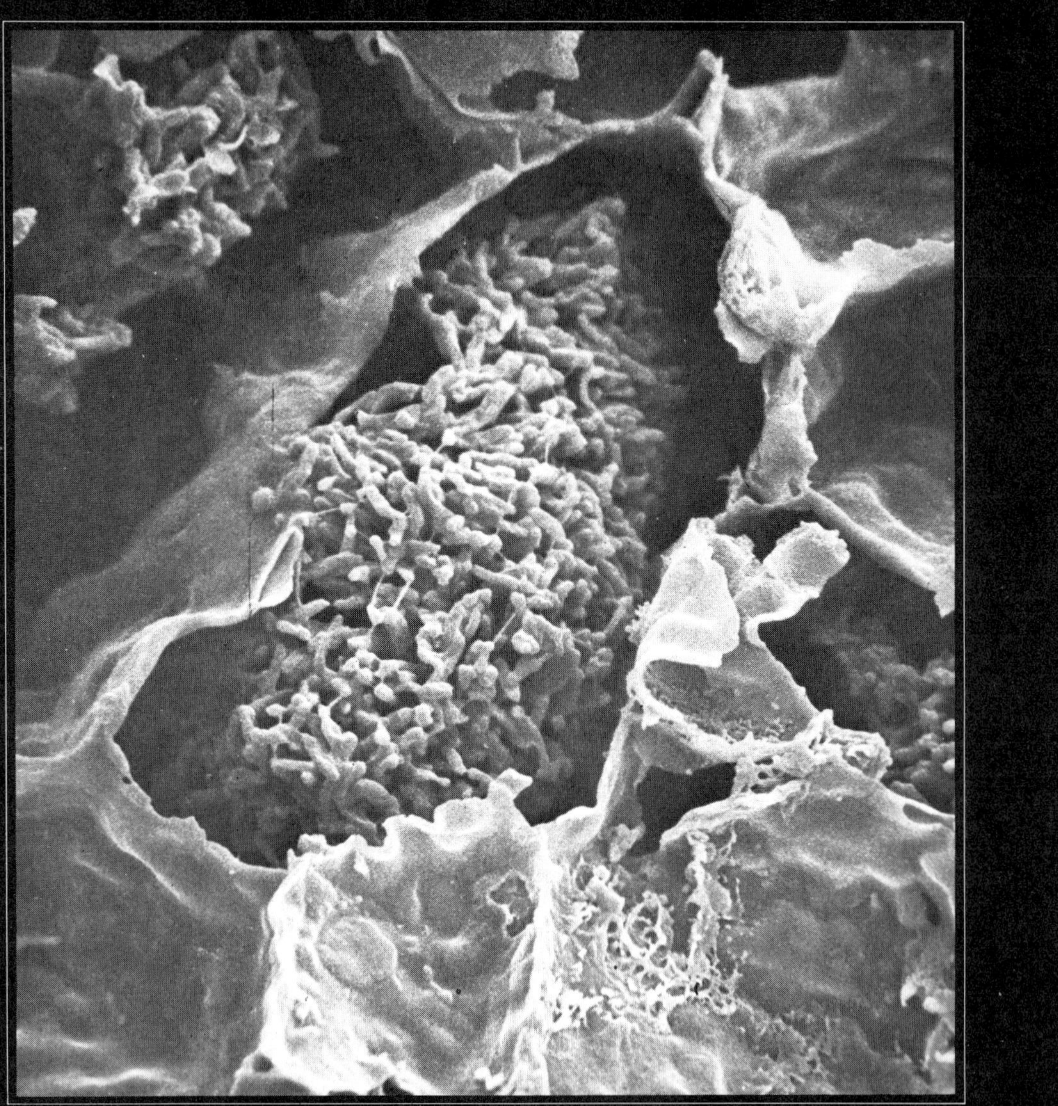

PLANTS IN ACTION

4

PHOTO OPPOSITE: NITROGEN-FIXING BACTERIA, *RHIZOBIUM TRIFOLII,* IN ROOT NODULE CELLS OF WHITE CLOVER, *TRIFOLIUM REPENS L.* (MAGNIFICATION 1600 $\times$)

11
MOLECULAR TRAFFIC

Living plants are the scene of endless activity. Numerous biochemical reactions are in progress, energy, is transformed and used, substances are transported from place to place, cells divide, enlarge and become specialized, cytoplasm streams around within the cells, and the plant grows and reproduces. Essential to these activities is a considerable traffic in substances between the plant and its environment and from cell to cell within the plant. Carbon dioxide, oxygen, water, and salts enter and leave the plants, and in addition food enters those plants that cannot make their own by photosynthesis. Our concern in this chapter is with the nature of this traffic into, out of, and through plants.

Much, though not all, of the movement of substances into and out of plant cells is by **diffusion,** a purely physical process requiring no expenditure of energy by the plant. Diffusion is a consequence of the ceaseless movement of all molecules and other small particles such as atoms and ions. When diffusion occurs the expended energy is utilized within the plant, but the energy comes from the kinetic energy of the moving molecules themselves rather than from any work done by the plant.

If we could see the molecules in any substance such as air or water, we would note that they are swarming about in every direction, bumping into each other or into solid surfaces and rebounding off in new directions without loss of velocity, and moving in straight lines until other collisions again result in changes of direction.

An increase in temperature brings about an increase in the rate of molecular movement. The smaller and lighter a molecule, the more rapidly it moves. The molecules of gases move more rapidly and freely than those of liquids, whereas the molecules of liquids generally have a greater velocity and freedom of movement than those of solids.

Despite their constant motion, the molecules of a substance attract each other. These attractive forces become greater as the distance between molecules decreases. In gases the attractive forces are slight, but in liquids they are great enough to hold the substance together and produce a surface. The molecules of solids are still closer together and the intermolecular attractions are great enough to hold the substance together in a more or less rigid shape. The movement of molecules in a solid is generally restricted to extremely short distances by the intermolecular forces, although occasionally the molecules of some solids, such as the naphthalene of moth balls, escape and move freely through the air as gases. Another solid that may "evaporate" is ice, the water molecules diffusing into the air as water vapor. Such "evaporation" of solids is called **sublimation.** Sublimation of ice in frozen plants often occurs in the winter.

Many substances can exist in solid, liquid, and gaseous states depending on the temperature and pressure. When ice (solid water) is heated, its molecules begin moving more rapidly and the water becomes a liquid. The more rapidly moving molecules in liquid water may, in turn, evaporate and so pass into the air as a gas **(water vapor).**

Of course, the melting and boiling temperatures of various substances differ greatly, carbon dioxide is a gas at ordinary temperatures but freezes into a solid (dry ice) that sublimates at −78.5°C. Lead melts at 327.43°C and boils at 1620°C.

DIFFUSION

Diffusion is a consequence of the movement of molecules (and also other small particles such as ions and colloidial particles) by their own kinetic energy, but by no means does all molecular movement result in diffusion. Diffusion is the *net* movement of the molecules of a particular substance from a region of higher molecular activity (free energy, or diffusion pressure) of this substance to a region of lower molecular activity of the substance. Thus, diffusion occurs only when more molecules of the substance are moving in one direction than another during any period of time. If the same number of molecules of a substance are moving in every direction within a system, there is no diffusion.

The **molecular activity** of a substance increases with its concentration (the number of molecules per unit volume), with temperature (because of the increased velocity of the molecules), and when the substance is subjected to an external imposed pressure (other than the prevailing atmospheric pressure). If the temperature throughout a system is uniform and there is no differential imposed pressure in any part of the system, then the gradient of molecular activity is determined by the differences in concentration of molecules. This is the simplest situation and the one we will consider first. It should be noted that each substance diffuses only on the basis of its *own* gradient of molecular activity, regardless of what the gradients of other substances present may be. Frequently two or more different substances occupying the same space will be diffusing in opposite directions at the same time.

Some examples will help make these facts clear. The diffusion of gases, solutes, and solvents (specifically water in biological systems) involve somewhat different factors and so will be considered separately. The diffusion of gases will be discussed first.

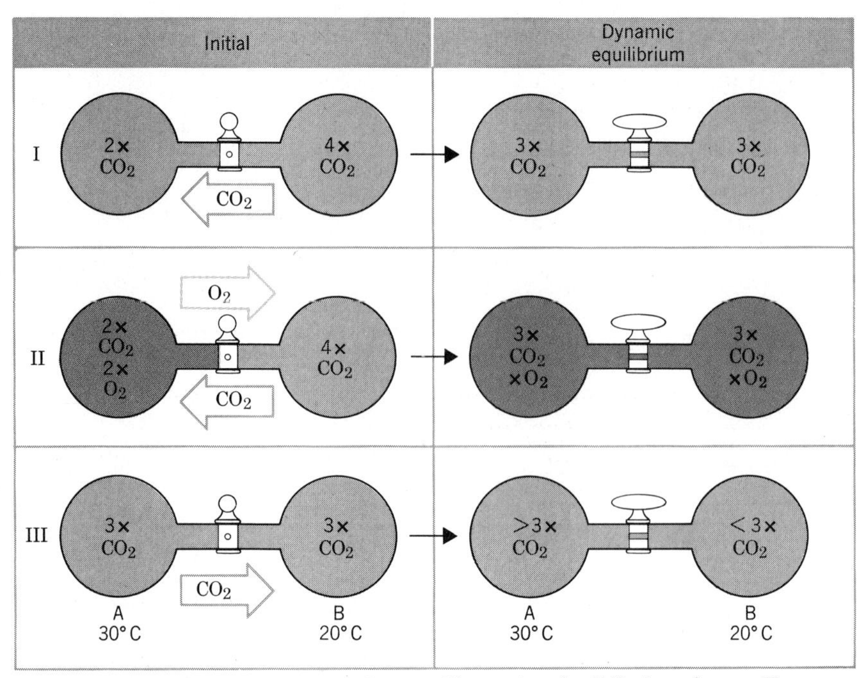

Fig. 11.1 A hypothetical experimental system illustrating the diffusion of gases. Two flasks of equal volume are connected by a stopcock. The flasks at the left give the initial concentrations of gas before the stopcock was opened. The flasks at the right give the concentration of the gases when a dynamic equilibrium was attained. I. Diffusion of CO_2 from a higher to lower concentration. II. Simultaneous diffusion of CO_2 and O_2 in opposite directions. The presence of O_2 does not affect the direction of diffusion of CO_2 nor its concentration at dynamic equilibrium. Note that CO_2 and O_2 diffused even though the total gas pressure was the same in flasks A *and* B *initially and when a dynamic equilibrium was established. III. When the temperature of flask* A *was maintained at 30°C and that of flask* B *at 20°C the CO_2 diffused from* A *to* B *despite the fact that its initial concentration was the same in* A *and* B.

Diffusion of Gases

A simple physical system consisting of two flasks with their necks joined by a stopper containing a stopcock valve (Fig. 11.1) can be used to demonstrate the basic concepts of diffusion. If flask A contains $2x$ molecules of carbon dioxide (CO_2) per unit volume and flask B contains $4x$ molecules of CO_2 per unit volume, diffusion from B to A will begin as soon as the stopcock is opened because CO_2 has a higher molecular activity in B (due to its higher concentration). Molecules of CO_2 will be moving through the open stopcock in both directions, but initially twice as many will be moving from B to A as in the reverse direction. As a result, the concentration of CO_2 in B continues to decrease as it increases in A, and the rate of diffusion progressively decreases. Finally, the concentration of CO_2 becomes the same in both flasks ($3x$) and diffusion ceases. A dynamic equilibrium has been reached. Molecules of CO_2 are still moving through the open stopcock in both directions, but there is no diffusion because just as many are going one way as another. Exactly the same diffusion of CO_2 would have occurred if, initially, flask *A* had contained $2x$ molecules of oxygen (O_2) in addition to the $2x$ molecules of CO_2, making the total initial gas pressure the same in both flasks.

Although there is no diffusion of CO_2 in the system after an equilibrium has been established ($3x$ molecules/unit volume in each flask), diffusion of CO_2 can be induced by heating flask *A* to 30°C while flask *B* is cooled to keep its temperature at 20°C. The CO_2 molecules in *A* would now be moving more rapidly than those in *B*, so more would move from *A* to *B* during a period of time than in the reverse direction, even though the initial concentration was the same. Eventually a new dynamic equilibrium of molecular activity will be established, but there will not be a concentration equilibrium. The CO_2 concentration in *A* at equilibrium will be less than $3x$ while in *B* it will be more than $3x$. However, because of their more rapid movement, just as many molecules of CO_2 will be moving from *A* to *B* as from *B* to *A* per unit of time, so an equilibrium of molecular activity exists.

Let us now turn to an example involving a plant and consider the diffusion of gases between the intercellular spaces of a leaf (Fig. 11.2) that is carrying on photosynthesis and the outside air adjacent to the leaf. Assume that the leaf is in the shade, and so at essentially the same temperature as the air. Here the only one of the three factors affecting the molecular activity gradient will be concentration. Since the cells of the leaf are producing oxygen by photosynthesis, the concentration of oxygen is greater in the intercellular spaces than in the outside air and, consequently, oxygen will diffuse out through the stomata. Of course, mole-

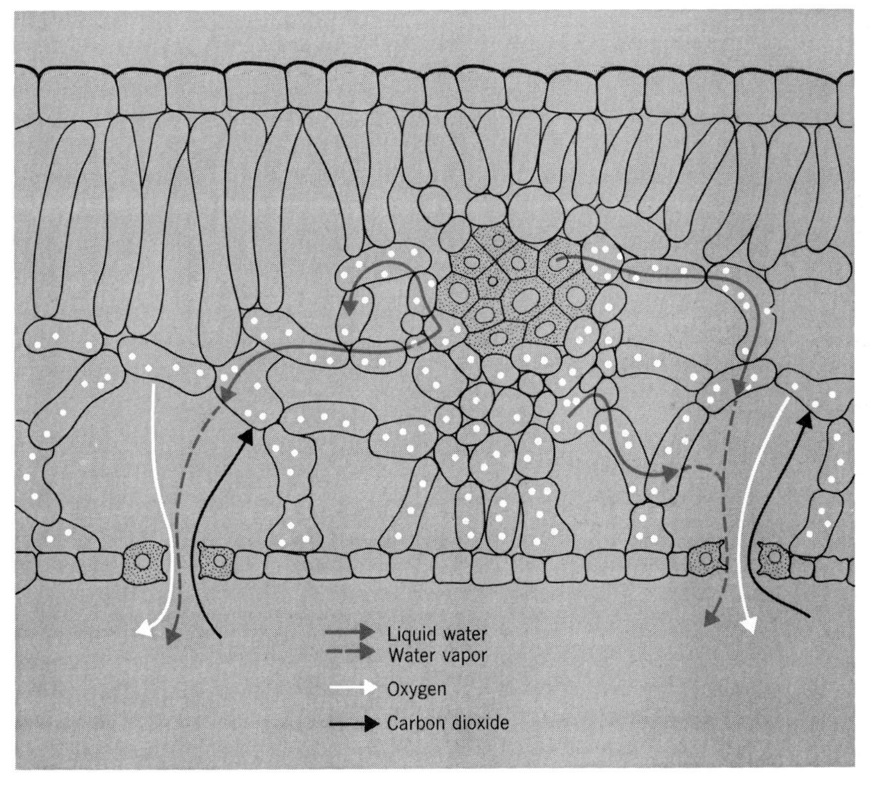

Fig. 11.2 Section through a leaf showing the simultaneous diffusion of water vapor and O_2 out of the leaf and CO_2 into the leaf through the stomata. This gas exchange occurs at all times when the rate of photosynthesis exceeds the rate of respiration.

cules of oxygen will be moving through the stomata in both directions, but since there are more molecules of oxygen per unit volume inside the leaf than outside, there will be a net movement (diffusion) out.

At the same time that oxygen molecules are diffusing out through the stomata, carbon dioxide molecules are diffusing in because they are being used in photosynthesis and so are less concentrated in the leaf than in the outside air. Note that two substances may diffuse in different directions at the same time and place, each one diffusing from the region of *its own* greater molecular activity to the region of *its own* lesser molecular activity. The presence of molecules of another substance affects the *rate* of diffusion (since they increase the number of molecular collisions) but not the *direction* of diffusion.

Now, by reducing the light intensity, let us reduce the rate of photosynthesis in the leaf until it is occurring just as fast as respiration. Under these conditions, all the oxygen produced by photosynthesis will be used in respiration, and all the carbon dioxide produced by respiration will be used in photosynthesis. Soon the concentration of both gases in the intercellular spaces will be the same as their concentration in the outside air; consequently diffusion will cease. Of course, molecules of both gases will still be moving through the stomata in both directions, but the number of molecules moving in during any time interval is the same as the number moving out, so there will be no net movement (diffusion) in either direction. Under such conditions, it is said that a dynamic equilibrium exists.

Since molecules of nitrogen gas are neither used nor produced by processes within a leaf, nitrogen does not diffuse into or out of a leaf as long as the leaf is the same temperature as the air. Let us next turn our attention to a leaf that has just been exposed to direct sunlight, and so becomes perhaps 10°C warmer than the outside air. Despite the fact that the concentration of nitrogen in the leaf is the same as outside, nitrogen will now diffuse out of the leaf because the molecular activity of the nitrogen molecules in the intercellular spaces has been increased by the rise in temperature. A new dynamic equilibrium will soon be established and diffusion of nitrogen will cease when the molecular activity of nitrogen is the same inside the leaf as out. At equilibrium the nitrogen will be less concentrated in the intercellular spaces than in the air, the greater velocity of the molecules inside the leaf compensating for their lower concentration. The product of velocity times concentration yields the same *total* molecular energy as that of the nitrogen outside the leaf where the molecules are more concentrated but have lower kinetic energy (velocity). The situation is comparable with the purely mathematical one of both 6×8 and 4×12 having 48 as the product.

Diffusion of Solutes

All principles of diffusion that apply to gases also apply to solute particles, although the rate of diffusion of solutes is much less rapid than that of gases because of the much greater density of the medium (the solvent) through which they are diffusing. When a few crystals of a colored substance such as potassium permanganate are placed in the bottom of a long thin glass tube filled with water (in which convection currents are reduced to a minimum) the substance diffuses through the water until a uniformly colored solution results, indicating that equilibrium has been attained. Several months are required for reaching an equilibrium in a tube a meter long.

Because of the short distances involved, diffusion provides an adequate means of movement of sugars and other solutes between adjacent cells in a plant. However, diffusion is too slow to account for the rates of solute movement through greater distances within

Table 11.1 **Influence of Selected Solutes on Osmotic Potential and Water Concentration**

Solute	Molality	Osmotic Potential, bars	Solution Volume, % of pure H_2O	H_2O Concentration, moles/liter
None (H_2O)	0	0	100	55.41
$MgSO_4$	0.08	5.80	99.87	55.44
NaCl	0.10	4.83	100.17	55.31
Sucrose	0.10	2.42	102.08	54.28
Sucrose	1.00	24.17	120.68	45.91

plants, as from the leaves to the roots. As we shall see in subsequent chapters, other processes are responsible for such long-distance transport.

Diffusion of Water (and Other Solvents)

Like solutes, water and other liquids diffuse much more slowly than gases. However, water vapor is a gas, and it diffuses at rates comparable with those of other gases.

An important factor in determining the molecular activity of liquid water is the presence of solute particles; the molecular activity of water decreases as the number of solute particles per unit volume increases. This has often been explained on the basis that solutes decrease the concentration of the water molecules, but this is not an adequate explanation. When most substances (e.g., sugar, sodium chloride, ammonium chloride) are dissolved in water, the volume of the resulting solution is greater than the volume of water used, and so the water is actually less concentrated than before. However, when some substances such as magnesium sulfate are dissolved in water, the volume of the solution is *less* than the volume of the water used, so the water is more concentrated than before (Table 11.1). Although different kinds of solutes have different effects on water concentration, the solute particles of *all* substances reduce the molecular activity of water by the same amount provided that the number of solute particules per unit volume is the same.

The kind, size, and weight of the solute particles are not significant in affecting the diffusion of water. The controlling factor is the total number of solute particles per unit volume. A glucose molecule reduces the molecular activity of water by the same amount as a molecule of sucrose, which is larger and heavier. A potassium ion or a nitrate ion each has the same effect as an intact molecule of potassium nitrate. It does not matter whether a solution contains solute particules of one substance, ten different substances, or a hundred. All that matters is the total number of solute particles per unit volume.

The molecular activity (free energy) of water is influenced by any imposed pressure, as well as by temperature and the concentration of solute particles. Imposed pressure is an important factor influencing the molecular activity of water within plant cells. Pressure is measured in various units such as pounds per square inch (lb/in^2), millimeters of mercury (mmHg), and atmospheres (14.7 lb/in^2), but we use the **bar,** a metric pressure unit. One bar equals 14.5 lb/in^2, 0.987 atmosphere, or 750 mmHg. Because of the high imposed pressures that often develop in plant cells, the bar (and atmosphere) are units of convenient magnitude.

The influence of imposed pressure on the molecular activity of water can be illustrated by considering a simple physical system. If water were placed in a strong steel cylinder fitted with a watertight piston and subjected to a pressure of 10 bars by means of a hydraulic press (Fig. 11.3) it would have a molecular activity 10 bars higher than water subjected only to the atmospheric air pressure (which is not considered as an imposed pressure). This is an example of increasing the molecular activity of water.

The molecular activity of water can be decreased by adding a solute. If enough solute were added to reduce the molecular activity by 10 bars and this solution were subjected to a pressure of 10 bars by the hydraulic press, this imposed pressure would increase the molecular activity of the water in the solution by 10 bars, completely offsetting the decrease caused by the solute. Consequently, the molecular activity of the water in the solution under pressure is the same as that of pure water not subjected to an imposed pressure (0 bars). If the solution had been subjected to an imposed pressure of only 4 bars, the water in it would have a molecular activity of −6 bars.

In plant cells water is often subjected to high imposed pressure (the wall pressure), and this factor must be considered when dealing with the diffusion of water into and out of cells. Wall pressure is the back pressure of the plant cell wall against the turgor pressure of the cell exerted against the wall. The wall pressure is equal in magnitude but opposite in direction from turgor pressure.

Diffusion Through Membranes

Any substance diffusing into or out of a plant cell must pass through the cell wall and the plasma membrane. If the substance enters the vacuole, it must also diffuse through the vacuolar membrane. Diffusion of particles through membranes is, then, a matter of considerable biological interest.

From the standpoint of diffusion a **membrane** may be regarded as any partition between two regions, so cell walls as well as the protoplasmic membranes could be considered as membranes. Most cell walls are **permeable,** that is, all molecules and ions can diffuse through them. The walls of cork cells, how-

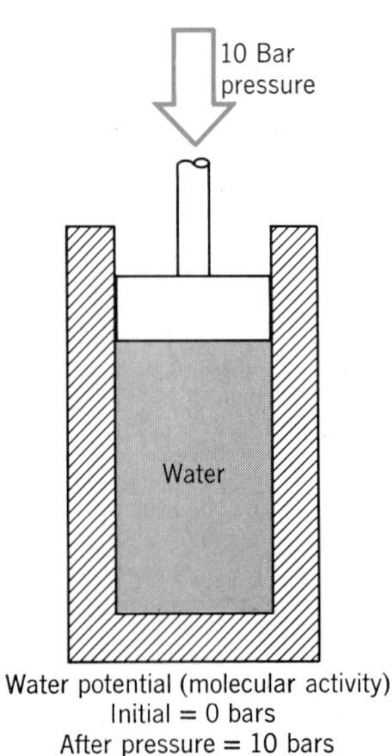

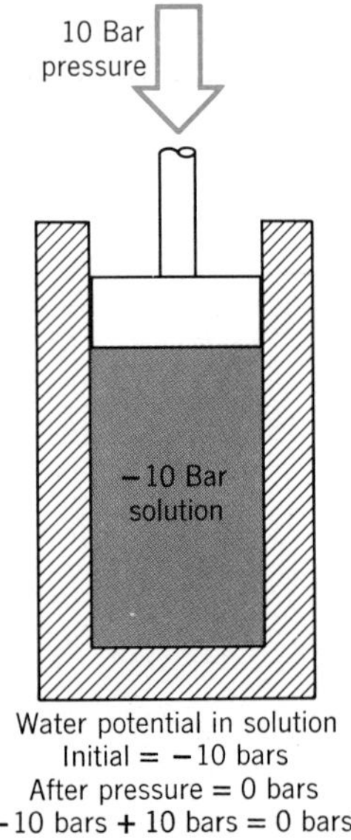

Fig. 11.3 When water or a water solution are subjected to pressure by the piston of a hydraulic press the water potential is increased by the amount of the applied pressure.

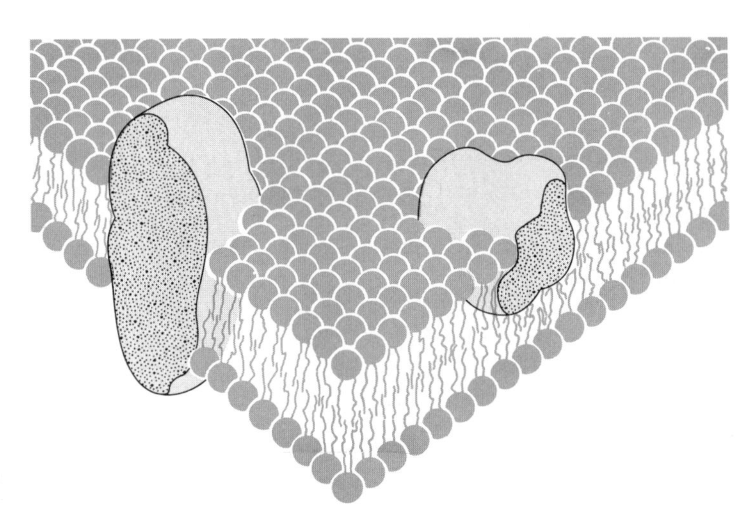

Fig. 11.4 The lipid-globular protein mosaic model of cell membrane structure. The spheres represent the polar (water soluble) portions of the phospholipid molecules and the wavy tails the nonpolar (fat soluble) fatty acid portions of the molecules. The phospholipid molecules are precisely oriented in two layers and provide the continuity of the membrane. The globular proteins occur at intervals and some extend through both the phospholipid layers. (From S. J. Singer and G. L. Nicholson, Science, 175, *723, February 18, 1972, with the permission of* Science *and the authors. Copyright 1972 by the American Association for the Advancement of Science.)*

ever, are **impermeable,** and diffusion cannot occur through them. The particles of some substances can pass through the various protoplasmic membranes whereas those of others cannot, so these membranes are said to be **differentially permeable.**

The protoplasmic membranes are generally very permeable to gases and water, less permeable to solutes such as salts and simple sugars, and impermeable to larger solute particles and colloidal particles. Fat-soluble substances, however, can penetrate these living cell membranes with considerable ease even though their molecules are relatively large.

It has long been known that the membranes of cells are composed of phospholipids and proteins, but it has not been easy to determine just how the molecules of these substances are organized into membranes. Various hypothetical models of membrane structure have been proposed. The most recent generally accepted model is of two layers of phospholipid molecules with protein molecules spaced at variable intervals (Fig. 11.4). The protein molecules may extend through one or both of the phospholipid layers, but in either case generally protrude beyond the surface of the phospholipid portion of the membrane. Sectional views of cell membranes in electronmicrographs appear as two dark lines with a light line between them, thus providing visible evidence for the two-layered structure of a membrane (Fig. 11.5).

Fat-soluble substances such as hydrocarbons, fatty acids, and lipids apparently diffuse through the phospholipid regions of membranes after dissolving in the phospholipid. Water and water-soluble substances such as sugars, salts, and gases probably diffuse through the protein portions of membrane. In either case the membranes are more permeable to certain kinds of molecules or ions than others.

Various nonbiological membranes are also differentially permeable. These are principally sheets or tubes of plastics, including cellophane, that have had their waterproofing components removed by immersion in alcohol. The differential permeability of such membranes is a considerably simpler matter than that of cell membranes. The plastic is perforated by submicroscopic pores that are too small to permit the diffusion of relatively large molecules (such as those of sugars) through them, but through which smaller molecules such as those of water and gases and the ions of salts can diffuse. Cigarette pack wrappers are differentially permeable after soaking in alcohol.

Fig. 11.5 An electron micrograph of a pair of unit membranes (44,000X).

Osmosis

Osmosis has commonly been defined as the diffusion of water through a differentially permeable membrane. However, a variety of evidence secured since 1953 indicates that a major portion of the water passes through a membrane by mass flow rather than by diffusion, thus explaining the fact that the rate of water movement through membranes is often greater than can be accounted for by the known rates of diffusion. However, the way in which water passes through a membrane has no influence on the direction of water movement through the membrane, or on osmotic factors such as turgor pressure or osmotic equilibria.

The term "osmosis" applies only to the diffusion of water (or other solvents) through a differentially permeable membrane, and not to the diffusion of solutes through such a membrane. A differentially permeable membrane is essential for osmosis, since an impermeable membrane permits no diffusion through it and a permeable membrane is essentially the same as no membrane at all, because all solute particles as well as water molecules can diffuse through it. One reason for considering osmosis as a special type of water diffusion is that it can result in the development of pressure. In cells the pressure is turgor pressure, which causes the cells to be turgid (inflated). In physical systems such as a U-tube osmometer (Fig. 11.6) the pressure results in the rise of water in the tube. The diffusion of water can generate such pressures only when it occurs through a differentially permeable membrane.

Various sets of terminology have been used in dealing with osmosis, but most plant physiologists now deal with osmosis in terms of water potential (rather than diffusion pressure, molecular activity, etc.). Water potential is the difference between the free energy (molecular activity) of water in a specific osmotic system and the free energy of pure water (that is not subjected to an imposed pressure) at the same temperature. The temperature specification makes it unnecessary to consider temperature in connection with osmosis, even though the free energy of water increases with temperature. The free energy of pure water (containing no solutes) not subjected to any imposed pressure is arbitrarily considered to be zero (0). (Actually, the free energy is thousands of bars, the precise value depending on the temperature of the water.) In any osmotic system in-

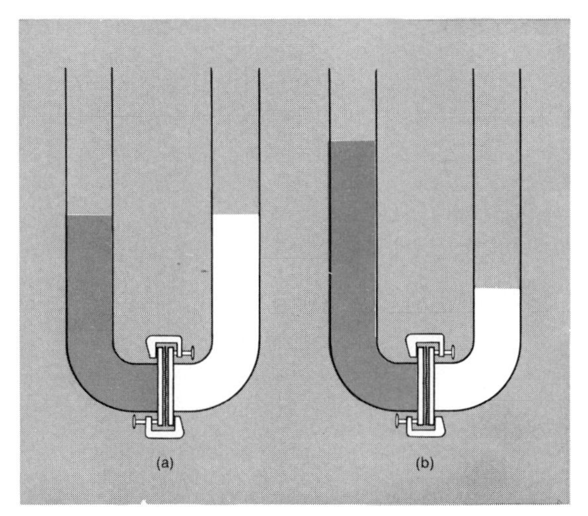

Fig. 11.6 A U-tube osmometer with the two arms separated by a membrane permeable to water but not solute molecules. Initially (A) *a solution was placed in the left arm and an equal volume of water in the right arm. Because of the reduction of water potential by the solutes water diffused from right to left. A dynamic equilibrium* (B) *was attained when the hydrostatic pressure was great enough to raise the water potential in the solution to that of the water in the right arm.*

volving plant cells the water potential is never more than zero and it is usually less than zero (a negative quantity), as we shall see shortly.

Since water potential involves the free energy (molecular activity) of water molecules, it should be considered in terms of energy units. The one used is joules/gram. However, this energy unit can be converted readily into pressure units (1 joule/g = 0.1 bar). Pressure units are more suitable for use in considering osmosis in plants, so we express osmotic quantities in bars.

Water potential in osmotic systems is influenced by two principal factors: it is reduced by the presence of solute particles (osmotic potential) and it is increased by any positive pressure (pressure potential). The positive pressure operating in plant cells is the wall pressure, exerted because the wall of a turgid cell soon reaches the limit of its elasticity and so presses back against the cell contents as the turgor pressure is exerted on the wall. The Greek letter psi (Ψ) is used as a symbol for water potential. For osmotic potential the symbol ψ_s will be used, the *s* indicating solute concentration. For pressure potential the symbol ψ_p will be used.

The relationships between the three osmotic quantities can be expressed by the simple mathematical equation given below, along with some random examples of what the water potential in various cells might be:

$$\text{water potential} = \text{osmotic potential} + \text{pressure potential (wall pressure)}$$

$$\Psi = \Psi_s + \Psi_p$$

$$0 \text{ bars} = -20 \text{ bars} + 20 \text{ bars}$$

$$-12 \text{ bars} = -25 \text{ bars} + 13 \text{ bars}$$

$$-25 \text{ bars} = -25 \text{ bars} + 0 \text{ bars}$$

The last situation would exist in a cell with no turgor pressure or wall pressure and also in a solution not subjected to an imposed pressure, such as a solution in a beaker or seawater. Note that the water potential and the osmotic potential are equal if there is no imposed pressure. The osmotic potential of pure water, as well as the water potential, is zero.

Osmosis in a Cell Model. An osmotic model of a cell can be constructed by tightly tying one end of a piece of differentially permeable plastic tubing, filling the tubing with a sugar solution, and then tightly tying the other end and rinsing off any sugar that was spilled on the outside of the tubing. Since the tubing is strong and has little elasticity, it corresponds to both the cell wall and the differentially permeable cytoplasmic membranes of a cell. The solution in the tube corresponds to the contents of a cell, both the cytoplasm and vacuole. If the cell model is placed in a beaker of pure water, the water molecules will diffuse into the cell model because the water potential is higher outside than inside, since inside it has been

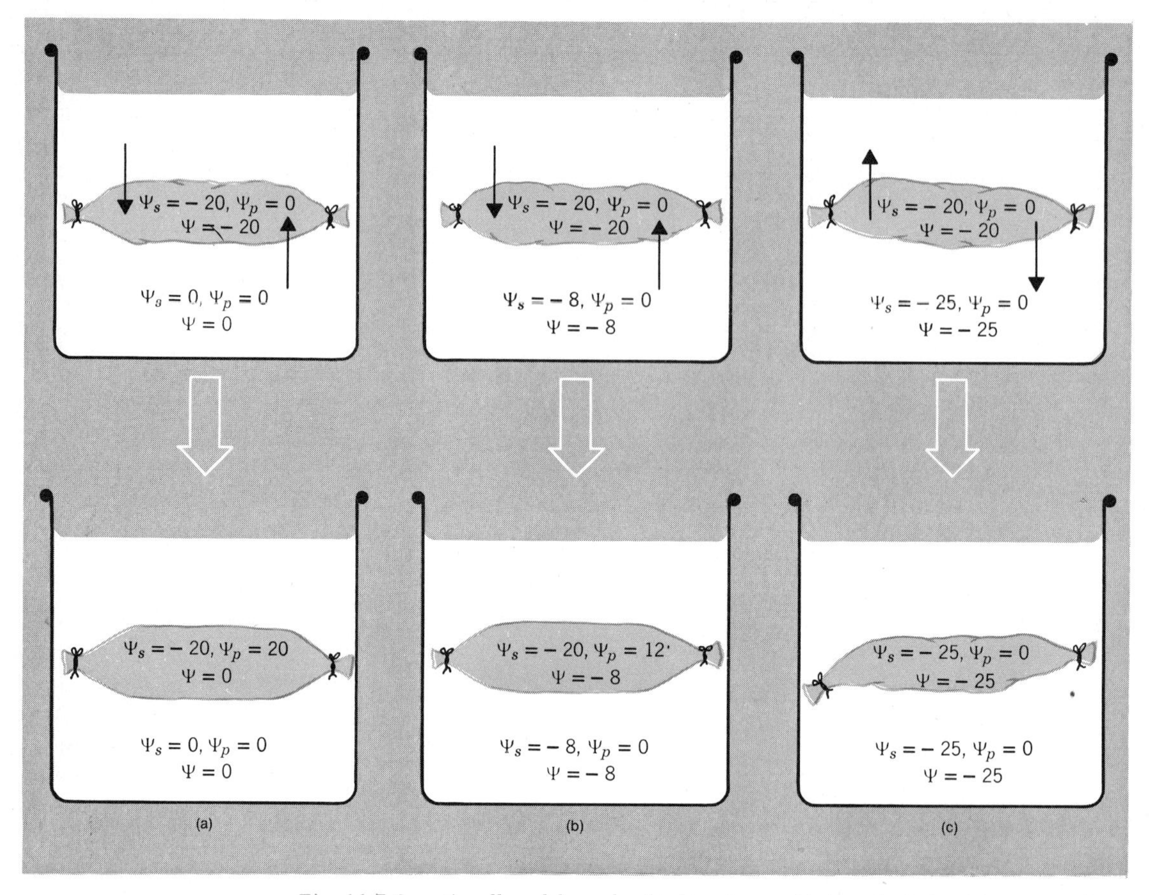

Fig. 11.7 Osmotic cell models made of tubing permeable to water but not to sucrose. In each of the three experiments the initial osmotic quantities are given in the upper drawing and the quantities at dynamic equilibrium in the lower drawing. (a) *A solution with an osmotic potential of −20 bars is placed in the cell model, which is immersed in water. Water diffuses into the cell model until the pressure potential reaches 20 bars, thus increasing the water potential in the cell model to 0 bars.* (b) *When the same cell model is placed in a solution with an osmotic potential of −8 bars equilibrium is again reached when the water potential becomes the same inside and outside (−8 bars). In this case the pressure potential (turgor pressure) at equilibrium is only 12 bars.* (c) *Here the cell model is placed in a solution with an osmotic potential of −25 bars, so the water potential is higher inside than out and water diffuses out. Since turgor pressure and wall pressure remain at 0 bars water will continue to diffuse out until the water potential (and osmotic potential) are the same inside and out (−25 bars).*

reduced by the presence of solutes.

Suppose that the sugar concentration is enough to reduce the water potential to −20 bars (Fig. 11.7*a*). Water will continue to diffuse inward until a dynamic equilibrium is attained, that is, until the water potential is the same inside as outside (0 bars). This occurs when the wall pressure becomes great enough to increase the internal water potential by the same amount that it is reduced by the presence of the solute particles.

As water diffuses into the cell model, it increases the internal water potential because it reduces the sugar concentration. However, if the cell model was essentially filled with solution at the beginning the increase in internal water potential would be very small, and we will ignore it in this and future examples for the sake of simplicity. If we considered this dilution factor, we would have to know exactly how much water diffused into the cell model or cell. Thus the situation inside the cell model (ignoring the dilution factor) would be:

	Ψ		Ψ_s		Ψ_p
Initially:	−20 bars	=	−20 bars	+	0 bars
At equilibrium:	0 bars	=	−20 bars	+	20 bars

The turgor pressure of the cell, as well as the wall pressure, is 20 bars at equilibrium. Note how important it is to consider the influence of the wall pressure, as well as the presence of solutes, on the water potential inside the cell. No matter how much water diffused in, the internal water potential would still be lower than that outside if only the solute factor were considered.

If, instead of placing the cell model in a beaker of pure water, we had placed it in a solution with a water potential of −8 bars (Fig. 11.7*b*) water would also have diffused in, since its potential was higher outside than inside. Again, diffusion would continue until the internal water potential became the same as the external water potential, but now the turgor pressure and the wall pressure at equilibrium would be less:

$$-8 \text{ bars} = -20 \text{ bars} + 12 \text{ bars}$$

If we had placed the cell model in a solution with a water potential of −25 bars, water would diffuse *out* of the cell model instead of in, since the water potential was higher inside (−20 bars) than outside (Fig. 11.7*c*). (The less negative the number, the higher the water potential.) As before, an equilibrium would occur when the water potential inside became the same as that outside:

$$-25 \text{ bars} = -25 \text{ bars} + 0 \text{ bars}$$

In this case we cannot ignore the change in the internal Ψ_s (osmotic potential), since the wall pressure becomes 0 bars and remains there no matter how much water diffuses out. (However, we *have* ignored the fact that as water diffuses out the external water potential increases slightly.) At equilibrium the internal and external osmotic potentials (as well as the water potentials) are the same, but this occurs *only* when the wall pressure is 0 bars.

Osmosis in Plant Cells. The preceding discussion applies to plant cells as well as to the cell model. However, there are several complications in plant cells that do not occur in the cell models. The cell membranes are not impermeable to all solutes, and solute particles as well as water molecules are almost continuously entering and leaving cells. Processes in cells are constantly using some solutes and producing others. For example, sugar may be produced by photosynthesis or by the digestion of starch and may be used in respiration or in the synthesis of cellulose. Thus the osmotic potential of plant cells fluctuates more or less continuously. Despite this, the osmotic potentials of cells have a characteristic range, depending on the species and organ in which they occur. Root cells usually have an osmotic potential of about −5 bars, leaf cells of herbaceous plants are in the −10 to −15 bar range while those of trees usually range between −20 and −30 bars. Salt marsh and ocean plants have osmotic potentials of −50 bars or less.

The turgor and wall pressures of plant cells also fluctuate greatly, since water is almost always diffusing into or out of a cell. The fluctuations in both of the factors that determine the water potential in a cell (osmotic potential and wall pressure) result in an almost endless diffusion of water into or out of a cell, and a dynamic equilibrium such as is attained in a cell model rarely if ever occurs. Nevertheless, the same quantitative relations apply to plant cells as to cell models.

Most of the cells of multicellular plants secure water from other cells adjacent to them, since only a small fraction of the cells are in contact with the external sources of water—the soil or some body of water. The same osmotic principles apply to the diffusion of water from one cell to another as to the diffusion of water from an external source into cells, with one exception. When the external water supply is large in volume as compared to the volume of the cell, equilibrium occurs by a change in the internal water potential until it becomes the same as that of the external water potential. If two adjacent cells are of the same volume, equilibrium occurs when their water potentials attain the average of their original water potentials. For example, if one cell had a water potential of -24 bars and its neighboring cell had a Ψ of -18 bars, at equilibrium (if it occurred) both would have a water potential of -21 bars. If the cells are of different volumes, a volume correction factor would be needed.

Plasmolysis. If a plant cell or tissue is in a solution that has a lower water potential (a more negative value) than that in the cell, water will diffuse out of the cell. Most of this water comes from the vacuole. As the vacuole decreases in size the elastic cytoplasm contracts and separates from the cell wall, which shrinks only slightly. This diffusion of water out of the vacuole and the shrinking of the cytoplasm away from the wall is **plasmolysis** (Fig. 11.8). The lower the water potential of the external solution, the greater the degree of plasmolysis. In a highly plasmolyzed cell the protoplast may completely separate from the wall and assume a spherical shape. If the water potential of the external solution is only slightly less than that in the cell, the cytoplasm may separate from the wall slightly in only one corner (incipient

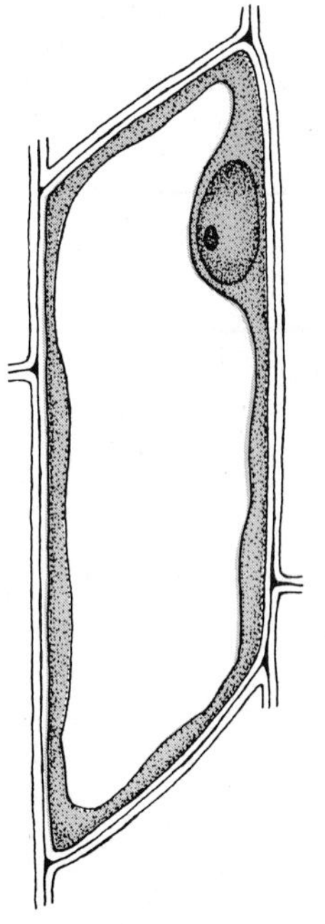

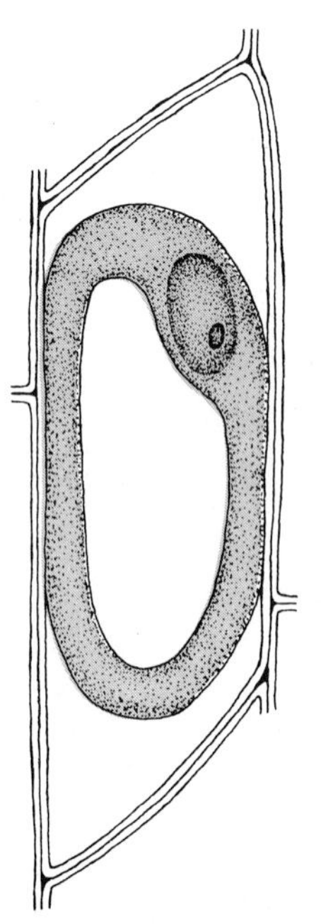

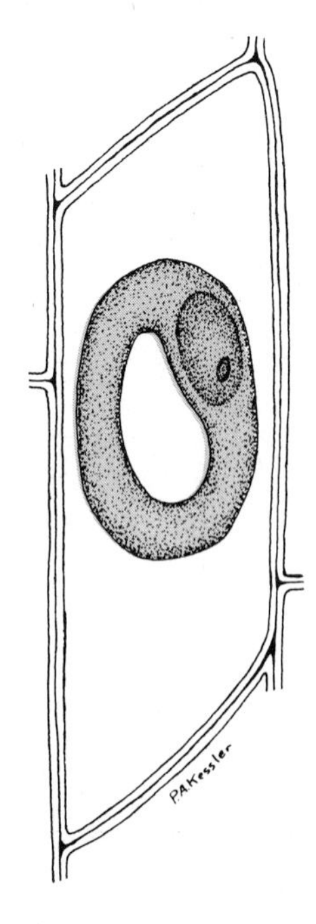

Fig. 11.8 Left: a turgid parenchyma cell. Center: after being placed in a solution with a water potential lower than the water potential in the cell water diffused out of the vacuole, plasmolyzing the cell. Right: when the cell was placed in a solution with a still lower water potential more extreme plasmolysis occurred as more water diffused out of the vacuole.

plasmolysis), but in either case the turgor pressure and wall pressure of the cell are 0 bars.

If a plasmolyzed cell is placed in water (or a solution with a higher water potential than the cell had originally), the cell will regain its turgidity and survive with no apparent damage, but continued plasmolysis results in the death of the cell.

Concentrated salt solutions kill plants by plasmolyzing their cells. This may happen unintentionally, as when salt, placed on streets or walks to melt ice, gets on lawns or when too much fertilizer is supplied to plants, or intentionally when salt is applied to a certain area to kill all the weeds. Freshwater plants are generally plasmolyzed when placed in a salt marsh or in seawater.

The Significance of Turgor Pressure. Plant cells must be turgid if the plant is to live and grow normally. Turgor pressure helps support non-woody tissues and loss of turgor results in wilting (Fig. 11.9). Stomata close when guard cells lose turgor and so gas exchange is hampered. The folding of the leaves of oxalis, bean, mimosa, and other plants at night and when the plants lack water results from reduced turgor, whereas the rapid movements of the leaves of the Venus's-flytrap (Fig. 9.29) and the sensitive plant (Fig. 11.10) result from a sudden diffusion of water from certain specialized cells. The leaves of some grasses roll into a tube when they begin to wilt because of the rapid loss of water from longitudinal rows of specialized cells in the upper epidermis and the consequent reduction in their turgor pressures (Fig. 11.11). Since most of the stomata of these grasses are in the upper epidermis, the rolling of the leaves results in a considerable reduction in the rate of water vapor diffusion out of the stomata. (The air in the tube soon becomes saturated with water vapor.) Although this may prevent excessive wilting of the leaves, the leaves do not roll up "in order to save water."

Fig. 11.9 *Multiple exposure (10 minute intervals) photograph of a young* Smilax *plant recovering from severe wilting after the soil was watered. Note the elevation of the leaves as well as the erection of the drooping stem.*

Since under natural conditions plant cells are rarely in contact with pure water, their turgor pressures are rarely at the possible maximum. Rainwater, however, is relatively pure, and during rainy periods the cells of fruits such as cherries may develop unusually high

Fig. 11.10 A sensitive plant (Mimosa pudica) *with expanded leaves (left) and the same plant 3 seconds after the stems had been tapped (right). The sudden folding and drooping of the leaves results from rapid loss of turgor pressure by the cells of the leaf pulvini. (Courtesy of the copyright holder, General Biological Supply House, Chicago.)*

turgor pressures, often resulting in stresses that burst the fruit. If rain falls directly on pollen, the turgor pressure of the pollen may become so high that the wall can no longer resist the pressure and then the pollen bursts. Because of this, and also because some of the wetted pollen germinates prematurely, prolonged rains while apple trees are blooming may seriously reduce the apple crop.

Other Special Types of Diffusion

Imbibition. The diffusion of the molecules of a liquid or a gas (usually water) in between the molecules of a solid, causing it to swell, is called **imbibition.** For example, the initial absorption of water by dry seeds is largely imbibitional. A board placed on moist soil imbibes water on its underside, causing it to swell and warp. Wooden doors and windows swell in humid weather because they have imbibed water vapor from the air. The imbibitional pressures resulting from imbibition are commonly very great, often being as high as 1000 bars.

Solids imbibe liquids or gases only when strong intermolecular attractive forces exist between them. Thus, cellulose imbibes water but not ether, and rubber imbibes ether but not water. The imbibed molecules are packed more closely together than they were in the liquid or gas state and their rate of movement or kinetic energy (and thus their diffusion pressure) is greatly reduced. The lost energy is converted into heat, so a release of heat always accompanies imbibition.

Evaporation. The diffusion of molecules of a liquid into the air is called **evaporation.** Molecules released by evaporation exist as gases. The gaseous form of water is called **water vapor.** Water is almost constantly evaporating from the wet walls of plant cells into the air of the intercellular spaces or of the outside atmosphere. The diffusion pressure of water vapor is commonly referred to as its **vapor pressure.** The vapor pressure of the air varies greatly with the concentration of water vapor and temperature, ranging from just a few millimeters of mercury to around 40 mm of Hg in hot, humid situations. Evaporation is slower when the vapor pressure of the air is high than when it is low.

ION ACCUMULATION

Although the ions of mineral salts are usually much more concentrated in plant cells than the same ions are in the surrounding soil or water, ions still continue to enter the cells and **accumulate,** principally in the vacuoles. The direction of ion movement during accumulation is just the opposite of what it would be in diffusion, so it is evident that some other factor is operating in ion accumulation.

Since ions are electrically charged particles, their movement is also influenced by electrical potential gradients, such as those that exist across the cytoplasmic membranes. However, even this is not adequate to explain the accumulation of ions.

Only living cells can accumulate ions. When a cell is killed the ions diffuse out until a diffusion pressure equilibrium is attained. Cells must expend energy in accumulating ions, the energy source being ATP from respiration or photosynthesis. As shown in Fig. 11.12, an increase in the rate of respiration of root cells is accompanied by an increase in accumulation. The way in which the energy from ATP is used in bringing about accumulation is not well understood, but apparently the membrane itself plays an active role. Enzymes are essential, and it is possible that the membrane proteins serve in this capacity. The membrane phospholipids may also play an active role, and there is some evidence that there may be a change from one kind of phospholipid to another during accumulation. In any event, it seems likely that the ions are carried across the membrane by some sort of ferryboat mechanism.

It may be a bit difficult to understand how

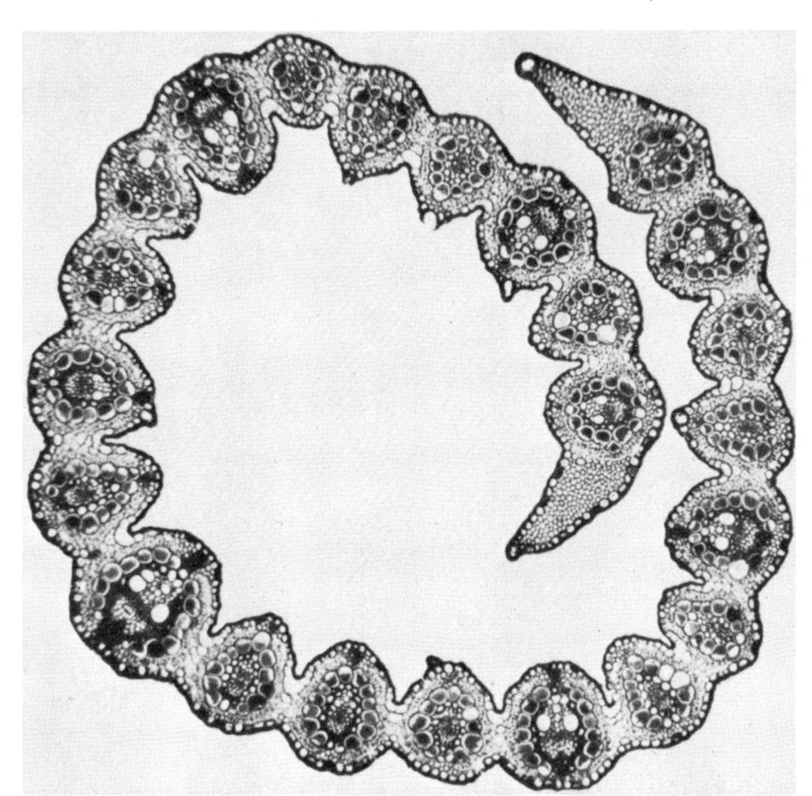

Fig. 11.11 Photomicrograph of a section cut across a rolled up grass leaf. The ridged upper surface is on the inside of the tube formed by the rolling. The large white cells in the depressions between ridges rapidly lose their turgor pressure when there is a water deficit and so cause the rolling. Since stomata are mostly in the upper epidermis, the increase in vapor pressure of the air in the almost-closed tube reduces the rate of transpiration. (Copyright by the General Biological Supply House, Chicago.)

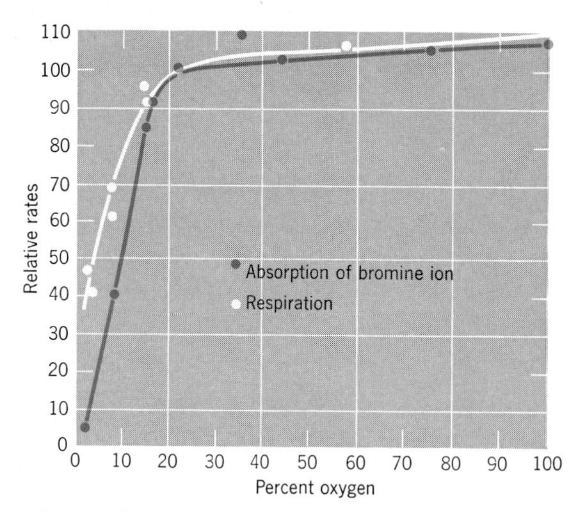

Fig. 11.12 Influence of the rate of respiration in roots on the rate of absorption of bromine ions. The rate of respiration was controlled by varying the oxygen content of the air bubbled through the solution in which the roots were growing. (Data of F. C. Steward, Protoplasma, 18, *208–242, 1933.)*

Table 11.2 **Ion Concentration in the Cells of *Nitella clavata* (an Alga) and in the Pond Water in Which It was Growing (Data of Hoagland & Davis, *Protoplasma* 6,611, 1929.)**

Ion	Ion, Concentration, Milliequivalents per Liter [a]	
	Pond Water	**Cells**
Ca^{++}	1.3	13.0
Mg^{++}	3.0	10.8
Na^{+}	1.2	49.9
K^{+}	0.5	49.3
Cl^{-}	1.0	101.1
SO_4^{--}	0.7	13.0
$H_2PO_4^{--}$	0.008	1.7

[a] A milliequivalent of an ion is 0.001 of its gram ionic weight divided by its valence.

a net movement of ions can occur in a direction exactly opposite from that in which they would diffuse, but an analogy may help. A basketball will roll down hill without the application of outside force, expending the kinetic energy stored in it when it was taken up the hill. This is comparable to diffusion. The ball will not roll up hill any more than a substance can diffuse from a region of lower to higher diffusion pressure. The ball can, however, be moved to the top of the hill by the expenditure of an outside supply of energy, just as ions can be moved from a lower to higher diffusion pressure by the expenditure of energy from respiration.

The "pumping" of ions against a diffusion pressure gradient does not always occur inward. At least in some marine plants sodium ions may be pumped out of the cells, with the result that they have a lower concentration in the cells than in the surrounding seawater. At the same time, the cells may be accumulating other kinds of ions. The various kinds of ions may be accumulated to different degrees (Table 11.2), indicating that somewhat different mechanisms of ion transport may be involved.

Accumulation of ions in vacuoles may be much less important in the general economy of plants than has been assumed. Ion accumulation may occur primarily at the vacuolar membrane, and the plasma membrane may be much more permeable than the vacuolar membrane, permitting relatively free diffusion in and out of the cytoplasm. However, mitochondria and perhaps other cytoplasmic structures also accumulate ions and their membranes are probably similar to the vacuolar membranes as regards permeability and accumulation.

The regions of a plant cell outside the vacuolar and mitochondrial membranes, through which ions may diffuse with considerable freedom, are referred to as **free space.** The cell walls constitute a part of the free space of a cell, but since the intercellular spaces are usually filled with air they are not considered to be a part of free space by some botanists.

Ions and water moving through a tissue, as through the cortex of a root on their way from the soil to the xylem, probably pass through the free space rather than through the vacuole of one cell after another. Movement through free space may be by diffusion, by cytoplasmic streaming, or by mass flow through the cell walls. The ions accumulated in the vacuoles of root cells are apparently held there more or less permanently and are not transported on up to the shoot of the plant (at least not promptly).

Ions and other solutes such as sugars that are in the free space of plant tissues may be washed out by rain. However, ions of essential mineral elements that are leached out are not necessarily wasted since they may again be absorbed by the roots of the plant.

PINOCYTOSIS

We shall briefly mention another way in which large particles to which the cytoplasmic membranes are not permeable may get into cells. This is **pinocytosis,** a process whereby the cell membrane invaginates and thus carries particles that are on its outer surface down into the cytoplasm. The membrane closes over the invaginated portion, which is thus pinched off as a spherical vesicle within the cytoplasm. Later on, the membrane of the vesicle may disintegrate, releasing the enclosed particles into the cytoplasm. Pinocytosis is known to occur in the cells of various animals and fungi, and in some of the cells of vascular plants.

RELATED READING

Bretscher, M. S. "Membrane structure: some general principles," *Science, 181,* 622–629, 1973.

Capaldi, R. A. "A dynamic model of cell membranes," *Scientific American, 230*(3), 26–33, March 1974.

Galston, A. W. "The membrane barrier," *Natural History, 83*(7), 86–90, 1974.

Lee, A. G. "Interactions within biological membranes," *Endeavour 34,* 67–71, 1975.

12
PLANTS AND WATER

Water is one of the essential requirements for the maintenance of plant life. Water constitutes from 80 to 90% or more of the fresh weight of active, living plant tissues. It is necessary for hydration of the protoplasts, the maintenance of cell turgor, as a solvent for most of the substances that participate in the biochemical processes of plants and for their transport, and as a reactant in various biochemical reactions including photosynthesis and digestion. Without an adequate water supply, plants cannot carry on their various life processes, grow, develop, or reproduce. Even desert plants cannot thrive without adequate water. Although seeds, spores, and even entire adult plants of some species can survive for some time with a very low water content, active life, growth and development cannot proceed without an adequate water supply.

In this chapter, we center our attention on the water relations of vascular land plants that absorb water from the soil through their roots, transport it throughout the plant through the xylem and by diffusion from cell to cell, and lose it into the atmosphere by transpiration (Fig. 12.1). For such plants there is a soil-plant-atmosphere continuum of water (SPAC), a fact that should be kept in mind in the following discussions. The water relations of plants that grow submerged in water are substantially less complex than those of land plants. There is no transpiration and water is generally absorbed by any surface of the plant. Of course, algae and other nonvascular plants have no xylem, and even in submerged vascular plants the xylem plays a less vital role than it does in plants with aerial shoots.

THE LOSS OF WATER

Despite the many essential roles that water plays in plants, most of the water absorbed by land plants evaporates from the cells into intercellular spaces and then diffuses through the stomata (and to a lesser degree through the cuticle and lenticels) into the outside air. This diffusion of water vapor from plants is called **transpiration.** Plants may also lose small quantities of liquid water.

The Loss of Liquid Water

When soil water is abundant and the atmosphere is essentially saturated with water vapor, drops of water are frequently found on the edges of leaves. These drops of water have been forced out of the ends of the xylem of the leaf veins under pressure. This process is known as **guttation** (Fig. 12.2). The drops of water exuded by guttation should not be confused with dew drops that are often present on the leaf surfaces at the same time. Dew is water that has condensed from moist air.

Liquid water may also exude from cut or injured stems, a process known as **bleeding.** Bleeding is most common under conditions of high soil moisture and atmospheric humidity, particularly in deciduous plants without leaves. Grapevines may bleed especially profusely under these conditions. A limited amount of water is **secreted** by plants, as in the nectar of some kinds of flowers.

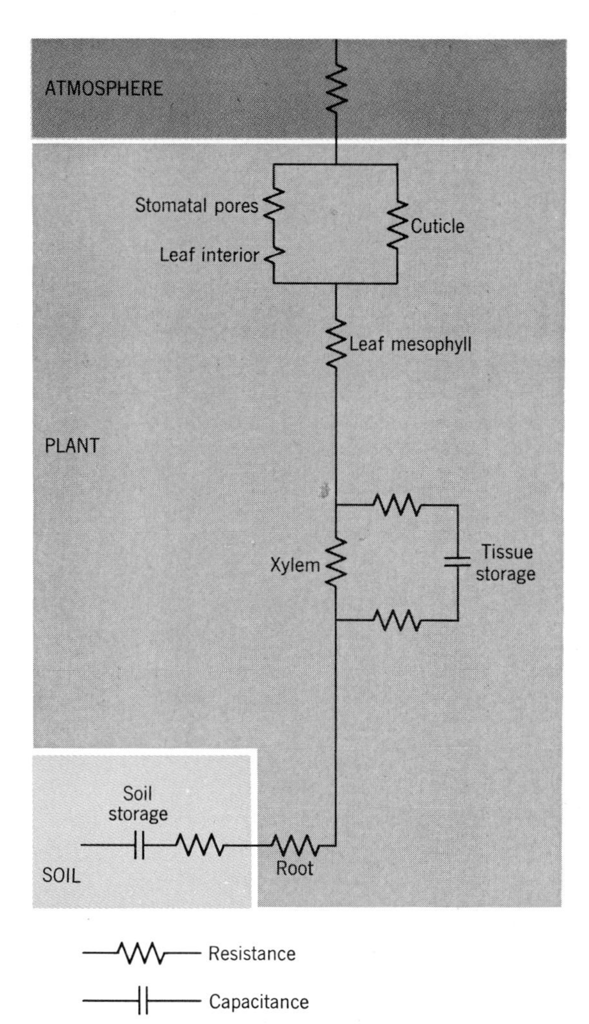

Fig. 12.1 *Diagram comparing the flow of water through the soil-plant-atmosphere continum (SPAC) with the flow of an electrical current along a potential gradient and through a series of resistances and capacitances. Standard electrical symbols are used. (After P. J. Kramer,* Plant and Soil Water Relations, *McGraw-Hill, 1969.)*

Transpiration: The Loss of Water Vapor

Most of the water lost by plants is transpired as water vapor. The quantities of water

Fig. 12.2 Guttation by a strawberry leaf.

lost by transpiration are immense. For example, in an acre of corn the plants transpire about 325,000 gallons of water during a growing season. This is equivalent to about 1300 tons of water, enough to cover an acre to a depth of 11 inches. It is estimated that the trees in an apple orchard transpire enough water during a growing season to cover the orchard to a depth of 9 inches, whereas a good stand of red maple trees may transpire enough water to cover the area under them to a depth of 23 inches.

Factors Affecting the Rate of Transpiration. Plants of two species growing side by side may lose quite different quantities of water by transpiration. This may result from the fact that one species has a greater total leaf area than the other. However, the rate of transpiration per unit area of leaf surface may also differ. Factors such as the diffusion pressure of water in the cells, thickness of the cuticle, size and numbers of stomata, and position of the stomata (whether flush with the epidermis or sunken below the general epidermal surface—Fig. 12.3) may influence the rate of transpiration. However, we are more concerned with the influence of environmental factors on the rate of transpiration.

Transpiration is generally very slow during the night, increases rapidly from early morning to mid-afternoon, and then decreases rapidly (Fig. 12.4). Transpiration is slower on cool, cloudy days than on warm, sunny days. Such variations in the rate of transpiration result from the interplay of two basic factors: (1) stomatal condition (whether the stomata are open or closed), and (2) the difference between the vapor pressure of the water vapor in the intercellular spaces and in the air.

Stomata are usually closed during the night and during the latter part of the afternoon on hot, sunny days, and are open at other times. However, succulent plants, many of which live in deserts and other dry environments, have stomata that are closed during the day and open at night (Fig. 12.9). This greatly reduces their transpiration and is an important factor in their survival, but it also impedes the diffusion of carbon dioxide into the plants at the time photosynthesis is occurring. The CO_2

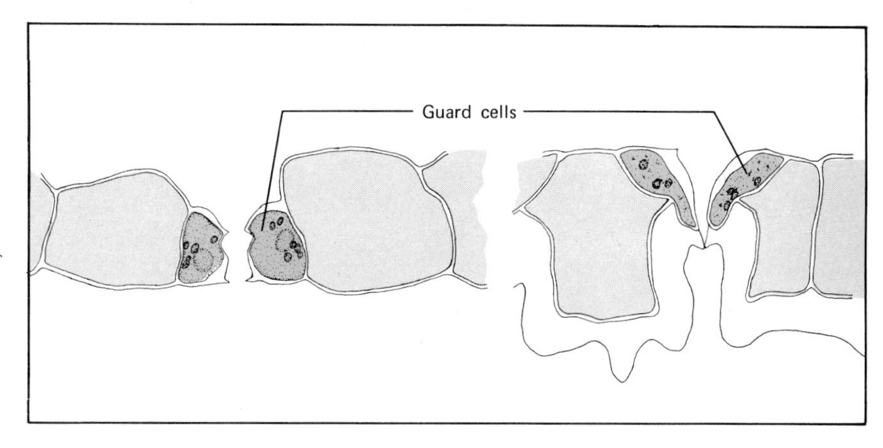

Fig. 12.3 Sections through a surface stoma (left) and a sunken stoma (right) of two different species. Both are from the lower epidermis. The rate of diffusion through a tube is inversely proportional to the length of the tube, so sunken stomata reduce the rate of diffusion of water vapor and other gases. Note also the thick cuticle at the right.

used in photosynthesis diffuses in mostly during the night, is incorporated in organic acids, and is then released from these acids as it is used in photosynthesis.

Stomata provide a very effective diffusion pathway. Stomatal openings generally occupy only 1 to 3% of the total leaf surface area, but from 50 to 75% as much water vapor diffuses through them as would diffuse from a water surface equal in area to the total leaf surface. This high diffusive capacity of stomata is particularly important in permitting substantial diffusion of carbon dioxide into leaves as it is being used in photosynthesis.

When stomata are closed the rate of transpiration is very low, since little transpiration generally occurs through the cuticle. When stomata are open the rate of transpiration is generally determined by the difference between the vapor pressure in the intercellular spaces of the leaf and in the outside air (the vapor pressure gradient, VPG). However, the degree of stomatal opening is a factor, especially with wind (Fig. 12.5). The vapor pressure in the intercellular spaces is always high (at or near the saturation pressure—100% relative humidity) because the wet cell walls provide a large evaporating surface in comparison with the volume of air in the intercellular spaces. The concentration of water vapor in the atmosphere (and so its vapor pressure) fluctuates more widely, but most of the time the VP of the atmosphere is lower than that in the intercellular spaces and so transpiration occurs.

The concentration of water vapor in the air is generally much higher in humid regions such as forests and areas near large bodies of water than in dry regions such as short grasslands and deserts, so the rate of transpiration is likely to be higher in the latter. Also, as is

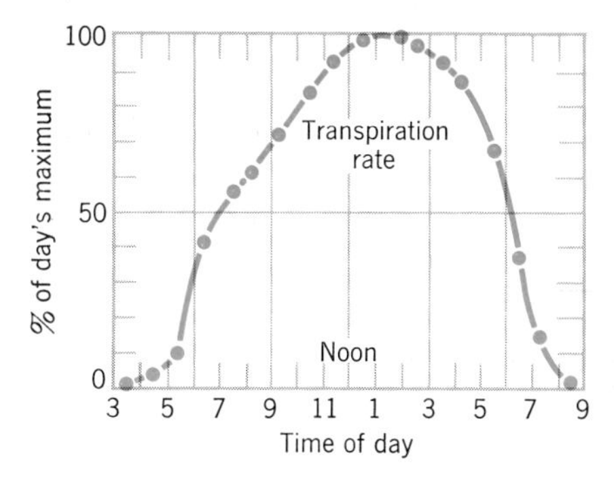

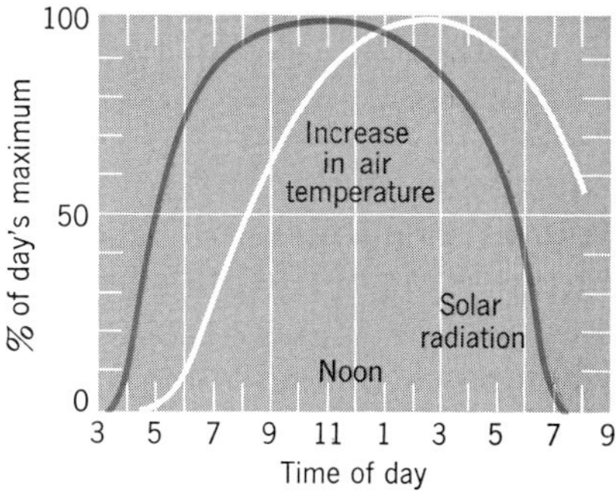

Fig. 12.4 Changes in the rate of transpiration of alfalfa plants in a field during an 18-hour period, compared with changes in temperature and the intensity of solar radiation. (Data of Briggs and Shants, Journal of Agricultural Research, 5, *583, 1916.)*

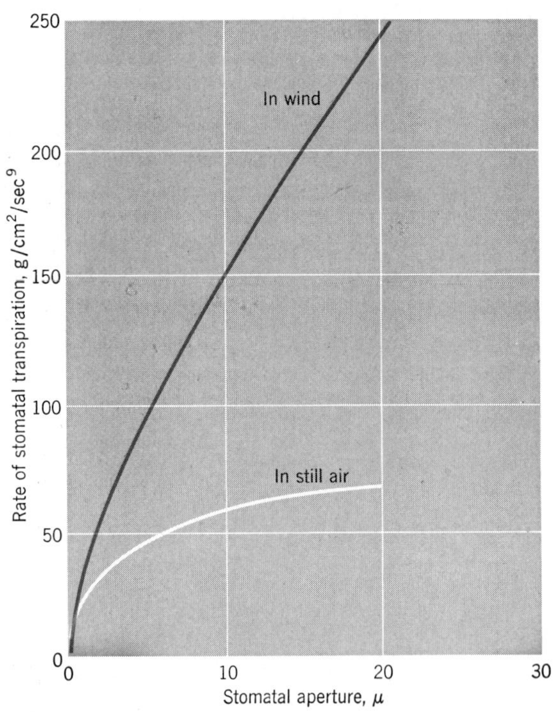

Fig. 12.5 Influence of the width of stomatal opening on the rate of transpiration of Zebrina *in still air and wind. (Data of G. G. J. Bange,* Acta. Bot. Neer., 2, *255, 1953.)*

widely known, the water vapor concentration of the air at any one location can fluctuate greatly from time to time, and so influence the VPG and the rate of transpiration. However, temperature has a much more important effect on the vapor pressure gradient than the concentration of water vapor in the atmosphere.

This is *not* because temperature increases the rate of movement of the water molecules, because this occurs both in the air and in the leaf. The increase in the VPG (and thus in transpiration) with a rise in temperature does result from the following: (1) The saturation vapor pressure (VP at 100% relative humidity) increases with temperature. (2) As the saturation VP in the air of the intercellular spaces increases, so does the water vapor concentration, since evaporation from the wet cell walls is adequate for keeping the air in the intercellular spaces at or near 100% relative humidity. (3) There is little or no increase in the water vapor concentration in the outside air with the increase in saturation vapor pressure because the atmosphere is expansible and the evaporating surfaces are small in comparison with the great volume of the atmosphere. (4) The marked increase in water vapor concentration in the intercellular spaces, and the lack of such an increase in the outside air, result in a marked increase in the VPG and so in the rate of transpiration (Fig. 12.6, Table 12.1). The steepness of the VPG is the factor that affects the rate of transpiration so greatly.

Stomatal Opening and Closing. Stomata are open when the guard cells are turgid, and closed when the guard cells have little or no turgor pressure. At first glance this situation

Fig. 12.6 Typical changes in the vapor pressure in the intercellular spaces of leaves and in the outside air as temperature increases. The leaves are in the shade and so have essentially the same temperature as the surrounding air. The marked increase in vapor pressure in the intercellular spaces with temperature results because of the large evaporating surface (wet cell walls) in comparison with the volume of air, so saturation vapor pressure is essentially maintained as it increases with temperature.

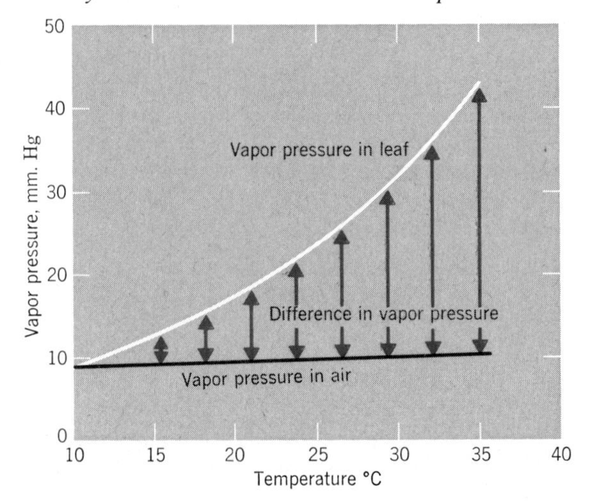

Table 12.1 **Random Examples Illustrating the Influence of an Increase in Temperature During a Day on the Increase in the Vapor Pressure Gradient Between the Intercellular Spaces of a Leaf and the Outside Air**

	Leaf in Shade				Leaf in Sun	
	Morning		Afternoon		Afternoon	
	Leaf	Air	Leaf	Air	Leaf	Air
Temperature °C	15	15	25	25	35	25
Relative humidity	100%	90%	100%	50%	100%	50%
VP, mm Hg[a]	12.79	11.51	23.76	11.88	42.18	11.88
VPG, mm Hg	1.28		11.88		30.30	

[a] The vapor pressures were secured from a table that gives the actual vapor pressures at various temperatures and relative humidities.

may seem to be just the reverse of what it should be, but guard cells of different types have a variety of structural features that provide the basis of this turgor mechanism. Many species of plants have guard cells with thickened cell walls adjacent to the stomatal opening (Fig. 12.7). As turgor pressure increases, the thinner parts of the cell wall are stretched more than the thickened parts, causing the thickened areas to cup inward and thus widen the stomatal pore.

The mid-afternoon closing of stomata on a hot, sunny day results from a water deficit in the plant and the consequent loss of turgor pressure in the guard cells. Once the stomata have closed, the reduction in transpiration may result in the guard cells regaining turgor and opening again, at least for a while.

The opening of stomata with the appearance of light and their closing in the dark, however, has a much less obvious explanation and has puzzled plant physiologists for a long time. Many investigators have worked on this problem during the past half century and one theory after another has been proposed, only to be found lacking upon further investigation. It does seem clear that the first event in the sequence is the beginning of photosynthesis in the guard cells with the appearance of light, and that in some way or another this results in an increase in the solute content of the guard cells, thus reducing the water potential and so causing water to enter the guard cells, increasing their turgor pressure. The first and most obvious suggestion was that the sugar produced by photosynthesis was responsible for

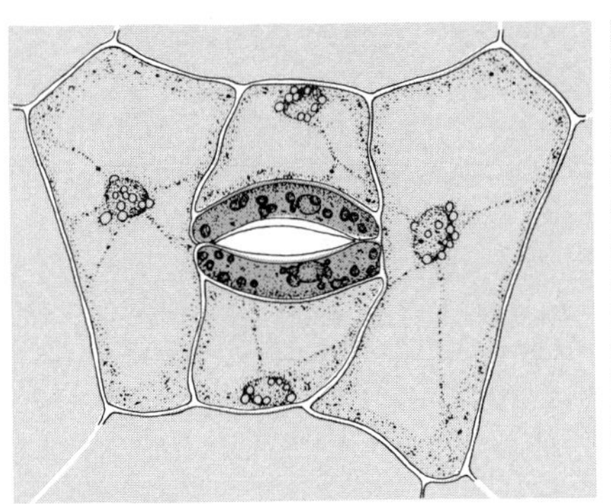

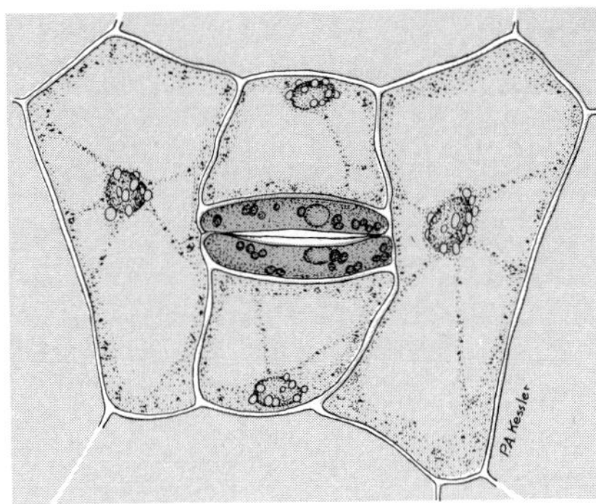

Fig. 12.7 Open and closed stomata of Zebrina. *Note the thicker walls of the guard cells adjacent to the stomatal pore.*

the increase in solute concentration, but this idea had to be discarded when it was found that too small an amount was produced too slowly to account for the rapid opening of stomata.

The fact that vapors of weak acids and high concentrations of CO_2 (which react spontaneously with water forming carbonic acid) cause stomates to close, while weak bases cause closed stomata to open, suggests that photosynthesis in the guard cells exerts its effect by reducing the CO_2 content and thus the acidity of the guard cells. Since the enzyme starch phosphorylase catalyzes the conversion of glucose-1-phosphate to starch at relatively high acidity (pH 5) and the reverse reaction at relatively low acidity (pH 7), it was proposed that the decrease in acidity brought about by photosynthesis caused starch to be converted to sugar, thus increasing the solute content of the guard cells. When it was realized that the phosphate used in making the glucose-1-phosphate from starch was as effective as the glucose-1-phosphate produced in reducing the diffusion pressure of water, this relatively simple and attractive theory had to be discarded.

It now appears that at least two additional enzymatic steps are essential for a workable theory — the conversion of the glucose-1-phosphate to glucose-6-phosphate and then the breaking down of this into glucose and phosphate, which would increase the total solute concentration of the guard cells. If this is actually the case, then the reverse reaction resulting in stomatal closure in the dark would require a supply of ATP from respiration or photosynthesis for producing glucose-1-phosphate from glucose and phosphate preliminary to the use of the glucose-1-phosphate in making starch. Whether this or some other system is involved is not known at present, but the point is that an apparently simple process such as the opening and closing of stomata can involve a complicated series of interrelated reactions (Fig. 12.8).

A series of cause and effect relations such as this illustrates the complexity of superficially simple plant responses and the dependence of an acceptable theory on the accumulation of a wide variety of research data. Contrast this *cause and effect* explanation with some of the *teleological* explanations of stomatal movement sometimes seen in print: that stomata open in the morning "so that the plant can secure carbon dioxide for photosynthesis" and that stomata close at night "in order to save water." Such teleological explanations, crediting the plant with intelligent and purposeful behavior, are easy to formulate but totally inadequate in explaining plant responses. Teleological explanations get the cart before the horse by converting a possible result into a cause. If plant physiologists were satisfied with teleological explanations for plant behavior, research aimed at discovery of the actual course of events would cease.

Many other teleological explanations of plant processes and behavior are often given, for example, "Plants bend toward the light in order to get more light for photosynthesis." "In the dark potato sprouts grow long because they are seeking the light." Such statements not only credit plants with intelligence, but also with considerable information about processes such as photosynthesis, and in the case of the potato sprouts, memory that there is such a thing as light. Teleological explanations of plant behavior are just as absurd as teleological explanations of physical phenomena would be, for example, "Water runs downhill in order to get to the ocean."

Wilting

The loss of water by transpiration may at times result in the loss of turgor by cells of the plant and the consequent **wilting** of the plant. Two types of wilting can be distinguished. **Permanent wilting** occurs when all soil water the plant can absorb has been exhausted. It is re-

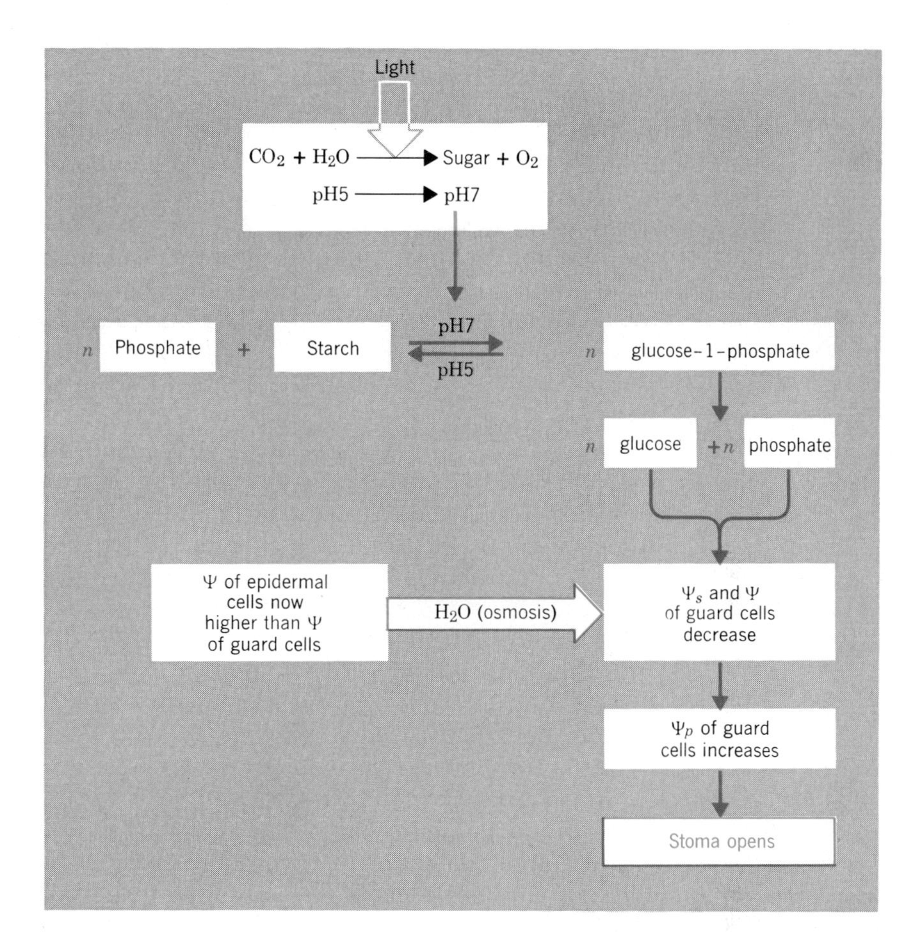

Fig. 12.8 Outline of one theory of how light causes stomatal opening. All the events occur in the guard cells, except for the osmotic movement of water into them from adjacent epidermal cells. See the text for further discussion.

ally permanent only insofar as the plant will not recover from wilting unless water is added to the soil. **Temporary wilting** occurs when there is adequate soil water but when the rate of transpiration exceeds the rate of water absorption (Fig. 12.9). Plants recover from temporary wilting when the rate of transpiration falls below the rate of absorption (Fig. 12.10). Temporary wilting is common on hot, sunny days. If a plastic bag is placed over a wilted plant it will recover from temporary wilting.

The Role of Transpiration

Botanists have long been interested in the question of whether or not transpiration plays an essential role in plants, particularly since most of the water absorbed by plants is lost by transpiration. Among the possible roles of transpiration that have been suggested are water and mineral salt transport through the xylem and cooling of leaves and other plant organs.

Although the rate of transpiration is the principal factor determining the rate of water flow through the xylem, both methods of water transport described in the following section could occur without transpiration. Water flow through the xylem would be much slower if transpiration were not occurring, but water would still be transported as fast as it was being *used* by the plant. Although some experimental evidence suggests that an increase in the rate of transpiration results in an increase in the rate of mineral salt absorption and transport, there is also evidence to the contrary. At any rate, it seems probable that mineral salts would be supplied to the cells of the plant rapidly enough and in adequate quantities even if there were no transpiration.

The evaporation of water during transpiration does, of course, contribute toward the heat loss from plants. A leaf exposed to the sun and carrying on transpiration may be as much as 5°C cooler than a similar leaf in which transpiration has been reduced by experimental closure of the stomata. However, the principal means of heat loss by plants is reradiation, and there is also some heat loss by convection. The heat loss by transpiration usually represents less than a third of the total heat loss from a plant, and is most marked in rela-

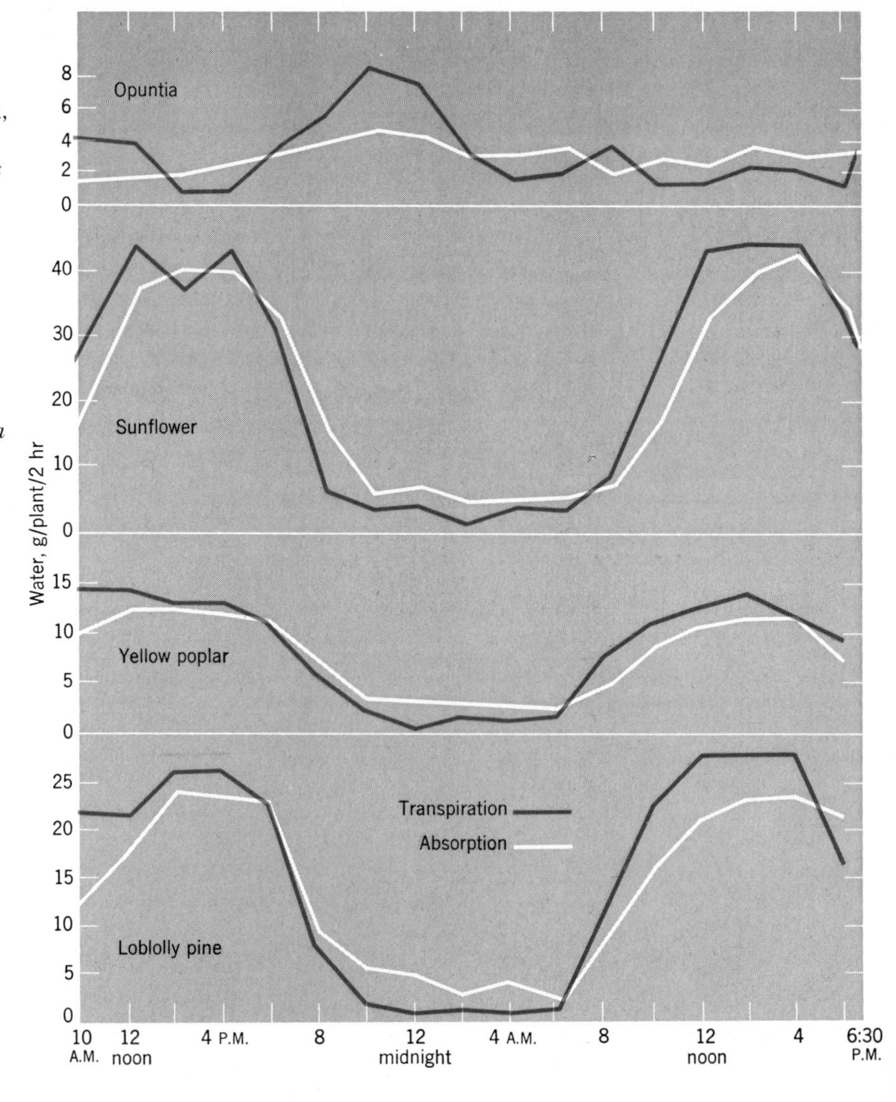

Fig. 12.9 Daily periodicity of transpiration and absorption of water in four species. In all but Opuntia, *note the marked increase in transpiration and the excess of transpiration over absorption during the day (resulting in temporary wilting), and excess of absorption over transpiration during the night. In* Opuntia *the peak in transpiration at night results from the opening of stomata at night and their closing during the day. Note also the low rates of both transpiration and absorption in Opuntia as compared with the other species. (Data of P. J. Kramer,* American Journal of Botany, 24, *12, 1937.)*

Fig. 12.10 Temporary wilting of squash leaves. (Left) Early morning. (Middle) Afternoon of the same day. (Right) The following morning. No water was supplied to the plants during this period.

tively still air. When the air is moving at more than 6 km/hr convection may become more important than transpiration. Whether or not the heat loss by transpiration ever plays an essential role in preventing the heat injury of plants is open to question, but it is quite probable that under conditions of high temperature and bright light it may play a significant role in holding temperatures nearer an optimal level.

It is not necessary, however, to evoke any possible useful role of transpiration to explain why it occurs. Transpiration occurs simply because water vapor has a higher diffusion pressure inside a plant than outside and because there are stomata and lenticels through which water can diffuse. Plants might possibly thrive better if there were no transpiration, since temporary wilting would never occur and permanent wilting would occur much less frequently. A plant without diffusion pathways such as stomata and lenticels would, however, be greatly handicapped, since exchanges of carbon dioxide and oxygen with the atmosphere during photosynthesis and respiration could not occur freely.

The evolution of stomata that could open and close was an important factor in enabling land plants to survive. The closure of stomata when plants are wilting prevents extreme desiccation and perhaps death, while the closing of stomata at night also reduces water loss. On the other hand, open stomata during much of the day permit rapid diffusion of CO_2 into plants as it is being used in photosynthesis. It might have been better if plants had evolved a cuticle permeable to CO_2 and O_2 but not to H_2O, with no stomata present, but evolution did not take this course.

However, man has accomplished something similar by spraying plants with antitranspirants. Some of these substances form a waxy or plastic film over the epidermis that retards water loss, while others bring about more or less permanent stomatal closure. Antitranspirants are used quite widely in dry regions, on nursery plants dug up for shipment, on recently planted trees and shrubs, and on young plants being transplanted.

ABSORPTION OF WATER

Water can enter plants through the leaves, stems, roots, flowers, or fruits, but most of the water absorbed by land plants enters through the roots. Water absorption occurs primarily through the young tissues of the root tip. The root hairs, in particular, provide a very large total absorbing surface in close contact with soil particles (Fig. 12.11). A single rye plant was found to have over 14 billion root hairs with a total surface area of about 370 square meters, and the absorbing surface of the remainder of the root system totaled about 230 m^2. In contrast, the total surface area of the stems and leaves was only 5 m^2. Although most land

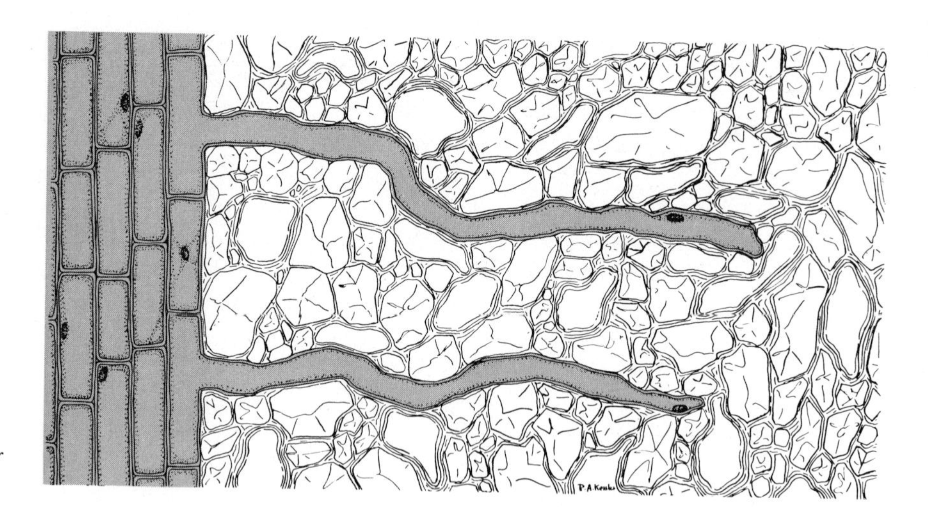

Fig. 12.11 Root hairs are in close contact with soil particles and have a large absorbing surface. Many root hairs are much longer than those shown here.

plants have abundant root hairs when growing in moist (but not flooded) soil, a good many species of trees have few or no root hairs.

It has generally been assumed that no absorption occurs through cracks in the suberized bark of woody plant roots, but Paul J. Kramer of Duke University has found that both water and mineral salts may be absorbed freely through root bark. Since suberized roots generally constitute over 95% of the total root surface of an older tree or shrub, the bark probably provides for a major part of the absorption.

The Method of Water Absorption

Water is absorbed from soils by at least two different methods. One is **active absorption,** which involves the movement of water by osmosis from the soil, through the root tissues, and into the xylem of the root. Active absorption requires a living root system. The other method is **passive absorption,** by which water flows through the free space of the root tissues as a result of tension generated in the shoot of the plant. Passive absorption does not require a living root system, or even any roots at all. (It operates when flowers or branches are placed in a container of water.) The two kinds of water absorption are associated with two different mechanisms of water flow through the xylem, and will be considered in greater detail when they are discussed toward the end of this chapter.

Soils. Since most land plants absorb the greater part of their water from moist soils, some basic information about the nature of soils and the ways they hold water is helpful in understanding water absorption. Most soils are composed principally of weathered rock particles. The size of these rock particles is important in determining the nature and properties of soils, including their water-holding capacities. American soil scientists have agreed to call particles less than 0.002 mm in diameter **clay,** particles 0.002 to 0.02 mm in diameter **silt,** and particles 0.02 to 2.0 mm in diameter **sand.** Although most soils contain particles of all three size ranges, soils are classified as clay, silt, or sand on the basis of the most abundant particle size range. Most soils also contain some **humus,** that is, finely divided plant and animal remains in various stages of decomposition. The gray and black colors of many soils result from their humus content. Although rock particles make up the greater part of most soils, **muck** soils contain more humus than rock particles.

The small particles in a clay soil tend to pack closely together, leaving only very small capillary spaces between them, whereas the much larger sand particles fit together more loosely, leaving larger capillary spaces and also still larger spaces filled with air. Unlike sandy soils, clay soils tend to be hard when dry and sticky when wet. These undesirable features of clay soils can be largely overcome by causing the small clay particles to cling together in irregularly shaped clusters known as **crumbs.** A clay soil with good crumb structure contains more air spaces than a packed clay soil and is much more porous. Good crumb structure may be obtained by adding lime or organic matter to the clay soil. The positively charged Ca^{++} ions from the lime become attached to the negatively charged soil particles, and since each ion can attach to two particles because of its double positive charge it binds the two particles together. Thus, a cluster of particles may be bound together into a crumb. Humus and commercial soil conditioners similarly bind clay particles together into crumbs. Compacting of a clay soil, as by frequent walking over it, results in a loss of crumb structure and thus of porosity of the soil.

The pore spaces between soil particles generally constitute a considerable percentage of the soil volume, from about 30% in clay soils to 60% or so in sandy soils. The smaller pores most commonly hold water by capillary forces and the larger pores are usually filled with air, but all the pores may contain air when the soil is very dry or water when it is very wet. The presence of adequate quantities of air in the soil is as important in the growth of most kinds of plants as the availability of sufficient water (Fig. 12.12), since the rate of respiration in roots decreases greatly when oxygen is defi-

Fig. 12.12 Influence of oxygen concentration on the growth of tomato roots. The plants were raised in separate solution cultures with different oxygen concentrations and then assembled together for the photograph. The percent of oxygen in the air above the solutions was (left to right): 1, 3, 5, 10, and 20.

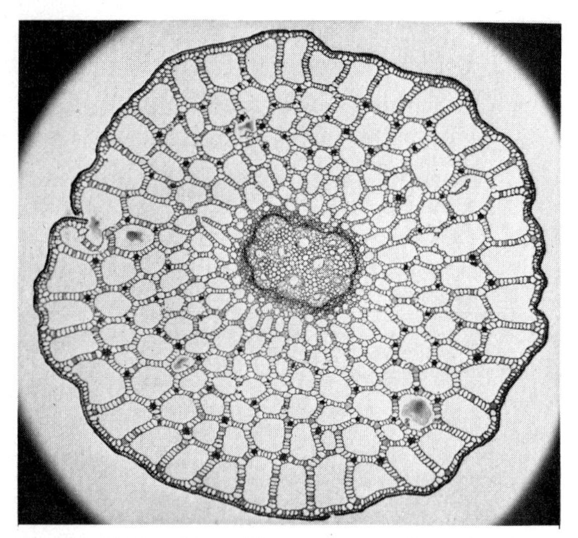

Fig. 12.13 The stems of Potamogeton, *shown in microscopic cross section, and many other water plants have large air spaces in the cortex of their stems through which gases can diffuse readily. (Copyright, General Biological Supply House, Chicago.)*

cient. Decreased respiration, in turn, may result in decreased water absorption, mineral salt absorption, and growth. A relatively few species of plants, such as paddy rice, cattails, and some willows, thrive in water-logged soils. However, most species of plants grow well only in well-aerated soil. Some species of plants that grow in wet situations have large air spaces in their stems and roots through which oxygen can diffuse (Fig. 12.13), whereas others have all their roots near the soil surface. Still others can survive even though the root cells are carrying on anaerobic respiration or a very low rate of aerobic respiration.

Because soil organisms as well as roots are constantly carrying on respiration, soil air generally has a lower concentration of oxygen and a higher concentration of carbon dioxide than the atmosphere.

Soil Water. The water held by soils consists of three principal fractions. First, each soil particle is covered by a film of water only a few molecules thick. This film of water is held by strong attractive forces between the molecules in the particle and the water molecules, and these forces greatly reduce the rate and freedom of movement of the water molecules. Thus the diffusion pressure of the water in the films is so low that it is unavailable to plants. These water films are present even in air dry soil but can be removed by heating the soil in an oven. Because clay particles are smaller and more numerous than sand particles, and so have a much greater total surface area, clay soils hold much more surface water than sandy soils. This fraction is called **hygroscopic water.**

In the smaller pores of a soil, water is held by capillary forces. These capillary forces reduce the diffusion pressure of water only slightly, and **capillary water** is readily available to plants (provided that its diffusion pressure has not been reduced too much by the presence of high concentrations of salts). In the larger pores of a soil the capillary forces are not great enough to hold water against the force of gravity. In a well-drained soil, these larger pores are only temporarily filled with water **(gravitational water)** after irrigation or a rain, becoming filled with air as the water drains out.

Field Capacity and Permanent Wilting Percentage. The water content of a soil is usually expressed as its percentage of the dry weight of the soil. The percentage water content of a soil at the time plants growing in it undergo permanent wilting is the **permanent wilting percentage.** At this point, the plants have absorbed all available capillary water. The permanent wilting percentage varies with the type of soil, ranging from almost 20% in clay soils to as low as 3% in sandy soils. These differences result from the fact that there is much more total particle surface area in a clay soil than in an equal volume of silty or sandy soil, and so more unavailable water is held in the surface films. Despite the great variation in permanent wilting percentage with soil type, permanent wilting occurs in all soils when the water potential of the soil water is reduced to about − 15 bars (Fig. 12.14), regardless of the species of plant used in making the determination.

The percentage water content of a soil when it holds all the capillary water it can, but no gravitational water, is known as the **field capacity** of the soil. In soils with water contents above field capacity, capillary movement of water to adjacent regions of drier soil takes place, but capillary movement essentially ceases once the water content is reduced to field capacity. Since clay soils have more numerous pores of capillary size than sandy soils, they have much higher field capacities than sandy soils. The field capacity of clay soils may be as high as 45%, whereas that of sandy soils may be as low as 5%.

Since field capacity represents the point at which maximal capillary water content is combined with adequate soil aeration, and the permanent wilting percentage represents the point at which the available soil water has been exhausted, these two levels of soil water content are important as far as plants are concerned. The difference between them may be called the **storage capacity** of a soil. Thus, a clay soil with a field capacity of 40% and a wilting percentage of 18% would have a storage capacity of 22%, whereas a sandy soil with a field capacity of 7% and a wilting percentage of 4% would have a storage capacity of only 3%.

Within two days or less after rain or irrigation, movement of water through a well-drained soil brings the water content to field capacity. If more water is then supplied, the moist layer of soil will only temporarily have a water content above field capacity, because the added water is moving through the layer already at field capacity and is bringing an additional layer of soil to field capacity. The final result of a heavy rain, as compared with a light one, is to bring a thicker layer of soil to field capacity, rather than to bring the soil to a higher percentage water content. The dividing line between soil at field capacity and the dry soil below it is quite pronounced, and although the dry soil may be at the wilting percentage, water will not move into it from the adjacent soil that is at field capacity, except for the diffusion of small quantities of water vapor.

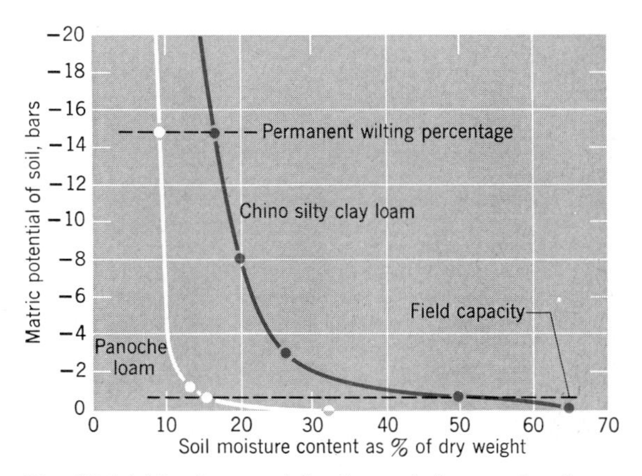

Fig. 12.14 Matric potentials of a sandy loam and a clay loam soil at different soil water percentages. Note that field capacity and the permanent wilting percentage of the two soils occur at different soil water contents but at the same matric potentials. (Data of Wadleigh et al. and of Richards and Weaver.)

As roots absorb water, the water content of the adjacent soil is reduced, perhaps even to the wilting percentage, and there will be no capillary movement of water into this dry soil if the soil next to it is at field capacity or less. However, as the root continues to grow, it enters new soil that may still contain available water—one reason why continued root growth is important in the economy of the plant. The roots are, of course, not searching for water. They are simply growing, and the chances are that they will grow into soil with a higher water content. In dry soil, root growth is greatly restricted (Fig. 12.15).

WATER TRANSPORT

The final aspect of the water economy of plants that we wish to consider is that of water transport in vascular plants. We are concerned here, not with the osmosis of water from one

Fig. 12.15 A box with one glass side was partially filled with dry soil at a slope and then soil at field capacity was added. Note that the roots of the corn seedlings grew throughout the moist soil, but only a very short distance into the dry soil. Note that there was no capillary movement of water from the moist to dry soil.

living cell to another or with the flow of water through the "free space" of plant tissues, but with the flow of water through the tracheids or the vessels of the xylem from the roots to the stems, leaves, and reproductive structures of a plant. Botanists have known for many years that long-distance transport of water through a plant **(translocation** or **conduction)** occurs in the tracheids or vessels of the xylem, but completely satisfactory explanations of the forces responsible for the flow of water have been elusive. Stephen Hales formulated the basic concepts of both the current explanations of water rise in plants early in the eighteenth century, but despite much subsequent research we still have only a slightly better understanding of translocation than he had.

It is obvious that some force must be applied to the water if it is to flow upward. That water flows through the xylem rather than diffusing through it is evident from the fact that water frequently moves as fast as 75 cm per minute in the xylem, much faster than the highest rates of diffusion of liquid water. Any acceptable explanation of water flow through the xylem must account for forces great enough to raise water to the tops of the tallest trees (Fig. 12.16), including redwoods and Douglas firs that are about 120 m high. One bar pressure can support a column of water 10 m high, so 12 bars pressure is required to support a 120 m water column. About 18 bars additional pressure is required to overcome the resistance to the movement of water (friction) through the tissues, so a total force of about 30 bars is essential for lifting water to the tops of the tallest trees.

Since air pressure provides a force of only 1 bar it is clearly inadequate for raising water in a tree of any great size. Furthermore, even in small plants there would have to be a vacuum at the top of the plant and the ends of the xylem in the roots would have to be exposed to the air. Although biology textbooks sometimes give capillary action as an explanation for the rise of water in plants, calculations show that in tubes the diameter of vessel elements or tracheids, the capillary rise is only a few inches to a few feet. The suggestion that water might rise much higher through the much finer capillary spaces within the cell walls was discarded because the bulk of the water flows through the cell cavities rather than through the walls. Several botanists have proposed various theories implicating the living cells of the xylem in translocation, but since water con-

Fig. 12.16 Any generally acceptable theory of the mechanism whereby water rises in the xylem of plants must be able to explain the rise of water to the tops of tall trees such as this one. Some trees are even taller.

tinues to flow through the xylem even when these cells are killed, these theories have not been considered acceptable. Of the various theories of water translocation proposed, only two appear to be even partially plausible. One is **root pressure.** The other is referred to by a variety of names such as the **cohesion theory, transpiration pull,** and the **transpiration-cohesion-tension theory,** but we shall refer to it here as **shoot tension** since this latter term is analogous with the term *root pressure.*

Root Pressure

If the stem of a well-watered plant of certain species, for example, a tomato, is cut off, then water will exude from the stump. When a piece of glass tubing is attached to the stump with a short length of rubber tubing, water may rise in the tube to a height of several feet. The force responsible for this rise of water is referred to as **root pressure.** Root pressure is a rather complicated phenomenon, and all its details are not thoroughly understood. In general, we may state that it results from the continuing movement of water from the soil, through the tissues of the root, and into the xylem, thus forcing the water already in the xylem upward. Root pressure involves more than just the osmotic movement of water from the soil, through the root tissues, and into the xylem. Energy from the ATP produced by respiration is needed, not to move the water, but in the translocation and accumulation of ions by root cells. This results in a gradient of water potential from the soil to the xylem, and so the osmotic movement of water into the xylem. There is active transport of ions into the xylem by adjacent cells. If roots are killed, or even just deprived of oxygen, root pressure ceases. Because living, respiring roots are essential for root pressure, the absorption of water associated with root pressure is known as active absorption.

Despite the fact that root pressure is a real force in some plant species, most plant physiologists do not believe that it offers a satisfactory general theory of water translocation. The magnitude of root pressure may be measured by attaching a manometer to the stump of a plant (Fig. 12.17). Root pressures in excess of 2 bars have rarely been measured and they are frequently much less. At the most, root pressure would seem to be able to support a column of water no more than 20 m high on the basis of these measurements, although Philip White has reported the development of root pressures of 6 bars or more by single isolated tomato roots. Not only does root pressure appear to be inadequate to explain the rise of water to the top of even a moderately tall tree, but also root pressure occurs only under conditions of high soil moisture and low transpiration—not when water is flowing the most rapidly through the xylem. Furthermore, root pressure has never been demonstrated in many species of plants. Finally, the rate of water flow resulting from root pressure is too slow to account for the volume of water frequently flowing through the xylem. Although root pressure probably provides the motive force for the rise of water in some plants part of the time, the root pressure theory apparently is inadequate as an explanation of the rise of water in the xylem of most plants most of the time.

Shoot Tension

The pulling up of water through the xylem by shoot tension is possible because of several basic phenomena. One is that water columns enclosed in tubes such as the vessels or tracheids of the xylem have a high degree of cohesion. A pull or tension of 30 bars or more is required to overcome the cohesive forces between the water molecules and so break the water column apart. In addition to being strongly cohesive, the water also adheres to the walls of the vessels or tracheids, and this is another factor essential to the shoot tension mechanism.

Another requirement for shoot tension is a motive force (or forces) that can exert a pull on the cohesive water columns in the xylem. This motive force results from the loss of water from (or the use of water by) cells of the leaves and other parts of the shoot and the consequent reduction in the diffusion pressure of water in these cells. As water evaporates from cell walls during transpiration, the water content of the cellulose and other colloidal cell wall components is reduced. Since these cell wall colloids have a great affinity for water, water now moves into the wall from the protoplasm and vacuole or from the walls of adjacent cells, reducing its diffusion pressure in these regions. The osmosis of water from one cell to another continues until water is moving out of the cells adjacent to the vessels or tracheids of the xylem. Water now diffuses from the xylem, and because of the cohesiveness of the water columns they are pulled up as water leaves the xylem. Transpiration is the principal factor creating a water deficit in cells, and the rate of movement of water through the xylem is determined largely by the rate of transpiration. Any use of water by a cell (as in photosynthesis), however, creates a water deficit and so may contribute toward the diffusion of water from the xylem. The pull of water through the xylem creates a negative pressure or tension in the water columns.

We have already indicated that forces of around 30 bars are essential for lifting water to the tops of the tallest trees. If shoot tension is

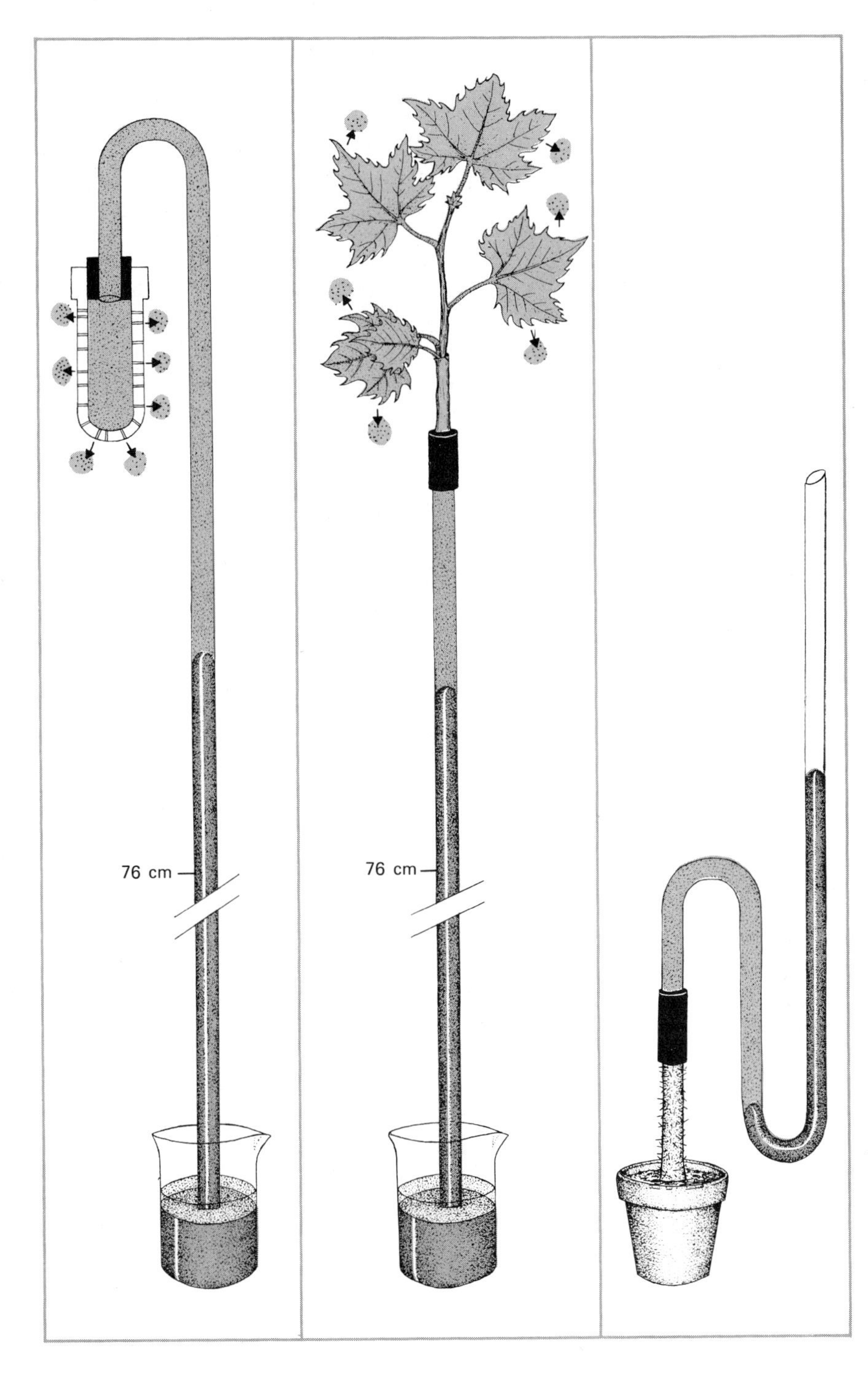

Fig. 12.17 Demonstrations of the cohesion-tension and root pressure mechanisms of water rise in plants. In the cohesion-tension apparatus (left), water is pulled up the tube as it evaporates from the porous clay tube or transpires from a branch. The mercury is pulled up because of the adhesive forces between it and the water. Air pressure contributes toward the rise of mercury to a height of 76 cm only, but in such demonstrations the mercury often rises to heights of 100 cm or more. In the root pressure demonstration the mercury level was originally the same in both arms of the U-tube manometer. The magnitude of the root pressure can be calculated from the difference in height of the mercury in the two arms. (Redrawn from "The Rise of Water in Plants," by Victor A. Greulach, Scientific American, 187(*4*), *78–82, 1952.* *Copyright 1952 by the Scientific American, Inc. All rights reserved.)*

to be considered a generally acceptable theory of the rise of water in plants, water potentials at least as low as −30 bars must exist in the cells of the tallest trees. Measurements show that such water deficits do occur. The water potential in leaf cells of herbaceous plants is often no lower than −10 bars, but even this would be sufficient to raise water 5 m if half the energy is used in overcoming friction. There is little doubt but that the motive forces involved in shoot tension are theoretically adequate to account for the lifting of water to the tops of even the tallest trees.

Shoot tension can be demonstrated by securing a branch of a tree to a long piece of glass tubing filled with water and having its lower end immersed in mercury (Fig. 12.17). As water is pulled up the tubing, it pulls the mercury up behind it. Since atmospheric pressure can support a column of mercury only about 76 cm high, mercury rise beyond that height can be ascribed only to shoot tension. When the apparatus is carefully and properly set up, mercury will rise to a height of 100 cm or more in the tube. The cohesion mechanism can also be demonstrated by a purely mechanical device, evaporation of water from a porous clay tube filled with water substituting for the loss of water from the branch (Fig. 12.17). In both demonstrations it is essential that all connections be airtight and that the dissolved gases in the water be reduced to a minimum (usually accomplished by boiling the water just before use) so that air bubbles will not form as the water is subjected to tension.

Although such demonstrations help support the shoot tension theory, perhaps the most convincing evidence that shoot tension does (as well as can) exist in plants comes from a variety of experiments showing that water in the xylem is commonly under tension. In one type of experiment, the tissues external to the xylem in the stem of a herbaceous plant such as bean are carefully removed, and a vessel element is punctured with a fine needle while it is being observed under a microscope. The water in the vessel will usually snap apart at the puncture. If the water were under pressure it would ooze or squirt out. That water in the xylem is under tension is also evident from the sinking of water into the stump when the trunk of a rapidly transpiring tree is cut down.

Water cannot be pulled up through the xylem below a point where air has entered a vessel and so broken the continuity of the column of water. This may occur as a result of the rupture of a vessel by injury, as a result of excessive tensions, or when flowers or branches are cut from a plant. If cut flowers or branches are recut under water several centimeters above the original cut, the vessels containing air are removed, the water columns are reestablished, and rapid recovery from wilting occurs (Fig. 12.18). Cuttings not recut under water may eventually regain continuity of the water columns (by absorption of the gases or other means) and so recover from wilting, but only after a considerable period of time.

The shoot tension or cohesion theory is not new. It was proposed in 1894 by the Irish scientists, Henry H. Dixon and John Joly, not to mention the fact that Stephen Hales had anticipated the theory in the early eighteenth century. Subsequent work by Dixon and Joly and many other investigators has added support to the theory, and most plant physiologists now agree that shoot tension is the mechanism responsible for the rise of water in most plants most of the time.

While shoot tension is operating, water may be pulled, not only through the xylem but also through the tissues of the roots from the soil. This type of absorption is referred to as **passive absorption,** and is probably the most common type of absorption. In contrast with active absorption, passive absorption does not depend on living root cells. Indeed, if a root system of a potted plant is killed by immersion in boiling water or by other means, the rate of passive absorption may increase markedly be-

Fig. 12.18 Two shoots of Amaranthus *(pigweed) were detached and allowed to wilt severely. The photograph shows them 30 min after they were placed in water. The difference in them resulted from cutting 3 cm from the end of the stem of the plant at the left under water, thus removing the portion of the xylem filled with air and so reestablishing the continuity of the water columns in the vessels.*

cause the dead tissues offer less resistance to the movement of water through them than do living tissues. Absorption ceases, however, when the dead root tissues disintegrate.

RELATED READING

Gates, D. M. "Heat transfer in plants," *Scientific American, 213*(6), 76–84, December 1965.

Sutcliff, J. *Plants and Water.* New York: St. Martin's Press, 1968.

Waggoner, P. E. and I. Zelitch. "Transpiration and the stomata of leaves," *Science, 150,* 1413–1420, 1965.

Zelitch, I. "Control of leaf stomata—their role in transpiration and photosynthesis," *American Scientist, 55* 472–486, 1967.

Zimmermann, M. H. "How sap moves in trees," *Scientific American, 208*(3), 133–142, March 1963.

13 PLANTS AND MINERALS

Many people have been intrigued with **hydroponics,** the cultivation of plants in solutions containing the essential mineral elements, either as a hobby or as a means of commercial crop production (Fig. 13.1). In some hydroponic facilities the plants are supported in one way or another so that their roots are growing in the solution, some means of aerating the water usually being provided. In other types of facilities the plants are raised in some substance such as gravel, cinders, sand or vermiculite and are watered with the mineral solution, thus simplifying the problem of both support and oxygen supply to the roots. Despite the claims of some hydroponic enthusiasts, crops raised by hydroponics are not necessarily better than those raised in good fertile soil, but hydroponics has proved to be a desirable cultural technique in greenhouses and in experiments where it is necessary to control the environment of plants as precisely as possible. During and after World War II the armed forces raised vegetables by hydroponics on some of the small coral islands where there was poor and scanty soil and in other areas where the soil was contaminated with worms and other human parasites.

As a cultural practice, hydroponics developed only during the second quarter of the present century, but plant physiologists

Fig. 13.1 Tomato plants raised in a greenhouse by the use of hydroponics.

have long used similar techniques in experiments designed to determine which mineral elements are essential for plants and in what quantities. These experiments have provided the information needed for formulating hydroponic solutions containing all essential elements, and more important, they have provided information of great value to farmers and gardeners as regards the use of fertilizers in maintaining soil fertility.

DEVELOPMENT OF CONCEPTS OF MINERAL NUTRITION

Up to the beginning of the nineteenth century, practically nothing was known about the mineral nutrition of plants. Van Helmont's conclusion, early in the seventeenth century (Chapter 14), that plants were composed entirely of water, was for some time accepted by those who did not continue to cling to the old Greek idea that plants derived all their substance from the soil. One of the first suggestions that mineral salts might be essential for plants came from J. R. Glauber, a German chemist, in 1656. He found that potassium nitrate applied to the soil increased plant growth and concluded that this substance was the "essential principle of vegetation." In 1699, John Woodward raised spearmint plants in rainwater, Thames River water, conduit water, and conduit water plus garden mold, and found that growth was proportional to the quantity of dissolved material present. He concluded that van Helmont was wrong, and that an unknown earthy substance as well as water entered into the composition of plants. At about the same time, Stephen Hales was conducting experiments on plant nutrition, but the real beginning of our modern knowledge of the mineral nutrition of plants was in the early nineteenth century. At the same time that increasing knowledge of photosynthesis was making it clear that the organic constituents of plants were derived from the carbon, oxygen, and hydrogen of carbon dioxide and water, our present-day knowledge of the mineral nutrition of plants was beginning to be pieced together.

In 1804, Theodore de Saussure provided some of the basic concepts of the mineral nutrition of plants, as well as of photosynthesis (Chapter 14). He found that the nitrogen of plants came from the nitrogen compounds in the soil rather than directly from the air, that good plant growth could not occur without mineral elements, that plants absorb mineral substances in ratios different from those in which they occurred in the soil, and that plants absorbed substances regardless of whether they were useful, useless, or even poisonous. It was not until 1842 that other investigators began reporting on the mineral nutrition of plants. By 1861, considerable information had been secured on what elements are essential for plant growth. The technique of raising plants in wax-coated pots containing well-washed sand or pulverized quartz and supplying the plants with solutions of mineral salts lacking only one of the elements suspected of being essential gradually made it evident that the essential elements included phosphorus, potassium, nitrogen, calcium, sulfur, and magnesium. These are now known as the **major elements,** since they are used by plants in relatively large quantities (Table 13.1). Whether or not still other elements were essential in plant nutrition could not be determined by the experimental techniques than available, and it was frequently assumed that only these elements were essential.

The first step toward a more complete understanding of the mineral nutrition of plants was taken by the German plant physiologist Julius von Sachs, who in 1860 introduced the method of raising the experimental plants with their roots immersed in the mineral salt solutions, thus eliminating the possibility of introducing unwanted elements as contaminants of the sand or quartz. It was not until the present century, however, that it became clearly evident that a number of elements—iron, boron, copper, manganese, molybdenum, chlorine, and zinc—are essential in very small quantities. These are referred to as the **minor** or **trace elements.** Demonstration of the essential nature of the trace elements could not be made until very pure chemical compounds were available (Fig. 13.2), and extreme precautions had to be taken to make sure that the culture solutions were not contaminated by substances from the pots and glassware used or even by dust from the air. The seeds sometimes contained enough of a trace element to permit normal growth to maturity, and it was necessary to use seeds from these experimental plants in a second experiment before it could be determined that the element was essential. Although we are quite certain that all the major elements are known, additional essential trace elements may still be discovered. At present there is doubt as to whether or not sodium and cobalt are essential trace elements for all plants. Some fungi have been reported to require traces of various rare elements such as gallium, scandium, and vanadium, and it is at least possible that these or other elements may eventually be found to be essential to higher plants.

Associated with the development of our fundamental knowledge of the mineral nutrition of plants has been the development of much practical information about fertilizers and soil fertility that has been of incalculable value to farmers and horiculturists. Much of the research in agricultural experiment stations deals with soil fertility and the development of better fertilizers for specific soils and crops. One of the first scientists to stress the importance of soil fertility was the German chemist Justus von Liebig, who between 1840 and 1873 devoted much effort to convincing agricultural scientists and farmers of the importance of replenishing the mineral elements of the soil, and whose researches along these lines have led to his designation as the father of agricultural chemistry.

MINERAL DEFICIENCY SYMPTOMS

Plants lacking adequate quantities of one or more of the essential elements usually develop characteristic **mineral deficiency symptoms** (Fig. 13.3) that can be used by experts to determine which element or elements are deficient (Table 13.1). Diagnosis is frequently complicated by the fact that certain plant diseases, particularly those caused by viruses, have symptoms similar to some mineral deficiency

Table 13.1 **Information about Elements Essential for Higher Plants**

Element	Chemical Symbol	Relative Quantity Required	Important Roles	Deficiency Symptoms
Nitrogen	N	1,000	Component of amino acids, proteins, nucleic acids, coenzymes, ATP, chlorophyll, etc.	Stunted growth, chlorosis, leaf fall, anthocyanin formation; older leaves most affected
Potassium	K	250	Enzyme activator, required for protein synthesis, stomatal function	Necrosis, mottled chlorosis, weak stems; older leaves most affected
Calcium	Ca	125	Middle lamella component, membrane structure and permeability, α-amylase component	Necrosis of root and shoot tips, stunted roots; younger leaves most affected
Magnesium	Mg	80	Chlorophyll component, enzyme activator, maintains ribosome structure	Mottled leaf chlorosis, leaf tips turned up; older leaves most affected
Phosphorus	P	60	Component of nucleic acids, nucleotides, phospholipids, sugar phosphates, ATP, coenzymes, etc.	Stunted growth, dark blue-green leaves, anthocyanin, maturity delayed; older leaves most affected
Sulfur	S	30	Component of proteins, thiamine, biotin, CoA	Chlorosis of young leaves between veins
Chlorine	Cl	3	Activates photosynthetic enzymes	Chlorosis, necrosis, and wilting of leaves, stunting and thickening of young roots
Iron	Fe	2	Component of cytochrome and ferredoxin, enzyme activator, chlorophyll synthesis	Chlorosis of young leaves between veins, short and slender stems
Boron	B	2	Carbohydrate translocation, prevents phenolic acid accumulation and toxicity	Black necrosis of stem and root tips, swollen roots, leaves twisted
Manganese	Mn	1	Enzyme activation, catalyst, electron carrier	Chlorosis except along smallest veins, necrosis between veins; younger leaves most affected
Zinc	Zn	0.3	Synthesis of IAA and chlorophyll, component of some enzymes	Chlorosis, stunted leaves and internodes, distorted leaf margins; older leaves most affected
Copper	Cu	0.1	Nitrate reduction, electron carrier, component of some coenzymes	Young leaves distorted, dark green, wilted
Molybdenum	Mo	0.001	Electron carrier, nitrogen fixation	Chlorosis, necrosis, and distortion of young leaves

symptoms. Toxic gases, such as those in smog, or the sulfur dioxide produced during some manufacturing processes, may also give rise to symptoms in plants resembling mineral deficiency symptoms. In general, the following are the most common and striking types of mineral deficiency symptoms:

1. Stunted growth. Although a deficiency of almost any essential element may stunt plant growth, growth is usually retarded the most by

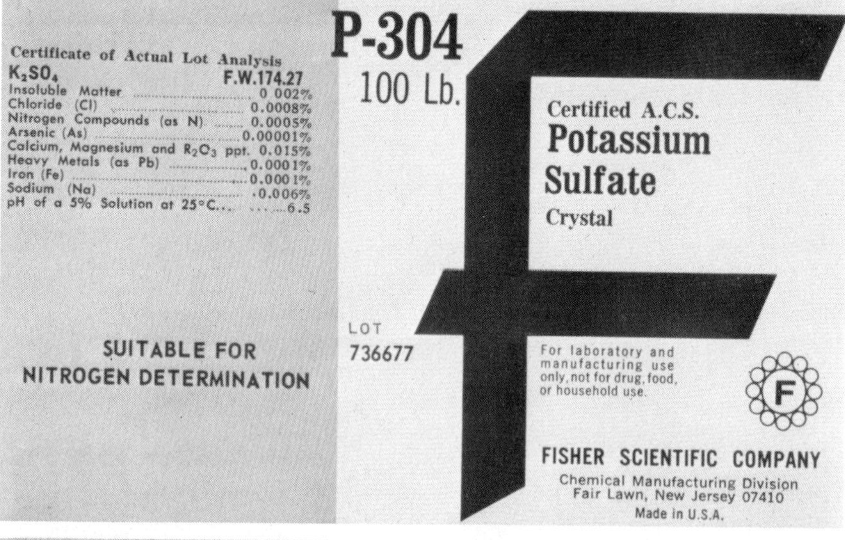

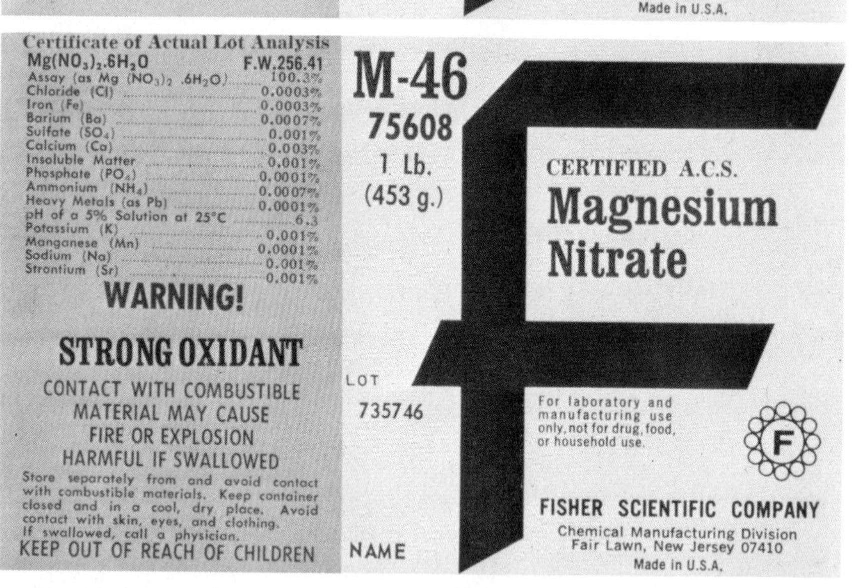

Fig. 13.2 Even high quality A.C.S. chemicals suitable for most laboratory uses contain enough impurities to make them unsuitable for determination of the trace elements required by plants, or even demonstration of many trace element deficiency symptoms.

a deficiency of nitrogen, phosphorus, potassium, calcium, magnesium, or boron (Fig. 13.3). Plants completely deprived of nitrogen will hardly grow at all.

2. Chlorosis. One of the most common mineral deficiency symptoms is **chlorosis,** the pale green or yellow color resulting from a reduced chlorophyll content. Chlorosis may be uniform over all the leaves, present in only the older or younger leaves, or present only along the veins or between the veins (Fig. 13.4). The pattern of chlorosis depends on which element is deficient and so provides one of the better diagnostic symptoms. Since both nitrogen and magnesium are components of the chlorophyll molecule it is obvious why deficiencies of these elements cause chlorosis. Other elements like iron, though not a part of chlorophyll, are essential in its synthesis and so chlorosis results when they are lacking. Plants lacking the

Fig. 13.3 Tobacco plants showing deficiency symptoms resulting from the lack of seven essential elements. The check plant (Ck) was supplied all the essential elements, and each of the others received all the essential elements except the one shown on the label. All the plants are the same age and variety. The plant lacking Ca had green, but small and distorted leaves. The plant lacking P had abnormally colored dark blue-green leaves, and only the older leaves were chlorotic.

element phosphorus do not develop chlorosis initially, but instead have an unusual dark bluish-green color.

3. Necrosis. The death of parts of a plant, **necrosis,** is another common mineral deficiency symptom. For example, scattered spots of dead tissue may occur in leaves if iron is deficient, leaf tips and edges may die when phosphorus is lacking, and roots and buds may die and turn black when boron is deficient (Fig. 13.5).

4. Anthocyanin formation. The formation of anthocyanin, and a consequent red color, in plant structures that usually do not produce this pigment may indicate a mineral deficiency, particularly N or P.

Fig. 13.4 Left: leaf from a sunflower plant deficient in iron. Center: leaf from a sunflower plant supplied with all essential elements. Right: leaf from a sunflower plant supplied with iron and with an excess of manganese, that made the iron unavailable.

Fig. 13.5 *Three redbeets deficient in boron have extensive black necrotic areas on their roots. The redbeet at the left was supplied with an adequate amount of boron.*

5. Stem symptoms. Unusually slender, woody stems may be developed by herbaceous plants lacking nitrogen, phosphorus, potassium, or magnesium.

6. Poor reproductive development. Mineral deficiencies, particularly of nitrogen, phosphorus, calcium, and potassium, may result in the development of unusually small and light seeds and fruits or even a complete failure to produce fruits and seeds.

Chemical Tests for Mineral Deficiencies

Both plants and soils may be analyzed chemically to determine how much of each essential mineral element they contain. The various state agricultural experiment stations will usually make such tests, at least for nitrogen, phosphorus, and potassium, for residents of their states at little or no cost. Accurate chemical analyses can be made only by experts, but kits for making approximate analyses of the N, P, and K content of plants and soils are available in the larger seed stores and may provide more reliable diagnostic information to the amateur than mineral deficiency symptoms.

ROLES OF MINERAL ELEMENTS IN PLANTS

The question now arises as to how mineral salts are used in plants, and as to how such uses are related to mineral deficiency symptoms. One important use of mineral elements is in the synthesis of various chemical compounds, many of them essential to the life of the plant. For example, nitrogen is an essential component of all amino acids, and so of proteins. Nitrogen is also a constituent of the molecules of chlorophyll, nucleic acids, and alkaloids as well as of many coenzymes, hormones, and vitamins. Calcium is a constituent of the calcium pectate of the middle lamella, and we have noted that magnesium is an essential part of the chlorophyll molecules. Proteins contain sulfur, as do vitamin B_1 and coenzyme A. Phosphorus is a constituent of many important organic compounds such as ATP, NAD, NADP, DNA, RNA, phosphorylated sugars, and phospholipids. Coenzymes involved in oxidation-reduction reactions have metallic elements as essential constituents of their molecules, with iron, magnesium, copper, zinc (Fig. 13.6), molybdenum, or cobalt being among the elements present in the molecules of one or more of the various coenzymes. For example, iron is a component of the various cytochromes. Some of these metallic elements have more than one valence state, so their ions can serve as electron acceptors and donors. Many of the other organic compounds synthesized by plants have mineral elements as essential constituents of their molecules. However, the abundant carbohydrates, fats, waxes, and some other compounds such as xanthophylls are composed only of carbon, hydrogen, and oxygen, and a few compounds such as carotenes and turpentine are composed entirely of carbon and hydrogen. Among the major elements, potassium is the only one that is not a stable component of the molecules of one or more organic compound.

The second major role of mineral elements is serving as enzyme activators. These are not stable components of the enzyme molecules, but they do form temporary complexes with the enzyme proteins. They do not function as electron acceptors and donors, but rather appear to be essential for maintaining the proper structural configuration of the enzyme proteins. Among the elements that serve as enzyme activators are calcium, magnesium, manganese (Fig. 13.7), chlorine, sodium, and potassium. Numerous enzymes require univalent cations as activators, and in most cases potassium ions are effective. In some cases, but by no means all, other univalent cations can substitute for potassium. The extensive use of potassium as an enzyme activator probably explains why it is essential in such large quantities, even though it is not a stable component of organic molecules.

Mineral salts play several other roles in plants. The ions and molecules of salts contribute substantially toward the osmotic potential of cells, as do sugars and other organic solutes. The ions formed when salts dissolve in water

Fig. 13.6 (Top) Zinc deficiency symptoms in tomato plants. Note the marked reduction of growth, even though zinc is a trace element required in only small amounts.

Fig. 13.7 (Bottom) Another example of severe mineral deficiency symptoms because of the lack of a trace element (manganese).

have many different effects on protoplasm. For example, they influence the amount of water that can be imbibed and affect the degree of permeability of cell membranes. Different ions may have opposing effects, for example sodium ions (Na^+) increase membrane permeability and calcium ions (Ca^{++}) decrease membrane permeability.

A. Wallace and D. Van Noort of UCLA have reported that soon after iron-deficient *Chlorella* plants were supplied with iron, considerable quantities of a new type of RNA were synthesized, and they suggested that this RNA might in turn code the production of an enzyme essential for chlorophyll synthesis. If this should be the case it would represent quite a different mode of mineral element action than had previously been suspected, in addition to clarifying the role of iron in chlorophyll synthesis.

TRANSLOCATION OF MINERAL SALTS

The absorption of mineral salts was discussed in Chapter 11, but we have not considered the translocation of mineral salts through plants. Like water and foods, mineral salts are translocated through the vascular tissues, but there has been considerable doubt as to whether salts are translocated through the xylem or the phloem or both. For many years it was considered that salts were translocated upward through the xylem and downward through the phloem, but during the second and third decades of this century at least one group of plant physiologists accumulated considerable data suggesting that most mineral transport might be through the phloem. When radioactive elements became available for use as tracers (Fig. 13.8), much research was conducted on the translocation of mineral salts, and as a result it now appears clear that the bulk of the salts are translocated upward through the xylem, although translocation in both directions occurs through the phloem. Tracer studies show that salts readily pass back and forth between the xylem and phloem. Most mineral elements can be translocated both upward and downward in plants; however, a few like calcium can be translocated only upward. Like water and foods, mineral salts are translocated much faster than they can diffuse. In both the phloem and the xylem the salts are dissolved in water and the forces involved in their transport are presumably those described for water (Chapter 12) and food (Chapter 14).

THE NITROGEN CYCLE

Nitrogen is in a somewhat different category from the other essential elements we have been discussing. Strictly speaking, it is not really a mineral element, although it is commonly considered along with the mineral salts since plants usually obtain their nitrogen through nitrogen salts absorbed with the other salts. The basic source of the true mineral elements is the disintegrated rock that constitutes the bulk of most soils, but the important basic source of nitrogen is the atmosphere, which is about 78% nitrogen (N_2). Neither animals nor most plants are able to tap this great reservoir of nitrogen directly, but a few species of bacteria, fungi, and blue-green algae can synthesize nitrogen compounds from atmospheric nitrogen by processes known as **nitrogen fixation.** These nitrogen compounds may then be used by plants, and in turn plants serve as the basic nitrogen source of animals. The flow of nitrogen and nitrogen compounds from one organism to another and between organisms and environment is called the **nitrogen cycle** (Fig. 13.9). Consideration of the nitrogen cycle may begin at any point, but we shall start with nitrogen fixation.

Nitrogen Fixation

Most species of nitrogen-fixing bacteria and blue-green algae are free-living soil or water organisms. Most of these nitrogen-fixing bacteria are aerobic, like the widely distributed genus *Azotobacter* (Fig. 13.10), but some like *Clostridium* are anaerobic. Nitrogen-fixing blue-green algae such as *Anabaena* are the most self-sufficient of all organisms, since they can carry on both photosynthesis and nitrogen fixation. Some of the nitrogen compounds synthesized by these nitrogen-fixing bacteria and blue-green algae diffuse out and may be absorbed by plants, whereas other compounds

Fig. 13.8 Autoradiographs of bean plants showing the translocation of ^{35}S (a) *and* ^{45}Ca (b) *at the indicated number of hours after the roots were supplied with solutions of the radioactive salts. The darkest regions of the organs have the highest radioisotope concentration. Note the differences in the distribution and redistribution of sulfur and calcium.*

(a)

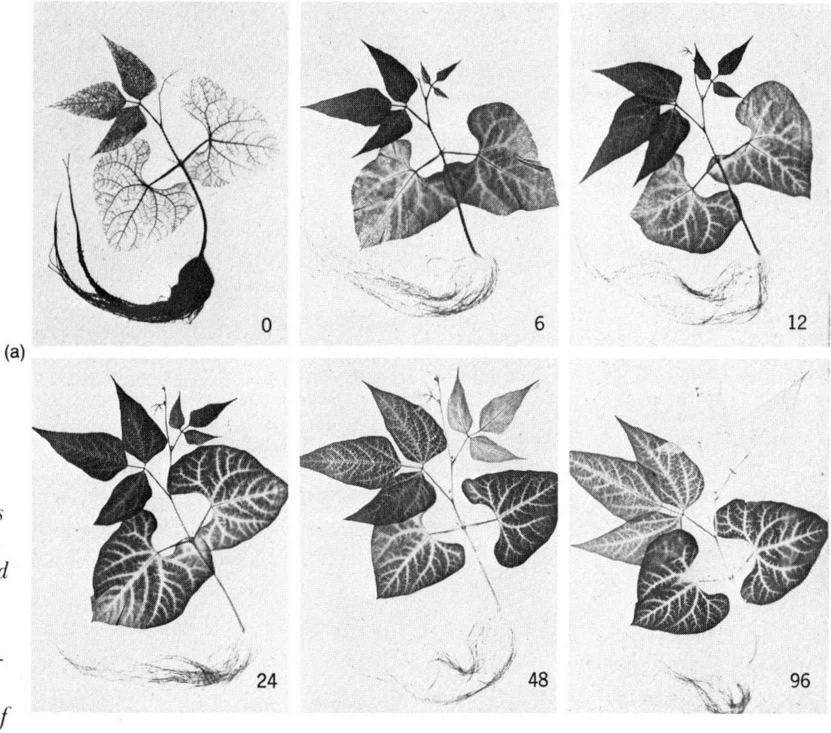

(b)

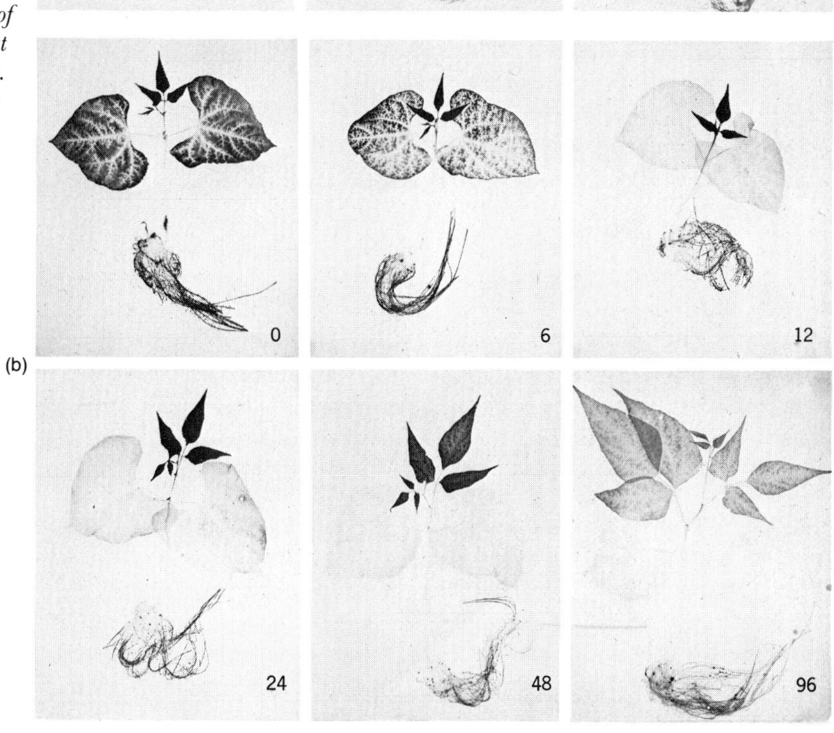

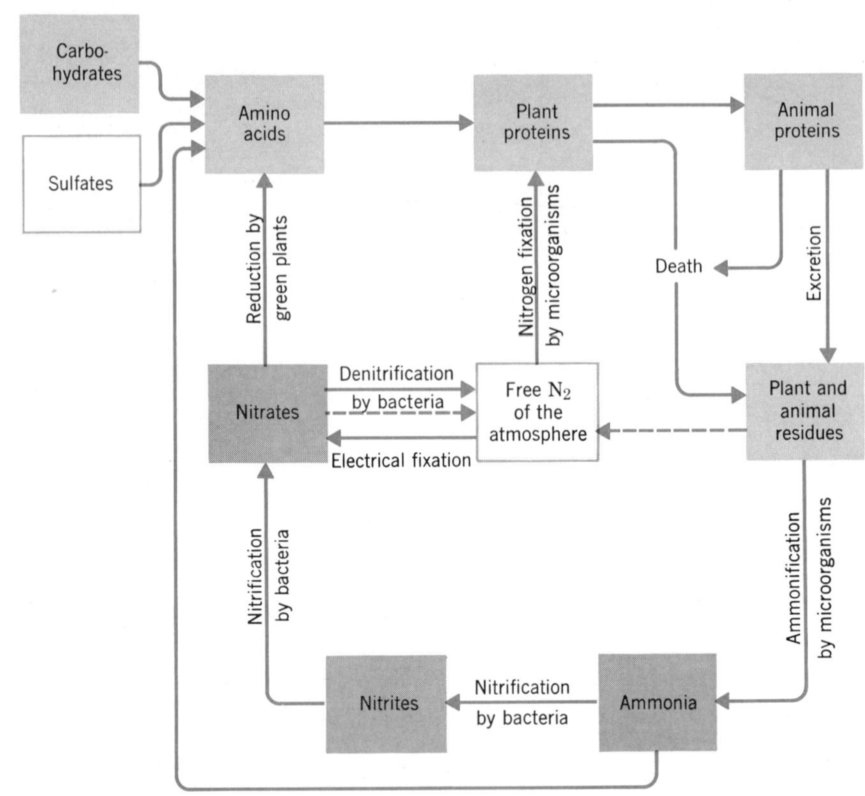

Fig. 13.9 Outline of the nitrogen cycle. (Redrawn from Plant Physiology *by Meyer and Anderson with the permission of the publisher, D. Van Nostrand.)*

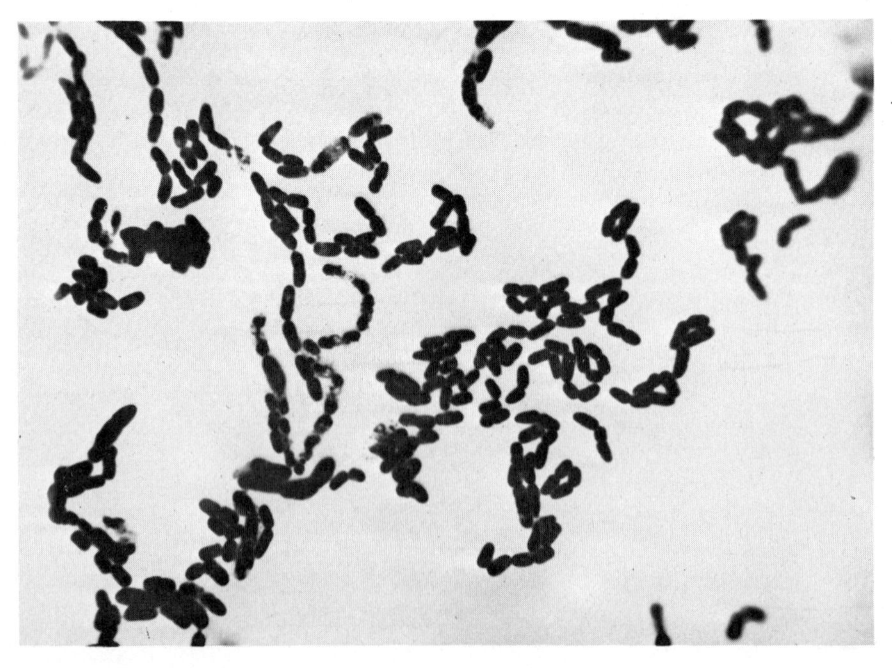

Fig. 13.10 A species of Azotobacter, *a nitrogen-fixing bacterium that lives in the soil and is not symbiotic. (Courtesy of the copyright holder, General Biological Supply House, Chicago.)*

may not be available until the nitrogen-fixing organisms die. The nitrogen-fixing organisms in an acre of typical midwestern soil fix about 7 to 12 kg of atmospheric nitrogen in a year. A second class of nitrogen-fixing bacteria invade the roots of leguminous plants (such as peas, beans, clover, and alfalfa), causing the formation of **root nodules** filled with the bacteria (Figs. 13.11 and 13.12). These bacteria can fix nitrogen only when in the nodules. Both the legumes and the bacteria benefit from this relationship. The legumes secure a supply of nitrogen (Fig. 13.13), and the bacteria secure food from the roots. Such relationships between two organisms benefiting both are known as **mutualism** or **symbiosis.** About 45 kg of nitrogen are commonly fixed in an acre of most leguminous crops per year, but as much as 115 kg may be fixed in an acre of alfalfa. There is generally an excess of nitrogen

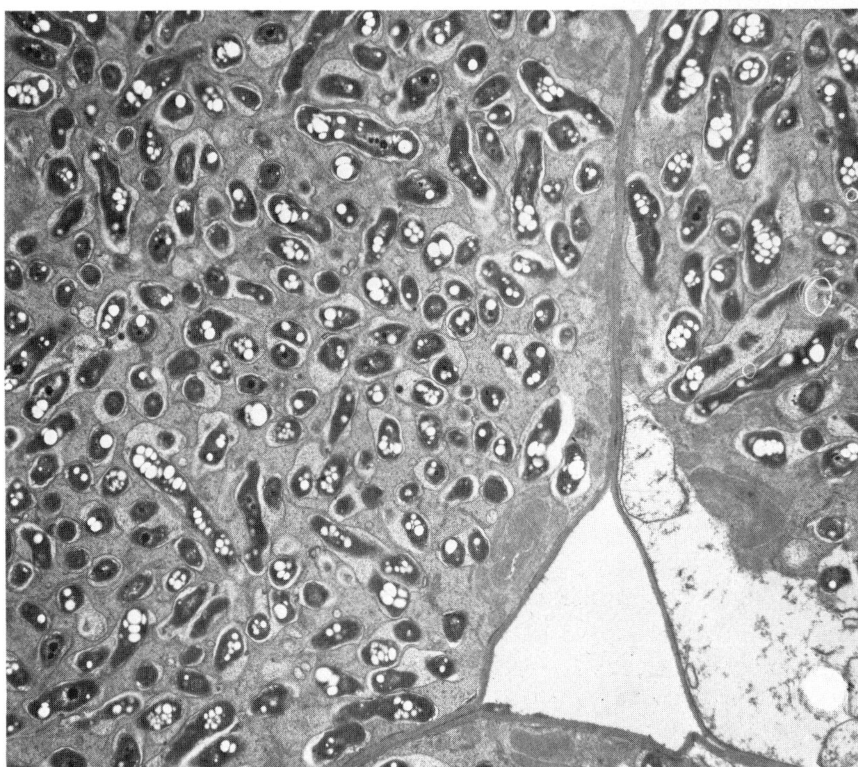

Fig. 13.11 (Top) Abundant root nodules containing nitrogen-fixing bacteria on Erythrina indica, *a legume.*

Fig. 13.12 Portions of three cells from a soybean root nodule containing numerous nitrogen-fixing bacteria (Rhizobium). *10,400X.*

Fig. 13.13 The effect of inoculating red clover with Rhizobium. *Both pots of plants were supplied with a minus nitrogen mineral solution.*

beyond that used by the legumes, and this may supply much of the nitrogen used by nonleguminous crops (about 28 kg per acre per year) planted in the same fields during the next season or two.

Symbiotic nitrogen fixation also occurs in a variety of nonleguminous vascular plants including alder and wax myrtle. In several genera of tropical plants the nitrogen-fixing bacteria are in leaf nodules rather than in root nodules.

The fixation of nitrogen (N_2) involves its reduction to ammonia (NH_3), which can then be used in the synthesis of amino acids (Chapter 14). The N_2 is first converted to intermediate compounds that are less highly reduced than ammonia, and a variety of intermediates such as hydroxylamine (NH_2OH) and hydrazine ($NH_2 \cdot NH_2$) have been suggested. Like all reductions, nitrogen fixation requires energy and large quantities of pyruvic acid are used. Both ATP and reduced hydrogen acceptors such as NADH are used. Molybdenum and probably cobalt and magnesium are required, presumably as components of essential coenzymes.

Lightning and other electrical discharges also fix atmospheric nitrogen, and the resulting nitrogen compounds are carried to the soil by rain. Electrical nitrogen fixation is a relatively minor source of nitrogen compounds compared with biological fixation, for only about 2.3 kg of nitrogen are added to an acre of typical midwestern soil by electrical fixation during a year.

Green Plants in the Nitrogen Cycle

The nitrogen compounds absorbed by green plants are derived, not only from nitrogen fixation, but also from decomposed plant or animal tissues and animal excretions, and under cultivation from fertilizers. Most nitrogen absorbed by plants is in the form of nitrate ($—NO_3$) salts, although plants also absorb ammonium ($NH_4—$) salts and other nitrogen compounds. The first step in the utilization of nitrates by plants is their reduction to ammonium compounds (Chapter 14), a process that requires the expenditure of considerable energy. Although plants synthesize many different nitrogen-containing compounds, the amino acids and proteins are of particular importance in the nitrogen cycle.

Animals and Nongreen Plants

The proteins synthesized by plants are the principal nitrogen source, either directly or indirectly, of all animals and of many species of fungi and bacteria. After digesting the proteins to amino acids they use the amino acids in synthesizing their own proteins. Some species of fungi and bacteria can, however, use inorganic nitrogen compounds in the synthesis of amino acids and from these they synthesize proteins. Amino acids that animals must get from their food are known as essential.

Return of Nitrogen to the Soil

Animal excretions contain urea and other nitrogen compounds. These may be absorbed by plants, but they are commonly converted first into ammonium compounds and then nitrates before absorption. When plants or animals die, bacteria and fungi may cause decay of the tissues, using the proteins and amino acids in them as a source of their own proteins, and in doing so also produce considerable quantities of ammonia (NH_3) and ammonium compounds. Although plants may absorb these ammonium salts directly, they are usually converted into nitrite salts (—NO_2) quite rapidly by soil bacteria commonly called **nitrifying bacteria.** Next, another type of nitrifying bacteria convert the nitrites into nitrate (—NO_3) salts that may then be absorbed by plants. Both the conversion of ammonium compounds to nitrites and the conversion of nitrites to nitrates are oxidation processes, and both kinds of bacteria use the energy released in the chemosynthesis of carbohydrate from carbon dioxide and water.

Loss of Soil Nitrogen

Nitrogen, once fixed, may continue to circulate from plants to animals to fungi and bacteria to the soil and back again to plants, but there are several ways nitrogen may be lost from the soil. In many soils there are **denitrifying bacteria** that break down nitrates and release gaseous nitrogen into the atmosphere. In contrast with the nitrifying and most kinds of nitrogen-fixing bacteria, the denitrifying bacteria thrive in situations where oxygen is deficient, and so are likely to be abundant and active in poorly drained soils. Like other salts, nitrogen salts may also be lost from the soil by leaching and erosion of the soil. Both leaching and erosion are likely to be much more severe in a cultivated field than in a natural plant habitat. An even greater loss of nitrogen from cultivated areas results from harvesting and removal of crops. Soils under cultivation show a continued decrease in nitrogen content (Fig. 13.14) unless the supply is replenished by planting leguminous crops or by using fertilizers. In most natural plant communities, soil nitrogen is maintained at a rather stable level that is frequently adequate for good plant growth. However, the nitrogen level is generally low in bogs, and it tends to be lower in dry than in humid regions and lower in hot than in cool climates. These differences result from effects of the environment on the operation of the nitrogen cycle. For example, where the temperature is frequently high, nitrogen-fixing bacteria are relatively inactive and decay is rapid.

Although denitrification by bacteria is generally undesirable in soils, denitrifying bacteria play a useful role in eutrophication, which begins with the introduction into lakes of large quantities of essential elements, especially nitrogen and phosphate, from sewage and heavily fertilized fields. This results in greatly increased growth of algae, which prevents fish and other algal herbivores from keeping the algal population under control.

Fig. 13.14 The decline of nitrogen in the soil of midwestern fields during a 70-year period of cultivation. (Redrawn from the 1938 Yearbook *of the U.S. Department of Agriculture.)*

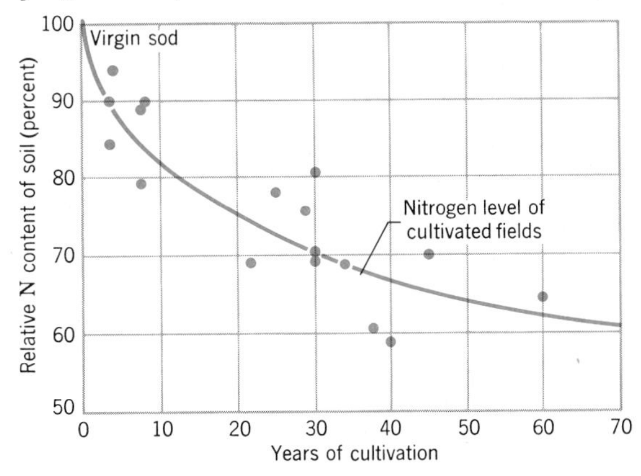

The decomposition of the excess algae by bacteria and fungi then seriously reduces the oxygen content of the water. The denitrifying bacteria thrive under these anaerobic conditions and their destruction of nitrates helps correct the eutrophication.

FERTILIZERS AND SOIL FERTILITY

Unless a farmer or gardener adds mineral elements to his soils, the fertility level will constantly decrease and his crop yields will decline. The amount and kind of fertilizer needed depends on a variety of factors such as the original fertility of the soil, the types of crop plants, the amount of leaching and erosion, and the soil structure. Positively-charged ions (cations) are held on the surface of soil particles where they are relatively resistent to leaching. Clay soils hold much greater quantities of cations than do sandy soils because of the much greater total particle surface area in clay soils. The colloidal clay particles bear multiple negative electrical charges. Different cations are held on the particle surfaces by electrical attraction with different degrees of firmness, the order from greatest to least for several common cations being H^+, Ca^{++}, Mg^{++}, K^+, NH_4^+, and Na^+. Any ion in the series can displace an ion to the right of it from the soil particles, a process known as **cation exchange.** Ions displaced from the surface of a particle can be absorbed by plants, but they are also subject to leaching from the soil. One reason for the low fertility of acid soils is that the H^+ ions have displaced most other ions from particle surfaces and these latter ions have been absorbed or leached away.

Types of Fertilizers

The term *fertilizer* may be applied to any material containing essential mineral elements that may be applied to the soil (or to bodies of water). Some fertilizers also improve soil structure. Although fertilizers supply mineral salts rather than food (Chapter 14), they are frequently referred to as plant foods. **Commercial fertilizers,** or **chemical fertilizers** as they are sometimes called, are made from a variety of things such as pulverized rock, bone meal, dried blood and other animal remains, cottonseed meal, and soybean meal. A so-called complete fertilizer contains the three elements that are most likely to be deficient in soils (N, P, K), the percentage of each present usually being indicated on the container. Thus, a 4-12-4 fertilizer contains 4% nitrogen, 12% phosphorus, and 4% potassium (potash). A few of the more expensive fertilizers contain all the known essential elements and so are really complete, although ordinary fertilizers may contain more than the three common elements as impurities. In recent years, mixtures of essential mineral salts have become available at garden supply stores. These are sometimes referred to as instant or concentrated fertilizers, and may be used in hydroponics, or dissolved in water and applied to the soil or to the leaves of plants as **foliar sprays.** The latter practice is becoming more widely used and is particularly valuable for elements such as iron that may become converted into insoluble and unavailable forms in the soil. Calcium, however, cannot be supplied to plants in foliar sprays since it is not translocated out of the leaves.

Lime is applied to soils principally to reduce soil acidity, but since it consists of calcium carbonate or other calcium compounds it also corrects calcium deficiencies. In addition, it helps release other elements from the soil particle surfaces and improves the crumb structure of clay soils (Chapter 12).

The **organic fertilizers** include barnyard manure, green manure (legumes or other plants plowed under), compost, and sludge from sewage disposal plants. In some countries it is a common practice to leach the solutes from manures with water and water plants with the solution. In addition to supplying mineral elements, organic fertilizers generally improve soil structure, making the soil easier to cultivate and increasing its capacity for holding air, water, and mineral salts. Some substances such as peat moss, vermiculite, and synthetic soil conditioners contain few or no mineral salts and are used primarily to improve soil structure.

Organic Gardening

Some amateur gardeners and farmers are enthusiasts for the cult of organic gardening. The promoters of organic gardening claim that only "natural" organic fertilizers, such as compost and manure, should be used, that plants can reach their maximum development only when supplied with such organic fertilizers, that plants provided with them are highly resistant to insects and diseases, and that both the plants and the animals or people who eat them are harmed by commercial fertilizers. Such ideas are based on incorrect, incomplete, and distorted information about plants and soils. Although the practice of organic gardening may be quite sound, some of the ideas behind it are not. Mineral elements from organic fertilizers are just the same as those from commercial fertilizers and either type of fertilizer can provide good soil fertility. Organic fertilizers do improve soil structure more than commercial fertilizers and are not as likely to burn plants when they are used in excess because of their lower solute content, but they also have some disadvantages. At times they may not contain adequate quantities of all the essential elements and they frequently do contain weed seeds, destructive insects, parasitic bacteria and fungi, and other pests. The burning of plants by commercial fertilizers results from plasmolysis of root cells rather than from a toxic effect, and may readily be avoided by using smaller quantities of the fertilizer. The beautiful crops that can be produced by hydroponics should make it evident that, not only organic fertilizers, but even soil itself are quite unessential for the excellent growth and development of plants.

However, the recycling of organic substances by adding to the soil composts, manures, and green manures (plowed under plants) is sound horticultural and agricultural practice if these do not contain plant pathogens. Legumes are particularly valuable as green manure because of the nitrogen fixed in their nodules.

PLANTS IN THE MINERAL NUTRITION OF ANIMALS

Man and other animals require about the same essential mineral elements as do plants, and they secure most of their supply either directly or indirectly from plants. Therefore, good mineral nutrition of our crop plants is important for man and his domestic animals as well as for the plants themselves. At times, crops in soils slightly deficient in one or more elements may grow reasonably well even though they are less nutritious for the animals or humans that eat them than might be desired. This situation is likely to be most widespread for elements such as iodine and sodium that are essential for animals but not for plants, except perhaps in minute traces. The deficiency of iodine in soils of the Great Lakes area resulted in extensive exophthalmic goiter among the residents of the region until iodized salt became available.

It should be stressed once more that plants fertilized organically are no better from the standpoint of human nutrition than those supplied with inorganic fertilizers. However, overfertilization with nitrates can result in problems for man and domestic animals, especially the young. If plants absorb more nitrates than they can reduce to ammonium compounds the nitrates accumulate in them. Although nitrates themselves are not toxic to animals, they have enzymes that convert nitrates to nitrites, which are toxic. The principal problems have arisen from baby foods that are high in nitrates.

PLANTS AS INDICATORS OF MINERAL DEPOSITS

Prospectors for deposits of valuable mineral ores are making increasing use of plants in their work in at least three principal ways. First, chemical analysis of plants growing in a certain region may reveal the location of underlying ore deposits by the presence of an unusually high concentration of the desired elements in the tissues of the plants. The sensitivity of this method of analysis is increased by the fact that plants may accumulate ions even though they are not essential. Second, there are certain indicator plants whose distribution is affected by ore deposits. For example, *Merceya latifolia* is an excellent indicator of copper deposits since it thrives in soils with an unusually high copper content. This plant is commonly called "copper moss." Finally, high concentrations of certain elements may influence the pattern of growth and development of plants. Thus, plants growing over certain iron ore deposits have stunted shoots, thickened roots, and changes in cell structure. High cobalt concentrations may result in white necrotic spots on leaves, and molybdenum may cause stunting and a yellow-orange chlorosis. Russian prospectors have reported that plants growing in soil rich in boron are two or three times their usual size and have a spherical shape and exceptionally large, dark green leaves. Plants growing above radioactive ores such as uranium may absorb radioactive ions if they are soluble and become radioactive.

TOXIC EFFECTS OF MINERAL ELEMENTS

The ions of a number of elements are quite toxic to plants, at least in excessive amounts. These elements include aluminum, arsenic, copper, boron, lead, magnesium, manganese molybdenum, nickel, selenium, silver, and zinc. Although some of these are essential elements, concentrations only slightly higher than the required ones may be quite toxic. Several examples of toxicity have just been given in the previous paragraph. Perhaps the most striking example of mineral toxicity is the copper basin of southeastern Tennessee (Fig. 13.15), where SO_2 and other toxic substances discharged into the air from copper smelters killed the vegetation over an area of many square miles, followed by severe erosion of the hilly terrain. True toxicity involves an adverse influence of the ions on metabolic processes. Neither the burning of plants by plasmolysis when too much fertilizer is applied, nor the unavailability of one element when another is present in the soil in too high a concentration (Fig. 13.4), is generally considered a toxic effect.

Plants may absorb mineral elements that are not particularly toxic to themselves but may be quite poisonous to the animals that eat the plants. The element selenium, present in many soils from Alberta to Arizona, is an outstanding example of this. Selenium accumulates in quantity in some plants, particularly the vetch. When animals graze on these plants they develop a serious poisoning known as alkali disease or blind staggers. Fortunately, most crop plants do not accumulate enough selenium to become toxic to animals or humans. Selenium is generally toxic to these crop plants, but it is not toxic to plants like vetch that accumulate it and may even be an essential element for them. The accumulation by plants of various kinds of radioactive ions from fall-

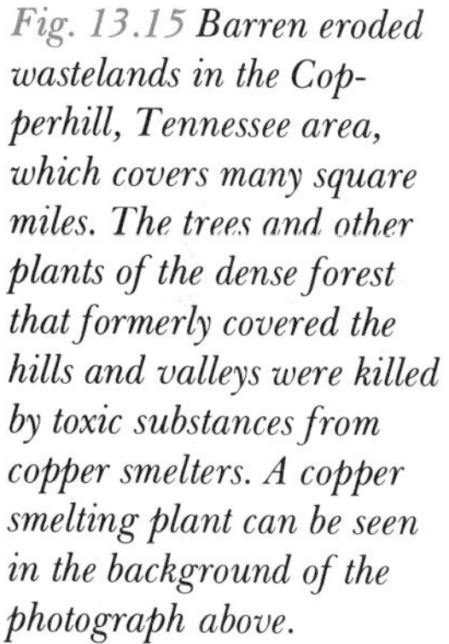

Fig. 13.15 Barren eroded wastelands in the Copperhill, Tennessee area, which covers many square miles. The trees and other plants of the dense forest that formerly covered the hills and valleys were killed by toxic substances from copper smelters. A copper smelting plant can be seen in the background of the photograph above.

out and the possible subsequent consumption of these radioactive substances by humans and other animals is a problem that is causing much concern and may well become more serious in the future. The problem is made more serious because each animal in a food chain, as well as the plants, further concentrate the radioactive ions by accumulation.

SOIL pH AND PLANT GROWTH

The degree of acidity or alkalinity of soils, as well as their mineral element content, may have marked effects on plant growth. Most plants grow best in soil with a pH of 6 to 7, although some species such as camellias, azaleas, and cranberries thrive only in an acid soil with a pH of 4 to 5.5, and a few species are restricted to alkaline soils of about 7.5 to 8. Plants may be able to survive in soils with an unsuitable pH, but their growth is severely restricted.

If the pH is extremely high or low (over 8 or below 4) there may be direct damage to the plants, but in general the influence of an unfavorable pH is indirect through one of the following factors: (1) If the pH is too low, the physical condition of clay soils becomes poor because of the loss of crumb structure. The separated colloidal clay particles then pack tightly together, and they may also be washed into the subsoil where they precipitate into an impervious layer known as hardpan. (2) The availability of the essential mineral elements is affected by pH, generally because of its influence on the formation of insoluble compounds. Phosphorus becomes less available when the pH drops below 6.5. Calcium and magnesium are also less available in acid soils. Iron, manganese, copper, zinc, and boron become less available when the pH rises above 7. (3) Low pH is unfavorable to most of the bacteria of the nitrogen cycle, resulting in a decrease in available nitrogen. (4) Certain fungus diseases develop only in acid soils, whereas other diseases occur only when the pH rises above a certain minimum. (5) The pH also influences the solubility of toxic substances. At low pH ranges, soluble aluminum and iron compounds become concentrated enough to be toxic. (6) In an acid soil, H^+ ions displace other cations from the soil particles by cation exchange, and these other cations may then be lost by leaching.

For agricultural and horticultural purposes the pH of a soil may be increased by the application of lime (calcium hydroxide or calcium carbonate), whereas the pH may be reduced by the use of ammonimum sulfate, aluminum sulfate, powdered sulfur, or peat moss.

The pH of the water that aquatic plants grow in is just as important as is the pH of the soil for land plants. Bacteria and fungi grow well only in limited pH ranges, and when they are cultured in laboratories it is important to adjust the pH to a suitable level.

RELATED READING

Anderson, A. J. and E. J. Underwood. "Trace element deserts," *Scientific American, 200*(1), 97–106, January 1959.

Burt, J. C. "Desert in the Appalachians," *Nature Magazine, 49,* 486–488, 1956.

Delwiche, C. C. "The nitrogen cycle," *Scientific American, 223*(3), 136–146, September 1970.

Ellfolk, N. "Leghaemoglobin, a plant haemoglobin," *Endeavour, 31,* 139–142, 1972.

Galston, A. W. "The blind staggers (Selenium), *Natural History, 83*(5), 38–45, 1974.

McElroy, W. D. and C. P. Swanson. "Trace elements," *Scientific American, 188*(1), 22–25, January 1953.

Pratt, C. J. "Chemical fertilizers," *Scientific American, 212*(6), 62–71, June 1965.

Throckmorton, R. I. "Organic farming—bunk," *Reader's Digest, 61*(4), 45–48, October 1952.

Wallace, T. "Mineral deficiencies in plants," *Endeavour, 5,* 58–64, 1946.

14 SYNTHESIS AND USE OF FOODS BY PLANTS

Most people are aware of the fact that human beings and other animals must have an adequate supply of foods such as carbohydrates, fats, and proteins as a source of energy and body-building material. Lack of an adequate food supply results in malnutrition, hampers growth, and may even lead to starvation. There is also general recognition that plants require food if they are to thrive and grow, but many people believe that the foods of plants are quite different from those of animals and that plants secure their foods from the soil. Specifically, plant foods are thought to be the essential mineral ements (or sometimes even organic substances) from the soil. This concept is strengthened by the fact that manufacturers of fertilizers often label and advertise these elements as plant foods and suggest proper feeding of plants.

Current differences in opinion as to the nature of plant foods stem entirely from different definitions of *food.* A few biologists prefer to define a food as "any substance which an organism obtains from its environment and utilizes, directly or indirectly, in its metabolism." From this definition, the foods of green plants would be water, carbon dioxide, mineral salts, and even oxygen, whereas the food of animals, bacteria, and fungi would include also carbohydrates, fats, proteins, and vitamins. Most botanists and many zoologists dislike such a broad, general definition of food and prefer to consider as a food any organic substance usable by a living organism as a source of energy or as cell-building material. In other words, **foods** are substances used in respiration and assimilation. On the basis of this definition, the foods of all organisms consist of the same groups of compounds: carbohydrates, lipids, proteins, and closely related substances such as organic acids. We use the term *food* in this sense throughout the book. Excluded by the definition are carbon dioxide, water, mineral salts, and oxygen. Important as these simple inorganic substances are in the life of plants, their roles are quite different from those of foods. Excluded also are the vitamins, important in plants as well as animals, but as regulatory substances rather than as energy sources.

Now that we have defined foods we can state the sources of the foods of plants. In photosynthetic plants, the food source is internal, all the food being synthesized by photosynthesis and subsequent synthetic processes. Some species of bacteria carry on photosynthesis, whereas a few other species synthesize their own food by processes similar to photosynthesis except that chemical energy rather than light energy drives the processes. Most bacteria, the vascular plants lacking chlorophyll (Fig. 14.1), and the fungi secure their foods from their environment, as do animals. The few species of insectivorous plants secure some food from the insects and other small animals they trap and digest, but most of their food is derived from photosynthesis. The most important food derived from the insects is protein, which provides a source of nitrogen.

Fig. 14.1 The Indian pipe (Monotropa uniflora) *is one of the relatively few species of vascular plants that lack chlorophyll and so must secure food from external sources. Such plants are saprophytes, parasites, or symbionts.*

DEVELOPMENT OF CONCEPTS ABOUT PLANT FOOD

The idea that plants secure their food from the soil is a very ancient one. Aristotle and other Greek scholars believed that the substance of plants was derived from the soil and this concept persisted into the Middle Ages. At a time when everything was thought to have originated from earth, air, fire, or water, this hypothesis was not illogical.

As far as is known, this idea was first challenged by Nicolaus of Cusa (Cusanius), a bishop with interest in the sciences. About 1450, he described experimental evidence that led him to conclude that the substance of plants was derived from water, not earth (soil). In 1648, the Dutch investigator Jean-Baptiste Van Helmont published a report of his experiment that led to the same conclusion. Whether or not he knew about the Cusanius report has not been determined, but the two experiments were essentially similar. Van Helmont's complete report follows.

That all vegetable matter immediately and materially arises from the element of water alone I learned from this experiment. I took an earthenware pot, placed in it 200 lb of earth dried in an oven, soaked this with water, and planted in it a willow shoot weighing 5 lb. After five years had passed, the tree grown therefrom weighed 169 lb and about 3 oz. But the earthenware pot was constantly wet only with rain or (when necessary) distilled water; and it was ample in size and imbedded in the ground; and, to prevent dust flying around from mixing with the earth, the rim of the pot was kept covered with an iron plate coated with tin and pierced with many holes. I did not compute the weight of the deciduous leaves of the four autumns. Finally, I again dried the earth of the pot, and it was found to be the same 200 lb minus about 2 oz. Therefore 164 lb of wood, bark, and root had arisen from water alone.

Of course, we now know that plants are not made from water alone, but the water hypothesis was an improvement on the soil hypothesis. Water does constitute the bulk of the fresh weight of living plant tissues and is one of the substances used in photosynthesis. In 1969, John Woodward, on the basis of his pioneering experiments on mineral nutrition (Chapter 13), concluded that Van Helmont was wrong and that an earthy substance as well as water contributed to the substance of plants. Van Helmont might have reached the same conclusion if he had not ignored the small decrease in the weight of the soil.

Perhaps the first suggestion that air was a source of plant substance was made in a 1687 report by Thomas Brotherton to the Royal Society of London. According to Robert Hooke, Brotherton's experiments showed that plants "are nourished and increased by a double food; the one an impregnated water, and the other an impregnated air. These do mutually mix and coalesce, and parts of the air convert to water, and parts of the water convert to air." At a time when modern chemical concepts had not yet developed this was a remarkable anticipation of photosynthesis. Between 1772 and 1860, a series of important investigations revealed the existence of photosynthesis and its basic nature. During the same period the basic facts about the mineral nutrition of plants were elucidated. Both were dependent on the concurrent development of modern chemistry.

Thus the old problem of the source of plant substances was finally solved. Plants are composed predominantly of water and of organic compounds derived from carbon dioxide and water by the process of photosynthesis, and to a small extent from mineral salts absorbed from the soil. Green plants are not dependent on an external source of food. They produce their own food by photosynthesis.

FOODS AND ENERGY

Although both the uses of food (as cell-building materials and as energy sources) are of great fundamental biological importance, the energy aspects of food require special consideration at this point. Life may be considered as a series of controlled, interrated, and coordinated energy transformations, and the result of their cessation is death. Because energy is important in life we should consider a few basic facts about the nature of energy.

Energy may be defined as the capacity for doing work. Energy has no mass and does not occupy space, and so is not a substance, although energy may be converted into matter and matter into energy, as in nuclear reactors. Energy exists in various forms such as heat, light and other radiation, sound, electricity, magnetism, and kinetic energy of movement (mechanical energy). Various compounds, incuding foods, contain considerable energy in the bonds holding the atoms of their molecules together and this energy is released when the compounds are converted into other substances of lower energy content by oxidation processes such as respiration and combustion.

Energy Transformations

One kind of energy can be converted into another, machines (as well as living organisms) being energy converters. Thus, a steam engine converts the chemical energy of coal or oil first into heat and then into mechanical energy that may be used to drive an electric generator, converting the energy into electricity. This electrical energy, in turn, may readily be converted into light, heat, sound, magnetism, mechanical energy, chemical energy, or other types of energy by appropriate devices.

Oxidation and Reduction. Energy transformations in plants will be considered later in this chapter, but it may be noted here that the energy transfers in biochemical **oxidation** and **reduction** reactions are of great biological significance. Fundamentally, oxidation is the removal of electrons from atoms or molecules while reduction is the addition of electrons, but many oxidation and reduction reactions also involve the transfer of oxygen or hydrogen. If oxygen is added to a molecule or hydrogen is removed, oxidation occurs, whereas removal of oxygen or addition of hydrogen to a molecule reduces it.

Whenever one substance is oxidized in a chemical reaction another is reduced, so it is proper to refer to such reactions as reduction-oxidation (or redox) reactions rather than as oxidations or reductions. However, if in a process there is a net increase in chemical bond energy (at the expense of some kind of energy such as heat or light) we commonly refer to it as a reduction process. If there is a net decrease in chemical bond energy (with the release of some kind of energy such as heat or light) we refer to it as an oxidation process. In any chemical reaction energy is required to break the chemical bonds of the reacting substances, and when the new chemical bonds of the products are formed energy is released. Whether there is a net input of energy or a net output of energy in a chemical reaction depends on which of these energy changes is greater.

The principal processes of plants involving oxidations and reductions are photosynthesis and respiration. In photosynthesis, carbon is reduced from a low energy level in CO_2 to a high energy level in sugar, the required energy being obtained indirectly from light energy absorbed by the chloroplast pigments. In respiration, sugars are oxidized to CO_2 and H_2O and the energy thus released is used in metabolic work.

Energy Transfer Through Coenzymes. Oxidation-reduction reactions in organisms occur in a stepwise fashion with only a relatively small amount of energy transfer in any one of the many individual reactions that make up the total process. Later in this chapter, we outline the ways in which glucose is converted into other compounds by a series of reactions as it is being oxidized in respiration. In a number of the individual reactions making up the process of respiration, oxidation occurs by the removal of hydrogen ions and electrons and the conversion of carbon to CO_2. As the hydrogen ions and electrons are transferred from one compound to another, each at a somewhat lower energy level, there is a con-

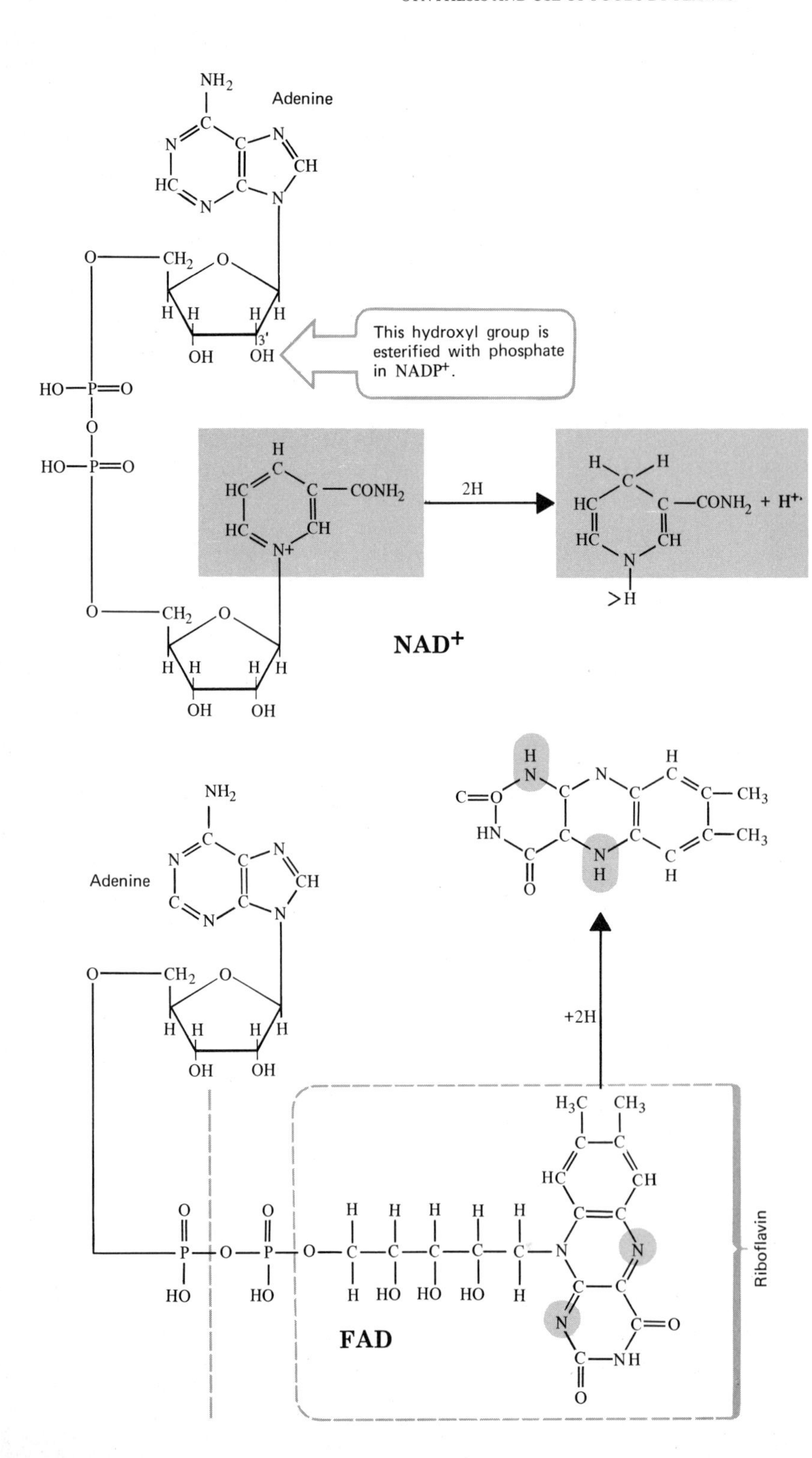

Fig. 14.2 The structural formulae of two H accepting and donating coenzymes. The complete formulae of the oxidized molecules of NAD^+ and FAD are shown at the left. $NADP^+$ differs from NAD^+ in that it has a phosphate group at the point shown. On the right side only the portion of each molecule to which the H is added in the reduced form is shown. Note that FADH contains both of the H atoms, but NADH only one. The electron of the other H neutralizes the N^+ of the molecule, leaving a proton (H^+).

trolled release of energy. The total energy release is the same that would occur in a rapid and uncontrolled way if an equal quantity of glucose were burned. The end products (CO_2 and H_2O) of both respiration and combustion are the same, but there is a great difference in the temperature at which the two processes will occur and in the usability of the energy released. Incorporation of energy into chemical bonds during photosynthesis also occurs in a stepwise fashion as hydrogen and electrons are transferred from one compound to another.

The stepwise oxidation-reduction reactions in organisms cannot occur unless the particular enzyme required to catalyze each specific reaction is present, but the protein enzymes are quite ineffective unless they are accompanied by particular kinds of nonprotein coenzymes. Some of the coenzymes pick up hydrogen (a proton and an electron) as it is removed from the substrate molecule, and then pass the hydrogen on to another coenzyme or a substrate molecule at a slightly different energy level. These coenzymes thus serve as hydrogen acceptors and donors. Other coenzymes accept and donate only electrons. In either case they make possible the stepwise energy transfer of biological oxidation-reduction reactions.

In our discussions of respiration and photosynthesis, we have to consider three principal hydrogen-transferring coenzymes. One is nicotinamide adenine dinucleotide, an organic compound with a molecular structure as complicated as its name. For obvious reasons it is commonly referred to simply as NAD^+. It should be noted that this is an acronym and that the letters are not symbols for chemical elements. Such acronyms are commonly used for complex compounds of biochemical importance. The reduced form is written as NADH[2] (the H *is* a chemical symbol) or, more properly, as $NADH + H^+$ since the compound is ionized (Fig. 14.2). A second hydrogen-transferring coenzyme, $NADP^+$, differs from NAD^+ in having an additional phosphate group in its molecule. Its reduced form is $NADPH + H^+$. Another hydrogen-transferring coenzyme is flavin adenine dinucletide (FAD), and there is the related flavin mononucleotide (FMN). These and other flavin coenzymes, along with their characteristic enzyme proteins, are often referred to as flavoproteins (FP). Their reduced forms are, respectively, $FADH_2$, $FMNH_2$, and FPH_2.

The cytochromes (Fig. 14.3) are electron acceptors and donors, but unlike the above compounds they do not accept and donate protons (and so not H). The cytochrome coenzymes have iron as a component of their molecules. When the iron picks up an electron, it is reduced from ferric iron (Fe^{3+}) to ferrous iron

Fig. 14.3 The proposed structure of cytochrome c. Note the iron atom (Fe) in the center of the porphyrin ring. This is characteristic of all cytochromes, but a magnesium atom is present inside the porphyrin ring of the chlorophylls.

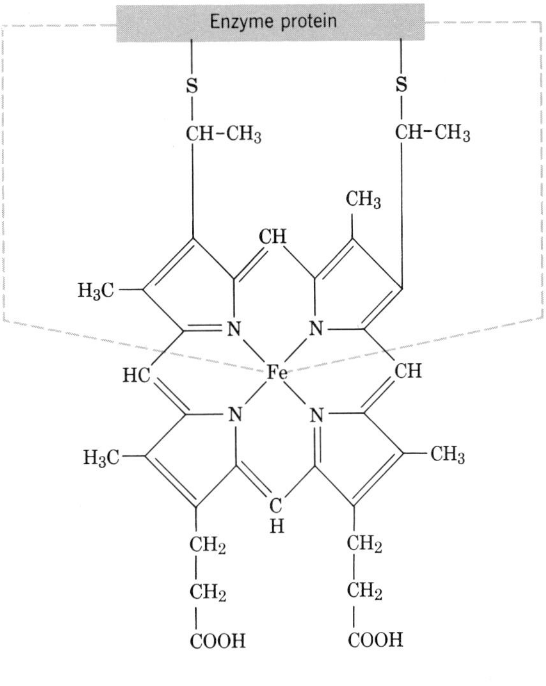

(Fe^{2+}), and when it then donates the electron, it is oxidized back to the ferric state. There is a series of cytochromes and as electrons are transferred from one to another at a lower energy level there is a release of energy, which can be used metabolically.

Although it is the coenzymes that accept and donate H or electrons, it should be noted that they would be quite ineffective without their protein apoenzyme. The terms we have been using actually apply to the complete enzyme, that is, the enzyme protein and the coenzyme.

Another substance that plays a very important role in energy transfer in organisms (in a different way from those already mentioned) is adenosine triphosphate (ATP). ATP is composed of adenine, ribose, and three phosphate groups (Fig. 14-4). In organisms it is usually derived from adenosine diphosphate (ADP) by the addition of a third phosphate group. This requires energy, which is usually supplied by hydrogen or electron donors such as NADH, NADPH, or reduced cytochrome. A still smaller molecule in the series is adenosine monophosphate (AMP), which has only one phosphate group. Addition of a phosphate group to AMP results in ADP, and also requires energy. Thus, ADP has a higher energy level than AMP but a lower energy level than ATP. ATP can be hydrolyzed to ADP (ATP + water → ADP + phosphate), and the energy released by this reaction can be used in doing a variety of metabolic work. The uses of ATP are outlined later in this chapter.

Quantity of Energy. Quantitative measurements of energy can be made, and quantity of energy can be expressed in various units. One of the most commonly used energy units in biological work is the **calorie** (cal), the amount of heat energy required to raise the temperature of one gram of water one degree centigrade. The **kilogram-calorie** (kcal) is 1000 times as great as a calorie. The calories used in nutritional work, and so familiar to those on diets, are kilogram-calories. Although calories are specifically units of heat energy, they can be used to express the quantity of any kind of energy, since one kind of energy can be converted into another. Thus, when we say that 180 grams of glucose (its gram molecular weight) contain 673 kcal (or 673,000 cal) we mean that if all the energy in this glucose were converted into heat, the resulting heat would be sufficient to raise the temperature of 673,000 grams of water 1° C.

Energy Sources. Not all the energy used by plants is freed from foods by respiratory processes. Light energy is utilized in photosynthesis and also in other processes including chlorophyll synthesis, the formation of some anthocyanins, and certain reactions affecting growth. The energy released from foods in respiration is, indeed, in the final anal-

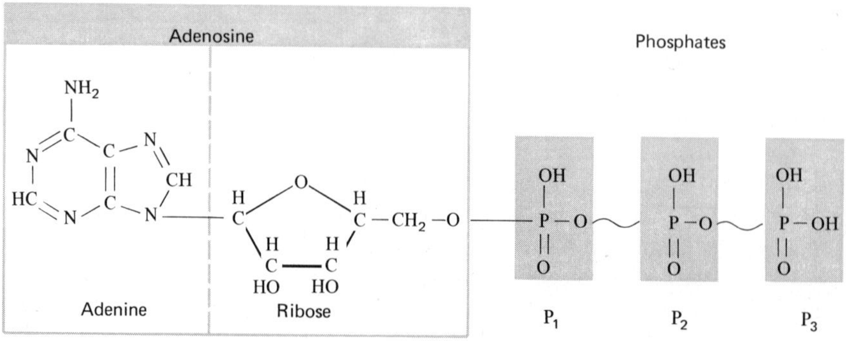

Fig. 14.4 The structure of ATP (adenosine triphosphate). ADP has only one phosphate group and AMP only one.

ysis derived from light energy utilized in photosynthesis and incorporated in the food produced. Molecular energy is expended, not only in diffusion but also in other processes, such as the flow of water through the xylem, deriving their energy from diffusion. Plants, then, secure energy from three principal sources: light direct, respiration (light indirect), and the kinetic energy of molecules. In contrast, animals secure most of their energy supply from foods through respiration, with relatively little molecular energy and even less direct light energy being used.

In the final analysis, the bulk of the energy expended by living organisms can be traced back to the light energy from the sun utilized in photosynthesis. Molecular energy, too, can be traced to the radiant energy of the sun. Indeed, all energy on earth comes directly or indirectly from the sun (except for a negligible quantity from other suns—the stars). Even atomic energy may be considered as having a solar origin, since the materials of the earth were probably originally derived from the sun.

Each year 13×10^{23} cal of radiant energy from the sun reach the earth. About a third of this energy is reflected immediately, so about 9×10^{23} cal are available for heating the earth (and so making it habitable) or for photosynthesis. Only about 0.1% of this available energy is used in photosynthesis, but this amounts to the still impressive sum of about 10^{21} cal per year. It is on this supply of energy, converted into chemical energy by plants during photosynthesis, that all living things depend for their existence.

The energy in fuels such as wood, peat, coal, petroleum, and gas, like the energy in food, is chemical energy derived from the light energy of the sun during photosynthesis. The fossil fuels—coal, petroleum, and gas—that provide the bulk of the energy used in our homes, industries, and transportation are all partially decayed remains of ancient plants and animals, and when we burn them we are releasing energy tied up by photosynthesis millions of years ago.

PHOTOSYNTHESIS

The Magnitude of Photosynthesis

Photosynthesis is by far the greatest production process on earth. A recent estimate places the annual photosynthetic production of organic matter by plants (in excess of that used by them in respiration) at about 87 billion tons. Thus, there is a net consumption of 128 billion tons of carbon dioxide and 52 billion tons of water and a net production of 93 billion tons of oxygen each year by photosynthesis. Earlier estimates were about five times as great because photosynthesis in the oceans was overestimated, but it is still thought that marine algae carry on about half of all photosynthesis.

The oxygen added to the air by photosynthesis is of great biological importance because of its use by both plants and animals in aerobic respiration. Although oxygen is continually being removed from the air by nonbiological processes such as combustion, rusting of metals, and weathering of rocks, as well as by respiration, photosynthesis is the only process that *adds* any substantial quantity of oxygen to the air. Indeed, if it were not for photosynthesis, the atmosphere would be devoid of oxygen.

The atmosphere contains only 0.03% carbon dioxide, but there is no danger of plants exhausting the supply by using it all in photosynthesis, for the atmosphere is so vast that this small percentage represents about 2200 billion tons of carbon dioxide. The oceans of the earth contain over 50 times this quantity of carbon dioxide in the form of dissolved gas or carbonates; this vast reservoir of the gas is in equilibrium with the carbon dioxide of the air and holds its concentration more or less stable.

The readily available carbon dioxide of the earth is enough to supply photosynthesis for about 250 years, even if none were added to the air. Actually, combustion, weathering of rocks, and other chemical processes, as well as respiration, add several hundred billion tons of carbon dioxide to the air each year. It is not generally realized that green plants contribute much more carbon dioxide to the air as a result of respiration than do animals, and that bacteria and fungi contribute even more than the green plants. The cycling of carbon dioxide through photosynthesis to organic compounds and the return of carbon dioxide to the air by respiration constitute the core of the carbon cycle (Fig. 14.5). During the two billion years or so that life has been present on earth, any particular carbon atom we might select for consideration has probably been through the carbon cycle billions of times and has been a constituent of innumerable and varied plants and animals.

Development of Concepts of Photosynthesis

Man's knowledge of the existence of photosynthesis and its basic nature was compiled by a number of different investigators over a period of almost 100 years. The first real clue was provided in 1772 by Joseph Priestley, an English clergyman and part-time scientist, who found that plants "restored the air injured by the burning of candles." Thus, he found that plants produced oxygen, although he did not announce his discovery of oxygen until two years later. Stimulated by Priestley's experiments, the Dutch physician Jan Ingenhousz conducted hundreds of experiments on the effects of plants on air. In his 1779 report, he confirmed Priestley's results, and also reported that green tissues and light were essential for purification of the air by plants. In 1782, a Swiss pastor, Jean Senebier, found that the air-restoring capacity of plants was dependent

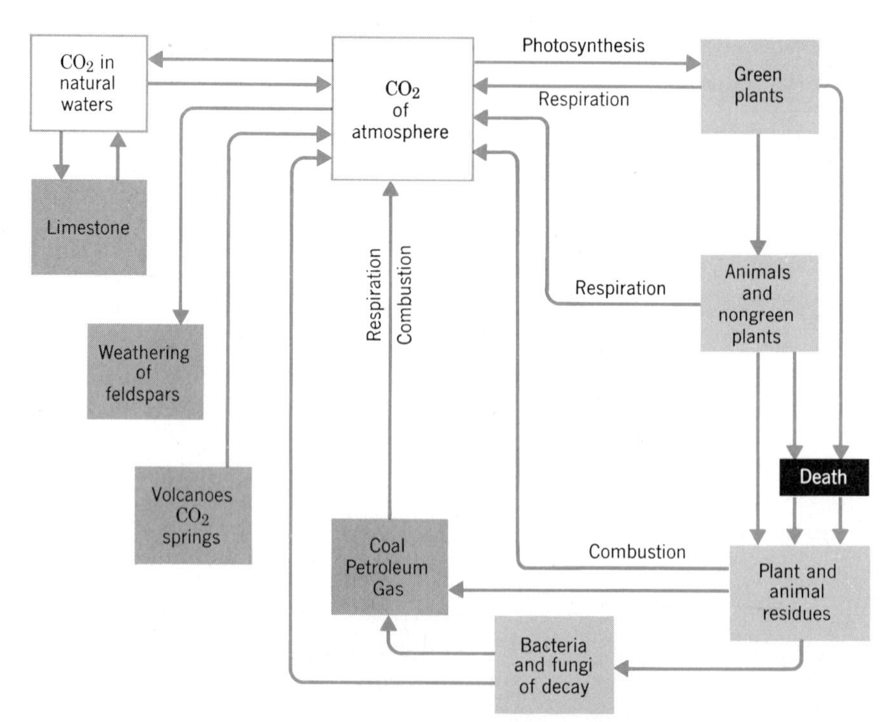

Fig. 14.5 The carbon cycle.

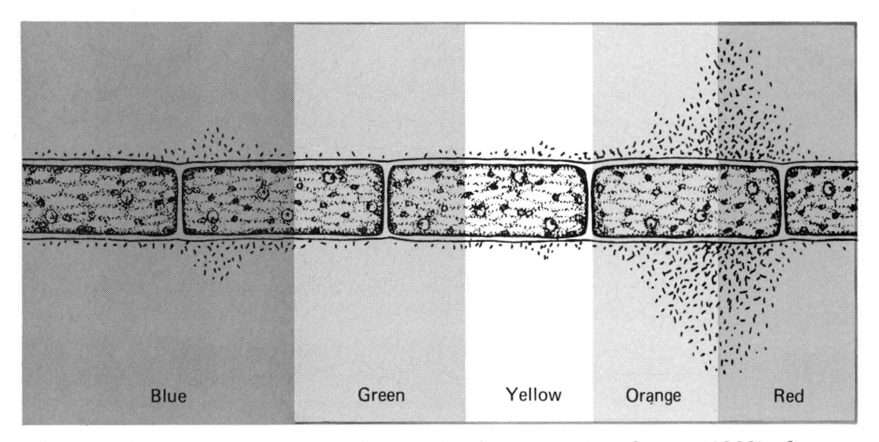

Fig. 14.6 Modified reproduction of a drawing by Th. W. Engelmann (1882) of an algal filament under a microscope when exposed to a microspectrum. The motile bacteria included in the mounted slide move toward regions of higher oxygen concentration. Since most of them clustered around the alga in the part exposed to red light, Engelmann could conclude that the red wavelengths were most effective in photosynthesis. The smaller cluster of bacteria in the blue region indicated that those wavelengths were next most effective.

upon the presence of "fixed air" (carbon dioxide), and suggested that it provided nourishment for the plants. In 1796, Ingenhousz also reported the use of carbon dioxide by plants and recognized that it was the source of carbon in plants. In 1804, the Swiss scientist Nicolas Theodore de Saussure published a book reporting his numerous experiments on photosynthesis. He confirmed and extended previous discoveries, distinguished the gas exchanges from those of plant respiration, and made the first quantitative measurements of photosynthesis. However, his most important contribution was that water, as well as carbon dioxide, was used in the process. This discovery was a product of his quantitative measurements.

The final important basic concept about photosynthesis was provided in 1845 by the German scientist Julius Robert Mayer, who was the first to recognize clearly that photosynthesis provided the energy used by organisms by converting light energy into the chemical energy of the organic compounds produced. Also in 1845, the German biochemist Liebig pointed out that *all* of the organic compounds of plants were derived from the carbon dioxide and water used in photosynthesis. In 1862, the German plant physiologist Sachs reported that the photosynthetic product was starch. (We now consider it as a secondary product derived from the sugar produced.)

By 1862, then, the main outlines of the process had been worked out, although the name *photosynthesis* was not used until 1898. In the 70 years between 1862 and 1932, the progress of research on photosynthesis was relatively slow and little was added to our understanding of the nature of the photosynthetic processes, although much information was assembled on the influence of environmental factors on the rate of photosynthesis. Perhaps the most fundamental discoveries during this period were those of the German botanist Th. W. Engelmann, who in the 1880s used ingenious experiments (Fig. 14.6) to determine the wavelengths of light most effective in photosynthesis; of F. F. Blackman, the English plant physiologist, who in 1905 found that at

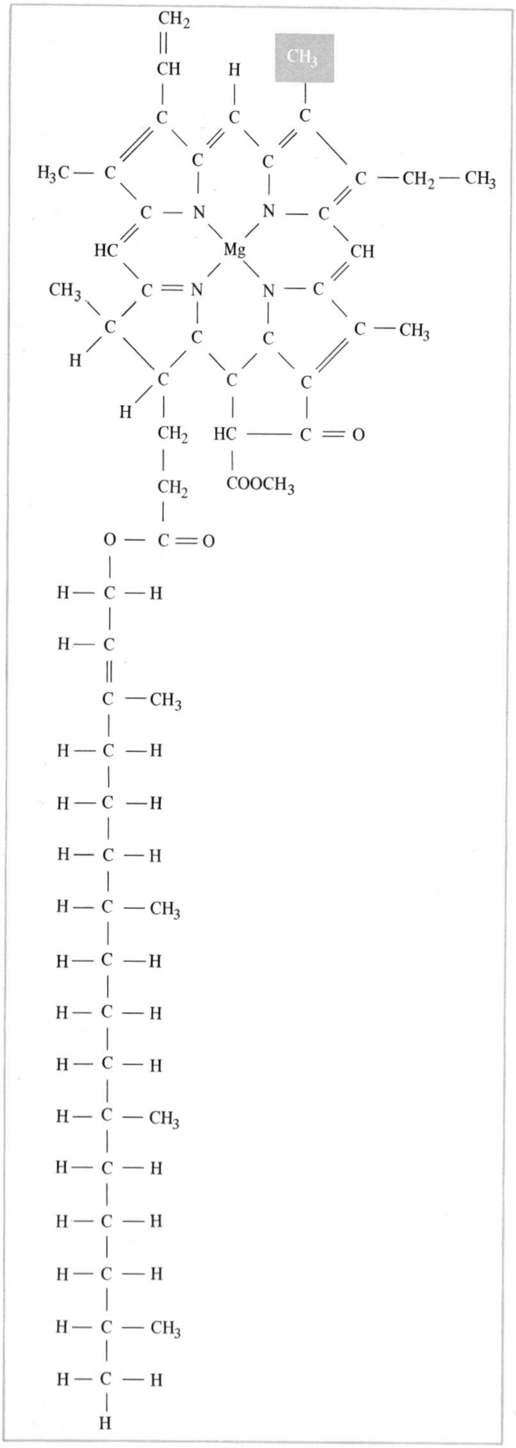

Fig. 14.7 Structural formula of a chlorophyll a molecule. Chlorophyll b *differs only in that it has an aldehyde (CHO) group in place of the circled methyl (CH_3) group of chlorophyll a. Note the Mg atom in the center of the porphyrin ring of the molecule.*

least one step in the photosynthetic process did not require light; and of the German chemist R. Willstätter and his associates, who between 1906 and 1922 worked on the chemical structure of chlorophyll (Fig. 14.7). Beginning in 1922, the German physiologist Otto Warburg and his associates conducted extensive experiments on the efficiency of photosynthesis in relation to the quantity of light energy absorbed.

In 1929, C. B. van Niel summarized research data that he and others had obtained about photosynthetic and chemosynthetic bacteria and pointed out the similarity of their processes to green plant photosynthesis. Bacterial photosynthesis differs from green plant photosynthesis in that different kinds of chlorophyll and other pigments absorb the light energy, and that substances such as hydrogen sulfide (H_2S) or hydrogen (H_2) rather than water serve as the hydrogen source:

$$2\ H_2O + CO_2 \longrightarrow [CH_2O] + H_2O + O_2$$
$$2\ H_2S + CO_2 \longrightarrow [CH_2O] + H_2O + 2\ S$$
$$2\ H_2 + CO_2 \longrightarrow [CH_2O] + H_2O$$

The overall synthetic reactions of bacterial chemosynthesis are the same as those of green plant photosynthesis: carbon dioxide and water are used and carbohydrate, water, and oxygen are produced. The difference between the processes lies in the energy sources, with chemosynthetic bacteria securing the energy that drives the process from the oxidation of inorganic compounds rather than from light.

C. B. van Niel pointed out that his data strongly suggested that the oxygen evolved by photosynthesis came from the water rather than from the carbon dioxide as was generally assumed, although he had no direct experimental evidence for this.

Despite such advances in our understanding of photosynthesis, we entered the 1930s with little more knowledge of the steps in the photosynthetic process than had been available at the time of the Civil War. The most generally held theory was that formaldehyde was the principal intermediate product, although there had never been any really satisfactory evidence to support this theory. In 1938, the English biochemist R. Hill reported that chloroplasts removed from plant cells and exposed to light produce substantial quantities of oxygen if they are provided with hydrogen-accepting compounds such as ferric salts, thus providing some evidence in support of both van Niel's proposal that the oxygen evolved is derived from the water and Blackman's concept of photosynthesis as at least a two-step process, only one of the steps being directly dependent on light. The Hill reaction, as it has come to be called, is only partial photosynthesis, since carbon dioxide is not used and no carbohydrate is produced.

A more complete understanding of the processes of photosynthesis awaited some means of labeling the carbon dioxide and water used in the process so that the labeled atoms could be traced through the various intermediate substances produced during the reactions.

The first attempt along this line was made in 1939 by the American biochemist Samuel Ruben and his associates, who used carbon dioxide containing a radioactive isotope of carbon, ^{11}C. The radioactivity of this isotope, however, was exhausted so rapidly that it provided little information. In 1941, Ruben and his co-workers, trying another approach, supplied the green alga *Chlorella* with water containing a heavy, nonradioactive isotope of oxygen, ^{18}O. All the carbon dioxide supplied contained the usual commonly occurring isotope of oxygen, ^{16}O. They found that some of the oxygen gas evolved was ^{18}O, proving that the oxygen released during photosynthesis is derived from the water used and substantiating the conclusions of both van Niel and Hill.

Beginning about 1940, two principal groups of investigators, one led by the German biochemist Otto Warburg and the other by Robert Emerson of the University of Illinois, initiated an extensive series of investigations designed to determine how many quanta of light energy were required in photosynthesis to reduce each carbon atom. Although this information is of considerable theoretical value in the elucidation of the photosynthetic processes, consideration of it is beyond the scope of this book. It may be noted, however, that in paper after paper the Warburg group presented evidence indicating that four quanta were necessary, while the Emerson group repeatedly presented evidence for a requirement of eight or more quanta. Each group criticized the other's conclusions, thus generating a classical biological feud. Most, but not all, of the neutral scientists qualified to judge have come to accept the results of the Emerson group.

Before he met an untimely death in an airplane accident, Emerson made other important contributions to a knowledge of photosynthesis, including the first evidence that two different pigments are essential for absorbing the light used in photosynthesis. These are now thought to be two kinds of chlorophyll *a* with different maximum absorption peaks.

Despite the continuing increase in our understanding of photosynthesis, at the close of World War II there was still essentially no information regarding the sequence of the reactions making up the process of photosynthesis, but in the years since 1950 the pathway of carbon from CO_2 to carbohydrate has been worked out in complete step-by-step detail, principally by Melvin Calvin and his associates at the University of California. These investigations and other similar ones were made possible by the availability after World War II of quantities of ^{14}C, a long-lived radioisotope which was much more suitable for tracing the pathway of carbon than ^{11}C, and by improved means of separating and identifying small quantities of substances by paper chromatography (Fig. 14.8).

Securing detailed information about the light reactions of photosynthesis has proved to be an even more difficult task than determining the nature of the dark reactions. Since 1957, Daniel Arnon of the University of California has pieced together one of the first comprehensive models of the sequence of events making up the light reactions, but his models and those of other investigators are subject to continuing modification and revision as more and more information is obtained. Physical processes, as well as biochemical and biological ones, are particularly important in the light reactions and the many investigators in this area include physicists as well as biochemists and plant physiologists.

A third major type of research that has been in progress since World War II and that is contributing to an understanding of photosynthesis is the clarification of the detailed structure of chloroplasts by electron microscopy and other techniques.

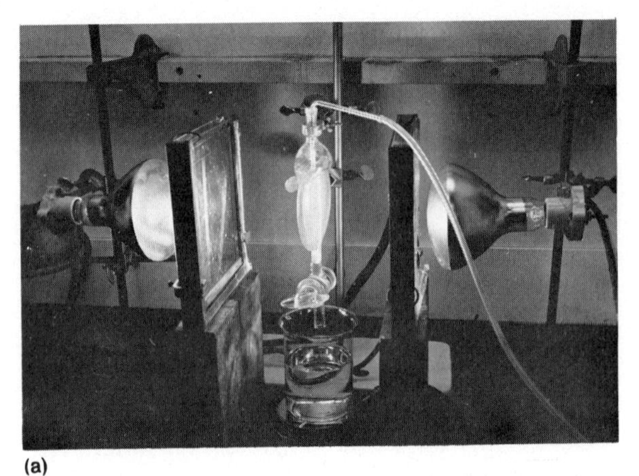

(a)

Fig. 14.8 (*Top*) *Apparatus used by Melvin Calvin and his co-workers in their pioneering studies of the pathways of carbon in photosynthesis. The flat closed flask between the two lamps and heat filters contained the unicellular algae. The* $^{14}CO_2$ *was introduced into the flask through the tube and after periods of time ranging from a few seconds to a minute or more the algae were suddenly killed by opening the stopcock, thus dropping the algae into the beaker of organic solvent. Extracts from the algae were then analyzed by separating the carbon compounds by two-dimensional paper chromatography and then making radioautograms of the chromatograms.* (*Bottom*) *Reproduction of a radioautogram made after 5 sec of photosynthesis, with the several labeled compounds identified. Note the high level of* ^{14}C *in the PGA and the sugar phosphates.*

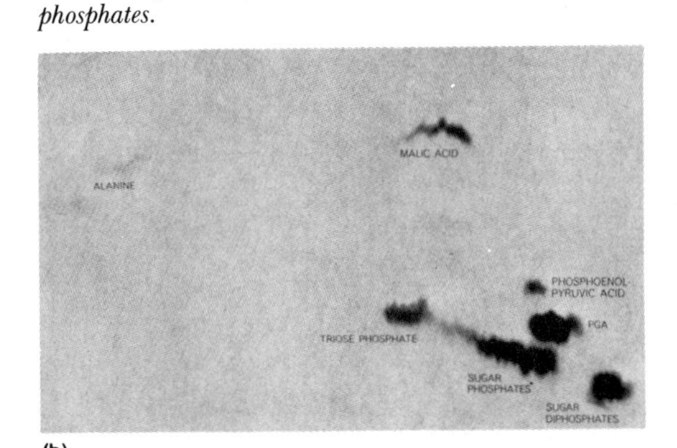

(b)

Chloroplasts: The Site of Photosynthesis

It has been known for a long time that chlorophyll is essential for photosynthesis and that at least the light reactions of photosynthesis occur within the chloroplasts, and it has now been established that the entire complex of the photosynthetic reactions occurs within the chloroplasts. If intact chloroplasts that have been isolated from plant cells are placed in a liquid medium containing the necessary reactants and exposed to light, carbohydrates will be synthesized and oxygen will be produced. If the membranes bounding chloroplasts have been broken and the stoma of the chloroplasts has been lost, the light reactions of photosynthesis may still occur with the production of oxygen, but no carbohydrate is synthesized. When the chlorophylls and carotenoids are extracted from chloroplasts by the use of suitable solvents, these pigments will still absorb light, but they are incapable of carrying on any of the photosynthetic reactions. The highly organized structure of the chloroplasts, the precise orientation of the pigment molecules in the chloroplast lamellae, and the presence of all the necessary enzymes and coenzymes are essential for photosynthesis.

The description of chloroplast structure in Chapter 8 should be reviewed at this time. All of the chloroplast pigments are located in the membranous lamellae that are more or less parallel with one another and extend from one end of a chloroplast to the other. The lamellae are composed of two membranes with a space between them and are closed at their ends, thus forming long and broad but thin, flattened sacs. These sacs are called **thylakoids,** especially where the lamellae branch or fold over on themselves and so form discs. The thylakoid discs usually occur in stacks, which are called **grana** (Fig. 14.9). The pigment molecules occur predominantly in the thylakoids of the grana. Like cell membranes in general, the membranes of the chloroplast lamellae are composed basically of phospholipids and proteins. The arrangement of these in the membrane is not definitely known, but it seems most likely that the membrane is basically a bimolecular layer of phospholipid molecules with globular protein molecules partially embedded in it at intervals. High-resolution electronmicrographs reveal such small spherical particles on both sides of the chloroplast membranes (Fig. 14.10). The protein component of the membranes is partly if not entirely enzyme protein. The chlorophylls and carotenoids are fat soluble and are located in the phospholipid part of the membrane. These pigment molecules are precisely oriented and spaced from one another, the long axis of the molecules being approximately perpendicular to the membrane surface. This precise orientation of the chloroplasts is essential for the light reactions of photosynthesis. The carbon fixation reactions occur in the stroma of the chloroplast, which is the matrix that surrounds the lamellae of the chloroplast.

The photosynthetic pigments include chlorophylls, carotenes, xanthophylls, and, in the red and blue-green algae, phycobilins. Chlorophyll *a* is the principal photosynthetic pigment, except in bacteria. A second kind of chlorophyll (chlorophyll *b* in higher plants and green algae) is also generally present. The other classes of pigments cannot carry on photosynthesis by themselves, but the light they absorb can be utilized by the chlorophyll. The green color of chlorophylls masks the yellow color of the carotenes and xanthophylls. However, fucoxanthol imparts a brown or yellow color to the brown algae and diatoms, phycoerythrin a red color to the red algae, and phycoxanthin a bluish color to the blue-green algae.

All these pigments absorb light energy that can be used in photosynthesis, but the energy absorbed by pigments other than chlorophyll *a* must all be passed on to chlorophyll *a* before it is usable in photosynthesis. Energy transfer from one chlorophyll molecule to another is much more efficient than energy transfer from carotenoids to chlorophyll. This can account for the difference between the absorption spectrum of chloroplasts and the action spectrum of photosynthesis in the regions of major absorption by the carotenoids (Fig. 14.11). The photosynthetic pigments absorb some light in most regions of the spectrum, but all have characteristic absorption peaks (Fig. 14.12). For example, the absorption peaks of the chlorophylls are in the red and blue regions with little absorption and much reflection in the green (which explains their green

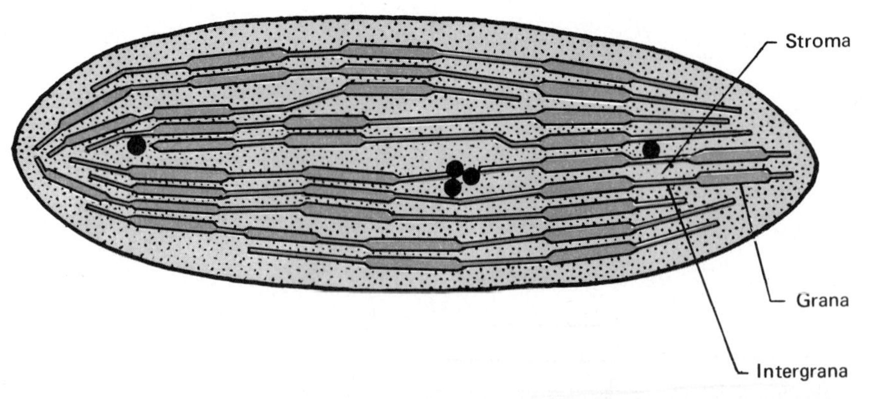

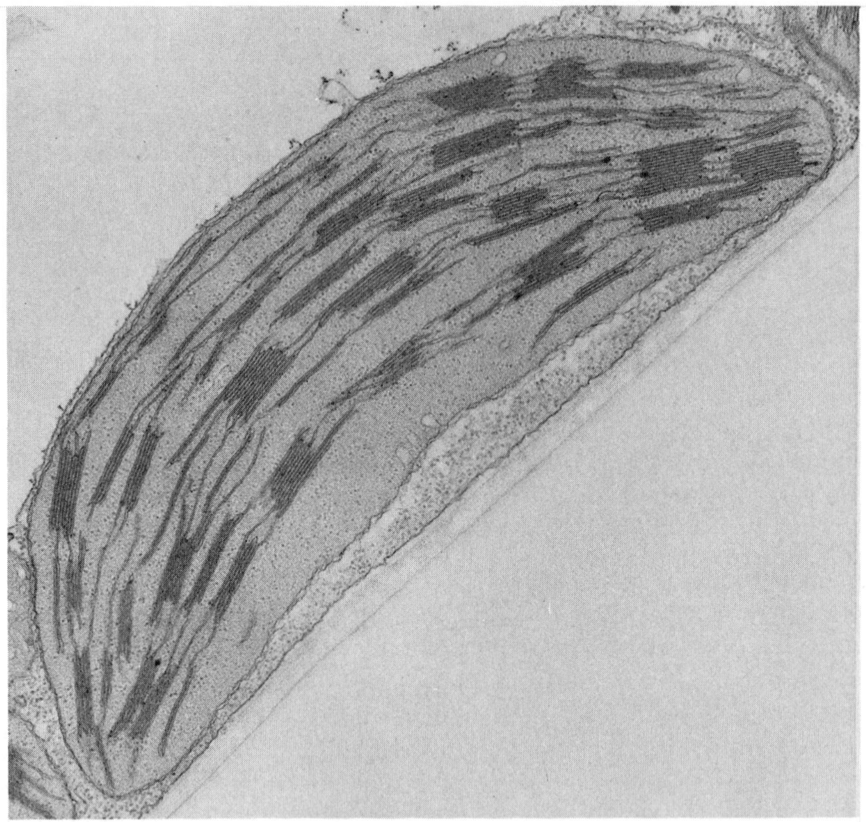

Fig. 14.9 (Top) Simplified drawing of a chloroplast. (Bottom) Electron micrograph of a tobacco chloroplast showing grana and membranes. Note that the lamellae are roughly parallel. (26,000X).

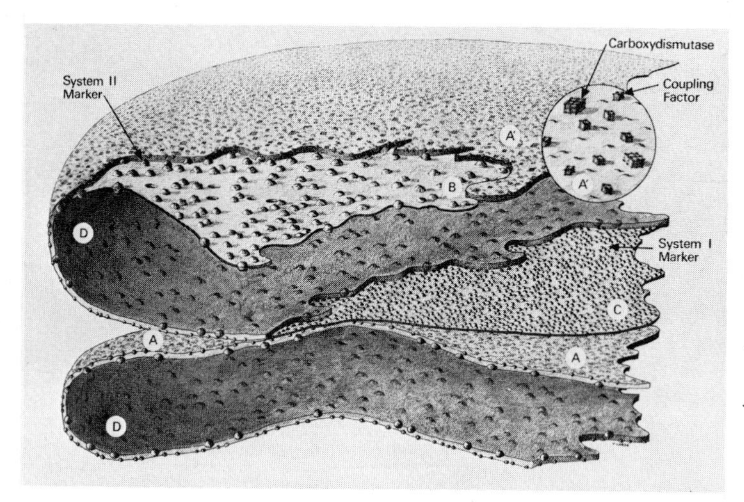

Fig. 14.10 Drawing of two granum discs showing particles in the membranes, based on electron micrographs about 28,000X. A, A′: *Carboxydismutase and phosphorylation coupling enzymes on the outer surface.* B, D: *System II markers.* C: *The smaller and more numerous System I markers.*

Fig. 14.11 Light absorption spectrum of Elodea *(solid line) and its photosynthetic action spectrum (broken line). Note that the spectra match closely in the region between 550 and 700 nm and around 430 nm, where most of the light absorption is by the chlorophylls, but deviate between 450 and 550 nm, where much of the light absorption is by the carotenoids. (Data of Max Hommersand and F. T. Haxo.)*

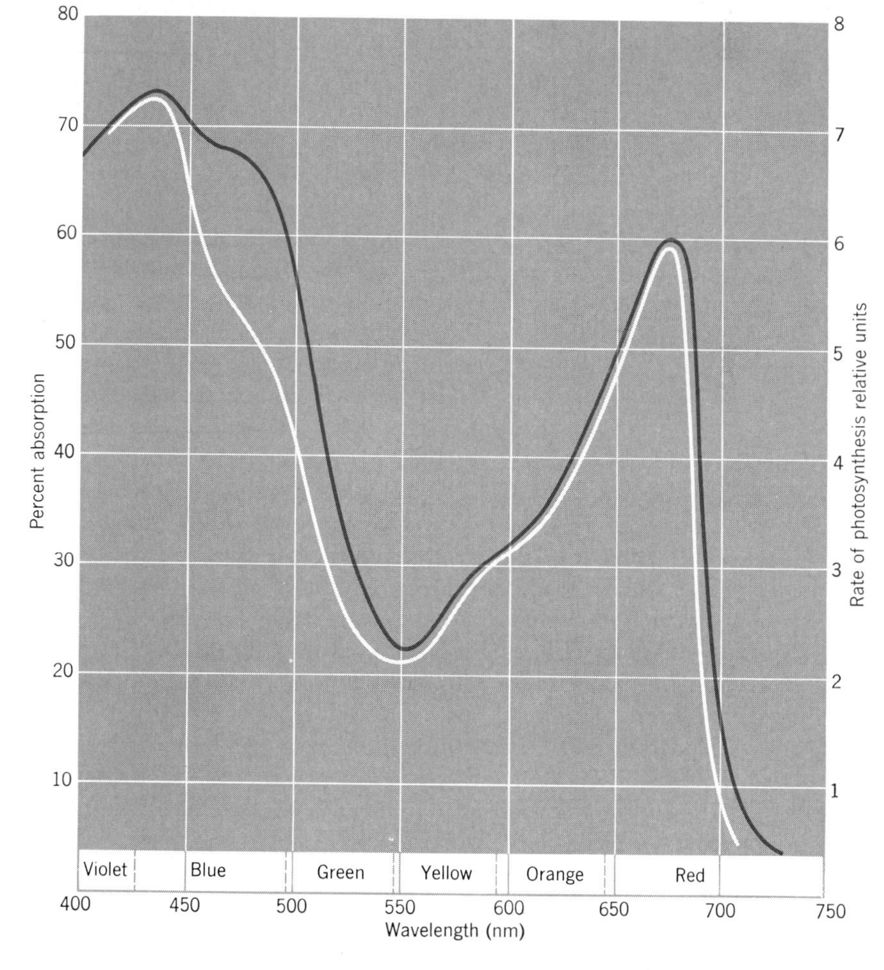

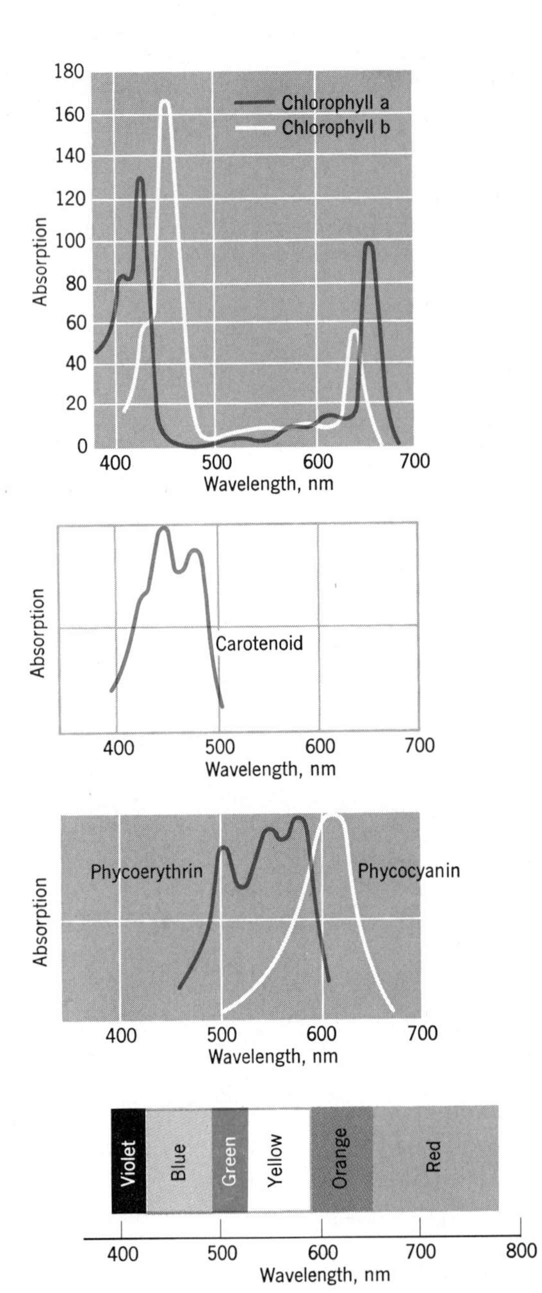

Fig. 14.12 The absorption spectra of several of the photosynthetic pigments.

color) while the carotenoids have absorption peaks in the blue region, resulting in their yellow or orange colors. The absorption peaks of phycoerythrin are in the green and yellow regions, the wavelengths least absorbed by water. This is why red algae can thrive much deeper in the ocean than other algae. In deep water most of the light absorption is by phycoerythrin, the energy then being transferred to chlorophyll *a*, but in shallow water the chlorophyll itself absorbs more of the light.

The photosynthetic pigments are organized into pigment systems, each one containing from 200 to 300 molecules of chlorophyll. Light absorbed by any pigment molecule in the system can eventually be transferred to a certain molecule of chlorophyll *a* in the system that is known as the exposed chlorophyll because it extends beyond the phospholipid layer. This concentration of energy in one molecule of the system probably is essential for the effective operation of the light reactions of photosynthesis. Two different pigment systems are involved in photosynthesis. These differ from one another in the proportions of the various pigments in them and have different absorption peaks. In System I the red absorption peak is at 700 nm (nanometer) while in System II it is between 682 and 690 nm. The roles of the two pigment systems will be described later.

The Energy Efficiency of Photosynthesis

Leaves commonly absorb about 83% of the light that strikes them, reflect about 12% and transmit about 5%. Of the 83% absorbed, only around 4% is absorbed by the chloroplast pigments. Most of the remainder is converted into heat that is lost by reradiation, conduction, or convection.

This may suggest that plants are very inefficient converters of solar energy into chemical energy. Furthermore, only part of the light falling on an area occupied by vegetation reaches leaves and other chlorophyll-bearing organs. About 1.6% of the light reaching a corn field during the growing season is used in photosynthesis. Corn plants use about a quarter of the sugar produced by photosynthesis in respiration, so the net efficiency is only 1.2%.

Since the corn grains contain about a third of the net photosynthetic product, efficiency from the standpoint of food usable by man is only 0.4%, and this is reduced to less than 0.13% when adjustment is made for the fact that the growing season is about one-third of a year. The 0.13% is the fraction of the solar energy that reaches the corn field during the season and that has been converted by plants into the chemical energy of foods readily usable by man. When photosynthetic efficiency is based on the light energy absorbed by chlorophyll, however, photosynthesis is a very efficient process, because about 90% of the absorbed light energy is converted to chemical energy.

The Light Reactions of Photosynthesis

The net result of the light reactions of photosynthesis is the production of oxygen gas (O_2) from the water used, the reduction of nicotine adenine dinucleotide phosphate ($NADP^+$) to NADPH, and the production of adenosinetriphosphate (ATP) from adenosinediphosphate (ADP) and phosphate. The ATP and NADPH, which contain chemical bond energy derived from the light energy, supply the energy that drives the reduction of carbon in the dark reactions, thus providing the connection between the two sets of photosynthetic reactions. The electrons (e^-) and protons of (H^+) of the hydrogen from the water are used in the course of the light reactions. The H^+ is used in the production of NADPH. The electrons are transferred through a series of electron acceptors at different energy levels and are finally incorporated in the NADPH.

There are two different kinds of photosynthetic light reactions (Fig. 14.13). The more basic one is noncyclic photophosphorylation, or noncyclic electron flow, which has been summarized above. The other one is cyclic photophosphorylation (or cyclic electron flow). Its only product is ATP. No NADPH is produced, and no water is used. The electrons are derived from chlorophyll rather than water, and after flowing through a series of electron acceptors they return to chlorophyll.

The Dark Reactions of Photosynthesis

The dark reactions are so-called not because they must occur in the dark but because they can occur in the dark if the product of the light reactions (ATP and NADPH) is available. Actually, the dark reactions are occurring in the light most of the time and continue only very briefly after a plant is placed in the dark since the ATP and NADPH produced in the light reactions are consumed very rapidly. The dark reactions are also referred to as carbon fixation, since the net result is the reduction of the carbon of CO_2 to a higher energy level of sugars. Since radioactive ^{14}C provides an excellent means of determining the series of compounds into which the carbon from CO_2 is incorporated, Calvin and others have been able to work out the sequence of events in the dark reactions in considerable detail.

In Fig. 14.14, only the most important steps in the dark reactions have been outlined, with emphasis on the number of carbon atoms in each of the compounds shown. More complete details are available elsewhere. The CO_2 is first picked up by a 5-carbon phosphorylated sugar, ribulose-1,5-diphosphate (RDP), with the production of two molecules of the 3-carbon compound phosphoglyceric acid (PGA). It should be noted that this type of carboxylation reaction, in which CO_2 reacts with some organic compound forming an acidic -COOH group, is by no means restricted to photosynthesis and occurs widely in both plants

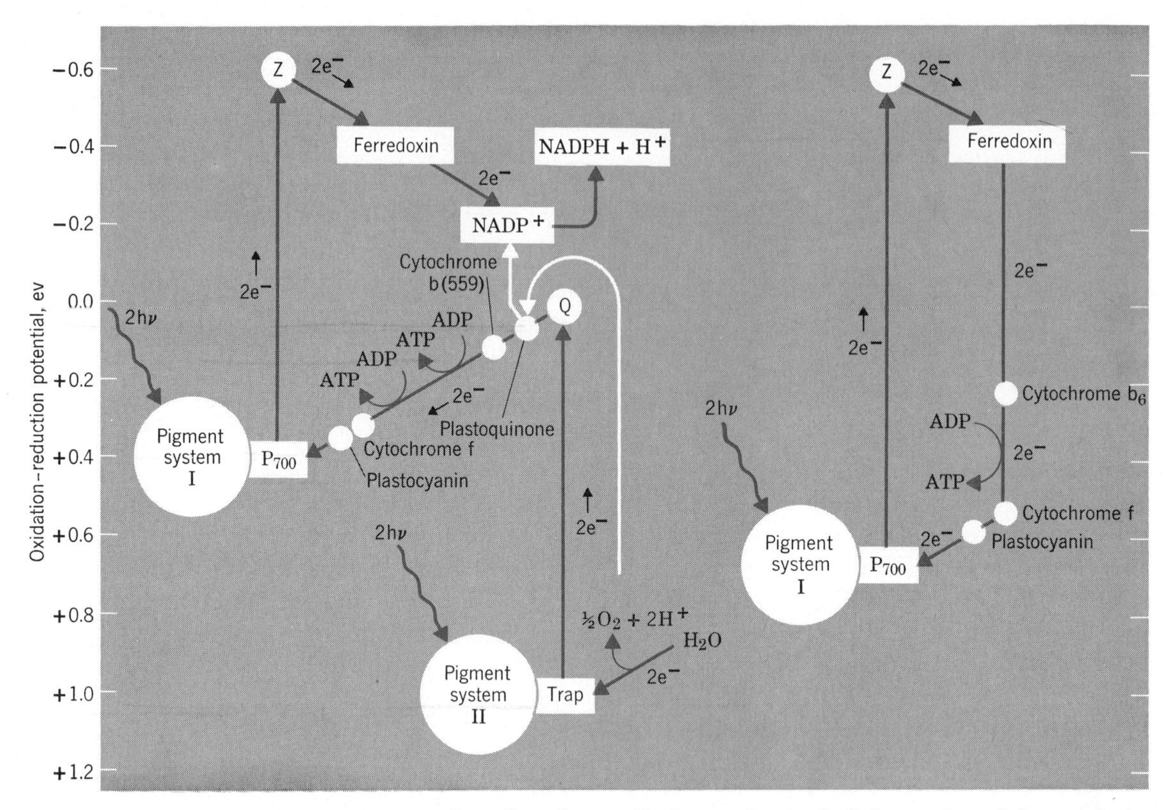

Fig. 14.13 Left: outline of noncyclic electron flow in the light reactions of photosynthesis. The first event is the absorption of two quanta (hν) of light energy by pigment system II. This raises the energy of two chlorophyll a *electrons (e^-) to a much higher level, resulting in their emission and transfer to electron acceptor Q. Two electrons from H_2O then fill the electron holes in the chlorophyll. Q donates the two electrons to a series of acceptors at progressively lower energy levels, the released energy being used in synthesizing ATP. When pigment system I absorbs two quanta two electrons from chlorophyll are raised to a much higher energy level and incorporated in electron acceptor Z. The pigment system I electron holes are filled by the two electrons originally emitted from pigment system II. From Z the two electrons from pigment system I are transferred to somewhat lower energy levels of ferredoxin and then $NADP^+$, reducing it to NADPH. The two protons (H^+) from the H_2O are not transferred along with the electrons, but rather reach the $NADP^+$ by way of plastoquinone. Thus the H of NADPH (electrons and proton) are derived from water, but the additional chemical bond energy comes from the light. The above process must occur twice to produce one molecule of O_2 and the two molecules of NADPH needed to reduce one molecule of CO_2. Right: cyclic electron flow, which is essentially a short-circuiting of noncyclic electron flow. Only pigment system I is involved and the electrons from chlorophyll return to it after flowing through Z, ferredoxin, and the other electron acceptors shown. The released energy is used in the synthesis of ATP. No e^- or H^+ from water is used and no NADPH is produced.*

and animals. A generalized equation for carboxylation is $RH + CO_2 \rightarrow R\text{-}COOH$. The symbol RH represents a variety of organic compounds and RCOOH the organic acids derived from them by carboxylation. However, the carboxylation of RDP occurs only during photosynthesis.

The PGA is next reduced to a phosphorylated triose sugar (phosphoglyceraldehyde, PGAL) by incorporation of hydrogen from the NADPH $+ H^+$ along with the incorporation of energy from ATP. It is at this point that the energy from the light reactions is introduced into the dark reactions and drives them. For every three molecules of CO_2 used, six molecules of triose are produced, but only one of these six is drained off as a net product. The other five are converted to three molecules of ribulose phosphate (RP). This conversion involves a considerable number of steps that we shall not attempt to describe here, with the production of a variety of C_4, C_5, C_6, and C_7 sugars as intermediates. The ribulose phosphate is then phosphorylated at the expense of the remaining ATP produced in the light reactions. This results in ribulose-1,5-diphosphate, which can now react with more CO_2 and so initiate another cycle. The ATP used in the dark reactions is converted back to ADP and the NADPH back to NADP, and so these are once more available for use in the light reactions.

It should be noted that for every molecule of CO_2 used in photosynthesis, two molecules of triose phosphate are produced. Since these two molecules contain a total of six carbon atoms it is evident that only one came from the CO_2, the remaining five being contributed by the ribulose-1,5-diphosphate. Of course, the carbons in the RDP were derived from CO_2 in previous photosynthetic cycles by way of the sugar interconversions. It should thus be evident that CO_2 is incorporated in carbohydrates in a piecemeal fashion, not all at one time.

The triose phosphate may be regarded as the product of photosynthesis since it is a sugar ester and the first substance produced after the

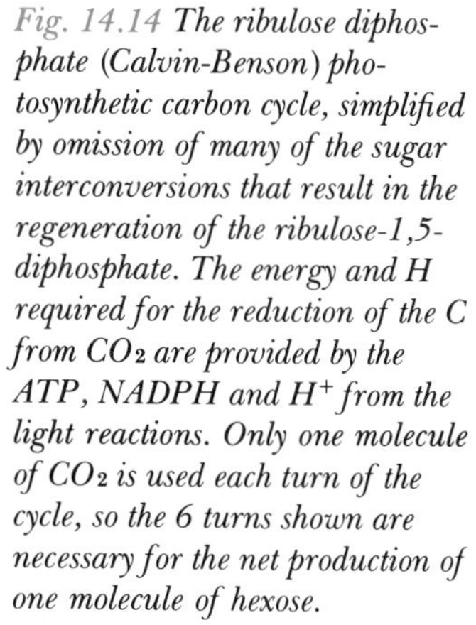

Fig. 14.14 The ribulose diphosphate (Calvin-Benson) photosynthetic carbon cycle, simplified by omission of many of the sugar interconversions that result in the regeneration of the ribulose-1,5-diphosphate. The energy and H required for the reduction of the C from CO_2 are provided by the ATP, NADPH and H^+ from the light reactions. Only one molecule of CO_2 is used each turn of the cycle, so the 6 turns shown are necessary for the net production of one molecule of hexose.

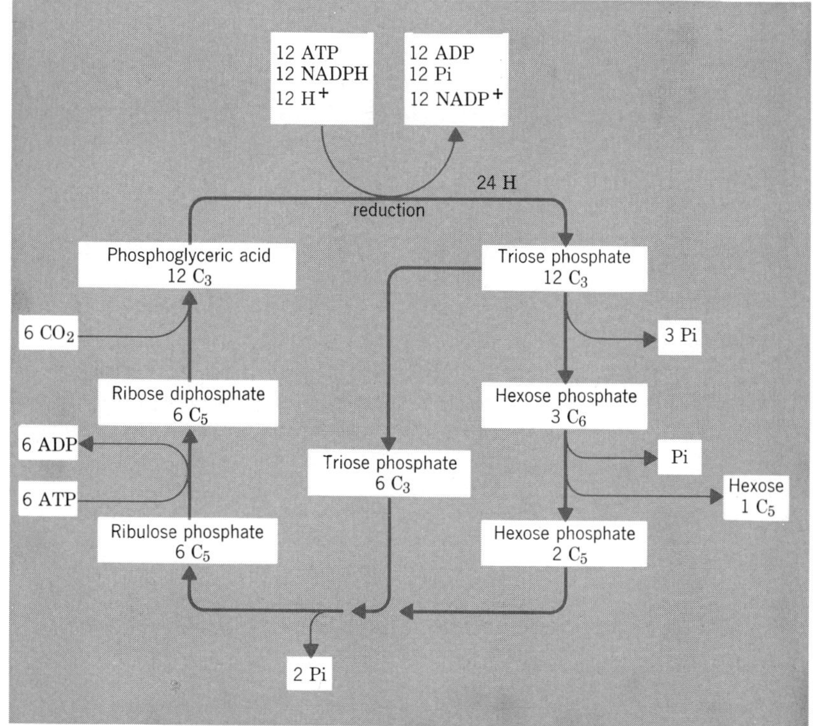

ATP and the NADPH from the light reactions have participated in the reduction of the carbon. However, it is fructose-1,6-diphosphate that is drained out of the cycle, and it is customary to consider the product of photosynthesis as being a hexose sugar. The fructose-1,6-diphosphate can be converted readily into other carbohydrates such as fructose, glucose, sucrose, and starch, and such conversions may occur within the chloroplasts, as well as in other parts of the cell. From the standpoint of a balanced summary equation for photosynthesis, it is most convenient to consider the product of photosynthesis as a nonphosphorylated hexose sugar.

However, sugar is not the only product of photosynthesis. The PGA may be used in the synthesis of various organic acids including amino acids and fatty acids rather than being converted to triose phosphate. Also, the triose phosphate may be converted into glycerol, that can then react with fatty acids and produce fats. The normal photosynthetic product of some algae is fat rather than sugar. Furthermore, the ATP and NADPH from the light reactions may be used in reductions such as the conversion of nitrates to ammonia. However, since sugar is the principal product of photosynthesis, we will not consider the alternative products further.

Summary Equations for Photosynthesis

Considering the product of photosynthesis to be a hexose sugar, we can now present summary equations for the light and dark reactions and for the process as a whole. The light reactions can be summarized as follows:

$$12H_2O + 12NADP^+ + 18ADP + 18H_3PO_4 \longrightarrow 12NADPH + 12H^+ + 18ATP + 6O_2$$

The dark, or carbon fixation and reduction reactions, can be summarized as follows:

$$6CO_2 + 12NADPH + 12H^+ + 18ATP \longrightarrow C_6H_{12}O_6 + 12NADP^+ + 18ADP + 18H_3PO_4 + 6H_2O$$

If the indicated number of molecules of each substance that appears on the left side of the above two equations is cancelled out by the number of its molecules that appear on the right side, we have the following traditional general summary equation for photosynthesis:

$$6CO_2 + 6H_2O \longrightarrow C_6H_{12}O_6 + 6O_2$$

A most important aspect of photosynthesis not shown in the above equations is that the chemical bond energy of the sugar is derived from light energy absorbed by the chloroplast pigments and used in driving the photosynthetic reactions.

An Alternative Pathway of Carbon Fixation in Photosynthesis

For some time the photosynthetic dark reactions described above (often referred to as the Calvin-Benson cycle) were regarded as the only pathway of photosynthetic carbon fixation and reduction. However, in 1965 George Burr and his colleagues, working in Hawaii, found that four-carbon acids rather than PGA were the first products of photosynthesis in sugarcane. Soon afterward, the Australians M. D. Hatch and C. R. Slack found that the carbon dioxide used in this alternate pathway of carbon fixation reacted with phosphoenol pyruvic acid rather than with ribulose-1,5-diphosphate. The product was oxaloacetic acid (Fig. 14.15). This then reacted with ribulose-1,5-diphosphate or ribulose-5-phosphate, with pyruvic acid and two molecules of PGA as the products. From PGA on the reactions are apparently the same as in the Calvin-Benson cycle.

The sugarcane type of photosynthesis (the

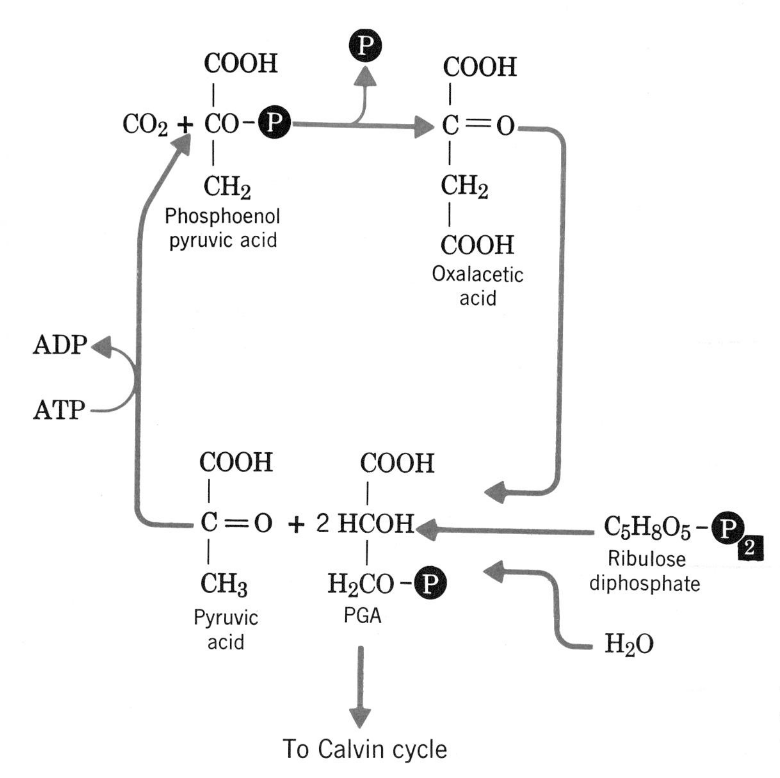

Fig. 14.15 In the sugarcane (Hatch-Slack) type of photosynthetic CO_2 fixation, the CO_2 reacts with phosphoenol pyruvic acid rather than ribulose diphosphate. However, the resulting oxaloacetic acid reacts with ribulose-1,5-diphosphate and the phosphoglyceric acid (PGA) then enters the Calvin-Benson pathway. However, the CO_2 acceptor (phosphoenol pyruvic acid) is regenerated from pyruvic acid and some of the ATP produced by the light reactions.

Hatch-Slack pathway) has been found to occur also in corn, various other tropical grasses, and a variety of dicotyledon plants including pigweed (*Amaranthus*) and *Atriplex*. It is likely to be found in many other species of plants. Photosynthesis by the Hatch-Slack pathway is much more efficient and productive than by the Calvin-Benson cycle. For one thing, it can reduce the available CO_2 to a much lower concentration, but perhaps most important is the fact that plants with the Hatch-Slack pathway do not carry on photorespiration, as do plants with the Calvin-Benson cycle. Photorespiration involves the oxidation of glycolic acid, an intermediate product of photosynthesis we have not mentioned previously. This greatly reduces photosynthetic productivity. Unlike ordinary respiration that occurs in the mitochondria, photorespiration is promoted by light and its rate increases with light intensity.

The high productivity of the sugarcane type of photosynthesis is of significance from a practical standpoint. More species with this photosynthetic pathway are likely to be considered for use as crop plants. Also, plant breeders may be able to introduce the pathway into species now lacking it or modify other traits of plants that do have the pathway so as to make them more desirable as crop plants.

Factors Affecting the Rate of Photosynthesis

Although most of the current research on photosynthesis is directed toward a better understanding of the nature of the process, our knowledge of the influence of various factors on the rate of photosynthesis is of greater practical value at present. Since photosynthesis is the ultimate source of food of practically all organisms, photosynthetic productivity is of general biological concern. To farmers and gardeners, whose income depends on the yield of their crops, it is a matter of monetary importance. The possibility of using algae as a source of both food and oxygen during prolonged space travel has resulted in extensive work directed toward increasing the photosynthetic productivity of algae.

Both hereditary and environmental factors affect the rate of photosynthesis. Plants of different species growing in the same environment may have quite different rates of photosynthesis. At the same time, the rate of photosynthesis in a particular plant varies greatly with the prevailing environmental conditions. Particularly important are light intensity, light quality (wavelength composition), carbon dioxide, water, temperature, and factors affecting chlorophyll content.

In any particular chloroplast and at any particular moment, only one factor limits the rate of photosynthesis, and the rate can be raised only by increasing this **limiting factor.** For example, if light intensity is limiting, an increase in light intensity will raise the rate, but supplying a higher concentration of carbon dioxide or a higher temperature would not. As light intensity is increased more and more, however, a point is reached where a further increase in light intensity is not accompanied by a rise in the rate of photosynthesis (Fig. 14.16); some other factor is now limiting. If we find that now only an increase in carbon dioxide will raise the rate of photosynthesis we know that this is the limiting factor.

Under natural conditions, deficient chlorophyll content is not commonly a limiting factor in photosynthesis unless a plant is diseased or is deficient in elements such as magnesium or nitrogen that are used in synthesizing chlorophyll molecules, or other elements such as iron, sulfur and manganese that affect chlorophyll synthesis. In most plants light is required

Fig. 14.16 Influence of CO_2 concentration on the rate of photosynthesis at four light intensities. CO_2 is limiting up to the point where the curves flatten out, where light becomes limiting. Note that even at these relatively low light intensities, photosynthesis increases with CO_2 concentrations substantially greater than the 0.03% in the atmosphere. Outdoor light intensity is on the order of 10,000 ft-c. (Data of W. H. Hoover et al., Smithsonian Institution Miscellaneous Collection, 87, *1, 1933.)*

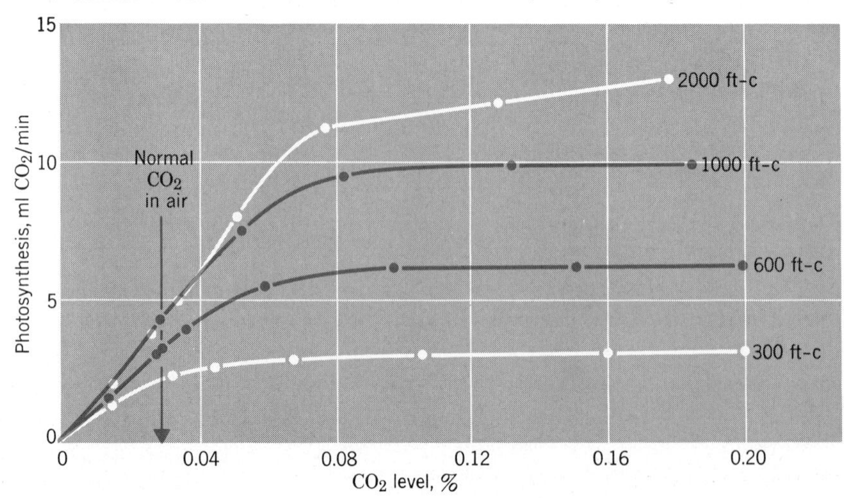

Fig. 14.17 Green and albino corn seedlings. Since they are unable to carry on photosynthesis, the albino seedlings starve to death when the food in the grain is exhausted.

in the final steps of chlorophyll synthesis. Thus, a plant that has grown from a seed in darkness cannot begin photosynthesis as soon as it is placed in the light, but must first synthesize chlorophyll. In many seedling beds, some young plants can be found that never synthesize chlorophyll because they lack the necessary hereditary potentialities. Such plants are called **albinos** and they are, of course, unable to carry on photosynthesis (Fig. 14.17).

Both light intensity and light quality may be limiting factors in photosynthesis. Light intensity is obviously the limiting factor at night, early morning, or late evening, and may also be a limiting factor on very cloudy days and in the shade. Although most plants can carry on some photosynthesis even in very dim light, a plant will survive and grow over an extended period only if the product of photosynthesis is at least equal to the amount of food used in respiration plus assimilation. Plants that naturally grow in the shade, as in a forest, carry on this basic minimum rate of photosynthesis at lower light intensities than those species found in sunny habitats. Although light quality is not a limiting factor in plants exposed to direct sunlight, the light received by the lower-growing plants in a thick forest is deficient in quality as well as intensity since wavelengths (red and blue) best absorbed by chlorophyll have largely been filtered out by the leaves of the tree canopy. Since water absorbs the longer wavelengths of light more than the shorter ones, water plants that live deep in lakes or the ocean receive light deficient in orange and red.

Carbon dioxide is very commonly a limiting factor in photosynthesis, particularly on clear summer days when plants are adequately supplied with water. Unfortunately, there is now no practical and economically feasible method of supplying field crops with more than the usual 0.03% of carbon dioxide present in the air, except perhaps by increasing the population of soil bacteria, fungi, and animals, thus increasing the total production of carbon dioxide by respiration. Such additional carbon dioxide may promote increased photosynthesis in low-growing plants, and it is conceivable that some of the beneficial effects of manures and other organic fertilizers might be related to an increased population of soil organisms. Carbon dioxide gas can be added to the air in greenhouses, and sometimes is.

In a wilted plant, or one that is just beginning to wilt, water is likely to be the limiting factor in photosynthesis, principally because the rate of most life processes, including photosynthesis, decreases when protoplasm has too low a water content. When a water deficiency results in stomatal closing, photosynthesis may cease almost entirely. Although brought about by a lack of water, this is really a situation where carbon dioxide is the limiting factor. Under natural conditions, water may be the limiting factor in photosynthesis, not only during prolonged droughts when the plants approach permanent wilting, but also amlost every afternoon during hot, summer weather when there is temporary wilting.

Under natural conditions, temperature is likely to be a limiting factor in photosynthesis only on very cool days. As the temperature rises, a point is reached where the rate of respiration rises more rapidly than the rate of photosynthesis (Fig. 14.18). Thus, a plant may have a greater net food production at a lower temperature, even though the process of photosynthesis itself is more rapid at a higher temperature. At high temperatures, the rate of photosynthesis may start falling while that of respiration is still rising. This results in an even lower net rate of photosynthesis, or even an excess of respiration over photosynthesis.

RESPIRATION

As essential as photosynthesis is for the continuation of life on earth, it is no more essential than respiration, since both plants and animals depend on energy made available from food by respiration for their continued existence. Photosynthesis converts light energy into the chemical energy of sugar, but respiration makes available the chemical energy of sugar (and other foods) and permits the use of this energy in a variety of useful ways within the organism. The end products of photosynthesis (sugar, oxygen, water) are the substances used in ordinary aerobic respiration, whereas the end products of respiration (carbon dioxide and water) are the substances used in photosynthesis.

We are using the term **respiration** in its basic biological sense, referring to the series of biochemical oxidations of foods occurring in living cells, the energy contained in these foods being transferred to a substance (ATP) that serves as a readily available energy supply. Unfortunately, the term *respiration* has been used in several other ways: as a synonym for *breath-*

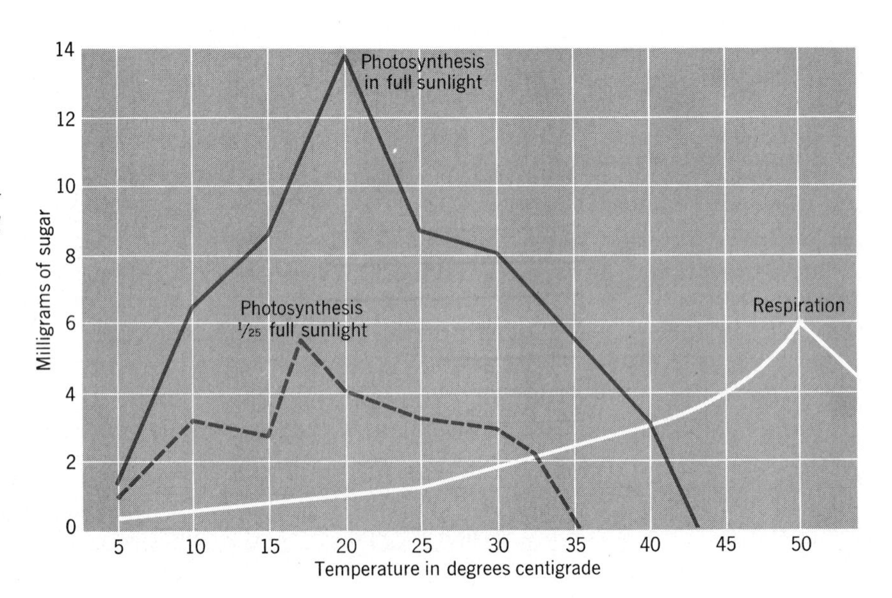

Fig. 14.18 Rates of photosynthesis and respiration in potato leaves at different temperatures. (Data of Lundegardh. Redrawn from Textbook of Botany *by Transeau, Sampson, and Tiffany with the permission of the publishers, Harper & Row.)*

ing, as a label for the diffusion of gases in and out of the lungs and gills or between the blood and tissues of animals, or even to refer to the diffusion of gases between a plant or animal and its environment. Although plants and many animals do not breathe, all living cells carry on respiration.

In animals much of the energy made available by respiration is used in muscular contraction and the generation of nerve impulses. Although plants do not use energy from respiration in these ways, they do use it (as animals do) in accumulating mineral ions (Chapter 11), in synthesizing fats, proteins, and certain other compounds, in cell division and other aspects of growth, in assimilation, in the maintenance of protoplasmic structure and differential permeability of membranes, in light production by some species (Fig. 14.19), and in other ways.

If plants used as much energy from foods in their activities as animals do in muscle contractions alone, the net food production by plants would be small and they could not serve as food suppliers to the entire living world. The extensive use of molecular energy in the transport of substances through plants, in contrast with the use of energy from foods in pumping blood through the body of an animal, represents one substantial saving of food energy in plants.

Aerobic Respiration

The oxidation of foods proceeds in various ways in different species, or even in one species under different conditions, but the majority of plants and animals have very similar if not identical pathways of **aerobic respiration,** that is, respiration dependent on a supply of oxygen as well as food. In plants, departures from this standard pattern of respiration are restricted almost entirely to some bacteria and fungi, although other plants may carry on anaerobic respiration for a while in the absence of oxygen. Aerobic respiration may be summarized as follows:

$$C_6H_{12}O_6 + 6O_2 \rightarrow 6CO_2 + 6H_2O$$

This summary equation, however, provides little information about the nature of respiration. The process occurs in many steps, just as photosynthesis does, and in recent years biochemists have worked out the series of reactions in great detail. It is not necessary or desirable to give the complete details in an introductory course, but Fig. 14.20 outlines the process in a general way.

Aerobic respiration can be regarded as consisting of four principal stages, or series of chemical reactions. These are, in the order of their occurrence, glycolysis, the citric acid cycle (Krebs cycle), oxidative phosphorylation, and

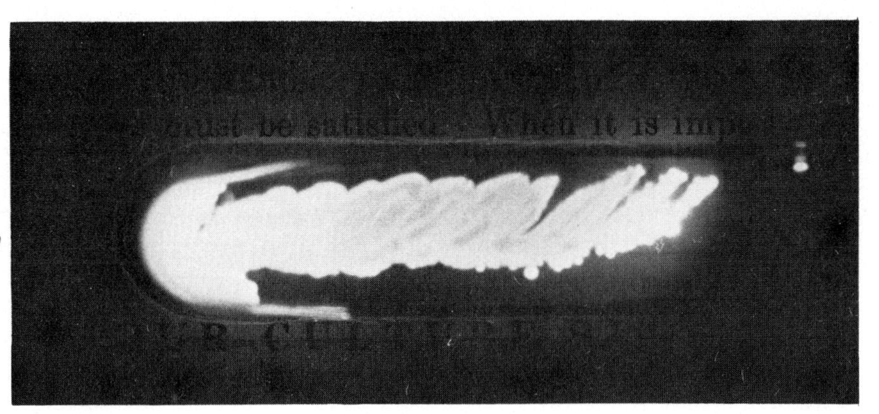

Fig. 14.19 Light production by a colony of Photobacterium fischeri, *illustrating one of the less universal uses of energy from respiration. Some other bacteria, fungi, and animals convert ATP energy into light. The photograph was made by light emitted by the bacteria (6 hours at f/4.5).* (Courtesy, Carolina Biological Supply Co.)

finally the release of energy that can be used in doing metabolic work. Each of the stages will be discussed below, and each one is also outlined in some (but not complete) detail in Fig. 14.20.

Glycolysis

Glycolysis consists of a series of reactions in which glucose is first phosphorylated and then converted to fructose phosphates, triose phosphates, phosphoglyceric acid, and finally pyruvic acid (pyruvate). Since pyruvate is a 3-carbon compound, two molecules are produced from each glucose (C_6) molecule. The two molecules of pyruvate contain a total of 8 atoms of H, as compared with the 12 atoms of H in glucose. The hydrogen removed in the oxidation of glucose to pyruvate is transferred to two molecules of NAD^+, reducing them to $NADH + H^+$. In addition, sufficient energy is released during glycolysis for a net production of two molecules of ATP. When we consider only the net input and the net output of glycolysis, the reactions shown in Fig. 14.20 may be summarized as follows:

$$C_6H_{12}O_6 + 2NAD^+ \rightarrow 2CH_3 \cdot CO \cdot COOH + 2NADH + 2H^+$$

$$2H_3PO_4 + 2ADP \rightarrow 2ATP + 2H_2O$$

During glycolysis, only a small portion of the chemical bond energy of glucose is converted to the readily usable chemical bond energy of NADH and ATP. Most of the energy originally present in glucose is now in the pyruvic acid. Note that the [H] acceptor in glycolysis is NAD^+, not $NADP^+$ as in photosynthesis. Glycolysis does not occur in the mitochondria (Chapter 8), but the remaining stages of respiration do.

The Citric Acid Cycle

The net result of the citric acid cycle is the breaking down of pyruvic acid into CO_2 and [H]. The [H] is transferred to hydrogen-acceptor coenzymes, principally NAD^+, thus reducing them. Most of the chemical bond energy originally present in the glucose is now in the $NADH + H^+$, although during the citric acid cycle some is used in making a molecule of ATP.

The decomposition of the pyruvate in the citric acid cycle occurs in the stepwise, cyclic fashion characteristic of many biochemical processes (Fig. 14.20). Before entering the cycle proper the pyruvate is converted to a 2-carbon compound by the loss of CO_2 and 2[H]. This compound, which is related to acetate, is bonded to coenzyme A, forming an activated substance known as acetyl-coenzyme A. The acetyl (C_2) then enters the cycle by reacting with oxalacetic acid (C_4) and thus forming

Fig. 14.20 Simplified outline of the sequence of reactions in aerobic respiration. (a) *Glycolosis. This results in the conversion of a molecule of glucose into two molecules of pyruvic acid and the incorporation of energy released by the oxidations into two molecules of NADH and a net of two molecules of ATP.* (b) *Citric acid cycle (Krebs cycle). The pyruvic acid is converted into acetyl-CoA, which enters the cycle. These reactions completely break down the pyruvic acid into CO_2 and H, which along with H from the water used in the cycle, reduces NAD^+ and other H acceptors. Not shown is 1 molecule of ATP generated during the cycle.* (c) *Oxidative phosphorylation. The H from $NADH + H^+$ is transfered to a flavoprotein (FP) at a lower energy level, and the FPH_2 then donates the electrons to a series of cytochromes at progressively lower energy levels. Oxygen is the final electron (e^-) and proton (H^+) acceptor, and so water is produced. The release of energy is sufficient to make three molecules of ATP per molecule of NADH oxidized, or two molecules of ATP per molecule of FPH_2. Some of the NADH produced during respiration is used in other ways.*

citric acid (C_6). The rest of the cycle consists in the conversion of one organic acid into another, most of the steps involving the loss of either CO_2 or [H] or both. The oxalacetic acid is finally regenerated and is then available to participate in another cycle. By the time this has occurred the atoms originally contributed by the pyruvate are all in the form of CO_2 and [H]. Each cycle releases 10[H], enough to reduce five molecules of hydrogen acceptor, but only four of the ten come from the pyruvate. The remainder come from the three molecules of water used in the cycle. Considering only the net input and the net output, the citric acid

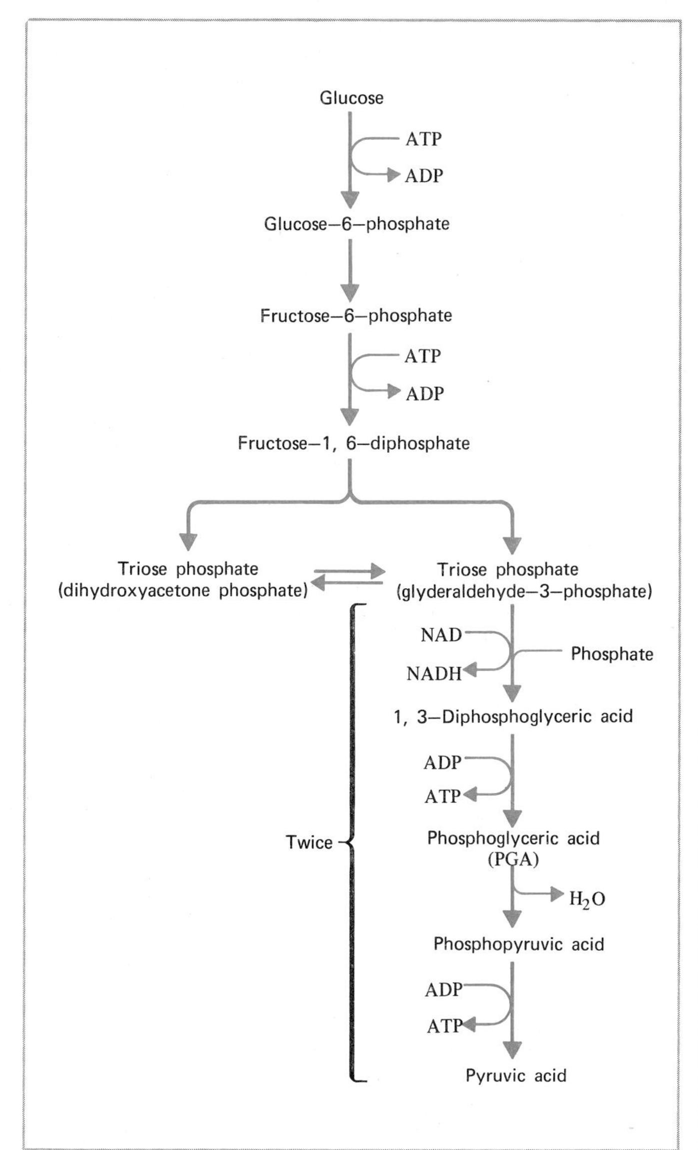

(a) Glycolysis

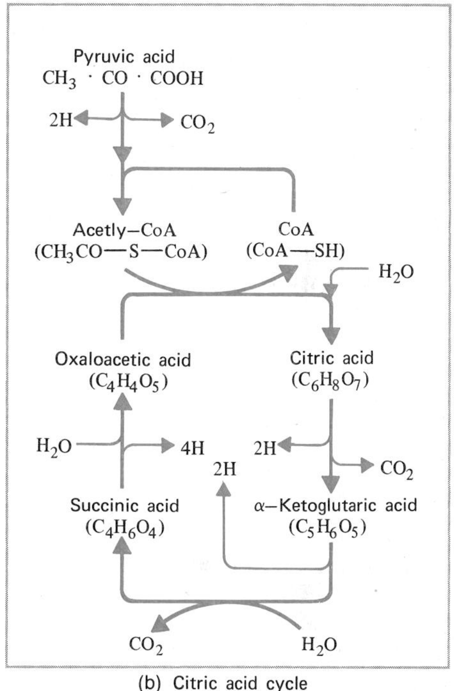

(b) Citric acid cycle

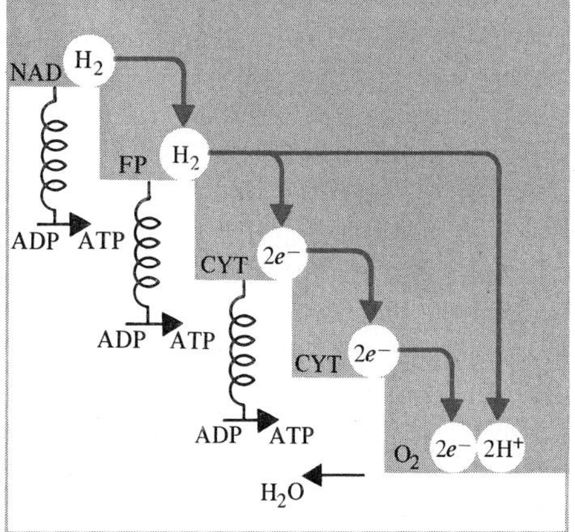

(c) Oxidative phosphorylation

cycle may be summarized as follows, the $NADH_2$ and other reduced H-acceptors being designated as AH_2:

$$\begin{matrix} CH_3 \cdot CO \cdot \\ COOH \end{matrix} + 3H_2O + 5A \rightarrow 3CO_2 + 5AH_2$$

$$H_3PO_4 + ADP \rightarrow ATP + H_2O$$

Since two molecules of pyruvate are derived from each molecule of glucose, the citric acid cycle turns twice for each glucose molecule used. Note that all six molecules of CO_2 that appear in the general summary equation for respiration are produced during the breakdown of the pyruvate. Note also that up to this point no O_2 has participated in the reactions.

Oxidative Phosphorylation

The NADH and other reduced H-acceptors produced during glycolysis and the citric acid cycle may participate in a variety of metabolic processes such as fat synthesis and nitrate reduction by serving as hydrogen donors, thus reducing the substrate. However, the NADH participating in the completion of the process of aerobic respiration enters into the sequence of reactions known as oxidative phosphorylation. First, its [H] is transferred to a flavoprotein (FP) coenzyme, reducing it to FPH_2 at a lower energy level than the NADH, and the energy released is used in synthesizing a molecule of ATP from ADP and phosphate. At this point it should be noted that the [H] actually consists of a proton (H^+) and an electron (e^-). The FPH_2 now transfers $2e^-$ to a cytochrome coenzyme, thus reducing the cytochrome, and the $2H^+$ are released. The specific component of the cytochrome that is reduced is iron, from Fe^{+++} to Fe^{++}. The reduced cytochrome is at a lower energy level than the FPH_2. Again, the energy is used in making a molecule of ATP. There is now a transfer of electrons from the reduced cytochrome to other cytochromes (only two of them shown in Fig. 14.20), each one being at a lower energy level than its predecessor. The energy released is used in making a third molecule of ATP. Finally, in the very last step of the series, $2H^+$ and $2e^-$ react with an atom of oxygen, producing a molecule of water. Thus oxygen serves as the final hydrogen acceptor. It is here that all the H_2O shown as a product of respiration in the general summary equation is formed.

The important result of oxidative phosphorylation is the transfer of energy from the NADH to the high energy bond holding the third phosphate of ATP. The energy of one NADH molecule is distributed among three ATP molecules, but it should be noted that about a third of the NADH bond energy is lost as heat rather than being incorporated in ATP bonds. It is important that the energy is released in a stepwise fashion in relatively small amounts, rather than in a single large surge, and it should be noted that this is characteristic of biological oxidation-reduction reactions.

Twelve molecules of hydrogen acceptor are reduced for each glucose molecule used in respiration (two in glycolysis plus ten in the two turns of the citric acid cycle). Ten of the 12 molecules of reduced hydrogen acceptor are NADH. As their H^+ and e^- are transferred through the flavoprotein and cytochromes three molecules of ATP are generated by the energy available from each molecule of NADH, thus producing 30 molecules of ATP. At one step in the citric acid cycle (succinate to fumarate) the [H] bypasses NAD^+ and is picked up directly by flavoprotein, thus giving rise to only two ATP in the terminal oxidations. Considering the two turns of the citric acid cycle per glucose molecule, this adds four more ATP produced during the terminal oxidations, for a total of 34. Oxidative phosphorylation may, then, be summarized as follows:

$$12AH_2 + 6O_2 \longrightarrow 12A + 12H_2O$$

$$34H_3PO_4 + 34ADP \longrightarrow 34ATP + 34H_2O$$

In addition, we must not forget the molecule of ATP generated directly in each of the two turns of the citric acid cycle and the net production of two ATP in glycolysis. Thus, a total of 38 molecules of ATP is produced per molecule of glucose used in respiration. One mol of glucose contains 673 kcal of energy, but the 38 mols of ATP contain only about 380 kcal. The remainder of the energy is lost as heat.

ATP Energy Release

Respiration is often considered to end with the terminal oxidations and the production of ATP. However, the actual use of the energy in useful work in a plant or animal does not occur until the ATP later breaks down into ADP and phosphate:

$$38ATP + 38H_2O \longrightarrow 38ADP + 38H_3PO_4$$
$$\hookrightarrow \text{energy} \longrightarrow \text{metabolic work}$$

Since the result of respiration is the transfer of chemical energy from sugar to ATP, and particularly since some of the energy from the sugar is dissipated as heat, the long and rather complicated series of reactions making up the process of respiration might appear to be lost motion. This is not the case. Sugars are quite stable and unreactive compounds with their energy scattered through many low energy bonds and do not provide a readily available energy source. The energy derived from the sugar during respiration is partitioned among many ATP molecules, each of them with the energy contained in a high-energy bond that is readily broken, thus releasing a burst of energy usable in a variety of cellular work. We might compare the transfer of energy from sugar to ATP during respiration with the refining and concentrating of a bulk product, and the packaging of the product in smaller containers of usable size. Or, we might compare respiration with the conversion of securities into readily negotiable bank notes of convenient size. Then the ATP would represent a cell's readily available cash supply of energy.

If you will now go back and write down the summary equations for each of the four stages of respiration and then cancel out each substance that appears on both the left-hand and right-hand sides of any of the equations, you should find that what is left constitutes the old general summary equation for respiration:

$$C_6H_{12}O_6 + 6O_2 \longrightarrow 6CO_2 + 6H_2O$$

Hopefully, this will now have more meaning to you than it did before. Although this summary equation is perfectly acceptable, you should now realize that it really provides little information about how respiration proceeds and how the chemical bond energy of glucose is made usable for metabolic work.

Enzymes and Respiration

Each of the many individual chemical reactions making up the process of respiration is activated by a specific enzyme, hence several dozen different enzymes are involved in the complete process. This is why it is impossible to give *the* respiratory enzyme in the summary equation of respiration. All the enzymes involved in the citric acid cycle and the terminal oxidations are found in the mitochondria of a cell, probably arranged in an orderly fashion. Thus the mitochondria appear to be the site of the major part of respiration, just as chloroplasts are the site of photosynthesis. As more and more is learned about the structure and chemistry of cells, it is becoming evident that many, if not all, biochemical processes are localized in specific cell structures. The organized proximity of the various enzymes involved in successive steps of processes such as respiration and photosynthesis undoubtedly greatly facilitates the reactions and eliminates the diffusion of the product of one step to another place in the cell before it reacts in the next step.

As has been noted in the discussion of the respiratory reactions, coenzymes such as NAD, NADP, coenzyme A, and the cytochromes play important roles in oxidation-reduction reactions, principally as hydrogen or electron carriers. However, it should be emphasized that such coenzymes are quite ineffective unless they are associated with the appropriate kind of enzyme protein.

Anaerobic Respiration

The widespread type of aerobic respiration just described and other less common types of aerobic respiration can proceed to completion only when oxygen is available. However, some plants can carry an **anaerobic respiration** in the absence of free oxygen. The most common type of anaerobic respiration is **alcoholic fermentation,** a process carried on by yeasts, a variety of other fungi, some bacteria, and even green plants when deprived of oxygen. In summary, alcoholic fermentation proceeds as follows:

$$C_6H_{12}O_6 \rightarrow 2C_2H_5OH + 2CO_2$$

When the process is considered in greater detail, we find that the first series of reactions in alcoholic fermentation is the same glycolysis series that occurs in aerobic respiration. The pyruvic acid resulting from glycolysis is then converted in two steps to alcohol and CO_2:

$$C_6H_{12}O_6 + 2NAD^+ \longrightarrow 2CH_3{\cdot}CO{\cdot}COOH + 2NADH + 2H^+$$

$$2CH_3{\cdot}CO{\cdot}COOH \longrightarrow 2CH_3{\cdot}CHO + 2CO_2$$

$$2CH_3{\cdot}CHO + 2NADH + 2H^+ \longrightarrow 2C_2H_5OH + 2NAD^+$$

During alcoholic fermentation there is a net production of only two molecules of ATP for each molecule of sugar used. The great bulk of the energy originally present in the sugar is now in the alcohol. Anaerobic respiration is a very inefficient energy-releasing process compared with aerobic respiration, which results in the formation of some 38 molecules of ATP from one molecule of sugar. On the other hand, anaerobic respiration does have the advantage of proceeding in the absence of oxygen.

The two principal commercial uses of yeasts—baking and brewing—are dependent on the two products of alcoholic fermentation—carbon dioxide and alcohol. When the alcohol concentration of the yeast culture medium reaches 12 to 14%, the alcohol kills the yeast cells. This is why unfortified wines and undistilled beverages never contain more than this concentration of alcohol. Yeast and some other microorganisms that carry on anaerobic respiration can also carry on ordinary aerobic respiration in the presence of an adequate supply of oxygen.

If higher plants are deprived of oxygen they may carry on alcoholic fermentation in the same way as yeast. Under natural conditions this is most likely to occur in roots when the soil is water-logged, thus reducing the available oxygen to a low level. Although the roots of some species of plants thrive in flooded soil, those of most species are seriously injured or killed after being deprived of oxygen for several days.

Respiration Versus Combustion

Although the summary equation for aerobic respiration could also be used for the complete combustion of sugar, and although both processes involve the oxidation of the sugar and the release of the same total quantity of energy from it, they are vastly different in detail and should not be confused. It is quite improper to refer to the burning of sugar within plant and animal cells. Whereas the oxidation of sugar during respiration is a step process activated by enzymes, releasing energy in small controlled quantities and transferring it from one compound to another, combustion is a rapid release of energy at high temperatures, the energy being converted into heat and light. If we compare respiratory energy release with the energy released as a person walks down a flight of stairs, combustion could be compared with the person jumping out of the window.

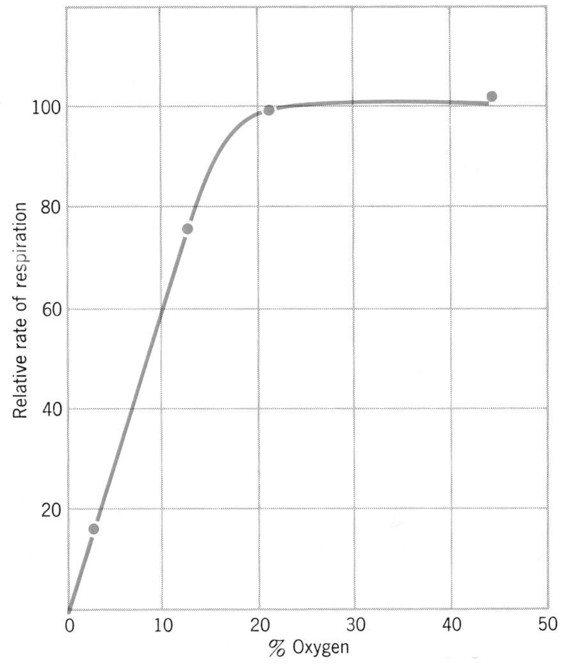

Fig. 14.22 Influence of oxygen concentration on the rate of plant respiration. (Data of Steward, Berry, and Broyer, 1936.)

Both methods of descent involve the loss of the same amount of potential energy, and both carry him to the same level, but there is certainly a marked difference in the methods of descent and energy expenditure, not to mention the effects on the individual.

Fig. 14.21 Influence of the water content of wheat grains on their rate of respiration. (Data of Bailey and Gurgar, 1918.)

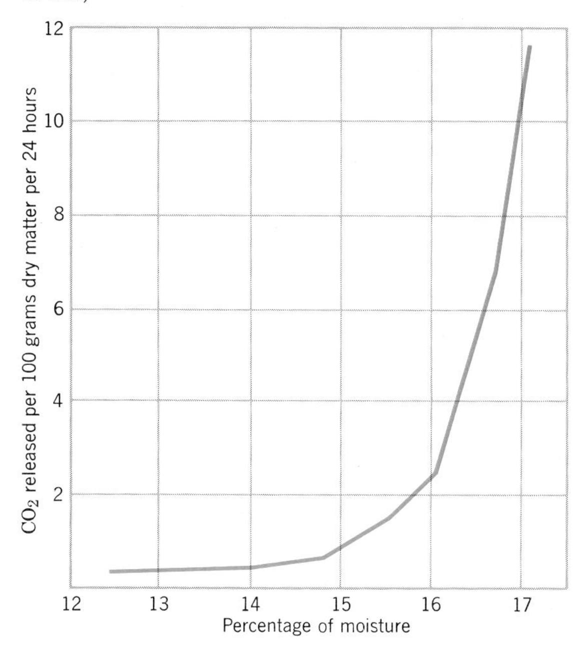

Factors Affecting the Rate of Respiration

Various environmental factors have a marked effect on the rate of respiration in plants and the principle of limiting factors operates just as it does in photosynthesis (and all other biological processes). The principal factors affecting the rate of respiration are temperature (Fig. 14.18), water content of the cells (Fig. 14.21), food availability, and oxygen concentration (Fig. 14.22). Adequate food supplies are usually present in a healthy plant, and oxygen is generally not a limiting factor unless its concentration is less than the 20% or so normally present in the air. Low water content of the protoplasm is generally the limiting factor in dry, mature seeds, and probably also in most wilted plants. The most common limiting factor in respiration is, however, temperature, and most of the time the rate of respiration in plants will increase with temperature up to the point (about 50°C) where enzymes are inactivated.

Photosynthetic/Respiratory Ratio

The rate of photosynthesis divided by the rate of respiration of a plant is known as the **photosynthetic/respiratory ratio** (P/R ratio). The P/R ratio may be based on a single cell, a leaf, the shoot of the plant, the entire plant, or even a community. The time period used may range from a few minutes to an entire 24-hour day or even a whole growing season. Perhaps the most useful base for the P/R ratio from the standpoint of the general economy of the plant is the entire plant for a period of at least 24 hours. On this base, a P/R ratio of 1 would mean that just as much food was used in respiration as was produced in photosynthesis, that no food would accumulate, and that any assimilation would occur at the expense of accumulated food. A P/R ratio of less than 1 would result in the use of accumulated food even in respiration, and if continued over a long period of time the plant would eventually starve. Actually, under ordinary favorable growing conditions plants usually have a P/R ratio of around 5 or even more. Thus there is an abundance of food that can be used in respiration and assimilation, with a surplus that accumulates.

Gas Exchanges of Plants

Because oxygen diffuses out of chlorophyll-bearing plants during the day and carbon dioxide diffuses into them, many people have the mistaken idea that plants carry on a type of respiration just the reverse of animal respiration, even though they may know that during the night oxygen diffuses into plants and carbon dioxide out. The misconception arises, of course, from the confusion of photosynthesis and respiration. Respiration occurs continuously in plants, both day and night, whereas photosynthesis occurs only in the presence of light. Since the rate of photosynthesis is generally much higher than the rate of respiration during the day, oxygen does diffuse out and carbon dioxide in, although respiration is still using oxygen and producing carbon dioxide. In the dark, and at all times in tissues lacking chloroplasts, only the respiratory gas exchange occurs. In dim light, or under other conditions severely limiting the rate of photosynthesis, the rates of photosynthesis and respiration may be the same and neither gas will be diffusing into or out of the leaf.

ASSIMILATION

Although our discussion of respiration has been quite lengthy, we have presented only a rather small fraction of the information that has been accumulated as a result of research on respiration. We cannot devote as much space to the other way organisms use food, that is, **assimilation,** simply because very little is really known about it. Whereas some biologists restrict the term *assimilation* to the making of living matter (protoplasm) from food and water, we shall use it in its broader sense, including the formation of all essential cell structures—cell walls as well as protoplasm.

Much is known about how plants synthesize cell wall substances like cellulose, lignin, and pectic compounds, but there is still much to be learned about how such components of cell walls are organized into the characteristic structural patterns of cell walls. Among the variety of useful bits of information that we do possess is the fact that the plant hormone, auxin, is essential in cell wall enlargement (Chapter 15), and the fact that in regions of cell wall formation there is an aggregation of rather distinct masses of actively moving protoplasm.

Even less is known about how chemical compounds are converted into protoplasm than about cell wall assimilation, although there is much information on the synthesis of the various constituents of protoplasm by cells. Assimilation itself is probably more a matter of physical organization of the various cell structures than of chemical change. The electron microscope has helped reveal the fact that organized cell structure extends to a far smaller level and a greater degree of complex and precise structural patterns than had been realized previously (Chapter 8).

FOOD ACCUMULATION

Whenever a green plant makes more food by photosynthesis than it uses in respiration and assimilation, or when the food intake of a nongreen plant or animal exceeds its use of food in respiration and assimilation, the excess food **accumulates.** This is frequently referred to as food *storage,* but the term *accumulation* is more descriptive of what actually happens.

Fats constitute the great bulk of food accumulated in animals. In plants, foods such as starch (Fig. 14.23), sucrose, and proteins, as well as fats, may accumulate. Foods can accumulate in almost any organ of a plant but they are usually the most abundant in seeds, fruits, fleshy roots, and tubers and other fleshy underground stems. Sugarcane plants (Fig. 14.24) have large quantities of sucrose in their stems, and there is considerable food in the twigs of trees and shrubs.

Surplus foods may be digested and used by plants at some later time. For example, the food in twigs is used in the initial stages of growth in the spring before the leaves provide much food from photosynthesis, and the food in seeds supports the respiration and assimila-

Fig. 14.23 (Top) Starch grains from potato tubers, about 675X. (Bottom) Some of the starch grains are darker in this photomicrograph because of treatment with iodine reagent, a test for starch.

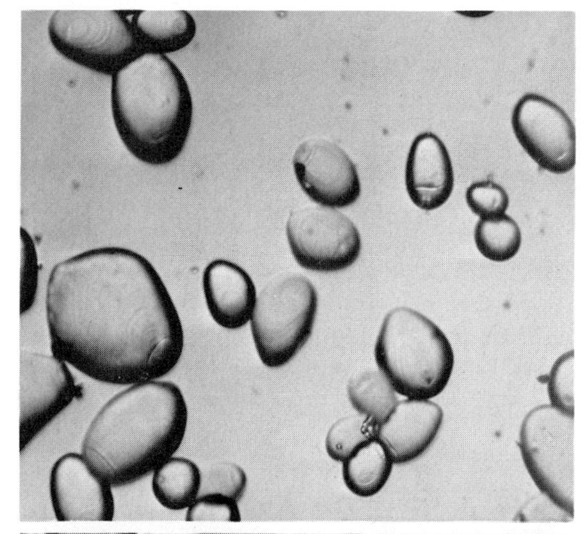

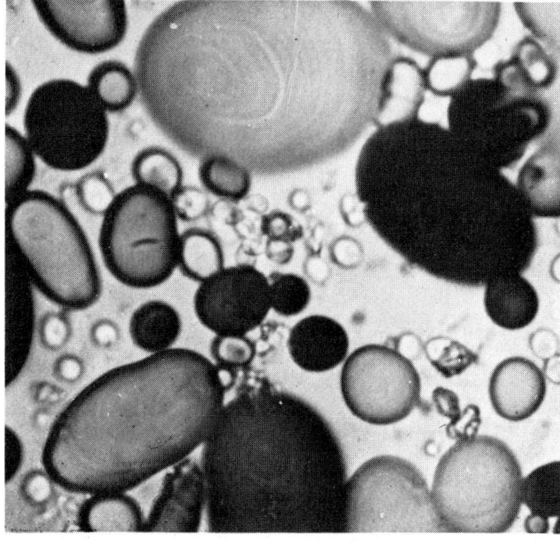

Fig. 14.24 The two plants that are the principal commercial sources of sucrose. (Left) Sugarcane being cut. (Right) A sugar beet plant showing the large taproot from which sucrose is obtained.

tion of the seedlings before they begin carrying on photosynthesis.

Some species of plants accumulate substantial quantities of substances other than foods, including rubber latex, turpentine, resins, alkaloids, and volatile oils like peppermint oil. Unlike foods, these substances apparently cannot be digested and used in respiration or assimilation and their roles (if any) in the life of the plants are generally not known. That they are not of general importance in plants is suggested by the fact that any one of these substances is produced by only a few species of plants. Although they may not play essential roles in the plants that produce them, many of them are among man's more important items of commerce. The alkaloids in particular are of great significance to human beings. Many such as atropine, quinine, colchicine, and morphine have important medical uses. Some such as heroin, cocaine, and LSD are addictive drugs.

FOOD SYNTHESES

From the sugar produced during photosynthesis, plants synthesize all of the numerous organic compounds present, including other carbohydrates, fats, proteins, a great variety of organic acids, hormones, vitamins, pectic compounds, lignin, volatile oils, carotenoids, chlorophylls, latex, alkaloids, purines, and pyrimidines, to mention only a few groups of compounds. Many of these organic compounds contain only the carbon, hydrogen, and oxygen derived from the sugar (or, in the final analysis, from carbon dioxide and water), whereas others also contain other elements derived from salts. Some of the synthetic pathways leading from the sugar are long and involved, with many intermediate substances produced along the way; other pathways are quite short and direct. Some of the synthetic pathways are well known, others are still imperfectly understood, but we can be certain that they lead back to the product of photosynthesis. A consideration of more than a few of these synthetic processes of plants is far beyond the scope of an introductory textbook. We discuss here only the synthesis of four foods—starch, sucrose, fats, and proteins. These synthetic pathways are also outlined in Fig. 14.25. In this figure note that respiration is important in synthesis as well as energy release.

Sucrose Synthesis

Sucrose, ordinary table sugar and one of the most abundant sugars synthesized by plants, is one of the **disaccharides.** Disaccharides are formed by the union of two molecules of simple sugars. The simple sugars used in the synthesis of sucrose are glucose phosphate and fructose, both of them being hexoses. We shall not consider the various reactions involved in the formation of these hexoses from the triose phosphate produced

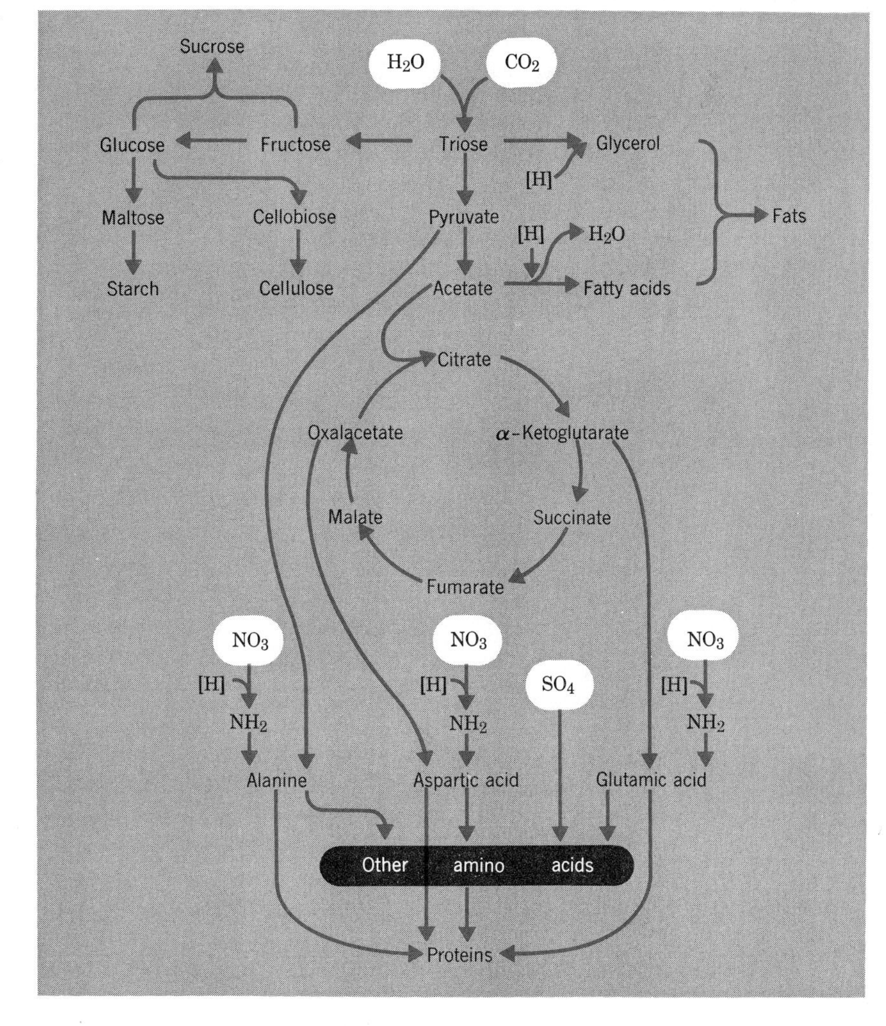

Fig. 14.25 Simplified pathways of synthesis of some of the foods by plants. The substances in the white ovals enter the plant from the environment. The trioses produced during photosynthesis (and glycolysis) can be considered as the basic carbohydrates from which all other foods are synthesized. All the numerous and varied organic compounds of plants can be traced back to the substances shown in this diagram. Note that fats are synthesized from products derived from glycolyses and that citric acid cycle intermediates can be converted into amino acids.

in photosynthesis, although it should be pointed out that it can be converted quite readily into all the other kinds of simple sugars such as pentoses as well as into the various hexose sugars. If glucose phosphate, fructose, and the enzyme sucrose phosphorylase are present the synthesis of sucrose can proceed as follows:

$$C_6H_{11}O_6 \cdot H_2PO_3 + C_6H_{12}O_6 \xrightarrow{\text{sucrose phosphorylase}} C_{12}H_{22}O_{11} + H_3PO_4$$

The synthesis of other disaccharides proceeds in a similar manner. For example, by substituting glucose for fructose and changing the enzyme, the preceding equation would represent the synthesis of the disaccharide *maltose*.

Starch Synthesis

Plants synthesize starch by linking hundreds to a thousand or more glucose molecules together, forming long spiral starch molecules that are sometimes branched (Chapter 7). The formula for starch may be written $(C_6H_{10}O_5)_n$, where n is the number of glucose molecules used in making a molecule of starch. If in a particular starch molecule n were 1000, we could write its formula $C_{6000}H_{10000}O_{5000}$. The formula of starch suggests that at each glucose linkage a molecule of water is produced as starch is being synthesized. Actually, phosphate is produced rather than water, since glucose can participate in starch synthesis only after it has been phosphorylated.

Starch synthesis can occur in most living plant cells, although some lower plants and a few seed plants such as onion and other members of the lily family are unable to synthesize starch. The use of the iodine test for starch to determine whether photosynthesis has been occuring in a leaf might suggest that starch is the product of photosynthesis. Actually, photosynthesis is completed with the production of sugar, but in most plants much of the sugar is almost immediately converted into starch.

Fat Synthesis

Animals are unable to synthesize either sucrose or true starch, but both plants and animals can synthesize fats from sugars. Fat synthesis occurs in three principal steps: (1) The synthesis of glycerol (glycerine) from phosphorylated triose, (2) the synthesis of various fatty acids from triose by way of pyruvic acid and acetyl-coenzyme A, and (3) the linking of glycerol and fatty acids into molecules of fat (Fig. 7.16).

We shall not consider the first two series of reactions in detail, but it is worth noting that glycerol $[C_3H_5(OH)_3]$ molecules contain the same number of carbon and oxygen atoms as triose $(C_3H_6O_3)$ molecules, but two more hydrogen atoms. The added hydrogen comes from reduced hydrogen acceptor (AH_2). Since the addition of hydrogen to a molecule results in its reduction, each glycerol molecule contains more energy than a triose molecule. The extra energy was provided by respiration.

For the synthesis of three molecules of palmitic acid $(C_{15}H_{31}COOH)$, a typical fatty acid, 16 molecules of triose are used. A little calculation will reveal that the three molecules of palmitic acid contain just the same number of C and H atoms as the 16 molecules of triose, but the palmitic acid contains only 6 oxygen atoms in comparison with 48 in the triose molecules. The fatty acid has been produced from the sugar by reduction processes and has a much higher energy content than the sugar, but here reduction has occurred by the removal of oxygen rather than by the addition of hydrogen. Again, the energy used in the reduction comes from ATP produced in respiration.

Thus, both glycerol and fatty acids have a

higher energy content per unit weight than the triose (or any other carbohydrate), and the fats formed by the union of fatty acids and glycerol also have this high energy content. A gram of fat contains about twice as many calories as a gram of sugar or starch. The final reaction in fat synthesis follows:

$$C_3H_5(OH)_3 + 3C_{15}H_{31}COOH \longrightarrow C_3H_5(C_{15}H_{31}COO)_3 + 3H_2O$$

Fats differ from one another chemically in the kinds of fatty acids used in their synthesis.

Protein Synthesis

All organisms can synthesize their own particular proteins by linking various amino acids together in specific and characteristic sequences. Although animals and some bacteria and fungi must depend on the proteins they absorb and digest for their basic supply of amino acids, green plants (and some bacteria and fungi) can synthesize their own amino acids from sugars, nitrates, and sulfates. The sugar used first passes through glycolysis to pyruvic acid and then part way through the citric acid to several of the keto acids of the cycle. These keto acids are converted to amino acids by substitution of an $-NH_2$ (amino) group derived from ammonium salts for the oxygen of the $>C=O$ group of the keto acid. The amino group is the characteristic structural feature of all amino acids, along with the $-COOH$ group common to all organic acids. Although the ammonia may come from ammonium salts absorbed by the plant, it is generally formed by the reduction of nitrate ($-NO_3$) salts. The energy used in this reduction is derived from respiration through AH_2. Since the keto acids are intermediate products formed in the course of the respiratory reactions, respiration is even more intimately associated with their synthesis. Respiration may, indeed, be considered as a synthetic process as well as an energy-releasing process, since glycerol, fatty acids, and other compounds, as well as amino acids are derived from substances produced during respiration.

Only a few of the amino acids are formed directly by the basic processes just outlined; however, these amino acids can be converted into others that are generally somewhat more complicated in structure. Three of the common amino acids contain sulfur, and most proteins include a few molecules of these sulfur-containing amino acids.

The next step in protein synthesis is the linking together of amino acid molecules into the long molecules of protein. The $-NH_2$ group of one amino acid is linked to the $-COOH$ group of another (a **peptide linkage**), as shown in Fig. 14.26. Note that one of the amino acid residues has a free NH_2 group and the other a free COOH group, so it is possible for an indefinite number of other amino acids to be added, thus forming a long chain.

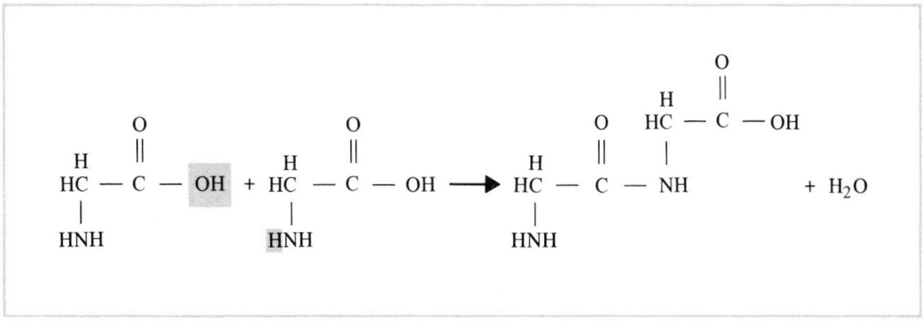

Fig. 14.26 The synthesis of a dipeptide from two molecules of an amino acid, showing the peptide linkage. Note that one amino acid still has an NH_2 group and the other a COOH group, so amino acids can be linked together in long polypeptide and protein chains.

Before the peptide linkages between amino acids can be formed, the amino acids must be activated. This is accomplished by the reaction of an amino acid with ATP and an activating enzyme. Two of the phosphate groups of the ATP are freed in this reaction, so the activated amino acid complex consists of the amino acid linked with AMP (adenosine monophosphate) and the enzyme.

The problem of protein synthesis is complicated by the fact that twenty different kinds of amino acids, can be used in making proteins and these must be linked together in a specific and precise sequence if a certain kind of protein is to result. We noted in Chapter 7 that each species of organism has its own particular kinds of protein, including certain kinds of enzymes, and it is extremely important that these particular kinds of proteins be synthesized. A single error in the amino acid sequence of an enzyme can completely alter its catalytic capabilities. It has been realized for many years that the kinds of proteins produced by an organism are under hereditary control, but until the last decade or two the way in which the genes controlled protein synthesis so precisely remained a mystery.

Following discovery of the fact that the genetic code is carried by the DNA (Chapter 23) of the chromosomes, the way in which this genetic code determines the sequence of amino acids in proteins has been clarified by a series of spectacular and important investigations. Although various details still remain unclarified or are subject to modification, the general outline of the processes is well established. The sequence of events involved is described in some detail in Chapter 23. Here we point out only that the genetic code of the DNA is transcribed to the complementary code of messenger RNA and that this then determines the sequence in which the various amino acids are added to the growing protein molecule (Fig. 14.27). This insures that each organism has the specific kinds of structural and enzyme proteins specified by its genetic code, and is of great importance because the kinds of enzymes an organism has determines what biochemical processes it can carry on.

DIGESTION

We have seen that plants synthesize more complex foods from simple foods: sucrose and starch from simple sugars, fats from fatty acids and glycerol, and proteins from amino acids, for example. Plants also break down these more complex foods into the simpler foods composing them. The two sets of processes might be compared with the building of a structure with an erector set and the subsequent dismantling of the structure. Even though plants break down the more complex foods through several different pathways, we shall restrict our discussion to the type of decomposition reactions involving the use of water in breaking the chemical bonds between the component units, a type of reaction referred to chemically as **hydrolysis** and biologically as **digestion.**

Green plants do not have digestive tracts as most animals do, and generally they do not secrete digestive enzymes into the surrounding medium as bacteria and fungi do, but the chemical reactions of digestion are essentially similar in all groups of organisms. In general, digestion in green plants occurs within cells **(intracellular digestion),** whereas in most animals and nongreen plants, the digestive enzymes are commonly secreted by the cells producing them and the digestion reactions occur outside the cells **(extracellular digestion),** either in a digestive tract or the surrounding food medium. Considerable intracellular digestion also occurs in fungi. The biological significance of digestion is similar in all organisms. Digestion converts large, complex molecules that are frequently insoluble into smaller, soluble molecules that can diffuse through cell

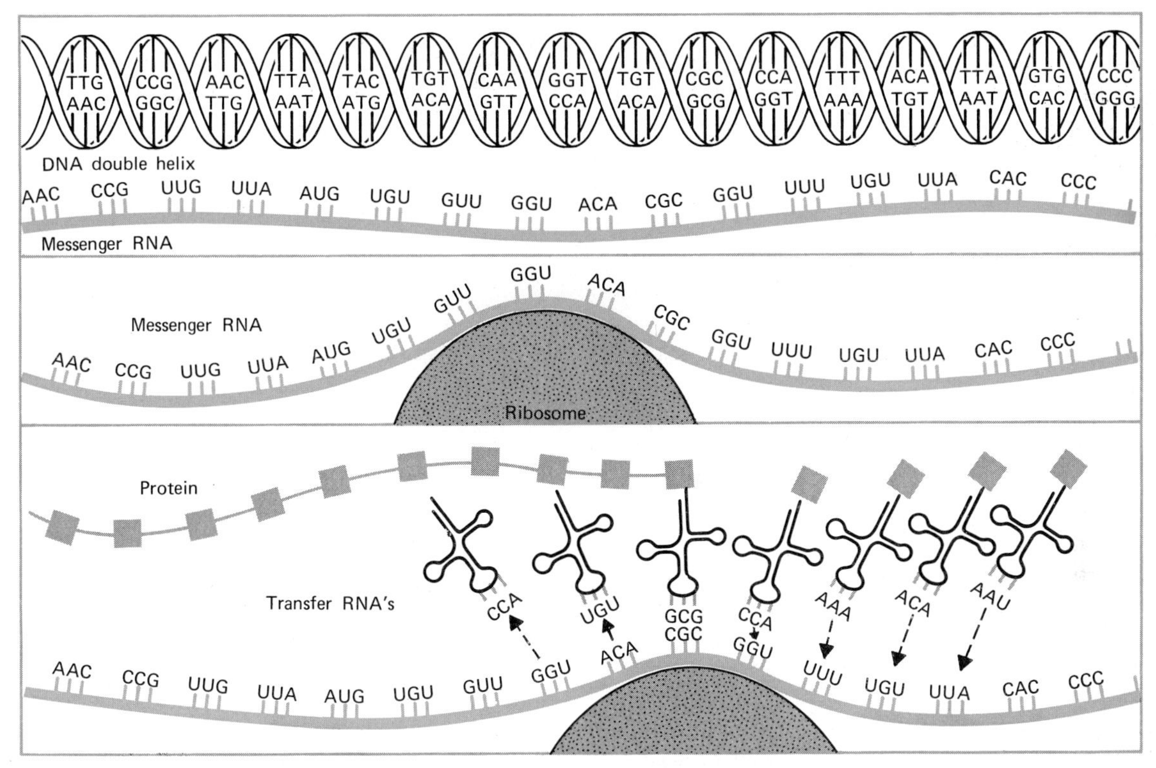

Fig. 14.27 The genetic control of the sequence of amino acids in proteins begins with the triplet codons of DNA. One of the DNA strands translates itself into the complimentary codons of mRNA, which in turn (with the assistance of the specific tRNA molecules and ribosomes) determines the sequence of amino acids in the growing polypeptide chain and also the length of the chain. Thus, each organism can synthesize only those proteins (including enzymes) specified by its DNA codes.

membranes, can be transported through the organism, and can be used in respiration. Neither a starch grain nor a single starch molecule could possibly leave the cell in which it was formed without disrupting the cell membranes. However, the glucose produced by digestion of the starch (Fig. 14.28) can diffuse through the cell membranes. Similarly, a fungus cannot absorb starch, cellulose, or proteins, but it can absorb the products of their digestion. Accumulated foods must first be digested before they can be used. Foods that do not require digestion (simple sugars, amino acids, fatty acids, glycerol) are not accumulated.

Digestive Processes

Since the various processes are relatively simple and in general quite similar, we can list several of them in a single group:

1. Sucrose digestion:

$$C_{12}H_{22}O_{11} + H_2O \xrightarrow{\text{sucrase}} \underset{\text{glucose}}{C_6H_{12}O_6} + \underset{\text{fructose}}{C_6H_{12}O_6}$$

2. Starch digestion:

$$(C_6H_{10}O_5)_{2n} + nH_2O \xrightarrow{\text{amylase}} \underset{\text{maltose}}{nC_{12}H_{22}O_{11}}$$

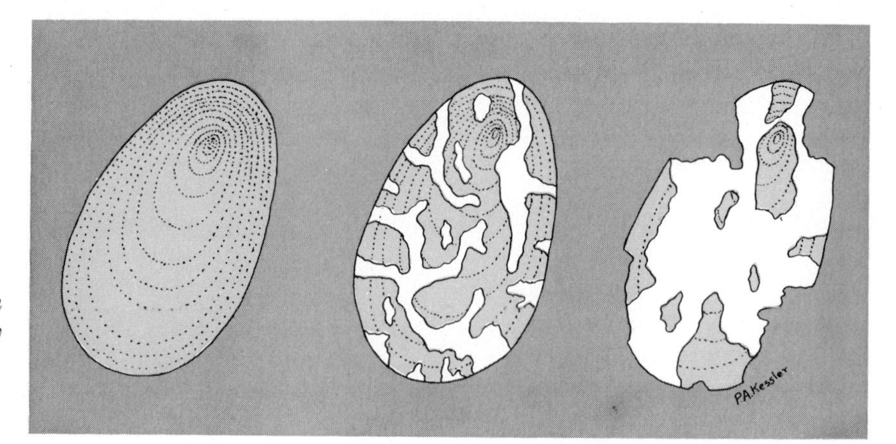

Fig. 14.28 An intact starch grain and two grains partly eroded as they were being hydrolyzed (digested) by amylase.

$$\underset{\text{maltose}}{C_{12}H_{22}O_{11}} + H_2O \xrightarrow{\text{maltase}} \underset{\text{glucose}}{C_6H_{12}O_6} + \underset{\text{glucose}}{C_6H_{12}O_6}$$

3. Fat digestion:

$$C_3H_5(C_{15}H_{31}COO)_3 + 3H_2O \xrightarrow{\text{lipase}} 3C_{15}H_{31}COOH + C_3H_5(OH)_3$$

4. Protein digestion:

$$\text{a protein} + \text{water} \xrightarrow{\text{proteases}} \text{amino acids}$$

The products of the various digestive reactions, then, are simple sugars, fatty acids, glycerol, and amino acids, all of them soluble and relatively simple substances. Little energy change of consequence occurs during digestion.

Digestive Enzymes

In writing the foregoing equations, we have departed from our usual practice of not indicating the enzymes that increase the rate of biochemical reactions. Since the digestion process represented (except for protein digestion) are specific reactions, rather than summary equations for a series of reactions as in many previous instances, listing the enzymes involved is a simple matter. It has been done for several reasons: to emphasize the fact that enzymes catalyze practically all biochemical reactions, to point out again the specificity of enzymes, to illustrate the current method of naming enzymes, and because digestive enzymes are the ones most likely to be encountered in the laboratory work of introductory courses.

Although sucrose and maltose are both disaccharide sugars, different enzymes are involved in their digestion. The digestion of cellulose would be represented by the same chemical formulae (but by different compound names) used for starch digestion, and even though both result in the production of glucose, amylase will not catalyze the digestion of cellulose. Another enzyme, **cellulase,** is essential. Whether or not cellulose is a food for a specific organism depends on whether it produces cellulase. Thus, some fungi and bacteria, as well as the protozoa that live in the digestive tracts of termites, produce cellulase and so can digest and use cellulose, whereas most species of plants and animals do not. At present, enzymes are generally named by suffixing **-ase** to the root of the name of the substrate or a term descriptive of the reaction. **Amylase** is so named because starch is technically called **amylose.** However, enzymes named before the present system was adopted generally ended in **in,** for example **pepsin, ptyalin,** and **papain.**

Several of the many digestive enzymes produced by plants are extracted and sold commercially. **Diastase** is a crude enzyme extract secured principally from germinating barley or from molds, and contains both amylase and maltase. It is available at many drug stores. Meat tenderizers contain a protease (papain) extracted from the fruits of the papaya or tropical paw-paw. Fresh pineapples also contain a protease and when added to gelatin desserts the gelatin (a protein) is digested and liquified instead of forming the usual semisolid gel.

FOOD TRANSLOCATION

One other aspect of the food economy of plants remains to be considered: the problem of food transport from one part of a plant to another, for example from leaves to roots or from tubers or other sites of accumulation to meristems (Fig. 14.29). Simple, soluble foods to which the cell membranes are permeable may move from one cell to another by simple diffusion. However, diffusion is far too slow a process to account for the observed rates of food transport through a distance of a few centimeters to many meters within a plant. Calculations show that sugars are transported to a developing pumpkin fruit at rates ranging from 100 to 500 cm/hr, and rates of 100 cm/hr or so are quite commonly found in many species of plants. In contrast, sugar diffusing from a 10% solution would take about two and half *years* to diffuse a distance of 100 cm. If diffusion were the only means of food transport in plants no plant more than a few centimeters in size could survive, unless its chlorenchyma tissues were distributed throughout the plant.

Food transport **(translocation)** from one part of a plant to another occurs within the sieve elements of the phloem. Although translocation is known to occur through mechanisms other than just simple diffusion, no completely satisfactory explanation of translocation has yet been proposed. Various theories have been proposed. Each of them has some degree of theoretical plausibility and some experimental support, but none is free from serious objections. We shall consider briefly three of the theories: interfacial flow, cytoplasmic streaming, and mass flow.

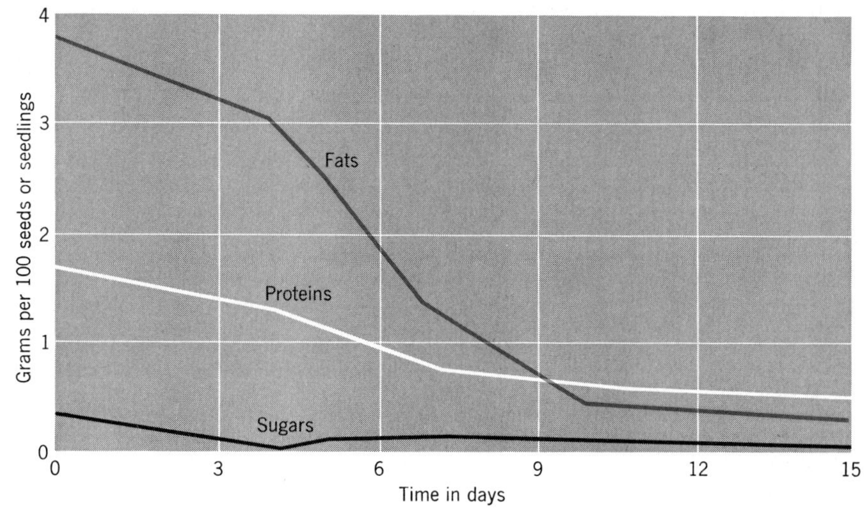

Fig. 14.29 The decrease in three classes of foods in the cotyledons of sunflower seedlings during germination and 15 days of growth. Most of the food originally present was digested and translocated to other parts of the plants. (Data of E. C. Miller.)

Interfacial Flow

At the interfaces between two immiscible liquids (such as oil and water) or between a liquid and a gas solute particles can move with great freedom and at high velocities that approximate those of diffusing gases. It has been proposed that phloem translocation occurs at such interfaces that are present in the cells. This theory has the advantage that it can account for the observed rates of phloem translocation. However, since the sugars and substances translocated in the phloem are highly soluble in water and since water is almost certain to be one of the substances at any interface within a cell, it appears quite unlikely that a sufficient portion of the solutes would be concentrated at the interfaces to account for the quantities of solutes known to be translocated per unit of time.

Cytoplasmic Streaming

This theory holds that foods and other solutes are transported from one end of a sieve element to another by the cyclic flow of the cytoplasm. This theory is supported by the fact that conditions known to retard or stop cytoplasmic streaming, such as reduced temperature or oxygen supply or death of the sieve elements, will retard or stop translocation, and by the reports that solutes may be transported simultaneously in opposite directions through sieve elements. There are, however, several serious objections to the cytoplasmic streaming theory. The observed rates of cytoplasmic streaming are too slow to account for many measured translocation rates, and besides, cytoplasmic streaming has rarely (if ever) been observed in mature sieve elements.

Mass Flow

This theory holds that the high turgor pressure of leaf cells produces a flow of the solution of sugar and other substances through the plasmodesmata to adjacent cells and then on to and through the sieve elements into cells with lower turgor pressures, particularly those of the roots. The evidence for mass flow includes exudation from cut or injured phloem and the fact that viruses, hormones, and other substances present in leaf cells in low concentration are translocated from the leaves only when sugar is being translocated out.

An interesting experimental technique that has provided evidence in support of the mass flow hypothesis involves the use of aphids (plant lice). When aphids are feeding on plants, they insert their stylets into sieve tubes. When a feeding aphid is cut from its stylet, contents of the sieve tube exude through the stylet, indicating that there is considerable pressure in the sieve tube. Aphids do not need to expend energy to suck their food from the plant.

Among the objections to the mass flow theory is the fact that the cytoplasm of sieve elements is rather viscous and would seem to offer very high resistance to the flow of a solution through it. Mass flow can account for translocation in only one direction at a time, and it cannot account for the fact that lowering the temperature or oxygen supply of the cells reduces the rate of translocation through them. Another objection is that the receiving cells may not always have a lower turgor pressure than the supplying cells. Like cytoplasmic streaming, mass flow appears to be too slow to explain some observed translocation rates.

Despite the fact that mass flow, like the other proposed mechanisms, has evidence against it as well as for it, most of the investigators of phloem translocation now believe that it is the most plausible hypothesis and that continuing research will clarify the situation and remove the current objections. However, it is possible that more than one of the proposed mechanisms operates or that modified or even entirely new hypotheses may be proposed.

Girdling and Translocation

Since the phloem of trees is in the bark, removal of a ring of bark from the entire circumference of a stem **(girdling)** will stop food translocation at the girdle (Fig. 14.30). When a tree trunk is girdled, the stem and roots below the girdle receive no more food from the leaves and will starve when all previously accumulated food is exhausted, although starvation may not occur for several years. Once the roots have starved, the shoot of the tree will die from lack of water. Pioneers frequently girdled trees a year or more before clearing land for cultivation. This practice led to little or no root sprouting. Orchard owners sometimes girdle branches of fruit trees to secure unusually large fruits for exhibition at fairs. Rabbits and similar herbivorous animals sometimes injure trees by eating their bark and at times girdling and killing them.

Fig. 14.30 Effect of girdling a cherry stem, shown two years after girdling. Secondary growth from the vascular cambium below the girdle was inhibited by a lack of food.

RELATED READING

Arnon, D. I. "The role of light in photosynthesis," *Scientific American, 203*(5), 104–118, November 1960.

Bassham, J. A. "The path of carbon in photosynthesis," *Scientific American, 206*(6), 88–100, June 1962.

Bjorkman, O. and J. Berry. "High efficiency photosynthesis," *Scientific American, 229*(4), 80–93, October 1973.

Bornman, C. H. and C. E. J. Botha. "The role of aphids in phloem control," *Endeavour, 32,* 129–133, 1973.

Gabriel, M. L. and S. Fogel. *Great Experiments in Biology*. Englewood Cliffs, N.J.: Prentice-Hall, 1955. (Enzymes, pp. 23–53, metabolism, pp. 84–101, photosynthesis, pp. 152–184.)

Haneda, Y. "Glow plants," *Natural History, 65,* 482–484, 1956.

Levine, R. P. "The mechanism of photosynthesis," *Scientific American, 221*(6), 58–70, December 1969.

Rabinowitch, E. I. "Photosynthesis," *Scientific American, 179*(2), 24–35, August 1948.

Rabinowitch, E. I. "Progress in photosynthesis," *Scientific American, 189*(5), 80–85, November 1953.

Rabinowitch, E. I. and Govindjee. "The role of chlorophyll in photosynthesis," *Scientific American, 213*(1), 74–83, July 1965.

Walker, D. A. "Carboxylation in plants," *Endeavour, 25,* 21–26, 1966.

Zimmermann, M. H. "Movement of organic substances in trees," *Science, 133*, 73–79, 1961.

15 PLANT GROWTH AND DEVELOPMENT

Plant and animal growth is one of those common, everyday things usually taken for granted, but to biologists growth is one of the most complex, puzzling, and interesting aspects of living organisms. Plant growth involves all the plant activities already discussed as well as many others, and its end products are the highly organized cells, tissues and organs of the plant. Much is known about plant growth and development, but many problems remain to be clarified and solved by future research. No one really knows just how the single cell of a fertilized egg gives rise to certain numbers of cells of specific kinds, at just the appropriate time and place with only a rare mistake, resulting in the development of a particular kind of plant or animal. One specific problem is presented by development of sieve tubes, companion cells, and other cells of the phloem from cells cut off to one side by the cambium, whereas cells cut off to the other side of this same meristematic tissue always develop into vessel elements and other cells of the xylem. It is relatively easy to describe such differences in cell development, but their causes have been quite elusive. The better understanding of growth that biologists are gradually accumulating is of great practical, as well as theoretical, importance. For example, the solution of the cancer problem depends on better knowledge of growth, and research on both normal and abnormal plant growth, as well as animal growth, is contributing to an understanding of cancer.

THE NATURE OF GROWTH

Definition of Growth

Growth is not easy to define. We usually think of growth as involving an increase in weight, but children may grow taller without gaining weight. The weight of a growing plant may fluctuate up and down as it gains and loses water during the course of a day. Plants may grow rapidly in the dark, but aside from the water in them, the weight of their component substances is constantly decreasing as they grow. Growth also usually involves an increase in size, but a frog embryo may have developed several hundred cells and yet be no larger than the fertilized egg from which it grew. As the turgidity of plant cells increases, plants get somewhat larger without really growing. Perhaps the best indication of growth is an increase in the quantity of protoplasm in an organism, but since this is difficult to measure it does not provide a very practical criterion of growth. Probably the best way to define **growth** is as an increase in the amount of protoplasm in an organism, usually accompanied by an irreversible increase in size and weight and involving the division, enlargement, and (usually) the differentiation of cells.

The differentiation of cells, tissues, and organs in a growing organism, resulting in its characteristic pattern of organization, is referred to as **development** or **morphogenesis.** Some biologists consider development as one aspect of growth; others regard the two as separate processes and restrict the term *growth* to purely quantitative increases in size, weight, or amount of protoplasm.

Measurements of Growth

In studies of growth, quantitative data are usually desired, so measurements of increase in size, fresh weight, or dry weight are obtained, and in some experiments all three are used. While size and fresh weight of an organism can be measured at intervals, dry weight can be determined only once for any individual since it is killed in the process of drying. Development is generally not subject to quantitative measurements, at least of any simple type, and so is usually described, illustrated, or represented by a system of graphic symbols.

Growth Rates

After a seed germinates, the seedling plant grows rather slowly for a while and then enters a period of much more rapid growth that is maintained until the plant approaches maturity, when growth slows down or even ceases. If the height of such a plant is measured at intervals and plotted on a graph, the growth curve has something of an S shape (Fig. 15.1). Such S-shaped growth curves are characteristic of animals as well as plants, and also of individual cells and tissues. The growth curve may, of course, be modified considerably if a plant is, for example, treated with a growth inhibitor or if growth is inhibited by the lack of some essential environmental factor such as a mineral element. Most plants grow less than 3 cm a day, but asparagus may grow as much as 30 cm per day and bamboo sometimes grows 60 cm in a day.

Plant growth is not steady but varies diurnally and seasonally. Seasonal growth variations are the most striking in trees, shrubs, and other perennials and in biennials. Many trees

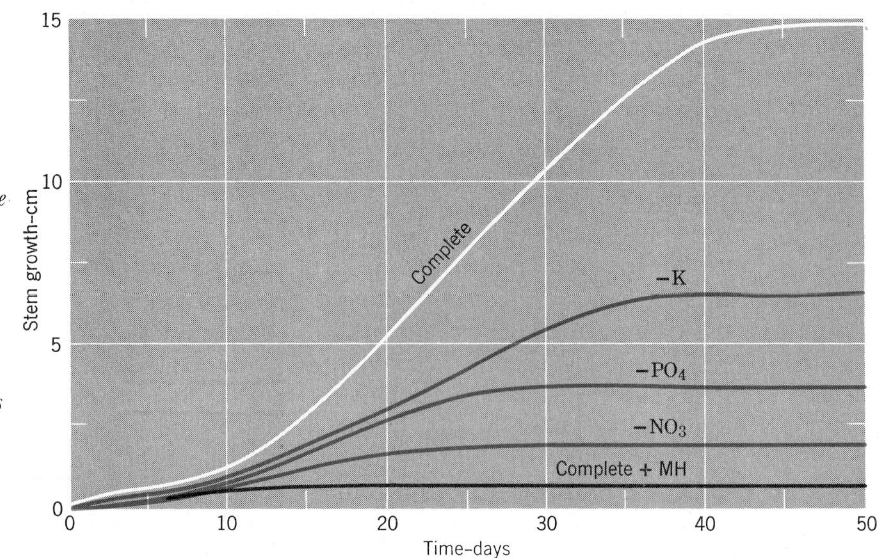

Fig. 15.1 When plants growing in vermiculite were supplied with a complete mineral nutrient solution and otherwise good environmental conditions, their growth curve had the typical shape. Plants lacking one of the essential elements or treated with the growth inhibitor maleic hydrazide grew much less and had flattened growth curves. (Data of V. A. Greulach.)

grow rapidly only during the spring and early summer, the growth of some species being limited to a few weeks during the spring. Sometimes the statement is made that plant growth continues as long as the plant lives, whereas animal growth ceases long before the animal dies of old age. Although this is generally true, it should not be construed to mean that plant growth is continuous without interruptions.

GROWTH AT THE CELLULAR LEVEL

All the structural changes we observe as a plant or animal grows depend on cell division, cell enlargement, and cell differentiation. The time, place, rate, plane, and pattern of these cellular activities determine the nature of the resulting tissues and organs of the organism.

Cell Division

All cells arise from the division of other cells. The billions or trillions of cells in a large plant or animal are all derived from a unicellular fertilized egg if the organism is a product of sexual reproduction. Cell division also occurs in unicellular organisms, but there it serves as a means of reproduction (population growth) rather than as a part of the growth of the individual. In vascular plants, cell division is mostly restricted to certain regions—the younger apical portions of roots and stems and the vascular cambium—although cell division also occurs in such regions as young leaves, fruits, and seeds. Cell divisions contributing to the growth in length of stems are frequently said to be restricted to the meristematic tissue of the buds, but several investigations suggest that a considerable amount of cell division may occur in the parenchyma cells of the younger internodes. In grasses and some other plants, continued cell division may occur at the bases of the stem internodes and the leaves, one consequence being the necessity for the periodic mowing of lawns.

The activities of nuclei during cell division (mitosis) have already been described (Chapter 8). Although the sequence of events during cell division has been observed and described in considerable detail, relatively little is known about the physiology of cell division. Some substances such as maleic hydrazide inhibit cell division, and consequently growth, and gibberellin and kinetin promote cell division in plants. A better understanding of how such substances affect cell division will probably eventually help clarify the nature of the physiological processes involved in cell division, and thus our understanding and control of such abnormal growths as cancers which involve the division of cells.

Cell Enlargement

Continued cell division without any cell enlargement would result in an increase in the number of cells but a progressive decrease in cell size and no increase in the size of the tissue or organism. In the meristematic tissues of plants, however, newly formed cells generally enlarge to about the size of the parent cell before they in turn divide, and so the plant increases some in size. The greatest increase in size of a growing plant, however, is the result of the very extensive enlargement of the derivatives of the older meristematic cells into parenchyma type cells that are many times the volume of meristematic cells (Fig. 15.2). In stems and roots cell enlargement is usually greater longitudinally than laterally.

The initial event in cell enlargement appears to be an increase in the area of the wall, a process that requires the presence of one of the group of plant hormones known as auxins as well as a supply of food that can be used in the synthesis of cellulose and other wall substances. As the wall increases in size, the turgor pressure decreases, resulting in a decreased water potential in the cell and thus inward diffusion of water. Much of this water enters the vacuoles, and the cell becomes more highly vacuolated as it increases in size. In a fully enlarged cell, most of the volume is occupied by the vacuole, the cytoplasm being restricted to a thin layer adjacent to the walls, although in some cells strands of cytoplasm extend through the vacuole. In addition to the great vacuolation during enlargement, there is also an increase in the quantity of protoplasm.

Cell Differentiation

Cell division and cell enlargement account for the increase in size of a plant as it grows, but a plant carrying on only these two phases

Fig. 15.2 *The enlargement of a meristematic cell into a parenchyma cell involves a marked increase in size, particularly in length. During cell enlargement, the most marked increase is in the amount of water (especially in the large vacuole), but there is also a substantial increase in the amount of cytoplasm (despite its restriction to the thin peripheral layer) and in the cell wall components.*

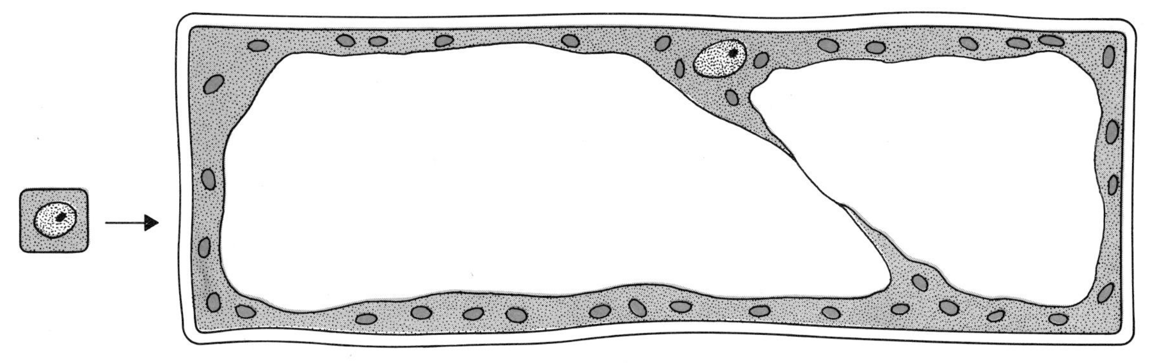

of cellular growth would consist of nothing but meristematic and parenchyma tissues. In some types of plant tissue cultures, growth of this kind may occur over extended periods, but development of plants with all their characteristic cell types, tissues, and organs cannot occur without cell differentiation. Cell differentiation involves various modifications of cells such as changes in shape, secondary thickening of the walls, incorporation of such substances as lignin and suberin in the walls, formation of chloroplasts or chromoplasts, death and decomposition of the protoplasm, loss of the end walls of vessel elements, and disintegration of the nuclei of the sieve tube elements.

Much is known about the sequence of microscopic structural changes during cell differentiation, but we are just beginning to acquire some understanding of the factors that cause cells to differentiate into certain types at specific times and places in a plant. Some of these factors will be discussed later on in this chapter.

THE INTERPLAY OF FACTORS IN PLANT GROWTH

The growth pattern and the resulting structural organization of a plant depend on both its hereditary potentialities and its environment. There should be no argument as to whether heredity or environment is more important, for every process involved in the behavior, growth, and development of a plant or animal depends on both hereditary potentialities (genes) and an environment that permits (or limits) expression of the potentialities. Heredity limits what an organism *can* do, and environment determines which of these things the organism *will* do. Some hereditary potentialities find expression in almost any environment suitable for survival, but others are expressed only in specific environments. The relation of heredity and environment to plant activities and the behavior, growth, and development of plants may be outlined as follows:

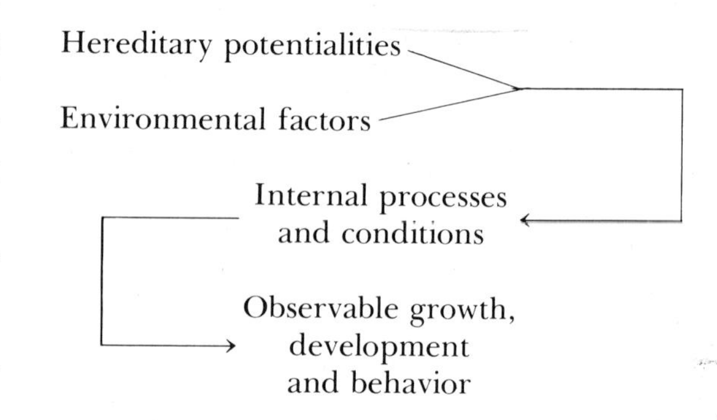

This outline emphasizes an important point: that heredity and environment exert their control over the pattern of behavior, growth, and development through their influence on the many and varied internal life processes and conditions of an organism such as photosynthesis, respiration, protein synthesis, chlorophyll synthesis, digestion, turgor pressure, and mitosis. Although it is relatively easy to observe, measure, and describe the growth and development of a plant with certain hereditary potentialities as it grows in a specific environment, securing even a rough understanding of the internal processes linking these is a difficult problem that generally requires years of research.

An understanding of the way in which hereditary potentialities (genes) exert their control over the internal processes and conditions has been acquired only during recent decades as a result of investigations that have demonstrated the relationship between genes and enzyme synthesis and other investigations that have clarified the chemical nature of genes, the nature of the genetic code, and the sequence of events connecting genes with enzyme synthesis (Chapters 7, 14, and 23). In brief, the genes are DNA and the DNA genetic code is transcribed to a messenger RNA code. The mRNA then moves from the nucleus to the ribosomes where, with the assistance of transfer RNA that is specific for certain amino acids, the particular kind of enzyme or protein that is coded for is synthesized against the mRNA matrix. The kinds of enzymes that an organism can synthesize are thus controlled by its genes, and the enzymes in turn determine what biochemical reactions the organism is capable of carrying on. An organism is a manifestation of its biochemical reactions.

Experimental confirmation of nuclear control of development has been provided by experiments involving nuclear transplantation in unicellular organisms. Classical experiments of this type have been conducted by Brachet and others on *Acetabularia,* a unicellular marine alga of unusual size and shape (Fig. 15.3). Each cell is 2 to 3 cm high and is umbrella shaped. If the umbrellalike cap is cut off, a new one characteristic of the species is regenerated. One species (*A. mediterranea*) has a rather flat disc-shaped cap, another (*A. crenulata*) has a thicker cap that is deeply notched and scalloped. It is possible to cut an *Acetabularia* cell into three pieces: a short basal portion that contains the nucleus, the long stalk, and the cap. If the base of one species is grafted to the stalk of the other, the cap that is regenerated is that of the species providing the nucleated base, not of

Fig. 15.3 Demonstration of nuclear (genetic) control of development by a regeneration experiment on Acetabularia, *a genus of large and complex unicellular green algae. Plants of two species with different shaped caps were cut into three pieces: cap, stalk, and basal portion containing the nucleus. The caps were discarded and the stalks were grafted to the bases of the other species. In each case the stalks regenerated caps of the species providing the base. However, in other experiments with a single species the caps were removed as was the part of the base of each plant containing the nucleus. Characteristic caps were regenerated despite lack of the nucleus. Evidently the cytoplasm contained controlling factors, presumably mRNA.*

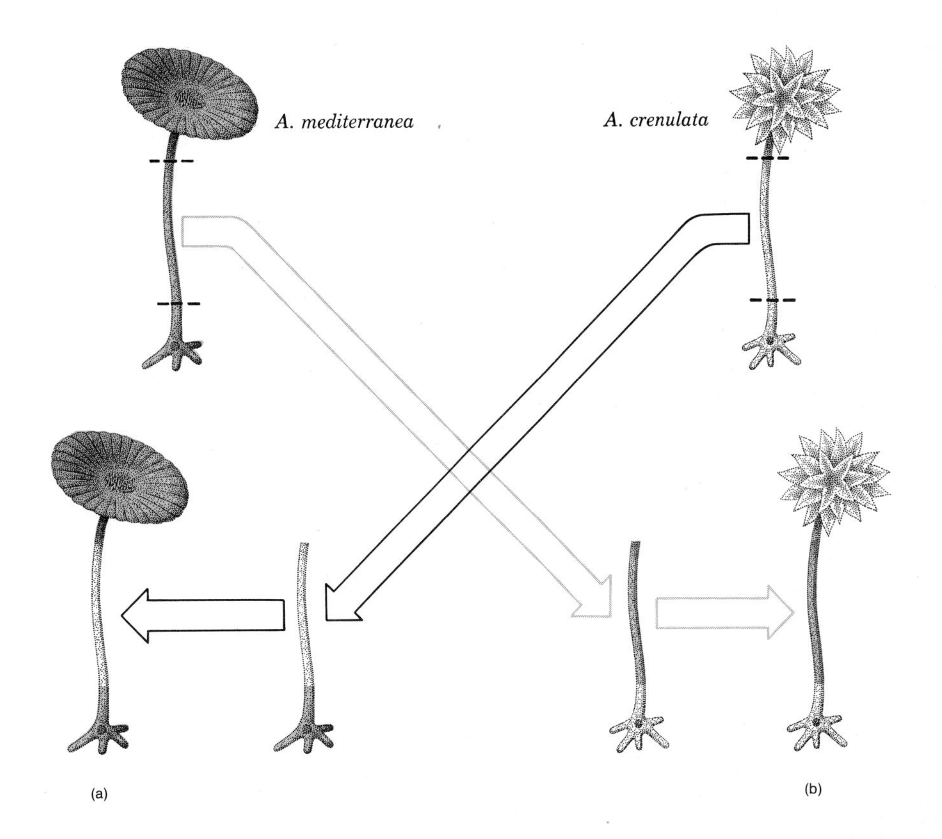

the one providing the stalk. However, the first cap regenerated may have some characteristics of the species providing the stalk, presumably because some mRNA of the stalk species is still present. If this cap is removed, the next one regenerated is definitely characteristic of the species providing the nucleus. Other experiments on *Acetabularia* are described in references listed at the end of this chapter.

Environment exerts its influence on the internal processes and conditions in at least two different ways. First, it determines which of the biochemical reactions that an organism *can* carry on actually *are* carried on at any particular time or place within the organism. For example, a plant may have all the enzymes essential for photosynthesis but cannot carry on its various reactions in the absence of CO_2 or light. Limiting quantities of these or other environmental factors will influence the rate of the process. Second, the environment (and in particular the local environment of specific cells) may determine which of the numerous enzymes the organism has a DNA code for and thus can produce, *are* actually synthesized by any particular cell. It is now apparent that at a given time a specific cell is synthesizing only a fraction of the mRNA for which it has DNA coding and that factors of the cell environment may determine which kinds of mRNA are to be produced at any particular time. This programming of mRNA production has great significance as regards differentiation and development, and it will be discussed in greater detail later on in this chapter.

Finally, the complex of internal processes and conditions of the organism determine the course of its activities, growth, development, and behavior. These are the structural features and functional activities of the organism that we can observe externally, or by suitable manipulations internally, either macroscopically or microscopically. The phenotype of any plant (how it looks and behaves) is determined by both its heredity and environment.

ENDOGENOUS RHYTHMS OF PLANTS

Many aspects of the metabolism, behavior, growth, and development of plants (as well as animals) occur in rhythmic cycles or sequences, generally over a period of a day (24 hours), a lunar tidal period ($29^1/_2$ days), or a year ($365^1/_4$ days). Although these rhythms are attuned to rhythmic environmental changes such as daily changes in temperature and light or seasonal changes in temperature or day length, the rhythmical responses generally continue even when plants are placed in a uniform environment with no fluctuation in any of the known environmental factors such as temperature, light, or humidity, although the rhythmic responses may gradually become less marked (Fig. 15.4). These are referred to as **endogenous rhythms** (self-contained rhythms), or sometimes as the biological clock. Biologists investigating these rhythms are of two schools of thought regarding their basic nature. Some biologists think the rhythms are not truly endogenous and that they really result from rhythmic changes in environmental factors such as atmospheric pressure, magnetic fields, or cosmic radiation that were not controlled in the experiments and are at present not known to have significant influences on organisms. However, others think the rhythms are truly endogenous and are not brought about by any known or unknown environmental factor, although they may be synchronized with such a factor.

The daily rhythms are often referred to as **circadian** (about a day) rhythms because the inherent rhythm is not exactly 24 hours. However, the rhythmic responses may become synchronized or set to the natural fluctuation of some environmental factor. For example, the leaves of bean plants and various other plants fold up in the evening and open again in the morning. This rhythm is endogenous because it continues even though the plants are kept

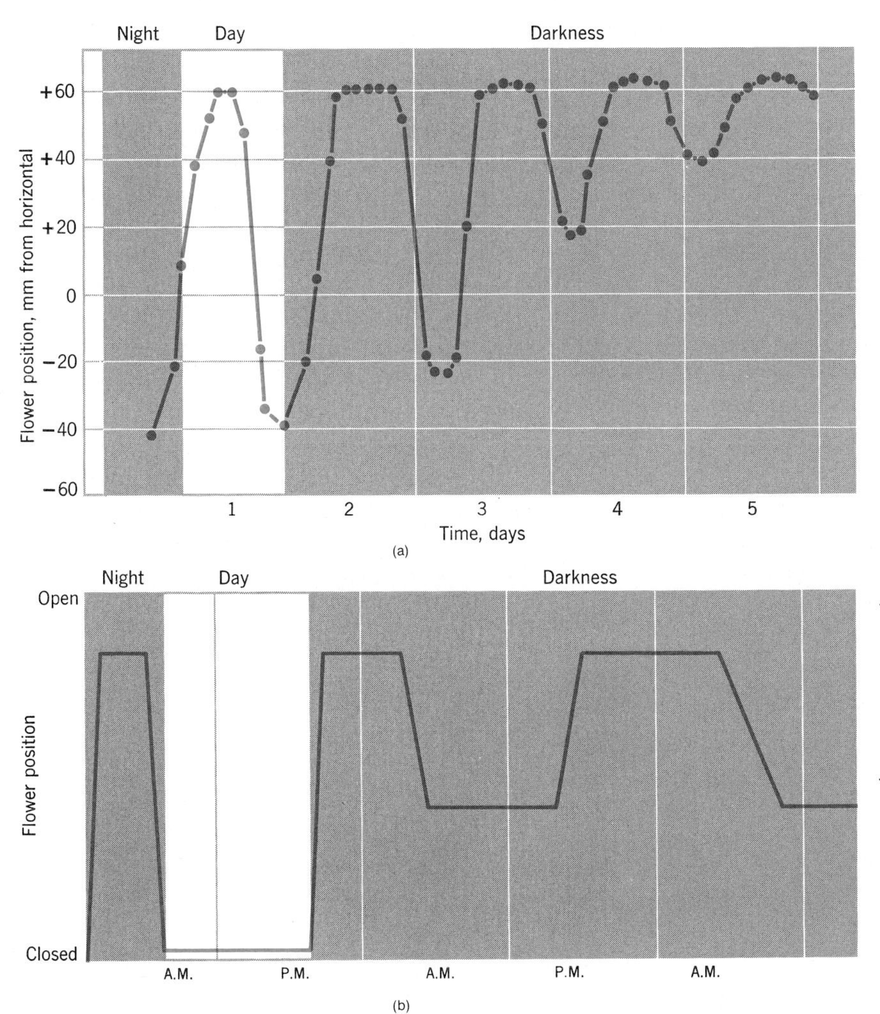

Fig. 15.4 Circadian rhythms in the flower movements of two species. (a) Kalanchoë *flowers elevate by day and droop at night, but the movements continue for several days even though the plants are placed in continuous darkness. (Data of R. Bünsow,* Biol. Zent., 72, *465, 1953.*) (b) Cestrum nocturnum *flowers close during the day and open at night, but even in continuous darkness there is periodic partial closing of the flowers. (Data of L. Overland,* American Journal of Botany, 47, *378, 1960.*)

under uniform conditions for a week or so. The synchronization occurred when the bean seedling was first exposed to light when it emerged from the soil. Other circadian rhythms of plants include the daily opening and closing of flowers, the daily changes in the rates of respiration and cell division, and perhaps some of the daily growth periodicity. However, some things such as the occurrence of photosynthesis only during the day may be somewhat rhythmical, but they are direct responses to environmental changes.

One example of a plant attuned to a lunar tidal period is the brown alga *Dictyota,* which produces eggs and sperm at about monthly intervals. The plants in one area may be synchronized with a different phase of the moon than those in another area. Among plants, such lunar rhythms are apparently restricted principally to marine algae.

Most seasonal or annual rhythms such as the time of blooming, the onset of dormancy, the breaking of dormancy, and seasonal growth periodicity are probably not endogenous since they can be controlled by suitable experimental alteration of the effective environmental factor. Thus, a short-day plant that normally blooms in the late summer or autumn can be made to bloom in midsummer by placing it under artificially shortened days. However, experiments in which the normal 24 hour day-night cycle is replaced by one of some other length (48, 36, or 72 hours, e.g.) indicate that there is probably an endogenous rhythm involved in the photoperiodic responses of plants.

It may appear that endogenous rhythms do not fit in with the interactions of heredity and environment in determining the internal processes and thus the observable growth, development, and behavior of plants, which we have just considered. However, it should be noted that endogenous rhythms are part of the complex of hereditary potentialities that an organism possesses, even though their interaction with the environment is different from the usual.

INTERNAL FACTORS RELATED TO PLANT GROWTH AND DEVELOPMENT

Practically all of the internal processes and conditions of plants that we have considered in previous chapters are involved in growth and development. An adequate water content is essential for hydration of the protoplast and cell walls, for maintenance of turgor pressure, as a solvent for chemical reactants and the transport of solutes, and as a reactant in various processes including photosynthesis and hydrolyses. Normal plant growth and development require an adequate supply of the essential mineral elements that are used in the synthesis of many important organic compounds, as enzyme activators, and in various other ways. Photosynthesis must provide an adequate supply of food that is used in respiration and assimilation and in the synthesis of the great variety of organic compounds such as carbohydrates, lipids, proteins, chlorophyll, nucleic acids, and coenzymes that are essential for the mere maintenance of life as well as for growth and development. Among the more important of these are the enzyme proteins that are specified by a plant's DNA codes and are essential for catalyzing practically all of the biochemical reactions that occur in plants. However, we have made only passing reference to still another group of substances synthesized by plants that are essential for their growth and development—the plant hormones (or phytohormones, growth regulators, or growth substances). We shall turn our attention to these in the following section.

PLANT GROWTH SUBSTANCES

That hormones play important roles in the metabolism and growth of humans and other animals is quite generally known, but many people are not aware of the fact that hormones are also essential in the growth of plants. The term *plant growth substances* includes both the natural plant growth hormones and various synthetic compounds not produced by plants but having hormone-like effects on plant growth. Botanists themselves have become aware of the existence of plant growth substances only during the present century. The **auxins,** the first type of plant growth hormone to be discovered, were definitely established as hormones by the work of F. W. Went in 1928, although somewhat earlier work by other investigators had provided important basic clues (Fig. 15.5). Since the 1920s plant physiologists have devoted much of their time to the study of plant growth substances and have published thousands of re-

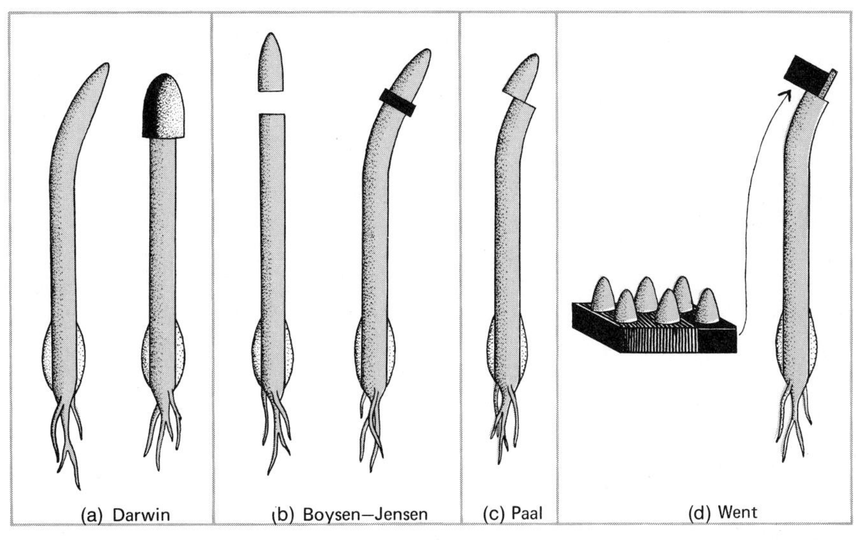

Fig. 15.5 Historical experiments leading to an understanding of auxin and the development of the Avena *(oat) coleoptile bioassay for auxin.* (a) *Charles Darwin (1880) found that coleoptiles did not bend toward light if the tip alone was covered with a light-proof cap.* (b) *Boysen-Jensen (1910) found that coleoptiles do not bend toward light if the tip is removed, but that bending resumes if the tip is replaced, even though separated from the rest of the coleoptile by a gelatin block.* (c) *Paal (1918) discovered that coleoptiles would bend even in complete darkness if the tip is cut off and replaced toward one side.* (d) *Went (1928) placed coleoptile tips on agar and after a time removed them, cut small blocks of agar, and placed them on one side of decapitated coleoptiles. The resulting bending confirmed the developing theory that a diffusable hormone produced by the tips caused the bending below the tips. (After K. V. Thimann,* American Journal of Botany, 44, *50, 1957.*)

search papers dealing with them, but our knowledge of this aspect of plant growth is still in its infancy.

The Nature of Plant Hormones

Hormones are generally considered to be substances produced in small quantities in one part of an organism and transported to other parts of the organism where they exert marked effects on the metabolism and growth of the organism. Unlike animal hormones, plant hormones are not produced in the special ductless glands known as endocrine glands. Both plant and animal hormones act at a very low concentration. For example, the optimal concentration of auxin for the growth of roots is commonly about 0.00000001 (1×10^{-8}) grams per liter (1000 grams) of water.

Most of the plant hormones are organic acids of one type or another, but they do not constitute a chemically related class of compounds as do the amino acids, proteins, carbohydrates, or lipids. Thus, they are grouped together because of similarities in their physiological roles and effects, not because they constitute a chemical class of compounds (Fig. 15.6).

All the plant hormones so far identified are involved in growth and development. Although many of the animal hormones play roles in growth and development, there are

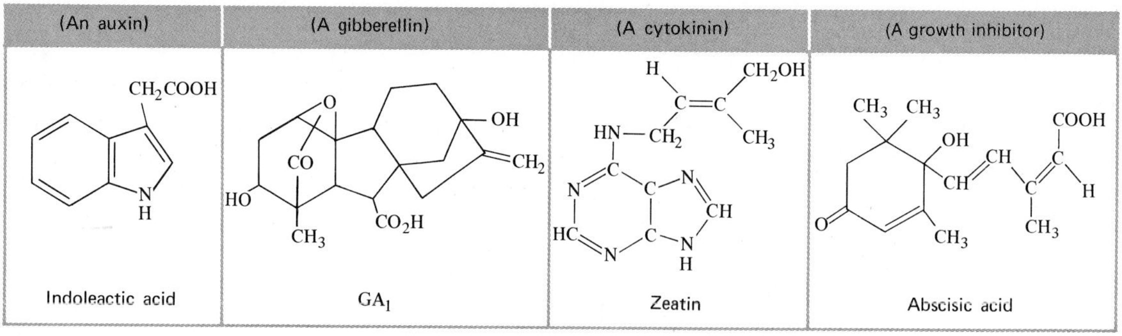

(An auxin)	(A gibberellin)	(A cytokinin)	(A growth inhibitor)
Indoleactic acid	GA_1	Zeatin	Abscisic acid

Fig. 15.6 Structural formulae of four plant hormones.

others like insulin that are primarily involved in metabolic processes only indirectly influencing growth and development.

Types of Plant Hormones

At the present time, five principal kinds of plant hormones have been identified. The chemical nature of at least some of the specific compounds in each of the five classes has been determined, much research has been conducted, and there is considerable information on their metabolic roles and their influences on cellular growth and development as well as on the growth and development of plants as a whole. However, much remains to be learned about each of them.

Three of the kinds of hormones—**auxins, gibberellins,** and **cytokinins**—are growth-promoting substances, although auxin inhibits growth at high concentrations. The fourth kind, **abscisic acid,** is a growth inhibitor and also promotes leaf abscission. Finally, there is **ethylene,** which inhibits growth and has other effects such as promotion of fruit ripening. Each of these five kinds of plant hormones are discussed below.

In addition to these, several dozen other plant hormones have been proposed at one time or another. Some of these are well established and others may prove to be more important than now realized. However, others may be found to be of little importance, to lack the qualifications for classification as hormones, or to be members of one of the established classes of hormones. Several remain hypothetical because they have never been extracted from plants and identified chemically. We will mention a few of these other hormones here, but will not discuss most of them further.

A large number of plant growth inhibitors in addition to abscisic acid and ethylene have been proposed. Some of these have proved to be abscisic acid, but there are probably other growth inhibitors among those that have not yet been isolated and identified chemically. Growth inhibitors are present in dormant buds and some dormant seeds and are responsible for their dormancy. These will be considered again later in the discussion of dormancy. Many more kinds of substances are known to cause seed dormancy than bud dormancy, and these are referred to as **germination inhibitors.**

Several of the proposed hormones are involved in reproductive development. One is **vernalin,** which has been reported to play a role in the low temperature preconditioning that makes biennials grow tall and bloom the second year of their life. Another is **antheridogen,** which promotes the development of antheridia on gametophytes. Several species of water molds have a complex system of reproductive hormones. The female plants produce a hormone that induces the formation of the sex organs of the male plants and then another hormone that attracts the sperm from the males to the female sex organs.

Perhaps the most extensively investigated reproductive hormone of plants is **florigen,** which was proposed by a Russian botanist in 1936 and which he suggested was produced in leaves and was transported to buds, causing them to develop into flower buds rather than leaf buds. If only the leaves of photoperiodic plants are subjected to a suitable day length for flower initiation they will bloom. The time required for the hormone to move from a leaf to a bud has been determined. If such a plant is grafted to another plant that has not been subjected to a suitable day length the second plant will also bloom. These and other experiments provide support for the florigen concept, but florigen has never been extracted definitively from plants and, of course, has not been identified chemically. It thus remains a hypothetical hormone. Florigen will be mentioned again in connection with photoperiodism.

Most of the vitamins are essential for plants as well as for animals. Some vitamins, including B_2 (riboflavin) and C (ascorbic acid), are synthesized throughout the plant. Others, such as B_1 (thiamine), nicotinic acid, and B_6 (pyridoxine), are synthesized primarily in leaves but are essential for root growth. Since they are translocated to the roots in low concentrations, they meet the requirements for classification as hormones. Also, young embryo plants cannot synthesize all the vitamins they need and these must be translocated to the embryos. However, vitamins are used primarily in the synthesis of various coenzymes, a role that has not been ascribed to the principal plant hormones. Vitamins play essentially the same roles in plants as in animals, but the true plant hormones are quite different from animal hormones.

The Auxins

None of the plant hormones have been investigated more thoroughly or have been found to have more diverse effects on plant growth and development than the auxins. The principal auxin produced by plants is indole-3-acetic acid (IAA) (Fig. 15.6). A number of related indole compounds are present in plants and have been reported to have auxin activity, but it may be that these are first converted to IAA. If so, IAA is the only natural auxin. However, a variety of synthetic compounds not produced by plants have auxin activity and have had wide commercial use.

Auxin synthesis is most active in meristematic tissues such as coleoptile tips, the young leaves of buds, and embryo plants. Young leaves that have emerged from buds produce considerable auxin, but as they get older their auxin synthesis progressively diminishes. Germinating pollen grains produce considerable auxin. Auxins can move from cell to cell, but are primarily translocated through the phloem. Their translocation is polar, in a downward direction from the stem tips toward the roots. This is not a matter of gravity. For example, in the hanging branches of weeping willow trees, auxin is still translocated from the stem tip to the stem base.

Basic Effects of Auxin. At the cellular level, auxin is essential for cell elongation, but it also plays roles in cell division and cell differentiation. For example, in trees and shrubs, resumption of the division of the cells of the vascular cambium in the spring does not occur until the terminal buds lose their dormancy and begin producing auxin again. Auxin is apparently essential also for the differentiation of the young cells of the xylem into vessel elements and fibers.

A cell cannot enlarge unless auxin is present. The fibrils of cellulose in a cell wall are cross-linked to pectic compounds and proteins, thus preventing the walls from stretching more than a small amount. In some way not yet determined, auxin brings about the breaking of these linkages, thus making the wall plastic or stretchable. The turgor pressure of the cell then causes the cell to elongate. Auxin probably also brings about the synthesis of additional cell wall materials. If an enlarging wall merely stretched, the wall would get progressively thinner, but actually it becomes thicker. Auxin cannot cause cells that have produced secondary cell walls to elongate further.

Auxin is incorporated in nuclei, as shown by radioactive tracers, and may well bring about the production of new kinds of mRNA and thus new enzymes that are involved in the auxin-influenced processes. However, many investigators question whether this is the way auxin promotes cell elongation, because auxin acts within just a few minutes whereas this mechanism requires at least 15 minutes and probably longer. They believe that auxin has its primary effect on the wall itself or on the plasma membrane.

Biological Tests for Auxins. Even the most sensitive chemical tests do not permit the accurate identification and measurement of the small quantities of auxins present in plants, but several much more sensitive biological tests have been devised. The *Avena* (oat) coleoptile test is the most widely used and is considered as a basic standard. The **coleoptile** is the cylindrical sheath covering the young leaves of grass seedlings. For the *Avena* test, oat seedlings are allowed to grow in the dark until the coleoptiles are about a centimeter long. Then, under red light (which does not cause phototropic bending), the coleoptile tips are cut off, the rolled up leaves inside are pulled loose and raised a little above the cut coleoptile tip, and a small block of agar containing the auxin to be measured is placed on the resulting ledge (Fig. 15.7). Plain agar is used for the controls. A solution of auxin may be included in the agar before it has gelled, but in testing for auxins in plant tissues the tissue is placed on solidified

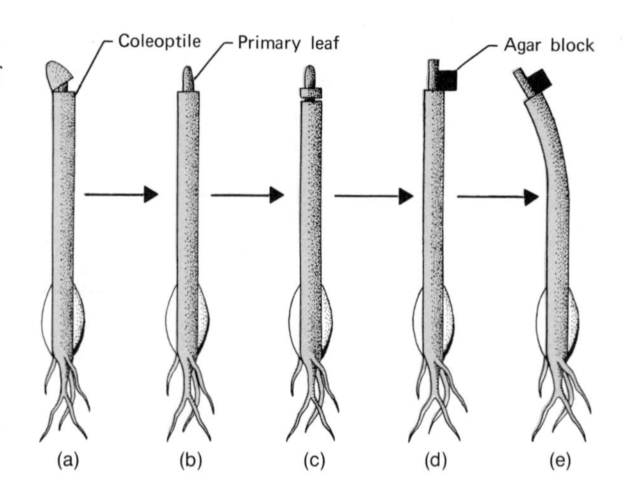

Fig. 15.7 Successive steps in the Avena *coleoptile bioassay for auxin.* (a) *Auxin-producing tip detached.* (b) *Three-hour interval.* (c) *Second decapitation. The rolled-up primary leaf is pulled loose so it will not raise the agar block as it grows.* (d) *Application of agar block containing auxin.* (e) *After 90 minutes the degree of curvature is measured. The degree of curvature is proportional to the auxin concentration. All operations are in a darkroom with only a dim red light that is not phototropically effective and the seedlings are kept vertical, to prevent geotropic bending. (From* Principles of Plant Physiology *by Bonner and Galston, W. H. Freeman & Co.)*

agar and the auxins diffuse into the agar. The more auxin contained in a standard sized agar block, the more the side of the coleoptile below the block will grow. Since the opposite side of the coleoptile is growing little if at all, the coleoptile bends, and the concentration of auxin present can be calculated from the degree of bending as compared with the bending produced by known auxin concentrations. As little as a hundred-billionth of a gram (10^{-11}g) of auxin can be measured by the *Avena* test.

Two other auxin tests are suitable for use with solutions containing auxins, but are not so well adapted to measurement of auxins from plant tissues. In the *Avena* straight growth test, 5 mm long pieces of oat coleoptile are placed in the solutions being tested, the amount of growth in length indicating the quantity of

Fig. 15.8 The slit pea bioassay for auxins. Above: pieces of pea seedling stems are slit longitudinally for about ¾ of their length. Lower left: when placed in water the half stems curve outward. When placed in solutions of auxins they curve inward, the degree of curvature being proportional to the auxin concentration.

auxins present. In the slit-pea test (Fig. 15.8), sections of pea stems from seedlings raised in the dark are slit about two-thirds of their length. The degree of curling of the cut portions of the stems is related to the concentration of auxins present in the solution.

Some Effects of Auxins on Development. Although the fundamental effect of auxins appears to be related principally to cell wall formation and cell enlargement, the auxins have numerous and rather varied observable effects on plant growth. For example, auxin is involved in **apical dominance,** the inhibition of lateral bud growth by terminal buds. If a terminal bud is removed, apical dominance is lost and the lateral buds grow into branches, but if a paste containing an auxin is placed on the stem tip, lateral bud growth is inhibited just as it was by the terminal bud (Fig. 15.9). Normal branching of plants usually occurs only some distance below the terminal bud.

Auxins are also involved in leaf abscission. The abscission of leaves of deciduous plants in the autumn, the abscission of the older leaves of evergreens, and the abscission of leaves during the growing season if they are diseased, extensively destroyed by insects, or subjected to prolonged drought are all results of a reduction in the quantity of auxins reaching the petiole base. This changes the ratio of auxin on the two sides of the petiole base and as a result the leaves abscise. If the blade of a coleus leaf is cut off, the petiole will abscise within a few days, but a dab of lanolin containing an auxin will prevent abscission if applied to the cut end of the petiole. Fruits may also abscise before maturity if they are producing too little auxin. Such premature fruit drop may be prevented by spraying the plants with solutions of auxins or auxin-like synthetic growth substances.

Growing stems exposed to brighter light on one side than another bend toward the brighter light, a response known as **phototropism** (Fig. 15.10). A common but erroneous and teleological explanation of phototropism is that the plants are trying to get more light for photosynthesis. Although this may be a result of the bending, the bending occurs because the light reduces the auxin concentration on the more brightly lighted side of the terminal bud and so less auxin reaches the elongating cells on the more brightly lighted side of the stem. Since the shaded side of the stem has a higher auxin content it grows faster, the unequal growth resulting in bending as the *Avena* coleoptile test. The more rapid growth of stems in the dark than in the light is also related to the effects of light on auxin concentration.

If a plant is placed in a horizontal position, its roots will bend and grow downward while its stem will bend upward until it is again grow-

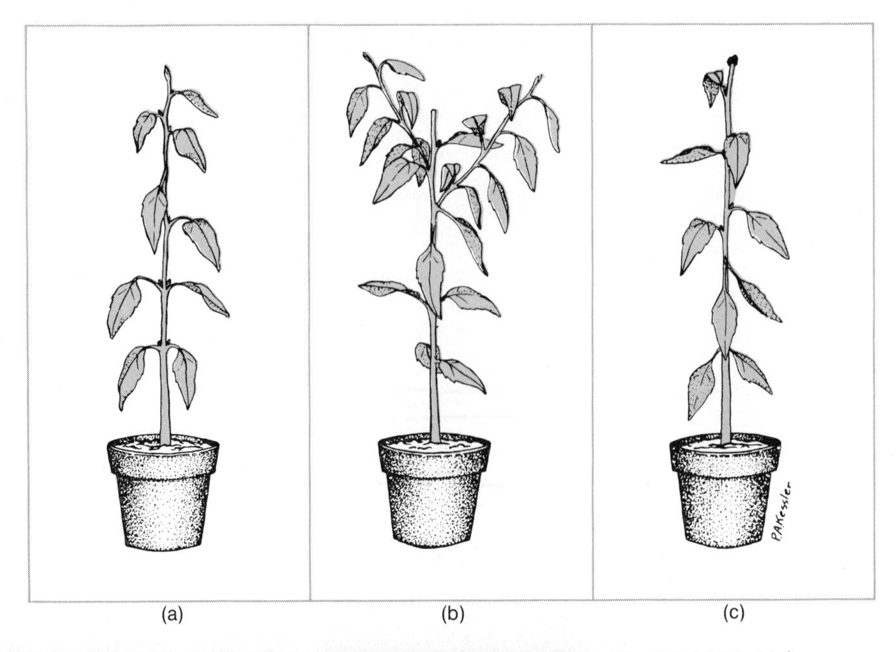

Fig. 15.9 *Removal of the terminal bud of an unbranched sunflower plant* (a) *results in the loss of apical dominance and the growth of two lateral buds into branches* (b). *When lanolin containing auxin was placed on the cut stem of another debudded plant growth of the lateral buds was inhibited, as in the intact plant.*

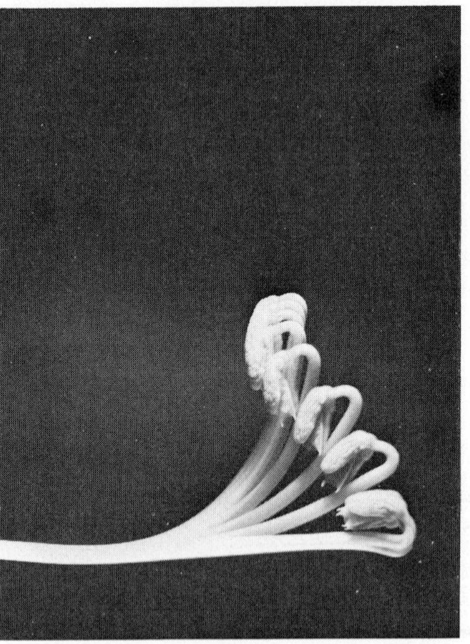

Fig. 15.10 *Multiple exposure photographs of phototropic (left) and geotropic (right) bending of bean seedlings (exposures at 40 and 50 min respectively).*

ing vertically (Figs. 15.10 and 15.11). This growth movement, known as **geotropism,** also involves unequal auxin distribution. Auxin is more concentrated on the lower sides of both stems and roots oriented horizontally. Whereas this higher auxin concentration promotes growth of the lower sides of the stems, it inhibits growth of the lower sides of the roots (roots being more sensitive to auxin), resulting in the two opposite directions of bending.

The auxins also participate in the formation of adventitious roots, particularly on stem

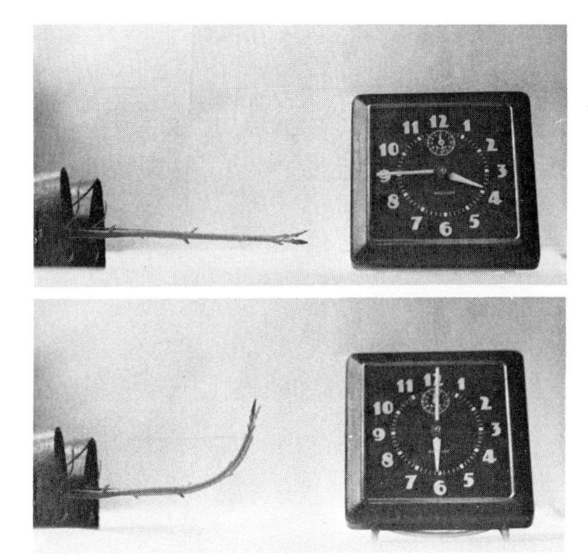

Fig. 15.11 (*Top*) *Geotropic response of two young tomato plants during a 135-minute period. The leaves were removed just before the first photograph to make the bending of the stems clearly evident.*

Fig. 15.12 (*Bottom*) *Cuttings of European silver fir stems six months after treatment with auxin. The cuttings at the right received twice the auxin concentration of the ones in the center. The untreated controls* (*left*) *developed callus on the stem ends, but no adventitious roots.*

cuttings. The cuttings of some plants such as willow and coleus apparently contain adequate quantities of auxins and root readily when placed in water or moist sand. Cuttings of some species, however, form few or no adventitious roots unless supplied with a suitable concentration of an auxin (Fig. 15.12). Cuttings with leaves generally root better than those with leaves removed. Auxins can induce root initiation in tissue cultures that would otherwise not root at all (Fig. 15.13).

The development of the ovulary (ovary) of a flower into a fruit is dependent on an adequate concentration of auxin (and perhaps other plant growth substances). In most plants the auxin comes from the germinating pollen and from the developing embryos within the ovules, and so fruits fail to develop unless pollination and often fertilization (resulting in a zygote and then an embryo) occur. However, in many species of plants unpollinated flowers will develop seedless fruits if they are treated with an auxin solution (Figs. 15.14 and 15.15). This is known as **artificial parthenocarpy.** However, some kinds of plants such as bananas, seedless grapes, and seedless citrus fruits have **natural parthenocarpy.** Their ovularies contain adequate auxin without fertiliza-

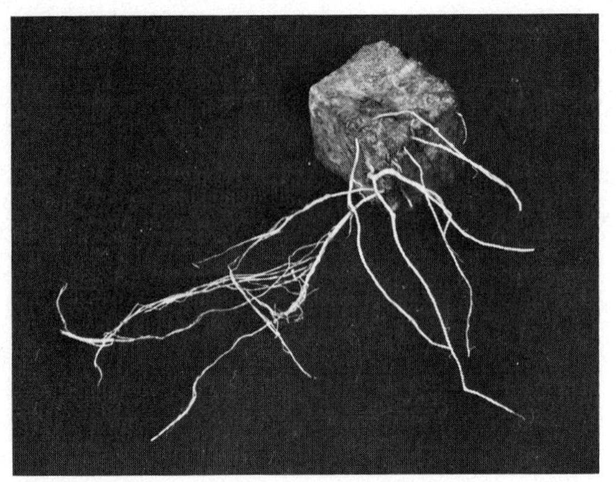

Fig. 15.13 Adventitious roots developing from a cube of potato tissue in a culture medium supplied with auxin.

flower initiation in pineapple plants. A pineapple plant can even be induced to flower by placing it on its side, since the auxin concentration becomes higher on the lower side.

Ethylene

Our consideration of ethylene follows that of auxin because of the intimate relationship between the two. The higher concentrations of auxin that inhibit growth promote ethylene production by plants, and it has been proposed that the growth inhibition is really brought about by ethylene. A number of the auxin effects such as the development of adventitious roots, inhibition of flowering, and promotion of pineapple flowering can be brought about by exposing plants to ethylene.

Ripening fleshy fruits increase their production of ethylene just prior to ripening, and it has been known for some time that treatment of fruits with ethylene will promote their

Fig. 15.14 Influence of pollination and auxin application on the development of strawberry receptacles into accessory fruits. (Left) The strawberry on the left was pollinated normally. The one in the center was from a flower bagged to prevent pollination and did not develop into a normal fruit. When pollination of the flower was prevented but the flower was treated with auxin, the result was the fruit on the right. Note that only the normally pollinated strawberry has true fruits ("seeds") developed from the ovularies of the numerous pistils. (Right) Enlarged photograph of a strawberry from a flower of which only three pistils were pollinated. The three true fruits that developed are clearly visible and the receptacle grew only below each of the fruits.

tion and embryo development, and often even without pollination.

The general influence of auxin on flower initiation, if any, is to inhibit it. Auxin concentration in buds generally decreases at the time of flower initiation. However, auxin promotes

Fig. 15.15 Artificial parthenocarpy of unpollinated holly plants sprayed with a weak solution of 2,4-D when the flower buds were beginning to open. The untreated control (left) failed to develop fruits.

ripening. Ethylene causes epinasty (downward growth of leaf petioles), which can also be produced by auxin application to the upper side of a petiole. Tomato plants are particularly sensitive, and have been used to check for slow leaks from gas pipes. Artificial gas contains considerable ethylene.

Ethylene is a simple hydrocarbon gas ($H_2C=CH_2$) and so is a rather strange substance to be a plant hormone, but there is no doubt about its marked effects on plants. Perhaps it would best be considered as an intermediate substance in certain auxin-induced reactions rather than a distinct plant hormone on its own.

The Gibberellins

A fungus (*Gibberella fujikuroi*) parasitic on rice causes the infected plants to grow much taller than healthy plants. In 1926, a Japanese scientist discovered that extracts from the fungus promoted plant growth, and by 1938 other Japanese scientists had isolated and identified the growth-promoting substance, **gibberellic acid.** However, it was not until 1955 that this growth substance became known in the western world. Since then the very marked promotion of plant growth by the **gibberellins** (the various kinds of gibberellic acids) has created much interest among botanists and horticulturists and hundreds of research papers on the gibberellins have been published. Since gibberellins have now been isolated from higher plants, it is evident that the gibberellins are naturally occurring plant hormones.

Plants treated with gibberellins generally grow at least two or three times as tall as untreated plants, and sometimes the growth promotion is much more marked than this (Fig. 15.16). The effect is principally on stem growth, promotion of leaf growth being less marked. Gibberellins have little effect on root growth at concentrations usually used, but higher concentrations inhibit root growth. Some genetic dwarf varieties of corn, peas, and other species grow as high as the tall varieties when supplied with gibberellins. At least several species of plants grow well over a broader

Fig. 15.16 (*Left*) *S. H. Wittwer of Michigan State University with cabbage plants he treated with gibberellic acid. Control plants of the same age are at the left. Cabbage is a biennial that develops tall stems and blooms only the second year of its life, but the GA caused these plants to bolt and bloom their first year.*

Fig. 15.17 (*Top*) *The spinach plant on the right was treated with gibberellic acid, and is the same age as the untreated plant on the left.*

temperature range when treated with gibberellins than they do otherwise.

One of the more striking effects of the gibberellins is the promotion of flowering, particularly in biennials and long-day plants (Fig. 15.17). Biennials ordinarily do not bolt (produce tall stems) and bloom until they have been exposed to a period of low temperature, but when treated with gibberellins they develop without the cold treatment. Some species that bloom only under long days will bloom even if kept under short days when treated with gibberellins, but gibberellins will not cause short-day plants to bloom if they are kept under long days. Among its other effects, gibberellin overcomes some kinds of seed dormancy and eliminates the light requirement for germination of some seeds such as those of lettuce. Gibberellins can also break bud dormancy.

The effects of gibberellins on plant growth and development are in general different from those of auxin. For example, auxin does not promote stem growth when applied to intact plants and does not induce the bolting and blooming of biennials. One exception is that both auxin and gibberellins may bring about artificial parthenocarpy. In some plants, such as tomatoes, either one is effective, while in others, such as apples and pears, gibberellins are effective while auxin is not. In still other species, auxin is effective but gibberellins are not.

At the cellular level gibberellins, like auxin, promote both cell elongation and cell division, but the principal effect of gibberellins is promotion of cell division in stems below the apical meristem. Their promotion of cell elongation may result from the increased auxin production they cause. Gibberellins cannot bring about cell elongation in the absence of auxin.

In germinating seeds the embryos produce gibberellins and these diffuse into the endosperm where they induce the synthesis of α-amylase and other hydrolytic enzymes such as proteases, thus bringing about digestion of the foods in the endosperm. If the embryo is excised from a seed, these enzymes are not produced unless gibberellin is supplied to the endosperm (Fig. 15.18). There is evidence that the gibberellin acts by inducing the production of new kinds of mRNA that code for the enzymes. Whether or not this is the way gibberellins act in all cases where they promote growth is not known. Excised endosperms of barley seeds are used as one kind of bioassay for gibberellins. Another bioassay for gibberellins involves the use of genetic dwarf plants known to be deficient in gibberellins.

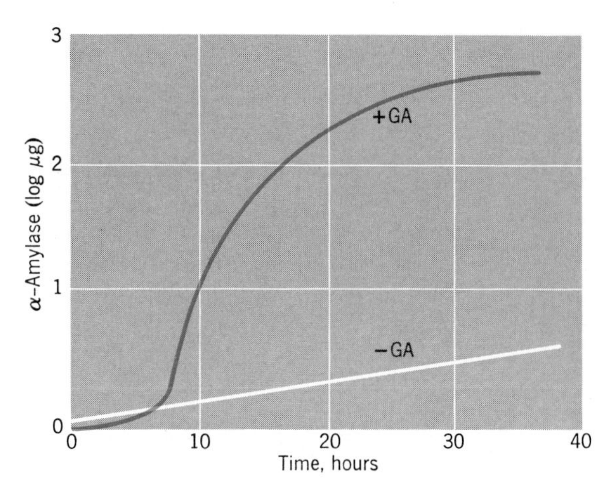

Fig. 15.18 Effect of gibberellin (GA) on α-amylase production by excised barley endosperm. (Redrawn from Varner et al., 1965.)

In contrast with auxins, where IAA is the principal if not the only natural one, several dozen different kinds of gibberellic acid have been isolated from plants. These are designated as GA_1, GA_2, GA_3, and so on. Their molecules have the same general structure as the gibberellin shown in Fig. 15.6 but differ from one another in various details. The different kinds of GA have variable effects on plant growth. Some promote growth while some are more effective in flower initiation.

The Cytokinins

In 1954, in the laboratory of Folke Skoog at the University of Wisconsin, a substance named **kinetin** that induced cell division in tobacco pith cultures was isolated from autoclaved herring sperm. Auxin promoted enlargement of the cells in the isolated pith, but not cell division. Kinetin proved to be a purine derivative. It was effective at the low concentration of 1 part per billion. Kinetin has not been isolated from plants, but various other purine derivatives have been (Fig. 15.19) and it is now clear that they are a distinct class of plant hormones, that are generally referred to as the **cytokinins.**

Cytokinins are widely distributed within plants, but roots are a particularly rich source and the cytokinins are evidently translocated to the shoots through the xylem. They may also be translocated downward through the phloem. The influence of cytokinins on cell division is basic, and although auxin and gibberellins also influence cell division, they cannot substitute for the cytokinins. The cytokinins have also been reported to promote cell enlargement in some cases, but principally in a lateral rather than lengthwise direction. Apparently the cytokinins exert their basic influences at the level of mRNA and enzyme production.

At the level of plant growth and development the cytokinins may promote the thickening of stems and roots and promote the growth of branch roots. They promote the development of buds by tissues in culture, whereas auxin promotes root development. The auxin/cytokinin concentration ratio is important in determining whether buds or roots will develop, and both do so at a medium ratio. Cytokinins also promote bud formation by detached leaves or pieces of leaves of many species. In addition, they release lateral buds on a stem from apical dominance even though the apical bud is not removed, and can break the dormancy of buds and some seeds. The cytokinins are apparently essential for the growth and development of embryos. In at least some species they bring about the conversion of staminate flowers into pistillate flowers, as auxin may also do. However, gibberellins can cause pistillate flowers to develop in place of staminate flowers.

Cytokinins bring about the mobilization of various solutes, including amino acids, auxin, and phosphorus, as has been shown by experiments in which cytokinins have been applied to only part of a leaf (Fig. 15.20). They also delay the senescence of leaves, particularly detached ones, when applied to the leaves. This is at least partly a result of their mobilizing effect. When leaves are detached there is extensive hydrol-

Fig. 15.19 The structural formulae of kinetin and three of the naturally occurring cytokinins of plants. Note that all are adenine derivatives but that they have different side chains.

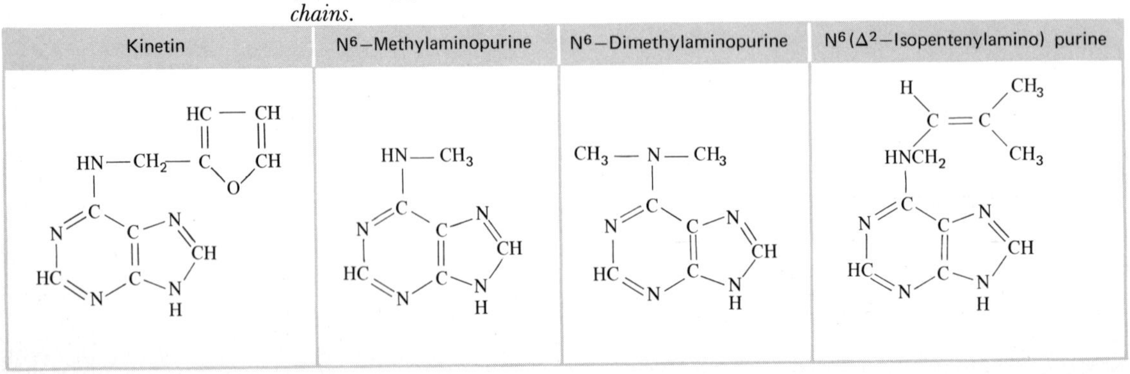

Kinetin	N^6–Methylaminopurine	N^6–Dimethylaminopurine	N^6(Δ^2–Isopentenylamino) purine

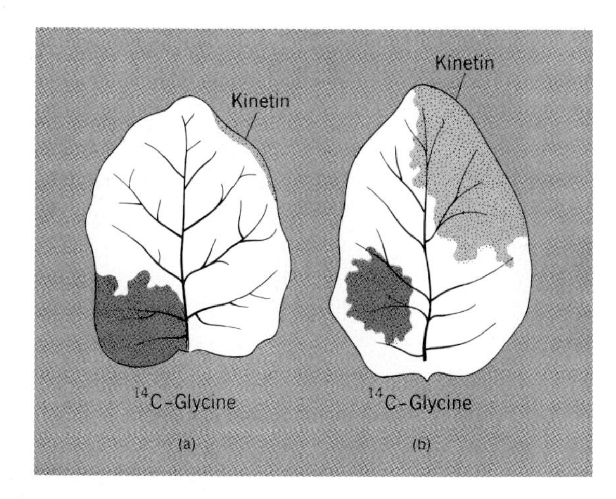

Fig. 15.20 Drawings of radioautographs of a young leaf (a) *and an old leaf* (b) *of* Nicotiana rustica. *Radioactive glycine was applied to the lower left of each leaf and kinetin to the upper right of each one. Later the leaves were removed and radioautographs were made. Note the strong mobilization of glycine by kinetin in the old leaf, but not in the young one. (After radioautographs in K. Mothes,* Naturwiss., 47, *337, 1960.*)

ysis of substances such as proteins and RNA and translocation of amino acids, mineral salts, and other solutes out of the leaves. Cytokinins apparently promote the synthesis of proteins, RNA and other essential substances, as well as reducing translocation of such solutes out of the leaves because of their mobilizing influence.

Abscisic Acid

During the early 1960s, chemical analysis of several plant growth inhibitors that had been called dormin, β-inhibitor, and abscisin revealed that they were all the same substance, which was named abscisic acid (ABA). ABA (Fig. 15.6) is a terpene derivative and is effective in as small amounts as 10^{-7}g. It brings about bud dormancy and promotes the abscission of leaves and also leaf senescence. When applied to seeds it makes them dormant, and is the natural cause of seed dormancy in some species. When a terminal bud is removed the ABA in lateral buds decreases, but the application of ABA keeps them from developing into branches. Thus ABA, as well as auxin, is evidently involved in apical dominance. ABA is also a general growth inhibitor. At the cellular level ABA inhibits both cell division and cell elongation.

Thus, ABA influences plant growth and development in a generally opposite way than auxin, gibberellins, and cytokinins. It should be stressed that normal plant growth and development is dependent on a suitable balance between the growth-promoting hormones and the growth-inhibiting hormones such as ABA and ethylene and subtle interactions among them.

Practical Applications of Plant Growth Substances

Synthetic auxins and auxin-type plant growth substances have a number of horticultural uses, particularly in the rooting of cuttings and in artificial parthenocarpy. Auxins have been especially useful in greenhouse tomato production, and have also been used to prevent preharvest drop of apples and other fruits. Auxins sprayed on pineapple plants bring about blooming and fruit production out of season, although application of auxins to most plants delays rather than promotes blooming. The pineapple industry has profited greatly from the use of plant growth substances that either promote or delay blooming as fruits can now be harvested around the year rather than during a limited period. The packing plants are not longer plagued with a rush season followed by a period of inactivity.

The most widely used plant growth substance is the synthetic compound 2,4-D (2,4-dichlorophenoxyacetic acid). The related compound 2,4,5-T (2,4,5-trichlorophenoxyacetic acid) is also used extensively. These and certain other phenoxy compounds have many of

Fig. 15.21 Leaf modifications of cotton plants resulting from treatment with nonlethal concentrations of 2,4-D. Right: a normal cotton leaf.

the effects of auxins when used in extremely low concentrations, and since they are considerably less expensive than the auxins, they are used in many of the ways described previously. When used in higher concentrations of around 0.1%, these phenoxy compounds act as **selective herbicides,** killing broad-leaved plants but not grasses and certain related plants. The use of 2,4-D in controlling weeds in lawns, pastures, and fields of various grains such as corn, wheat, oats, and rice has been of great practical value. Woody weeds such as poison ivy and Japanese honeysuckle are controlled more effectively by 2,4,5-T than by 2,4-D. Attempts have been made to find a selective herbicide that would kill grasses but not broad-leaved plants, and although several substances of this type have been discovered, they are not as effective as might be desired. Relatively low concentrations of 2,4-D that do not kill broad-leaved plants may, however, cause stem bending and leaf deformation (Fig. 15.21).

The widespread use of 2,4-D and 2,4,5-T as herbicides and defoliants in the Viet Nam war (especially on mangroves and rubber trees) has been severely criticized by botanists as well as others for two reasons. One is the serious disturbance of the natural ecosystems that they caused. The other is their possible adverse effects on human beings and other animals. Experiments have shown that these substances can cause birth defects and other damage at the concentrations used in the war. However, their less extensive and more controlled horticultural and agricultural uses should probably not cause any great concern.

Synthetic gibberellin is available commercially but has had limited practical application. It has been used to increase fruit size and to bring biennials into bloom their first year, thus speeding seed production, but its promotion of plant height is of little practical value. In general, compact bushy plants are more desirable from a horticultural standpoint than tall spindly ones with long internodes. The cytokinins have had little practical application so far. As mentioned earlier, ethylene is widely used to promote fruit ripening or is removed from fruit storage houses to delay ripening.

Abscisic acid has not been used much

practically as a growth inhibitor or to promote leaf abscission, but less expensive synthetic substances have been. Defoliants have been used on mature cotton plants to facilitate the use of mechanical harvesters. A variety of synthetic growth inhibitors not synthesized by plants have had horticultural uses. These include Amo-1618, CCC, B995, Phosfon-D, and maleic hydrazide (MH). MH has had the most extensive practical use. It causes cessation of cell division in plants (but not in animals and some fungi) and so the plants stop growing although they remain alive. Maleic hydrazide is available commercially as MH-30 and MH-40 and has been used extensively to prevent the growth of suckers (branches) on tobacco plants that have been topped and so have lost apical dominance. Maleic hydrazide is also used to prevent sprouting of potatoes and onions in storage, to control the growth of hedges after pruning, and to reduce the labor of mowing plants along highways. Although maleic hydrazide inhibits the growth of lawn grass, it is not extensively used for this purpose since a slight overdose may cause the grass to turn brown.

DORMANCY

Earlier in this chapter it was noted that there are daily and seasonal variations in the rate of plant growth. Plants or plant parts, particularly seeds, buds, and the vascular cambium may cease growing and enter into periods of **dormancy.** The striking thing about a dormant bud or seed is that it will not resume growth, even though the environmental factors may be suitable for growth, until the dormancy is broken. The buds of trees and shrubs in temperature regions enter dormancy in the fall and remain dormant through at least the early part of the winter; potato tubers and onion bulbs are dormant for some time after harvesting. Among the species with seeds that are dormant for one to many years are clover, water lotus, cocklebur, pigweed, holly, orchid, vetch, and honey locust. The seeds of many plants such as beans, corn, peas, and radish never become dormant. A seed is not necessarily dormant simply because it is not germinating. Dormancy *does* exist if a viable seed fails to germinate when provided with the environmental factors generally necessary for germination, that is, water, oxygen, and a suitable temperature.

The short days of autumn appear to be the principal environmental factor initiating bud dormancy, but the breaking of dormancy usually occurs only after a period of low temperature. For example, a dormant lilac bush kept over the winter in a warm greenhouse will not start growing, even at the usual time in the spring, but a branch of the plant projecting through a hole in the glass and so exposed to the low winter temperatures will begin to grow as soon as the weather becomes warm enough in the spring. Bud dormancy can also be broken by a variety of chemicals including thiourea and ethylene chlorhydrin, and there is some evidence that the gibberellins may do so, too. Dormant plants are much more resistant to freezing injury than active plants, damage by freezing being the most common in the spring after dormancy is broken. Dormancy keeps buds from opening in brief warm periods during the winter, and so probably has considerable survival value.

Seed dormancy also has survival value, since seeds maturing during the summer or fall will not germinate before at least the following spring if dormant. Dormant weed seeds may remain in the soil for several years before they finally break dormancy and germinate. Although the dormancy of the seeds of some species results from the presence of germination inhibitors, in other species, it results from the fact that the seed coats are impermeable to either water or oxygen or are so hard and resistant that germination cannot occur. The latter types of dormancy are generally broken naturally either by the decay of the seed coats or their cracking by freezing and thawing, but can be overcome also by scarifying the seed coats, treating them with strong acid, or simply removing them. Dormancy resulting from germination inhibitors or similar physiological conditions is more comparable with bud dormancy, and is generally broken by periods of low temperature. It can also be broken by treatment with chemicals such as those used on buds.

PLANT MORPHOGENESIS

The development of the characteristic tissues and organs of an organism from a single undifferentiated cell or from a group of cells separated from the parent organism is one of the most complex, interesting, and puzzling aspects of life. Growth and development are, of course, the product of the cell division, cell enlargement, and cell differentiation going on in the developing organism. But what is it that determines how many cell divisions will occur in various parts of a leaf primordium and in what planes, or how much the cells will enlarge, thus producing a leaf of a certain characteristic size and shape? What causes the most cell divisions and the greatest cell enlargement in root and stem tips to occur in one direction, thus producing elongated organs, whereas in a developing spherical fruit the cell divisions occur almost equally in all planes? Why do certain cells differentiate into components of xylem and phloem, in a pattern characteristic of the species, whereas nearby cells remain essentially undifferentiated? What causes the first true leaves of a bean plant to be opposite and simple while the subsequently developed leaves are alternate and compound? Why does the spore of a fern develop into a small heart-shaped prothallus and the other unicellular stage in the life cycle, the fertilized egg, develop into a large plant with roots, stems, and leaves? These are just a few examples of the problems that face a student of plant **morphogensis** (the origin of form or structure).

The sequence of structural changes, both external and internal, have been observed and described in detail for many specific examples of development. It is obvious that heredity plays an important controlling role in development, since the development of an individual follows the pattern characteristic of the species. Many specific instances of the influence of environment on development have been identified, and it is possible to alter many aspects of development predictably by suitable experimental control of the environment. Development may also be altered and controlled by the application of various plant growth substances. Yet, with all this information about development, biologists have just begun to make progress toward understanding the series of internal processes and conditions linking hereditary potentialities and environmental influences with observable development. It is this aspect of morphogenesis that provides one of the greatest challenges to biologists during the coming years.

The older investigations of development were largely descriptive, and continuing research of this type is still needed, but most of the current work is experimental and it appears that only this type of approach can contribute toward an increased understanding of morphogenesis. The experimental techniques are varied, including manipulation of the environment, use of growth substances and the identification of natural growth substances and organizers, elucidation of metabolic pathways, and appropriate surgical manipulation of the developing tissues. The isolation of cells, tissues, or organs from an individual and their cultivation in sterile cultures has provided considerable information about development and will undoubtedly provide much more in the future. Tissue cultures provide a somewhat simpler system that is more subject to experimental control than the intact organism. The extensive use of lower plants including fungi, algae, mosses, slime molds, and fern gametophytes in the study of development is another approach toward securing as simple a system as possible for elucidation of some of the basic problems. Even the vascular plants provide a less complex situation than the higher animals, and we can expect that many of the basic advances in an understanding of morphogenesis will come from work with plants. One experimental advantage of plants over animals is that plants have continuing development from the stem and root tips and in many cases from the cambium, whereas animal development occurs mostly in the embryonic stages. It is for this reason that the study of animal development has been much more closely identified with the science of embryology than has the study of plant development.

Among the aspects of plant morphogenesis are correlations, polarity, symmetry, differentiation, regeneration, and abnormal development. We shall be able to consider each of these only briefly.

Correlations

The influence of one part of an organism on the growth and development of other parts is referred to as **correlation.** Although correlations are basically under hereditary control, in plants they operate through at least two main sets of internal factors: food and hormones. We have already considered several examples of hormonal correlations brought about by auxins: apical dominance, the development of ovularies into fruits following fertilization, and the influence of auxins on root growth. As an example of a relatively simple type of nutritional correlation, we may take the ratio between the number of leaves and fruits on an apple tree as it influences fruit size. If a tree bears many fruits, the available food is distributed among them and a limited amount reaches each one, resulting in small fruits, but if the same quantity of food is available to fewer fruits each one will grow larger. There are also correlations between fruit development and vegetative growth. For example, tomato plants bearing numerous fruits have reduced vegetative growth, but removal of the fruits results in increased growth.

The ratio between the sizes of the shoots and roots of a plant (the shoot/root ratio) is a correlation controlled by a variety of factors, some of them nutritional. If a plant is supplied with a large quantity of nitrogen in proportion to the carbohydrate produced by the shoot (a low carbohydrate/nitrogen or C/N ratio), a high shoot/root ratio results because the large supply of nitrogen available to the shoots permits the synthesis of much protein and nucleic acid. Consequently, extensive shoot growth occurs. This consumes most of the available carbohydrate and little is translocated to the roots, so root growth is limited by lack of carbohydrate. On the other hand, a high C/N ratio results in a low shoot/root ratio, since the roots use most of the limited supply of nitrogen, leaving relatively little to be translocated to the shoots. However, there is a surplus of unused carbohydrate that is translocated from the shoots to the roots.

Many of the correlations occurring during the growth and development of a plant are more subtle and less well understood than the examples given, but they play essential roles in keeping the pace of growth and development of one part in step with others and so in the production of a characteristic structural pattern.

Polarity

Most organisms, both plants and animals, have a longitudinal axis that is **polarized,** that is, one end differs from the other. Thus, a fish has a head end and a tail end whereas one end of the axis in vascular plants develops into the roots and the other becomes the shoot. Some filamentous algae have an axial structure without obvious polarity, since one end looks just like the other, but others develop a specialized holdfast cell at the lower end and so have a polar structure. Some coccus bacteria and some unicellular algae lack an obvious axial or polar organization. However, even unicellular organisms may be axial and polar as regards their shape, the location of their cell structures, and the attachment of their flagella.

Indeed, it appears probable that all cells and organisms have physiological polarity even though they may have no obvious structural polarity. This physiological polarity involves gradients in such factors as electrical potential, pH, rate of respiration, hormone concentration, and osmotic pressure. The translocation of auxins is polar, occurring only in a morphologically downward direction even if a stem or root is inverted so that the flow is upward in relation to the earth. The translocation of Ca^{++} ions is also polar, occurring only in a morphologically upward direction. There is an electrical potential difference (a difference in voltage) between the two ends of a cell and also between two points along the axis of a plant. Although the possible role of such electrical potential differences in development is little understood, they may play an important part.

That stems have physiological polarity can easily be demonstrated by the fact that stem cuttings develop roots only at the morphologically lower end, whereas the buds at the upper end grow into branches. This occurs even if the pieces of stem are inverted (Fig. 15.22). Transverse polarity is evident in the fleshy roots of the yam. If discs are cut across the root and then sliced in half, roots always develop from the circumference of the half-disc and shoots originate from the cut diameter (Fig. 15.23).

Much work has been done on the development of polarity in the fertilized egg of the brown alga *Fucus*. Shortly after its fertilization the originally nonpolar egg falls to the bottom of the ocean and within a day it develops a protuberance on its lower side, followed by a cell division at right angles to the protuberance (Fig. 15.24). The lower cell forms the rhizoid that anchors the plant, and the upper rounded cell develops into the main part of the plant. Experiments have shown that the lower light intensity on the under side (rather than gravity

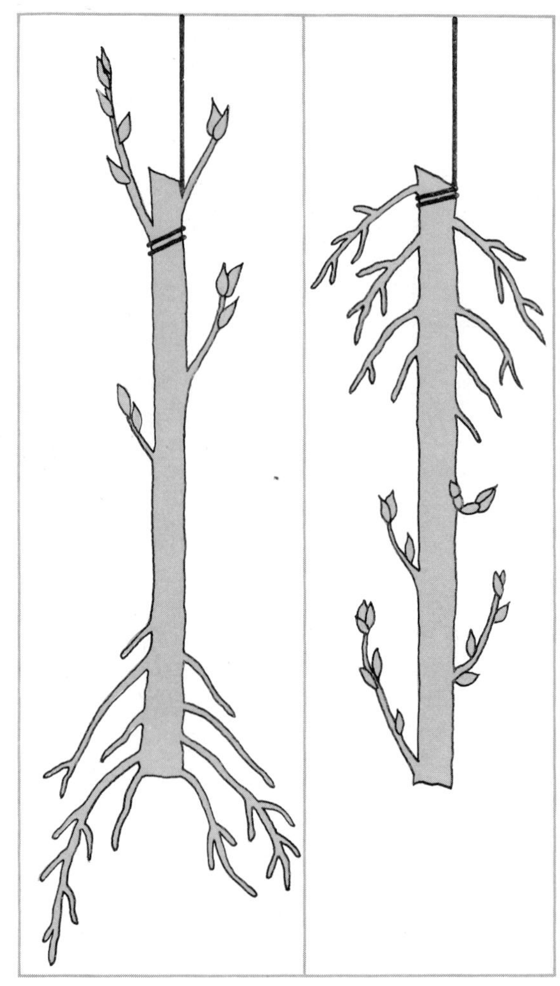

Fig. 15.22 Polarity of bud growth and adventitious root formation of willow cuttings. Left: a cutting suspended in the usual upright position. Right: a cutting suspended in an inverted position. (After W. Pfeffer.)

as might have been expected) is responsible for the polarization of the egg and the appearance of the protuberance on only the lower side. In the dark, the protuberance may appear on any side. It has also been found that the protuberance will grow toward the positive pole in an electric current, toward a warmer environment, toward a lower pH, toward the centrifugal pole when centrifuged, toward the side to which auxin has been applied, and, in a group of eggs, toward the center of the group. These results suggest some of the internal processes involved in the natural induction of polarity by light and also indicate the types of data needed for a better understanding of other and more complex examples of polarity.

Symmetry

One of the striking features of development in most organisms is that it results in a symmetrically organized individual (Fig. 15.25). The symmetry may be **radial,** as in

Fig. 15.23 Transverse polarity of regeneration from a half of a slice through a Dioscorea *(yam) tuber. (After K. Gobel.)*

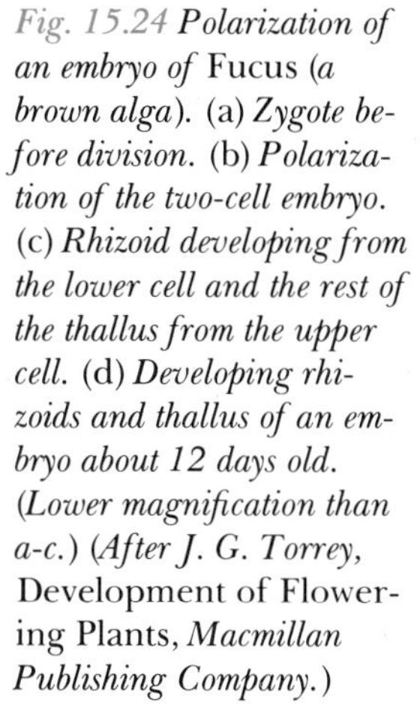

Fig. 15.24 Polarization of an embryo of Fucus *(a brown alga).* (a) *Zygote before division.* (b) *Polarization of the two-cell embryo.* (c) *Rhizoid developing from the lower cell and the rest of the thallus from the upper cell.* (d) *Developing rhizoids and thallus of an embryo about 12 days old. (Lower magnification than a-c.) (After J. G. Torrey,* Development of Flowering Plants, *Macmillan Publishing Company.)*

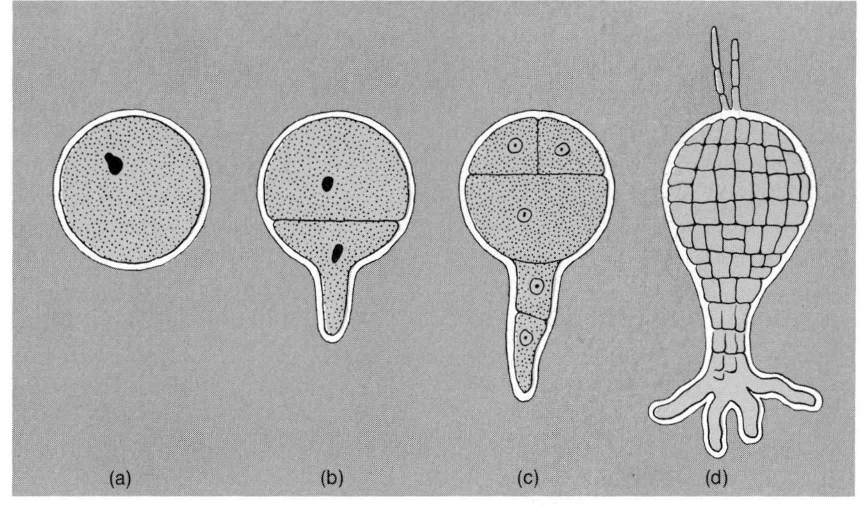

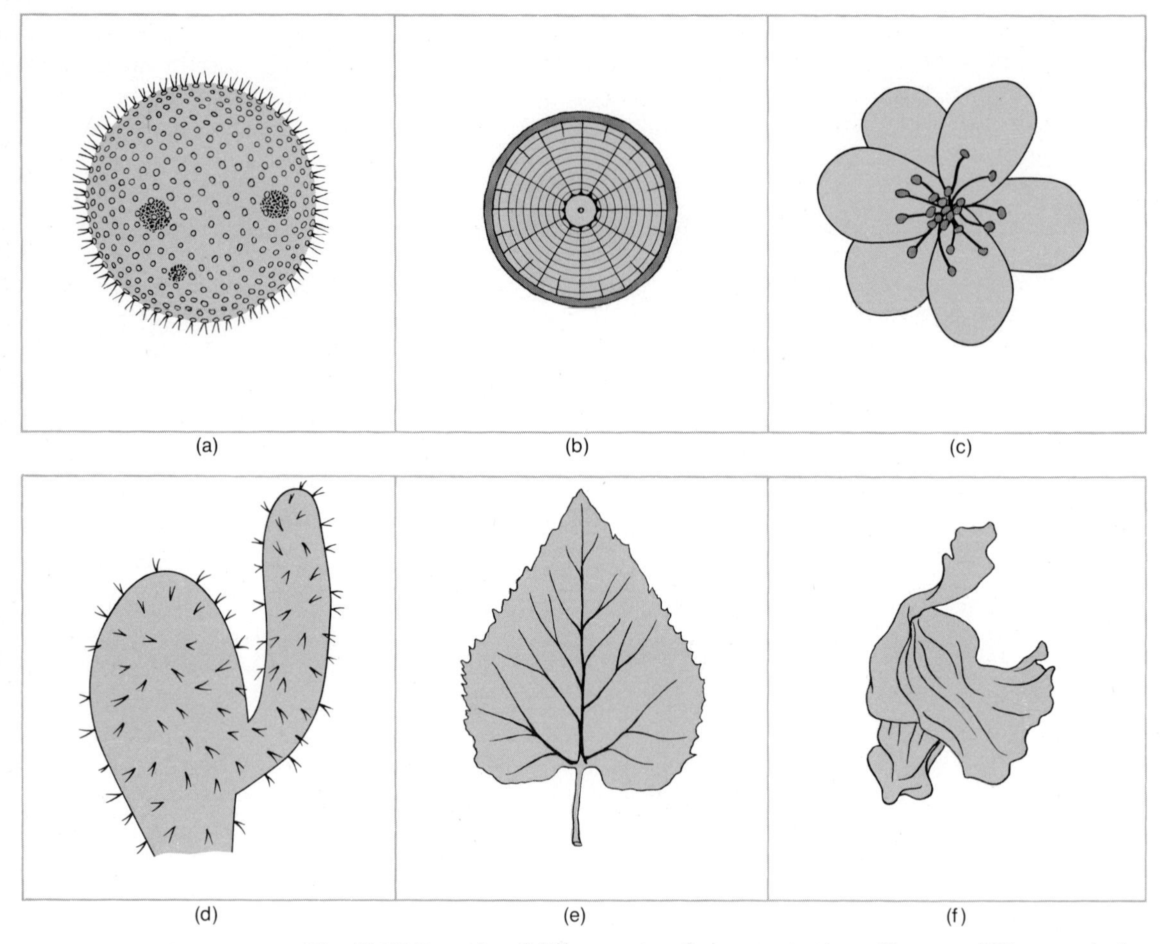

Fig. 15.25 Examples of different types of symmetry in plants (drawn to different scales.) (a) *Spherical,* (b, c) *radial,* (d) *bilateral,* (e, f) *dorsiventral.*

most stems and roots, in many fruits, in flowers such as those of apple or strawberry, in sunflower flower heads, in the umbrella-shaped sporophores of mushrooms, or in *Volvox.* If a radially symmetrical structure is cut along any radius (i.e., diameter) the resulting halves will be essentially the same. The symmetry generally extends to internal as well as external structures. In a **bilaterally symmetrical** structure, the two sides and also the front and back are similar. Thus, essentially equal halves can be obtained by cutting the structure through the center along either of two, and only two, diameters. The flattened stems of some cacti and the leaves of the iris are good examples of this rather rare type of symmetry. **Dorsiventral symmetry** is sometimes considered as bilateral, but differs from true bilateral symmetry in that the front and back sides are different, so that the structure can be cut along only one plane to get similar halves. Most leaves have dorsiventral symmetry, although the leaves of some plants such as elm and beech are unsymmetrical. Many flowers such as those of sweet pea and snapdragon have dorsiventral symmetry. Horizontal stems are sometimes dorsiventral rather than radial, and the branches of several conifers in turn branch only in a horizontal plane, producing a flattened dorsiventral spray.

Symmetry is found, not only in individual organs of plants, but also frequently in the shoot and root systems as a whole. Trees growing in the open where they are subjected to reasonably uniform environmental conditions on all sides show a general radial symmetry, although factors such as shading on one side may prevent this symmetrical development. Nonuniform environments may also modify the natural symmetry of individual organs.

Leaves may appear to be attached to stems in a random pattern, but closer observation reveals that they are symmetrically arranged in a definite pattern characteristic of the species. Some species have opposite leaves, the leaves of one node generally being at a 90° angle from those at the next node. More species have alternate leaves arranged on the stem in a spiral pattern, the steepness of the spiral being a characteristic of the species (Fig. 15.26). In the simplest type of alternate arrangement, the successive leaves are 180° apart around the stem and so every second leaf is directly above another leaf. In going from a leaf to the one directly above it through the intervening leaf one spiral is made around the stem. This situation may be represented by the fraction $^1/_2$, the numerator representing the number of turns around the stem and the denominator the number of leaves from the starting point. We can describe the **phyllotaxy,** or pattern of leaf arrangement, of such a plant merely with the fraction $^1/_2$. Another common kind of phyllotaxy is $^1/_3$, the third leaf being directly above the first one and one spiral turn around the stem being required to reach it. Only a limited number of phyllotaxies are found, the series being $^1/_2$, $^1/_3$, $^2/_5$, $^3/_8$, $^5/_{13}$, $^8/_{21}$, $^{13}/_{34}$, $^{21}/_{55}$, $^{34}/_{89}$, and so on, although the phyllotaxies from $^8/_{21}$ on are found only in cones and other compact structures with modified leaves. It is interesting that any phyllotaxy in the series represents the sum of the numerators over the sum of the denominators of the previous two phyllotaxies.

Fig. 15.26 Diagram of the leaves of a shoot with ⅜ phyllotaxy, viewed from above. Every eighth leaf is directly above another leaf (e.g., 9 over 1) and three spirals around the stem are required to reach the eighth leaf. (After Sinnott and Wilson, Botany: Principles and Problems, *by courtesy of the publishers, McGraw-Hill Book Co.)*

Differentiation

The differentiation of the cells, tissues, and organs of a developing organism constitutes the core of most morphogenetic problems. We have already considered cell differentiation briefly, and the differentiation of tissues and organs involves the differentiation of certain kinds of cells at specific places and times. However, the differentiation of organs cannot be accounted for entirely by cell differentiation. The location and shape of an organ are determined primarily by the pattern of cell division. Thus, cell divisions at precisely spaced points on the stem apex result in outgrowths that are the leaf primordia. Further cell divisions, largely in one plane, result in the formation of a young leaf in its characteristically flat form. Whether the leaf will be entire or lobed, simple or compound, as is characteristic of the species, also depends on the pattern of cell division. The size of the leaf depends largely on how long cell division continues before it stops, and so depends on the number of cells present rather than on their size. This is also generally true of other organs, although the size to which cells enlarge also influences organ size to some degree. Thus the pattern of cell division determines the phyllotaxy, size, and shape of leaves, whereas the differentiation of certain types of cells at specific places determines the character of its tissues. Protuberances also grow from the stem apex in the axils of the leaf primordia (Fig. 9.1), and through their own characteristic pattern of cell division and differentiation they develop into the axillary buds. In contrast with the stem apex, the root apex is smooth and lacks outgrowths and so does not produce leaves and buds. This basic difference in development is just one of the many morphogenetic problems without a satisfactory solution.

The difference between woody and herbaceous stems is basically a matter of how long cell divisions in the cambium continue. Although some herbaceous annuals have no cambium at all, others have a cambium but it remains active only one year, whereas woody stems have a cambium that remains active year after year. Of course, some plants considered as herbaceous annuals in temperate climates are woody perennials in tropical climates.

The pattern of differentiation may depend on the age of the plant. Thus the first leaves (the **juvenile** leaves) of a plant may be quite different in size and shape from the ones developed later (Fig. 15.27). What appears to be an extreme example of this occurs in the English ivy, which may grow for long periods of time as a vine with relatively large lobed leaves and no flowers. When flowering branches finally appear they have smaller, entire leaves (Fig. 15.28) and the branches grow out in a rather rigid, shrubby pattern. Cuttings taken from such branches produce bushy plants with entire leaves, although seeds from these plants grow into plants with lobed juvenile leaves. The discovery that gibberellic acid causes flowering branches to produce lobed juvenile leaves provides a clue to some of the internal factors controlling leaf form. Some plants such as the morning glory have a series of leaves at successive nodes with shapes gradually progressing from the juvenile to the adult form.

Fig. 15.27 *A seedling of* Acacia melanoxylon *with juvenile pinnately compound leaves (characteristic of the genus) and the adult foliage consisting of phyllodes (flattened petioles only). Note the intergrading forms. (After Velenovsky.)*

The age of a plant may influence flower, as well as leaf, development. Thus, in the acorn squash the first flowers are all underdeveloped staminate flowers. Normal staminate flowers develop at the next several nodes, then for some time a mixture of staminate and pistillate flowers appears, followed by a mixture of inhibited staminate flowers and giant pistillate flowers. Finally, only pistillate flowers that produce fruits parthenocarpically are formed. The sequence of development seems fixed, but the length of each stage is influenced by day length and temperature.

The conversion of buds from a vegetative to a reproductive condition is an interesting example of differentiation. As has been pointed out earlier, the controlling factor appears to be a flowering hormone translocated from the leaves, and this is produced by some plants only during certain lengths of day and night. The transformation from a vegetative to

Fig. 15.28 Left: a shoot from a mature plant of English ivy (Hedera helix). *The mature plants are upright and shrubby and bear flowers. Right: a shoot of a juvenile plant of English ivy. The juvenile plants are trailing or climbing vines and never bloom. Note the differences in the leaves of the two stages. Most English ivy plants remain in the juvenile stage indefinitely.*

a flower bud involves several basic changes in development. (1) The leaves are modified into flower parts (sepals, petals, stamens, and carpels in that order); (2) lateral bud formation is suppressed; (3) the internodes do not elongate; and (4) continued growth of the stem apex ceases. The result is a bud that develops into a flower rather than a leafy branch. One of the striking things about the development of a flower is the switch from the formation of sepals to petals and then to stamens and carpels as development progresses along the minute distance from onc modified leaf to the next.

Little is know about the complex of internal factors providing such precise and specific control of differentiation, even though it is known that environmental factors such as length of day may determine whether an imperfect flower will be staminate or pistillate. In some flowers, especially those with numerous parts, there may be an intergrading series between petals and stamens (Fig. 15.29), suggesting a gradual change in developmental factors. In some plants, age, as well as environmental factors, influence flower differentiation. Trees that do not bloom until they are several years old are extreme examples.

There is growing evidence that an important factor in differentiation, and perhaps the most basic one, is the production of new kinds of messenger RNA by a cell just prior to the time that it begins differentiating. As a consequence, new types of enzymes are produced and these presumably catalyze the specific biochemical processes involved in differentiation. It has been shown, for example, that in the epidermal cells of roots which will later develop root hairs (but not the other epidermal cells), new enzymes appear before there is any evidence of a developing hair. In buds of plants that have been induced to bloom by suitable photoperiodic treatments, there is an increase in both RNA and proteins before any visible differentiation into flower buds occurs.

Fig. 15.29 Intergradations between petals and stamens of Nymphaea. *(Courtesy of the copyright holder, General Biological Supply House, Chicago.)*

Furthermore, chemicals that inhibit RNA synthesis also prevent the development of flower buds. While it has been known for some time that not all cells of an organism produce all the enzymes that the organism is capable of producing (i.e., for which it has a DNA code), it is now becoming evident that this is probably the result of the synthesis of certain kinds of messenger RNA only by certain cells, at particular stages of development, and under specific conditions.

A certain cell at a certain time is likely to be producing only a small fraction of the mRNA for which it and the other cells of the organism have a DNA code. The reason is that in some way or another most of the DNA (genes) is prevented from making mRNA, possibly by being linked to or jacketed by certain proteins. Hormones and certain other substances are apparently able to bring about or remove the DNA inhibitor selectively, so that at certain stages in the development of the cell or under certain environmental conditions the production of specific kinds of mRNA (and consequently enzymes) is stopped or started. This programming of mRNA and enzyme production could well be the basic controlling factor in differentiation.

For example, a plant with genes for red flowers may be producing the enzymes essential for anthocyanin synthesis only in the cells of the petals, even though all the cells of the plant contain the necessary DNA code for producing the enzymes. On the other hand, the cells of the petals do not produce chloroplasts even though their chromosomes contain all necessary hereditary potentialities for doing so. Evidently, differentiation of the petals involved turning on the production of some kinds of mRNA and turning off the production of other kinds. That every cell of a plant still contains all the hereditary potentialities of the zygote from which it developed (except after meiosis) is indicated by the fact that certain isolated vegetative cells can develop into complete new plants under suitable conditions, even though this does not occur when they are in place in the tissues of the plant. The necessary conditions include substances that apparently activate the production of the kinds of mRNA required for development. Conversely, while the cell was in the intact tissue it seems probable that neighboring cells may have been producing substances that inhibited the production of these kinds of mRNA.

Visual evidence that only certain parts of chromosomes are actively producing mRNA has been provided by microscopic examination of the giant chromosomes of *Drosophila* (the fruit fly) at various stages of development of the organism. Regions of the chromosome actively producing mRNA appear as enlarged puffs, and mRNA production has been pinpointed on these puffs by the use of radioactive nucleotides and autoradiography.

Regeneration

The restoration of lost parts by an organism is known as **regeneration.** Plants and certain invertebrate animals such as hydra, planaria, and starfish can regenerate entire organs but regeneration in higher animals is limited to the healing of wounds.

The restoration of bark that has been scraped from a tree trunk is an example of regeneration in plants. If a portion of a herbaceous stem is cut out so as to interrupt one or more vascular bundles, the bundles may be regenerated by the differentiation of a series of pith cells into what are essentially short, vascular elements (Fig. 15.30). If a stem tip is cut vertically into several segments, each segment will regenerate the missing portions, thus forming several stems in place of one.

There are many examples of more extensive regeneration in plants. If a terminal bud is removed, not only do lateral buds begin growing because of the loss of apical dominance, but the uppermost bud often begins growing vertically, assumes dominance, and thus essentially replaces the missing terminal bud. A piece of a stem cut from many kinds of plants will develop adventitious roots at its lower end, thus regenerating an entire plant. It has been found that even a small piece of a stem tip, as short as 0.25 mm, from some plants will grow on a suitable nutrient medium and the resulting stem may form adventitious roots, thus regenerating an entire plant. Pieces cut from the roots or leaves of some species develop adventitious roots and buds, regenerating an entire plant, although the leaves of many species can produce neither roots or buds, or perhaps only roots, and so fall short of complete regeneration. Stems bearing lemon fruits can regenerate roots (Fig. 15.31) but not buds. Cuttings that can regenerate all missing organs are widely used in the vegetative propagation of plants, and all means of vegetative propagation depend on the capacity for regeneration. Regeneration in relation to plant propagation is discussed in Chapter 20.

The fact that applied auxins promote the formation of roots on cuttings of many species that do not otherwise root readily (Fig. 15.12) provides some information about the internal factors involved. Presumably species that root

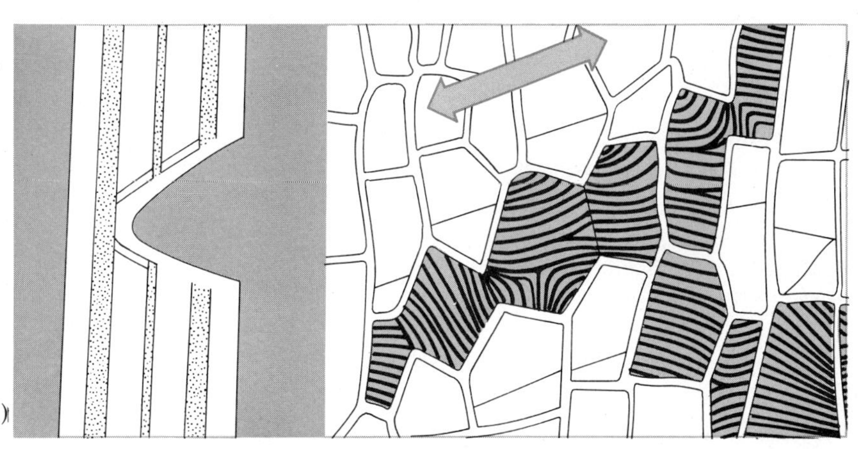

Fig. 15.30 Left: sketch showing regeneration of vascular tissue in a Coleus *stem after some of the vascular bundles were severed by cutting a wedge from the stem. Right: drawing of a microscopic view of vessel elements regenerated from parenchyma cells. As shown by the arrows, regeneration began in the center and progressed both ways. (After E. W. Sinnott and R. Bloch.)*

Fig. 15.31 Development of adventitious roots on a lemon stem with attached fruit. Rooting occurs without application of auxin, but added auxin increases the number of roots.

readily contain an adequate supply of auxin. Cuttings from some species, however, will not initiate adventitious roots even when auxin is supplied so there are evidently other factors to be considered.

Probably the most extreme and spectacular example of regeneration is the development of isolated vegetative cells (that had stopped dividing, elongating, or differentiating before their separation from other cells of a tissue) into embryo-like structures and then into mature plants. Pioneering work of this type has been done since 1955 by Professor F. C. Steward of Cornell University with cells from the phloem of carrot roots and from carrot embryos, and subsequently he and other investigators have also used other species of plants. When small cylinders cut from the phloem were placed in special tissue culture flasks which were slowly rotated, some of the cells became detached from the tissue and these cells then began enlarging and dividing, producing irregularly shaped clusters of cells. Some of these clusters later acquired polarity and developed into embryolike structures. When these were transferred from the liquid medium to an agar culture medium they developed into young plants, which could later be transplanted to pots, where they grew to maturity and produced flowers and seeds (Fig. 15.32).

This development occurred only when the culture medium contained coconut milk (the liquid endosperm of coconut seeds) or liquid endosperm from young corn or buckeye seeds in addition to the usual components of plant tissue culture media such as foods, mineral salts, vitamins and auxin. The endosperms evidently contain substances essential for growth stimulation, and it has been determined that there are a good many such substances including various growth substances that stimulate cell division and the type of alcohols called inositols that play supporting roles. Some tissues, including those of potato tubers, contain growth inhibitors whose influence must be counteracted before even a complete medium with endosperm is effective.

Professor A. C. Hildebrandt and his associates at the University of Wisconsin have secured the development of mature tobacco plants from isolated pith cells on a completely defined nutrient medium which contained kinetin, IAA, and inositol as well as mineral salts, sugars, amino acids, and vitamins (Figs, 15.33 and 15.34). Walter Halperin has secured development of embryos from cell clusters from wild carrot roots in an even simpler medium, containing only sucrose and mineral salts, but only when the concentration of cells is quite high.

Fig. 15.32 Development of carrot plants from cells isolated from embryos. (Similar development occurs from isolated phloem cells.) (a) *Isolated cells in liquid medium.* (b) *A petri dish containing about 100,000 embryoids derived from cells isolated from one embryo.* (c) *Higher magnification of part of* (b). (d) *Embryoid and cell clusters.* (e–i) *Successive stages in embryoid development.* (j, k) *Later stages in development of plants derived from isolated cells.* (l) *Detail of the inflorescence of one of these plants.*

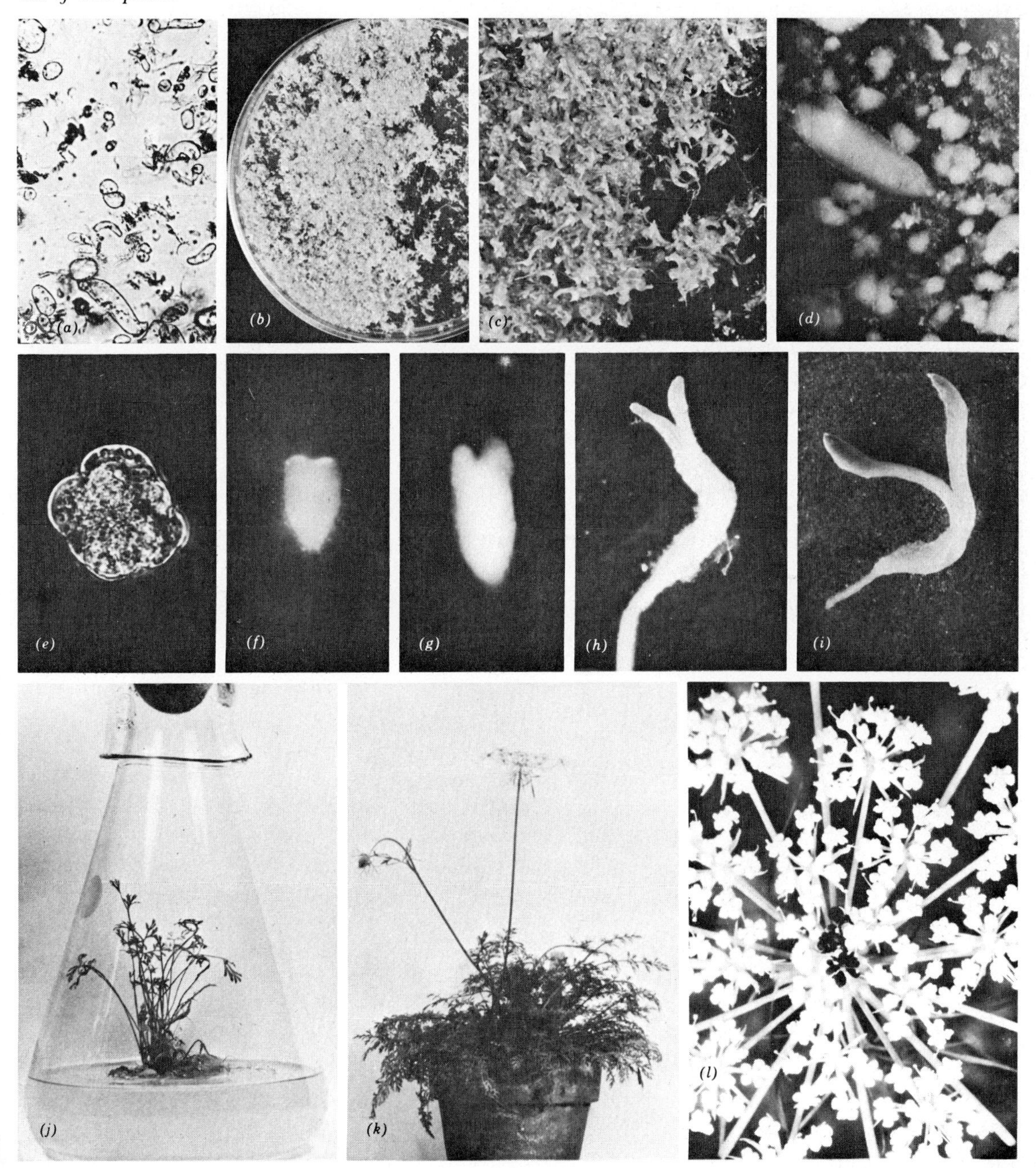

Fig. 15.33 Cells isolated from cultures of tobacco pith callus and clusters of cells developed from isolated cells. (a) *A single isolated cell.* (b–h) *Successive stages in the growth of a cell mass from the cell in* (a). (i–k) *Successive divisions of another isolated cell.* (l) *First division of an isolated cell. Note that the pattern of cell division in the three examples shown is not the same, and also that the cell clusters are not symmetrical or differentiated.*

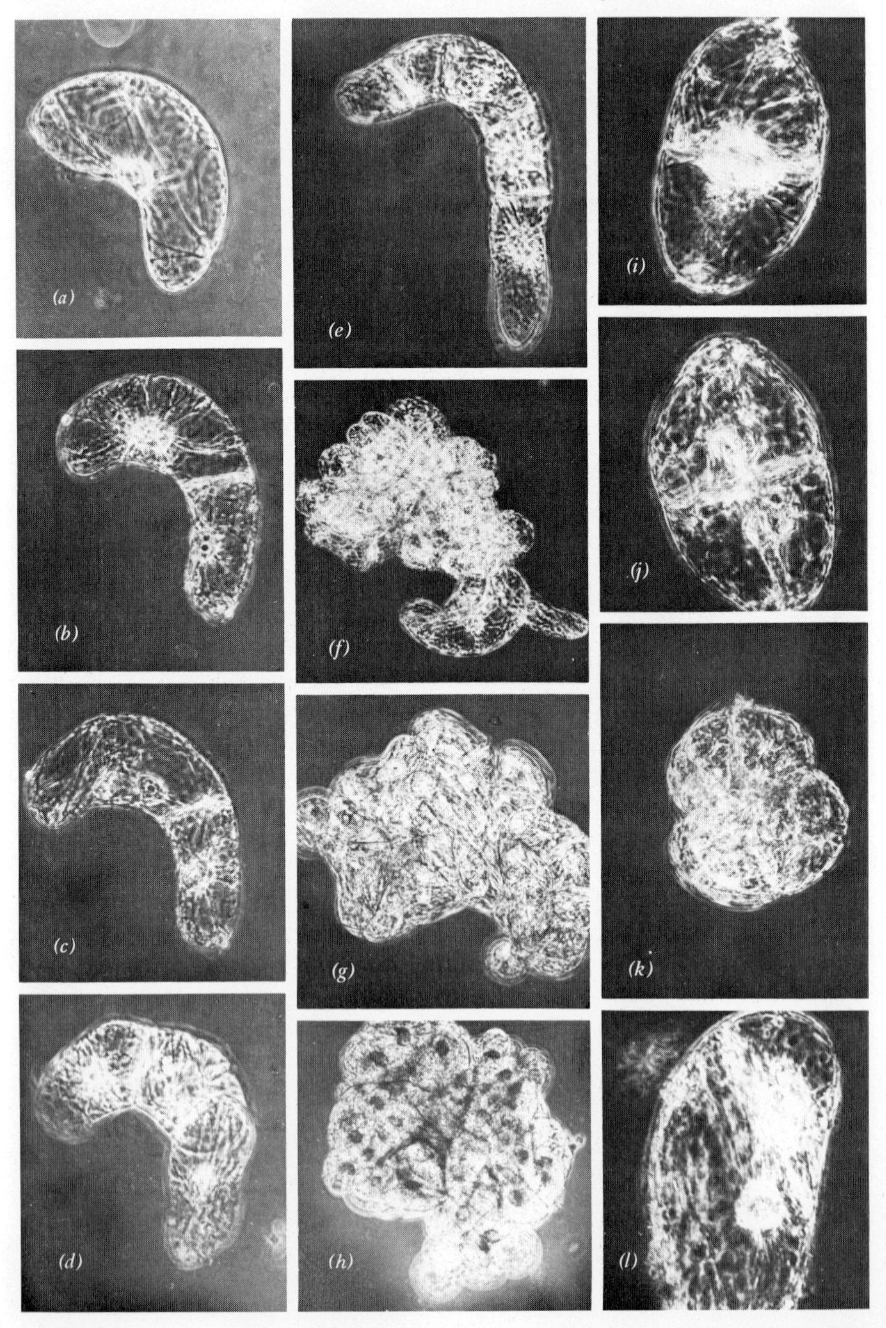

Fig. 15.34 Later stages in the development of tobacco plants from isolated pith cells. Unlike carrot and some other species, development of tobacco plants from isolated cells does not involve embryoid formation. (a) *Undifferentiated callus from a single isolated cell.* (b, c) *Plantlets developing from callus.* (d) *Two plantlets isolated from callus.* (e–h) *Development of plants after transfer from sterile culture to pots.*

Fig. 15.35 A tobacco plant exposed to gamma radiation from cobalt-60 for 11 weeks (right) compared with a control plant of the same age (left). The treated plant was not only extensively malformed but also much smaller than the control plant (Note the two rulers for difference in scale).

Experiments of this type demonstrate that certain mature vegetative cells have all the potentialities of a zygote for developing into a complete mature plant provided that they are freed from the growth inhibiting influences of intact tissue and placed in a suitable environment that provides all the substances essential for growth. Such experiments are one of the most promising means of determining what conditions are essential for development.

Abnormal Development

Development usually proceeds in a highly organized manner, resulting in an individual with a structural pattern characteristic of the species. However, the normal course of development is sometimes disrupted, resulting in an abnormal individual. The amazing thing is not that such deviations from normal development occur occasionally, but that they do not occur more frequently. Since there is a considerable range of hereditary variation in most species, and since environment can modify development markedly, it is a little difficult to determine just where the line can be drawn between normal and abnormal development. For example, an etiolated plant that has been growing in the dark differs considerably in structure from a plant of the same species growing in the light, but we would not consider it abnormal. Indeed, its development has been normal for that particular environment. Genetic mutations sometimes result in the development of structures that can be described as abnormal, but they could also be regarded as normal features of a new hereditary strain.

One class of abnormalities is the product of known environmental factors. Thus radiation may cause extremely distorted development (Fig. 15.35), insects, fungi, or other organisms may cause the development of galls (Fig. 17.15), or applications of growth substances may cause callus formation (Fig. 15.36), or may greatly modify leaf form (Fig. 15.14) or stem form (Fig. 15.37). Other abnormalities may appear to occur spontaneously, although they are undoubtedly the result of changes in internal processes or conditions brought about by an unusual combination of environmental factors or localized hereditary changes. For example, sepals may occasionally develop into what are essentially foliage leaves, the stem bearing a flower may not stop growing as it normally does, resulting in a flower located around a stem rather than at the end of it, or tendrils may develop into leaflike structures. Stems and other radially symmetrical structures occasionally become fused and flattened into broad ribbons, an abnormal development known as **fasciation.** A better knowledge of the causes of such types of abnormal development can also contribute to an understanding of normal development. Much has already been learned by studies of the growth of calluses in tissue culture, and these unorganized and unsymmetrical masses of tissue can certainly be considered as examples of abnormal growth.

PHOTOMORPHOGENESIS

Although all light-influenced processes of plants are related to growth in some way or another, we should distinguish between the **photomorphogenic** processes, which usually have a relatively direct and specific influence on growth and development, and a second

Fig. 15.36 A callus produced by treating the cut end of a Cleome *stem with a synthetic auxin, photographed 20 days after treatment.*

Fig. 15.37 Normal Kalanchoë *plant (left) and one treated with parachlorophenoxyacetic acid (a chemical relative of 2,4-D). The goblet-shaped structure with plantlets at its top developed from the two uppermost leaves. Note the normal plantlets on the leaf margins of the untreated plant (left).*

group of processes that have a more indirect and less specific influence on growth and development. Among the latter are photosynthesis, which is related to growth through the production of food, and stomatal opening and closing, which affects the diffusion of gases into and out of plants and so in turn a variety of processes involved in growth. Phototropism is an example of a photomorphogenic process, and as we have already noted the light effective in phototropism is absorbed by a yellow pigment (probably riboflavin). However, the light effective in most photomorphogenic processes is absorbed by a pale blue pigment (phytochrome) that has been discovered and isolated only in recent years. Phytochrome-mediated processes will be discussed later.

To appreciate the marked and varied influences of light on the growth and development it is only necessary to compare plants that have grown in the dark with plants that have grown under ordinary conditions of day and night. In the dark, plants cannot generally synthesize chlorophyll and so are a pale yellow color. The stems and hypocotyls are unusually tall and spindly, with poorly developed vascular tissue (Fig. 15.38). The hypocotyls and plumules remain bent into a hook and fail to straighten out. The leaf blades fail to enlarge and may remain folded together as they were in the bud. This complex of characteristics is known as **etiolation** (Fig. 15.39). Less extreme etiolation occurs when plants are growing in dim light. As we shall point out later on, light has other photomorphogenic influences in addition to those made evident by comparing plants that have grown in light with etiolated plants.

Photoreceptor Pigments

Before light energy can be used in any biological process it must be absorbed by some substance within the organism. Such substances are known as **photoreceptors.** All photoreceptors of visible light are colored because of the unequal reflection of the various wavelengths making up white light and so are

Fig. 15.38 Bean seedlings of the same age that had grown in bright light (a), *dim light* (b), *and darkness* (c).

Fig. 15.39 Influence of red and far-red light on the development of bean seedlings. Left to right: plant kept in dark ten days. Plant exposed to 2 minute red light during the ten days. Plant exposed to 2 minute red +2 minute far-red. Plant exposed to 5 minute far-red.

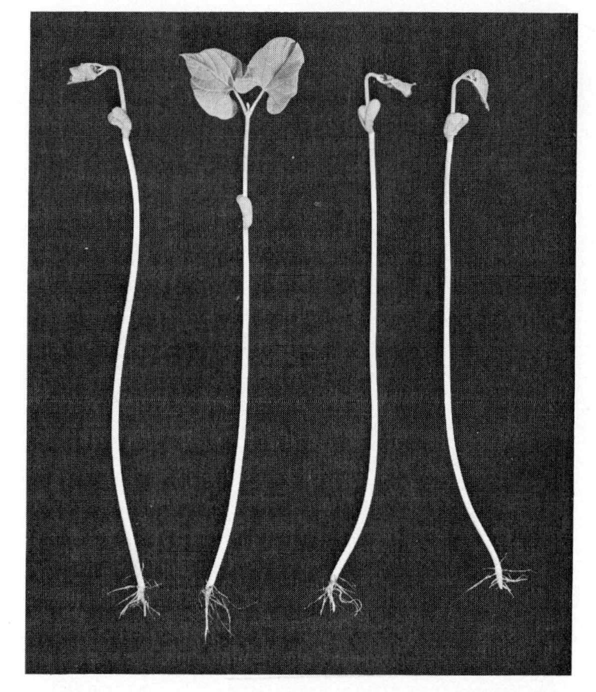

frequently referred to as **pigments.** Some photoreceptor pigments like the chlorophylls and carotenoids are so abundant that they impart a color to the tissues containing them, but other pigments are so dilute that we cannot detect their influence on the color of the tissue. Each pigment has a characteristic **absorption spectrum** in which the peaks of maximum absorption of light energy generally correspond closely with peaks of maximum efficiency of the light-requiring process under various wave-lengths of light (the **action spectrum)** (Fig. 15.40). An action spectrum for a process is determined by exposing a series of plants to light of different colors (i.e., different narrow wave-length bands) by the use of either colored filters or a projected spectrum and then measuring the rate of the process under each wave-length band. The relationship of a photoreceptor pigment to a certain photochemical process cannot be considered as definitely established until the absorption spectrum of the pigment is known to correspond with the action spectrum

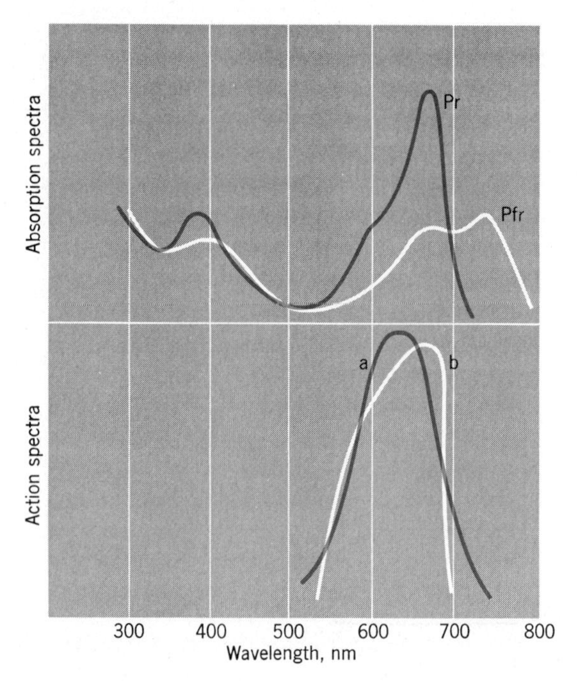

Fig. 15.40 Above: absorption spectra of red-absorbing (P_r) *and far-red-absorbing* (P_{fr}) *forms of phytochrome. Below: action spectra of inhibition of flowering of soybean* (a) *and of promotion of lettuce seed germination* (b) (*Absorption spectra from the data of Butler et al. Action spectra from data of Parker* (a) *and Hendricks* (b).

of the process. Among the more important photoreceptor pigments of plants are chlorophyll, protochlorophyll, the carotenoids, flavin compounds such as riboflavin, and phytochrome.

Besides its wavelength composition **(quality),** light has two other aspects that are biologically important: its **intensity** (measured in **foot candles** or other units) and its **duration.**

Phytochrome and the Red, Far-red Reaction

More than half of the photomorphogenic influences of light on plants have essentially similar action spectra, with the highest peak in the red region of the spectrum. These include the various photoperiodic responses of plants, the light requirement for the germination of seeds and fern spores, the opening of hypocotyl hooks (Fig. 10.3) and plumule hooks (Fig. 15.39) in light, the enlargement and unfolding of young leaves (Fig. 15.39), the synthesis of some kinds of anthocyanin, and the influence of light on the angle between a petiole and the stem. It has been found that the photoreceptor pigment for all these various processes is **phytochrome,** a pale blue substance which is widely (if not universally) distributed in plant tissues but which is present in too low concentration to impart any visible color to them.

Phytochrome was discovered and isolated by Sterling B. Hendricks and his associates at the USDA research laboratories in Beltsville, Md. They have identified it as a protein with an attached pigment molecule, the latter apparently being an open-chain tetrapyrrole chemically related to the bile pigments and to the phycocyanin and phycoerythrin found in red and blue-green algae. An interesting and important property of phytochrome is that it exists in two forms that can readily be converted from one to the other. One form has an absorption peak in the red at 660 nm and is designated as P_r. The other form has an absorption peak at a longer wavelength (730 nm) of red and is designated as P_{fr}, for far-red. Light at 730 nm is at the far end of the visible spectrum, almost at the beginning of the infrared band, and is sometimes even referred to as infra red. It is just barely visible, and some people cannot see it at all. When P_{fr} absorbs light at 730 nm it is converted to P_r. When P_r absorbs light at 660 nm it is converted to P_{fr}. In addition to these interconversions brought about by light, which are very rapid, there is a very slow conversion of P_{fr} into P_r in the dark. The result is that during the night all the phytochrome is converted to the P_r form. In sunlight and broad spectrum artificial light containing both the red and far-red wavelengths, the conversion of P_r to P_{fr} predominates, so their influence is the same as that of red light at 660 nm. The phytochrome interconversions are summarized in Fig. 15.41.

The significance of these P_r and P_{fr} interconversions is that P_{fr} is active as an enzyme and so is the biologically effective form, while P_r lacks enzyme activity. It should be noted that under natural conditions sunlight converts phytochrome to the active P_{fr} form, while in darkness the P_{fr} is slowly reconverted back to P_r. It is only under experimental conditions that plants are exposed to narrow bands of red or far-red light. In most cases only a short exposure of a few minutes to light is necessary for the production of effective quantities of P_{fr}, and the more intense the light the shorter is the time required. If an exposure to red light is followed by an exposure to far-red light before the P_{fr} has been able to exert its enzymatic influences it is converted back to P_r and the effect of the light exposure is canceled out. Alternating exposures to red and far-red wavelengths can be repeated for some time, the last one in the series always being the effective one.

One of the more interesting red, far-red reactions mediated by phytochrome is the influence of light on seed germination. The seeds of many plants will germinate either in light or darkness, but the seeds of a good many other species become sensitized to light after they have imbibed water. The germination of some seeds, including those of primrose (*Primula spectabilis*), *Phacelia,* and many members of the Lily family, is prevented by light, whereas the germination of other seeds such as those of tomato and Jimson weed is retarded by light. On the other hand, the seeds of many species, including some varieties of tobacco and lettuce, peppergrass (*Lepidium*), *Lobelia,* mullein, and *Primula obconica,* will not germinate unless they have been exposed at least briefly to light after they have imbibed water. Seeds of carrot, some figs, and many grasses are among those that germinate better after exposure to light, but will germinate in darkness. If the light intensity is great enough, the soaked seeds require an exposure to light of only a few seconds. The light requirement for the germination of lettuce and some other seeds has been reduced or eliminated by treating the seeds with gibberellins, suggesting that these or similar growth substances may be involved at some point in the internal processes associated with germination.

Fig. 15.41 Interconversions between the red (P_r) and far-red (P_{fr}) absorbing forms of phytochrome. The conversions induced by light are rapid, but the others are slow.

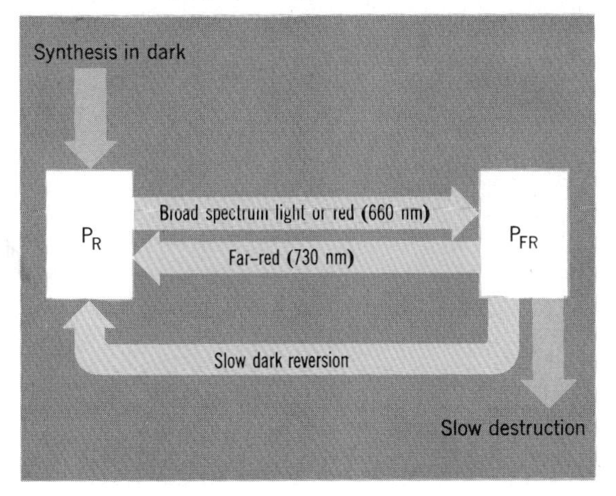

Photoperiodism

That photoperiodism is a red, far-red response has been learned only in recent years, as is also true of the other red, far-red reactions. **Photoperiodism** is the influence of the *duration* of the daily periods of light and darkness on the growth, development, and reproduction of plants and animals. The discovery just after World War I of this previously unsuspected environmental influence on organisms ranks as one of the major biological advances of the present century. Photoperiodism was more or less accidentally discovered during the course of an experiment being conducted for other reasons in the Washington,

D.C. area by W. W. Garner and H. A. Allard of the U.S. Department of Agriculture. Maryland Mammoth, a new variety of tobacco, failed to bloom and produce seeds at the usual time but instead continued to elongate. The plants were eventually moved into a greenhouse and finally began to bloom. The investigators became interested in this unusual flowering behavior and turned from their original experiments in an effort to determine the reasons for the delayed flower formation by Maryland Mammoth. After painstakingly eliminating all possible known environmental factors such as temperature and light intensity by a series of experiments, they came to the rather startling conclusion that blooming had been induced by the short days of late autumn. Their conclusion was substantiated when they found that Maryland Mammoth could be induced to bloom even in midsummer if artificially short days were provided by placing the plants in the dark for part of each day.

Before publishing their results, Garner and Allard investigated the influence of day length on the reproductive development of many other species of plants and found that plants could be classified into at least three groups as regards their response to day length: (1) **Short-day** plants that initiate flowers only when the day length (photoperiod) is shorter than a certain critical number of hours (Fig. 15.42), (2) long-day plants that initiate flowers only when the day length is longer than a certain critical number of hours (Fig. 15.43), and (3) day-neutral plants that bloom under either long or short days. The report on their research, published in 1920, contained such extensive and convincing evidence for photoperiodism and provided such a surprising explanation of the previously puzzling seasonal blooming of plants that it created a sensation among botanists. Garner and Allard conducted many more experiments on photoperiodism and other botanists soon began to research on the subject. Not long

Fig. 15.42 Chrysanthemum plants of the same age kept under 8-hour days (left) and 16-hour days (right).

after the discovery of photoperiodism in plants it was found that aphids (plant lice) have a photoperiodically controlled reproductive cycle. Since then the seasonal breeding of many other invertebrate animals and of birds and other vertebrates, as well as such seasonal responses as changes in fur colors of mammals and possibly bird migrations, have been found to be photoperiodic.

By 1940 the photoperiodic responses of hundreds of species and varieties of plants had been determined, a few of them being listed in Table 15.1. In addition to its influence on

Table 15.1 **Photoperiodic Classification of Selected Plants As Regards Flower Initiation, with Approximate Critical Day Length in Hours if Known**

Short-Day Plants	Less than	Long-Day Plants	Greater than	Day-Neutral Plants
Bryophyllum	12	*Chrysanthemum frutescens*		Bean, string
Chrysanthemum indicum	15	Clover, red	12	Buckwheat
Cocklebur	15	Dill	11	Celery
Cosmos, Klondyke	13	*Hibiscus syriacus* (Althea)	12	Corn (most)
Goldenrod		Larkspur		Cotton
Orchid		Radish		Cucumber
Poinsettia	12	*Rudbeckia hirta*	12	Geranium
Ragweed		Ryegrass, Italian	11	Pansy
Rye, winter	12	*Sedum*	13	Snapdragon
Soybean, Biloxi	14	Spinach	13	Strawberry (everbearing)
Strawberry (most)	10	Sugar beet		Tobacco (most)
Tobacco, Maryland Mammoth	14	Wheat, winter (most)	12	Tomato

Fig. 15.43 Dill plants of the same age kept under 8 hour days (left) and 16 hour days (right). Note the promotion of internode elongation as well as flowering by long days.

blooming, day length was found to influence many other aspects of plant development. For example, long days promote onion bulb development and the vegetative growth of most species of plants, whereas short days favor tuber formation and promote the onset of both dormancy and leaf abscission. In some plants such as *Rudbeckia* and mullein, long days not only initiate blooming but also bring about elongation of the internodes of the stems. Under short days such plants have virtually no internodes and their leaves are all close to the ground in a pattern known as a **rosette.**

Temperature has been found to have a modifying influence on the photoperiodic responses of some species, some plants being day neutral at one temperature range and either long or short day at another. Much has also been learned about the influence of alternating light and dark cycles of unusual length on the photoperiodic responses of plants. One of the more important discoveries is that interruption of the long dark period accompanying short days by only a brief period of light will prevent blooming in short-day plants (Fig. 15.44), provided that the interruption occurs somewhere near the middle of the dark period. The few minutes to an hour of light required, depending on its intensity, are not enough to make the total light period equal a long day. These results indicate that slow reactions requiring a long dark period for completion are involved in photoperiodism, and that short-day plants are really long-night plants whereas long-day plants should be called short-night plants.

Despite much information like this about

Fig. 15.44 Photoperiodic responses of cocklebur plants. (1) *Long nights induce blooming.* (2) *Short nights prevent blooming.* (3) *Interruption of long night by short bright light period prevents blooming.* (4) *Low intensity light is effective in providing a long night.* (5) *One long night and short day cycle is enough to induce blooming.* (6) *Long nights for only one leaf induce blooming.* (7) *Long nights induce blooming.* (8) *Plant 7 grafted to plant kept in short nights induces it to bloom.*

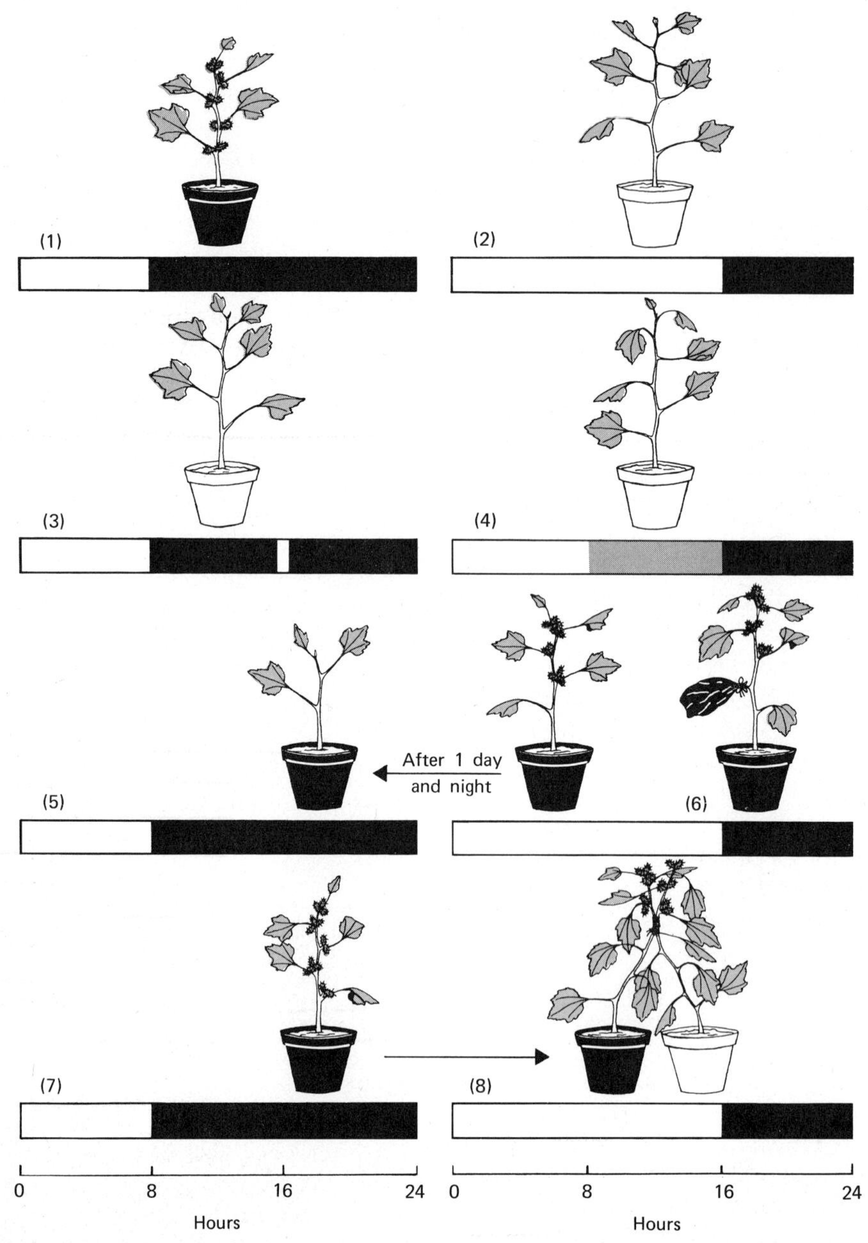

the photoperiodic responses of plants under different environmental conditions, and some information about the inheritance of photoperiodic potentialities of plants, little is known yet about the internal processes and conditions associated with the photoperiodic responses. Since interruption of the long dark period can be effected by red light at 660 nm as well as by white light, and since the effect of red light can be canceled out by far-red light supplied within a half hour, it appears that the enzymatically active P_{fr} catalyzes some process that prevents the production of the floral stimulus and that the long dark period is required for reduction of P_{fr} concentration to a low enough level to permit the floral stimulus to be produced during the remainder of the dark period. Experiments such as the one illustrated in Fig. 15.44(6) have shown that the leaves are the organs that receive the photoperiodic stimulus and a variety of experiments (e.g., Fig. 15.44(8)) have shown that a diffusable stimulus, presumably the hypothetical hormone florigen, is translocated from the leaves to the bud primordia, causing them to develop into flower buds rather than leaf buds. However, very little is known about the presumably numerous and complicated steps between phytochrome and florigen synthesis in the leaves, or about the reasons for the different photoperiodic responses of long-day, short-day, and day-neutral plants. Most of the current research on photoperiodism is directed toward clarification of the series of metabolic processes that occur in the leaves and also those that occur in the bud primordia and we can anticipate an increasingly better understanding of them.

Our scanty knowledge of the internal mechanisms of photoperiodism has not hampered practical application of the discovery. Soon after Garner and Allard discovered photoperiodism, its possible practical applications became evident, and for many years greenhouse plants have been brought into bloom out of season by either extending or reducing the natural day length. During the fall, winter, and early spring, long days may be provided by extending the day length with ordinary incandescent light bulbs. Less than one foot-candle of light is generally adequate, although the light intensity suppled is usually greater than this. During the summer, short days can be supplied by placing the plants in the dark for part of each day (Fig. 15.45). Florists can count on having Easter lilies even for an early Easter, poinsettias by Christmas, and chrysanthemums for supermarket openings even in mid-summer. Controlled day lengths have also been used to bring different varieties of a species into bloom simultaneously, thus permitting them to be crossed for the first time. Pho-

Fig. 15.45 A Japanese morning glory (Pharbitis nil) *plant (short day) that was kept under short days from the beginning. The first leaves and all but one bud were removed, but the cotyledons remain. The bud developed into a flower, rather than into the tall stem characteristic of morning glories, which would have developed if the plant had been kept under long days. Some other species of plants do not develop flowers until several nodes have been produced, even though they have been under suitable photoperiods from the beginning.*

toperiodic control of garden and field crops is less feasible, but floodlights have been used experimentally to prevent the blooming of sugarcane plants. This results in a greater yield of sugar, since none is used in the growth and development of flowers.

MORPHOGENETIC EFFECTS OF TEMPERATURE

Except for purely photochemical reactions, such as the light reaction of photosynthesis, all of the many biochemical processes of plants and all physical processes such as diffusion and water transport are influenced by temperature. The rates of both physical and chemical processes in plants increase with temperature to the point where it becomes high enough to inactivate enzymes or injure protoplasm, unless some other factor is limiting. Plant growth, however, is generally the most rapid at moderate temperature ranges, most temperate zone plants growing best between 70°F and 80°F. Decreased growth at higher temperatures results from such factors as the more rapid increase of respiration than of photosynthesis with a rise in temperature and the marked increase in transpiration and the subsequent wilting and stomatal closure.

In addition to the varied influences of temperature on the rate of plant growth, temperature has some marked effects on the pattern of plant development. These morphogenetic influences of temperature are of two principal types: low-temperature preconditioning and thermoperiodicity. In addition to influencing plant development both of them also influence rates of growth.

Low-Temperature Preconditioning

Relatively low temperatures are effective in breaking both bud dormancy and some kinds of seed dormancy. Since the effects of the low temperature do not become evident until growth is resumed or germination occurs, this breaking of dormancy is an example of **low-temperature preconditioning** (Fig. 15.46). Similar low-temperature preconditioning is a natural prerequisite for the formation of elongated internodes **(bolting)** and blooming of winter wheat, celery (Fig. 15.47), cabbage, foxglove, and all other **biennials,** that is, plants that bloom the second year after planting and then produce seeds and die. Some biennials also require a long photoperiod in addition to the low-temperature preconditioning. Before the importance of low-temperature preconditioning was discovered, truck gardeners sometimes had a complete failure of biennial crops such as celery because the plants bolted and bloomed the first year, making them unsuitable for marketing. After it was learned that this resulted from transplanting the young plants outdoors too early, thus subjecting them to low-temperature preconditioning, crop failures have been avoided simply by waiting to transplant until all danger of cold weather is past. Although the preconditioning requires temperatures a few degrees above freezing, it does not require freezing. Low-temperature preconditioning can be applied to seeds soaked in just enough water to permit germination to start but not to continue. This process is called **vernalization.** Vernalization is used mostly with winter wheat in Russia, where the winters are so cold that winter wheat planted in the fall is killed by freezing, and where suitable varieties of spring wheat are not available. Spring wheat varieties are annuals rather than biennials and, like annuals generally, do not require low-temperature preconditioning. The term *vernalization* is sometimes extended to include all low-temperature preconditioning and also the high-temperature preconditioning of seeds, for example cotton seeds. However, the term is more generally and more properly used in its stricter sense.

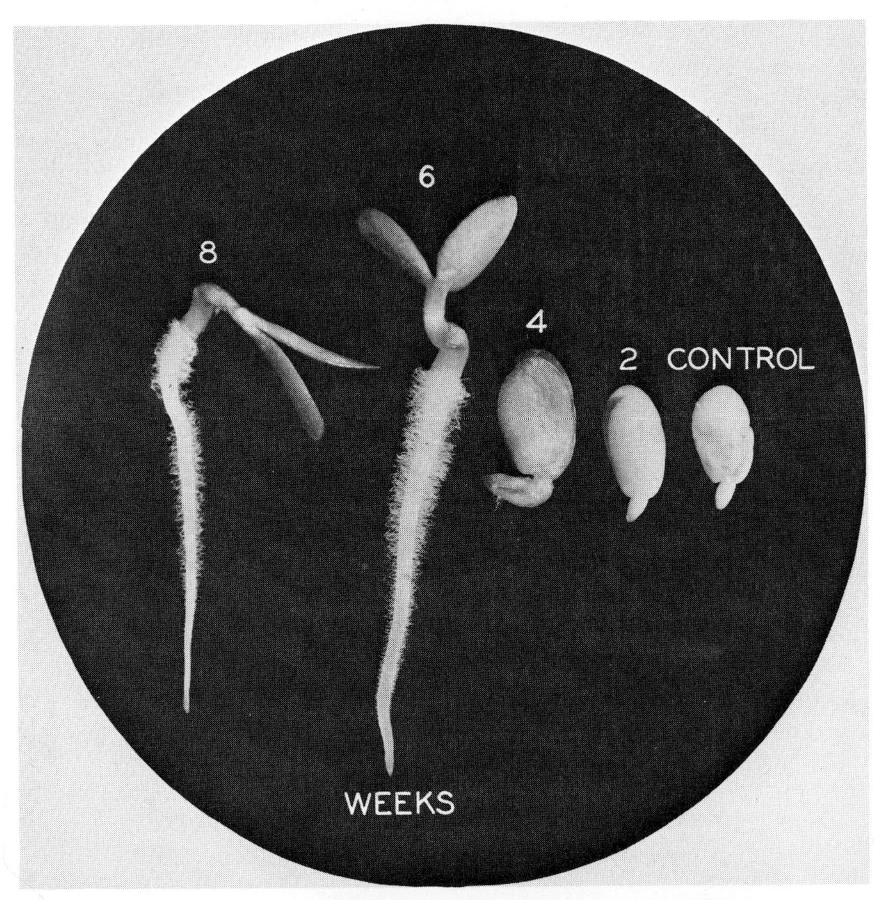

Fig. 15.46 (Left) Embryos of European mountain ash (Sorbus aucuparia) *kept at 1°C for the indicated number of weeks. Those given six and eight weeks of low temperature preconditioning have broken dormancy and are germinating.*

Fig. 15.47 (Right) The celery plants on the left were given low temperature preconditioning (10–16°C) while those at the right were kept at 17–21°C.

Thermoperiodicity

The day to night temperature fluctuations have marked effects on the growth of most species of plants **(thermoperiodicity).** Most kinds of plants grow best when the night temperatures are lower than the day temperatures, but a few species such as the African violet flourish best when it is somewhat warmer at night than during the days (Fig. 15.48). Night temperatures are particularly important in their effects on growth; a difference of only a few degrees frequently produces very different growth patterns. The optimal day and night temperatures vary from species to species, and in addition the optima may change as a plant grows older. Tomatoes are particularly sensitive thermoperiodically, the reduced fruiting of tomatoes during hot summer weather resulting from too high night temperatures. Young tomato plants grow best when the day temperature is about 26°C and the night temperature around 20°C, but mature plants grow and set fruit best at night temperatures of 15°C and no fruit at all develops if the night temperature is over 25°C, even though the plants are blooming.

Fig. 15.48 The solid line represents average day and night temperatures at Pasadena, California during a year. The thermoperiodic requirements of three species are shown by the dotted circles. Zinnia (Z) and China aster (C) thrive outdoors in Pasadena but African violet (Saintpaulia, *SP*) *does not. (Data of F. W. Went,* American Scientist, 44, *378–398, 1956.*)

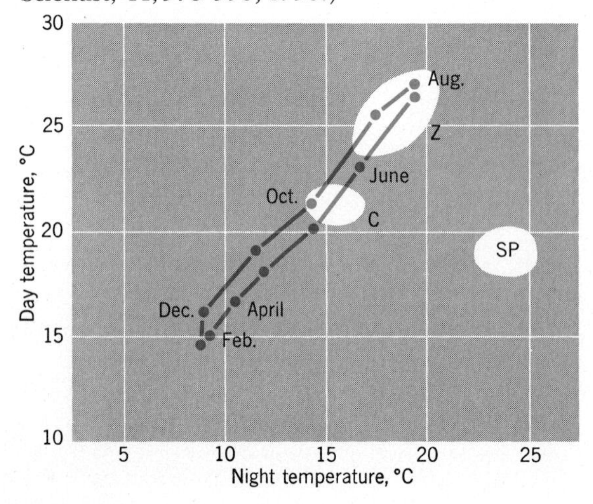

In addition to the diurnal thermoperiodicity we have been describing there is a seasonal or annual thermoperiodicity that is important in the lives of certain species of plants, particularly bulb plants of temperate latitudes such as tulip, hyacinth, and narcissus. Not surprisingly, most of the research on this type of thermoperiodicity has been done by Dutch botanists. In these species of plants each of several critical stages of development has a different optimal temperature, and the plants flourish only in regions where the seasonal temperature changes provide these optimal temperatures in proper sequence. In the tulip, for example, the bulbs must be kept at near 20°C for about three weeks to permit initiation of flowers and differentiation of all flower parts. Further flower development within the bud will not occur until the temperature drops to about 9°C and remains there or lower for about 14 weeks. By the end of this period the flowers within the bulb are fully developed. Gradually increasing temperatures, up to about 20°C, are then required for emergence of the leaves and flower from the bulb and for full opening of the flower. These thermoperiodic requirements make it evident that tulips cannot be successfully grown in the south, even though bulbs purchased from a northern nursery may bloom well the first year. Tulip bulbs can be held completely dormant during any stage of development without injury if kept at about 35°C.

The thermoperiodic requirements of most species have still not been determined, but it is already obvious that there is a close relationship between the optimal thermoperiods for a species and the prevailing temperature fluctuations during the growing season in regions where the plants either grow naturally or are cultivated successfully. Apparently most species of plants flourish only when both the thermoperiod and photoperiod are optimal for the species, even though the plants may be day neutral as far as reproduction is concerned. A better knowledge of the thermoperiodic and photoperiodic requirements of economic species promises to be of great value in the selection of regions for the cultivation of specific crops and in the control of greenhouse environments.

RELATED READING

Amen, R. D. "The concept of seed dormancy," *American Scientist, 51,* 408–424, 1963.

Bell, P. R. "Physical interactions of nucleus and cytoplasm in plant cells," *Endeavour, 34,* 19–22, 1975.

Bonner, J. T. "The growth of mushrooms," *Scientific American, 194*(5), 97–106, May 1956.

Bonner, J. T. "How slime molds communicate," *Scientific American, 209*(2), 84–93, August 1963.

Brachet, J. L. A. "*Acetabularia.*" *Endeavour, 24,* 155–161, 1965.

Braun, A. C. "Plant cancer," *Scientific American, 186*(6), 66–72, June 1952.

Braun, A. C. "The reversal of tumor growth," *Scientific American, 213*(5), 75–83, November 1965.

Brown, F. A., Jr. "The 'clocks' timing biological rhythms," *American Scientist, 60,* 756–766, 1972.

Davidson, E. H. "Hormones and genes," *Scientific American, 212*(6), 36–45, June 1965.

Galston, A. W. "Which end is up? (Geotropism)," *Natural History, 83*(5), 20–23, 1974.

Galston, A. W. and P. J. Davies. "Hormonal regulation in higher plants," *Science, 163,* 1288–1297, 1969.

Gibor, A. "*Acetabularia:* a useful giant cell," *Scientific American, 215*(5), 118–124, November 1966.

Hillman, W. S. "Light, time, and the signals of the year," *BioScience, 23,* 81–86, 1973.

Khudairi, A. K. "The ripening of tomatoes." *American Scientist, 60,* 696–707, 1972.

Salisbury, F. B. "The initiation of flowering," *Endeavour, 24,* 74–80, 1965.

Shen-Miller, J. and R. R. Hinchman. "Gravity sensing in plants: a critique of the statolith theory," *BioScience, 24,* 643–651, 1974.

Steward, F. C. "Totipotency, variation, and clonal development of cultured cells," *Endeavour, 29,* 117–124, 1970.

van Overbeek, J. "The control of plant growth," *Scientific American, 219*(1), 75–81, July 1968.

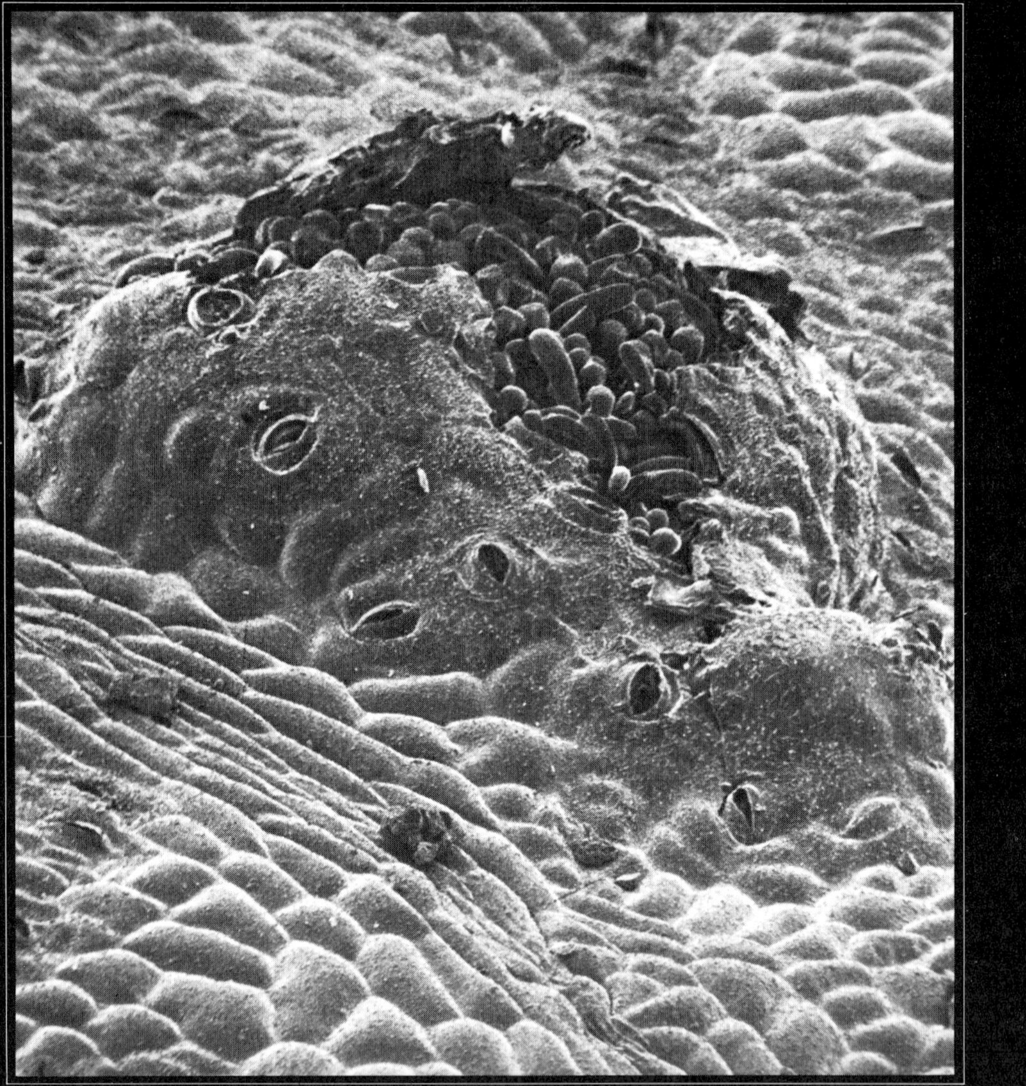

PLANTS AND THE ECOSYSTEM

5

PHOTO OPPOSITE: RUST INFECTION, *PHRAGMIDIUM MUCRONATUM,* ON ROSE LEAF. (MAGNIFICATION 415 ×)

16 THE PHYSICAL ENVIRONMENT

Most of the Earth's surface is occupied by one or another of a variety of plant communities: forests of various kinds, grasslands, arctic tundras, desert communities, marshes, bogs, or aquatic communities that inhabit streams, lakes, or oceans. It is by no means a matter of chance that certain kinds of plant communities are found in specific locations. Of primary importance in determining where various kinds of plants can thrive is the complex of factors that make up the physical environment of plants: temperature, light, water availability, kinds of soil, topography, and even fires. The factors of the physical environment are sometimes classified as climatic and edaphic, the latter being soil factors, but many plants including epiphytes and some aquatics do not have roots in the soil. The physical environment, in addition to determining what kinds of plants can live in a certain region, has in the course of plant evolution played an important role in natural selection. The plants with hereditary potentialities that enabled them to live and thrive in the particular complex of physical environmental factors in a certain region survived and now populate the area, while those that could not thrive in such an environment were eliminated. Thus the plant communities in a region are well adapted to its physical environment.

In the course of geological history, there have been many changes in the physical environment in particular regions of the Earth, including the submersion or emergence of land, the building or erosion of mountains, and marked changes in climate. Such changes have often made the environment unsuitable for the plant communities that occupied the area so these plants were unable to survive and were replaced by plant communities that were adapted to the new environment and invaded it from other areas. For example, the region now occupied by the southwestern desert was once much more humid than it is now and was occupied by forests (Fig. 16.1). The arctic regions were once much warmer than at present and were occupied by palms and other plants characteristic of tropical or temperate plant communities (Fig. 16.2).

A region with a complex of physical envi-

Fig. 16.1 Petrified tree trunks in the Petrified Forest National Park, a desert region in Arizona. There has evidently been a marked change in the physical environment since the time when forests flourished in this region.

Fig. 16.2 A fossil leaf of Glossopteris *found in central Antarctica. This genus thrives in temperate climates in Peru. Such fossils indicate that the regions around the poles were once much warmer than they are now.*

ronmental factors suitable for a particular type of plant community such as a grassland does not contain all the species of plants that could thrive in the region. There are grasslands in various parts of the world occupied by different species or genera than those in the grasslands of the United States, but they have been unable to become established here because of barriers such as oceans and mountains. Man has sometimes deliberately or accidentally transported plants from one such community to another and these generally thrive, because the physical environment of any type of community such as a grassland is similar wherever it is found. Indeed, the introduced plants may thrive to the extent that they become weeds.

The physical environment can be subdivided into the **macroenvironment** and the **microenvironment.** The macroenvironment is similar throughout a large area occupied by a specific type of plant community such as the eastern deciduous forest as regards amount of rainfall and evaporation, temperature ranges, and other general climatic factors. These are influenced by latitude and altitude. The soils of the region may also be of generally similar types. The microenvironment is the immediate environment of a plant and may be quite different from the prevailing macroclimate. Herbaceous plants growing in the shade of the trees of a forest have a microenvironment that is more humid and has lower temperatures and light intensity than that of the trees. The microenvironment of a rock outcrop is quite different from the surrounding environment, not only in regard to the scanty soil but also in such factors as temperature and water availability. Even minor differences in topography may have marked influences on the microenvironment. Plants in even shallow depressions or valleys are subject to more frosts than those in the higher areas (because of the downward drainage of cold air) and there is also likely to be a higher soil water content. Plants growing on a south-facing hill have a different microenvironment than those on a nearby north-facing hill, primarily because of the degree of exposure to the sun.

The plant communities we have been talking about consist of all the populations of plants that occupy a particular habitat. Actually, we should probably have been talking about communities in terms of both plants and animals because each plant community is closely associated with a specific and characteristic type of animal community. If we consider both a community of living organisms and its physical environment we are dealing with an **ecosystem.** The term ecosystem can be applied at various levels: to the ecosystem of the Earth as a whole (**ecosphere** or **biosphere**), to a biome such as an evergreen forest, or to a community in a particular microenvironment. It can even be applied to artificial environments created by man such as cultivated fields, aquaria, terraria, and space capsules. In any event, the physical environment of an ecosystem must be suitable for the organisms and the communities in it if it is to be viable.

TEMPERATURE

In previous chapters we have considered the influence of temperature on the rates of plant processes such as respiration, photosynthesis, and transpiration, on the rate of plant growth, and on the pattern of plant development. Here we are concerned primarily with the influences of temperature on plant distribution, although these involve genetic differences between species as regards the effects of temperature on their processes, growth, and development.

An important influence of temperature on plant distribution is the length of the frostfree growing season (Fig. 16.3). Most tropical plants are injured if not killed by even brief periods of freezing weather, even though the temperature may be only a few degrees below 0°C, so their distribution is restricted to the tropical zones. The plants of temperate zones, if biennials or perennials, are able to survive the lowest winter temperatures in the region. However, sudden temperature drops, especially in the fall before the plants are fully dormant or in the late spring after their dormancy is broken, may kill plants that could ordinarily survive much lower temperatures. Even if low temperatures that kill plants of a certain species occur only once every five or ten years the plants cannot remain a viable part of the natural ecosystem. The temperate zone annuals must have a long enough frostfree growing season so they can complete their life cycle by producing seeds. The annuals of the frigid zones have a very short growing season and must be able to complete their life cycles quickly. The perennials must be able to survive the extremely low winter temperatures of the region.

Even if plants of a certain species are not killed by freezing in a particular region, they may not be able to thrive and compete with other species because the prevailing temperature ranges are not optimal for their growth. The native plants of a region have minimal, optimal, and maximal temperature requirements for growth similar to the prevailing temperatures of the region. For tropical plants, these temperatures are generally on the order

Fig. 16.3 Map showing the last month of killing frosts in the various regions of the United States. The first killing frosts in the autumn come in the reverse order, so the map also represents the length of the frost-free growing seasons.

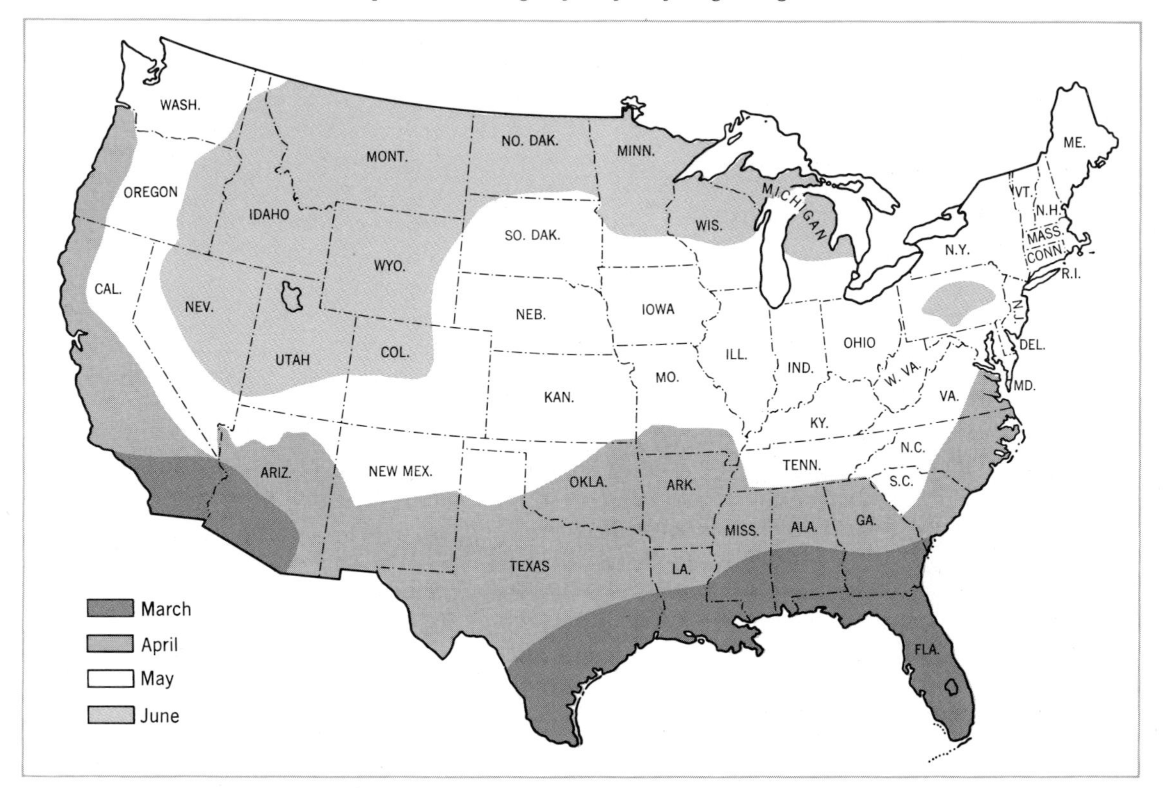

of 15, 30, and 45°C. For temperate zone plants, they are around 5, 20, and 35°C. For arctic plants, they may be as low as 0, 5, and 15°C. The prevailing temperatures in a region are determined by altitude as well as by latitude. As is well known, there is a temperature decrease from the base to the top of a mountain. Because of the thermoperiodic requirement of plants, the night temperatures as well as the day temperatures must be near optimal if the plant is to thrive. Plants with seasonal thermoperiodicity must have a suitable seasonal change in temperature.

Temperate zone plants may not be able to survive in tropical climates because the temperatures that prevail are too much above their optimum for growth. Also, plants that require low temperature preconditioning for breaking bud or seed dormancy or inducing bolting and blooming (if they are biennials) cannot survive in regions where the winter temperatures are too high.

The temperatures in microenvironments may be quite different from those prevailing in the region generally. We have mentioned the temperature differences on north- and south-facing slopes. Also, cold air drains down slopes into valleys or even slight depressions so that plants in these habitats may have a shorter frost-free growing season than those at somewhat higher elevations. Rock outcrops and areas of bare soil absorb more heat than areas covered with plants and so have higher temperatures.

Among the reasons why plant growth decreases at temperatures above the optimal is that the photosynthetic/respiratory ratio decreases with a rise in temperature and that a rise in temperature brings about a marked increase in the rate of transpiration. The rate of water evaporation from the soil also increases. Thus, a rise in temperature can influence plant growth by reducing available water. In addition, dry soil heats more rapidly than moist soil. The influence of temperature on the rate of transpiration can influence plants in another way. If there is a warm period during the winter, the rate of transpiration of evergreens increases, while the rate of water absorption does not because the soil is still cold if not frozen. Thus the plant becomes desiccated and may die. This is called **winter killing.**

LIGHT

In previous chapters, we have considered various influences of light on plant processes and plant growth and development. Light is essential for photosynthesis and photorespiration (Chapter 14) and is responsible for the phototropic responses of plants. It is involved in the opening and closing of stomata and in anthocyanin synthesis. Light absorbed by phytochrome has a variety of photomorphogenetic influences including promotion of germination of some kinds of seeds and the photoperiodic responses of plants. Here we are concerned with the ecological influences of light including plant distribution and the composition of plant communities. At least three aspects of light are of importance: its **intensity,** its **quality** (wavelength composition), and its **daily duration.**

The intensity of light reaching plants at various places on the Earth's surface is influenced by several factors. The intensity of full sunlight increases with altitude because the gases of the atmosphere absorb and scatter light. Light intensity at sea level is about 83% of that at the top of an 11,000-foot mountain. The water content of the atmosphere is even a more important factor. A dense cloud cover can reduce light intensity to about 4% of full sunlight. Also, smog or dust in the air can reduce light intensity by as much as 90%. Latitude influences light intensity because of the angle of incidence of the sunlight on the Earth's surface. If the angle is small, the light must pass through a thicker layer of the atmosphere. Thus, light intensity is highest at the equator and decreases with increasing latitude. The daily change in light intensity and the seasonal changes in light intensity are also the result of changes in the angle of light incidence and so the thickness of the atmosphere through which the sunlight passes. In the temperate zones the winter light intensity may be only 10% of that in summer. Such differences in light intensity from time to time and place to place have definite influences on the rates of plant processes and plant growth, but are less important in determining the distribution of the major plant communities of the Earth than other factors of the physical environment such as temperature and water availability. For one thing, photosynthesis and photorespiration are the principal processes influenced by high light intensities and, except on very cloudy days, carbon dioxide rather than light is likely to be the limiting factor in photosynthesis. A low rate of photorespiration is beneficial to the plant.

However, at the microclimate level, light intensity is likely to be a more important factor in determining the composition of the plant communities. In a dense forest, the light intensity may be as little as 1% of full sunlight, barely enough for photosynthesis of even the shade plants that live on the forest floor. Some herbaceous forest plants can carry on adequate rates of photosynthesis at somewhat higher light intensities while others can do so only when the leaves of the trees have abscissed. The low light intensity in forests influences tree species principally in regard to their seedlings and young plants. Whether or not a particular species of forest tree can reproduce itself depends to a considerable degree on whether their offspring can thrive and grow under low light intensities. For example, hemlock seedlings can thrive, but aspen seedlings cannot. The light in a dense forest is not only of low intensity but also of low quality, since the leaves of the trees have almost completely

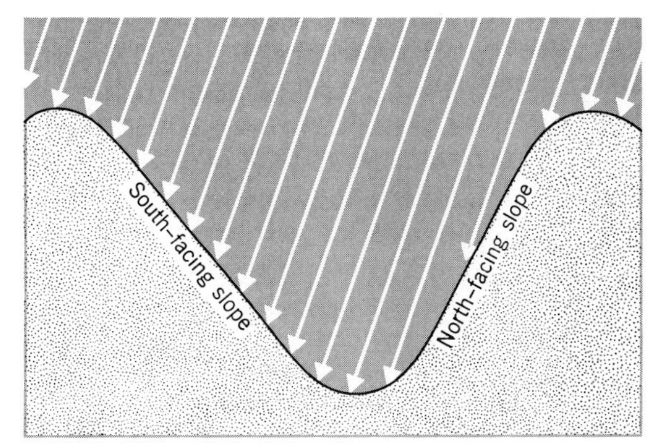

Fig. 16.4 Diagram illustrating the fact that (except in midsummer) the south-facing slopes of hills receive more radiation from the sun per unit area than the north-facing slopes. The parallel lines represent the direction of the sun's rays, and the closer together they are on a hillside the greater the intensity of light and heat radiation per unit area.

filtered out the wavelengths (red and blue) most effective in photosynthesis. Topography also influences light intensity. The light intensity on a north-facing hill, as well as the temperature, is considerably lower than on a nearby south-facing hill (Fig. 16.4). (Of course, the situation is reversed in the southern hemisphere.)

Light intensity is greatly reduced in bodies of water of any depth, including deep lakes as well as the oceans. Water surfaces reflect considerable light, particularly in the blue and green regions of the spectrum. Water also absorbs considerable light of all wavelengths, but predominantly in the blue and red portions of the spectrum. These are the wavelengths best absorbed by chlorophyll. At a depth of 18 meters, about 50% of the light not reflected by the water has been absorbed, and at a depth of 50 meters, practically all the light has been absorbed. These figures are for clear water. Of course, much less light penetrates muddy water or water that has abundant algae in its upper layers. Note that with increasing depth from the surface, there is a decrease in light quality as well as light intensity, further reducing the light available for photosynthesis. Green plants cannot survive at a level lower than that where low light intensity has reduced the rate of photosynthesis to a point where it is no faster than the rate of respiration (a P/R ratio of 1). Red algae can grow deeper in the oceans than other photosynthetic plants because of their accessory red photosynthetic pigment phycoerythrin, which absorbs light primarily in the green and yellow regions of the spectrum. These wavelengths (especially the green) penetrate deeper into the water than the red and blue wavelengths absorbed most completely by chlorophyll.

The daily duration of light plays an important role in the latitudinal distribution of plants, since day length varies with latitude as well as seasonally. At the equator the day length is about 12 hours throughout the year, but going from the equator to either pole the day length is progressively longer in the summer and shorter in the winter (Fig. 16.5). The photoperiodically effective day length is greater than the almanac time from sunrise to sunset because the light intensity during most of the twilight periods is great enough to affect photoperiodism. The length of the twilight periods also varies with latitude, being very short at the equator and becoming progressively longer toward the poles. Short-day plants could not reproduce in arctic regions because when the days are short the temperature is too low for active growth. At or near the equator the day length is never long enough for the reproduction of long-day plants. Of course, as far as the photoperiod is concerned, day-neutral plants can reproduce at any latitude. Some photoperiodic species of plants do have a wide latitudinal distribution, but these have a number of genetic strains that have different critical photoperiods. Going from the equator toward the poles the strains have progressively longer critical photoperiods.

OTHER RADIATION

Light (radiation visible to man) constitutes only a minor portion of the spectrum of elec-

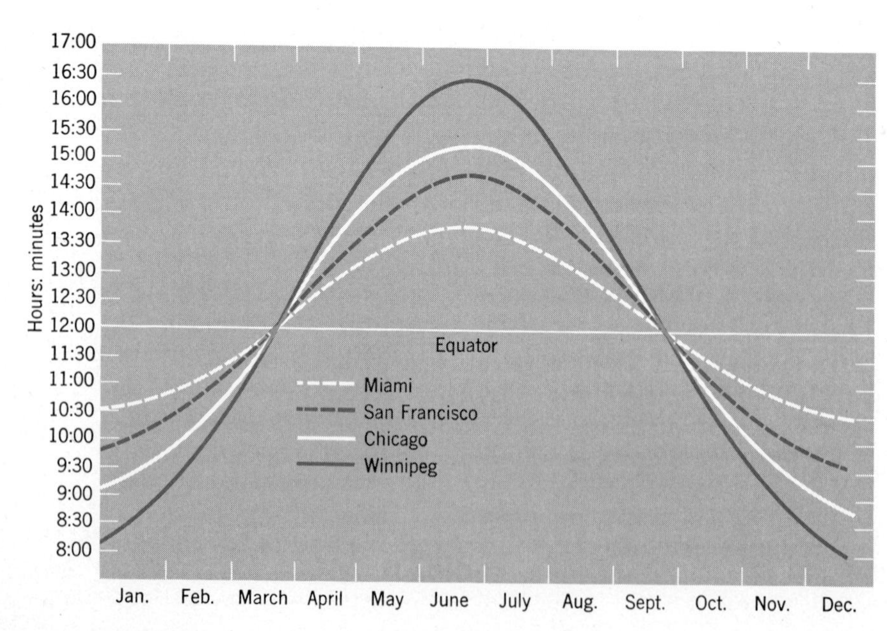

Fig. 16.5 The seasonal variation in day length (sunrise to sunset) at five different latitudes. Note that the seasonal differences increase with latitude. Because of the daily twilight periods the photoperiodically effective day lengths are somewhat longer than those shown. The length of the twilight periods also increases with increased latitude.

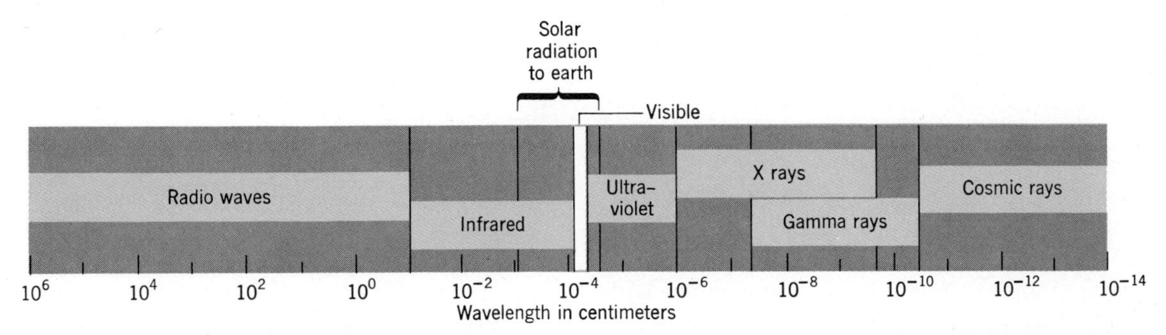

Fig. 16.6 The spectrum of electromagnetic radiation. For an expanded spectrum of light (visible radiation), see Fig. 14.11. The wavelength scale here is in centimeters, but for light the unit of length usually used is the nanometer (nm). One nanometer = 10^{-7} cm.

tromagnetic radiation (Fig. 16.6). Radio waves and infrared or heat radiation have longer wavelengths than light. The wavelengths shorter than those of light are, in order, ultraviolet, X-rays, gamma rays, and cosmic rays. The radiation from the sun that reaches the Earth's surface consists of only the longer ultraviolet wavelengths, light, and about the shorter third of the infrared wavelengths. Radiation from the sun is broader than this, but the atmosphere filters out the other wavelengths. Other types of radiation coming from space include the very short cosmic rays and some radio waves, although most radio waves are generated by man.

Fortunately, the widely used radio waves apparently do not influence organisms. Infrared radiation influences organisms largely by heating the Earth and so increasing its temperature. It is the principal natural source of heat. Ultraviolet radiation can be lethal, but the shorter wavelengths that are most lethal are filtered out by the atmosphere. However, even the narrow band of ultraviolet that reaches the Earth's surface can be lethal to microorganisms. The wavelengths are apparently not injurious to higher plants because they do not penetrate deeply and are mostly reflected or absorbed by the cuticle of the epidermis. Ultraviolet radiation is not essential for plants. Ordinary glass filters out almost all ultraviolet, but plants still grow well in greenhouses. X-rays, gamma rays, and cosmic rays are ionizing radiation. They promote the ionization of molecules, including many that do not ordinarily ionize, and so disrupt metabolic processes and induce abnormal growth and development (Fig. 16.7). Another way in which they may be injurious is by promoting mutations and chromosome breakage, which may affect future generations. Cosmic rays are present throughout the Earth, although their intensity varies somewhat from place to place. Thus, they are part of the natural environment or organisms and apparently are not particularly injurious. X-rays are mostly generated by man and gamma rays come principally from some radioactive isotopes.

Ionizing radiation includes particles emitted by radioactive isotopes as well as electromagnetic radiation. The principal particles emitted by radioactive atoms are alpha particles (helium nuclei), beta particles (high speed electrons), and neutrons. Although neutrons are not charged, they can cause ionization by hitting molecules.

Radiation other than light has little influence on the distribution of plant communities at the macroenvironment level, except for the effect of infrared radiation on temperature, but it may have definite effects at the microenvironment level. Plants growing near rock outcrops containing naturally occurring radioisotopes such as uranium, radium, or thorium may be injured or have high rates of mutation. But the greatest interest to ecologists and geneticists is the contamination of the environment by man-made radioisotopes. The fallout from atomic bomb explosions may cover wide areas. Although the radioactivity may not be great enough, except near the point of explosion, to kill or injure plants, the plants generally accumulate the radioactive salts just as they do salts in general. Further concentration of the radioactive atoms may occur at subsequent steps in the food chain and so the radiation level ultimately may be much higher than it was in the soil. Particular concern has been expressed by ecologists and geneticists in regard to the concentration of radioisotopes such as strontium (^{90}Sr), especially from the standpoint of people who consume milk or meat from animals that have eaten strontium-containing plants. The disposal of wastes from nuclear reactors and nuclear generating plants for electricity may contaminate both streams and soils with radioisotopes, even when the best precautions are taken. Radiation ecolo-

Fig. 16.7 *A tobacco plant exposed to 300 roentgens of gamma radiation from cobalt-60 for six weeks. The photograph was taken 50 days after treatment was terminated. Plants in nature are not exposed to ionizing radiation as intense as this, so they rarely have tumors, other malformations, or growth inhibitions as severe as this. However, man is responsible for radioactive contamination that does inhibit growth, produce less severe morphogenetic changes, and induces mutations.*

gists have conducted numerous experiments to determine the effects of various levels of radiations on plant and animal communities as well as on individual plants and animals. Since various species differ in their sensitivity to radiation, it can have an influence on the composition of a community.

WATER

We have already considered (Chapter 12) the water relations of plants, including the absorption, translocation, and transpiration of water, and the ways in which water is held in the soil. Here we are concerned with the influences of water availability on the composition and distribution of plant communities. No factor of the physical environment, except possibly temperature, has a greater influence than water in determining the distribution of the major plant communities (biomes) such as forests, grasslands, and deserts over the face of the Earth (Fig. 16.8). Of course, bodies of water such as ponds, lakes, streams, marshes, and oceans have entirely different plant communities than those in nearby land areas. Water is also an important factor at the microenvironment level. Water availability in nearby habitats may vary considerably as a result of such things as rock outcrops, shallow soil, topography, and shading by larger plants.

There is a continual and rather rapid circulation of water from the atmosphere to the land or sea and back into the atmosphere again. Most of the water vapor in the atmosphere evaporates from the oceans, but some also evaporates from lakes, streams, marshes, or moist soils, or is lost by transpiration. The water molecules may aggregate into droplets that form clouds, and from these, precipitation occurs in the form of rain, snow, sleet, or hail. Some of the precipitation returns water directly to lakes or oceans. Some of the rain that

Fig. 16.8 *Plant communities as diverse as these result because of differences in the physical environment. (Left) A salt marsh. (Right) An Arizona desert.*

falls on land runs off the surface of the soil into streams and then into lakes and oceans. The amount of surface runoff depends on various factors such as topography, type of soil, amount of plant cover, and intensity of the rain. The water that infiltrates the soil provides the water absorbed by the roots of land plants, or if there is enough it may percolate through the soil to the water table. If the water table extends to the lower layer of soils, this ground water may be absorbed by the roots of plants, but often the water table is very deep and may drain into cavities in the underlying rocks. Some of the ground water emerges at lower elevations as springs and some is pumped up by man, but much of the deeper ground water may remain out of circulation for long periods.

It is well known that the amount of rainfall varies greatly from one part of the Earth to another. A few regions receive several hundred inches of precipitation a year, forested regions generally receive about 40 inches, while deserts receive little or almost none. There are various causes for this unequal precipitation. Regions distant from any ocean are generally quite dry. However, one of the more important causes is that prevailing winds blowing from oceans rise, expand, and cool when they reach a mountain range. Thus the moisture-laden air is no longer capable of holding much of its water vapor, which condenses into the water droplets of clouds and is then precipitated as rain or snow. The region between the coast and mountains thus receives abundant precipitation. Once the moving air has passed over the mountain range it has lost most of its water vapor and the region beyond the mountains receives little precipitation. There may also be a distinct seasonal difference in the amount of rainfall in various places. In northern India the heavy monsoon rains fall in the summer, the rest of the year being quite dry, but in California most of the precipitation is during the cool seasons, with very little during the summer. In other regions, such as the eastern states, the rainfall is more evenly distributed throughout the year. The plant communities of a region are well adapted to the prevailing amount of precipitation and its seasonal distribution, but other species of pants not subjected to this kind of natural selection are not likely to thrive or even survive in the region.

The water available to plants in a particular region depends, not only on the amount and seasonal distribution of precipitation but also on the amount of water lost from the soil during a year by evaporation and transpiration. The rates of evaporation and transpiration are influenced by various factors including the humidity of the atmosphere, but the most important factor is temperature. The rate of evaporation increases greatly as temperature increases, and the rate of transpiration increases even more with a rise in temperature. The amount of annual rainfall may be about the same in a grassland as in a desert adjacent to it but nearer the equator; however, the higher temperatures and consequently the higher rates of evaporation and transpiration in the desert reduce the available soil water below that in the grasslands. This is perhaps the most important factor in determining which kinds of plant communities are present.

Plants have been classified ecologically in regard to their water relations as hydrophytes, mesophytes, and xerophytes. Each group contains plants belonging to a considerable number of different families and genera. **Hydrophytes** are plants that grow in wet locations such swamps, bogs, and marshes as well as plants that grow in bodies of water such as streams, ponds, lakes, and oceans. Most of the vascular hydrophytes are anchored in the soil by their roots, but root hairs are often absent. Some, such as *Elodea,* tape grass (*Vallisneria*), eelgrass, and various pondweeds have entirely submerged shoots. Some, such as water lilies, have floating leaves with all the stomata on the upper surface. Others have shoots that extend well above the water surface and are only partially submerged if at all. These are often called **amphibious plants** and include rice, bullrush, cattail, water willow, some sedges, bald cypress (*Taxodium*), and mangroves. Several genera of amphibious plants such as *Proserpinaca* have leaves both below and above the water surface, the submerged leaves being quite different in shape and other characteristics from the aerial ones (Fig. 16.9). The submerged leaves are usually much more highly divided. Multicellular algae are often anchored to rocks or other substrates by their holdfasts and are submerged all or part of the time. These include many species of green algae that grow in streams or lakes and various species of brown and red algae that grow on rocky coastlines. Finally, many hydrophytes have no connection with the soil or rocks and float in the water. Some, like duckweeds, water hyacinths, and *Salvinia* (all vascular plants) float on the surface and so are exposed to the air. Others, such as many algae in both lakes and oceans, float under the surface of the water and are exposed to neither soil nor air.

The principal problem facing hydrophytes is a supply of oxygen adequate for a suitable rate of respiration, in view of the low solubility of oxygen in water. Most vascular hydrophytes have unusually large intercellular spaces in their stems and roots and often in their leaves. Indeed, there may be large air chambers (Fig. 12.13). These large air spaces facilitate the diffusion of gases from the aerial parts of the plants and also serve as reservoirs where oxygen produced by photosynthesis can accumulate. Spongy tissues also provide the bouyancy that enables leaves (as of water lilies) or entire plants (as water hyacinths) to float. Many brown algae have air bladders that help make them bouyant.

Mesophytes are plants that grow in regions where the capillary water of the soil is usually adequate but where the soil is not waterlogged. Practically all of the trees, shrubs, and herbs of forested regions are mesophytes, and the tall grasses of the more humid grasslands are also generally mesophytic. Even in dry regions the soil near streams and lakes may contain enough water to support mesophytes.

Xerophytes are plants that can survive and thrive in dry regions and constitute the plant communities of deserts. In deserts the soils are generally depleted of capillary water to a depth of at least 2 decimeters over considerable periods of time. Some xerophytes are also found in more humid regions occupied by grasslands or forests where there are dry microenvironments such as rock outcrops, shallow sandy soils, and dry ridge tops. Algae, mosses, and lichens that grow on tree trunks or rocks are often subjected to long dry periods and can survive despite considerable desiccation.

The xerophytes of deserts are of four different types. The desert annuals should perhaps be considered as **drought evaders** rather than true xerophytes, since they complete their life cycles during the rainy seasons, which may be as brief as a month. During this time there is adequate soil water. Their seeds remain dormant and viable for even several years where the periods between substantial rains are that long. The seeds of many desert annuals contain water-soluble germination inhibitors, but light rains are not sufficient for leaching out these inhibitors, so the seeds remain dormant. A second type of desert plant (generally perennials) has deep roots which absorb underground water that flows from surrounding mountains far below the desert surface and so secure an adequate supply of water during the long dry periods. These are also drought evaders.

Succulents such as cacti are able to survive in the dry environment because much water accumulates in their tissues during the rainy periods and is available during the dry periods. Succulents also have low rates of transpiration, partly because of their thick and compact cu-

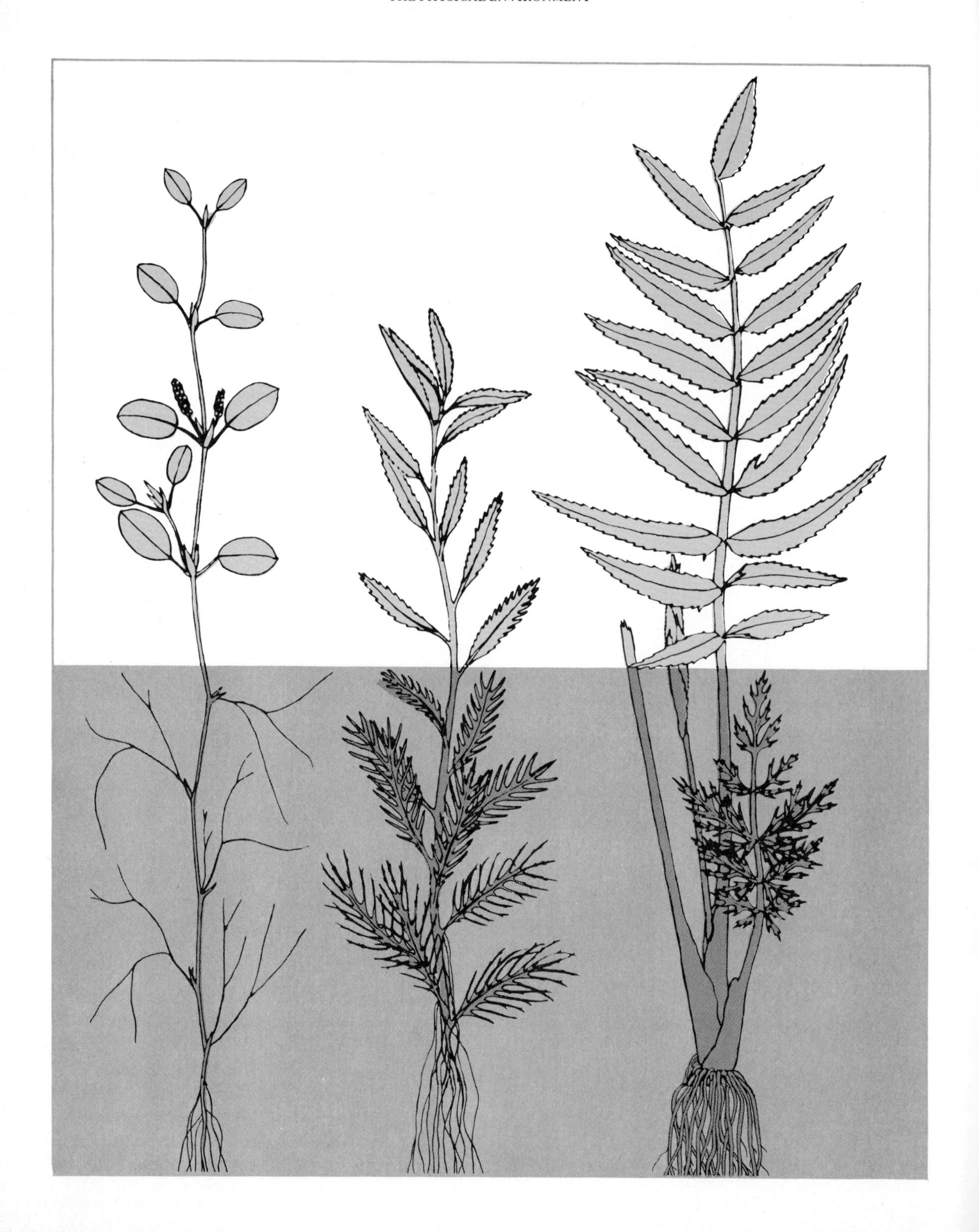

Fig. 16.9 Three species of amphibious plants, showing the influence of submersion on leaf development.

ticles and partly because their stomata are closed during the day and open at night. During the night they use carbon dioxide in making organic acids, and the following day these are decarboxylated, providing CO_2 that enables photosynthesis to continue. Also, succulents generally have shallow root systems that enable them to absorb water after even moderate rains that wet the soil to field capacity to a depth of only a few centimeters.

The fourth type of xerophytes is the perennial which can survive prolonged periods of desiccation and permanent wilting. These are **drought resistant** plants, and are considered by some ecologists to be the only true xerophytes. They have a variety of morphological and physiological features that make them drought resistant. They generally have very low osmotic potentials that enable them to absorb water from relatively dry soil. Their root systems are extensive and their shoots are generally limited in size (a low shoot/root ratio). Although they usually have high rates of transpiration during the wet seasons, the water loss by transpiration is very low during the long dry seasons. The leaves are generally small, and leaf surface may be reduced still further during dry periods by abscission or death of the leaves. Or, the cuticle may be thick and compact and the stomata close when the plants are wilted. Sunken stomata are common. In some species the leaves become more vertical (by elevation or drooping) when wilting begins, thus reducing the amount of solar radiation they receive. The leaves of xerophytic grasses commonly roll into a tube when desiccated.

Conifers and other evergreens that live in subalpine regions are essentially xerophytes, not because they live where the annual precipitation is low, but rather because the soil water is frozen and thus unavailable during a substantial part of the year while atomospheric conditions may be such as to permit continuing transpiration from the persistent leaves. Among the xerophytic features of such plants are reduced needlelike leaves, leaves with thick and compact cuticles, and sunken stomata.

THE ATMOSPHERE

The atmosphere is an essential and important factor in the physical environment of plants, but many of its influences are by way of its effects on other factors such as temperature, light, water, soil, or fire. The day-to-day temperature fluctuations result to a great degree from winds that bring warmer or cooler air into a region. Light intensity is influenced greatly by the presence of clouds, fogs, or smog in the atmosphere, and much of the sun's radiation is filtered out by the atmosphere. Water availability to plants is influenced by the flow of humid air from the oceans over the land, the humidity of the atmosphere, and the effects of breezes and winds on the rates of evaporation and transpiration. The atmosphere extends into soils, unless they are waterlogged, and winds may erode dry soil and cause dust storms. The spread of fires is greatly influenced by the wind velocity.

However, there are other more direct influences of the atmosphere on plants. Some relate to the gaseous composition of the atmosphere. Because of the almost continuous churning of the atmosphere by breezes and winds the percentage composition of the atmosphere does not vary greatly from one place to another. However, the concentration of each of the gases decreases with increased elevation, and so on high mountains the low concentrations of carbon dioxide and oxygen can become limiting factors in photosynthesis and respiration and so influence the kinds of plants making up the communities there. In places with dense vegetation, such as corn fields and forests, photosynthesis may reduce the carbon dioxide much below the usual 0.03%, thus limiting further photosynthesis. Because of the respiration of roots and soil organisms the soil atmosphere may have a considerably reduced concentration of oxygen. The reduced oxygen can limit the rate of respiration and also promote the growth of anaerobic bacteria which may produce toxic substances or carry on such processes as dentrification. However, the accumulation of carbon dioxide may be more serious, since 10% is toxic and 30% to 50% is fatal. Man has been responsible for polluting the atmosphere with gases not naturally present (except in some cases from volcanic eruptions), and such gases are generally toxic to plants (Fig. 16.10). For example, smelters emit sulfur dioxide and hydrogen fluoride, coal smoke contains considerable sulfur dioxide, and automobile exhausts pour out carbon monoxide and toxic hydrocarbons.

Winds also have effects on plants other than those mentioned above. The strong prevailing winds on high mountains play an important role in determining the location of the timberline (above which there are no trees). and in causing the stunted and lopsided growth of trees near the timberline (Fig. 16.11). Winds blowing landward from the ocean have similar influences, but primarily because of the salt spray they carry. Many species of plants that could otherwise thrive along the coasts are killed and eliminated by the salt spray. Hurricanes and tornadoes uproot, break, or otherwise destroy trees and even some smaller plants. On the other hand, wind plays a useful and important role in plant reproduction and dispersal. It is essential for the reproduction of wind-pollinated species, and it disperses plants by carrying spores, winged or tufted seeds and fruits, entire plants such as bacteria and algae, and the detached shoots of tumbleweeds with their seeds.

Fig. 16.10 Damage of plants by air pollutants. The tobacco plants at center and right grew in Yonkers, New York air. The plant at the left grew in a chamber supplied with air filtered through charcoal to remove the pollutants.

TOPOGRAPHY

Like the atmosphere, topography (the configuration of the Earth's surface) is an important aspect of the physical environment, but it has no direct influences on plants and affects them through its influences on the other factors of the physical environment. Topography determines where streams, ponds, marshes, lakes, and oceans are located. It is well known that temperature range decreases as elevation increases, and we have already pointed out that with an increase in elevation there is increased light intensity, decreased concentrations of the atmospheric gases, and often increased wind velocity. The important role of mountains in determining the amount of precipitation has also been mentioned. The steepness of a hillside influences the amount of water that will be absorbed by the soil in comparison with what will flow into a stream by surface runoff. The

Fig. 16.11 (Top) The timberline on a high mountain in the Rocky Mountain National Park. (Bottom) Dwarfed and distorted trees of Pinus aristata *just below the timberline of a mountain in Arapaho National Forest, Colorado.*

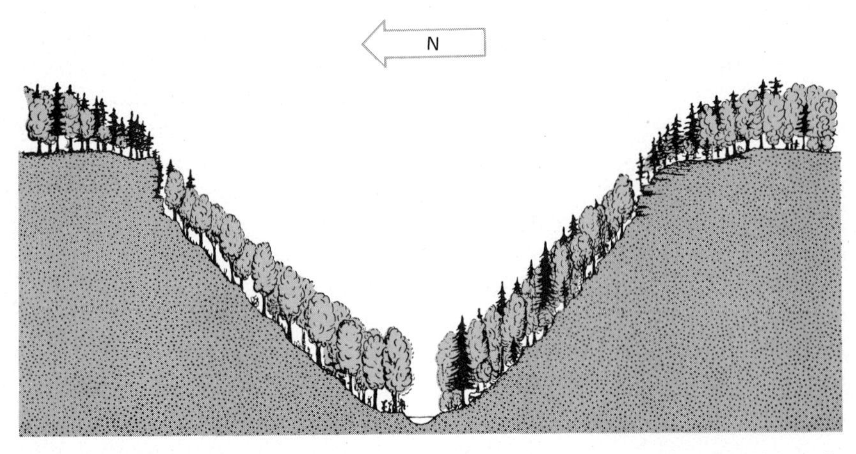

Fig. 16.12 Influence of topography on microclimates that determine the composition of plant communities, as shown by a section through a gorge in Pennsylvania. The upland trees are pines, oaks, and hickories. On the north-facing slope are hemlocks, beeches, birches, maples, and tulip poplars. On the south-facing slope are chestnuts, oaks, and hickories. (Data for the sketch were secured before most of the chestnut trees were killed by the blight.) (Redrawn from Transeau, Sampson, and Tiffany, Textbook of Botany, *with the permission of the publishers, Harper & Row.)*

differences in temperature, light intensity, and water availability on the north- and south-facing slopes of hills has been mentioned (Fig. 16.12), as has the fact that cold air drains into even rather shallow depressions. Thus, topography influences the other factors of the environment in many ways, including ones we have not mentioned.

FIRE

Every year we read in the newspapers about disastrous forest, chaparral, and grassland fires that were started by people careless with cigarettes, campfires, or matches. Yet man is by no means the cause of all fires. Lightning causes many fires, and was responsible for fires even before man appeared on earth. Also, man sometimes starts fires in forests and grasslands deliberately, instead of accidentally, for practical reasons that we will mention later. Although fires in any particular region rarely occur annually and often not for decades or centuries they constitute a factor of the physical environment that is of great ecological importance, and not all of the results of fires are adverse.

Fires, in particular forest fires, are of three types. **Ground fires** occur within the soil itself, burning the humus and other organic matter in the soil, and are generally flameless. They may occur even in moist soil since they burn slowly and dry out adjacent moist soil and so the fire spreads. Ground fires kill the smaller plants with roots in the upper layers of soil, but may not injure trees and other plants with deeper roots. **Surface fires** burn the herbaceous plants, shrubs, and seedling trees, but the large trees may escape injury. **Crown fires** burn through the treetops as well as on the ground and are the most serious type, often killing most of the plants and animals of the forest (Fig. 16.13).

The effects of fires are both direct and indirect. The direct effects are the killing and burning of plants because of lethal high temperatures and the burning of the organic matter of the soil and on the surface of the soil to ash. The indirect effects are more numerous and varied, but these also have important effects on the nature and composition of plant communities. They may favor the growth of surviving species by killing plants that were competing with them. For example, larch trees (*Larix occidentalis*) are not a major component of the coniferous forests of the northern Rocky Mountains, but after one or several fires they become the principal trees of the community because they are more fire-resistant than the other species. Fireweed (*Epilobium angustifolium*) is rare and stunted in forests where it occurs, but after a serious fire it may take over the entire area and grow much taller. Quaking aspen (*Populus tremuloides*) is a minor component of northern forests but may become abundant after a fire by sending up sprouts from its wide, spreading roots. Animals as well as plants are killed by forest fires and this may influence the plants that repopulate the burned-over area. Squirrels and other rodents that eat nuts may be killed, thus leaving a larger supply of seeds that can germinate and grow into trees. Birds that ordinarily live in open areas rather than forests may invade them after a fire and scatter seeds of blackberries, raspberries, currants, and similar fruits. Such plants may then become abundant in the burned-over area.

Fires (particularly crown fires) also result in marked changes in the physical environment that in turn influence the nature of the plant communities that reoccupy the area. There follows a marked increase in light intensity. The soil temperature increases, both because of the increased radiation and the presence of charred black organic matter. The soil moisture may increase initially because of greater exposure to rain, but the loss of humus from the soil can reduce the soil's water-holding capacity resulting in increased runoff of water. This and the lack of the plant cover can result in substantial soil erosion. The oxides of elements such as calcium, phosphorus, and potassium may be more soluble and thus more available to plants than their previous com-

Fig. 16.13 A severe crown forest fire in the Boise National Forest, Idaho.

pounds, but they also leach out of the soil more rapidly. The oxidized elements also bring about a higher soil pH (increased alkalinity). Nitrogen is lost to the atmosphere.

Some species of plants are much more fire resistant than others, and so are likely to repopulate the burned-over area. Certain species of plants, including several kinds of sumac, have seeds with hard seed coats that remain dormant for years. A moderate fire does not kill them, but does crack the seed coats, thus breaking dormancy and permitting germination. The cones of several species of conifers (including *Pinus contorta, P. banksiana, Cupressus sargenti,* and *Picea mariana*) remain closed for many years and still contain viable seeds. The heat from a surface fire causes them to open and release their seeds, which then germinate. *Pinus contorta* cones have been known to remain closed as long as 75 years and still contain viable seeds. Trees that lack oils and resins in their stems and leaves are more resistant to fire than those that have them. Trees with thick bark, such as longleaf pine, larch, and some oaks, are resistant to surface fires. Some species of trees have dormant lateral buds that may not be injured by fire or are capable of developing adventitious buds after a fire, and so can recover from fire injury.

We have mentioned that man sometimes starts fires deliberately in an effort to improve the composition of the plant communities from a practical standpoint. In semideserts occupied by grasses and sagebrush, continuous grazing reduces the grasses and favors the growth of the sagebrush. Fires that are deliberately started kill the sagebrush but not the grasses, thus permitting the grasses to occupy most of the area and improving it for grazing. In the southeastern states surface fires are sometimes started (especially in longleaf pine forests) to kill the trees and seedlings of oaks and other hardwoods that have little economic value and are beginning to replace the more valuable pine trees. This permits the growth of more pine seedlings and also promotes the growth of grasses that can be used for grazing. Such fires are generally carefully watched and

Fig. 16.14 *Deliberate burning of litter and undergrowth in a Florida pine forest. Such fires are carefully watched to keep them under control, and are repeated at four- to six-year intervals.*

are extinguished if they start getting out of control (Fig. 16.14). In some forests where man has prevented fires for many years, serious damage to trees has been caused by insects and parasites. Burning of the surface litter destroys these insects and also destroys ticks and other arthropods that attack animals. The controlled burning of surface litter may also enable the roots of seedlings to reach the soil easier, and can prevent the accumulation of excessive litter which, if accidentally set on fire, might result in a serious crown fire. However, the economic benefits of deliberately set fires are outweighed by the destruction of valuable forest trees by fires started accidentally by man or by lightning.

SOILS

The soil is one of the most complex factors of the physical environment, and one of the most important environmental factors for land plants. It is the source of practically all their water and essential mineral elements. It provides anchorage for plants. The degree of acidity or alkalinity (pH) of the soil has a marked influence on the solubility and availability of mineral salts, on the growth of plants, and on the kinds of plants that can thrive in it. The soil atmosphere must be extensive enough to provide an adequate supply of oxygen for the respiration of plant roots and soil organisms. The temperature of the soil has important influences on the growth of roots and on the rate of water absorption by roots. The soil is partly a biological factor in the plant's environment because of the numerous organisms that live in it—bacteria, fungi, algae, protozoa, and a variety of animals such as moles, earthworms, insects, and nematodes. These organisms affect plants in many ways. The nitrogen cycle bacteria (Chapter 13) help determine the amount of nitrogen in the soil, the bacteria and fungi that are decomposers determine the amount of humus in the soil, and earthworms loosen the soil and provide better soil aeration. We have already considered a number of these soil factors, particularly the role of soil texture (sand, silt, clay) on the water-holding capacity of soils (Chapter 12). Here we are concerned primarily with soil structure, the different structural types of soils found in various regions, and the factors involved in the formation of these different types.

Soil is composed principally of small particles derived from rocks. The disintegration of rocks into soil is brought about by mechanical factors such as the splitting action of ice in crevices or by growing roots, temperature changes, abrasion by glaciers and sand or other particles carried by wind or flowing water. Rock decomposition is also brought about by chemical processes such as oxidation, carbonation, hydration, and most extensively by the action of acids on the rocks. To a considerable extent the acid is carbonic acid resulting from the respiration of roots, soil organisms, or plants such as lichens, mosses, and algae that grow on rock outcrops, but organic acids produced by plants also play a role. In addition, plants aid in soil accumulation by reducing the rate of soil erosion, which, without a plant cover, would be much more rapid than the slow process of soil formation (Fig. 16.15). The upper layers of soils are also composed of varying amounts of humus, the partially decomposed remains of plants and animals. The most important factors determining the particular structural type of soil formed in a region are temperature, amount of rainfall and evaporation, and type of vegetation. Perhaps strangely, the chemical nature of the underlying rock is less important. Furthermore, soils are not always derived from their underlying rocks but may have been transported by wind, flowing water, gravity, or glaciers.

Examination of the soil at a highway cut or a deep trench makes it evident that soil profile is composed of distinct layers, which are called

Fig. 16.15 Severe erosion on a hillside denuded of vegetation.

Fig. 16.16 (Right) Profile of a chernozem soil in South Dakota. The marker is graduated in feet.

horizons (Fig. 16.16). The uppermost horizon is designated as the **A horizon** (topsoil), and this is composed of several subhorizons, from the litter and leaf mold on the soil surface, to the humus layer of the soil, and then often a lighter colored layer that has been leached of many of its solutes. The **B horizon** is composed primarily of mineral soil. The minerals are partly derived from the parent rock, partly by the complete decomposition of humus in the A horizon, and partly from minerals from the A horizon, the last two having leached down into the B horizon. The **C horizon** consists of essentially unmodified material from the parent rock, and often contains stones and boulders of considerable size. The thickness, composition, and texture of the soil in these three horizons varies considerably from one structural type of soil to another. (Fig. 16.17).

We will describe briefly four major structural types of soils: podzols, laterities, chernozems, and sierozems. The distribution of these types correlates well with the distribution of major types of plant communities such as forests, grasslands, and deserts. This results primarily because temperature and humidity in a region favor the type of soil formed and the plants that occur there independently, and because the plants present influence soil formation, the suitability of the soil type for the plants being a secondary factor.

Podzols develop in cold, wet regions oc-

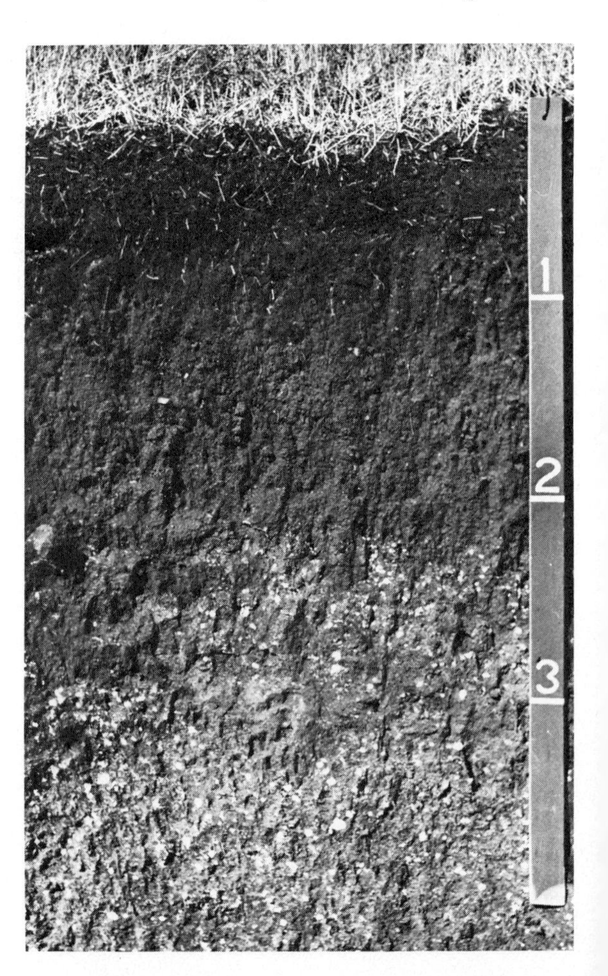

cupied by northern evergreen forests, but modified podzols also occur in relatively cool forested regions. There is extensive leaching of minerals and other solutes from the soil because of the high moisture, so the soils are poor from the standpoint of mineral nutrition. The soils are acid (pH 3 to 5), to a large extent because of acids produced during the decay of fallen pine, spruce, and fir needles. Because of the rather low temperatures, decomposition of the leaf litter into humus by decomposers is rather slow, but complete decomposition of the humus is even slower, so the A horizon has a substantial layer of humus (Fig. 16.17). Below this the A horizon consists of ash-gray sandy soil that is well leached. The fact that at low temperatures silicates break down more slowly than other rock minerals favors sand formation. The upper part of the B horizon is darker because of humus leached from the A horizon, while the lower part of the B horizon consists of sand, gravel, and rocks stained brown by the clay and aluminum and iron compounds leached from the A horizon.

Chernozems develop in grassland regions, which generally have only moderate rainfall, hot summers, and cold winters. There is not enough rainfall to leach the soil severely, so it is quite fertile. The extensive fibrous root systems of the grasses extend throughout the A horizon and when some of the roots die they are converted into humus. Thus the entire A horizon has a high humus content (about 10% by weight). The A horizon is usually from 0.6 to 1.0 meter deep (Fig. 16.16). In the B horizon, there is a high concentration of calcium carbonate that has leached down from above. The soils are neutral or slightly alkaline (pH 7 to 8), have a high field capacity, but are well drained and are often almost black. Because of their fertility and ease of plowing and cultivation, chernozems are the best agricultural soils,

Fig. 16.17 Sketches of the profiles of a podzol and a chernozem soil. Laterite and sierozem soils generally have little profile.

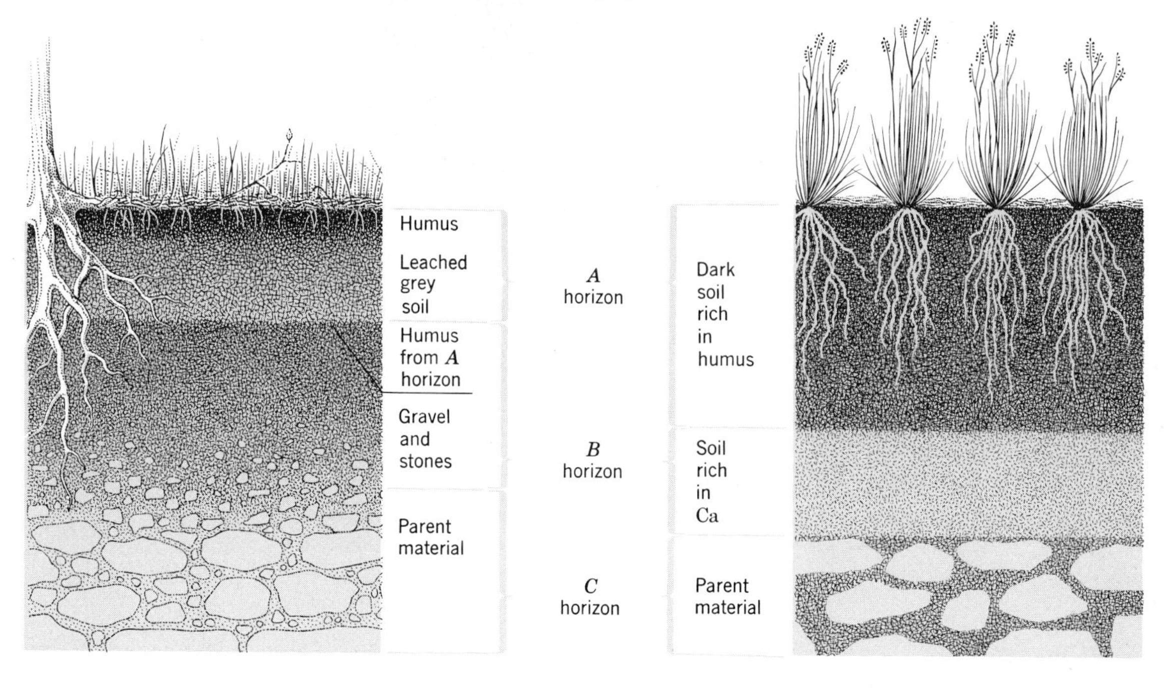

and the bulk of the food crops of the western world grow in them.

Laterite soils develop in regions with high rainfall and warm or hot temperatures through most of the year. They are typical of tropical rain forests but are also found in the more moderate climate of the southeastern states. They are somewhat acid to neutral (pH 5 to 7). Because of the high temperatures, complete decay of organic matter is very rapid, so there is very little humus in the soil. The high temperatures also cause silicates, which have high concentrations of aluminum and iron compounds, to break down more rapidly than other soil minerals. The abundant iron oxides give the soil a bright brownish-red color or even a definitely red color. The soil is mostly clay, with very little sand, and is poorly aerated. Leaching is extensive because of the heavy rainfall, but less so than if the soils were sandy. The A horizon is deep (often as much as 15 meters) except where there is solid rock near the surface, and there is no definite B horizon.

Sierozems are soils that develop in warm, dry regions, and are the soils of deserts and semideserts. They contain little humus both because of the sparse vegetation and because the high temperatures promote rapid and complete decomposition of the organic matter during the rainy seasons. There is very little leaching of the mineral salts from the soil because of the scanty rainfall. As a result the soils are fertile, but they are quite alkaline and the soil water has a low osmotic potential. This means that the plants must have an even lower osmotic potential if they are to absorb the scanty available water. When deserts are irrigated the crop plants generally thrive because of the abundant mineral elements, the high light intensity, and other factors. However, after irrigation has been practiced over several decades the salt content of the soil may become so high that further cultivation of the crop plants is impossible. The salts in the irrigation water are added to those already present in the soil and, since there is little leaching, the osmotic potential of the soil water becomes so low that the plants are no longer able to absorb water. There is no practical way to correct this situation. Sierozem soils have little structure beyond what can be called the A horizon, which is often very shallow and is in direct contact with the parent rock material.

THE COMPLEX OF ENVIRONMENTAL FACTORS

Although we have considered the factors of the physical environment separately, it should be stressed that it is the complex of these factors that is important in influencing plant processes, plant growth and development, and the composition and distribution of plant communities. In the preceding discussion, we have mentioned some of the many ways in which a particular factor is influenced by other factors of the physical environment and also by plants and other organisms. It has perhaps not been made clear that plants are subjected simultaneously to practically all of the factors we have discussed and that it is the entire complex of the physical factors that influences the growth and development of plants and the composition and distribution of plant communities. This is one of the things that makes ecology a complex, difficult, but challenging science. However, the environment of a plant consists of more than the physical factors. It also includes the biological environment, which consists of all the organisms that in any way influence a particular species of plant or animal. We turn our attention to this biological environment in the next chapter.

RELATED READING

Caborn, J. M. "Microclimates," *Endeavour, 32,* 30–33, 1973.

Claiborne, R. "When it is nature's way, forest fires are not bad, but in fact, a benefit," *Smithsonian, 3*(2), 16–23, May 1972.

Clark, J. R. "Thermal pollution of aquatic life," *Scientific American, 220* (3), 18–27, March 1969.

Cooper, C. F. "The ecology of fire," *Scientific American, 204*(4), 150–160, April 1961.

Enos, P. "Photosynthesis and atmospheric oxygen," *Science, 180,* 515–516, 1973.

Udall, S. L. "A message for biologists," *BioScience, 14*(11), 17–18, 1964. (This issue contains five other articles on the pesticide problem.)

Went, F. W. "Air pollution," *Scientific American, 192*(5), 62–73, May 1955.

Went, F. W. "The role of environment in plant growth," *American Scientist, 44,* 378–398, 1956.

Went, F. W. "Climate and agriculture," *Scientific American, 196*(6), 82–94, June 1957.

Woodwell, G. M. "The ecological effects of radiation," *Scientific American, 208*(6), 40–49, 1963.

17
INTERACTIONS AMONG ORGANISMS

In the laboratory at least some kinds of plants and animals can be cultivated in pure culture, completely isolated from all other species of living organisms. But in nature, plants and animals are surrounded by and affected by other organisms of many different species that constitute their **biological environment.** The biological environment of a plant may have effects on its processes, growth and development, and even its reproduction and survival that are quite as important and striking as the effects of the physical environment. The varied interactions among organisms in nature may be classified as **social** and **nutritive.** In a social interaction neither organism secures food from the other, whereas in a nutritive interaction one of the organisms does so. The community level of plant organization is a product of the interactions of the varied factors of both the physical and biological environments on the organisms constituting the community.

Although this botany textbook deals primarily with plants, we have from time to time mentioned interactions between plants and animals and in this chapter will do so even more extensively. In nature the biological interactions are among *organisms,* and there is no artificial distinction between plants, animals, or microorganisms.

SOCIAL INTERACTIONS

Although the organisms in some social plant interactions are in physical contact with each other part or all of their lives, perhaps the more important kind of social interactions are those that occur even though the organisms are not in actual physical contact. We shall consider two classes of interactions of the first type (epiphytes and vines) and two of the second type (antibiosis and the effects of organisms on factors of the physical environment).

Epiphytes

Plants such as Spanish moss (Fig. 17.1) and many species of orchids that grow perched on another plant but do not secure food from it are called **epiphytes.** Epiphytes are particularly abundant in the tropical rain forests (Fig. 17.2). The epiphytes may benefit by the relationship since they frequently receive more light than they would if growing on the ground. Epiphytes have no contact with the soil and they secure their water directly from rain or from the water vapor of the air and their mineral salts from airborne dust particles. Epiphytes do not take food or water from the supporting trees; however they may affect the trees in other ways. For example, Spanish moss may become so thick that it shades the leaves of the tree and reduces their photosynthetic productivity. Spanish moss may also be heavy enough, especially after a rain, to break some branches off the tree. Most species

Fig. 17.1 *Spanish moss, an epiphyte on oaks and other trees along the south Atlantic and Gulf coasts. The leaves and stems have a gray-green color. Spanish moss is a member of the pineapple family.*

Fig. 17.2 *Numerous epiphytes of several species on a Bucare tree in tropical Venezuela. Most of the epiphytes shown also belong to the pineapple family* (*Bromeliaceae*).

of the pineapple family, of which Spanish moss is a member, are epiphytes, the common pineapple being one of the exceptions. Many species of orchids are epiphytes. Lichens and algae growing on the trunks or branches of trees are also epiphytes.

The strangler fig and a number of other epiphytes of tropical forests are unusual in that they have long aerial roots that eventually grow from near the top of a tall tree down to the soil. After penetrating the soil, they branch and begin absorbing water and minerals from the soil. The numerous roots grow together, forming a network that encloses the tree trunk (Fig. 17.3). This crushes the phloem as the tree grows in diameter, preventing translocation, and killing the tree. After the dead tree rots away the strangler fig remains as an independent plant, with a hollow "trunk" composed of roots!

Fig. 17.3 A strangler fig (Ficus aurea) *that has developed a network of roots around the trunk of the host tree. By this stage the roots have entered the soil and are absorbing water and mineral salts from it, so it has changed from an epiphyte to a vine. After it has killed the host tree, it becomes an independent plant. Note the stems and leaves of the fig.*

Vines

Vines may climb on and over other plants (Fig. 17.4), but they differ from epiphytes in that they have roots in the soil throughout their lives. Like epiphytes, vines may benefit from the relationship by securing more light than they would trailing on the ground, and they may also affect the supporting plant by shading its leaves. In addition, some woody vines such as *Wisteria* may coil around the trunk of a tree and constrict the trunk as it grows in diameter. This may result in crushing the phloem and either partial or complete girdling that can injure or even kill the tree.

Antibiosis

Although the use of **antibiotics** from fungi and other microorganisms in the treatment of diseases is now a well-known practice, it is not so generally recognized that antibiotics play an important role in biological interactions in nature. Antibiotic substances diffusing from a mold or other plant may inhibit the growth of nearby bacteria and other organisms, thus reducing competition. Antibiotics may also protect the organisms producing them against certain parasites. At least some vascular plants such as the black walnut tree and the desert shrub *Encelia* are known to produce antibiotics that inhibit the growth of other plants in their immediate neighborhood. Antibiosis may play a more important role in the composition and structure of plant communities than is generally recognized (Fig. 17.5).

Fig. 17.4 (Left) A poison ivy (Rhus toxicodendron) *growing on a tree. Note the compound leaves with three leaflets. (Right) An older stem of poison ivy with numerous adventitious roots that hold it on the tree. Such roots are also present on the younger stems, but do not show clearly in the photograph at the left.*

Influences on the Physical Environment

Among the more important social interrelations of plants and animals are those in which one organism affects another by modifying the ordinary factors of its physical environment. The trees in a forest have a marked effect on the physical environment of the other forest plants through shading, temperature reduction, increased atmospheric humidity, absorption of water and mineral salts, and contribution of organic matter to the soil. The restriction of certain animals to specific types of plant communities results, not only from the availability of suitable food supplies, but also from the influence of the plants on the physical environment of the animals. Animals may also be responsible for marked effects on the physical environment of plants. For example, dams built by beavers (not to mention men) may result in the complete alteration of the physical environment of a limited area. Overgrazing by animals can result in soil erosion (Fig. 17.6). Of all organisms, man has perhaps had the greatest influence in altering the natural physical and biological environments, as he has removed forests, cultivated land, and constructed roads and buildings. Such activities are an essential aspect of modern civilization, but they are creating increasing concern. Even the natural communities not directly destroyed

Fig. 17.5 (Left) Antibiotic action of aromatic shrubs (Salvia leucophylla) *that have invaded a grassland in the Santa Inez Valley, California. Volitile substances from the shrubs have kept a two meter zone (a–b) essentially free of other plants, and there is less marked inhibition in the zone from b to c.*

Fig. 17.6 (Right) Overgrazing by cattle was a factor in the erosion of the soil from this Indiana woods and the consequent exposure of roots of the beech trees.

may be seriously affected by changes in drainage patterns and water availability, soil erosion and the silting of streams. Man has just begun altering the environment in another way that may have marked effects on life on earth: dispersal of radioisotopes and a variety of chemicals that may have adverse biological effects.

Considered from a broad and long-range standpoint, however, no group of organisms has had as marked an influence on the physical environment as photosynthetic plants. Without photosynthesis the atmosphere would probably be devoid of oxygen and most kinds of organisms would be unable to survive. Some biologists now believe that during the early history of life on earth there were no photosynthetic plants and that the atmosphere was not only devoid of oxygen but also contained high concentrations of carbon dioxide and also of other gases such as hydrogen, ammonia, and methane that would not last long if oxygen were present. The change from such an atmosphere to our present atmosphere can be credited to the appearance and spread of photosynthetic plants. If it were not for land plants there would be little or no soil since the rate of erosion would generally exceed the rate of soil formation. It is obvious that organisms may have marked effects on the physical environment, just as the physical environment has marked effects on organisms.

NUTRITIVE INTERACTIONS

In considering nutritive interactions between organisms, we are concerned only with the plants and animals that secure their food from living organisms, not with those that synthesize their food by photosynthesis or chemosynthesis **(autotrophic** organisms), that grow on nonliving organic matter **(saprophytic** bacteria and fungi), or that consume dead plants or animals **(scavengers).**

Ingestion of Other Organisms

Most species of animals secure their food by *ingestion* (i.e., by eating all or parts of other organisms and then digesting the food in their digestive tracts). Carnivorous animals ingest other animals, herbivorous animals ingest plants, and omnivorous animals ingest both plants and animals. The rather few species of carnivorous plants should probably also be placed in this category, but it should be noted that carnivorous vascular plants contain chlorophyll and in most cases probably make much of their food by photosynthesis.

Organisms that are consumed by others are certainly not benefited by the relationship, but, as we shall note later on, both herbivorous and carnivorous organisms play important roles in the scheme of biological checks and balances. The kinds and numbers of herbivorous animals present in a biological community may have a marked influence on the nature of the vegetation. For example, in England, hillside pastures grazed by sheep are dominated by bilberry plants and a grass called sheep's fescue, whereas similar ungrazed hillsides are covered with a member of the heath family called ling. Overgrazing in forests may result in the destruction of most of the young trees, and so in the gradual disappearance of the forest as the older trees die or are cut down (Fig. 17.6). Many insects and other animals consume only portions of plants and the effects on the plant may be minor, but at times swarms of insects consume such a large portion of the leaves or other organs of a plant as to interfere seriously with its growth and development (Fig. 17.7).

Parasitism

An organism that lives in or on another organism part or all of its life, securing its food from this host organism but not benefiting the host in any way, is referred to as a **parasite.** Many, but not all parasites are **pathogens,** that is, disease-producing organisms. Even a single species of parasite may be pathogenic in some hosts and not in others. For example, typhoid bacteria cause typhoid fever when they become established in most people, but typhoid carriers harbor large numbers of the bacteria and still have no symptoms of the disease. Generally the principal damage to the host by the parasite results from the destruction of tissue (Fig. 17.8) or the production of toxic substances by the parasites. Other factors may also be involved such as the plugging of the xylem

Fig. 17.7 Two examples of extensive damage to plants by insects. (Left) Eastern tent caterpillars consumed most of the leaves of this tree. (Right) Corn plants stripped of most of their leaves by a swarm of grasshoppers.

vessels by certain kinds of wilt bacteria and fungi.

The majority of the plants that are parasitic either on other plants or on animals are bacteria or fungi, but a few species of plants in other groups are completely or partially parasitic. The common mistletoe (Fig. 17.9) contains chlorophyll and probably synthesizes much of its own food, but it at least obtains water and mineral salts from the host tree. Another chlorophyll-containing seed plant that is probably only a partial parasite is the witchweed, a plant parasitic on the roots of corn and some other grasses (Fig. 17.10). Witchweed has only recently been introduced accidentally into the Carolinas from Africa and is causing serious damage to corn. Every effort is being made to control this pest and to prevent its spread, but this is difficult because of the numerous and very small seeds produced by the witchweed. One of the most common and destructive parasitic seed plants is dodder (Fig. 17.11). Although dodder contains small quantities of chlorophyll, it probably obtains most of its food from the host plants. The various species of dodder parasitize numerous plants, including many crop plants.

The state flower of Wyoming is the Indian paintbrush (*Castilleja linariaefolia*), which is apparently parasitic on big sagebrush (*Artemisia tridentata*). The repeated failure of residents of the state to cultivate their state flower in their gardens is apparently related to its obligate parasitism on sagebrush.

Several species of seed plants lack chlorophyll entirely and are complete parasites. These include a number of different genera of the broomrape family parasitic on the roots of plants, among them the beechdrops that parasitize beech trees. *Rafflesia* (Fig. 17.12), a genus of Malaysian plants parasitic on grape roots and a member of the Aristolochia family, is one of the most unusual of all seed plants. The vegetative organs grow entirely within the tissues of the host roots and have been reduced and modified until they resemble a fungal my-

Fig. 17.9 (Right) A young mistletoe (Phoradendron flavescens) *showing its haustoria embedded in a branch of the host tree. (Courtesy, Carolina Biological Supply Co.)*

Fig. 17.8 (Bottom left) Extensive destruction of the stems of a cactus by parasitic bacteria.

Fig. 17.10 (Bottom right) Several witchweed plants parasitic on the roots of a corn plant.

Fig. 17.11 Stages in the development of parasitism by dodder (Cuscuta). (a) *A dodder seed germinating near a plant that will become its host.* (b) *The dodder stem coils around the stem of the host plant.* (c) *Dodder haustoria have penetrated the host stem and the lower part of the dodder stem has severed.* (d) *An older dodder plant that has grown extensively and is blooming.*

celium. The only organs visible above the ground are the large, malodorous flowers, that are about a meter in diameter in one species. The white Indian pipe (Fig. 14.1) is frequently given as an example of a saprophytic seed plant, but it is apparently parasitic on a fungus, reversing the more usual parasitic relationships between fungi and seed plants.

Bacteria parasitize a wide variety of hosts, including practically all kinds of plants and animals. Fungi are perhaps best known as parasites of vascular plants, but they also parasitize animals, algae, and even other fungi. Among the fungi parasitic on humans are those causing ringworm and athlete's foot and a variety of species causing ear infections and pulmonary diseases. One group of fungi is parasitic on mosquito larvae and may prove to be valuable in mosquito control. Another group of fungi parasitic on nematode worms has a noose that contracts when a nematode passes through it, thus trapping the nematode (Fig. 17.13) which is then killed and digested by the fungus.

When plants are considered as hosts rather than as parasites, we find that almost every species of plant has parasites of one kind or another. It has been reported that the

ered to be living organisms or not, but at any rate viruses are among the more important causes of plant disease (Fig. 17.14).

Naturally man is distressed when he or his domestic plants or animals are attacked by parasites, but parasitism is just as normal a biological interaction as any other and is an important component of the general system of biological checks and balances.

Galls. Some parasites of plants, particularly insects, fungi, and bacteria, cause the de-

Fig. 17.12 The immense flower (compare with camera lens cover) of Rafflesia hasseltii *and its peduncle. This plant is a root parasite on grape vines in Malaysia and only the flower develops above ground.*

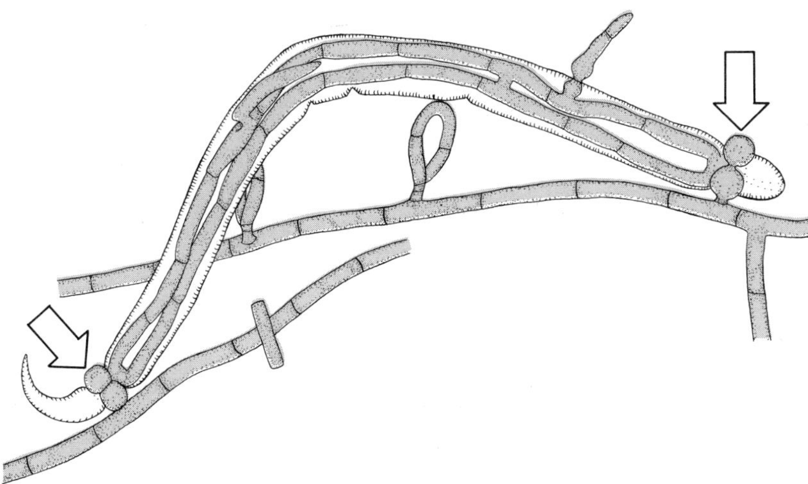

ginkgo tree is one species relatively free of parasites. Types of organisms parasitic on plants include not only other plants such as bacteria, fungi, and vascular parasites, but also nematode worms, scale insects, and other animals. Biologists are just beginning to realize how extensively nematode parasites damage plants and reduce crop yields. Whether viruses are parasites depends on whether they are consid-

Fig. 17.13 A nematode-trapping fungus with a nematode caught in the loops (arrows) of two hyphae. Note the hyphae growing inside the nematode. (After a drawing by John N. Couch, University of North Carolina.)

Fig. 17.14 A tobacco leaf infected with tobacco mosaic virus, showing damage by the virus.

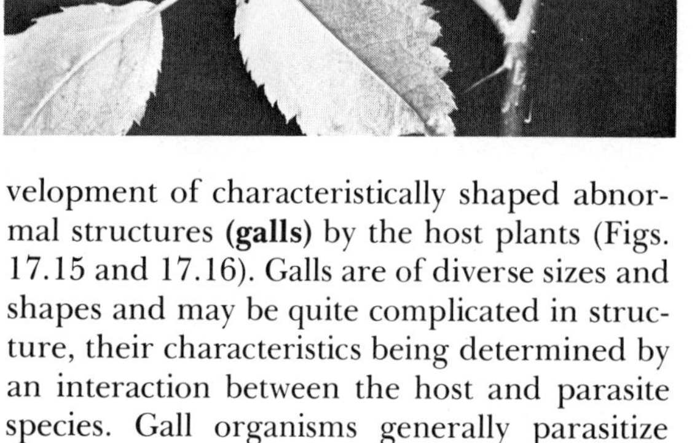

Fig. 17.15 (*Left*) *Spiny rose gall on a blackberry plant.*

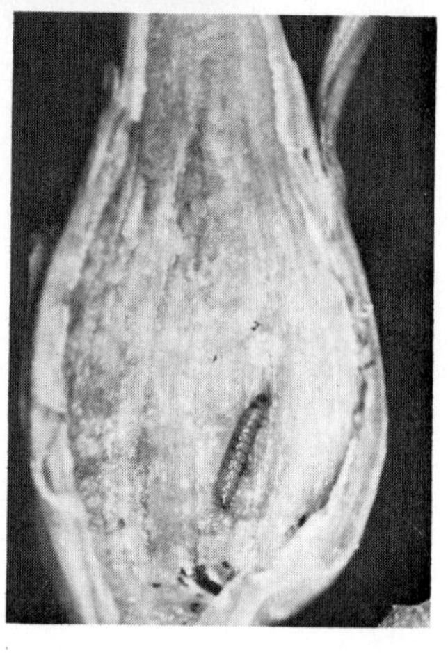

Fig. 17.16 (*Top*) *A gall on the stem of a goldenrod plant, caused by a moth.* (*Right*) *A sectioned gall showing a larva of the moth.*

velopment of characteristically shaped abnormal structures **(galls)** by the host plants (Figs. 17.15 and 17.16). Galls are of diverse sizes and shapes and may be quite complicated in structure, their characteristics being determined by an interaction between the host and parasite species. Gall organisms generally parasitize only one species of plant, or at most a few closely related species.

Wasps are probably the most common gall-inducing organisms. Ther lay their eggs in the tissues of the host plant, and as their larvae develop, the galls develop from the surrounding tissues. The larvae use some of the tissues of the gall as food. Oak trees are particularly susceptible to attack by gall wasps of various species.

Although it appears certain that gall organisms produce growth substances that cause the plant tissues to differentiate into characteristic galls, little is known about the nature or mode of action of most of these growth substances. It is known, however, that the galls caused by the crown-gall bacteria (Fig. 17.17) contain an unusually high concentration of auxin. The bacteria are apparently responsible for the initiation of the auxin production, but once a crown gall is established in a plant it may give rise to bacteria-free galls in other parts of the plant. These bacteria-free galls still have the capacity for synthesizing high auxin concentrations. If part of one of these bacteria-free galls is grafted to healthy plant tissue, it will cause the development of an auxin-producing tumor (Fig. 17.18). In contrast, the apparently similar gall-like growths produced

Fig. 17.17 Tumors on a Kalanchoë *caused by a crown gall bacterium.*

by treating plants with auxins (Fig. 15.36) do not spread and do not have the capacity for producing auxin themselves.

Alternate Host Parasites

Some parasites (the **alternate host parasite)** complete certain stages of their life cycle in one host, but can complete other essential stages only in another host species that is generally not closely related to the first host. A well-known example of an alternate host parasite is the protozoan that causes malaria and is parasitic on man and mosquito. The most common alternate host parasites of plants are the rust fungi. Of the many species of rusts, three of the more important are the black stem rust of wheat (with wild barberry bushes as its alternate host), the white pine blister rust (with currant or gooseberry bushes as alternate host), and the cedar apple rust of apple trees and cedar trees (Figs. 17.19 and 17.20). The diseases caused by these and other rusts have resulted in serious economic losses, particu-

Fig. 17.18 A small cube of tissue from a crown gall (left) continues growth in tissue culture without added auxin. If a bit of crown gall tissue from the culture is grafted to a stem segment from an uninfected plant (right) and cultured the tumor continues to grow and also induces tumor growth from the stem section. (Redrawn from Principles of Plant Physiology *by Bonner and Galston, with the permission of the publisher, W. H. Freeman & Co.)*

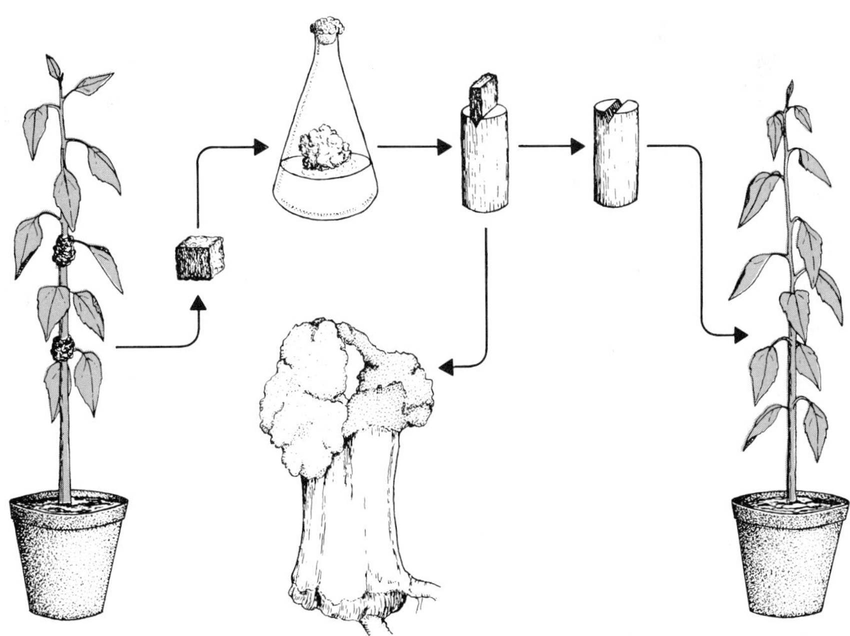

larly of wheat and white pine.

The life cycle of the black stem rust of wheat is diagrammed and described in Fig. 17.21. For our present discussion this life cycle is less important for its technical details than for emphasizing the fact that certain specific stages of development occur in each of the two hosts and that the full life cycle can be completed only when both host species are present. The diagram also illustrates the complexity of a rust life cycle.

Fig. 17.19 *A cedar apple gall on a cedar tree. Teliospores produced in the gelatinous horns of the gall give rise to basidia that produce basidiospores. The basidiospores are carried by wind to apple trees, the alternate host of the rust fungus.*

Fig. 17.20 *Aecia of the cedar apple rust fungus on leaves and a fruit of an infected apple tree. The aeciospores produced by the aecia infect only cedar trees, whereas the basidiospores produced on cedar trees infect only apple trees and their close relatives.*

Fig. 17.21 *(Opposite) Life cycle of the black stem rust of wheat. The teliospores* (a, b) *produced by the fungus on wheat germinate into basidia* (c) *that produce basidiospores* (d). *The basidiospores infect only wild barberry bushes* (e), *the alternate host. The aeciospores* (f) *produced by the fungus in barberry infect only wheat. The wheat stage of the fungus also produces uredospores* (g) *that can affect other wheat plants and so spread the disease during the growing season. (Redrawn from* Botany *by Sinnott and Wilson, with the permission of the publisher, McGraw-Hill.)*

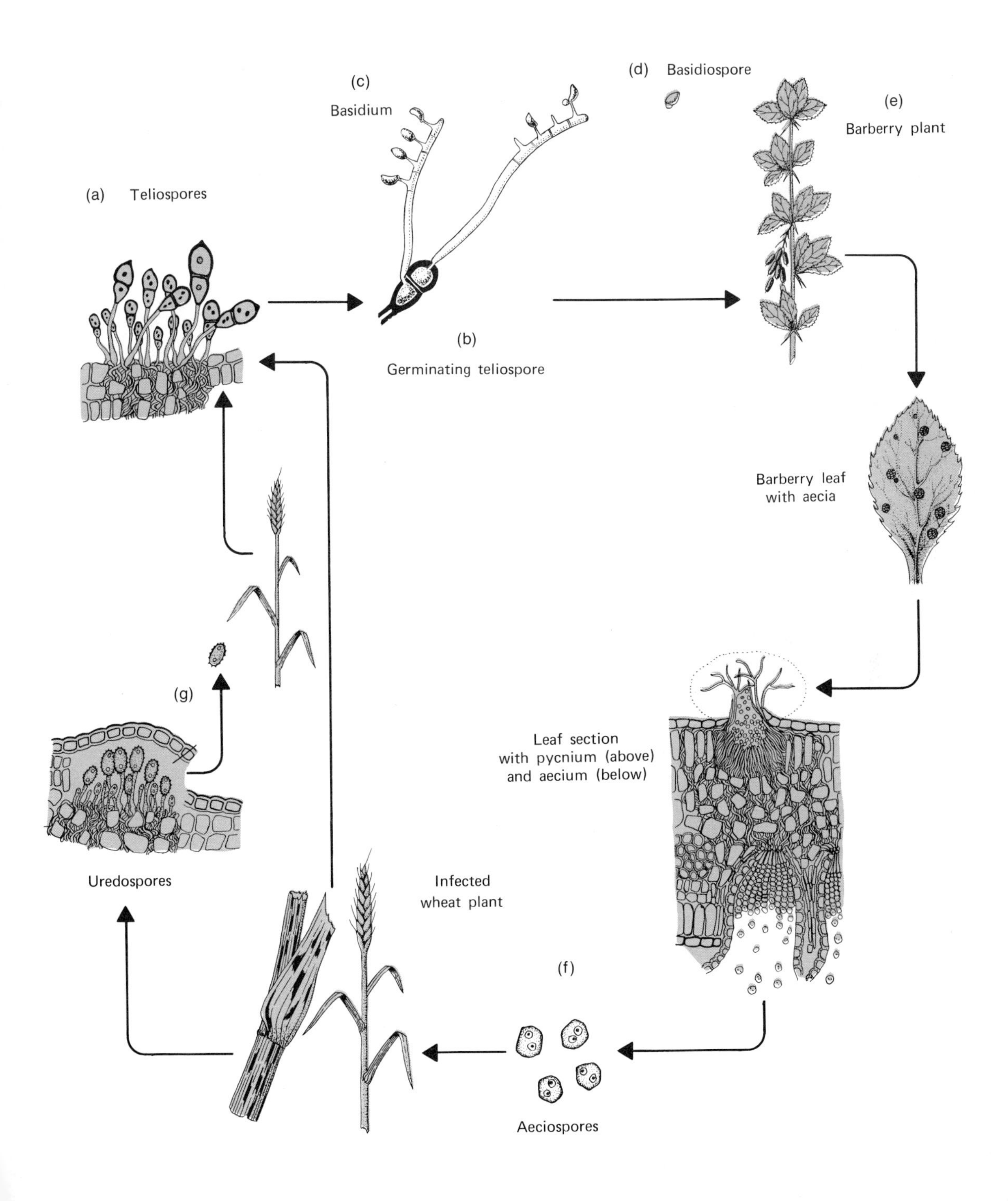
(a) Teliospores
(c)
Basidium
(b)
Germinating teliospore
(d) Basidiospore
(e)
Barberry plant
Barberry leaf
with aecia
Leaf section
with pycnium (above)
and aecium (below)
(f)
Aeciospores
Infected
wheat plant
Uredospores
(g)

Eradication of the less important host species provides a means of disease control limited to alternate host parasites. Thus, white pine blister rust has been controlled to some degree by eradication of as many wild currant and gooseberry plants as possible. Wild barberry bushes have been killed to control wheat rust, and cedar trees in the neighborhood of apple orchards have been cut down. This method of control is, however, not always effective against wheat rust, since the rust can spread directly from one wheat plant to another by means of the uredospores (Fig. 17.21), and since in regions with mild winters the uredospores can survive until the next spring and infect the new crop. In cooler climates only the barberry stage can survive the winter. The breeding of rust-resistant varieties of wheat has provided a more effective means of disease control, but even this is not a permanent solution since new strains of rust to which the wheat is not resistant keep arising by mutations and by genetic recombinations occurring during sexual reproduction of the rust.

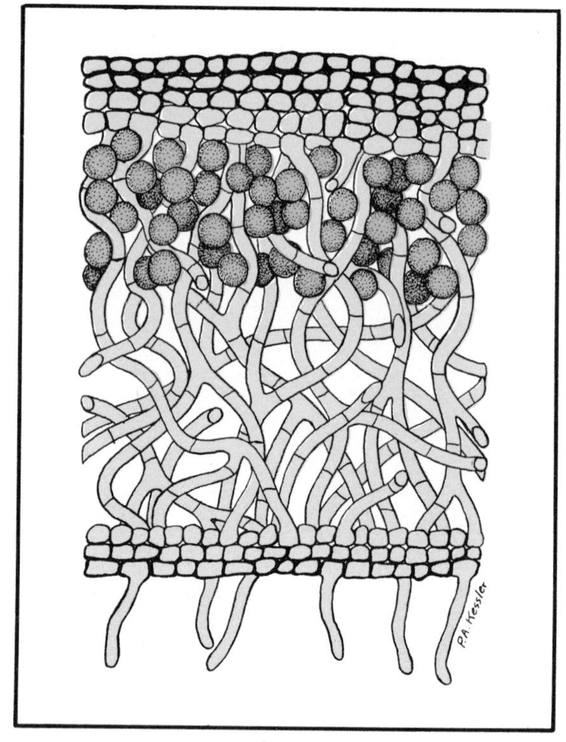

Fig. 17.22 Section through a portion of a lichen showing spherical unicellular algae among the hyphae of the fungus.

Symbiosis

The word **symbiosis** is used in at least two different ways: (1) To cover all interactions between two different species (although sometimes limited to interactions involving physical contact), and (2) to include only nutritive interactions beneficial to both organisms. We shall use the term in the second sense, although the term **mutualism** is preferred by some biologists for this type of biological interaction. It is sometimes difficult to determine whether an interaction is parasitism or symbiosis. Lichens, the composite plants consisting of algae within the mycelial mass of fungi (Figs. 17.22 and 20.4), are frequently cited as examples of symbiosis because the fungus obtains food from the alga and the fungus in turn is said to protect the alga from desiccation and other hazards. However, it appears probable that at least in some lichens the algae gain no benefits and the fungi are simply parasitic on the algae.

Certain species of fungi grow in or on the roots of many species of forest trees, orchids, members of the heath family and other plants (Fig. 17.23). The rather stubby root-fungus complexes are called **mycorhiza** (Fig. 17.24). The fungi obtain food from the roots, and since in many cases, at least, the mycorhiza appear to be important or even essential in the absorption of mineral salts by the plant the relationship is generally regarded as symbiotic. Mycorhiza may also participate in water absorption, and may benefit the plant in other ways. Many common forest mushrooms are the reproductive structures of mycorhizal fungi. Some plants do not thrive without mycorhiza, but others seem to grow as well without mycorhiza as with them, and in such instances it is possible that the mycorhizal fungi are simply parasites.

The association of nitrogen-fixing bacteria with leguminous plants (Chapter 13) is almost certainly beneficial to both organisms and so should be classed as symbiotic. An unusual habitat of one species of unicellular alga is among the cells of a hydra (a simple relative of the jellyfish), thus producing a green hydra.

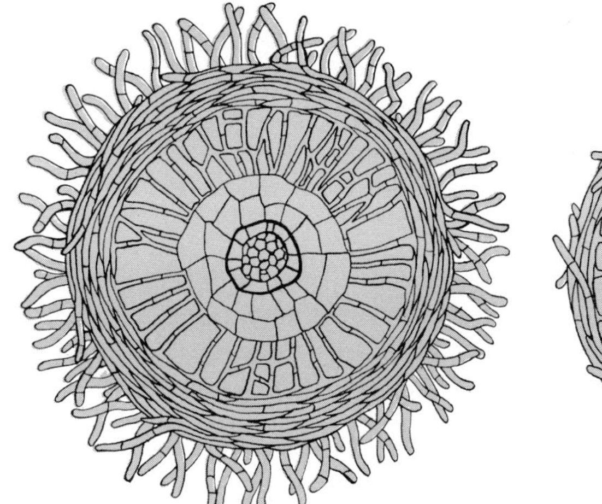

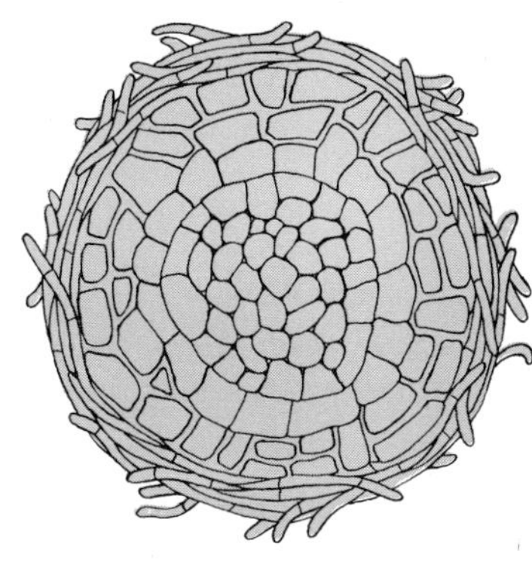

Fig. 17.23 Cross sections of two mycorrhizal roots. Left: hickory. Right: pine. (Redrawn W. B. McDougall, Plant Ecology, *with the permission of the publisher, Lea & Febiger.)*

Fig. 17.24 Mycorrhizal roots of Pinus virginiana *showing the fungal sheath covering the roots.*

This composite organism is considered a symbiont by some biologists. The hydra probably obtains food and oxygen from the alga, and the alga obtains carbon dioxide, mobility, and possibly protection from the hydra. Green hydra, unlike the usual kinds, are attracted toward regions of brighter light.

Symbiosis does not necessarily involve continued close contact between the organisms. Thus, the relationship between bees and other pollinating animals and plants is symbiotic. The bees use both the nectar and pollen they collect as food, but the plants are pollinated in the process. Birds and other animals that eat fruits and seeds play an important role in plant dispersal, since at least some of the seeds pass through their digestive tracts without being digested or injured. The relationship between man and the plants and animals he raises for food may be considered as symbiotic, since most domesticated plants and animals could not survive without man's care, having been selected for desirable commercial qualities rather than for characteristics contributing toward survival in nature.

A few other species of animals similarly care for plants or animals from which they secure food. For example, the agricultural ants chew up leaves into a substrate suitable for the growth of certain fungi that they place on the substrate. Later, they harvest and eat the fungi. Ants carry aphids (plant lice) from one host plant to another (to the detriment of the plants). When stroked, the aphids secrete a drop of honeydew (which contains a high concentration of sugars), and the ants consume it. The situation is not too different from our caring for cows from which we secure milk.

RELATED READING

Batra, S. W. T. and L. R. Batra. "The fungus gardens of insects," *Scientific American, 217*(5), 112–120, November 1967.

Bonner, J. "Chemical warfare among the plants," *Scientific American, 180*(3), 48–51, 1949.

Bonner, J. "Chemistry in plant societies," *Natural History, 68,* 508–513, 1959.

Denison, W. C. "Life in tall trees," *Scientific American, 228*(6), 74–80, June 1973.

Dobzhansky, T. and J. Murca-Pires. "Strangler trees," *Scientific American, 190*(1), 78–80, January 1954.

Ehrlich, P. R. and P. H. Raven. "Butterflies and plants," *Scientific American, 216*(6), 104–113, June 1967.

Eleazer, J. M. "Witchweed invades America," *Science Digest, 43*(6), 60–73, June 1958.

George, J. "Plants that eat insects," *Reader's Digest, 80*(2), 221–226, February 1962.

Grave, E. V. "A plant that captures animals under water," *Natural History, 66,* 74–77, 1957.

Halligan, J. P. "Bare areas associated with shrub stands in grasslands: the case of *Artemisia californica," BioScience, 23,* 429–432, 1973.

Hovanitz, W. "Insects and plant galls," *Scientific American, 201*(5), 151–162, November 1959.

Hutchins, R. E. "Dodder—a vampire plant," *Natural History, 64,* 378–381, 1955.

Hutchins, R. E. "Ants that grow mushrooms," *Natural History, 65,* 476–481, 1956.

Lamb, I. M. "The remarkable lichens," *Natural History, 67,* 86–93, 1958.

Lamb, I. M. "Lichens," *Scientific American, 201*(4), 144–156, October 1959.

Last, F. T. and R. C. Warren. "Non-parasitic microbes colonizing green leaves: their form and functions," *Endeavour, 31,* 143–150, 1972.

Maio, J. J. "Predatory fungi," *Scientific American, 199*(1), 67–72, July 1958.

18
ECOSYSTEM DYNAMICS

In any ecosystem the interactions of its organisms with one another and with the physical environment constitute a dynamic complex in which there is a constant flow of matter and energy. There is a tendency for the ecosystem to reach a dynamic balance, but the balance is delicate and may be disrupted quite easily. We will now examine several aspects of ecosystem dynamics: food chains and webs, energy flow, cycling of matter, checks and balances, disturbance of balances, and succession.

FOOD CHAINS AND WEBS

The basic food supply of any natural ecosystem is provided by its autotrophic organisms, which are predominantly photosynthetic, although there are a few species of chemosynthetic bacteria. These are called the **producer organisms** and they occupy the basic **trophic** (nutritional) level. All the other organisms in the ecosystem are consumers. At the second trophic level are the **primary consumers,** including the herbivorous animals and the various parasites and symbionts of plants. The third trophic level is occupied by carnivorous animals **(secondary consumers)** that consume herbivorous animals and by parasites of the herbivorous animals. In land ecosystems, a fourth or subsequent trophic level is rare, except for parasites, but in marine ecosystems several successive trophic levels of carnivores are common. A certain species may occupy more than one trophic level, as in omnivorous animals such as rats and human beings.

In any ecosystem there are numerous **food chains,** made up of the eaten and the eaters at each trophic level. One such food chain consists of plants, rabbits, and foxes. An aquatic food chain that is somewhat longer consists of algae, small crustaceans, dragonfly nymphs, minnows, and bass. However, in ecosystems herbivores generally consume more than one kind of plant, omnivores consume a variety of plants and animals, and carnivores consume a number of different animals. Furthermore, parasites, and symbionts secure their food from plants and from animals at the various trophic levels. Thus the nutritional interactions among the various species of a community constitute a complex **food web,** rather than a series of isolated food chains. The food web outlined in Fig. 18.1 is actually a substantial simplification of a food web as it exists in nature.

In general, the biomass at each trophic level (producers, primary consumers, secondary consumers, and tertiary consumers) decreases by a factor of about ten (Fig. 1.1), although when actual calculations are made the decreases may be somewhat larger or smaller than this (Fig. 18.2). **Biomass** is the total dry weight of the organisms at each trophic level, and consists primarily of organic compounds derived from the product of photosynthesis.

Fig. 18.1 A greatly simplified food web in a deciduous forest community.

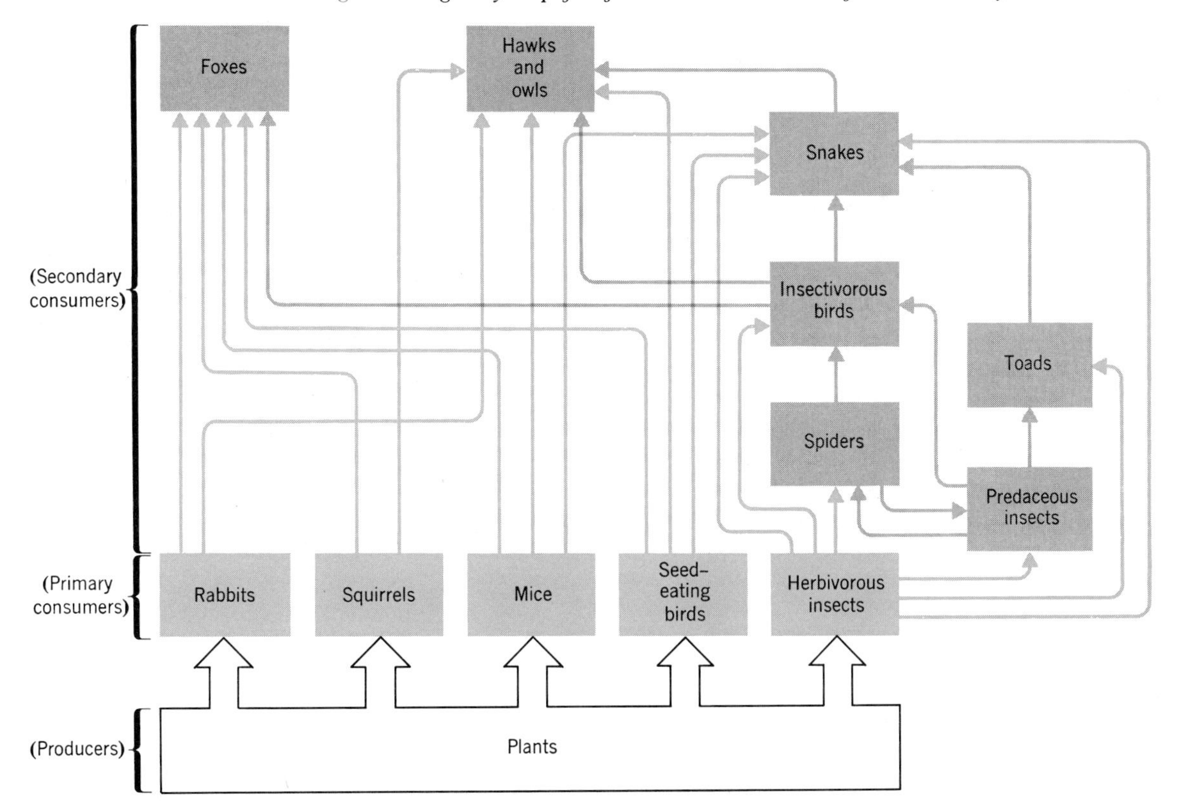

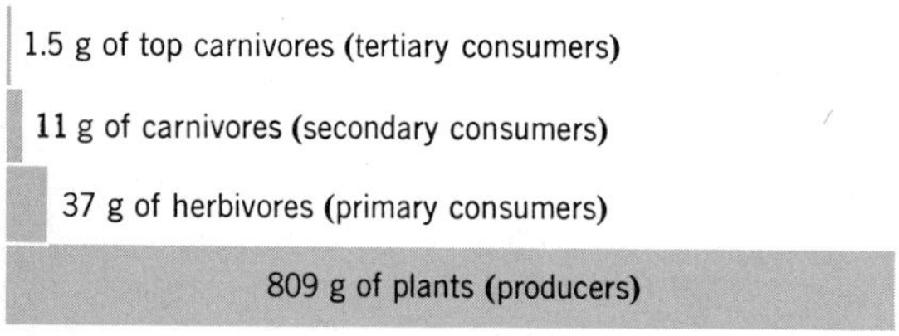

1.5 g of top carnivores (tertiary consumers)
11 g of carnivores (secondary consumers)
37 g of herbivores (primary consumers)
809 g of plants (producers)

Fig. 18.2 Biomass at successive trophic levels in an aquatic community at Silver Springs, Florida, g/m² dry weight.

Fig. 18.3 Decrease in numbers of individuals at each trophic level in a grassland community. The situation is typical of most ecosystems except when food chains involving small organisms such as insects and parasitic bacteria are considered. These greatly outnumber the individuals from which they obtain their food in most cases.

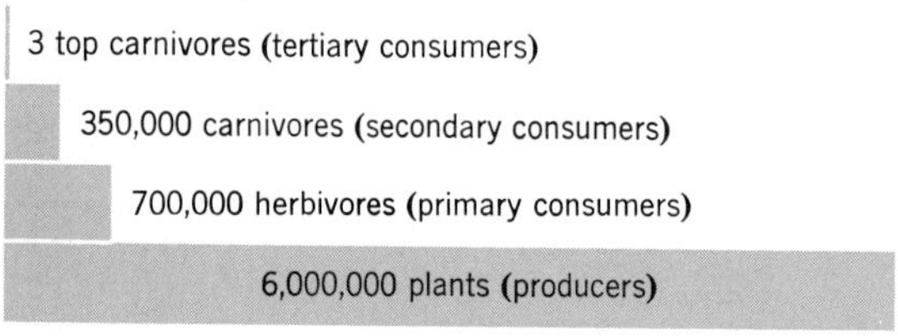

3 top carnivores (tertiary consumers)
350,000 carnivores (secondary consumers)
700,000 herbivores (primary consumers)
6,000,000 plants (producers)

This progressive decrease in biomass is unavoidable, since the organisms at each trophic level use much of their food in respiration. Also, an ecosystem would not be viable if the organisms at a trophic level used all or even most of the organisms at the next trophic level as food.

There is also generally a pyramid of numbers of individuals at each trophic level from plants to herbivores, carnivores, and top carnivores (Fig. 18.3). However, host to parasite food chains, insect infestations of plants, and similar food chains lead from a large organism to numerous small or very small organisms.

ENERGY FLOW THROUGH ECOSYSTEMS

Energy is introduced into the biosphere when plants convert light energy from the sun into the chemical bond energy of foods during photosynthesis (Fig. 18.4). Some of this energy is used by the plants in respiration, and the rest is retained in the chemical bonds of foods used in assimilation or accumulated in seeds, fruits, or other organs. The assimilated and accumulated foods are both available as food for herbivores, and these too use part of the food in respiration. The use of food in respiration is repeated by each level of carnivore and by parasites and decomposers. The energy provided by respiration is essential to each organism for carrying on various life processes, but represents a loss of energy and also biomass at each trophic level. Eventually all the energy released during respiration is converted to heat energy and dissipated, whether the heat is produced during the respiratory reactions themselves or after the ATP and NADH from respiration have been used in doing metabolic work. By the time decomposers (and their parasites) have used as food the remains of dead plants, animals, and animal wastes and carried on their respiration, all the chemical bond energy trapped during photosynthesis has been used in respiration, and then lost as heat. In Fig. 18.4, parasites have been included with the decomposers for simplification, even though they secure their food from living rather than dead tissues.

The energy flow outlined in Fig. 18.4 is characteristic of stable, mature (climax) communities. During a period such as a year essentially the same amount of energy is used by respiration (and eventually lost as heat) as is trapped from light by photosynthesis. As a result, there is neither gain nor loss in biomass (organic compounds) of the ecosystem during the period. However, in developing (successional communities, photosynthesis exceeds the total respiration of the organisms making up the community, and so there is a continuing increase in biomass until succession results in a climax community. This is discussed at the end of this chapter.

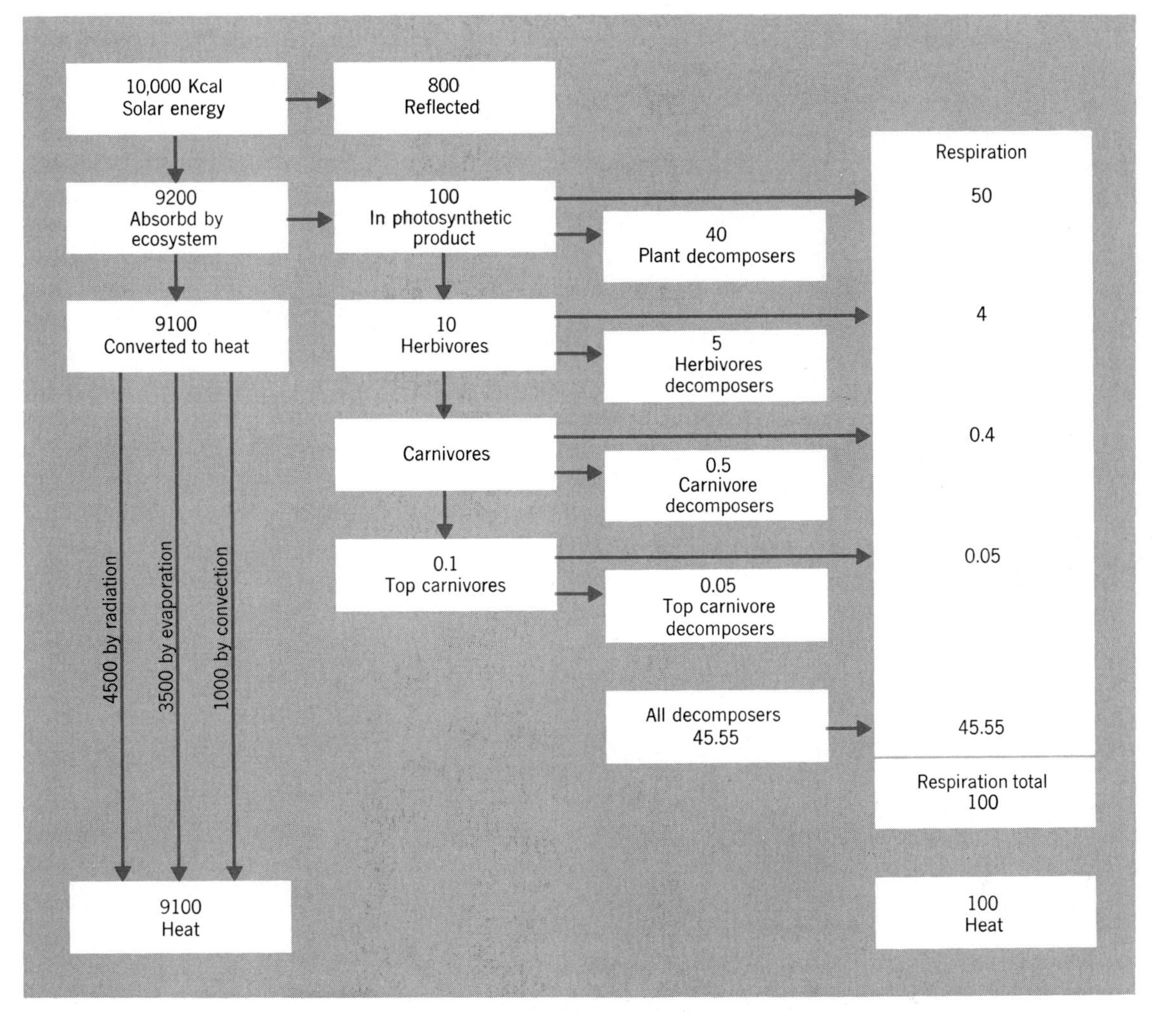

Fig. 18.4 Energy flow through a typical mature (climax) ecosystem per 10,000 kcal of solar radiation. In any time period such as a year total respiration essentially equals total photosynthesis, but the photosynthetic products used by consumers was often produced in previous years. Much of the food used by decomposers, such as the wood of logs, was produced by photosynthesis many years earlier. All the energy absorbed is eventually converted into heat and dissipated.

Note the relatively small role played by animals in the flow of energy through the ecosystem, and the very small fraction (about 1%) of the solar energy that reaches the ecosystem that is incorporated in the organic compounds produced by photosynthesis. Although the vast majority of the solar radiation absorbed by an ecosystem is converted to heat and lost, it is of great importance in providing a range of temperatures suitable for life.

Human beings have used very little wasted solar radiation as a source of energy, but because of the current energy shortage efforts are being made to use solar energy in heating.

THE CYCLING OF MATTER IN ECOSYSTEMS

In contrast with the one-way flow of energy through ecosystems, matter is cycled and recycled in ecosystems. The nitrogen cycle has already been described in Chapter 13. Another important cycle involves the cycling of carbon through ecosystems (Fig. 14.5). In photosynthesis the carbon of carbon dioxide is incorporated in sugars and other organic compounds, and then into the wide variety of organic compounds synthesized by plants and other organisms. The carbon flows through the food webs as described previously, and at each trophic level foods are oxidized in respiration with the production of carbon dioxide.

The cycling of water (Fig. 18.5) has already been referred to in Chapter 16, but will be summarized briefly here. We can begin with the precipitation of water from clouds as rain or snow, the soaking of water into soils, the absorption of water by plants, its use by plants, and its loss into the atmosphere by transpiration as water vapor. Another part of the biological cycling of water involves its consumption and use by animals and then its loss by excretion, perspiration, and as water vapor during breathing. Of course, much of the precipitation runs off into streams and lakes and eventually into the oceans. Evaporation of water from these and from the soil, as well as the loss of water from plants by transpiration, contributes to the water vapor of the air which is again precipitated. It should be noted that the use of water in photosynthesis and its pro-

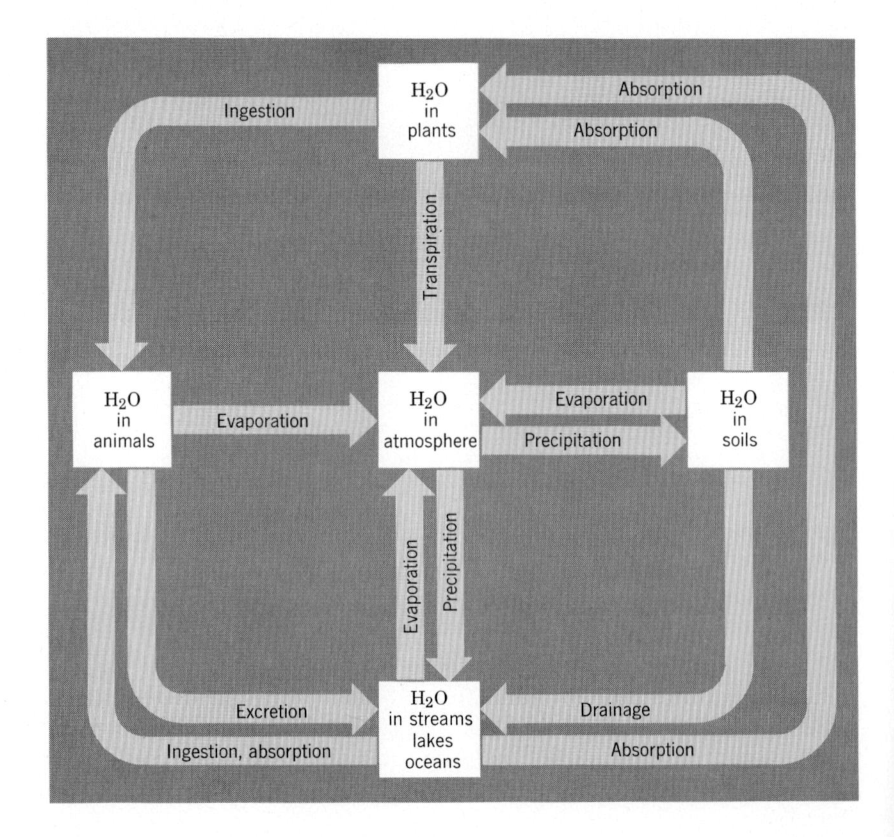

Fig. 18.5 The water cycle of an ecosystem.

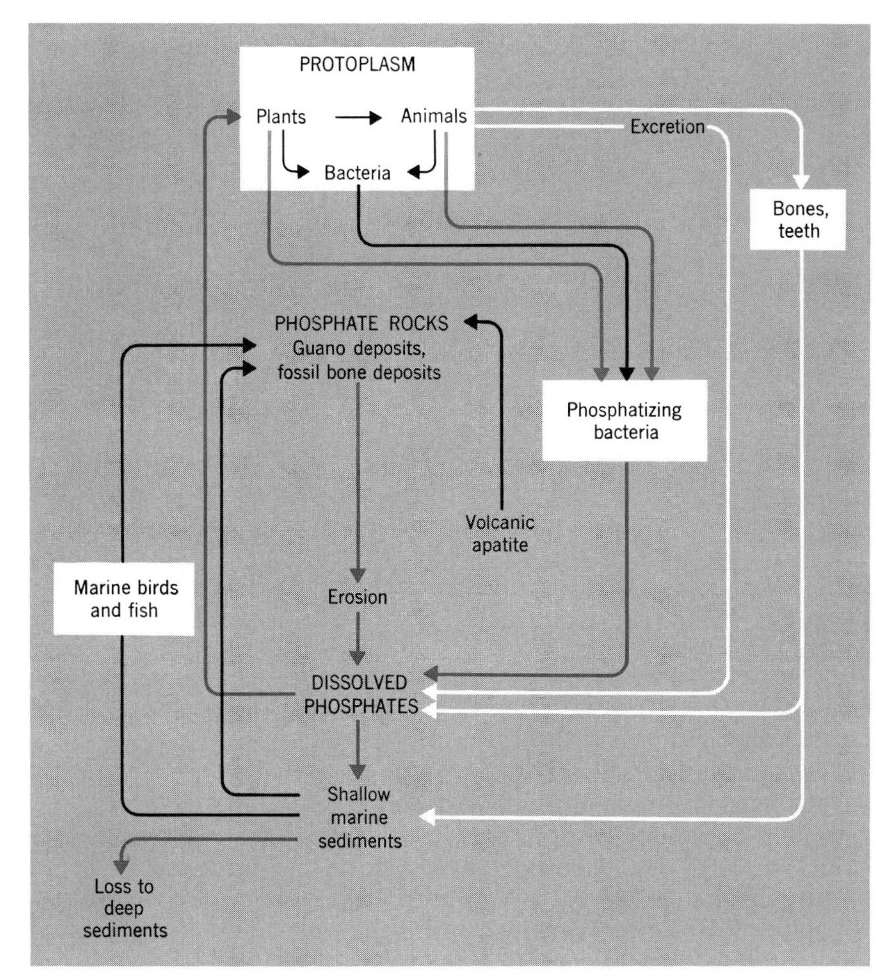

Fig. 18.6 The phosphorus cycle of the biosphere.

duction in respiration is also part of the water cycle.

All of the above cycles have involved gases of the atmosphere, but the circulation of mineral elements through the ecosystem involves a reservoir of compounds in the soils and waters of the Earth (Fig. 18.6). In general, the mineral element cycles involve their absorption by plants from the soil (or from the surrounding water by aquatic plants), the transfer of the minerals to animals and microorganisms through the food chains, and the return of the minerals to the soils or waters by decay of fallen leaves, by decomposition of other plant and animal remains, by leaching from plants, or by excretion by animals. In an undisturbed ecosystem the only significant loss of mineral elements is from runoff of dissolved salts into streams, lakes, and eventually the oceans. However, when the vegetation cover of an area is removed in clearing land for agriculture or for construction, or in cutting forest trees for lumber, extensive soil erosion may occur and mineral ions held on soil particles are lost. Runoff is also usually increased greatly. In cultivated fields the natural cycling of mineral elements is reduced when crops are harvested and removed from the area, and so it is necessary to supply fertilizers to maintain adequate quantities of the essential mineral elements.

BIOLOGICAL CHECKS AND BALANCES

There is a tendency of mature ecosystems to attain a steady-state condition in which the species of plants and animals that make up the community and the number of individuals of each species remain relatively constant over a period of hundreds of years. This dynamic biological balance is a result of the interactions among the organisms that make up the community. Every species of plant and animal has a much greater reproductive potential than is ever realized in a balanced ecosystem. The food web plays a most important role in preventing population explosions in an ecosystem, as herbivorous animals consume plants or their seeds, carnivorous animals consume the herbivorous ones, and parasites attack organisms at each of the trophic levels. Competition among individuals of a species or between individuals of different species that thrive best in the same environmental complex also tends to limit the population that occupies each of the microenvironments of the ecosystem. Since the kinds of organisms and the numbers of individuals of each species remain relatively constant from generation to generation, so do the food webs, the cycling of matter, and the flow of energy through the ecosystem.

DISTURBED BIOLOGICAL BALANCES

The biological balance of a plant and animal community is delicate, and may be upset by any substantial change in the physical environment, the partial or complete removal of one or more of the species making up the community, or the addition of species from other regions into the community. Any disturbance of a balanced community results in the initiation of a succession that eventually leads to the establishment of a new balanced climax community, although hundreds of years may be required for completion of the succession. The problems of conservation of our biological resources result largely from the disturbance of natural biological balances by man.

Upsetting Biological Balances by Altering the Physical Environment

We have already mentioned several ways that man may alter the physical environment, and all of these may result in some degree of disruption of the natural biological balances. A specific example may help to emphasize the degree to which man may upset a natural community by a relatively minor disturbance of the physical environment. Previous to 1918, Currituck Sound (Fig. 18.7) on the North Carolina coast was a sportsman's paradise. It teemed with fish and served as a winter feeding ground for ducks and geese, both the fish and the birds living on the numerous submerged water plants. In 1918 the sea-level locks of the canal connecting the Sound with the waters around Norfolk, Virginia, were removed. This made it possible for the north winds to blow brackish, sewage-laden water into the Sound from the Norfolk area. This brackish water also contained numerous small jellyfish that covered and smothered the water plants of the Sound. As the plants died, soil from the bottom of the Sound that had been held in place by the roots became stirred up and made the water muddy. The mud and sewage cut off much of the light from the few remaining plants, reducing their rate of photosynthesis so greatly that the plants starved. With their food supply gone and the oxygen content of the water reduced, the fish died and the ducks and

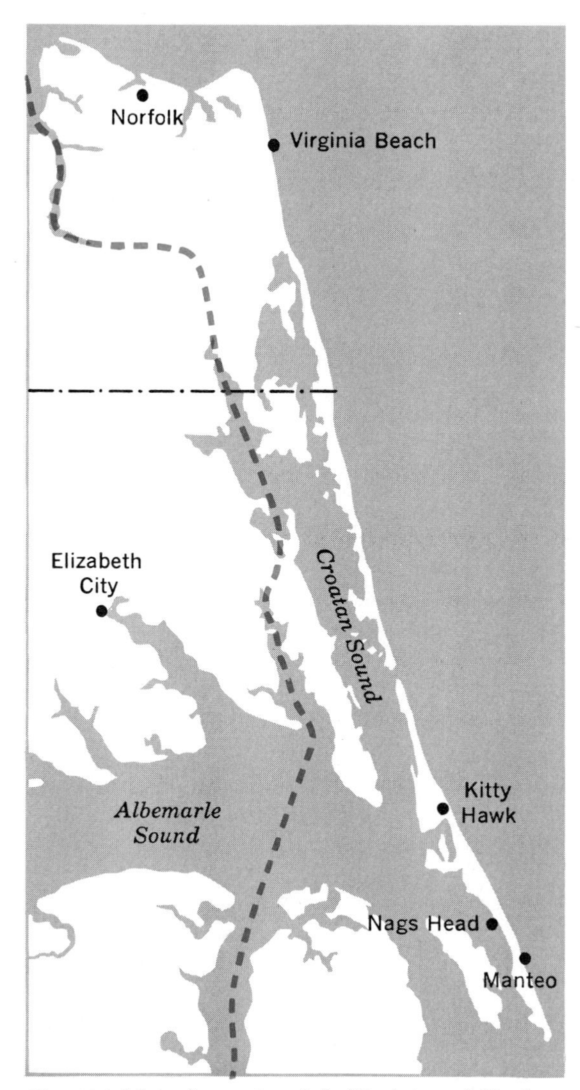

Fig. 18.7 Map of a portion of the Virginia and North Carolina coastal area including Currituck Sound. Note the course of the Intercoastal Waterway from the Norfolk region into Currituck Sound.

geese went to other feeding grounds. By the mid-twenties, disruption of the balance of life in the Sound was complete, ruining an investment of over $5 million and the means of livelihood of around 10,000 people in a region that had been a favorite hunting and fishing area for thousands of sportsmen each year. Years of painstaking biological study were required to clarify the series of causes and effects that resulted in the disappearance of the water plants, fish, and game birds. Once the situation was understood the Army Engineers were persuaded to restore the locks, and by 1943 the water life of the Sound was again flourishing.

For some time biologists have been concerned about the contamination of the environment with a variety of potentially harmful chemicals by man. These include toxic dusts and gases discharged into the atmosphere by factories and automobiles, detergents in streams from sewage disposal systems, chemical wastes from factories that pollute streams and often kill all or most of the organisms in them, and a variety of pesticides including insecticides, fungicides, and herbicides that are being used in agriculture in increasing quantities and are contaminating natural biological communities, not to mention radioactive fallout from testing of atomic and hydrogen bombs. Following the publication of Rachel Carson's *Silent Spring* in 1962, the general public for the first time became concerned about the effects of such contamination, particularly from pesticides, and there has been considerable discussion of the problem. Agricultural scientists and representatives of agricultural chemical companies have pointed out that the use of pesticides is essential and have tried to minimize their harmful effects. However, there is little doubt but that chemical contamination of the environment is harmful to plants and animals as well as humans and the only question is to what degree.

This disruption of the biological balances in streams, lakes, and even the oceans can occur not only through the introduction of toxic substances and oil spills but also in other ways. One way is thermal pollution, that is, the discharge of warm water (or sometimes hot water) from power plants or factories into bodies of water. An increase of the temperature of the water injures or even kills some of the plants and animals of the aquatic community while it may promote the growth of others, thus completely upsetting the natural balance. Streams and lakes may also have their biological balances completely disrupted by eutrophication (Chapter 13). This results, not from toxic substances, but from excessive quantities of essential mineral elements, particularly nitrogen and phosphorus. This results in a marked increase in the algae and other aquatic plants, far in excess of the quantity that can be kept under control by the primary consumers. When the plants die the decomposers (bacteria and fungi) increase greatly as they thrive on the dead plants, and they use so much oxygen in their respiration that the oxygen content of the water becomes too low to permit the survival of many of the aquatic animals. Thus the entire natural balance of the lake or stream is disrupted. The mineral elements come from excess fertilizers applied to fields as well as from sewage (which contains among other things phosphate-containing detergents).

Natural geological processes such as erosion, deposition, and submergence or emergence of land masses, as well as major long range changes in climate may have even more extensive disruptive effects on biological communities than the activities of man. For example, the forest trees occupying the southwestern part of what is now the United States several thousand years ago died when the region became more arid and were replaced by desert vegetation. About 3000 years ago the area now occupied by Ohio, Indiana, and Illinois became much drier than it had been and as the forests died out they were replaced by extensions of the western grasslands. When the climate once again became more humid, the grasslands in turn gave way to forests, except in isolated areas where the soil factors and topography favored the survival of the grasses. These prairie patches were a striking feature of the natural landscape before most of the region was placed under cultivation.

Upsetting the Biological Balance by Subtraction

We shall cite only a few of the many examples of disturbed biological balances resulting from removal of one or more species from a community. In Sumatra a good many years ago tigers were killing rather substantial numbers of people and domestic animals, so a systematic program of tiger extermination was begun. The extermination was so successful that wild pigs, an important source of food for the tigers, became very numerous and consumed most of the young palm trees. As the older trees died the palm tree population declined so greatly that the natives, who relied heavily on the palms for both food and shelter, became impoverished.

Another incident occurred in a London park, where wasps were exterminated because they had become a nuisance to park visitors. Soon the grass in the park began to die. At first there seemed to be no connection between the two events, but investigation by biologists revealed that the two events were related. The wasps had been keeping crane flies under control, and since the crane fly larvae lived on grass roots the increased crane fly population injured the grass seriously.

In our own country many upset balances, perhaps less spectacular than the ones just mentioned but at least as serious, have resulted from the destruction of native organisms. The killing of carnivorous animals such as wolves and foxes has resulted in increased numbers of rabbits, squirrels, and mice and consequent extensive damage to both native and cultivated plants. Destruction of birds and snakes results in an increase in plant-destroying insects. Although most native birds are now protected by law, birds may be poisoned by eating poisoned insects. Widespread use of insecticides may also kill useful insects such as bees and insects parasitic on the insects that attack plants, thus further upsetting the natural balance. This may result in the use of even larger quantities of insecticides to control the insect pests of plants and so the still further reduction in the organisms exerting biological control over the harmful insects.

Upsetting Biological Balances by Addition

In the fifteenth century, Portuguese settlers brought goats to the island of St. Helena, then covered with dense forests. The goats flourished to such an extent that they were soon consuming all the young trees as well as other low growing plants. As the older trees died or were cut down the forests gradually disappeared, the fertile soil eroded into the ocean, and the entire biological balance was destroyed. Similar situations exist in our own country where forests are too heavily grazed, and many people are concerned over the probable eventual results of granting grazing rights in our national forests.

In Jamaica, rats became serious pests when they invaded the island from ships, and so the mongoose was imported from India to destroy the rats. The mongoose did such a thorough job of this that it soon had to turn to poultry, native game birds, and snakes for food. The decrease in the populations of these insectivorous animals in turn resulted in a great increase in insects and extensive destruction of both native and cultivated plants. Within a period of about 20 years, both the biological balance and economy of the island were seriously affected by the introduction of a pest and a well-meaning but misdirected attempt at biological control of the pest.

Australia has been particularly unfortunate in having its biological balances upset by the introduction of foreign plants and animals. In 1840 a Dr. Carlyle took some prickly pear cactus plants to Australia and planted them in his yard. Since the environment was favorable and since none of the natural enemies of the cactus had been brought with it, the cactus plants flourished and spread rapidly. By 1916 cactus had overrun some 23 million acres of land, making them unsuitable for cultivation or grazing, and was spreading at the rate of over a million acres a year. Biologists have now been able to bring the cactus under control by introducing an insect that in its larval stage lives in and destroys the cactus (Fig. 18.8). Here, biological control has been effective, as is

Fig. 18.8 *(Top) One small part of the large area of Australia occupied by the introduced prickly pear cactus. (Bottom) This area was once as thickly covered by the cactus as the one above, but the plants were killed by the parasitic cactus moth, which was introduced as a means of biological control.*

biological control in general if it is well planned in advance.

Watercress, introduced into Australia from England, has also become a pest and is so dense in even the larger streams that it hinders navigation. In England, watercress flourishes in only the smaller streams, as it is kept under control by its natural enemies. Still another introduced pest in Australia is a species of rabbit that, despite extensive hunting and long fences designed to restrict its extension, has spread over the country in immense numbers, causing extensive damage to crops and native plants. Biological control by the introduction of a parasite causing a serious disease of the rabbits finally seemed to be solving the problem, but the few rabbits immune to the disease are now repopulating the country.

Many of our serious plant and animal pests, including weeds, plant diseases, insects, and undesirable birds such as starlings and English sparrows are introductions from other countries where the natural biological controls keep them from being serious pests. Efforts at biological control are not always successful in their new environment. In Japan, one of the most effective enemies of the Japanese beetle is a wasp, but efforts to introduce the wasp into the northeastern United States to control the beetle have failed since the wasp cannot survive our cold winters as it can the relatively mild Japanese winters. The Japanese beetle is particularly destructive since it attacks so many species of plants; the adults eat leaves and fruits and the larvae attack the roots of plants.

In the southern states, and particularly Florida, a very attractive floating water plant, water hyacinth (*Eichhornia crassipes*), with large hyacinthlike inflorescences was introduced from the tropics for cultivation. It escaped from cultivation and began spreading from one stream, pond, or lake to another. It thrived and multiplied rapidly and has become a serious pest, forming thick mats (Fig. 18.9) that usually cover all or much of the water surface and completely disrupt the aquatic ecosystem. Furthermore, the mats of water hyacinth make the bodies of water unusable for navigation and water sports. Mechanical removal, the use of herbicides, and other attempted control measures have all been ineffective.

Fig. 18.9 *Dense mats of water hyacinth in a Florida lake. Ponds, small lakes, and streams are often entirely covered by the plants.*

Another serious disruption of a Florida ecosystem involves the thick coastal stands of red mangrove trees. It has been investigated so recently that it is not yet known whether the problem arises from the addition or deletion of one or more members of the community or from a change in the physical environment. In the affected areas the prop roots of the mangroves are bored into by numerous small (1 centimeter long) crustaceans (*Sphaeroma*). They hollow out the roots, which then die and decay. The root destruction occurs up to the maximum intertidal level at which *Sphaeroma* can survive and work. With the loss of the basal portion of the prop roots (Fig. 18.10) the peat into which they had been growing begins eroding, the shoreline recedes, and the mangroves topple over. The process then begins again farther back from the shore. The recession of the shoreline is in contrast with the extension of

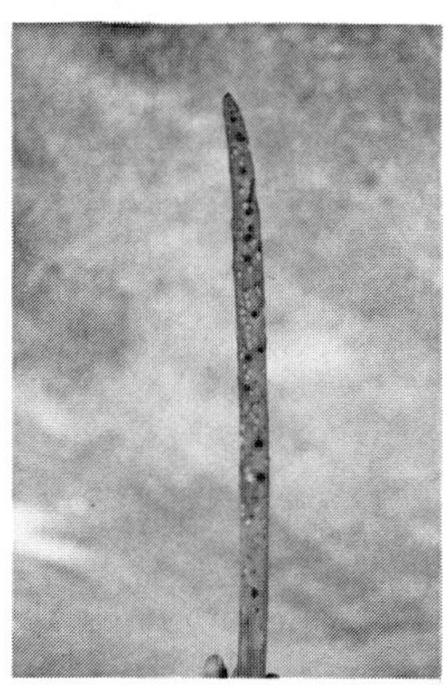

Fig. 18.10 (Top left) A mangrove tree with its adventitious prop roots intact. (Top right) Destruction of the roots of mangroves by the parasitic crustacean, Sphaeroma. *(Bottom right) A root showing the holes drilled by the parasites as they entered the root.*

the shoreline when it is occupied by uninfected mangroves. The *Sphaeroma* infestation is extensive along the southwestern coast of the Florida peninsula and the numerous offshore islands in the region. The full extent of the damage to the ecosystem is still not known, but it is likely that extensive disruption of the Everglades will occur as the coastline recedes toward that region.

The problems resulting from disturbed biological balances are undoubtedly among the most important of all practical biological problems, and are being given much consideration by ecologists as well as by conservationists and agricultural scientists. Our fundamental knowledge of the complex interrelations existing in biological communities is gradually increasing, but it is still far from complete, particularly in regard to specific situations. Unfortunately, much of the knowledge we do have is not applied and man is still disturbing biological balances with serious results. Further education of the public, and in particular of governmental officials, is badly needed.

PLANT SUCCESSION

If a natural plant community, biologically balanced and so rather stable, is disturbed, either by man or by major natural changes in the physical environment, many years are usually required before a new balance is achieved. The plants that first become established in the disturbed region are replaced by other species of plants, these by still others, and so on until a stable community has developed. Each kind of plant changes the environmental conditions in such a way that they are more favorable for the survival and reproduction of other species than for themselves. Such a series of changes in the species of plants composing the community is known as **succession.** Each group of species making up the community at any stage in the succession is known as an **association,** and the final stable association is known as the **climax association.** Such a climax association will maintain itself until it is disturbed by major changes in the environment, or by man. Whether the climax association is a certain kind of forest, grassland, or desert depends on the nature of the physical environment in the region, whereas the specific kinds of plants making up the association depend also on whether or not they are present in nearby regions and so can migrate into the disturbed region.

Old Field Succession

What happens when a field is withdrawn from cultivation in the Piedmont section of a southeastern state such as North Carolina provides a good example of a rather rapid succession following disturbance of the natural vegetation by man. The climax association in a certain region may be an oak-hickory forest, but when cultivation of the field is discontinued the vegetation does not immediately revert to an oak-hickory forest, even though there are many oak and hickory trees nearby to provide seeds. The environmental conditions in the abandoned field are not favorable to the survival of oak and hickory seedlings. However, seedlings of crabgrass flourish even though the soil is poor and the field is hot and dry in summer and subject to early freezes in the fall. Within a year or so, seedlings of tall weedy plants such as ragweed, aster, and goldenrod become established and attain dominance within two or three years. These are gradually succeeded by a dominant stand of broomsedge over a period of three or four years (Fig. 18.11).

About six to eight years after abandonment of the field, conditions have become favorable for the survival of pine seedlings and they will be found growing sparsely among the broomsedge. As the pine seedlings grow into trees they shade the low plants, which are able to carry on an adequate rate of photosynthesis only in bright light. The pines otherwise alter the environment so as to make it unfavorable for the broomsedge. The result is that these pioneer plants gradually die out. Within 25 years the pine forest is quite tall and dense, and it continues to flourish for the next 50 years or so, more or less unaltered except for the increasing size of the trees. The pine seedlings cannot survive in the dense shade produced by the parent trees, and so the pine association is not able to reproduce itself. However, the shady and relatively moist forest provides favorable conditions for the growth of the seedlings of oak, hickory and other hardwood trees (Fig. 18.11). Within 75 years or so after abandonment, these young hardwood trees make up a conspicuous part of the community, and as the pine trees die, the hardwood trees take their places in the forest. The result is that about 100 years after abandonment there is a mixed pine-hardwood forest, and after 200 years or so all the pines have died and the climax oak-hickory association has succeeded the pines. Since the seedlings of these hard-

Fig. 18.11 Old field succession in the Piedmont of the Southeastern United States. (a) *Crabgrass* (Digitaria) *pioneering in an abandoned corn field.* (b) *Tall weeds (mostly* Aster) *have replaced the crabgrass.* (c) *Broomsedge grass* (Andropogon) *has replaced the tall weeds.* (d) *Young loblolly pines* (Pinus taeda) *are becoming established in a field of broomsedge.* (e) *The broomsedge has been unable to survive under the well-established pine trees.* (f) *The climax oak-hickory forest has become established, and few pines remain.*

wood trees can survive in the forest, the association can reproduce itself and will continue to occupy the region until there is some other major natural or manmade disturbance. Although the oak-hickory association is made up of many different kinds of hardwood trees and shrubs, it is named for the two dominant kinds of trees in the association.

In some areas, the natural succession may be stopped at the stage just preceding the climax and held at that stage for long periods of time by the recurrence of major disturbances. Recurring fires often have this effect, as in the southeastern coastal plain where some forests of relatively fire-resistant pines are maintained as a **subclimax** by the constant destruction of the young hardwoods.

Fig. 18.12 This swampy marsh in New Hampshire was formerly the site of a lake. The recently established trees were destroyed by fire, but seeds from the surrounding forest trees will repopulate the trees.

Bare Rock Succession

Most natural successions require several thousand years instead of several hundred years to reach a climax. One such succession in the Eastern deciduous forest formation begins on bare rock, on which little else but lichens can survive. These are followed by mosses, which continue the process of rock disintegration begun by the lichens, and by herbs that can grow in the cracks in the rocks. As the rocks weather into soil, other kinds of herbs and shrubs can become established, and these are succeeded by pines and finally by an oak-hickory forest. In some places this may be the climax association, whereas in others it is succeeded by a climax beech-maple forest association.

Pond Succession

Another type of natural succession begins with a pond or small lake in which aquatic plants are growing. As eroded soil and the remains of dead plants accumulate on the lake bottom and make it shallower, plants such as cattails and rushes become established and these further promote the filling-in process until there remains only a swamp occupied by a sedge association (Fig. 18.12). The filling-in process continues until conditions become favorable for trees such as willows and alders, and these in turn are succeeded by an elm-ash-soft maple association. Eventually, as the soil becomes more abundant, swampy conditions entirely disappear and so the environment has become favorable for the establishment of the climax beech-maple forest association. In some regions, the same sort of pond or lake succession may lead to a conifer association as the climax vegetation.

Climatic Changes and Succession

The successions we have described, leading to the establishment of climax formations, occur under the fairly stable climatic conditions characteristic of regional or zonal climates. The character of the vegetational climax is, of course, controlled by the climatic tolerances of the native species of plants. Thus, forest, grassland, desert, or tundra are really expressions of species-tolerance of prevailing climate. Major climatic changes do occur, however, and those changes may result in the replacement of one major plant formation by another major formation. For example, in prehistoric times, dense forests covered much of the region now occupied by desert in southwestern North America. About 3000 years ago, the region including Ohio, Indiana, and Illinois became much drier that it had been, and prairie grasslands from farther west replaced the deciduous forests. Later, the climate again became moister and forests once more replaced the grasslands, except in small, isolated patches where local conditions made it impossible for the trees to become established. Fossil palm trees have been found in arctic regions, indicating that in one of the interglacial periods the climate there was warm enough for these tropical trees. At the present time, we are apparently still recovering from the last glacial period, and within the memory of living men the northern evergreen forest has made definite advances into the regions formerly occupied only by tundra vegetation. Indeed, it may be that the current advance of desert conditions southward from the Sahara in central Africa, with the attendant severe effects upon vegetation and upon human occupancy, is related to a major climatic change in progress.

Energy Aspects of Succession

As noted previously, the series of developing communities in a succession differs from a climax community in that there is an excess of photosynthesis over the total respiration of the organisms making up the community, and so an increase in biomass of the community. For example, in an old field succession, it is evident that there is a marked increase in biomass as succession proceeds from weedy annuals to grasses and then from a young to an older pine forest. The increase in biomass through these earlier stages of succession is quite rapid, but the increase becomes less as the climax community is approached. When the climax oak-hickory forest finally develops, total respiration in the community becomes essentially equal to the photosynthetic production, and so there is no further increase in the biomass of the community.

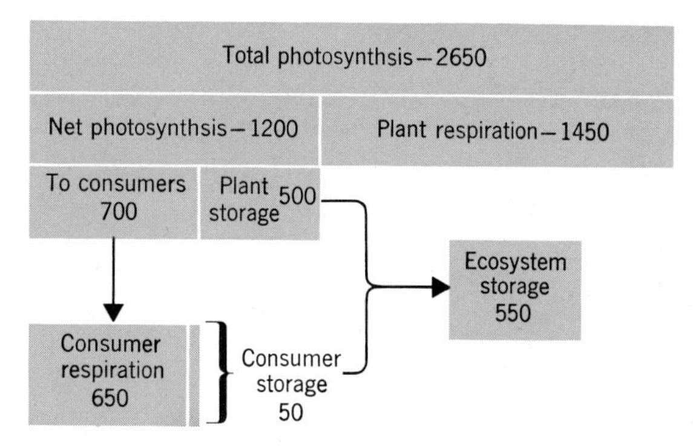

Fig. 18.13 Increase in the biomass of a late successional stage deciduous forest (g/m²) because of the excess of photosynthesis over respiration in the community. Contrast this with the balance of photosynthesis and respiration in a climax community (Fig. 18.4). The consumer respiration includes that of animals, decomposers, and other microorganisms. (From the data of Woodwell)

Most of the increase in biomass during succession results from an increase in plant tissues (Fig. 18.13), although there is also some increase in the biomass of the consumers. The data in Fig. 18.13 were obtained experimentally in a late stage successional forest on the property of the Brookhaven National Laboratory. The figures are given as grams per square meter (g/m²), but could also have been given in kilocalories as in Fig. 18.4. Total respiration was about 79% of photosynthetic productivity, the remaining 21% being accumulated as additional biomass (ecosystem storage). This is characteristic of late successional stages. In the earlier stages of succession, biomass accumulation constitutes a substantially greater percentage of the photosynthetic productivity. As succession approaches climax the P/R ratio becomes lower.

RELATED READING

Bolin, B. "The carbon cycle," *Scientific American, 223*(3), 124–135, September 1970.

Bormann, F. H. and G. E. Lukens. "The nutrient cycles of an ecosystem," *Scientific American, 223*(4), 92–101, October 1970.

Cloud, P. and A. Gibor. "The oxygen cycle," *Scientific American, 223*(3), 110–123, 1970.

Deevey, E. S., Jr. "Mineral cycles," *Scientific American, 223*(3), 148–158, September 1970.

Dolan, R., P. J. Godfrey, and W. E. Odum. "Man's impact on the barrier islands of North Carolina," *American Scientist, 61,* 152–162, 1973.

Gates, D. M. "The flow of energy in the biosphere," *Scientific American, 225*(3), 88–103, September 1971.

Hitchcock, S. W. "Fragile nurseries of the sea: can we save our salt marshes?" *National Geographic, 141,* 729–765, 1972.

Hutchinson, G. E. "Eutrophication," *American Scientist, 61,* 265–279, 1973.

Hutchinson, G. E. "The biosphere," *Scientific American, 223*(3), 44–53, September 1970.

Intersoll, J. M. "The Australian rabbit," *American Scientist, 52,* 265–273, 1964.

Isaacs, J. D. "The nature of oceanic life," *Scientific American, 221*(3), 146–162, September 1969.

Kellogg, W. W. et al. "The sulfur cycle," *Science, 175,* 587–596, 1972.

Miller, J. N. "The nibbling away of the west," *Reader's Digest, 101*(6), 107–111, December 1972.

Milne, L. J. and M. Milne. "The eelgrass catastrophe," *Scientific American, 184*(1), 52–55, January 1951.

Penman, H. L. "The water cycle," *Scientific American, 223*(3), 98–108, September 1970.

Rehm, A. and H. J. Humm. "*Sphaeroma terebans* a threat to the mangroves of southwestern Florida," *Science, 182,* 173–174, 1973.

Starbird, E. A. "A river restored: Oregon's Willamette," *National Geographic, 141,* 816–835, June 1972.

Woodwell, G. M. "Toxic substances and ecological cycles," *Scientific American, 216*(3), 24–31, March 1967.

Woodwell, G. M. "The energy cycle of the biosphere," *Scientific American, 223*(3), 67–74, September 1970.

19 MAJOR PLANT COMMUNITIES

At first glance it may appear that the complete organism represents the highest and final level of biological organization, but organization extends beyond the individual to **biological communities.** Like cells, tissues, organs, and organisms, the communities of plants and animals that occupy most areas of Earth have a specific and characteristic structural organization. Communities also engage in characteristic activities, and although these activities are the product of the activities of the individual members of the community, they involve a new level of interaction and interrelation beyond that of the individual organism. Communities even pass through stages of development from what might be considered birth through youth and on to maturity in a definite and predictable sequence. Some biologists have considered communities as being superorganisms, whereas others feel that community organization is hardly compact and consistent enough to warrant such a label. There is no doubt, though, that biological organization does extend beyond the individual to the community.

Certain social animals such as some bees, ants, termites, and man are organized into compact societies with considerable specialization and division of labor among the individuals in the

Table 19.1 **Major Communities (Biomes) of the World, with Dominant Types of Plants and Photosynthetic Productivity** [a]

Biome	Dominants	Area (10^6 km²)	Photosynthetic Productivity (g/m²/year)	Annual World Net Photosynthesis (10^9 metric tons)
Aquatic				
Oceans	Floating algae	359	2–600	51
Estuaries, shores	Algae, grasses	2	500–4000	4
Lakes, streams	Algae, angiosperms	2	100–1500	1
Swamps, marshes, bogs	Algae, sedges, rushes, sphagnum moss, etc.	2	800–4000	4
Terrestrial				
Forests				
Tropical forests	Broadleaved trees, evergreen and deciduous	20	1000–5000	40
Temperate forests	Evergreen or deciduous trees	18	600–2500	23
Boreal forests (taiga)	Evergreen trees (conifers)	12	400–2000	10
Scrubland (reduced forests)	Shrubs, small trees, grasses	7	200–1200	4
Grasslands				
Tropical savanna	Grasses, scattered trees	15	200–2000	11
Temperate grasslands	Grasses	9	150–1500	5
Deserts	Succulents, shrubs, sagebrush, herbs	18	10–250	2
Tundra	Dwarf shrubs, grasses, sedges, lichens, moss	8	10–400	1
Agricultural lands		14	100–4000	9
Totals		510		164

[a] Data from R. H. Whittaker, *Communities and Ecosystems,* Toronto: Macmillan, 1970. Modified and used by permission.

community. We are not concerned with such community organization here, particularly since it is limited to certain species of animals, but rather with the larger and more comprehensive types of biological communities of which practically all organisms are members. Each of these communities is composed of a considerable variety of plant and animal species tied together by interactions among them and the factors of their physical environment. Although every natural community of these broader types is composed of both plants and animals, the plants are always the basic structural members of a community and are responsible for its general character, appearance, and usually even its name. The reason for this is that photosynthetic plants always provide the basic food supply of any community and constitute the bulk of living tissue in it (Table 19.1). We shall first consider the general nature of plant communities and then turn our attention to the major plant communities, giving the characteristic dominants, aspect, and major geographic locations.

TYPES OF PLANT COMMUNITIES

Almost everyone is aware of the fact that the vegetation of the Earth is not uniform. In one place it consists of forests, whereas in others, there are grasslands, desert, or tundra. Still other types occur in bodies of water. These major types of vegetation generally cover large areas and are referred to as **biomes** or **formations** (Fig. 19.1). The principal plant formations are described in the following pages.

It is not mere chance that forests, grasslands, or deserts are found at certain places. The factors of the physical environment determine what types of plants can survive in a certain region and so determine the character of the community. Availability of water is of particular importance. Only desert communities composed of plants that can thrive under conditions of low moisture **(xerophytes)** can exist in extremely dry regions. Grasslands occupy regions where water availability is somewhat greater, whereas forest formations occur in still moister regions. The plants of forest communities and some grassland plants are **mesophytes,** in contrast with the xerophytes of deserts and the drier grasslands, and the **hydrophytes** that grow partially or completely submerged in bodies of fresh water. Although temperature influences the occurrence of the plant formations we have mentioned largely through its effects on water availability, low temperature is a primary factor in determining the location of tundra formations. Temperature may, of course, determine what particular

Fig. 19.1 Map showing generalized world distribution of major plant formations. Unmarked forest regions are chiefly of dry scrub or broadleaf evergreen. Based on Aitoff's Equal Area Projection, condensed. (Adapted from Polunin, Introduction to Plant Geography, *McGraw-Hill.)*

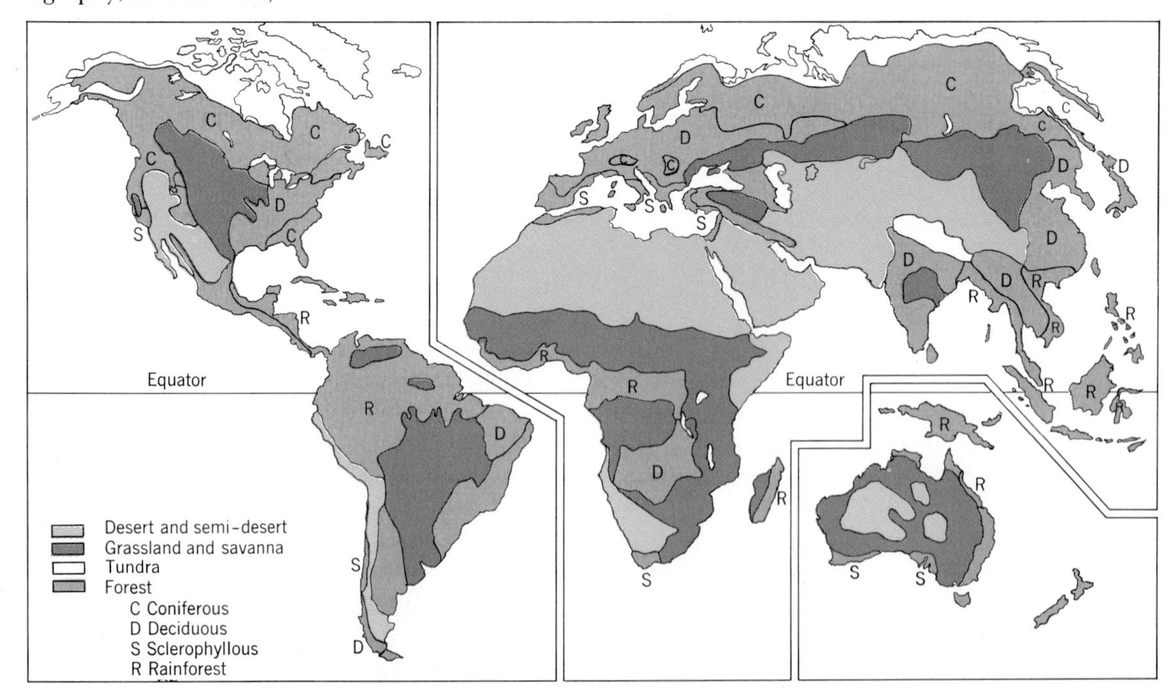

species make up a formation. Thus, trees that are not resistant to freezing cannot be a component of a temperate zone forest formation, whereas trees that grow well only in a cool climate are not found in tropical forests. Some species require low temperatures to break the dormancy of flower buds, or even leaf buds. Also, the seeds of some species require cold for breaking dormancy.

Each plant formation is composed of several different subdivisions known as **associations.** For example, in the Eastern Deciduous Forest formation there are several associations such as the swamp forest association, the oak-hickory association and the beech-maple association. Associations are commonly named for their **dominant** species, that is, those larger plants that are the most abundant in the community. Local factors such as topography, differences in water availability, soil structure, fertility, and acidity, or perhaps even just the stage of successional development are important in determining which association of a formation will occur in a certain locality, even though these factors have little influence on the extent of the formation as a whole. Associations are still further subdivided into smaller, more compact and more homogeneous **communities,** but we shall not be concerned with these smaller units of community organization here.

Although the physical environment determines the general character of the formations and associations that occupy any particular area of the Earth, the more detailed structural organization and the dynamics of any community are also influenced to a considerable degree by the interactions among the members of the community. Thus the trees of a forest produce shade that influences light intensity, temperature, and humidity and so they restrict the species of plants that can grow under them. The species of animals present in a community depend on the food available, that is, on the kinds of plants present, and despite their mobility, animals are just as restricted to particular formations or even associations as are plants. Once a community has reached maturity there is a remarkable stability in the proportion of individuals of each species in it year after year, unless there is a marked disturbance of the community by man or major changes in the physical environment. This balanced situation results from the checking of the great reproductive potential of any particular species by other members of the community that use it as food, parasitize it, or produce conditions unfavorable to the survival of individual members of the species.

The subdivision of biology dealing with communities, with the influence of the physical environment on the communities, and with the interactions between the members of a community is called **ecology.** A study of the distribution of plant communities over the face of the Earth is a special part of ecology sometimes referred to as **plant geography.** In earlier chapters we have considered some of the influences of the physical and biological environment on plants. The remainder of this chapter is devoted to plant geography.

TERRESTRIAL

Forests

Tree species are the dominant plants in the forest formations, and these are frequently associated with one or more shade-tolerant understories consisting of small trees, shrubs, herbs, and, rarely, grasses. Forests occur generally in regions of relatively high water availability. A few of the several important types of forest are described.

The **Tropical Rainforest** occurs in an area of high temperatures and abundant rainfall. The forest is dense and consists of tall, broadleaf evergreen trees of large diameter, representing many species. In the shade on the forest floor a rank understory of shade-tolerant ferns and herbs is found, along with some moderately high shrubs and trees. In some areas, the shade of the tall trees is so dense as to exclude understory associates. Many woody climbers ascend the tall tree trunks and display their foliage in the crowns of the trees. In reclaimed clearings and along stream banks where light is more abundant the classical tangle of the jungle prevails. The principal ranges of the rainforest are the drainage basins of the Amazon and Congo, the East Indies, the Philippines and coastal southeastern Asia, and Central America (Figs. 19.1 and 19.2).

The **Coniferous Forest** is indigenous chiefly to regions with cool temperate climate, as in the middle and higher latitudes of the temperate zones and higher elevations in the tropics. The trees are largely needleleaf gymnosperms, species of spruce, fir, hemlock, pine, and larch. Locally, in the northern regions, an admixture of some birch and poplar is common, but understory associates are few except in areas having considerable rainfall (Fig. 19.3). In North America the coniferous forest characterizes a broad area in the southeast, most of the region west of the plains except the very arid southwest, and much of Canada. In Eurasia, it extends from the Alps northward to Scandinavia and eastward to the Pacific Ocean (Fig. 19.1).

The **Deciduous Hardwood Forest** occurs chiefly in temperate regions with marked seasonal cycles. Typically the trees lose their leaves in a dry or in a cold season. The forest contains many species of broad-leaf trees in relatively open stands, with many understory associates, including small trees, shrubs, annuals, and few grasses. Oaks, hickories, tulip trees, gum, beeches, maples, and ash are common dominants (Fig. 19.4). Dogwood, sourwood, buckeye, poison ivy, huckleberry, and weedy annuals frequently compose the understory. Extensive areas in east-central United States, western Europe, and southern European Russia support typical deciduous forests, locally and regionally diversified as distinctive associations(Fig. 19.1).

Fig. 19.2 Rain forest along shore of Amazon River, Brazil. Note mixture of broad-leaf species and presence of woody lianas (climbers) which display their foliage in crowns of supporting trees. Dense undergrowth and sparse distribution of specimens yielding valuable cabinet woods make harvesting difficult and costly.

The **Sclerophyll Forest** is a distinctive type made up chiefly of small trees and coarse shrubs with broad, leathery, evergreen leaves in open stands. Coarse, short grasses sparsely cover the soil between the dominants. Occurring in semiarid warm regions, the sclerophyll forest is characteristic of many regions in the lands surrounding the Mediterranean Sea, coastal south-western United States, and areas in Chile, South Africa, and southern Australia (Fig. 19.1).

The **Savanna** is a forest type of a distinctive parklike aspect, with open or scattered stands of small broad-leaf evergreen or deciduous trees and with the space between occupied by coarse grasses. It is essentially a transitional type between forest and grassland, or forest and desert formations. Typically it is characteristic of high-temperature, low-rainfall climates. Broad east-west zones on the northern and southern flanks of the tropical rainforests in South America. Africa, much of India, and northern Australia are covered by this type of forest (Fig. 19.1).

Grasslands

Grasses are the dominant and often the only growth form in grassland. Small trees are rare and limited to the land immediately ad-

Fig. 19.3 Coniferous forests. (Top) Redwood forest in coastal northern California. Note the dense understory of ferns and shrubs. (Bottom) Forest of Engelmann spruce and Lodgepole pine in Colorado. This forest is practically free of understory.

Fig. 19.4 Oak-hickory association of eastern Deciduous Hardwood Forest, in early spring condition.

jacent to water courses (Fig. 19.5). Typically, grassland is a formation characteristic of regions of relatively low water supply. In its very dry expressions it consists chiefly of low-growing, shallow-rooted grasses in open formation, as in the plains grassland in western United States, Canada, and southern Asiatic Russia (Fig. 19.5). In cool areas, short-lived annuals and cactus or cactuslike plants may be present. In its moister expressions, as in the prairies of east-central United States and large areas of Asia and Africa, the grasses are taller, are sod-forming species, and are commonly associated with many annuals.

Deserts

Extreme aridity and commonly high daytime temperatures characterize the desert. Vegetation may be sparse, consisting of small-leaf deciduous or leafless shrubs of spiny habit, such as greasewood, or perennials with succulent stems or leaves, such as cacti or yuccas. The intervening space may be bare soil, or, in areas receiving moderate rainfall in an occasional year, short-lived annuals and sparse bunch grass may occur (Fig. 19.6). Large areas of western United States and adjacent parts of Mexico, western South America, the Sahara and Arabian deserts, and large areas of central Asia and Australia support this formation. Some extreme desert areas are, of course, destitute of vegetation (Fig. 19.1).

Tundra

The tundra formation occurs in regions characterized by freezing temperatures during all but a few weeks of the year. During a very brief growing season the soil thaws to a depth of several centimeters and the sparse vegetation, consisting of small annuals, hardy grasses, dwarf willows, depressed junipers, mosses, and lichens, complete their vegetative and reproductive activities. Tundra vegetation is characteristic of northern Canada, Alaska, northern Europe, and Asia, extending from the upper limits of tree growth to the region of permanent ice (Figs. 19.1, 19.7).

Of course, the limits of such plant assemblages are not precisely definable, for the formations blend into one another where they meet, in zones of transition, just as the environmental factors which they depend on vary gradually over a distance. As critical environmental conditions change at the margins of a formation, so will the range of the formation expand or retract.

Fig. 19.5 Grasslands. (Top) A northern prairie near Vermillion, Alberta, with groves of aspen along water courses and in moist depressions. (Bottom) Shortgrass plains of eastern Montana.

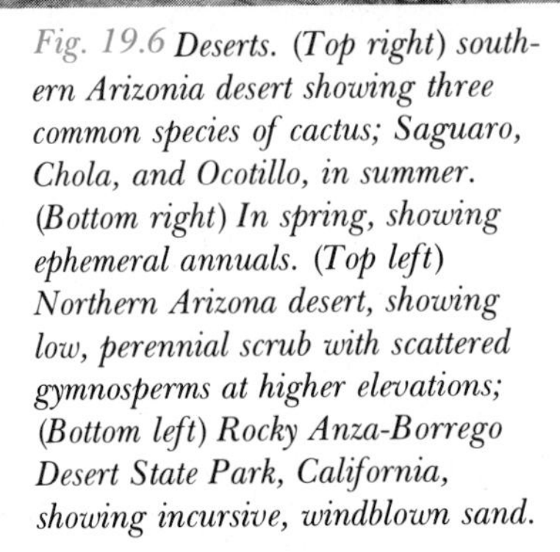

Fig. 19.6 Deserts. *(Top right) southern Arizonia desert showing three common species of cactus; Saguaro, Chola, and Ocotillo, in summer. (Bottom right) In spring, showing ephemeral annuals. (Top left) Northern Arizona desert, showing low, perennial scrub with scattered gymnosperms at higher elevations; (Bottom left) Rocky Anza-Borrego Desert State Park, California, showing incursive, windblown sand.*

Fig. 19.7 Alaskan tundra, well north of the northern limit of tree growth. Vegetation consists mostly of hardy grasses with a few small annuals in the spring.

AQUATIC BIOMES

Oceans

The oceans occupy about 62% of the surface area of the Earth and account for over 30% of its annual photosynthesis, but productivity per unit area is low and ranges from only 2 g/m²/year to 600 g/m²/year. The deep central areas of the ocean have low populations of plants and animals and thus are the areas of lowest productivity, but because of their extent, they contribute substantially to the total photosynthetic product of the oceans. The producer organisms of the oceans are primarily microscopic algae of various kinds that float in the surface layers of the water. The larger algae, including the brown, red, and some green algae, are found principally on or near the shores and do not contribute significantly to the food supply of the open oceans.

The shores and estuaries of the oceans are occupied by large populations of organisms that are not found either in the open oceans or on land and make up a variety of distinct biomes that may differ in species composition but have the same general characteristics (Fig. 19.8). The shore includes the intertidal zone and also the shallow water near the shore. The producer organisms of shores and estuaries include both phytoplankton (small floating algae) and algae attached to sand, mud, seashells, or rocks. Rocky coastlines are usually well populated by brown and red algae and some of the larger green algae, particularly in the intertidal zone, and these algae are also frequently found floating in the water near the shore (Fig. 19.9).

Estuaries are river mouths, bays, or sounds subject to tidal fluctuation with water that is intermediate in salinity between seawater and freshwater. In addition to algae, estuaries usually have extensive populations of grasses and other vascular plants, primarily in the intertidal zones (Fig. 19.10). These constitute salt marshes, often composed of a single species (*Spartina* sp.) of marsh grass. Some-

Fig. 19.8 *(Top left) Vegetation of coastal sand dunes. Grasses and other plants stabilize the dunes.*

Fig. 19.9 *(Right) Brown algae in the intertidal zone along a rocky coast.*

Fig. 19.10 *(Top right) A coastal salt marsh in Rhode Island.*

what higher areas may be occupied by sedge (*Juncus* sp.). These salt marsh plants and others contribute substantially toward the high productivity of the estuaries, although they serve as food for animals mainly after they have been broken by tides and partially decomposed by bacteria and fungi. The estuaries and their salt marshes provide favorable spawning areas for a variety of fish, shellfish, and crustaceans, many of great commercial value. Man has endangered the present life of the estuaries by silting and pollution of rivers, dredging, and the construction of recreational communities in unsuitable locations.

Freshwater Communities

Freshwater communities occur in lakes, ponds, streams, swamps, marshes, and bogs. Ponds and lakes generally have a littoral zone where there are rooted vascular plants and a deeper limnetic zone where the producer organisms are phytoplankton. Large lakes also have a still deeper central area where the phytoplankton is scarce and the fish and other animals secure their food primarily from the limnetic zone. Nearest the shore the littoral zone is occupied by plants with mostly aerial shoots such as cattails, arrowheads (*Sagittaria*), bur reeds, pickerelweeds, and rushes (Fig. 19.11). Somewhat farther from the shore are water lilies or similar plants with floating leaves. In still deeper water, completely submerged vascular plants are often abundant. These include pondweeds (*Potamogeton*), water milfoils, waterweed (*Elodea*), and *Vallisneria*. Green algae, diatoms, and blue-green algae are also found in the littoral zone. Sometimes filamentous green algae such as *Spirogyra* and *Oedogonium* form thick mats that float on the water. The small vascular plants known as duckweeds also float on the surface and may form dense masses that shade out submerged plants underneath them. Large lakes may have sandy beaches or rocky shores similar to those of the seashore which differ in their vegetation

Fig. 19.11 Vegetation on shore of fresh water lake.

from the mud-bottomed littoral zones we have described.

In streams, phytoplankton plays a less important role in primary production than it does in lakes and ponds, and is limited to large rivers and pools in the slowly flowing regions of smaller streams. In more rapidly flowing streams the producers are principally plants attached to rocks, logs, or other solid substrates. These include various species of filamentous green algae with holdfasts, aquatic mosses, and diatoms that may form extensive attached crusts. Of course, in a stream a substantial amount of food may be carried by the current from upstream, lakes, or adjacent land areas.

Freshwater marshes, swamps, and bogs constitute transitions between ponds and lakes and the truly terrestrial adjacent communities (Fig. 19.12). Indeed, they are commonly successional stages that occur as a lake or pond silts in and lead eventually to a forest biome. The three terms are overlapping and are to some extent used interchangeably, but in general, marshes are wet areas without trees, whereas swamps include trees and shrubs. Swamp trees include swamp red maples, alders, willows, and in some regions, tamarack or bald cypress. Bogs may also have shrubs and trees, but the most characteristic plant of bogs is sphagnum moss, which forms dense, thick mats that are often floating on water (Fig. 19.13). As the lower parts of the sphagnum plants die they sink to the bottom, forming a thicker and thicker layer of peat. Insectivorous plants such as sundews and pitcher plants are also characteristic of bogs.

Fig. 19.12 *A bald cypress,* Taxodium distichum, *swamp in Florida. Note the characteristic cypress "knees" which provide aeration for tree roots.*

Fig. 19.13 *A* Sphagnum *bog at a lake margin in Minnesota. As drier land develops through deposit of peat, the conifers, tamarack and spruce, will invade the area.*

RELATED READING

Amos, W. H. "The life of a sand dune," *Scientific American, 201* (1), 91–99, July 1959.

Billings, W. D. *"Plants, Man and the Ecosystem,* 2nd ed. Belmont, Calif.: Wadsworth, 1969.

Deevey, E. S., Jr. "Bogs," *Scientific American, 199*(4), October 1958.

Edwards, W. M. "Abundant life in a desert land," *National Geographic, 144,* 424–436, 1973.

Platt, R. "Flowers in the Arctic," *Scientific American, 194*(2), 88–98, February 1956.

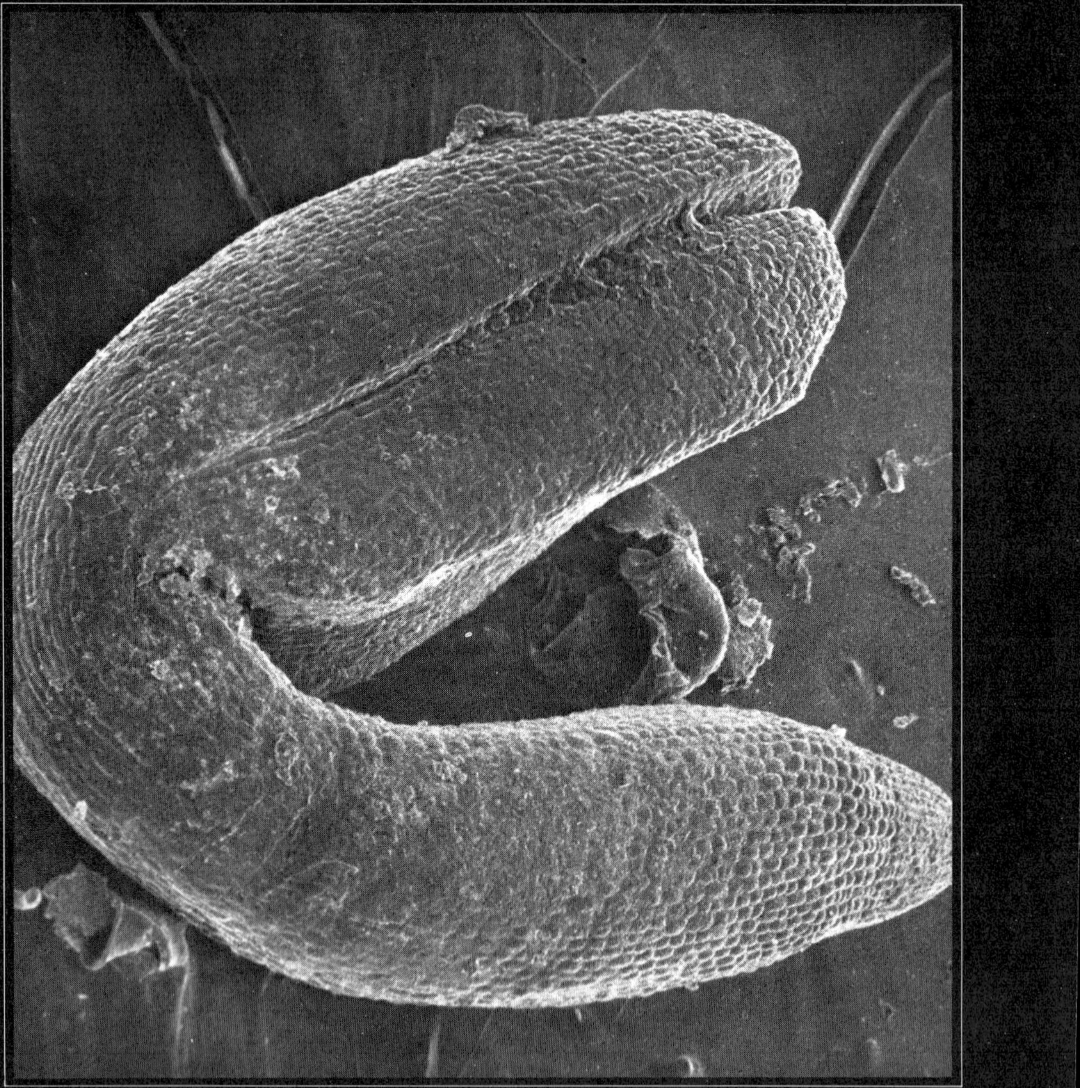

FROM GENERATION TO GENERATION

6

PHOTO OPPOSITE: OLDER EMBRYO FROM THE SEED OF SHEPHERD'S PURSE, *CAPSELLA BURSAPASTORIS*. (MAGNIFICATION 175 ×)

20 ASEXUAL REPRODUCTION

Perhaps no other single feature distinguishes living organisms from nonliving things better and more completely than the capacity of every species for reproduction. Without it life would continue for only a single generation since every living individual is faced with eventual death, whether after a life of only a few days or thousands of years. Reproduction is also the means of increasing the population of a species, and plants are generally dependent on dispersal of reproductive structures such as seeds or spores for movement from one favorable habitat to another. Reproduction plays at least one other important role: the hereditary variations associated with reproduction have played a major role in the evolution of the numerous and diverse plant and animal species inhabiting the Earth.

In this section we focus our attention on the reproductive continuity of plant life from generation to generation, the associated transmission of hereditary potentialities from parent to offspring, and the evolution through the ages of new plant species as a result of changes in these hereditary potentialities.

SEXUAL AND ASEXUAL REPRODUCTION

Basically, reproduction is detachment of a cell or group of cells from the parent and their development into a new individual. If reproduction is **sexual,** one detached cell must unite with another before development of the new individual proceeds. In **asexual** reproduction there is no fusion of cells; the detached cells, tissues, or organs develop directly into new individuals. Like most species of animals, the great majority of plant species can reproduce sexually, and we shall consider the sexual reproduction of plants in the following chapter. Unlike most animals, many plants can reproduce asexually as well as sexually, whereas many species of plants that do not have asexual reproduction in nature are propagated asexually by man. Only a few kinds of plants, including the bacteria, blue-green algae, and some fungi, are propagated entirely by asexual reproduction, and now appears that even these structurally simple organisms may possess some rudiments of sexual reproduction. Several species of vascular plants, mostly cultivated ones, are sterile and do not produce seeds even though they have flowers. Such plants, including cultivated bananas, seedless grapes, seedless citrus fruits, and most sweet potatoes are propagated vegetatively (asexually).

Sexual and asexual reproduction differ greatly in the degree of hereditary variation associated with them. Offspring reproduced sexually rarely have the same assortment of hereditary potentialities (genes) as either parent, and so they differ from the parents in many hereditary traits even though they possess the general characteristics of the species. On the other hand, plants reproduced asexually are really just a detached portion of the parent and, except for rather rare mutations, have exactly the same hereditary potentialities as the parent. The descendants of an individual produced asexually, even though they may number in the thousands, may be regarded as extensions of the single individual and are referred to as a **clone.** Thus, since apple trees are propagated vegetatively by grafting, every Winesap apple tree on Earth is really just a part of the original Winesap tree. Winesap apple trees constitute one clone, Delicious apple trees another.

ASEXUAL REPRODUCTION OF NONVASCULAR PLANTS

The algae, fungi, bacteria, and bryophytes all include species that reproduce asexually by a variety of methods. we shall consider some of the more important ones here.

Fission

Bacteria and unicellular algae and fungi, as well as protozoa in the animal kingdom, generally reproduce by a process known as **fission** (Figs. 20.1 and 20.2). Fission is the simplest possible type of reproduction, consisting merely of cell division, the two cells produced by the division separating instead of remaining attached as do dividing cells in a multicellular plant or animal. Each cell resulting from fission is a new individual, the only growth being the enlargement of the new cells to the size of the parent cell. Since the parent cell now exists

Fig. 20.1 *Electron micrograph of a bacterium* (Bacillus subtilis) *undergoing fission, 9200X.*

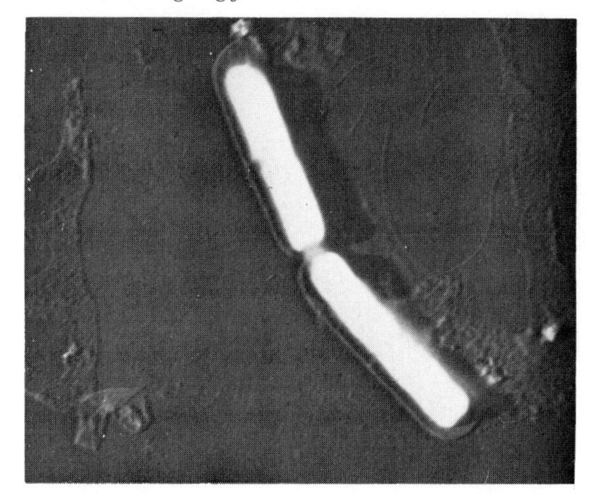

only as part of each of the offspring, all individuals produced by fission may be considered to be "orphans." On the other hand, since there is no such thing as old age or death from old age, organisms reproducing by fission may be considered to have a sort of immortality.

Reproduction by fission may be quite rapid. Under good conditions some bacteria divide every 20 minutes. If this rate were maintained for 12 hours, a single cell would have almost 68 billion descendants. However, such rapid increases in population are only theoretically possible, since many individuals die or fail to divide as food and water become exhausted and toxic waste products of metabolism accumulate. Despite this, the actual population increases are frequently very rapid.

Budding

Yeasts reproduce by a modified type of fission known as **budding.** In budding the offspring cells are smaller than the parent cell and may remain attached to the parent cell for some time (Fig. 20.2). The daughter cells may in turn produce a still smaller cell before becoming separated, thus forming a short chain of cells. Eventually all the cells separate and enlarge to the size of the parent cell. In budding, the parent cells thus retain their identity, unlike parent cells reproducing by fission.

Sporulation

True asexual **spores** are found only among the algae and fungi; the spores of the common bread mold (Fig. 20.2) are a good example. The spores of the bread mold and a good many other fungi are borne within a **sporangium;** other fungi such as some species of *Penicillium* and *Aspergillus* (Fig. 20.2) bear their spores in chains of cells that become detached as they mature. Such spores are called **conidiospores** or simply **conidia.** Aquatic algae and fungi generally produce motile spores **(zoospores)** that swim through the water by means of very slender filaments called **flagella** (Fig. 20.2). The flagella contain fibers that apparently contract in a rhythmical sequence, causing the flagellum to move in a whiplike fashion. Flagellar movement requires energy from respiration, and the energy transfer mechanism may be quite similar to that of muscle contraction. The few kinds of flagella that have been investigated all have eleven contractile fibers, and it is rather interesting that the flagella of plant sperm, protozoa, and other flagellated animal cells, as well as the tails of animal sperm, all seem to have the same number and arrangement of contractile fibers as the flagella of zoospores.

An individual plant usually produces from hundreds to millions of spores, aerial spores generally being produced in greater numbers than zoospores. Although most of the zoospores produced by a plant may germinate and grow into a new individual, only a small portion of the aerial spores are generally deposited by air currents in environments suitable for the germination and growth of the spores. Because of the immense numbers of aerial spores produced, however, they provide an effective reproductive mechanism and in view of the small quantity of protoplasm in each spore the failure of most of the spores to germinate is not particularly wasteful. Most spores are single cells, although a few species of fungi have spores composed of two or more cells.

The term *spore* has, unfortunately, been used for several different kinds of plant cells playing various roles in life cycles, including the thick-walled resistant spores that permit survival of some bacteria, algae, and fungi through periods of unfavorable environment such as extreme desiccation or high temperature, and certain cells **(zygospores)** resulting from sexual fusion. The most widespread kind of spores in the plant kingdom are the **meiospores,** which are produced by reduction divisions (Chapter 21) and which are really a stage

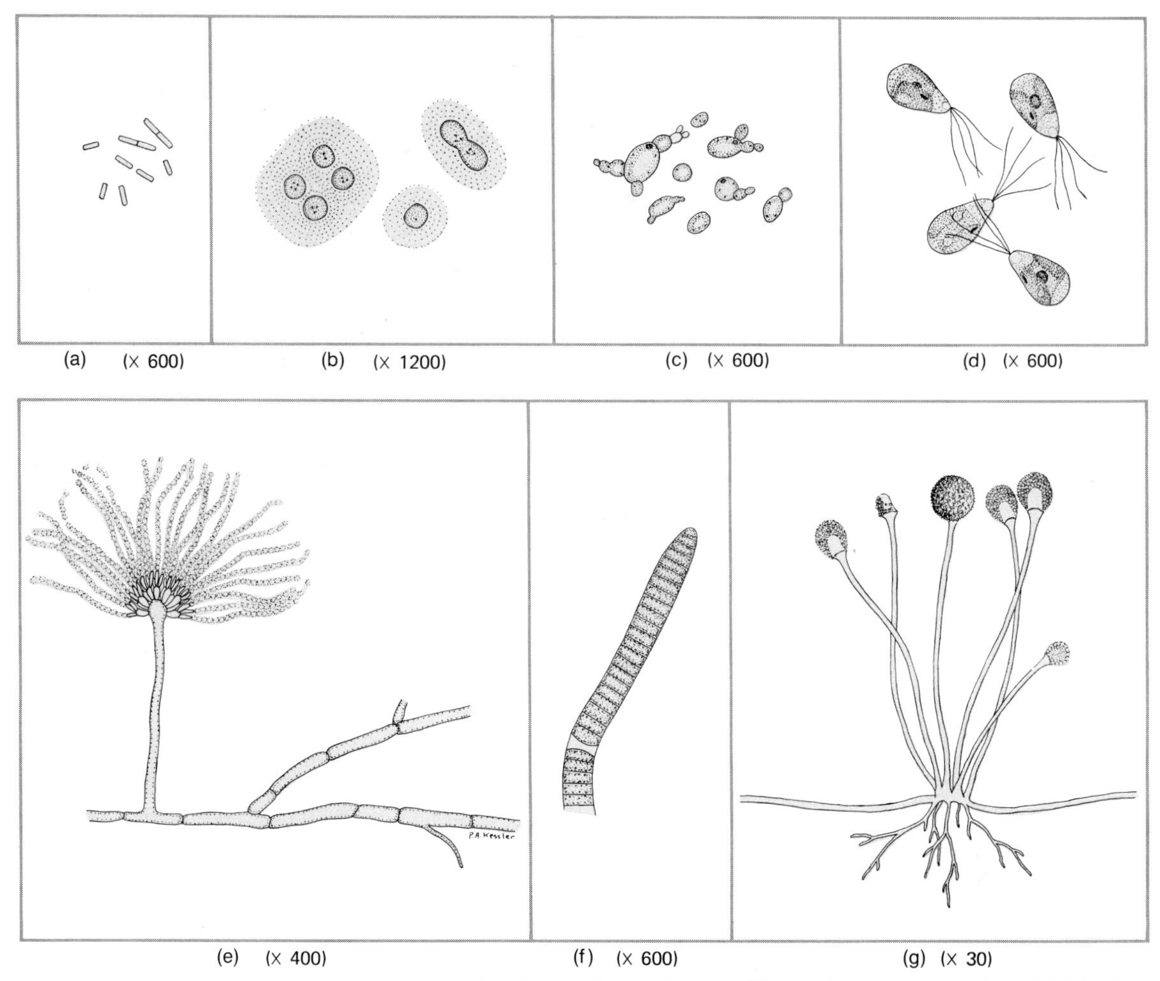

Fig. 20.2 *Examples of asexual reproduction of lower plants.* (a) *Bacteria multiplying by fission.* (b) Gleocapsa, *a unicellular blue-green alga, reproducing by fission.* (c) *Yeast cells budding.* (d) *Zoospores of* Ulothrix, *a green alga.* (e) *Conidia being produced by* Aspergillus, *a mold.* (f) *Fragmentation of* Oscillatoria, *a blue-green alga. The clear heterocyst has resulted in the formation of two filaments from the parent filament.* (g) *Sporangia and spores of* Rhizopus, *black bread mold.*

in the sexual reproduction of plants. The spores of bryophytes and of the ferns and other vascular plants are all meiospores, as are some of the spores produced by algae and fungi. None of these various types of spores are the true asexual spores that have been considered here. It is important to distinguish between the various uses of the term *spore.*

Vegetative Propagation

The term *vegetative propagation* is sometimes used more or less synonymously with *asexual reproduction;* however, we shall use it here to include those types of asexual reproduction involving the detachment of a multicellular mass from the parent plant. **Frag-**

Fig. 20.3 Marchantia (a liverwort) with gemmae cups containing gemmae. (Courtesy, Carolina Biological Supply Co.)

Fig. 20.4 Soredia of a lichen. Each soredium consists of a few algal cells and some fungal hyphae, and when detached propagates the lichen as a unit.

mentation by multicellular algae or fungi, particularly filamentous types, into two or more individuals may be considered as vegetative propagation. Although fragmentation is frequently accidental, some filamentous blue-green algae have specialized enlarged, thick-walled dead cells **(heterocysts)** at intervals along their filaments (Fig. 20.2) that are associated with fragmentation. The filaments separate at the heterocysts, thus propagating the species. Many fungi, particularly species that grow saprophytically in the soil, spread over large areas through the continued growth of their hyphae and any separation of hyphae from the parent plant results in a new individual, although one that is still a member of the clone. Cultivated mushrooms are generally propagated, not by spores, but by portions of the culture medium containing hyphae of the mushroom, referred to as **spawn.**

Several kinds of nonvascular plants have specialized structures involved in vegetative propagation. For example, liverworts may produce small budlike structures called **gemmae** within cup-shaped organs (Fig. 20.3). The gemmae eventually break loose from their stalks and may be washed by rains to locations where they may grow into new plants. Lichens produce reproductive bodies known as **soredia**

(Fig. 20.4), which consist of a small cluster of both the fungal hyphae and the algal cells that constitute the lichen. This results in the reproduction of the composite "organism" as a unit.

ASEXUAL REPRODUCTION OF VASCULAR PLANTS

All the various kinds of asexual reproduction found among the vascular plants may be considered as forms of vegetative propagation. Any of the organs of a vascular plant—stems, leaves, roots, and even flowers—may be involved in vegetative propagation. The most convenient way of classifying the types of vegetative propagation found among the vascular plants is on the basis of the organ concerned.

Propagation by Stems and Buds

Perhaps the most common type of vegetative propagation occurs in plants with horizontal stems growing either above ground **(runners** or **stolons)** or underground **rhizomes.** Plants with stolons include the strawberry (Fig. 20.5), the creeping buttercup, Kudzu vines, and St. Augustine grass. Many different species of plants have rhizomes; among them are most ferns, iris (Fig. 20.5), lilies of the valley, wild morning-glories, cattails, sedges, water hyacinth, blueberries, bananas, and numerous species of grasses. Both stolons and runners

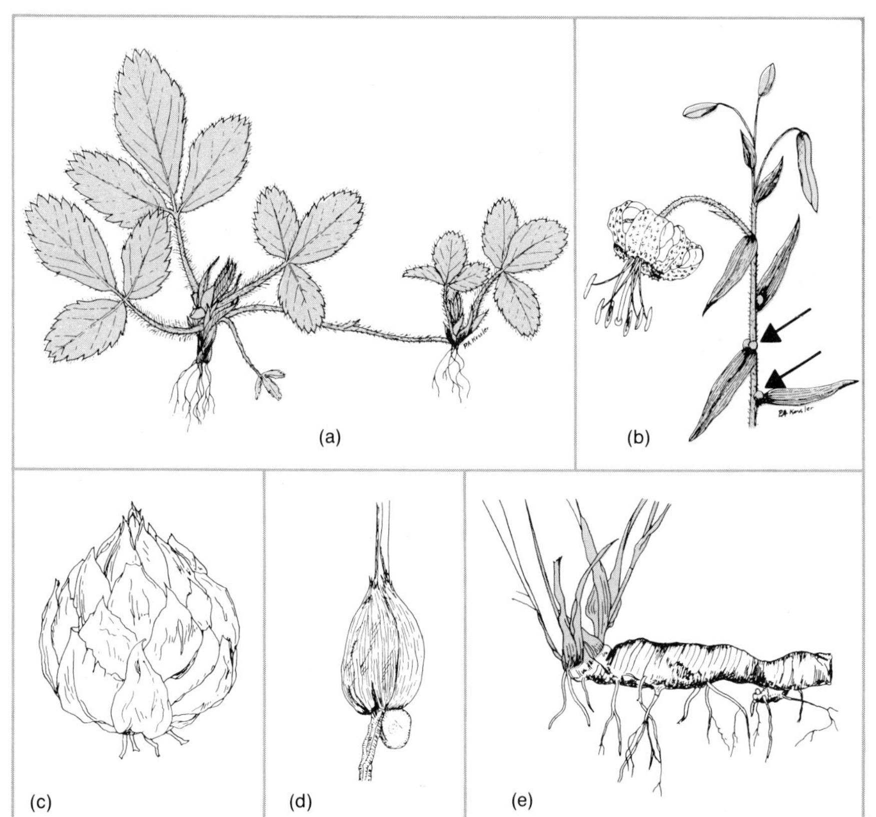

Fig. 20.5 (a) *A stolon (runner) of a strawberry plant.* (b) *Modified axillary buds (bulbils) of a tiger lily. The bulbils abscise and develop into plants when they reach the soil.* (c) *A lily bulb.* (d) *A* Freesia *corm.* (e) *An iris rhizome.*

produce aerial shoots and adventitious roots at their nodes, or sometimes at alternate nodes as in the strawberry, and so the plants spread over extensive areas. Many of the dominant grasses of our natural grasslands have rhizomes that undoubtedly contribute to the rapid spread and dominant position of these species in the community. Many of the beach grasses also have stolons or rhizomes. As long as the stolons or rhizomes remain intact, the entire colony is really still only one individual plant, even though it may cover a considerable area and include many spaced and apparently independent shoots. When, however, some of the connecting stolons or rhizomes die or are accidentally severed, we may consider that a new individual has been added to the clone and that vegetative propagation, rather than just the spread of an individual plant, has occurred.

Several plants including the Irish potato and the Jerusalem artichoke are propagated vegetatively by **tubers** (Fig. 20.6), the thickened, fleshy ends of rhizomes. Since the aerial shoots, the roots, and the rhizomes of tuber-bearing plants usually die at the end of the growing season, the isolated tubers may each give rise to a new individual the following season and so serve as a true means of vegetative propagation.

Some plants such as raspberries, dewberries, wild roses, currants, and gooseberries have long slender stems that may droop and touch the ground. When this occurs, adventitious roots develop from one of the nodes near the stem tip and the bud grows into an upright stem (Fig. 20.7). Although this is called **tip layering** it is not basically different from the development of roots and shoots at the node of a stolon. Almost impenetrable thickets of rasp-

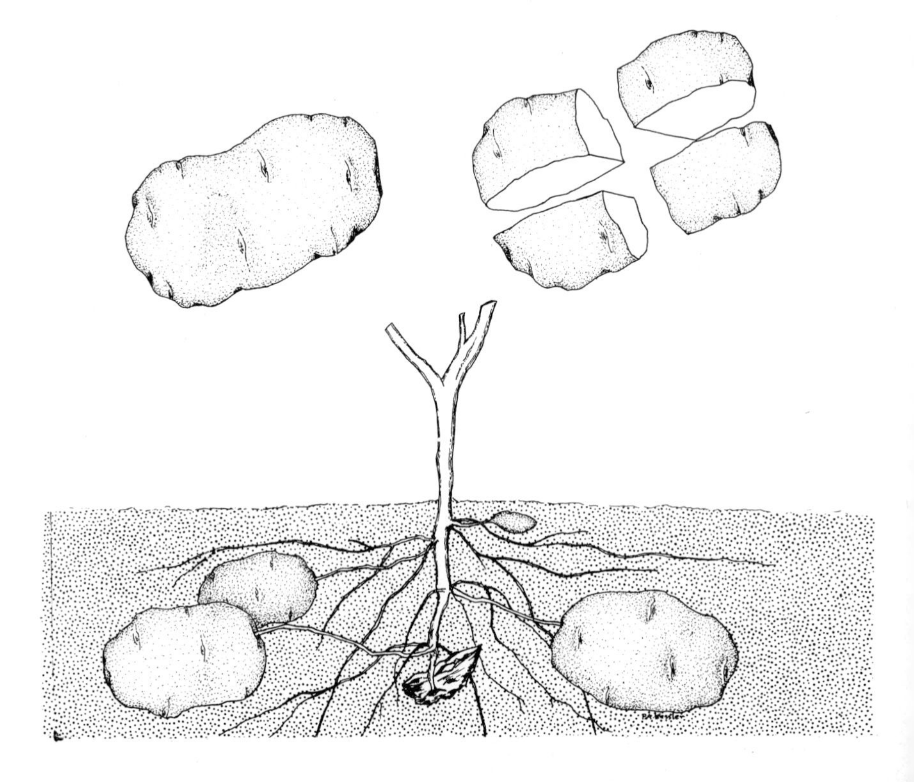

Fig. 20.6 *A portion of a potato plant showing roots, young tubers, and the piece of tuber from which the plant grew. Above: tubers used in propagating potato plants are cut into pieces, each with at least one bud ("eye"), so that more plants can be propagated from a tuber.*

Fig. 20.7 *Propagation of a raspberry plant by tip layering.*

berries or dewberries may arise from tip layering.

The submerged water plant, *Elodea,* is propagated vegetatively by the fragmentation of its fragile stems. Fragmentation of the stems of cacti is rather common, the broken-off stem segments falling to the ground and rooting. Stems may also be broken off willow trees growing along stream banks, and if the end of the stem happens to become immersed in the soft mud of the banks, roots will usually develop and the twig may grow into a tree.

Tiger lilies and some other species of lilies produce compact, spherical axillary buds superficially resembling seeds (Fig. 20.5). These buds abscise and fall to the ground, where they may develop into new plants. **Bulbs** (Fig. 20.5), such as those of onions, tulips, and lilies, are essentially large buds and provide a means of vegetative propagation. Bulbs form small branch bulbs that eventually develop to full size and grow into new individuals. Although bulbs are generally underground structures, onions may develop aerial bulbs in addition to their usual underground bulbs. **Corms,** the short, fleshy vertical underground stems of a number of species including gladiolus and crocus, differ from bulbs in that they have more stem tissue and small scale leaves rather than large fleshy leaves (Fig. 20.5). Like bulbs, corms form branches that give rise to young plants.

Propagation by Roots

The horizontal roots of various plants, including the Canada thistle, milkweed, and many other weeds as well as a number of trees and shrubs such as beech, Osage orange, sumach, and elderberry, serve as a means of spread and vegetative propagation much as do stolons and rhizomes. The horizontal roots, however, form adventitious buds whereas the stolons and rhizomes form adventitious roots. Plants may spread over considerable areas with numerous shoots growing from their horizontal roots. When most hardwood trees are cut down, numerous sprouts develop from the stump and roots even though the species may not otherwise form adventitious shoots from the roots. The fleshy, clustered roots of plants such as sweet potatoes (Fig. 20.8), dahlias, and peonies also provide a means of vegetative propagation. Sweet potato roots form numerous adventitious buds that grow into new plants. Dahlia and peony roots, however, do not normally produce adventitious buds and so are effective in propagating the plants only when a portion of the stem is attached to the roots.

Propagation by Leaves

Leaves are not as common a means of vegetative propagation in nature as are stems or even roots. However, several species of plants propagate vegetatively by means of their leaves. For example, when the tips of the long leaves of the walking fern come in contact with the ground they form adventitious buds and roots that grow into a new plant (Fig. 20.9). This may be considered as tip layering, comparable with the tip layering of stems. The fleshy leaves of the African violet, some species of begonia, and a number of other plants form adventitious buds and roots when removed from the plant and brought in contact with the soil. It is possible that, in nature, leaves accidentally broken from the plants may come in contact with the soil and give rise to plants.

Fig. 20.8 *The shoots that develop from adventitious buds of a sweet potato root when supplied with water produce adventitious roots. The individual plantlets are separated from one another and the fleshy parent root and then planted.*

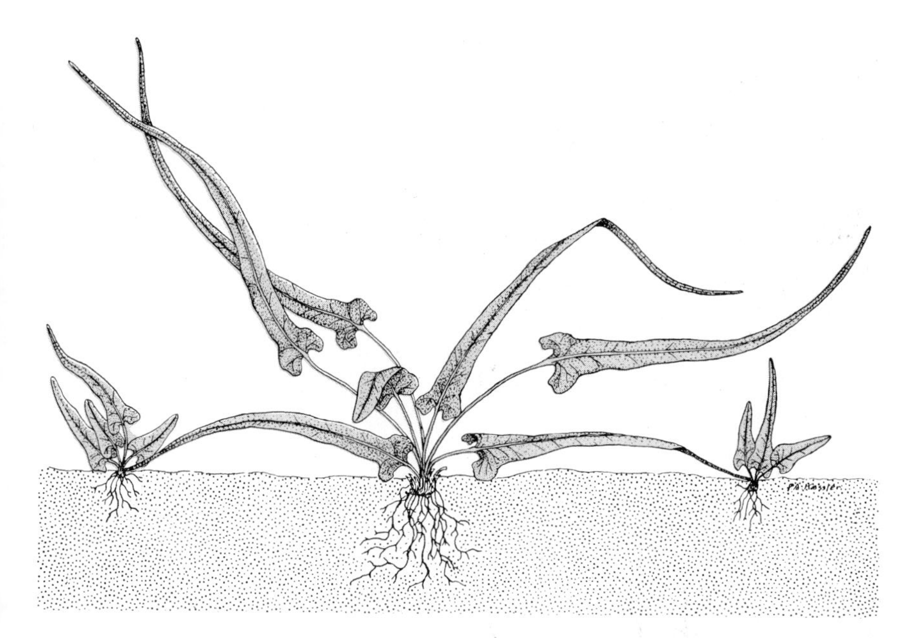

Fig. 20.9 Propagation of a walking fern by tip layering of the leaves. Extensive propagation of the plant in this way occurs in nature.

Fig. 20.10 A leaf of Bryophyllum calycinum (Kalanchoë pinnata) *with plantlets in notches of the leaf margin. The latent primordia of the adventitious buds and roots are present from the beginning, but develop only under long days. As regards flower initiation it is a short-day plant.*

The most striking examples of leaf propagation are found in the genus *Kalanchoë* (*Bryophyllum*). In some species of this genus, well-formed tiny plants develop in the notches of the fleshy leaves even while the leaves are attached to the plant (Fig. 20.10). These plantlets abscise from the leaves and fall to the ground, where they take root and continue to grow into mature plants. These plantlets develop under long days, and flowers form only under short days.

The very small floating seed plants known as duckweeds commonly propagate vegetatively by the development of new individuals from their leaves. In the autumn, small bulblets form on the leaves and sink to the bottom of the water. The following spring these bulblets rise to the surface and develop into mature plants. Duckweeds rarely produce flowers and seeds; some species practically never bloom.

Vegetative Propagation by Flowers

Although flowers are primarily associated with the sexual reproduction of plants, various tissues of flowers may occasionally be involved in vegetative propagation. The clustered aerial bulbs of onions (Fig. 20.11) actually arise from flower buds, and the flowers of some grasses may be replaced by vegetative propagules. In some species of plants, among them the dandelion, the eggs develop into embryo plants without fertilization. This is known as **parthenogenesis.** Since no fusion of sex cells is

Fig. 20.11 Aerial bulbs of onion that developed in the inflorescence in place of some of the flowers.

involved, parthenogenesis is really a form of asexual reproduction even though it is a modification of the normal pattern of sexual reproduction. At times embryo plants may arise from cells of the ovule other than an egg cell. The development of such cells into embryos is clearly a case of vegetative propagation. These embryos may occur in the seed along with an ordinary sexually produced embryo or may replace it.

Regeneration in Relation to Vegetative Propagation

In general, the vegetative propagation of plants is possible only because of the capacity of the propagating tissues or organs for forming either adventitious buds (that include, of course, stems and leaves) or adventitious roots or both. The formation of adventitious buds and roots, particularly in portions of plants lacking them, may be considered as examples of **regeneration.** Thus, when a detached twig develops adventitious roots it is regenerating the only lacking organ of an otherwise complete vegetative plant. The dependence of vegetative propagation on the regeneration of buds and roots is shown most clearly by the differing regenerative capacities of leaves of various species of plants. The leaves of a good many species of plants normally cannot form either adventitious buds or roots, whereas detached leaves of a good many other species readily form adventitious roots but no adventitious buds. Only leaves of the third type that can regenerate both buds and roots can, of course, serve as a means of vegetative propagation. Regeneration is not always associated with vegetative propagation. For example, the growth of sprouts from the stump of a tree is regeneration, but there is no propagation.

The capacity for regeneration and vegetative propagation are widespread throughout the plant kingdom, but only lower animals such as starfish, sponges, and some kinds of jellyfish and worms have great enough regenerative capacities to permit a portion of the animal to develop into a complete new individual and so undergo what is essentially "vegetative" propagation. Although some higher animals such as lobsters and salamanders can regenerate missing legs, most higher animals can regenerate nothing more than limited areas of skin and certain other tissues.

VEGETATIVE PROPAGATION AS A HORTICULTURAL PRACTICE

The propagation of our cultivated plants, along with the associated efforts to create and maintain desirable varieties, constitutes one of the most important aspects of agriculture and horticulture. Most annual and biennial plants, including practically all farm crops and garden plants, are propagated sexually through seeds, but a very large portion of both herbaceous and woody perennials are commonly propagated vegetatively. For example, strawberries are propagated by runners, iris by division of the rhizomes, Irish potatoes by tubers, sweet potatoes by roots, tulips by bulbs, gladioli by corms, and African violets by leaves. Most of our ornamental shrubs are propagated by stem cuttings, and our orchard trees and roses are usually propagated by grafts. All these plants are propagated by seeds only as a means of developing new varieties.

Advantages of Vegetative Propagation

The most obvious advantage of vegetative propagation as a horticultural practice is that it makes possible the propagation of plants such as bananas, seedless grapes, and citrus fruits, and the cultivated hydrangea and snowball bush that have lost their capacity for sexual reproduction. Other plants, including some kinds of Bermuda and Zoysia grasses, are propagated vegetatively because they produce only small quantities of seed. Another advantage of vegetative propagation is that some plants including lilies, tulips, and gladioli require several years to reach maturity when propagated by seeds, whereas they mature in a season or two when propagated vegetatively. Perhaps the most important advantage of vegetative propagation, however, is that all offspring propagated vegetatively have exactly the same hereditary potentialities as the parent plant, except in the rare cases of mutations, whereas great hereditary variation generally accompanies sexual reproduction. For example, if a thousand seeds from Winesap apples were planted, there would be only a slight chance of securing even a single tree bearing apples with all the characteristics of the Winesap variety. It might be possible to produce a variety of apples that would come true to form from seeds by means of an extensive breeding program designed to produce a considerable amount of hereditary uniformity (as has been done with beans, peas, wheat and other plants propagated by seeds), but the procedure would be lengthy, time-consuming, and highly impractical. Whenever cultivated plants can conveniently and practicably be propagated vegetatively, this means of reproduction is used in preference to seeds because of the ease of maintaining desirable combinations of hereditary traits, even though the vegetative propagation possesses no other advantage.

Use of Natural Methods of Vegetative Propagation

Man makes use of all types of natural vegetative propagation already discussed, such as tubers, bulbs, rhizomes, runners, and fleshy roots. Frequently, procedures are employed that increase the effectiveness of the natural method. Thus, a potato tuber simply left in the ground will give rise to only one potato plant, but if the tuber is cut into pieces (Fig. 20.6) each piece will produce a plant providing that it contains a bud (an "eye"). In tulips and other bulbous plants the conical stem can be cut out from the under side of a bulb. Each scale then gives rise to a new bulb at its base, thus increasing the rate of propagation. The scale leaves are frequently removed from lily bulbs and planted separately, a method that is really comparable with the leaf propagation of African violets and begonias. In addition to using natural methods of vegetative propagation, man has devised several extensively used methods that rarely, if ever, provide natural means of vegetative propagation. The most important of these artificial methods are **cuttings,** modified types of **layering,** and **grafting.**

Propagation by Cuttings

Although the division of any plant organ—roots, stems, or leaves—into pieces for propagation may be considered as cutting, stem cuttings are the type most commonly used. Branches cut from many species of both woody and herbaceous plants will readily develop adventitious roots when placed in water or in a rooting medium such as moist sand or soil. Species that do not root readily can frequently be induced to root by treating the cuttings with dilute solutions of auxins (Fig. 15.12) or dusting the cut ends with powders containing auxins. Most ornamental shrubs, sugar cane, pineapple, and many greenhouse plants including chrysanthemum, coleus, and geranium are propagated by stem cuttings.

Propagation by Layering

Layering differs from cutting in that the formation of adventitious roots or buds occurs before, rather than after, separation of the propagule from the parent plant. Some species that cannot be propagated successfully by cuttings are readily propagated by layering. In addition to utilizing natural tip layering in propagating plants such as raspberries (Fig. 20.7), man has devised several modified types of layering such as **mound layering** and **air layering** (Fig. 20.12). In mound layering, the bottoms of the lower branches of a plant are covered with soil, and after adventitious roots have developed on the branches the rooted branches are cut off and planted. In air layering, a branch is first cut about half way through and from this cut a longitudinal slit is made about 3 inches upward through the center of the stem. The cut surfaces are then generally dusted with a powder containing auxin-like growth substances. Moist sphagnum moss is then inserted between the cut surfaces and wrapped around the cut region of the branch Finally, the ball of moss is wrapped with a sheet of polyethylene plastic. The plastic is permeable to oxygen but not to water, and so retains moisture. As in other types of layering, the branch is cut off and planted after adventitious roots have developed.

Propagation by Grafts

Grafting is essentially just another modified method of propagation by cuttings. The cuttings, instead of being rooted, are grafted to another plant of the same species or genus, but generally of a different variety. The cutting is known as **scion,** and the plant to which it is grafted is known as the **stock.** The stock is generally secured by planting seeds and using the young plants that grow from the seeds or in some species the stocks may be produced by vegetative propagation. The stock provides the root system and the lower portion of the stem of the composite plant produced by grafting, and the scion generally provides the upper part of the main stem and all the branches. However, sometimes a portion of the shoot system of the stock is retained and scions of one or more other species or varieties are grafted to it, producing a plant bearing the flowers and fruits of several different varieties or species. Such composite plants are generally produced

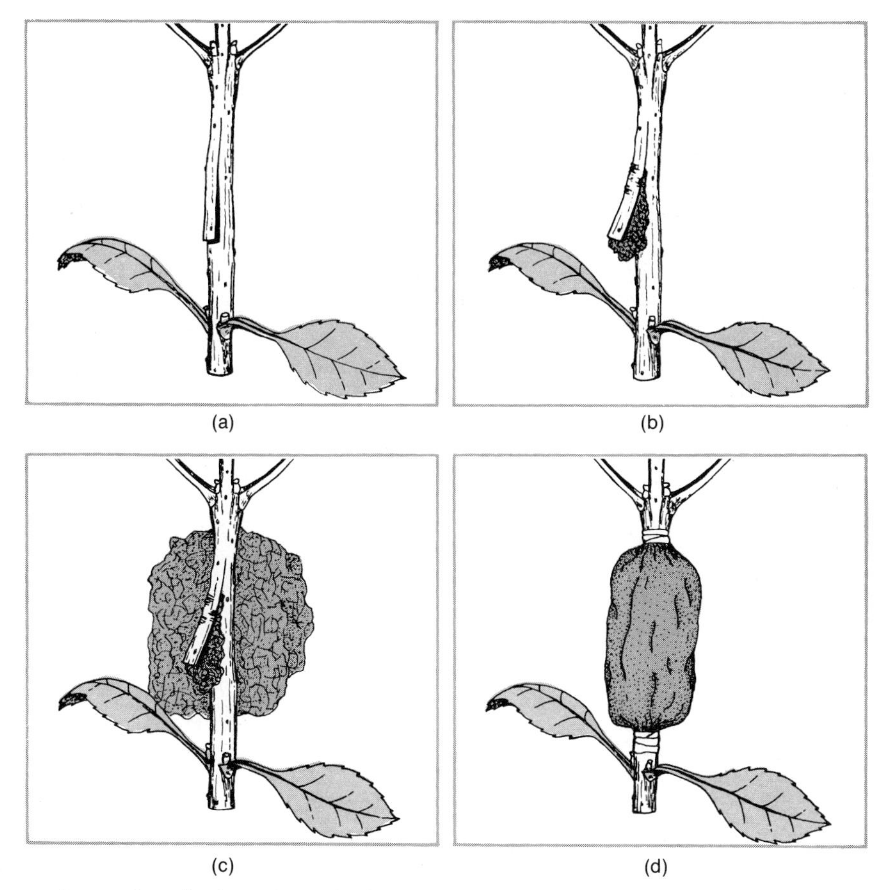

Fig. 20.12 Air layering is accomplished by cutting a flap of bark (a), *dusting the cut surfaces with a powder containing auxin, inserting moist sphagnum moss under the flap* (b), *and covering this region of the stem with more sphagnum moss* (c) *and moisture-proof plastic film* (d). *After adventitious roots have developed the stem is cut off and planted. (Modified from* Introduction to Plant Science *by Northen, with permission of the publisher, The Ronald Press.)*

only as horticultural curiosities, although an apple or pear tree bearing several different varieties may be useful to a person with a small lot who wishes to produce an assortment of fruits in limited quantities for home consumption.

One of the principal advantages of grafts over cuttings is that a variety or species with a more vigorous, more rapidly growing, and more disease-resistant root system than that of the plant used as the scion may be selected for use as the stock. Thus, roses are usually grafted on wild rose stocks, and lilacs on California privet stocks.

In making a graft, the stems of the stock and scion are cut so that they will closely dovetail into each other. A variety of cuts is employed (Fig. 20.13). The stems of the stock and scion need not be the same diameter at the point of union, but the cambia of the two must be in close contact. After the stock and scion have been fitted together, the graft union is tied with waxed string and covered with wax to prevent drying out and the entrance of bacteria, fungi, or insects. Grafts are generally successful when the two plants used are in the same species or genus and when their rate of growth in diameter from the cambium is about the same. Successful grafts have been made, however, between different genera of the same family, as between pear and quince, tomato and potato, and tomato and tobacco, and a few grafts have been produced between members of different but related families. Grafting may be done either early in the growing season or in the winter while the plants are dormant.

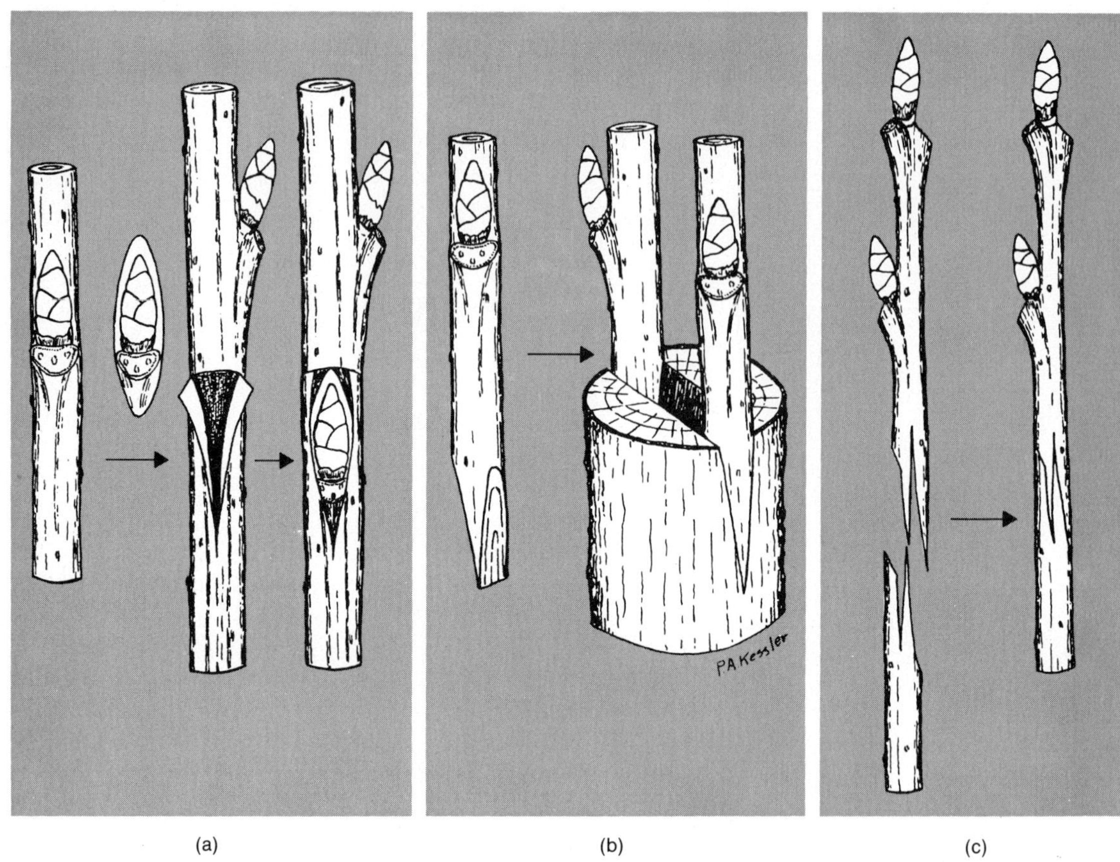

Fig. 20.13 *Budding and two kinds of grafting.* (a) *A bud with a small piece of bark is removed from the tree being propagated and inserted under the bark of a seedling tree stem. All the buds of the seedling tree itself are removed.* (b) *In cleft grating small scions are grafted onto a larger stock.* (c) *In tongue grafting the stock and scion are the same size and are cut so they will fit together tightly. In all three procedures the cuts are covered with grafting wax and tied together securely with waxed string.*

The dwarf, early bearing fruit trees sold by nurseries are products of grafting. Certain kinds of stocks have a dwarfing effect on the scions. Quince stocks cause the dwarfing of pear scions, apricot stocks the dwarfing of peach scions, and certain varieties of apples the dwarfing of apple scions. Apple trees may also be dwarfed by grafting only a short piece of the stem of a dwarfing variety between the scion and a nondwarfing stock.

A modified type of grafting known as **budding** (Fig. 20.13) is used extensively by nurserymen, budding being used almost exclusively for some species such as pecan. Budding differs from other types of grafting in that a bud rather than a developed branch is used as the scion. In budding a piece of bark containing the bud and at least some of the cambium is cut from the plant being propagated, inserted under a cut made in the bark of the stock, and tied and waxed. Branches growing from the buds of the stock must be cut off until the grafted bud assumes apical dominance and so prevents further development of buds of the stock.

Grafting is employed largely as a means of

propagating horticultural plants, but it is sometimes useful as an experimental tool. For example, much information about the transmission of the floral stimulus in photoperiodism has been obtained by grafting photoperiodically induced plants to noninduced plants, whereas grafts between tobacco and tomato plants have revealed the fact that nicotine is synthesized in the roots of tobacco plants even though it accumulates in the leaves.

Like other methods of vegetative propagation of plants, grafting is an old horticultural practice, and many misconceptions about grafting have become prevalent. Perhaps the most common misconception is that grafting is a means of hybridizing plants, when as a matter of fact hybridization can occur only in the course of sexual reproduction. Except in the rare cases of mutations that may occur in the buds of plants whether they have been grafted or not, the hereditary potentialities of both the stock and scion remain exactly the same as they were before grafting. If it were not for this, grafting would be unsuitable as a means of propagating a certain species and variety of plant. The stock may, of course, influence the scion physiologically as in dwarfing or in the absence of nicotine in the leaves of tobacco scions grafted on tomato stocks.

Some of the misconceptions about grafting as a means of hybridization may have resulted from the fact that occasionally a bud developing at the graft union may be composed partly of cells from the stock and partly of cells from the scion. The resulting branch will then contain tissues derived from both plants. These composite shoots are known as **graft chimeras.** Chimeras may be of varied composition—the scion may contribute the epidermis and the stock the other tissues, or vice versa, or one side of a stem may be composed of stock cells and the other side may be composed of scion cells. Some of the flowers and fruits on a chimeral branch may develop from stock tissue, some from scion tissue, or perhaps from both, resulting in the presence of different kinds of flowers and fruits on the same branch. However, each cell of a chimera still retains exactly the same hereditary potentialities as the stock or scion that it came from.

Propagation by Cell and Tissue Culture

Cell and tissue culture techniques described in Chapter 15 have been used primarily for experimental purposes, but they have great potential for use in plant propagation on a practical basis. In fact, some such practical use of cell and tissue cultures for plant propagation has already been made (Chapter 15). One advantage is that a very large number of plants can be propagated from a single plant that has desirable hereditary potentialities. As in other types of vegetative propagation, the offspring constitute a clone, each member of the clone having the same hereditary potentialities as the parent plant.

RELATED READING

Brodie, H. J. "Nature's splash guns," *Natural History, 61,* 403–407, 1952.

Gottleib, D. "Germination of fungus spores," *Endeavour, 23,* 85–89, 1964.

Mahlstede, J. P. and E. S. Haber. *Plant Propagation.* New York: Wiley, 1957.

Ingold, C. T. "Spore liberation in the higher fungi," *Endeavour, 16,* 78–83, 1957.

21
SEXUAL REPRODUCTION

It was shown in Chapter 20 that some species in all the divisions of the Plant Kingdom are capable of *asexual* reproduction. In the asexual method, essentially, a portion of the parent plant becomes detached and grows directly into an individual like the parent. The *sexual* method is somewhat more complex and indirect. In the common view, it involves merely the fusion of two specialized cells into a single cell that ultimately develops into a new individual. However, this simple statement is quite inadequate to describe the nature of the sexual process. The theory that plants reproduce sexually was suspected in ancient times, when the practice of hanging clusters of "male flowers" in the date groves assured an abundant crop of fruit. The details of the process, however, have been known for only a little over a hundred years.

Most species of plants reproduce sexually. The sexual method had simple beginnings in some of the very early members of the Plant Kingdom, and has become more elaborate through evolutionary changes in many succeeding species. However, the elaboration entailed no essential changes in the basic nature of the process, but consisted principally of the development of structural refinements to surround the essential action. This chapter traces the story of plant change as shown by the details of the sexual reproductive process in selected species.

SELECTED ALGAE

The unicellular, motile green alga, *Chlamydomonas,* is an example of a primitive sexually reproducing form, which, like many others, also reproduces asexually. *Asexual* reproduction is effected by the division of the protoplast into two or four motile **spores** (called **zoospores,** in allusion to their animal-like motility). These *asexual* spores are released by the breakdown of the parent cell wall. Structurally they resemble the parent unicellular plant. After discharge (Fig. 21.1*a, b, c*), they enlarge to adult size and become typical *Chlamy-*

Fig. 21.1 Green alga, Chlamydomonas. *Asexual and sexual reproduction. Explanation in text.*

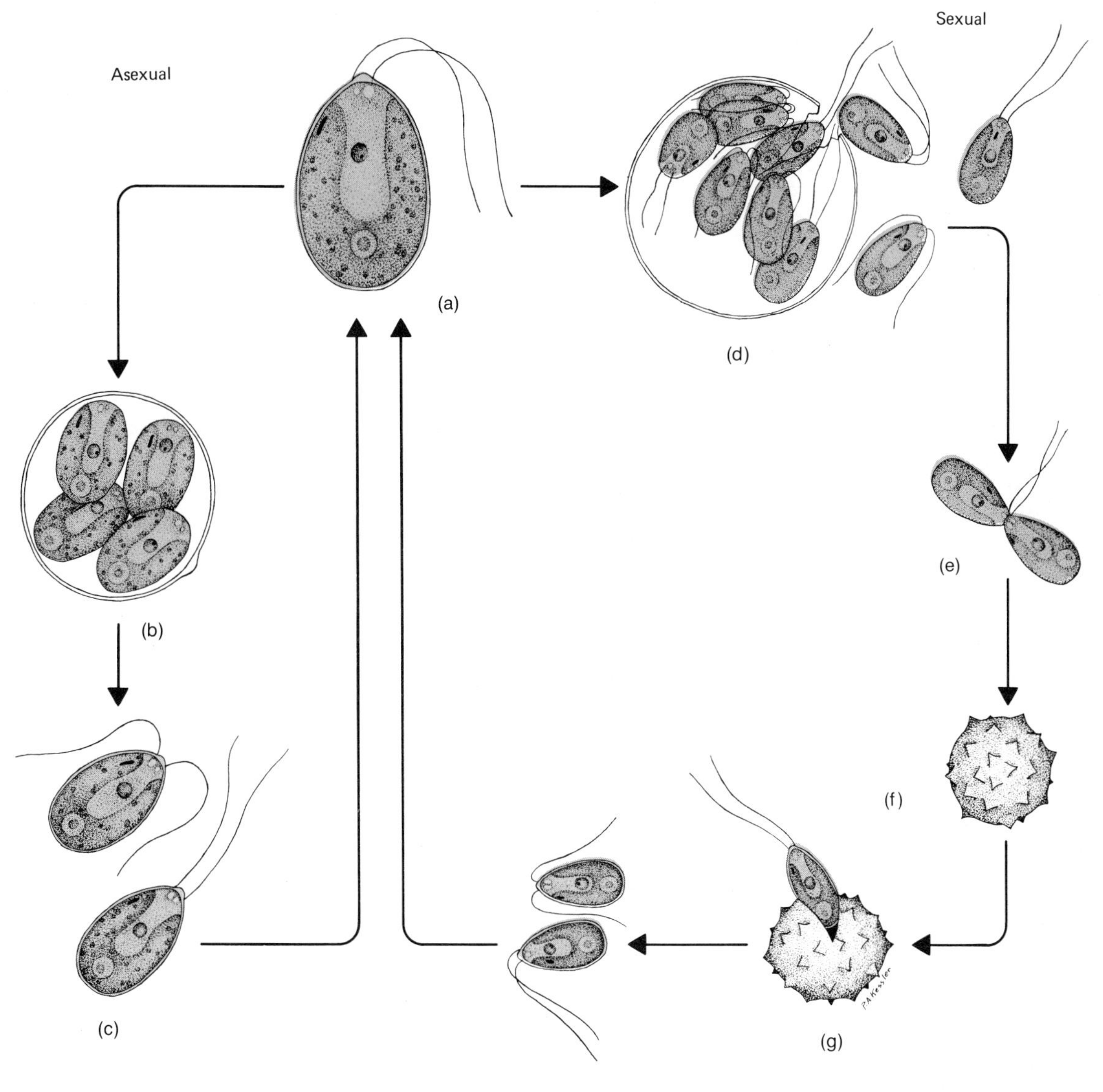

domonas plants. *Sexual* reproduction in *Chlamydomonas* begins with the division of the parent protoplast into eight cells which resemble the asexual spores except for their smaller size. These smaller motile cells, called **gametes,** are discharged, swim freely for a time and then unite in pairs. The resultant product may swim freely for a short time, during which two nuclei can be discerned in the cell. Finally it settles down, the two nuclei fuse, and a thick wall is secreted. This fusion product, called the **zygote,** ultimately germinates, releasing four motile cells that are structurally like the asexual spores. We designate these motile cells derived from the zygote by the special name **meiospores.** The reason for the qualifying prefix *meio-* will become clear later on. The meiospores grow to become adult *Chlamydomonas* plants like the original parent (Fig. 21.1*a, d, e, f, g*). Occasionally, two adult plants may themselves act as gametes; that is, they fuse, producing a zygote that behaves as described above.

Thus, it will be seen that the sexual cycle is more complex than the asexual, involving certain intermediate steps between two generations of adult *Chlamydomonas* plants. The intermediate steps are: formation of *gametes, fusion* of gametes to establish the *zygote,* germination of the zygote to yield *meiospores.* Although the meiospores and the asexual zoospores resemble each other structurally and behave similarly in giving rise directly to new *Chlamydomonas* plants, they are really different. Meiospores, it must be remembered, are part of the *sexual* reproductive cycle. Discussion of further detail and the significance of the sexual cycle is deferred until we consider a few additional algae.

The filamentous alga *Ulothrix* is a multicellular organism with simple structural differentiation (Fig. 21.2*a*). The filament, consisting of many cells, grows attached to the bottom of a stream by means of a specialized anchoring cell, or **holdfast.** The specialization of this cell is expressed not only in its form and function, but also in the fact that it is regularly without a chloroplast, and does not contribute to growth or reproduction of the filament. The free-living, self-maintaining organism here is clearly the whole filament, not any single cell. It is therefore not a colony of independent cells, but truly a multicellular plant.

Ulothrix may reproduce asexually by the formation of four or eight zoospores in any cell of the plant, except the basal one. Each zoospore has four anterior flagella. After a period of swimming, they settle down and divide transversely, the bottom cell becoming the holdfast. Further divisions in the upper cell give rise to the remainder of the filament.

In sexual reproduction, a cell of the filament may produce and discharge as many as 64 gametes (Fig 21.2*b*). The gametes are strikingly similar to the asexual spores, except that they are smaller and have only two flagella (Fig. 21.2*c*). If two gametes from different filaments meet, they promptly fuse. The resultant zygote develops a hard, thick wall and settles to the bottom of the stream where it may remain for a period of time (Fig. 21.2*d*). The zygote may be subjected to very low temperature and drying without evident injury. Under favorable conditions it may germinate, yielding 4 meiospores, each with 4 flagella (Fig. 21.2*e*). New filaments arise from these.

In *Ulothrix* we again see the structural similarity of gametes and asexual spores. The zygote is of special interest in view of its capacity to carry the plants over periods unfavorable to growth.

The marine alga *Ulva,* the sea lettuce, occurs along the ocean shores in the intertidal zone. The plant body consists of a bright green blade up to 30 cm long, somewhat resembling a leaf of garden lettuce (Fig. 21.3*a*). The plant is anchored to the bottom by a holdfast similar to that of *Ulothrix.* In sexual reproduction many cells at the margin of the blade produce by mitotic divisions 32 or 64 biflagellate ga-

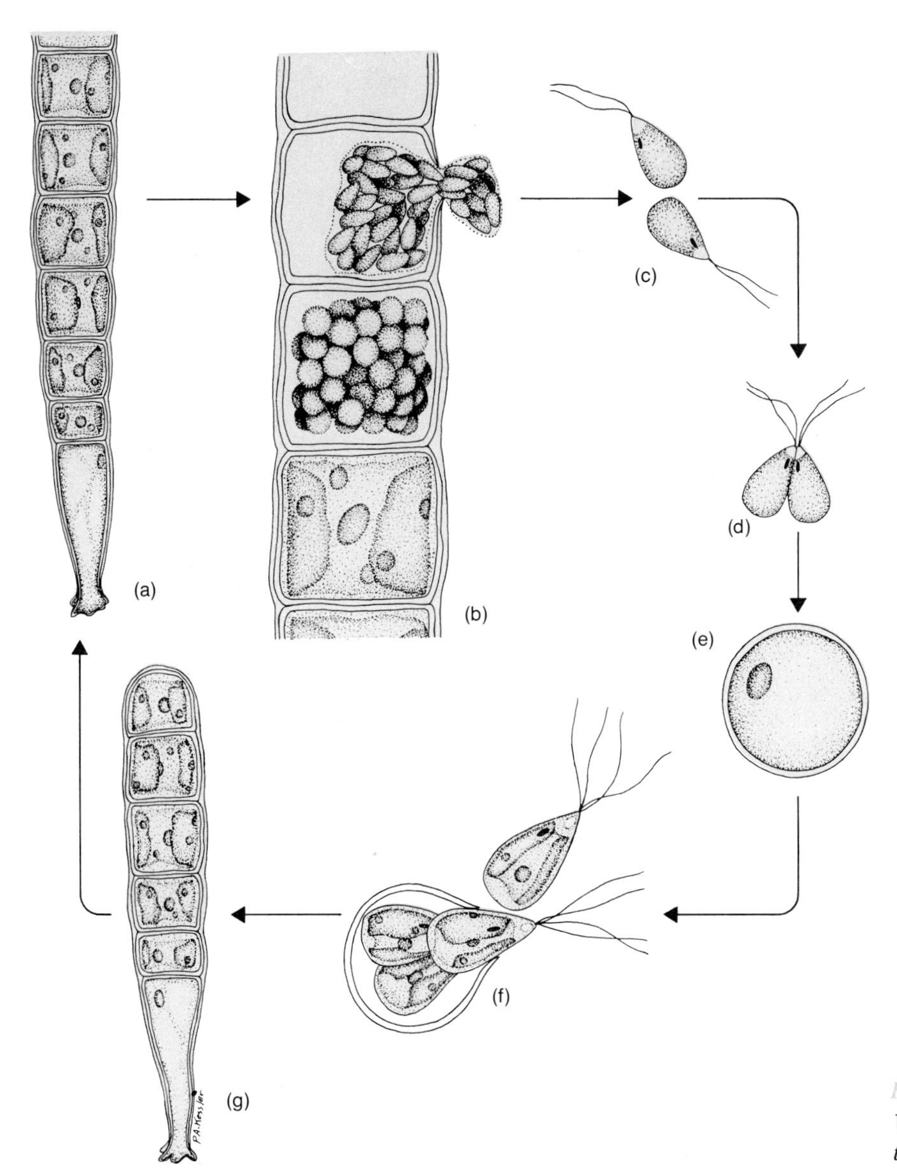

Fig. 21.2 Green alga, Ulothrix. *Sexual reproduction. Explanation in text.*

metes which are liberated through a small pore developing in each cell (Fig. 21.3*b, c*). The fusion of gametes from different plants occurs in the surrounding water (Fig. 21.3*d*). The resultant zygote germinates after a few hours, giving rise by repeated cell divisions to a filament of cells with a basal holdfast (Fig. 21.3 *e–g*). Through further growth, the filament becomes a broad blade indistinguishable in form from the parent plants (Fig. 21.3*h*). However, this plant produces meiospores in many of its marginal cells. Each meiospore, after a period of free swimming, settles down, germinates, and grows to become a leaflike plant similar to the original parent, which in its turn will produce gametes (Fig.21.3*a*).

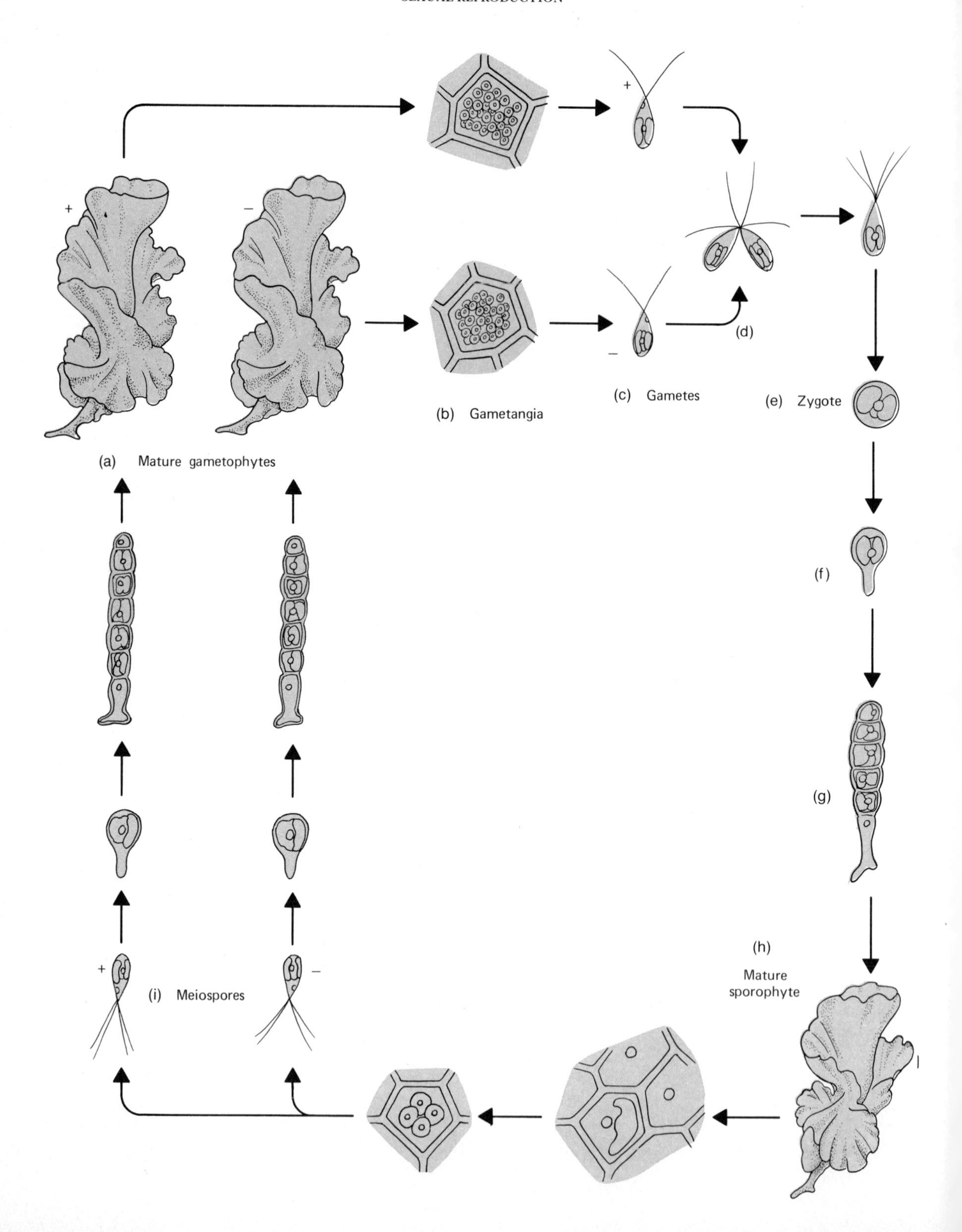
+
−
(a) Mature gametophytes
(b) Gametangia
+
−
(c) Gametes
(d)
(e) Zygote
(f)
(g)
(h)
Mature
sporophyte
+
(i) Meiospores
−

It is noteworthy that in *Ulva* the zygote yields upon germination, not meiospores directly as in *Ulothrix,* but a large multicellular plant. In effect, the production of meispores is delayed for a time during which a considerable amount of vegetative growth takes place.

Spirogyra, a common filamentous green alga of streams and ponds (Fig. 21.4) displays a number of interesting features in its sexual reproductive cycle. The cells of two different filaments lying side by side develop small protuberances (Fig. 21.4*e*) on their adjacent walls and as these enlarge the two filaments are pushed apart, the whole configuration somewhat resembling a ladder (Fig. 21.4*b*). The protoplasts of two opposing cells shrink and become gametes and the abutting walls of the protuberances disappear, forming a tubular connection between the cells. The gamete of one cell flows through the **conjugation tube** and fuses with the gamete of the other cell (Fig. 21.4*c*). The resultant zygote (Fig. 21.4*d*)

Fig. 21.4 Sexual reproduction in the green alga, Spirogyra. (a) *Vegetative filament.* (b) *Conjugating filaments; conjugation tubes formed, but gametes not yet organized.* (c) *Conjugation tubes open and gametes passing across.* (d) *Conjugation complete in three upper cells on right.* (e) *Small protuberance on cell is earliest stage in formation of conjugation tube; normally two such cells would be opposed. (Carolina Biological Supply Co.)*

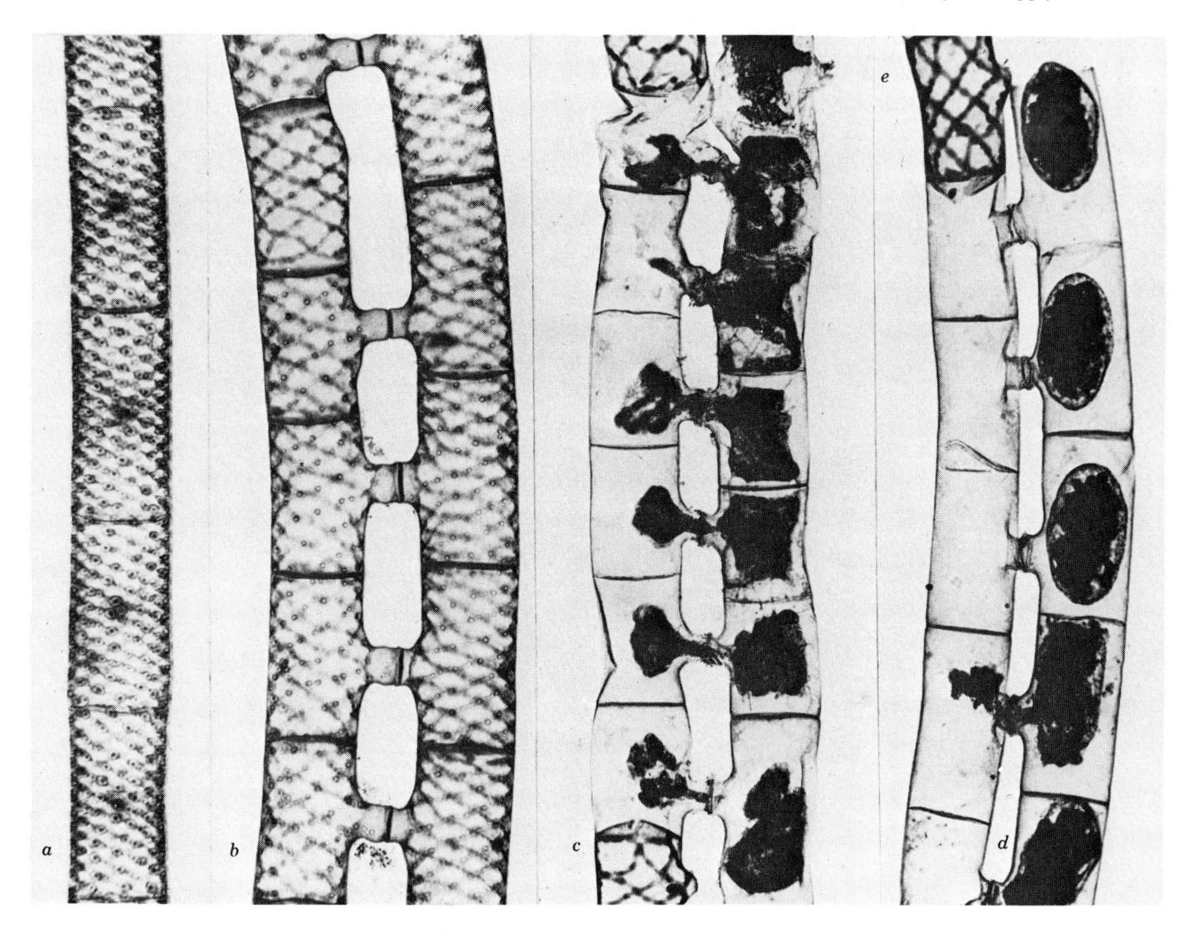

develops a thick wall and carries over the winter in a resting condition. Germination of the zygote in spring gives rise to four meiospores which are retained within the zygote wall. Three of the meiospores abort and the remaining one germinates, yielding a new *Spirogyra* filament.

In *Spirogyra,* as in *Chlamydomonas, Ulothrix,* and *Ulva,* the fusing gametes are structurally alike, although those of *Spirogyra* are not perfectly so. Gametes that are morphologically indistinguishable are commonly called **isogametes.** In most species, however, gametes derived from the same plant are incompatible and do not fuse with each other. Species in which such incompatibility obtains are said to be **heterothallic,** that is, gametic fusion may occur only when the two gametes are derived from different plants. This implies an innate sexual difference between the two fusing gametes and between the parent plants from which they came, even though morphological distinction cannot be made. To express this difference in heterothallic isogamous species, the terms plus (+) strain and minus (−) strain are commonly used. In *Spirogyra* the difference in behavior of the two fusing isogametes suggests a tentative beginning of structural differentiation.

The fresh-water green alga, *Oedogonium* (Fig. 21.5*a*), in general structure and asexual reproduction resembles *Ulothrix.* In sexual reproduction, the protoplast of a cell of *Oedogonium* may in its entirety become organized as a single large gamete **(egg)** which remains within the cell **(oogonium)** (Fig. 21.5*c*). A group of two to several short cells **(antheridia)** are produced by division of other cells on the same or a different filament. Antheridia each yield two small motile gametes **(sperms)** which escape by the opening of the antheridial wall. The sperms are very similar in form to the zoospores, but smaller (Fig. 21.5*b*). One of the sperms gains entrance into the oogonium through a pore, and fuses with the egg. The resultant zygote (Fig. 21.5*d*) develops a thick wall, is freed by rupture of the oogonial wall, and falls to the bottom where it remains dormant. Four meiospores, resembling the zoospores, are produced at germination of the zygote and after a period of swimming grow into new filaments (Fig. 21.5*e, f*).

In the sexual cycle of *Oedogonium* some important and interesting innovations appear, even though the basic pattern of the reproductive process is unaltered. First, the fusing gametes of *Oedogonium* are morphologically different, whereas those of the previously described genera are not. Morphologically dissimilar gametes are called **heterogametes.** *Oedogonium,* therefore, demonstrates an early and easily recognizable step in sexual differentiation in the development of a large nonmotile egg **(megagamete)** and small, motile sperms **(microgametes).** The union of gametes occurs within the egg-producing cell, not in the surrounding water as in *Ulothrix.* Some species of *Oedogonium* are heterothallic whereas others are **homothallic,** that is, gametes from the same plant may fuse with each other.

The plants we have described were selected from many because they probably reflect the origins and some of the early modifications of the sexual reproductive process in plants. It is important to keep in mind the basic steps of the process and to recognize their essential invariability. The apparent differences in the process among species are, as we have previously noted, for the most part accessory structural refinements to surround the essential action. A fact of special interest in the life cycles is the striking similarity, except in size, of one or both of the gametes and zoospores in the same species. In isogamous species, both gametes resemble the asexual zoospores; in the heterogamous species, the motile microgamete bears this resemblance. This fact gives rise to the theory that gametes were evolved from spores, and that sexual reproduction probably originated with the fusion of two "small-sized

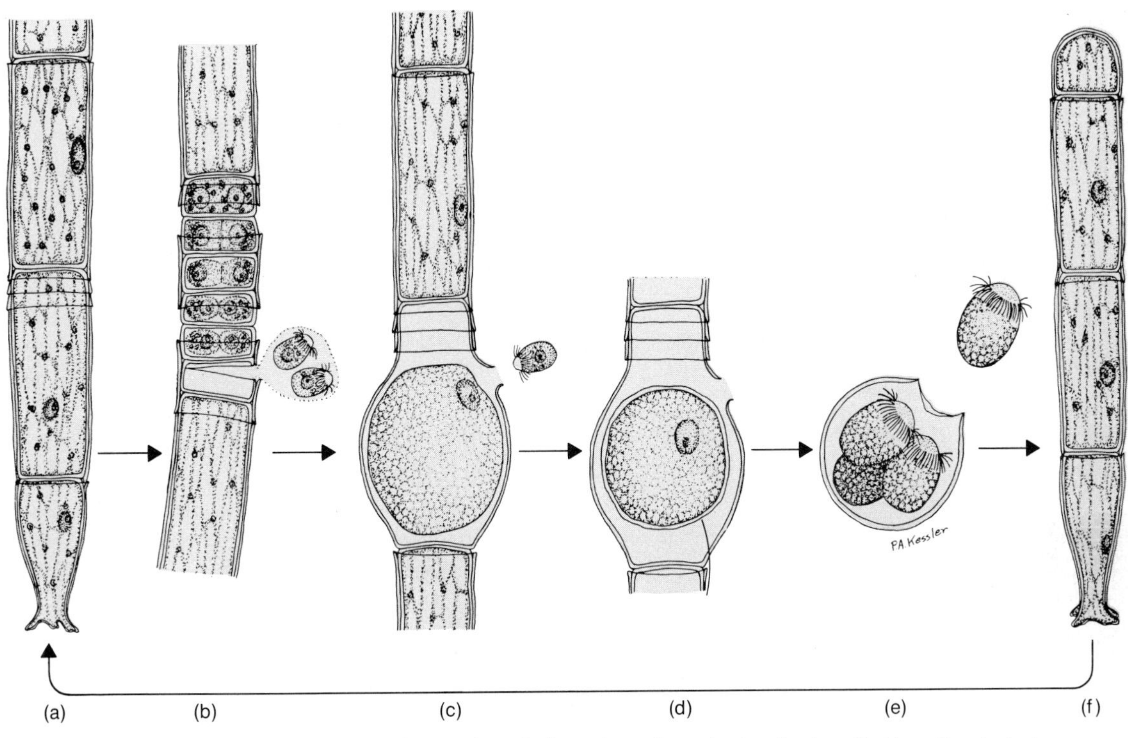

Fig. 21.5 *Green alga, Oedogonium. Sexual reproduction. Explanation in text.*

spores." The isogamous condition would be considered to have preceded the heterogamous in evolution.

If we pause for a moment to think about the sexual life cycles of the plants we have described, we realize that they are essentially alike with respect to the major events leading up to production of the next generation. The differences observed are mainly structural or operational in detail. The sexual reproductive cycle really consists of two recognizable *phases*. These phases are characterized by their specific contribution toward completion of the sexual cycle. One of these phases, the adult parent filament, as in *Oedogonium* or *Ulothrix*, or the parent free-swimming cell, as in *Chlamydomonas*, is expressly given over to the production of *gametes*. The other phase, the *zygote*, contributes *meiospores*. It is obvious that these two distinctive structures are not unrelated, but are recognizable parts of an uninterrupted process in which the protoplasm of the plant temporarily exists in two different forms. For this reason we call these manifestations *phases* of a life cycle. Gametes are produced by a "gamete plant," and meiospores by a "meiospore plant." To translate this into more usual botanical terminology, the gamete plant is the **gametophyte** (*phyton* = plant) and the meiospore plant is the **meiosporophyte.** The latter term is, in common practice, shortened to **sporophyte.** Thus, in a sexual cycle a **gametophytic phase** is followed by a **sporophytic phase,** and as the countless new generations of plants appear, the regular unvarying alternation of these two phases continues.

MEIOSIS

Study has revealed that there are differences between the gametophytic phase and sporophytic phase more fundamental than external appearance. We recall from Chapter 8 that, among other things, a species of plant is characterized by the number of chromosomes in its cells. Chromosome counts of both the sporophytic and gametophytic phases have been made for many species of plants. Typically, with but few exceptions, the sporophytic number is twice the gametophytic or, expressed symbolically, the sporophytic is **2n** or **diploid** (double) and the gametophytic **n** or **haploid** (single). In many species the chromosomes making up the gametophytic complement are distinguishable, so that a definite assortment or set of chromosomes can be recognized. Thus the expression $2n$ is to be taken to indicate not merely twice the n number, but *two complete sets.* That is, each type (size, shape) is represented by a *pair.*

The haploid gametophyte produces its gametes by ordinary mitoses, that is, by equational nuclear divisions. Therefore, the gametes will contain the n number of chromosomes. Sometimes the basic n number of chromosomes is called the **gametic number.** When, as a result of gametic fusion, the zygote is formed, it obviously will receive an n set from each gamete, and will therefore be $2n$ (n *pairs* of chromosomes). Since the meiospores produced by the zygote directly yield filaments with n chromosomes, the reduction of chromosome number from $2n$ to n obviously occurs at meiospore formation. This is indeed the case, and of the two divisions required to produce four meiospores, the first accomplishes the reduction. This operation is performed in an interesting manner (Fig. 21.6). Thus, when the chromosomes of the zygote assume the metaphase position on the spindle, the members of each pair have become closely associated physically. The physical association of members of a pair is spoken of as the pairing or **synapsis** of **homologous** (comparable) chromosomes, and suggests that the members of a pair are highly compatible and are mutually attractive because of a qualitative similarity. When the spindle becomes organized, the fibers attach to the members of the several pairs. During the ensuing anaphase, members of the several pairs separate and move to opposite poles. Thus, at telophase, each of the two daughter nuclei receive one complete n set. Two haploid cells are thus derived from a diploid cell (Fig. 21.6*a–e*). The nuclear division that effects this reduction from $2n$ to n is quite properly called a **reductional division.** In general performance it clearly resembles mitosis, except that whole chromosomes, members of pairs, are separated, rather than chromatids of a single chromosome.

Reductional division can, of course, occur only in cells with pairs of chromosomes ($2n$), whereas mitosis or equational nuclear division may occur in cells with either n or $2n$ chromosomes. As previously stated, haploid meiospores are regularly produced in groups of four. Thus a further division of each of the haploid cells resulting from the reductional division of the diploid cell occurs. This division is of the equational or mitotic type (Fig. 21.6*f–h*). The diploid cell that undergoes the successive reductional and equational divisions yielding the four haploid meiopores is a **meiospore parent cell.** The combination of reductional and equational divisions is called **meiosis** (to make smaller), or the **meiotic divisions.** Because meiospores are produced by the special method described, it is necessary to distinguish them from the asexual spores produced directly by the haploid plant, hence the name *meio*spore.

Since homologous chromosomes are separated at the reductional division of meiosis, it follows that the chromosome of any one pair contributed to the zygote by the sperm will be separated from its homologue contributed by the egg. However, since the arrangement of the pairs at metaphase of the first division (reductional division) is random, the haploid set of each daughter cell will have various assort-

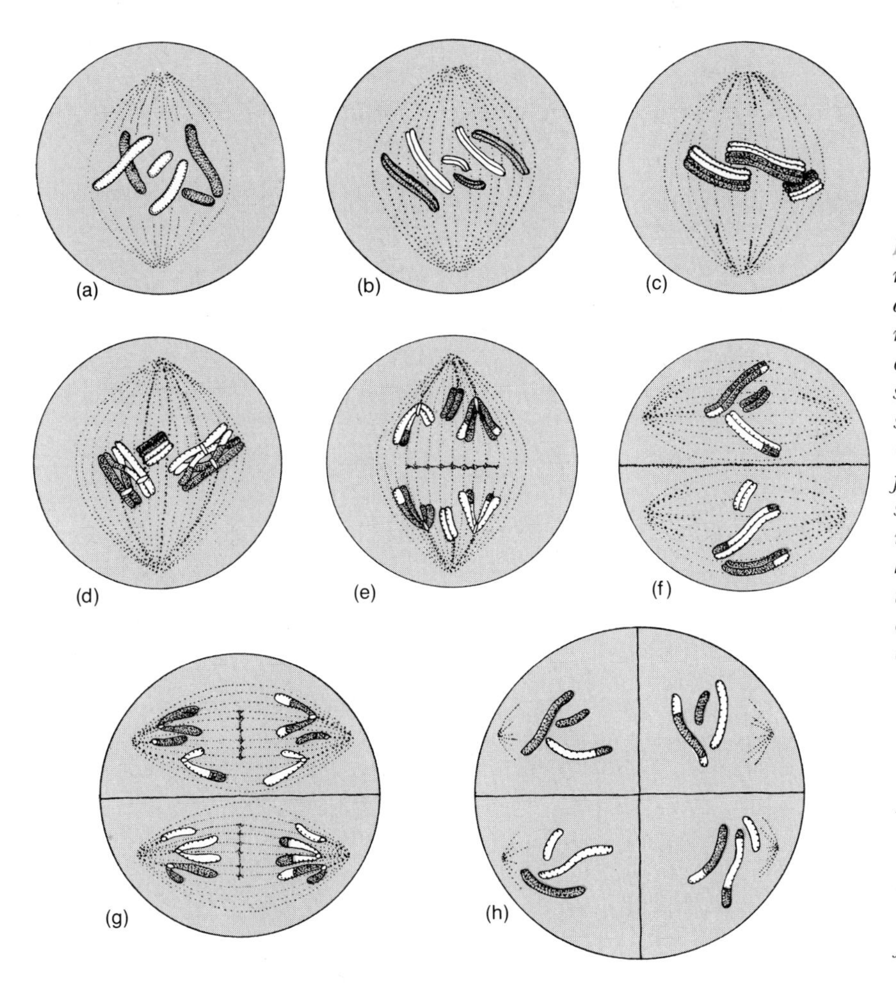

Fig. 21.6 *Diagrammatic representation of the meiotic divisions of a meiospore mother cell with three pairs of homologous chromosomes. One set of the three is shaded, one is unshaded.* (a) *Late prophase of the first or* reductional *division.* (b) *Homologous chromosomes appear double and have begun to pair.* (c) *Metaphase; note that each chromosome appears as two chromatids.* (d) *Anaphase.* (e) *Telophase of the reductional division and beginning of the second or* equational *division.* (f) *Metaphase of the second division.* (g) *Anaphase; sister chromosomes (Chromatids) are separating.* (h) *Meiosis complete with formation of four meiospores.*

ments of sperm-contributed and egg-contributed chromosomes. The significance of this fact will be fully dealt with in Chapter 22.

SELECTED FUNGI: A PHYCOMYCETE

The black molds of the genus *Rhizopus* to which the common bread mold belongs are mostly saprophytes, occurring on many kinds of organic substrates, such as bakery goods, other foodstuffs, and animal dung. Cultures are easily obtained by exposing a piece of moistened bread to the air for a few hours. Spores settling on the substrate germinate to produce white cottony masses of nonseptate, multinucleate (coenocytic) hyphae (Fig. 21.7*a*). If growing hyphae of mycelia derived from spores of compatible (plus and minus) sexual strains come in contact, the hyphal tips enlarge (Fig. 21.7*b*). Septa are promptly formed behind the enlarging tip, thus delimiting two multinucleate **gametangia** (Fig. 21.7*c*), each subtended by a swollen hypha called a **suspensor.** The walls between the abutting gametangia break down and the nuclei and cytoplasm of the gametangia commingle (Fig.

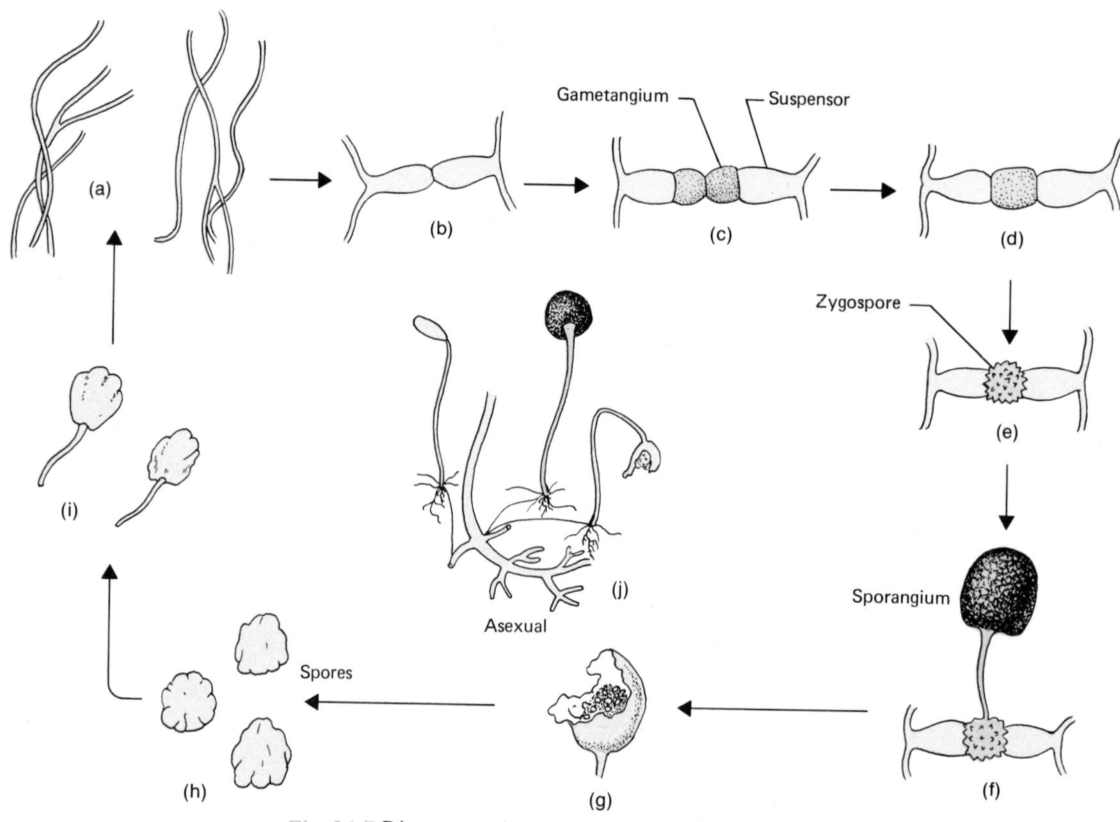

Fig. 21.7 Diagrammatic representation of the life cycle of the bread mold, Rhizopus nigricans. (a) *Mycelia of two compatible strains.* (b, c, d) *Stages in conjugation.* (e) *Mature zygospore.* (f) *Germination of zygospore with formation of sporangium.* (g, h) *Dehiscence of sporangium and release of spores.* (i) *Germination of spores, yielding hyphae of two sexual strains.* (j) *Asexual reproduction: formation of spores without sexual fusion.*

21.7*d*). The gametangia are sometimes referred to as *multinucleate gametes.* The product of fusion is a zygote called a **zygospore** (Fig. 21.7*e*). During its development the number of haploid nuclei increases. Many of the nuclei fuse, giving rise to diploid nuclei; the remaining unpaired nuclei disintegrate, leaving only diploid nuclei in the zygote. The zygote develops a thick rough wall and enters a dormant stage that may last for several months (Figs. 21.8 and 21.9). Germination of the zygote yields a short, multinucleate hypha whose tip enlarges to produce a spherical, thin-walled sporangium. Many cells are formed within the sporangium by a progressive cleavage of the cytoplasm into small protoplasts containing two to several nuclei. These multinucleate protoplasts round up, secrete a wall and become meiospores. Rupture of the delicate sporangial wall liberates the meiospores for wind dispersal (Fig. 21.7*g*).

Apparently, some of the details of zygote germination in the black molds are not perfectly understood. In some species all the diploid nuclei undergo meiosis, yielding haploid nuclei. These nuclei increase in number as ger-

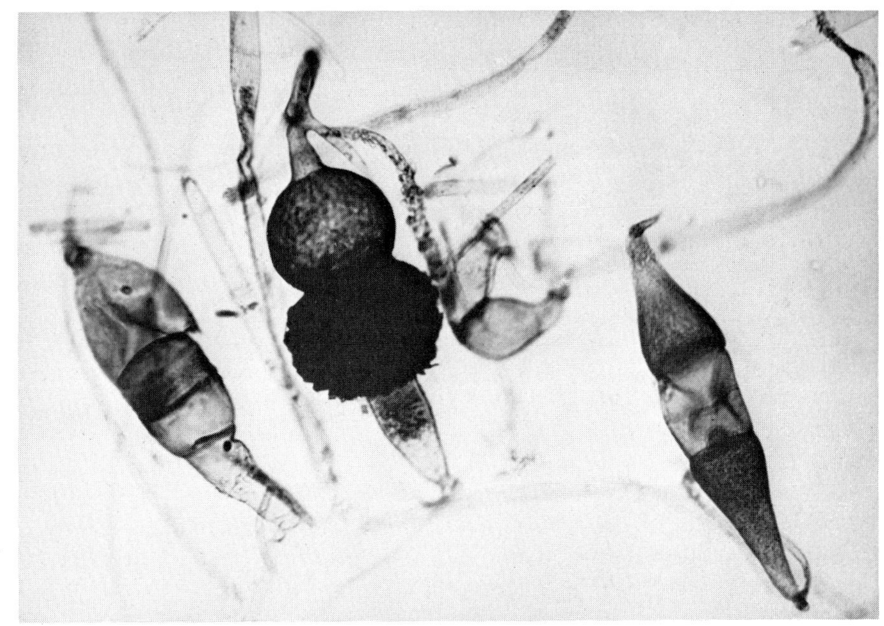

Fig. 21.8 Rhizopus nigricans. *Center: a mature zygospore (zygote) and its two suspensors, one greatly enlarged. Left and right: young developing zygospores shortly after conjugation. (Courtesy, Carolina Biological Supply Co.)*

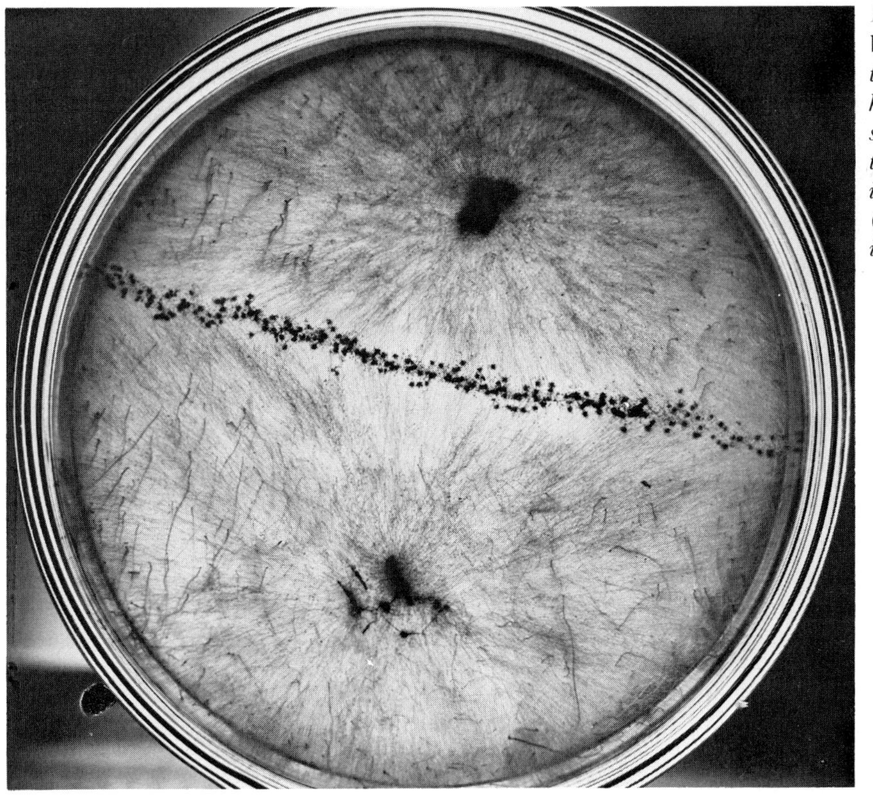

Fig. 21.9 Phycomyces blakesleeanus, *a heterothallic phycomycete. The hyphae of plus and minus strains have conjugated in the region of contact, forming zygospores (zygotes). (Courtesy, Carolina Biological Supply Co.)*

mination proceeds, many of them coming to reside within the developing sporangium. This behavior may account for the fact that in some species the spores yield mycelia of both sexual strains. In other heterothallic species all the diploid nuclei disintegrate except one. The surviving diploid nucleus undergoes meiosis yielding four haploid nuclei of which three disintegrate. The remaining functional haploid nucleus gives rise to all the haploid nuclei of the new developing sporangium by rapid successive divisions. Meiospores formed in this way would be all of one sexual strain, either all plus or all minus.

Some species are homothallic, that is, zygospores may be formed by conjugation of hyphae of a single mycelium. Most species are also capable of asexual reproduction. In asexual reproduction sporangia that are similar to those arising from the zygospores are produced directly on the vegetative mycelium without conjugation (Fig. 21.7*j*). Asexual spores so derived would, of course, be of the same sexual strain as that of the parent mycelium.

SELECTED FUNGI: A MUSHROOM

The common mushrooms and toadstools, fungi of the class Basidomycetes, display an interesting life cycle. They are predominantly saprophytic, living in moist soil rich in organic matter or on rotting wood. A few of them are parasitic, especially on the roots of certain species of trees.

The familiar mushroom at maturity is an umbrella-shaped structure consisting of a stalk or **stipe** bearing at its apex a flattened, conical, or hemispheric cap or **pileus.** On the underside of the pileus many delicate, membranous **gills** radiate out from the stipe to the margin (Fig. 21.10*a*). Under microscopic examination a gill is seen to consist of a sheet of interwoven **hyphae.** Some of the hyphae turn outward toward the surface and terminate in a broad club-shaped cell, a **basidium.** The densely crowded basidia form a distinctive layer, the **hymenium.** Usually each basidium bears four uninucleate **basidiospores,** each at the tip of a small extension or **sterigma** (Fig. 21.10*b, c*). When mature, the basidiospores are forcibly discharged from the sterigmata. The basidiospores fall through the space between the gills and may be widely dispersed by air currents. Falling upon a suitable substrate a basidiospore germinates, giving rise to a septate mycelium whose cells characteristically contain a *single* nucleus (monocaryotic). The mycelium may grow to become an extensive web beneath the surface of the substrate. In growth, some of its cells may come into contact with the hyphae of another mycelium. If the mycelia are of compatible (plus and minus) strains, contacting cells may **conjugate,** giving rise to a *binucleate* cell (dicaryotic) in which one nucleus is of the minus strain and one of the plus strain. The nuclei do not fuse. The binucleate cells give rise to a rapidly growing dicaryotic mycelium that may occupy a considerable area just below the substrate surface (Fig. 21.10*e*). Growth of the dicaryotic mycelium and maintenance of the binucleate condition is accomplished by a complex and unique process of nuclear and cell division in which distinctive intercellular migration of nuclei is a characteristic feature. At the periphery of the growing mycelium tangled masses of dicaryotic hyphae emerge and grow to produce a "button" which rapidly expands into a mushroom (Figs. 21.10*f, g* and 21.11).The mushroom is anchored in the substrate by a tangle of coarse mycelial threads **(rhizomorphs)** from which it grew. The mycelium is perennial, growing radially in all directions and producing seasonally a new crop of mushrooms at its periphery as it grows. In favorable situations, as in lawns and rich meadows, this characteristic growth produces a seasonal succession of ever-widening circular

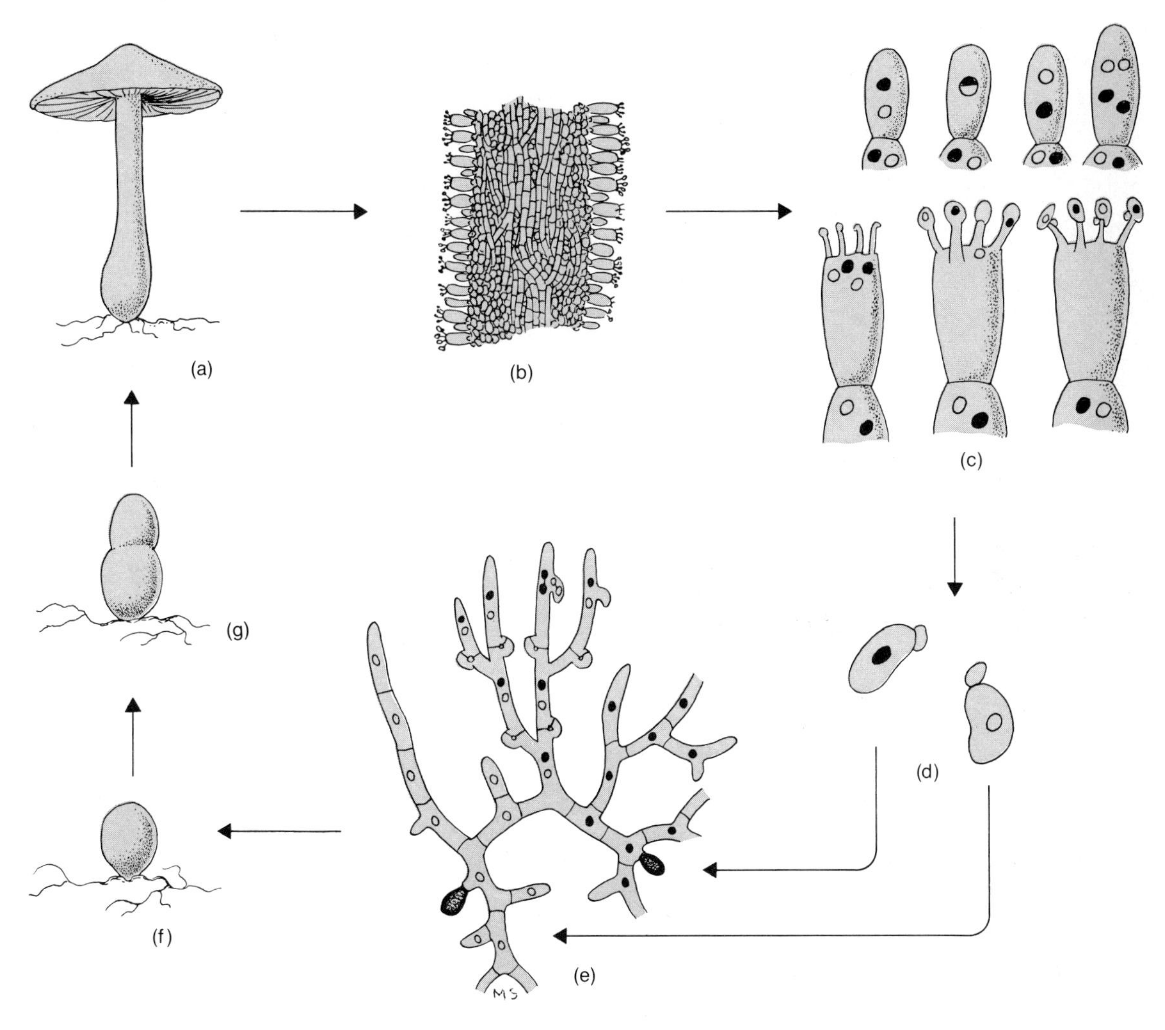

Fig. 21.10 Diagrammatic representation of the life cycle of a mushroom. (a) *Mature basidiocarp.* (b) *Magnified cross section of a gill, showing basidia of various ages.* (c) *Stages in the formation of the basidium, showing nuclear fusion and development of haploid spores* (d). (e) *Mycelia resulting from germination of spores have conjugated, producing a dicaryotic mycelium.* (f, g) *"Button" stages of sporocarp.* (e *after Buller.*)

stands of mushrooms. These striking circular colonies are the familiar "fairy rings" of childhood fancy. Growth of the basidiocarp is very rapid, and this accounts for the startling "overnight" appearance of mushrooms following a warm rain. All cells of the mushroom, including the young basidia, are binucleate. As the basidium matures the two nuclei fuse, creating a *diploid* nucleus. The diploid nucleus undergoes meiosis and the resulting four *haploid* nuclei migrate into the developing sterigmata (Fig. 21.10*c*). When the spores are mature they are shed and germinate as previously described. The basidia do not mature all at one time (Fig. 21.10*b*), thus the process of spore production may extend over a period of a few

Fig. 21.11 *A common poisonous mushroom,* Amanita chlorinosma, *showing sporocarps of three stages. See also Figs. 4.23 and 4.24. (Courtesy, Carolina Biological Supply Co.)*

days. When spore production has been completed the basidiocarp collapses and dies.

In the commercial culture of the common edible mushroom, *Agaricus campestris,* small blocks of substrate containing living mycelium are used as "spawn" to start fresh cultures.

Several features of the reproductice cycle of the mushroom are noteworthy. Specialized sex organs are lacking. The conjugation of uninucleate cells of compatible strains and the establishment of the dicaryotic condition brings together the essential elements of a sexual union, yet the actual fusion of nuclei does not occur until maturation of the basidium. Meanwhile the plus and minus nuclei coexist in all cells of the fruiting mycelium and the basidiocarp. The fusing nuclei of the young basidium function as gametes. The diploid zygotic structure is represented by the fusion nucleus which promptly undergoes meiosis. The basidiospores are, therefore, meiospores which germinate to give haploid mycelia, as expected. But it is difficult here to recognize perfect homologies and to apply the customary terms "gametophyte" and "sporophyte." However, the essential steps of the sexual cycle as we have observed it elsewhere are identifiable. In effect, the actual sexual fusion is delayed and the essential requirements of the "sporophytic" phase are met by the establishment of the dicaryotic condition. Mycologists generally regard this distinctive sexual cycle as the result of extreme evolutionary reduction.

Some species of basidiomycetes display a sexual cycle more complex than that of the mushroom. An important example of this is the life cycle of the black stem rust of wheat portrayed in Figure 17.13.

SELECTED BRYOPHYTES: A MOSS

We now attempt to apply our analysis of the sexual reproductive method to the bryophytes, a group judged to be more advanced in evolution than the algae. Of the three classes of bryophytes, liverworts, horned liverworts, and mosses, the better-known mosses serve our purpose.

A consideration of the moss may well start with its conspicuous phase, the green "leafy" plant. At the apex of a moss axis, surrounded by the terminal leaves, is a cluster of slender, multicellular, delicately proportioned, vase-shaped female reproductive structures, the **archegonia** (Fig. 21.12*a, c*). In the enlarged basal portion of each is a large cell, the megagamete.

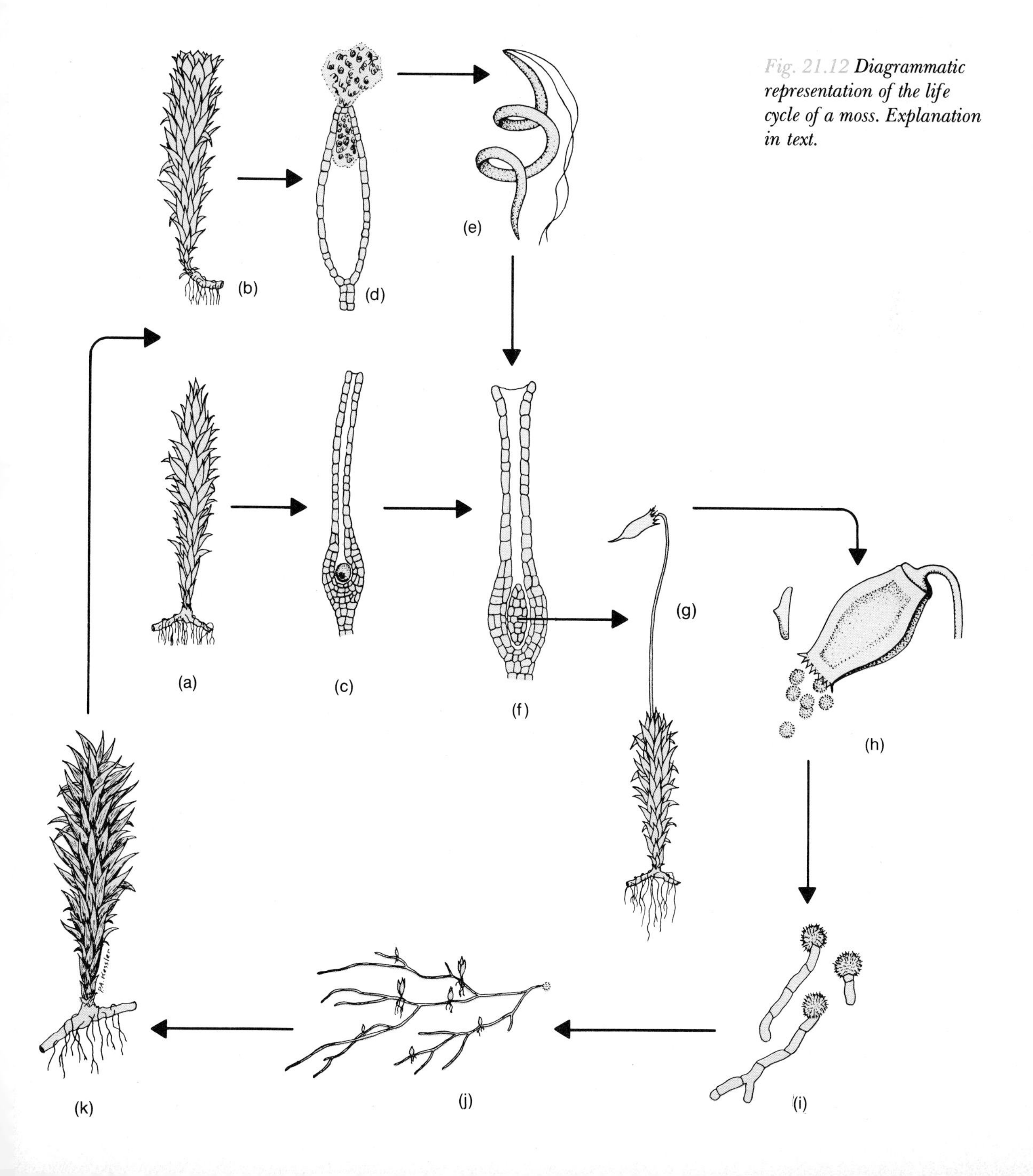

Fig. 21.12 Diagrammatic representation of the life cycle of a moss. Explanation in text.

On the same or another moss axis, the **antheridia** (the male reproductive organs) occur in the form of a group of club-shaped sacs, along with some hair-like filaments (Fig. 21.12*b, d*). The antheridia contain microgametes (Fig. 21.12*e*). The leafy moss plant will thus be recognized as the gametophyte, bearing multicellular sex organs, the egg-producing archegonia and the sperm-producing antheridia. All of these structures are, typically, haploid.

The microgametes, released from the antheridium through a rupture at the tip, are rapidly dispersed over the surfaces of the plants by splashing rain drops, or they may swim through the thin film of water usually covering the plant surfaces (Fig 21.12*d*). A microgamete swims down the slender neck canal of the archegonium and fuses with the megagamete (Fig. 21.12*e*). The product of the fusion is the diploid zygote, the first cell of the sporophytic phase. From this cell, through rapid cell divisions, a simple embryo-like structure is produced with a tapering absorbing **foot** deeply buried in the tissue of the gametophyte (Fig. 21.12*f*). Food is absorbed mainly from the green gametophyte, growth of the embryo occurs, and a slender **stalk,** surmounted by an enlarging **capsule** is produced (Fig. 21.12*g*). The foot, stalk, and capsule constitute the sporophyte. During growth, the sporophyte is green with chlorophyll and thus is at least partly self-sustaining. At maturity, however, the chlorophyll is lost and the whole structure becomes brown. The capsule, along with its stalk, is a multicellular structure and contains many diploid meiospore parent cells. The structural complexity of the moss sporophyte has been brought about by a period of vegetative growth of the zygote.

Meiosis, occurring in the meiospore parent cells within the capsule, produces a large number of meiospores (Fig. 21.12*h*). These are produced in fours, as meiospores always are, by meiosis. Meiospore formation marks the beginning of the haploid phase. The capsule is a specialized meiospore case called a **sporangium.** The meiospores are released from the capsule when the air is dry and are scattered by air currents. Germination of a meiospore on the moist soil yields a green, branched filament of cells. Specialized nongreen branches of this filament **(rhizoids)** penetrate the soil and anchor it firmly, and some of the green branches grow vertically and become the "leafy" moss axis bearing the antheridia and archegonia (Fig. 21.12*i, j, k*).

Certain features of the moss deserve special mention. The entire organization of the plant in both gametophyte and sporophyte, seems pointed toward a terrestrial mode of life. Note especially that the meiospores are produced in great numbers from elevated structures, and that they are air dispersed. Only a relatively small proportion of the total meiospores produced ever germinate. The sporophyte, although attached to the gametophyte, contains some chlorophyll and so has a limited measure of nutritional independence. It probably secures most of its food from the gametophyte, which has considerable photosynthetic tissue in its leaflike outgrowths. Although these features are judged to be land-plant characteristics, complete release from aquatic habitats has not been achieved. The moss has no vascular tissue nor specialized absorbing structures, and thus is essentially restricted to wet habitats. The wetness of plant surfaces is also very important in the reproductive cycle, for the union of gametes depends on the motility of the microgamete.

The evolutionary position of the bryophytes is difficult to assess, for factual evidence in the form of fossil remains is scanty. However, both the structural features and the life cycles of bryophytes represent a condition intermediate between those of the green algae and vascular plants. It seems probable that the bryophytes are derived from algal forms, probably some of which had developed independent multicellular sporophytes. The modern species of moss have probably undergone a reduction of the sporophyte to a relationship dependent on the gametophyte. The filamentous early stage of the moss gametophyte is quite alga-like. Bryophytes are believed to occupy a kind of evolutionary dead end. Although the tracheophytes are a higher group in the evolutionary scale, the most generally accepted view at present is that they were derived directly from an algal ancestor, probably one of the green algae, rather than from a bryophyte.

SELECTED TRACHEOPHYTES: A FERN

The acquisition of vascular tissue by the tracheophytes represents a significant step toward the development of a terrestrial mode of life, for it permits water absorption from the deeper layers of the soil by specialized absorbing roots and the rapid conduction of water through the plant. The great group of plants that have acquired this advantage have come to be ecologically more diverse and wider spread than those that have not. A small segment of the vascular plants, including the ferns, lycopods, and horsetails, however, have retained a feature (to be described later) that binds them closely to their aquatic ancestors, although in other respects they have evolved an effective land habit.

The plant body of a fern, as we encounter it in nature, consists of a vascular axis bearing highly differentiated leaves and roots (Fig. 21.13*a*). The internal organization of all these parts bears a striking resemblance to the corresponding structures in the seed plant: The fern leaves, sometimes called fronds, bear on their under surfaces a very large number of **sporangia,** variously grouped, each producing numerous meiospores (Figs. 21.13*b* and 21.14). The fern plant, therefore, is a sporophyte. The spore-bearing leaves are **sporophylls.** The sporangia are minute, multicellular, stalked sacs (Figs. 21.13*c* and 21.15). A special ring of thick-walled cells called the **annulus** causes the sporangium to break open, under dry conditions, in such a way that the contained meiospores are thrown free and carried off by air currents (Figs. 21.13*c, d* and 21.16). In the familiar species of fern, as in the mosses and in the algae discussed, all the meiospores are similar in appearance. This condition is call **homospory,** and the species are described as **homosporous.** A few species of highly specialized ferns of aquatic habitats produce meiospores of two sorts and are thus **heterosporous.**

A meiospore germinates on suitable moist soil, producing a short filament of green cells, the young gametophyte (Fig. 21.13*e*). Some of the filament cells develop slender protuberances, called **rhizoids,** which penetrate the soil surface and provide anchorage. The terminal cells of a filament undergo repeated divisions yielding a thin flat sheet of cells that finally, by differential growth, attains a heart shape (Figs. 21.13*f* and 21.17). This structure, called a **prothallus,** commonly less than 1 cm wide and closely appressed to the moist soil, is the mature gametophyte. On its underside it develops multicellular antheridia and archegonia (Fig. 21.13*g, h*). The motile microgametes are released from the antheridia through a terminal pore and swim free in the film of water (Fig. 21.18). Meanwhile, the disorganization of cells within the neck of the archegonium creates a passage through which several microgametes swim (Fig. 21.13*i, j*). A single microgamete enters the megagamete, fuses with the nucleus, and thus initiates the sporophytic stage.

The zygote, or fertilized egg, gives rise to a simple embryo that grows into a young sporophyte, consisting of foot, root, stem, and leaf (Fig. 21.13*k, l, m*). When the sporophyte has become well established, the gametophyte

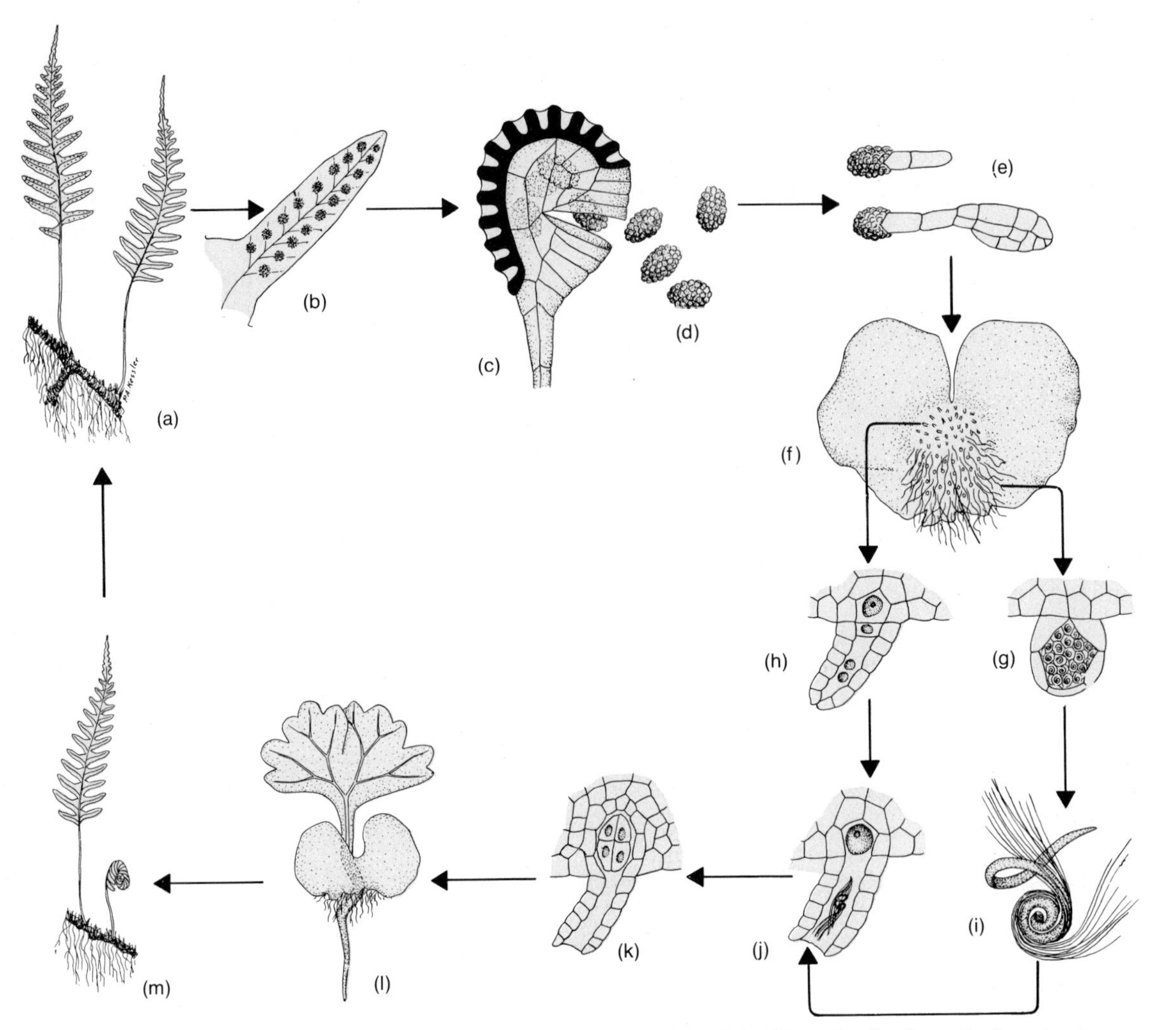

Fig. 21.13 Diagrammatic representation of the life cycle of a fern. Explanation in text.

withers and disappears. The sporophyte of some fern species produces an erect stem that may attain a height of up to 10 or 12 meters, as in the tree ferns of the tropics (Fig. 5.8). The stems of most ferns, however, are relatively short, and many are subterranean.

The conspicuous phase of the fern life cycle is the sporophyte, which is perennial, whereas the gametophyte is ephemeral and relatively inconspicuous. Both phases contain chlorophyll and so are autotrophic. The fern may properly be judged a terrestrial plant by virtue of its vascular axis and air-dispersed meiospores. However, as in the moss, the motile microgamete is the primitive ancestral feature that reflects the fern's ancient aquatic origins.

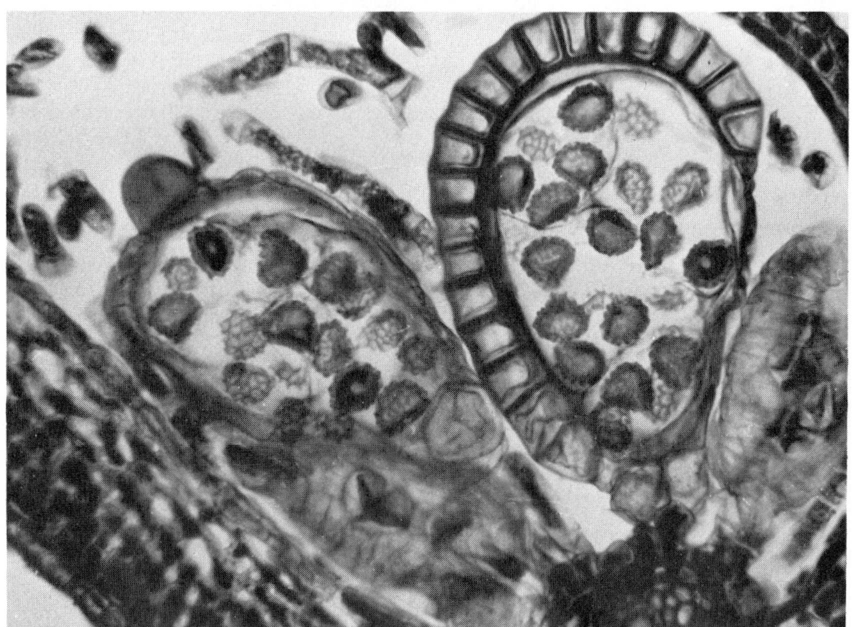

Fig. 21.14 (*Top left*) Pteris. *Lower surface of leaflet showing clusters (sori) of sporangia.* (*Top right*) Dryopteris. *Lower surface of leaf segment showing sori protected by indusia (membranous outgrowths of the leaf surface.)*

Fig. 21.15 Pteris. *Photomicrograph of sporangia and meiospores. (Courtesy, Carolina Biological Supply Co.)*

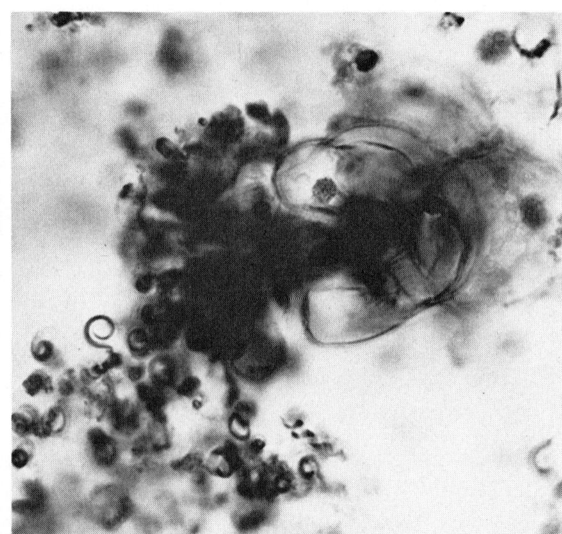

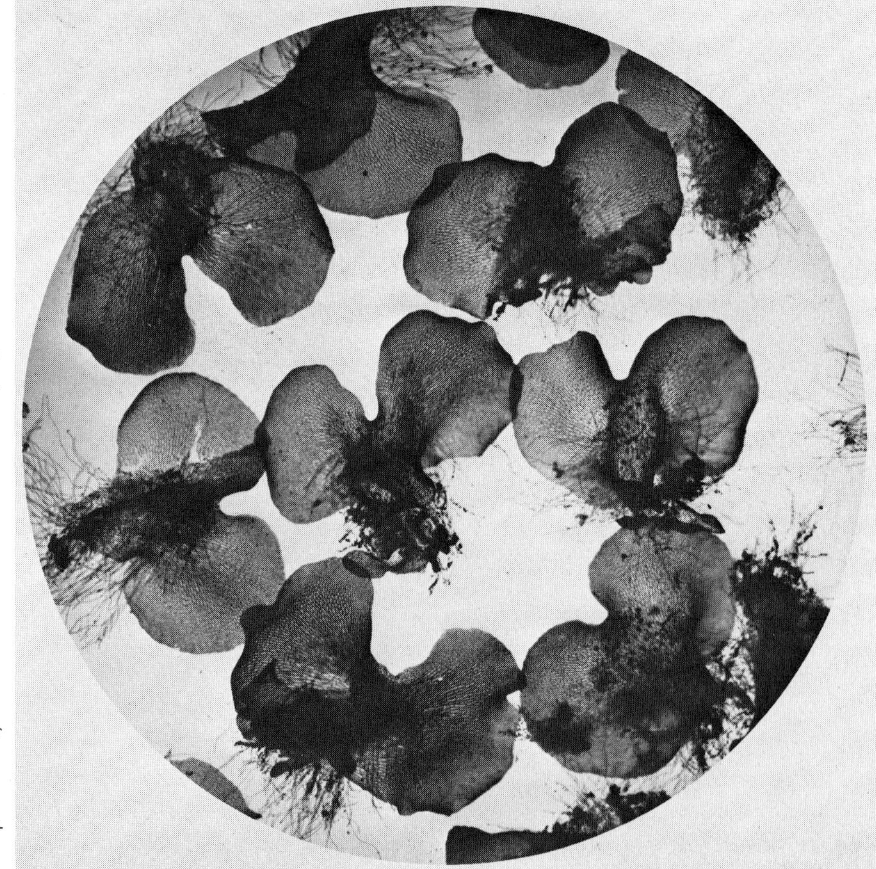

Fig. 21.16 (Top left) Ripe sporangia of Osmunda *discharging a cloud of meiospores. (Courtesy, Carolina Biological Supply Co.)*

Fig. 21.17 (Right) Fern prothalli (gametophytes), view of lower surface. Note the sex organs in broad median zone in prothalli at lower right and abundant growth of rhizoids. (Copyright, General Biological Supply House, Inc., Chicago.)

Fig. 21.18 (Top right) Swimming microgametes of fern in the vicinity of a receptive archegonium. (Courtesy, Carolina Biological Supply Co.)

SELECTED TRACHEOPHYTES: A GYMNOSPERM

The gymnosperms (naked seed) are a relatively small (about 700 species) but botanically and economically very important group. As representative of the gymnosperms, we shall select the familiar pine.

In general structure of the vegetative plant body, the pine is similar to that of the flowering plant described in Chapter 9. It will be recalled that the secondary xylem is composed principally of pitted tracheids rather than vessels and fibers. The pine tree is the sporophytic phase of the life cycle and produces two sorts of meiospores, **microspores** (micromeiospores) and **megaspores** (megameiospores). The pine, therefore, is **heterosporous.** The microspores are formed in elongate **microsporangia** that are borne in pairs on the underside of **microsporophylls.** The microsporophylls are arranged spirally in a cone or strobile, the **microsporangiate cone** (microstrobile, or male cone) (Fig. 21.19*a, b, c*). The microsporangiate cones are borne in clusters at the tips of last season's growth on most branches (Fig. 21.20). As the terminal bud begins growth in early spring the cones enlarge, often assuming a reddish or purplish color. During enlargement of the cones, the microspore parent cells undergo meiosis and the resultant haploid microspores (four from each parent cell) develop a characteristic pair of wings (Fig. 21.19*d*) through separation of two layers of the wall. While still within the microsporangium, the haploid nucleus of each winged microspore undergoes two successive mitotic divisions, giving two **prothallial cells** and an **antheridial cell** (Fig. 21.19*d, e*). The antheridial cell now divides to form a **tube cell** and a **generative cell** (Fig. 21.19*f*); meanwhile the prothallial cells are degenerating. These divisions of the microspore constitute the germination of the meiospore and thus the beginning of the gametophytic phase. The two-winged, four-celled structure is a young *micro*gametophyte. Dehiscence of the microsporangial wall liberates the young microgametophytes **(pollen).** Dense, yellow clouds of pollen can be seen drifting from the pines on a dry, breezy spring day.

The **megasporangiate cones** (megastrobiles, or female cones) are borne on short lateral branches arising from the current season's growth (Fig. 21.20). A female cone consists of an elongate axis bearing broad **ovuliferous scales** spirally arranged, each standing in the axil of a small and inconspicuous **bract** (Fig. 21.19*i, j*). Two ovules are borne upon the upper surface of each ovuliferous scale. The ovuliferous scale is commonly referred to as a **megasporophyll,** that is, a megasporebearing *leaf,* but its axillary position requires it to be interpreted as a modified dwarf *shoot.* The megastrobile, therefore, is *compound,* whereas the microstrobile is *simple.* Each ovule consists of a massive **nucellus,** or **megasporangium** with a single enveloping **integument** (Fig. 21.19*k*). The **micropyle,** a circular opening in the integument, at the apex of the ovule, is directed inward toward the cone axis (Fig. 21.19*k*). A **megaspore parent cell** differentiates from the nucellus (Fig. 21.19*k*). At this stage the tips of the ovuliferous scales are slightly spread apart and the airborne pollen falls among the scales, sifts downward, and comes to rest in the space near the micropyle (Fig. 21.19*k*). The transfer of pollen from the microstrobile to the megastrobile is **pollination.** Following pollination the ovuliferous scales thicken considerably at the tips, bringing the scales into contact with each other and thus closing the cone.

About one month after pollination the megaspore parent cell undergoes meiosis, giving rise to a row of four haploid megaspores (Fig. 21.19*l*). Three of the megaspores distintegrate and the remaining functional megaspore, after a delay of about five months,

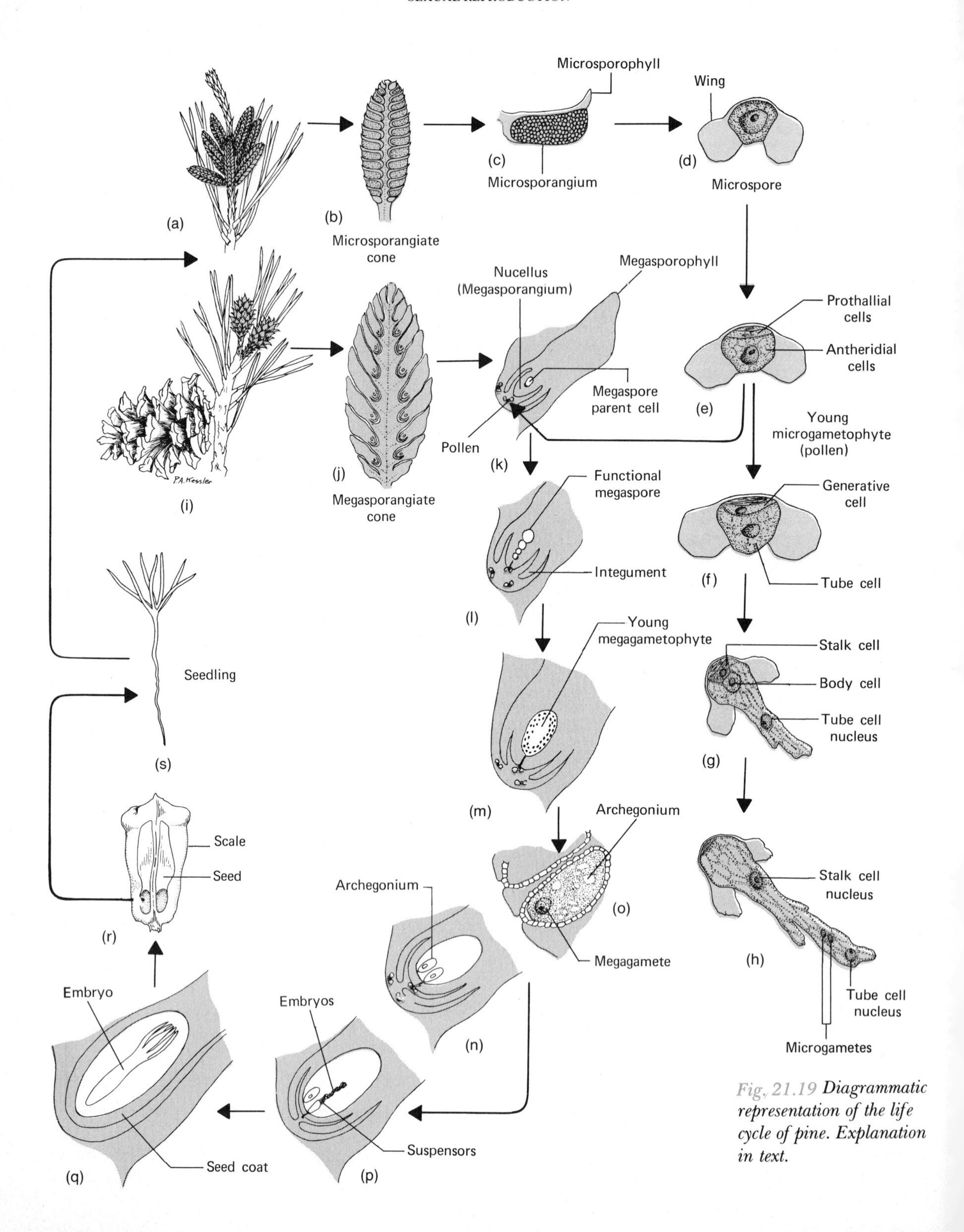

Fig. 21.19 Diagrammatic representation of the life cycle of pine. Explanation in text.

Fig. 21.20 Pine. *(Top) Mature microsporangiate cones shortly before release of pollen. (Middle) Megasporangiate cone at time of pollination in spring. Note the open, reflexed cone scales. Small buds on main branch will become clusters of needle leaves. (Bottom) Mature megasporangiate cones at time of seed dispersal. A few winged seeds are still present on some of the cone scales. (Top and middle: courtesy, Carolina Biological Supply Co.)*

begins development of the *mega*gametophyte by a long, slow series of nuclear divisions. These divisions continue for about six months, the ovule and entire megastrobile meanwhile increasing in size (Fig. 21.19*m*). Enlargement of the growing megagametophyte is accompanied by progressive disintegration of adjacent cells of the nucellus. In the second spring, 13 months after pollination, the multinucleate megagametophyte becomes cellular by the development of cell walls among the nuclei. Three or four of these cells near the micropylar end differentiate as **archegonial initials.** An archegonial initial divides to form an initial neck cell at the surface of the megagametophyte and a central cell. The neck initial by a few additional divisions forms a neck. The central cell enlarges greatly, its nucleus moving to a position just below the neck where it divides to produce an **egg nucleus (megagamete)** and a promptly disintegrating ventral canal nucleus. As the **archegonium** is thus maturing, adjacent cells of the megagametophyte become modified to form a layer of jacket cells around it (Fig. 21.19*o*). Usually from two to four archegonia develop. Development of the megagametophyte is now complete (Fig. 21.19*n*), the process having taken a little more than 13 months following pollination.

The pollen (four-celled *micro*gametophytes), it will be recalled, had reached the micropyle of the ovule. Assisted by a droplet of fluid exuded by the micropyle, the pollen grain passes through the opening and comes in contact with the megasporangium. Growth of the tube cell promptly ensues and the resultant

pollen tube begins a slow penetration of the megasporangial tissue. As the tube grows, it branches somewhat and the tube nucleus remains close behind the advancing tip. In the spring, one year after pollination, the generative cell divides to yield a **stalk cell** and a **body cell** (Fig. 21.19*g*). Shortly before the growing pollen tube completes penetration of the megasporangium and reaches the megagametophyte, the nucleus of the body cell divides to form two **microgametes** or **sperms** which move forward in the pollen tube (Fig. 21.19*h*). When the pollen tube makes contact with the archegonium the microgametes and other nuclei of the tube are discharged into the egg cell. One of the sperms fuses with the egg nucleus. The egg, thus fertilized, is the diploid zygote, the beginning of the sporophytic phase. Fertilization occurs in late spring or early summer, 13 to 14 months after pollination.

Shortly after fertilization the diploid nucleus of the zygote undergoes two successive mitotic divisions, yielding four diploid nuclei of the **proembryo.** These migrate to the base of the archegonium, arranging themselves in a tier of four. Further nuclear and cell divisions result in a 16-celled, 4-tiered arrangement. **Embryos** develop from the cells of the lowest tier. Cells of the second tier elongate greatly **(primary suspensors),** pushing the cells of the first tier into the tissue of the megagametophyte which meanwhile has become charged with reserve food materials. **Secondary suspensors,** cut off from the embryonal cells of the first tier, increase the thrust upon the developing embryos, forcing them deeper into the megagametophytic tissue (Fig. 21.19*p*). The nutritive gametophytic tissue is called the **endosperm,** and in the mature seed it completely envelops the embryo. Although four embryos may begin development, usually only one goes on to maturity.

The completely developed embryo (sporophyte) consists of a **hypocotyl,** an **epicotyl,** and several **cotyledons** (Fig. 21.19*q*). The hard, bony *seed coat* of the mature *seed* is derived from the integument by modifications of the cells in its middle layer.

The whole megastrobile meanwhile has enlarged greatly and the cone scales have thickened and become woody, to form the familiar ripe pine cone (Fig. 21.19*r*). From the time of fertilization to the maturation of the cone requires about four months. The ripe cone opens by a reflexing of the scales, the winged seeds fall out and are dispersed by the autumn winds (Fig. 21.19*i* and 21.20).

Certain features of the life cycle of the pine should be reëmphasized and comparisons should be made with the life cycle of the common fern. Many of these features are clearly derived from those of the more primitive ferns and at the same time foreshadow conditions to be found in the life cycle of angiosperms. Pine is heterosporous, producing its micro- and megaspores upon micro- and megasporophylls borne in separate cones. Appropriately, micro- and megagametophytes result from the germination of the two kinds of meiospores, and these are much reduced in size and complexity and are dependent on the sporophyte. The microgametes are not motile, but are carried to the archegonium by growth of the pollen tube. Thus the pine is not dependent on the presence of an external film of water for the movement of the microgamete. The embryo is relatively highly developed, encased in a copious food reserve (the endosperm), and well protected by the seed coat. In this situation the embryo may remain dormant for a period of time.

SELECTED TRACHEOPHYTES: AN ANGIOSPERM

The organization of the flower, the specialized reproductive structure of the flowering plant, or angiosperm, was described in Chapter 9. In this chapter we deal with important details of the sexual reproductive process omitted in that chapter. It may be helpful to review flower structure at this point.

The tree, shrub, or herb bearing flowers is the sporophytic phase of the life cycle of angiosperms. The micro- and megasporophylls (stamens and carpels, respectively) are borne upon the torus of the flower, along with the associated corolla and calyx. The flower is essentially a cone of sporophylls, somewhat like the cones of pine, except that typically, both micro- and megasporophylls are in the same cone (Fig. 21.21*a*).

The anther (microsporangia) of a stamen in a young flower bud, as seen in a cross-sectional view, consists usually of two or four chambers that contain a mass of large cells, the microspore parent cells (Fig. 21.21*c*). Each microspore parent cell by meiosis produces four microspores (Fig. 21.21*e*). The microspore nucleus divides and the daughter nuclei become incorporated into two cells of unequal size by the formation of separating membranes. The larger is called the **tube cell,** the smaller, the **generative cell.** This division is accompanied by thickening of the microspore wall and concomitant formation of ridges, spines, or grooves on the surface. The division of the microspore and its wall modification have transformed it into a young microgametophyte or pollen grain. At this time, in most flowering plants, the walls between adjacent pollen chambers break down, thus creating two **pollen sacs.** The pollen sacs open by pores or splits as the flower matures, freeing the pollen (Fig. 21.21*f*).

During organization of the pollen, as described, other events are taking place in the pistil. Within the ovulary of a young developing pistil, one or more small protuberances arise on a **placenta,** commonly where carpel margins meet. The protuberance is a hemispheric mass of cells, the megasporangium, or nucellus, in which a single large megaspore parent cell develops (Fig. 21.21*h*). After a slight elongation at the base, the megasporangium becomes almost completely enveloped by one or two collars of cells originating at the base. The enveloping layers are the **integuments.** The small opening left at the rim of the integuments is the **micropyle.** Differential growth at the base usually begins at this time so that the micropyle becomes turned toward the placenta.

By meiosis the megaspore parent cell produces a linear series of four **megaspores.** Usually, the three megaspores nearer the micropyle disintegrate, leaving the basal one functional (Fig. 21.21*i*). The functional megaspore enlarges rapidly, its nucleus undergoes three successive divisions yielding eight nuclei, and the cytoplasm develops a large central vacuole (Fig. 21.21*j, k, l*). Four of the eight nuclei are situated at each end. One nucleus from each group of four moves to the approximate center and each of the remaining six becomes incorporated into a cell by the formation of separating membranes. Thus three cells stand near the micropyle and three at the basal end. This structure, consisting of six cells and two free nuclei and contained within the much extended megaspore wall, is the mature megagametophyte, or **embryo sac** (Fig. 21.21*m*). In most flowering plants the enlargement associated with formation of the embryo sac results in the destruction of tissue of the megasporangium, so that the megaspore wall (embryo sac wall) comes to lie against the integument. The entire structure, consisting of embryo sac plus integuments, constitutes a mature **ovule.**

The position of the cells and nuclei of the embryo sac carries special significance, for their behavior and the role they play in subsequent events is characteristic. The two free nuclei that migrate to the center of the sac are called **polar nuclei.** The three cells farthest from the micropyle are the **antipodal cells.** One of the three cells near the micropyle is the

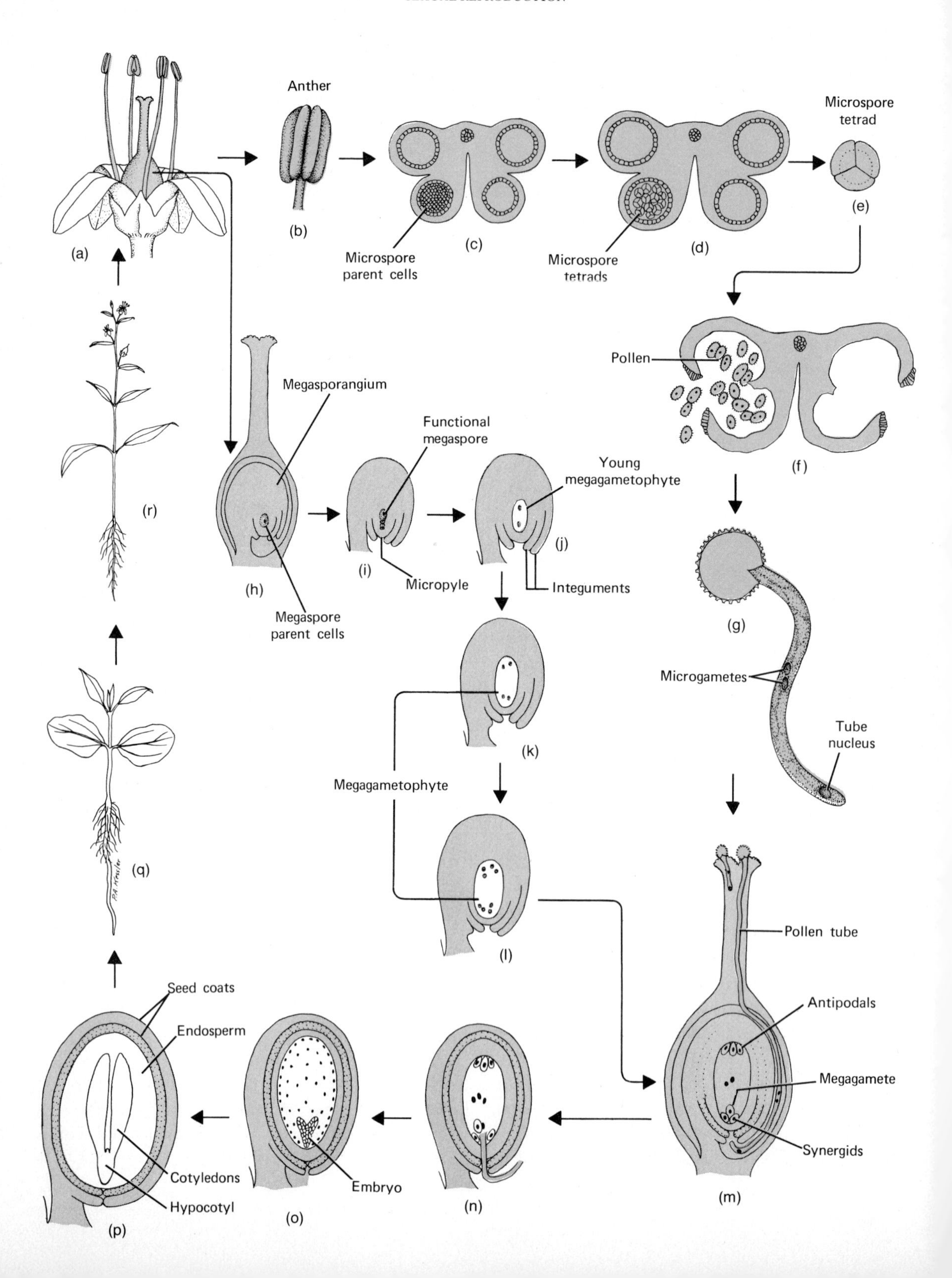
Anther
Microspore tetrad
(a)
(b)
(c)
(d)
(e)
Microspore parent cells
Microspore tetrads
Pollen
(f)
Megasporangium
Functional megaspore
Young megagametophyte
(r)
(h)
(i)
(j)
Micropyle
Integuments
Megaspore parent cells
(g)
Microgametes
Tube nucleus
(k)
Megagametophyte
(q)
(l)
Pollen tube
Seed coats
Endosperm
Antipodals
Megagamete
Synergids
Cotyledons
Hypocotyl
Embryo
(p)
(o)
(n)
(m)

Fig. 21.21 Diagrammatic representation of the life cycle of a flowering plant. Explanation in text.

megagamete, or egg, and the remaining two are called **synergids** (Fig. 21.21*m*). The antipodal cells and synergids are apparently functionless in embryo production, for they take no part in subsequent development and commonly soon disintegrate.

The type of embryo sac development described is one of the commonest, but among the angiosperms much variation in detail is observed from species to species, and ten or more distinct types have been described. These types differ with respect to such features as the number of megaspores participating in embryo sac formation and the number and arrangement of cells in the mature structure.

Now let us turn our attention again to the pollen grains in the open pollen sacs of the anther. By one means or another, as we shall discuss later in this chapter, pollen grains are transferred to the stigma of the same or another flower where they become firmly fixed by contact with a viscid stigmatic secretion. The transfer of pollen from anther to stigma is called **pollination.** Under the stimulating influence of the stigmatic fluid, the pollen grain germinates and sends downward through the stigma and style a slender pollen tube. As the tube emerges from the pollen grain and grows forward, the tube nucleus usually assumes a position just behind the advancing tip. Shortly, the generative cell moves into the pollen tube where it divides to produce two microgametes, or sperms (Fig. 21.21*g, m*). The pollen tube is the mature microgametophyte. The tip of the pollen tube enters the micropyle of the ovule, penetrates the wall of the embryo sac, and discharges the tube nucleus and two sperms into the embryo sac. The tube nucleus probably functions, as its name implies, as the nucleus for the growing pollen tube and after discharge into the embryo sac it disintegrates. One of the sperms moves deep into the sac and takes part in a **triple fusion** that involves also the two polar nuclei. The product of the triple fusion is called the **primary endosperm nucleus.** The remaining sperm fuses with the egg. The fusion of sperm and egg constitutes fertilization (Fig. 21.21*n*). The fertilized egg is the zygote, the first cell of the embryo, and thus the beginning of the new sporophytic phase. It is usually at this time that the synergids and antipodals disintegrate.

From this point on, events move rapidly and culminate in the development of a seed. The fertilized egg, lying close to the micropyle, undergoes several cell divisions to yield a row of cells extending a little way into the embryo sac. The terminal cell of the row now divides transversely and longitudinally, producing a spherical cluster of eight cells. This group of cells will contribute the major portion of the embryo, whereas the lower part of the row serves as a suspensor and through continued growth pushes the developing embryo more deeply into the embryo sac.

Meanwhile successive divisions of the primary endosperm nucleus produce a large number of free **endosperm nuclei** which congregate around the developing embryo and along the walls of the embryo sac (Fig. 21.21*o*). The endosperm nuclei gradually become incorporated into cells by the formation of separating membranes, and the cells so formed (the endosperm) grow rapidly and accumulate food in the form of starch, protein, or oil. The accumulated food comes, of course, from the parent plant. In some seeds the embryo soon reaches the limit of its growth, and is enveloped by a large mass of **endosperm** (Fig. 21.21*p*). This condition obtains in such seeds as those of castor bean and corn. In other species such as pea and bean, the developing embryo continues to grow until all of the endosperm is absorbed and its food substance is transferred to the cotyledons, which become very fleshy.

The one or two integuments, meanwhile, have undergone modification of the outer cell layer and now are the seed coats. The modification may take many forms. For example, in the bean the outer coat is heavily cutinized and rather tough, in the peanut it is thin, brown, and papery. In cotton, some of the cells of the outer integument extend outward like fine hairs, producing the long cotton fibers. In some species, such as trumpet vine, the seed coat develops membranous, wing-like extensions, or as in milkweed, feathery tufts of hairs (Fig. 9.44) which assist in dispersal by wind currents.

In the bean and many other seeds the micropyle may still be visible as a small depressed dot just above the **hilum** or scar where the seed was attached to its stalk. In seeds with outer coats of low permeability, much of the imbibition of water, which initiates germination, takes place through the hilum.

POLLINATION

Something more should be said about pollination. It is a fascinating subject which brings into view some of the most interesting aspects of plant form and behavior. The germination of pollen upon the stigma, the production of sperm, and the subsequent fertilization of the egg are indispensable to the usual development of viable seed. Pollination also plays an important role in the natural development of the fruit. Each seed produced involves the action of a separate pollen grain. A tomato with 100 seeds had at least 100 pollen tubes that reached 100 ovules, all at about the same time.

Air currents and insects are the principal agents for pollen transfer. Plants pollinated by air currents often have small and inconspicuous flowers borne well exposed on the plant. Frequently the corolla is much reduced in size, or lacking. Such plants produce a great abundance of buoyant pollen and have large feather-like stigmas. Many of these plants are among the early spring blooming species. Corn, the common grasses, oaks, hickories, elms, and ragweed are examples of wind-pollinated species. Hay fever sufferers are especially aware of the windborne pollen of certain species.

Insect-pollinated species usually produce conspicuous and brightly colored or nectar-bearing flowers with sticky pollen. The transference of pollen by the insects is, of course, incidental to the insects' search for food. Pollen adheres to the body and legs of insects chiefly of the bee order (Hymenoptera), although some moths and butterflies (Lepidoptera) are effective pollen carriers. Elaborate flower modifications are known that seem designed to assure pollen collection by insect visitors. In such flowers the positions of stigma, stamens, and nectaries are such that insects in search of nectar cannot avoid collecting and distributing pollen.

Based on the source of the pollen, two types of pollination may be recognized. **Self-pollination** is the transfer of pollen from anther to stigma of the same flower or of another flower on the same plant. **Cross-pollination** occurs when pollen is carried to the stigma on another plant of the same or related species. Many species are regularly self-pollinated. This is usually brought about by pollination before the flower opens fully, or by growth of the receptive stigma among the ripe stamens. Self-pollination may occur by either insects or air currents, but these agents, as a rule, are more significant in effecting cross-pollination. Cross-pollination is probably more general than self-pollination. In many instances, it has been shown that plants of greater vigor and productivity result from cross-pollination than from self-pollination.

Many species possess structural and functional modifications that tend to prevent or reduce the possibility of self-pollination. For example, liberation of pollen in some flowers occurs at a time when the stigma of the same flower is nonreceptive. Although this tends to favor cross-pollination, it does not preclude self-pollination involving other flowers on the same plant. In some species a condition of **self-sterility** is encountered, in which the plant's pollen is ineffective upon its own stigmas. Insect-pollinated species, however, possess the most complex mechanisms for ensuring cross pollination. In sage, a plant of the mint family, a bee may light upon the lip of the corolla and push into the flower for nectar, and in doing so cause the stamens to trip forward and dust his back with pollen. When the insect subsequently visits another, older flower in which the stigma has grown receptive, his pollen-laden back brushes against the stigma, transferring the pollen (Fig. 21.22). Several species produce, on different plants, flowers of two types in which the relative positions of stigma and anthers are reversed, so that the part of the insect's body bearing pollen picked up in one flower will deliver it to the stigma of the second flower (Fig. 21.23).

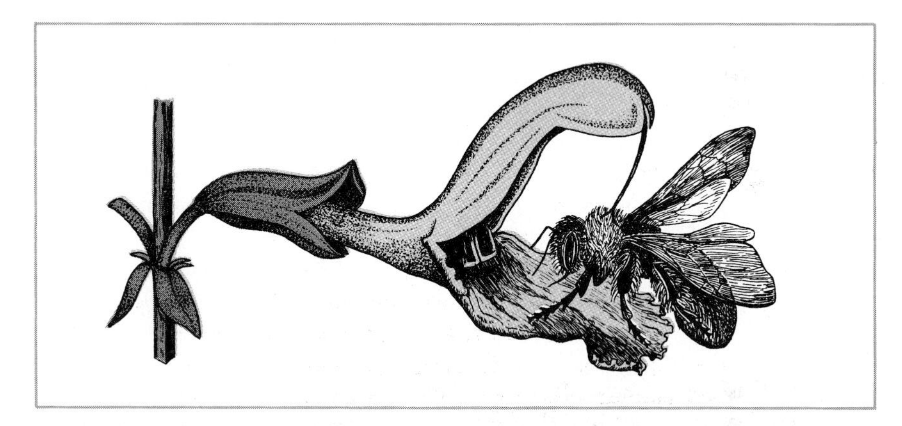

Fig. 21.22 Flower of sage, Salvia, *showing specialized stigma contacting pollen-laden back of bee. (After Kerner,* Natural History of Plants.)

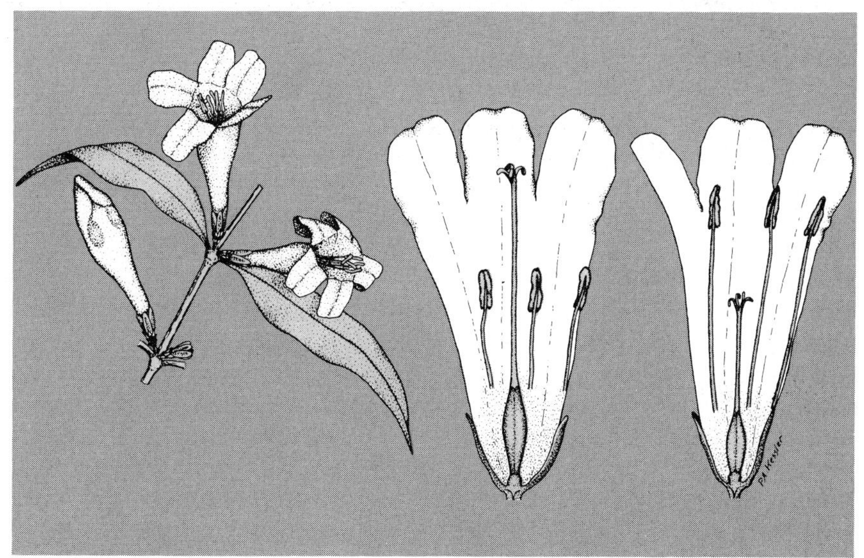

Fig. 21.23 Gelsemium *flower, showing dimorphism that ensures cross-pollination by insects.*

Knowledge of the pollination characteristics of some cultivated plants may be of great importance to orchardists. For example, although many apple varieties are **self-fertile** and set fruit when self-pollinated, they often do better when other varieties are available as a pollen source. Some varieties are quite self-sterile and do not set fruit unless they are cross-pollinated. Thus commercial orchards are often planted so as to provide compatible cross-pollinating varieties whose periods of flowering coincide. Some varieties of cherry require cross-pollination. Peaches, on the other hand, are all self-fertile.

The pollen of seed plants (and the spores of lower forms) are interesting objects for study. In size, shape, patterns of surface sculpture, and in other features, they present a great variety. Many of them are objects of great beauty in the intricacies of their design, and for very many species and larger categories they are distinctive. The outer walls are commonly quite resistant and thus they readily become fossilized in sedimentary formations. Wind-dispersed pollen becomes widely distributed and much of it falls into bogs and swamps where it accumulates over long periods of time. Microscopic examination of such situations frequently yields information about the kinds of plants that grew in or near ancient sites. This history of forest migrations and the dating of coal deposits have been worked out by pollen and spore analyses. The petroleum industry is using such data on an increasing scale in the exploration for oil. The study of pollen and spores, called **palynology,** is an important division of botany.

COMPARISON OF FLOWERING PLANTS AND GYMNOSPERMS

Comparison of a flowering plant and a gymnosperm, as represented by pine, should now be made with respect to certain features of their life cycles. They are similar, of course, in that both are heterosporous, the mature microgametophyte of both consists of a pollen tube, and both produce a seed that consists of an embryo plant, protective seed coats, and a food reserve. They are different, however, in important ways. The gymnospermous ovules are borne exposed on open, flat megasporophylls, whereas those of angiosperms are typically enclosed within an **ovulary** that develops into a true fruit (Chapter 9). The gametophytic phases of the flowering plant are somewhat simpler. The angiospermous microgametophyte consists of a single cell (the pollen tube and its nucleus) containing two sperms, and the megagametophyte (embryo sac) consists usually of only six cells and two free nuclei contained within the megaspore wall. Archegonia are not present in the megagametophyte of angiosperms. However, it may be interesting to interpret the embryo sac as a single, much reduced archegonium in which the synergids may be regarded as the remains of a more extensive archegonial wall and the antipodals as vestiges of a bulkier gametophyte. The endosperm of pine is the food-enriched tissue of the megagametophyte, whereas the endosperm of the flowering plant is the product of the special triple fusion of one sperm with two polar nuclei.

Seed plants are distributed in nature by the dispersal of young sporophytes (embryos). In the lower forms of plants (ferns, and the lower tracheophytes, bryophytes, and thallophytes) the dispersal structures are spores or meiospores, the gametophytes of these forms being capable of independent existence apart from the sporophyte. With various modifications of seeds and fruits, seed plants may effectively utilize a variety of agents for dispersal.

The seed, with its well-advanced embryo, efficient protective seed coat, and food supply is, perhaps, chiefly responsible for the great success of the seed plants. The seed confers on the species tremendous survival potential through improved chances for successful establishment.

SUMMARY FOR SEXUAL REPRODUCTION IN PLANTS

A number of interesting and important themes may be discerned in a comparative study of sexual reproduction in plants, from the most primitive to the most advanced. One of the most striking is a progressive reduction in the comparative size and complexity of the gameophytic phase as it progresses from the thallophytes to the higher tracheophytes. From a condition of structural dominance and independence in the sexually reproducing algae, the gametophyte becomes a dependent, microscopic structure of simple organization in the angiosperms. Concomitantly, the sporophytic phase becomes progressively larger and more complex and ultimately the dominant, independent phase.

Other general trends involve evolutionary changes in the form and behavior of various structures. Thus, motile isogametes are succeeded by nonmotile heterogametes; motile isospores are succeeded by nonmotile heterospores. It must be remembered, however, that these evolutionary changes do not in any way alter the biological significance of any structure, within its own cycle. Although homologous structures may change in appearance in the course, of evolutionary development, their basic functions and their position in the sexual cycle remain unchanged. Thus we may state this generalization:

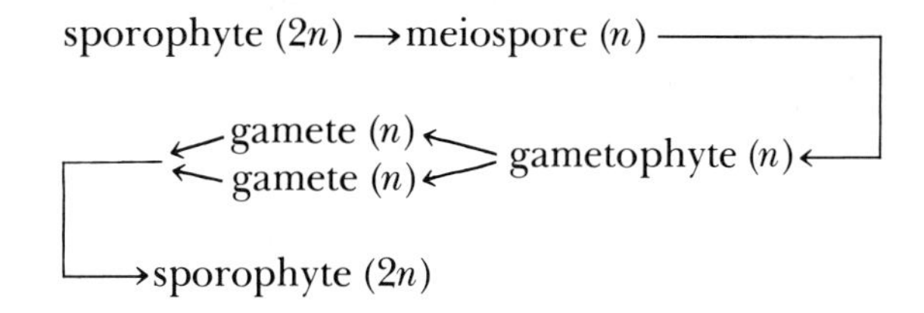

and observe that there occurs in a sexual life cycle a regular alternation of sporophytic and gametophytic phases.

RELATED READING

Echlin, P. "Pollen," *Scientific American, 218*(4), 80–90, April 1968.

Hutchins, R. E. "Flight secrets of a jungle seed," *Natural History, 62,* 416–419, 1953.

Jensen, W. A. "Fertilization in flowering plants," *BioScience, 23,* 21–27, 1973.

Mather, K. "Mating discrimination in plants," *Endeavour, 2,* 17–21, 1943.

Raghavan, G. "Control of differentiation in the fern gametophyte," *American Scientist, 62,* 465–475, 1974.

22 PLANT GENETICS

In Chapter 15 we pointed out that the hereditary potentialities of an organism, interacting with the various factors of the environment, determine the nature and course of the processes and conditions within the organism, and that these in turn dictate the pattern of the organism's behavior, growth, and development. We have already considered some of the influences of environment on plants, and now that we have described the reproductive processes of plants we are in a position to discuss the nature and mode of action of hereditary potentialities, their transmission from generation to generation, and their role in development and hereditary variation.

Although in this, as in other considerations of inheritance, hereditary variations will receive the greater part of our attention, hereditary stability is more universal and of at least equal importance in the biological world. Pea plants may differ from one another in height, flower color, seed color, and many other hereditary traits, but they all have in common a multitude of traits that set them apart as a distinct species of plants and that have been transmitted unchanged through countless generations. The fossil record reveals that many species of plants and animals have remained essentially unchanged through millions of years, a fact that is sometimes neglected as we center our attention on the evolution of new species through the natural selection of hereditary variations. Such hereditary stability generation after generation is even more remarkable than hereditary variation, especially when one considers the complex and delicate nature of living substance and the narrow protoplasmic bridge—commonly only unicellular spores or zygotes—that link one generation with another.

THE DEVELOPMENT OF OUR CONCEPTS OF INHERITANCE

Early Ideas About Heredity

For centuries man has recognized that plants and animals beget their own kind but that offspring may differ from parents in many details; however, even these basic and obvious concepts have not always been universally accepted. During ancient and medieval times people commonly believed that one kind of living organism could give rise to another (e.g., that geese developed from goose barnacles) and that worms, molds, and other lower forms of life arose spontaneously from slime and decaying matter. It was not until about 1650 that experiments discrediting such beliefs in spontaneous generation were first conducted, and until Pasteur's experiments in 1864 there was still much controversy as to whether microorganisms arose by spontaneous generation.

Even in those instances where organisms were recognized as being produced by parents of their own kind, the transmission of hereditary traits and hereditary variation remained a complete mystery for centuries. One barrier to any understanding of heredity was an almost complete lack of knowledge of sexual reproduction. It was not until 1676 that Nehemiah Grew suggested the true role of ovules and pollen, and it was in 1694 that Rudolph Camerarius first presented conclusive evidence that plants reproduced sexually. In 1761 Joseph Koelreuter announced his discovery of fertilization in plants and during the next several years he engaged in extensive hybridization of plants. He made over 500 different crosses and raised first and second generation hybrids and backcrosses. He had a strange concept of fertilization as being a sort of chemical reaction between two germinal fluids. In the 1830s and 1840s Carl Gaertner conducted some 10,000 hybridization experiments, repeating many of Koelreuter's, and made a number of advances, including the observation that first-generation hybrids were not always intermediate between the two parents, as Koelreuter believed, but might be identical with one parent or the other for certain traits. Even earlier John Goss (in 1822) observed segregation of traits in peas and in 1823 Andrew Knight reported dominance and recessiveness in peas as well as segregation. Other botanists who experimented with plant hybridization were Charles Naudin and G. B. Amici. Many of these investigators had experimental results that might have enabled them to formulate the basic principles of heredity, but they failed to do so, probably because they did not assemble adequate quantitative data that could be analyzed statistically. This has occurred repeatedly in the history of science. The man who makes a major breakthrough has usually been preceded by other investigators who might well have done so themselves if only their data had been a little more adequate or more carefully analyzed or if they had possessed a little more scientific creativity in drawing critical conclusions from their data.

The Work of Gregor Mendel

In 1866 Gregor Mendel (Fig. 22.1), an Austrian monk who had done careful and extensive experiments on inheritance in peas, published his *Experiments in Plant Hybridization,* a report now recognized as being one of the

Table 22.1 **Data of Mendel on Ratios of Dominants and Recessives in the Second Generation (F_2) Pea Plants Following Single Factor Crosses**

Experiment	Dominant Character	Recessive Character	Number of Plants	Ratio of Dominants to Recessives
1	Smooth seeds	Wrinkled seeds	7324	2.96:1
2	Yellow seeds	Green seeds	8023	3.01:1
3	Red flowers	White flowers	929	3.15:1
4	Inflated pods	Constricted pods	1181	2.95:1
5	Green pods	Yellow pods	580	2.82:1
6	Axial flowers	Terminal flowers	858	3.14:1
7	Tall stems	Dwarf stems	1064	2.84:1

most important of all contributions to biology and providing the foundation of modern genetics, the science of hereditary variation.

Actually, it turns out that peas are a very good choice of material for genetic studies such as those conducted by Mendel. They are naturally self-pollinated (and thus self-fertilized), resulting in genetically "pure" strains. They can, of course, be cross-pollinated manually to produce hybrids.

Mendel chose to work with seven sharply defined characters of peas (Table 22.1), each of which exhibited two alternate forms. For example, one character he studied was seed color. The seeds were either yellow or green. For each of these seven characters, he found that when he cross-pollinated (hybridized) a plant having one form of a character with a plant having the other form, all of the offspring were exactly like one of the parents as regards the trait, rather than being intermediate.

Thus, when a plant bearing smooth seeds was crossed with a plant bearing wrinkled seeds all the offspring (the **F_1 generation**) produced smooth seeds (Fig. 22.2), regardless of whether the smooth-seeded parent provided the pollen or received it. Mendel concluded that the trait for smooth seeds was **dominant** over the **recessive** trait for wrinkled seeds. The dominant and recessive characters discovered by Mendel in his seven experiments are listed in Table 22.1.

Fig. 22.1 *Gregor Mendel.*

Mendel's next step was to cross the F_1 hybrids among themselves. In each of the seven experiments he secured several hundred (or even several thousand) offspring (the F_2 generation). In the experiment with smooth and wrinkled seeds, 7324 F_2 plants were obtained. Although 5475 of them produced smooth

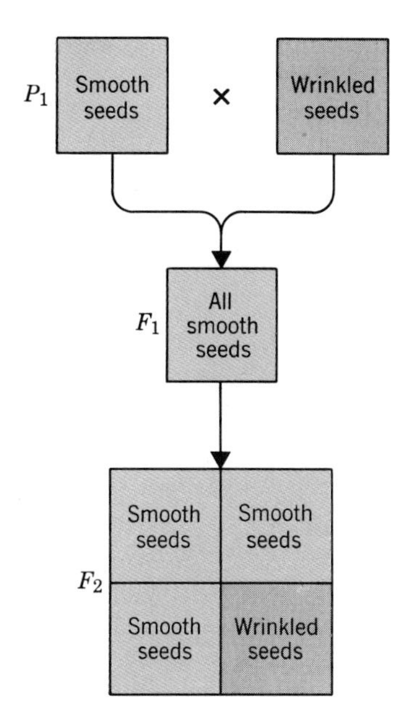

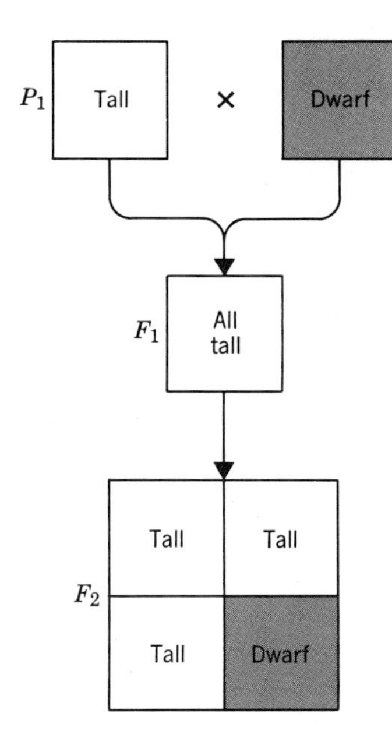

Fig. 22.2 Phenotypes of the F_1 *generation secured by crossing a homozygous dominant and a homozygous recessive* P_1 *and the phenotypes of the* F_2 *generation secured by self- or cross-pollination of the* F_1 *plants, for two different characters of pea plants.*

seeds (as did both of the parent plants), the other 1849 plants produced wrinkled seeds, a ration of 2.96:1 (Fig. 22.2). In each of the other six experiments the ratio of plants exhibiting the dominant form of the character to those exhibiting the recessive form of the character was also essentially 3:1 (Table 22.1).

Mendel also conducted experiments in which he followed the inheritance of two sets of contrasting traits at the same time (**dihybrid** crosses). For example, when he crossed pure tall, smooth-seeded plants with pure dwarf, wrinkled-seeded plants, all the F_1 plants were tall and had smooth seeds. When he then crossed the F_1 plants among themselves he found that about $^9/_{16}$ of the F_2 plants were tall and had smooth seeds, $^3/_{16}$ were tall and had wrinkled seeds, $^3/_{16}$ were dwarf and had smooth seeds, and $^1/_{16}$ were dwarf and had wrinkled seeds, a 9:3:3:1 ratio. In the first group dominance was exhibited in both traits, in the second and third groups dominance was exhibited in one trait, and in the fourth group in neither trait. The 9:3:3:1 ratio is actually the product of two 3:1 ratios (3:1 × 3:1 = 9:3:3:1).

Mendel's great contribution to our understanding of genetics was that: (1) characters are determined by genes (Mendel's unit factors) that occur in pairs in the body cells of organisms; (2) one form of a gene may be dominant over the other (may mask its expression); (3) the two genes of a pair separate from each other in the formation of the gametes (although Mendel thought in terms of pollen and ovules rather than gametes—sperm and egg), each gamete therefore receiving only one gene of each pair; (4) different gene pairs assort (segregate) independently of the other gene pairs, resulting in all possible combinations of the genes in the gametes, which then combine at random at fertilization, thus restoring the double number of genes; and (5) paired dominant and recessive genes segregate with purity, that is, neither gene is modified by their association.

Mendel was fortunate in that all the traits he selected for study showed dominance, that they all assorted independently of one another, that they all interacted to produce

9:3:3:1 F_2 ratios, and that he was working with plants that are usually self-pollinated, thus facilitating the production of the original pure parental strains (the P_1 plants). If he had happened to select traits that showed incomplete dominance, or were linked or interacted to produce other ratios, he might well have become confused. Yet it was not all luck, and Mendel must receive great credit for selecting specific traits for study, for designing suitable experiments, and for using enough plants to provide significant ratios that he could analyze and use in formulating hypotheses and drawing conclusions.

From 1866 to 1900

Mendel's excellent work might have been expected to create a sensation among biologists, but actually it was almost completely ignored. Apparently, few outstanding biologists even read the paper, and those who did read it missed its significance. Thus, biology in general still lacked any real understanding of inheritance. One reason for the neglect of Mendel's work was that, at the time, most biologists were greatly concerned with Charles Darwin's work on evolution. Another reason may have been that most biological work was still descriptive, and biologists who did read Mendel's paper may not have been prepared to appreciate his analytical and quantitative approach.

Furthermore, Mendel's proposed hereditary unit factors were purely hypothetical and neither Mendel nor any other biologist was aware of the nature and significance of mitosis, meiosis, chromosomes, and genes and their relation to heredity. Between 1875 and 1890 the work of many biologists gradually clarified the nature of mitosis and meiosis and the behavior of chromosomes during these two types of nuclear division as well as during fertilization of eggs by sperm. This newly accumulated understanding of the cellular basis of heredity set the stage for a real appreciation of Mendel's work, and in 1900 three botanists—Hugo DeVries of Holland, Carl Correns of Germany, and Erich von Tschermak-Seysenegg of Austria—working independently established the basic principles of heredity in plants, only to discover that Mendel had anticipated them by 34 years. The triple announcement of the rediscovery and verification of Mendel's work created great interest among biologists, and only a few failed to realize its importance in establishing the basic principles of inheritance. The science of genetics was now on a firm footing that made possible its rapid development during the twentieth century. Within a few years after the rediscovery of Mendel's principles of inheritance, they were found to apply to animals as well as to peas and other plants. The relationship between chromosomes and heredity was discovered, and Mendel's unit factors were identified as **genes.**

THE CELLULAR BASIS OF HEREDITY

It has already been pointed out that during mitosis (Chapter 8) each chromosome is duplicated and that consequently each of the resulting cells has the same number and kinds of chromosomes as the parent cell, whereas during meiosis (Chapter 21) the chromosomes are reduced from the $2n$ or diploid number of the spore parent cell (and other cells of the sporophyte) to the n or haploid number of the meiospores (and so of the gametophytes and gametes). Since both meiosis and fertilization play critical roles in the recombination of genes during sexual reproduction, we shall consider their genetic aspects briefly.

Meiosis and Hereditary Variations

Each diploid cell contains two complete sets of chromosomes, one contributed by the egg and one by the sperm, that gave rise to the zygote from which the plant developed. Any particular chromosome in one set has a homologous mate in the second set, comparable with it in gene content as well as in size and shape. Since only homologous chromosomes pair and then separate, each meiospore receives one complete set of chromosomes, and thus one complete set of genes. However, since it is just a matter of chance as to which chromosome of a homologous pair happens to be on one side of the equator or the other (Fig. 22.3) at metaphase I of meiosis (reductional division), the set of chromosomes in a particular meiospore can consist of any possible combination of the chromosomes of the two original sets, and thus of any possible combination of the genes located in different chromosomes (Mendel's principle of **independent assortment**). It is this shuffling of the two sets of chromosomes during meiosis that makes possible the extensive recombination of genes during sexual reproduction.

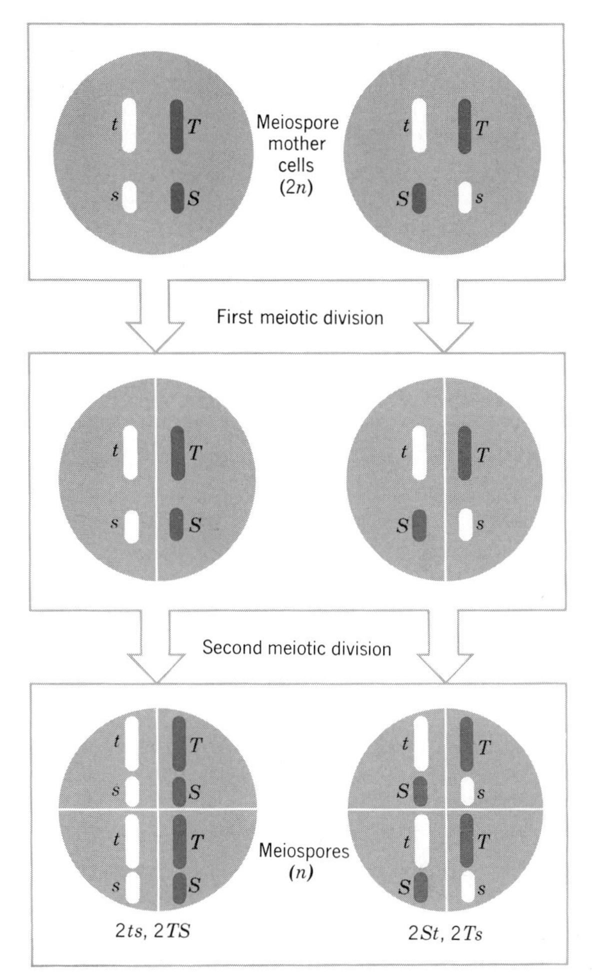

Fig. 22.3 *Independent assortment of two pairs of homologous chromosomes of a plant heterozygous for genes in each pair* (Tt *in one and* Ss *in the other*). *The left and right diagrams illustrate the two possible ways the chromosomes can line up at the cell equator at the first metophase. The four different kinds of meiospores (and so gametes) that result are produced in equal numbers since the probability of one type of assortment is the same as that of the other.*

Fertilization and Hereditary Variation

Any sporophyte individual produces many genetic kinds of meiospores (and thus of gametophytes and gametes) as a result of the random (chance) assortment of chromosomes during meiosis. It is also just a matter of chance which sperm will fertilize a particular egg, so fertilization is the second point in a sexual life cycle where hereditary variations by recombination arise. Once an egg has been fertilized and so becomes a zygote the hereditary potentialities of the new individual are established.

HEREDITARY VARIATIONS THROUGH RECOMBINATIONS

We now clarify the results of Mendel's hybridizations by first centering our attention on one pair of Mendel's "unit factors" (genes), a **monohybrid** difference, and later considering simultaneously two pairs of genes (a **dihybrid** difference). As our first example we will use one of the traits that Mendel studied: height.

A Monohybrid with Dominance

One of the two parent generations (P_1) used by Mendel was pure for tallness, that is, each of the two chromosomes of the pair in the $2n$ body cells contained the gene for tallness that we can represent by the capital letter T, the genetic makeup **(genotype)** of the plant as regards height being TT. The other parent was pure for dwarfness and its genotype can be represented as tt (the lowercase letters being used to represent the recessive genes for dwarf stems). As a result of meiosis, each meiospore, gametophyte, and gamete of the tall plants contained only one gene for tallness, represented by a single T, while those of the dwarf plants contained one gene for dwarfness, represented by a single t. Thus the zygotes (the beginning of the F_1 generation plants) resulting from the crossing of tall and dwarf plants all had the genotype Tt (Fig. 22.4). Because tallness is dominant, the F_1 plants are all as tall as their tall parent, that is, they have the same appearance or **phenotype.** However, unlike the tall parent, they have one gene for dwarfness and so are heterozygous tall (Tt) rather than homozygous tall (TT) like the tall parent. Of course, the dwarf parent was homozygous (tt).

When meiosis occurs in the F_1 plants, half of the meiospores receive a gene for tallness (T) and the other half a gene for dwarfness (t). In a large number of fertilizations we can expect all of the four possible fertilizations (T sperm $\times$ T egg, T sperm $\times$ t egg, t sperm $\times$ T egg, and t sperm $\times$ t egg) to occur with essentially equal frequency. The result is an F_2 generation consisting of $^1/_4$ homozygous tall plants (TT), $^2/_4$ heterozygous tall plants (Tt), and $^1/_4$ homozygous dwarf plants. The genotype ratio is 1:2:1, but the phenotype ratio is Mendel's 3 tall: 1 dwarf. Exactly the same ratios would be obtained with any other pair of genes with dominance, e.g. smooth or wrinkled seeds.

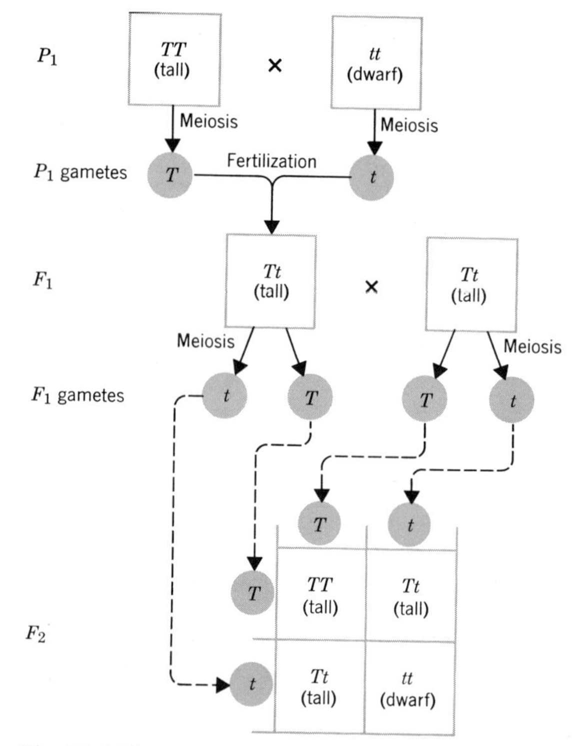

Fig. 22.4 The genotypes and phenotypes of the P_1, F_1, *and* F_2 *generations in a monohybrid cross. The* F_2 *phenotypic ratio of 3:1 is characteristic of dominance.*

Monohybrid Backcross and Test Cross

When an F_1 plant (Tt) is crossed with its recessive parent (tt), half of the offspring would be expected to be tall and the other half dwarf (Fig. 22.5). In such a backcross, both the genotype and phenotype ratios are 1:1. If a geneticist needs to know whether a particular tall plant is homozygous or heterozygous, he can make a similar cross with any dwarf plant. This is known as a test cross. If about half of the test cross offspring are tall and half dwarf, it is evident that the genotype of the tall parent is heterozygous (Tt), since a homozygous tall plant (TT) would have all tall offspring. Geneticists can observe phenotypes but not genotypes, yet they are able to deduce the genotype of an individual from its own phenotype and the phenotypes of its offspring in suitable test crosses (or even in natural populations).

A Monohybrid with Incomplete Dominance

Mendel worked only with genes that were dominant or recessive, but there are some genes that exhibit **incomplete dominance.** The heterozygous genotype results in a phenotype intermediate between the two homozygous genotypes. For example, in snapdragons and a number of other species the genotype RR results in red flowers, the genotype $R'R'$ in white flowers, and the genotype $R'R$ in pink flowers (Fig. 22.6). If a red-flowered plant (RR) is crossed with a white-flowered plant ($R'R'$), all the F_1 plants will have pink flowers ($R'R$), a phenotype different from that of either parent. Note that the assortment and recombination of the genes is the same as when dominance exists and that the genotype ratio is the same in both cases (1:2:1). However, with incomplete dominance the F_2 phenotype ratio is 1:2:1 rather than 3:1.

Fig. 22.5 A backcross of an F_1 *plant with its recessive* P_1 *parent. Note the 1:1 phenotypic ratio. In a test cross, a plant with the dominant phenotype but of unknown genotype (whether* TT *or* Tt*) is crossed with a dwarf (homozygous recessive) plant. If the phenotypic ratio of the offspring is about 1:1 (or indeed if only a few offspring are dwarf), the genotype of the tall parent was evidently* Tt*. If all of many test cross offspring are tall, it is safe to conclude that the genotype of the tall parent was* TT.

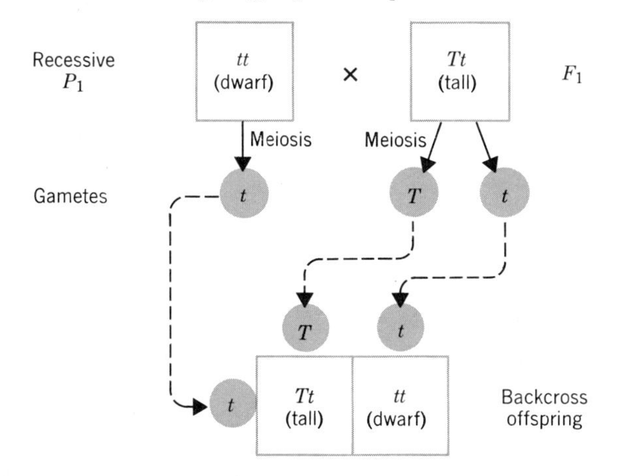

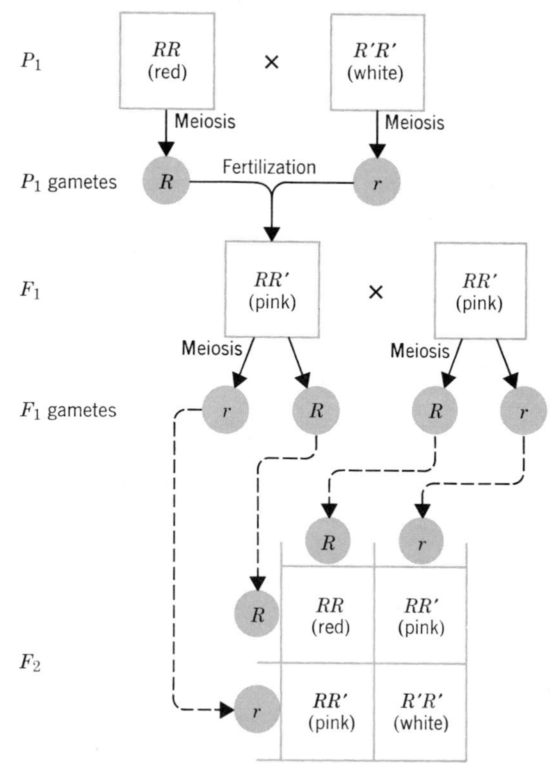

Fig. 22.6 A monohybrid cross with incomplete dominance (snapdragon or four-o'clock flower color). The procedures and the genotypic ratios are the same as in dominance, but the F_1 *phenotype is intermediate between the parents and the* F_2 *phenotypic ratio is 1:2:1 rather than 3:1.*

A Dihybrid

Mendel's experiments involving the simultaneous consideration of two sets of characters (two pairs of genes), that is, dihybrids, provided important information not available from monohybrids. If a homozygous tall, red-flowering pea plant ($TTRR$) is crossed with a dwarf, white-flowering pea plant ($ttrr$), all the F_1 plants will be tall with red flowers. (In peas, red is dominant over white, rather than incompletely dominant.) Their genotypes are *TtRr*. The two pairs of genes assort independently because they are in two different pairs of chromosomes. Thus the F_1 plants can produce four different kinds of meiospores, gametophytes, and gametes (*TR, Tr, tR,* and *tr*) in equal numbers.

These four types of eggs and four types of sperm produced by the F_1 plants can unite in 16 possible combinations (Fig. 22.7), with any one of the combinations as likely to occur as any other. There are, however, only nine *different* combinations (different genotypes) and four different phenotypes:

Genotypes	**Phenotypes**
1/16 *TTRR* 2/16 *TtRR* 2/16 *TTRr* 4/16 *TtRr*	9/16 tall, red
1/16 *TTrr* 2/16 *Ttrr*	3/16 tall, white
1/16 *ttRR* 2/16 *ttRr*	3/16 dwarf, red
1/16 *ttrr*	1/16 dwarf, white

Here we have the 9:3:3:1 phenotypic ratio observed by Mendel and found by many other investigators when they were studying any two pairs of genes that showed dominance and were located on different pairs of chromosomes. Genes may, however, show incomplete dominance or may interact with one another in a variety of other ways, so dihybrid F_2 phenotypic ratios other than the common 9:3:3:1 ratio are frequently encountered by geneticists.

An important result of the preceding dihybrid cross was the production of two new combinations of hereditary traits that did not exist in either the P_1 or F_1 generations: tall, white-flowering plants and dwarf red-flowering plants. These might be considered as new varieties of peas. Plant and animal breeders make extensive use of such recombinations of hereditary potentialities in the development of new and improved varieties. Thus, a high-yielding disease-resistant variety of wheat may be produced by crossing a high-yielding but susceptible variety with a variety that is disease-resistant but has low yields.

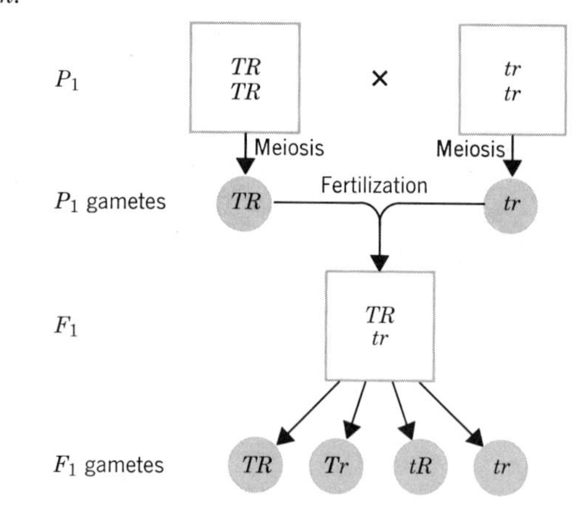

F_2

	TR	*Tr*	*tR*	*tr*
TR	*TR* *TR*	*Tr* *TR*	*tR* *TR*	*tr* *TR*
Tr	*TR* *Tr*	*Tr* *Tr*	*tR* *Tr*	*tr* *Tr*
tR	*TR* *tR*	*Tr* *tR*	*tR* *tR*	*tr* *tR*
tr	*TR* *tr*	*Tr* *tr*	*tR* *tr*	*tr* *tr*

Fig. 22.7 *In a dihybrid cross, two different traits are investigated simultaneously. The* F_1 *plants can produce four kinds of genetically different gametes (Fig. 22.3) and these can unite in 16 possible combinations at fertilization.*

Continuous Variations

So far we have considered only **discontinuous** hereditary variations. Pea plants are either tall or dwarf and the seeds are either green or yellow, with no intergrading heights or colors between these extremes. Although heterozygous snapdragons (RR′) have pink flowers, there is still no series of shades between pink and red or pink and white. Many hereditary characteristics such as height in humans or in some plants, the length of ears of corn, the color of wheat grains, some flower colors, and the size and weight of fruits of tomatoes, squashes, and many other plants show, however, **continuous** variations with a graded series of phenotypes connecting the two extremes.

Mendel studied only discontinuous variations, and this is probably another reason for his success in establishing the basic principles of heredity. Many of the earlier students of heredity had centered their attention on continuous variations and were unable to discover any logical explanation for them. We now know, however, that continuous variations, like the discontinuous ones, are influenced by genes assorting and recombining on the basis of Mendelian principles. The difference is that continuous variations are controlled by two or more pairs of genes that influence the same hereditary trait. As an example we shall consider the simplest possible type of such **multiple gene** heredity, where two pairs of genes (*Aa* and *Bb*) located on different pairs of chromosomes influence fruit weight. We shall assume that the genes of the *A* set have the same influence as the genes of the *B* set, although this may not always be the case.

Each effective (dominant) gene adds 125 grams to the basic 500 g weight of a fruit, whereas ineffective (recessive) genes do not add to the basic fruit weight. Thus, a plant with the genotype *aabb* will bear 500 g fruits while a plant with the genotype *AABB* will bear 1000 g fruits (500 g + 4 × 125 g). An F_1 plant or any other plant with two dominant genes (*AaBb, AAbb,* or *aaBB*) will have fruits weighing 750 g (500 + 2 × 125 g). Plants with any three dominants will bear 875 g fruits, while those with only one dominant will bear 625 g fruits. There is, then, a graded series of fruit weight (and size) phenotypes: 500, 625, 750, 875, and 1000 g.

If an *AABB* plant is crossed with an *aabb* plant the F_1 plants (*AaBb*) will bear 750 g fruits. At this point it would be impossible to determine (if the nature of the heredity were not known) whether fruit weight was a matter of monohybrid incomplete dominance or multiple gene heredity, but when the F_2 generation is analyzed (Fig. 22.8), there is a phenotype ratio of 1:4:6:4:1 in order of increasing (or decreasing) fruit weight, so we know that two pairs of independently assorting genes are involved. Although this ratio differs from the 9:3:3:1 phenotype ratio in the dihybrid previously considered because of a different kind of gene interaction, the assortment of genes during meiosis, the gene recombinations during fertilization, and the F_2 genotype ratios are just the same as in any dihybrid cross involving independent assortment.

One point worth noting is that $^5/_{16}$ of the F_2 plants bear fruits heavier than those of the F_1 parents, whereas another $^5/_{16}$ have lighter fruits than the parents. Even in natural populations it is quite common for offspring to be more extreme in continuous variations than either parent, since heterozygosity is quite prevalent in nature.

Another thing that should be stressed again is that genes are only hereditary potentialities and are expressed as phenotypic traits only when the environment is suitable for their expression. A plant with the potentiality for developing 1000 g fruits (*AABB*) might produce fruits weighing less in an environment unsuitable for optimal growth—for example, if some of the essential mineral elements were

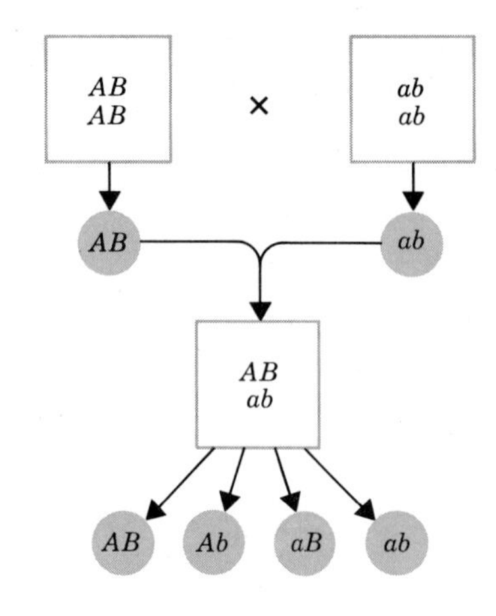

	AB	Ab	aB	ab
AB	AB AB (1000 g)	Ab AB (875 g)	aB AB (875 g)	ab AB (750 g)
Ab	AB Ab (875 g)	Ab Ab (750 g)	aB Ab (750 g)	ab Ab (625 g)
aB	AB aB (875 g)	Ab aB (750 g)	aB aB (750 g)	ab aB (625 g)
ab	AB ab (750 g)	Ab ab (625 g)	aB ab (625 g)	ab ab (500 g)

Fig. 22.8 Multiple gene heredity controlled by two independently assorting alleles (gene pairs). Note that the mechanics of this operation and the genotypes are the same as in a dihybrid cross, but the F_2 *phenotypic ratios of the two kinds of heredity are quite different.*

deficient, if there was a lack of sufficient food or water, or if the plants had been treated with a growth inhibitor such as maleic hydrazide. Phenotypes are a product of the interaction of heredity and environment, and geneticists must always be on the alert to make certain that the phenotypic expression of the hereditary potentialities is not being limited by an unfavorable environment. Of course, plants with an *aabb* genotype cannot produce fruits weighing more than 500 g regardless of how favorable the environment is, except that supplying an internal factor that cannot be synthesized by the *aabb* genotype might make possible the development of heavier fruits. The influence of gibberellic acid in causing genetic dwarf plants to grow as tall as those that carry genes for tallness is an example of this, the gene for tallness apparently providing the potentiality for synthesis of a gibberellin.

If the 1:4:6:4:1 ratio of F_2 phenotypes is plotted on a graph against fruit weight, an inverted *V*-shaped distribution curve results. If a continuous variation is influenced by three pairs of genes instead of two, the F_2 phenotype ratio will approximate 1:6:15:20:15:6:1 and the distribution curve will assume more of a bell shape. When still larger numbers of genes are involved, the curve becomes a typical bell-shaped normal distribution curve. In any natural population of plants or animals, continuous hereditary variations that have a normal frequency distribution, such as human height and intelligence or the weight of bean seeds, are controlled by multiple genes involving a substantial number of gene pairs.

Gene Linkage

Mendel studied only traits that assorted independently in his dihybrid crosses, and so far we have considered only this type of dihybrid. We now know that independent assortment occurs only when the two pairs of genes being considered in a dihybrid cross are located on two different chromosome pairs. Each chromosome, however, contains many genes, and the genes of any particular chromosome are said to be **linked** with one another, forming a linkage group. Linked genes do not assort independently in meiosis. For example, in tomatoes the genes for height (*T* and *t*) are located in the same chromosome as the genes for smooth (*S*) or hairy (*s*) stems. If a homo-

Table 22.2 **Results of a Test Cross of F_1 Tomato Plants (TS ts) with Recessive Plants (ts ts) Compared with the Expected Results If the Genes Were Either Completely Linked or Independently Assorting**

Phenotypes	Genotypes	Number of Test Cross Plants	Percent of Test Cross Plants	Percent Expected If the Genes Were: Completely Linked	Percent Expected If the Genes Were: Independently Assorted
Tall smooth	*TS ts*	96	48.5	50.0	25.0
Dwarf hairy	*ts ts*	95	48.0	50.0	25.0
Tall hairy	*Ts ts*	4	2.0	0.0	25.0
Dwarf smooth	*tS ts*	3	1.5	0.0	25.0
Totals		198	100.0	100.0	100.0

zygous tall smooth plant (*TS TS*) is crossed with a dwarf hairy plant (*ts ts*) all the F_1 plants will be tall and smooth just as if the genes were not linked. (The bars under the genes for height and hairs indicate that they are linked.) Since we are dealing with only one pair of chromosomes rather than two, as in previous crosses involving two traits, we can expect the F_1 plants to produce only *TS* and *ts* gametes rather than the four types (*TS*, *Ts*, *tS*, *ts*) that would result from independent assortment if the two pairs of genes were in separate chromosome pairs. We are dealing essentially with a monohybrid situation and so would expect an F_2 phenotype ratio of 3 tall smooth to 1 dwarf hairy, or a backcross ratio of 1 tall smooth to 1 dwarf hairy.

Crossing Over

The results of a test cross experiment with tomatoes involving F_1 plants (*TS ts*) are given in Table 22.2. Note that although the phenotype ratio of the backcross offspring was close to the 1:1 ratio expected if linkage were complete and far from the 1:1:1:1 ratio that would result if the genes assorted independently, the few test cross offspring that were tall hairy and dwarf smooth introduced a complication. The explanation for these two unexpected phenotypes is that in meiosis resulting in a few of the meiospores (and so gametes) the two pairs of genes changed linkages. Linkage changes such as these result from the crossing over between two of the four chromatids of the homologous pair of chromosomes. During synapsis, two of the chromatids sometimes twist around each other and may then break apart at the twist and exchange parts (Fig. 22.9). If this exchange of parts occurs, some of the old gene linkages are broken and new linkages are established. For example, in tomatoes, a crossover between the genes for height and hairs would result in one chromatid with a *Ts* linkage and another with the *tS* linkage. The data in Table 22.2 indicate that 3.5% of the F_1 gametes contained chromosomes resulting from crossovers someplace between the genes for height and for hairs. A crossover at some other point along the chromatids would not change the *TS* and *ts* linkages, but it would result in linkage changes for some of the other genes on the chromosomes.

Chromosome Maps

Since crossovers can occur at almost any point along a chromosome, the farther apart two genes are in a chromosome, the greater the chance that crossing over will occur between them. A third pair of genes present in the tomato chromosome pair we have been

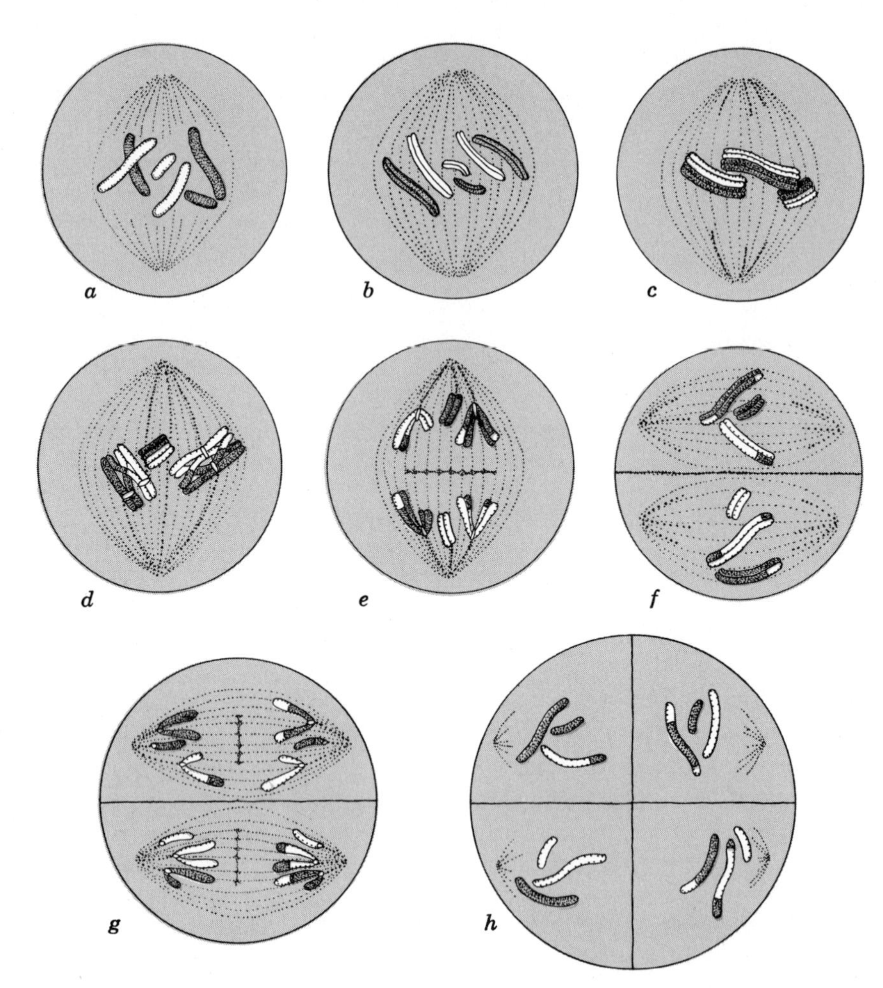

Fig. 22.9 Crossing over during meiosis. The meiotic divisions of the spore mother cells proceed in the same way as shown in the meiosis diagram in Chapter 20, but here crossing over is occurring in two of the three pairs of homologous chromosomes (d). *There is a double crossover in the long pair and a single crossover in the pair of medium length. As a result of the crossovers all four meiospores derived from the meiospore mother cell are genetically different. Of course, other assortments of the chromosomes can occur.*

considering influences fruit shape, the dominant gene *O* resulting in oblate (slightly flattened spherical) fruits and the gene *o* resulting in pear-shaped fruits when homozygous. The percent of crossovers between *T* and *O* is 17.4, indicating that *T* and *O* are much farther apart in the chromosome than are *T* and *S*. The percent of crossover between *S* and *O* is 13.9. On the basis of this information we know, not only the relative distances between the three genes, but also that they must occur in the chromosome in the order *T*, *S*, and *O* (Fig. 22.10). We have thus made a beginning toward mapping the location of the genes of the chromosome.

Geneticists have made rather extensive

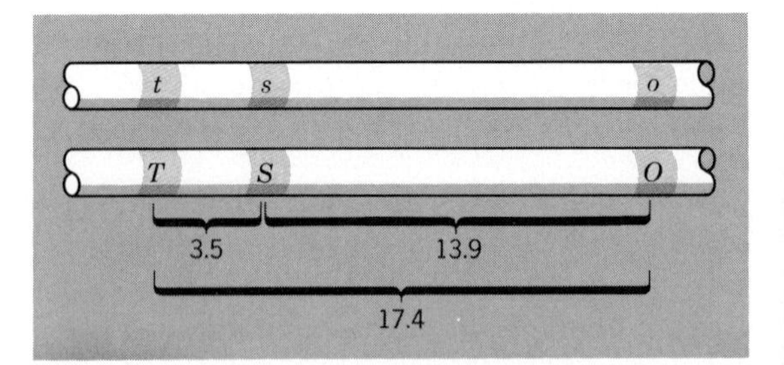

Fig. 22.10 Portions of two homologous chromosomes of a tomato plant, showing the location of three pairs of genes on the chromosomes. Chromosome maps such as this (and much more complete ones) can be made if the percentage of crossing over involving at least three pairs of genes is determined. The numbers are crossover percentages, but they also indicate the spacing of the genes.

Table 22.3 **Possible Hereditary Variation in F_2 Generations When Various Numbers (*g*) of Heterozygous, Independently Assorting Gene Pairs with Dominance Are Considered**

Number of Heterozygous Loci (Gene Pairs) That Assort Independently	Number of Different F_1 Gametes (Also Number of Different F_2 Phenotypes)	Number of Possible Combinations (Fertilizations of F_1 Gametes) (Also Number of Checkerboard Squares)	Number of Different F_2 Genotypes
1	2	4	3
2	4	16	9
3	8	64	27
4	16	256	81
5	32	1,024	243
6	64	4,096	729
7	128	16,384	2,187
8	256	65,536	6,561
9	512	262,144	19,683
10	1,024	1,048,576	59,049
17	131,072	17 billion	70 billion
23	8,388,608	70 trillion	94 billion
g	2^g	$(2^g)^2$	3^g

maps of all of the chromosomes of a few plants and animals, particularly corn and the fruit fly (*Drosophila*). Chromosome mapping is a striking example of the extensive information about genes that geneticists have been able to assemble by the analysis of phenotypic ratios.

Crosses Involving More than Two Gene Pairs

In a monohybrid cross there are two genetic types of F_1 gametes that can unite in four possible combinations, resulting in three different genotypes and, if dominance is involved, two different phenotypes in the F_2 generation. In dihybrid crosses there are four types of gametes, sixteen possible combinations, nine different F_2 genotypes, and four different phenotypes (if dominance is present in both pairs of genes). If a trihybrid cross were made, the types of gametes would increase to eight, the possible combinations (i.e., the number of squares in the F_2 checkerboard) would be 64, the number of different F_2 genotypes would be 27, and the number of different phenotypes in the F_2 generation (with dominance in all three pairs) would be 8. Similar data of crosses involving more than three gene pairs is given in Table 22.3, which makes it clear that the hereditary variability associated with gene recombinations increases rapidly as additional gene pairs are considered. Furthermore, the increases occur in a simple mathematical pattern. If we let g represent the number of heterozygous gene pairs being considered, then 2^g is the number of types of F_1 gametes (and also the number of F_2 phenotypes if dominance is found in all gene pairs), whereas $(2^g)^2$ represents the number of possible gamete combinations (fertilizations) and so the number of squares in the F_2 checkerboard, and 3^g is the number of different F_2 genotypes.

The number of pairs of independently assorting genes (g) is limited by the number of chromosome pairs in a species, or in other words the n chromosome number. Thus, in a species where $2n = 16$ and $n = 8$, g cannot be larger than 8. The data in the table do not include the added variability resulting from crossovers, and column two represents the number of different F_2 phenotypes only when all gene pairs exhibit dominance. The number of F_2 phenotypes would be quite different, and impossible to determine by any simple formula, if some of the gene pairs showed incomplete dominance, were involved in multiple gene heredity, or interacted in any of the various possible ways other than simple dominance.

Variation by Recombination in Natural Populations

In nature, most plants and animals have a rather high degree of heterozygosity, and will probably have at least one pair of heterozygous genes on each chromosome pair, so the data in Table 22.3 also represent the minimum amount of hereditary variation that may be expected within a species. Since apple trees have 17 chromosome pairs ($2n = 34$), there would be at least 131,072 phenotypes possible when apple trees were sexually propagated. Even this large number does not include the additional variability resulting from crossing over, incomplete dominance, and other kinds of gene interactions. The important point here is not the exact amount of variation possible, but the fact that an immense amount of hereditary variation can occur in a natural population simply as a result of gene recombinations during sexual reproduction. It should now be obvious why we cannot expect to get Winesap apple trees by planting seeds from Winesap apples.

Since human beings have 23 pairs of chromosomes, there is a much greater chance of hereditary variations than there is among apple trees, as shown in Table 22.3. Since there is a possibility of 8,388,608 phenotypes in any family (not even considering additional variation resulting from such things as incomplete dominance, multiple factor heredity, and mutations), it is not surprising that no two children in a family have identical hereditary characteristics, unless they are identical twins (which are derived from the same zygote).

Among cultivated plants propagated by seeds the story is quite different from that in apple trees and other plants propagated vegetatively. Plants such as peas, beans, and tomatos propagated by seeds have been bred and selected for a high degree of pureness or homozygosity, particularly as regards the traits important commercially and characteristic of the variety. Consequently, as long as the pure lines are kept from crossing with other varieties, there is little hereditary variation, and one individual of a variety has essentially the same phenotype as another. Cultivated ornamental flowers are, however, frequently less homozygous and so more variable than garden vegetables or farm crops. The extensive use of hybrids of corn and other crops will be considered later.

Despite the great hereditary variation within most species of plants, each species has many hereditary traits that ordinarily do not vary from one individual to another and are presumably essentially homozygous throughout the species. Thus, all common garden beans (*Phaseolus vulgaris*) regardless of varietal differences have the same type of fleshy cotyledons, lack endosperm in their seeds, have one pair of simple, opposite cordate leaves whereas the remaining leaves are alternate and compound, lack a cambium, and have flowers of a characteristic shape. Such hereditary traits that are generally the same in one member of a species as another are the ones used by taxonomists in describing and identifying members of the species and in distinguishing the species from related ones. Some hereditary traits are even more fundamental and widespread, occurring throughout a family or even a larger taxonomic group such as a class. Thus, all members of the legume family have pods rather than berries, achenes, or other kinds of fruits, and all tracheophytes have vascular tissue—a hereditary potentiality not present among the fungi, for example.

Cytoplasmic Inheritance

Although most genes are located in chromosomes some are in the cytoplasm, specifically in the chloroplasts and mitochondria. These organelles are self-replicating and contain DNA, and so carry some of their own hereditary potentialities, although they are also partly under the control of ordinary chromosomal genes. Since the sperm does not contribute to the cytoplasmic component of a zygote, cytoplasmic inheritance is through the female parent only and does not follow Mendelian principles of segregation and recombination. For example, four-o'clocks may have green, pale, or variegated leaves. Regardless of the phenotype of the male parent, the offspring always inherit their phenotype from the female parent. Female parents with green leaves always have offspring with green leaves; those with pale leaves always have offspring with pale leaves. However, female parents with variegated leaves can have green, pale, or variegated-leaved offspring in no predictable ratio. The reason is that variegated plants have some green chloroplasts and some pale chloroplasts. It is a matter of chance as to whether a particular zygote receives all green, all pale, or both kinds of chloroplasts. If both, the resulting plant will have variegated leaves.

HEREDITARY VARIATION THROUGH MUTATIONS

Hereditary variation may result, not only through the recombination of genes, but also from **mutations** or changes in genes or chromosomes. **Gene mutations** result from alteration in the chemical structure of a gene; **chromosome mutations** (or chromosomal changes) may result from either changes in chromosome structure or altered chromosome numbers.

Gene Mutations

Mutations, or chemical changes, of genes may occur in any living cell of a plant or animal at any time. Although mutations may occur rather frequently, the great majority of new genes resulting from mutation probably are never expressed phenotypically. There are several reasons for this. For one thing, only one gene of the pair in a diploid cell ordinarily mutates and the mutant gene is usually recessive. Even if a cell containing the mutant gene has many cell descendants the recessive mutant trait will not be expressed in the individual in which it occurred or in any asexually produced offspring. The mutation will be present only in those cells derived from the cell where it occurred, and the mutant gene can be transmitted to offspring sexually only if the gametes happen to be among these cells. Even then it may be several generations before the mutation is expressed phenotypically, since the recessive mutation must be present in both the egg and the sperm. At least one other factor may prevent expression of the mutant gene, at least in adult individuals. Some mutations are lethal because they result in such marked changes in essential metabolic processes that the individual cannot survive beyond an early stage of development. The albino mutant of corn is lethal because the plants cannot carry on photosynthesis and so starve when the food in the grain is exhausted. However, the albino gene can be transmitted from one generation to another because it is a recessive.

In haploid individuals such as the independent gametophytes of lower plants, recessive mutations are expressed immediately, since there is only one set of chromosomes and thus no possibility of masking by the corresponding dominant. If such a recessive mutation is lethal to the gametophyte, it cannot be passed on to offspring. Similarly, mutations lethal to the gametophytes of higher plants cannot be passed on, and we are rarely aware that they have occurred.

Even in a diploid individual a dominant mutation may appear phenotypically in first generation offspring, or in the individual in which the mutation occurred, provided that the trait affected is expressed in the cells containing the mutant gene. Thus, a mutation from purple to white flowers would be expressed immediately if present in the cells of a petal (Fig. 22.11) but not if present only in leaf cells. When a dominant mutation occurs in a very young bud, some or all of the cells in the branch developing from the bud will contain the mutation and the branch may have an obviously different phenotype from the rest of the plant. Such **bud mutations** or **bud sports** have occasionally given rise to new varieties of commercial value, such as the navel orange and the Golden Delicious apple.

Fig. 22.11 A mutation in snapdragon induced by gamma radiation from cobalt-60. This variety has purple petals, but the flowers derived from the cell where the mutation occurred have white petals.

Some mutations may be desirable economically or advantageous to the individual, but most mutations result in phenotypes less adapted for survival than the parent phenotypes (Fig. 22.12). Even when not lethal, mutations may result in rather marked changes in structure or metabolism that can handicap the individual, or they may result in some type of sterility, that is, the inability to reproduce sexually. Of course, in plants that can reproduce asexually, the latter would not necessarily prevent transmission of the mutation to offspring.

Since mutations apparently occur at random, the question arises as to why most mutant traits are less desirable than the previously existing hereditary potentialities. The answer seems to be that the gene pool of any species has been subjected to millions of years of natural selection, and so any phenotypes not adapted to the environment have failed to survive or at least to produce offspring and thus propagate the undesirable traits. Selection has, then, resulted in a complex of generally desirable hereditary potentialities and any change in these is likely to be for the worse. The undesirable mutations will, in turn, be subject to elimination by natural selection. Whereas dominant genes with negative survival value may be eliminated quite promptly, undesirable recessive genes may be transmitted through many generations in the heterozygous condition and some of the poorly adapted phenotypes may appear in each generation. However, the natural selection against these will tend to keep the frequency of the undesirable gene in the population at a relatively low level.

Fig. 22.12 Corn plants with a mutation referred to as Lazy Maize, compared with one row of normal corn plants.

Man has found some mutations in plants to be desirable commercially, even though they may have no survival value or even negative survival value. Thus, a mutation resulting in seedless fruits might make a plant unable to survive in nature, but man could propagate the plant vegetatively. A few of the desirable mutations that man has selected for propagation include pure red sweet pea flowers (as contrasted with the previous purple-red), the large, undulating petals of Spencer sweet peas, and the nectarines (smooth-skinned peaches). Nectarines originated as a bud mutation on a peach tree.

The rate of mutation is greatly increased by mustard gas and some other chemicals and by the various kinds of ionizing radiations including X rays, gamma rays, some ultraviolet rays, and alpha and beta particle radiation. Recent reports have implicated LSD (lysergic acid diethylamide), certain cyclamates, alcohol, and sulfur dioxide (a major component of air pollution) as mutagenic agents, capable of causing gene mutations or chromosome breaks. One should be cautious about accepting these early reports as conclusive and about extrapolating results of tests on one organism to another. Also, most of the tests have used unusually high concentrations of the substance being tested. Nevertheless, many geneticists are concerned and many commonly encountered substances are being tested for possible mutagenic effects.

Since radioactive isotopes and radiation have become available for use as tools, biologists have subjected many plants and animals to irradiation, both to learn more about the nature of mutations and the effects of radiation and in the hope that a few desirable mutations might result. New high-yielding strains of *Penicillium* and several new varieties of ornamental flowers have resulted from such artificially induced mutations, and extensive work is being done with peanuts and other crop plants. Biologists are greatly concerned, however, about the probable large increase in mutations in man as well as in plants and animals as a result of radioactive fallout as well as over the direct harmful effects of the radiation on individuals. Since mutations induced by chemicals or radiation apparently do not differ in character from those that occur naturally, we can expect most of the induced mutations to be undesirable.

Altered Chromosome Structure

Chromosomes may, from time to time, undergo structural changes that result in changes in hereditary potentialities. For example, a portion of a chromosome may break off, and if it lacks a centromere (kinetochore) it will not be included in either of the two daughter nuclei at the next division. The result of such a deletion is the loss of a group of genes. Another chromosomal change is the attachment of a detached piece of chromosome to a chromosome of another pair (translocation). Also, a chromosome may double back on itself and interchange ends at the junction point (Fig. 22.13). This is an inversion. For example, if the original order of the genes in a chromosome was ABCDEFGHIJ, the order after an inversion might be ABCGFEDHIJ, the inverted portion being underlined. Although inversions and translocations do not result in a change in the gene content of a cell, they do result in new hereditary potentialities since the action of a gene may be modified by the adjacent genes (position effect). Chromosomal changes may give rise to gene mutations in the next generation. Like gene mutations, chromosomal changes may be induced by radiation or chemicals.

Altered Chromosome Number

The cells of animals and sporophytes usually contain the diploid ($2n$) number of chromosomes, but occasionally individuals may have only the haploid (n) number. Haploid sporophytes may originate in several different ways, principally by the development of an unfertilized egg into an individual **(parthenogenesis).** Haploids may be sterile, or more rarely the life cycle may proceed without either reduction division or fertilization. Dandelions have such a parthenogenetic life cycle, except that both the sporophytic and gametophytic generations are diploid. Since parthenogenesis is essentially a type of asexual reproduction, the hereditary variation associated with sexual reproduction is lacking, but as we have seen, mutations can be expected to express them-

Fig. 22.13 Three types of chromosomal alterations. (a) *Inversion. A chromosome forms a loop and where its ends cross they break off and exchange positions, resulting on a new, and reversed, sequence of genes.* (b) *Deletion. A portion of a chromosome breaks off and is not incorporated in one of the daughter nuclei because it lacks a centromere.* (c) *Translocation. Two nonhomologous chromosomes exchange parts. Note that this differs from the crossing over of homologous chromosomes.*

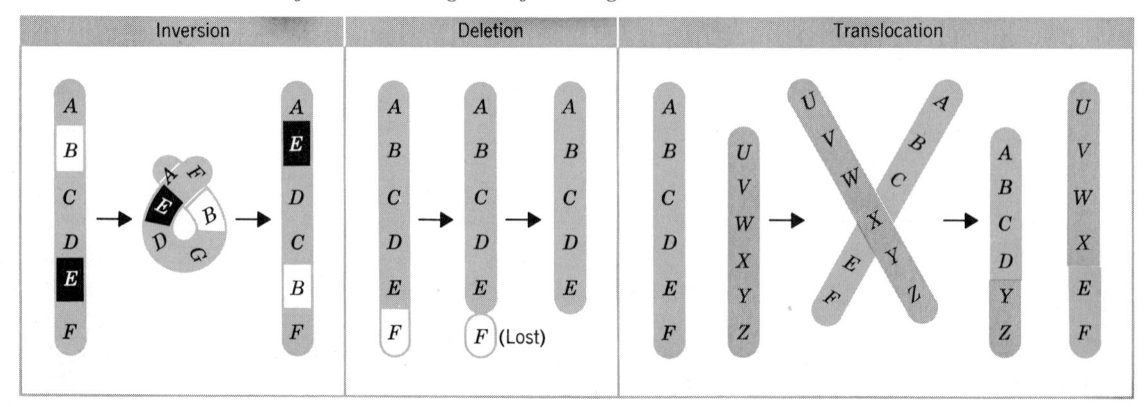

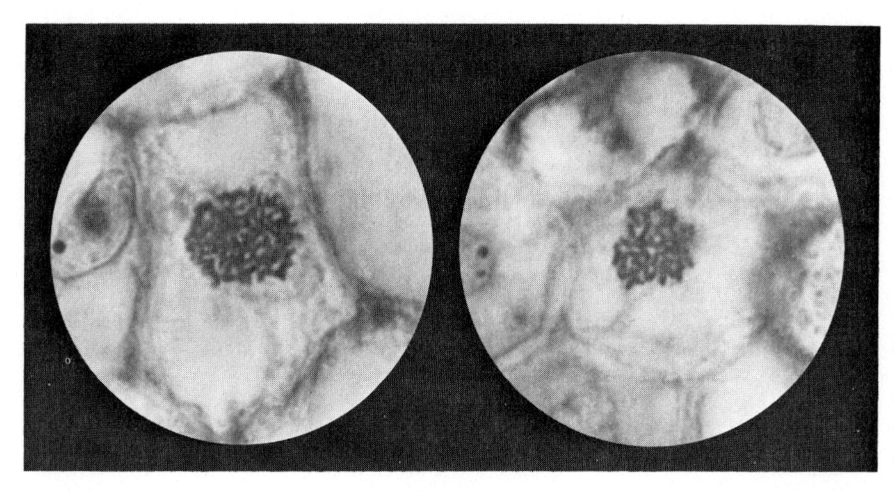

Fig. 22.14 Cells from the pith of a diploid apple tree (left) with 34 chromosomes and a tetraploid with 68 chromosomes. Note also the difference in cell size.

selves more frequently in haploids than in diploids. Of course, recessive genes present in haploids will all be expressed phenotypically, provided that the environment is suitable, and haploids will be likely to have fewer dominant genes in multiple gene series controlling continuous variations than will a diploid.

The gametophytes of some ferns and bryophytes (and apparently to a lesser extent in other plants) give rise to sporophytes without fertilization of the eggs **(apogamy).** The sporophytes usually develop from vegetative cells rather than by parthenogenesis. Such sporophytes will be haploid unless the gametophyte was derived from its parent sporophyte by **apospory** (a lack of meiosis and so meiospores). The diploid gametophytes arise from vegetative cells, or perhaps at times from the spore mother cells. In some species of ferns with apospory or apogamy both the gametophytes and sporophytes are diploid, while in other species the sporophytes as well as the gametophytes are haploid.

More common than haploid sporophytes are **polyploids** which contain more than two sets of chromosomes per cell (Fig. 22.14). Tetraploids ($4n$) are perhaps the most common polyploids, but there are also triploids ($3n$), hexaploids ($6n$), octaploids ($8n$), and even higher polyploids. Polyploidy is much more common among plants than among animals. It may come from failure of meiosis or by mitotic division of a nucleus without subsequent cell division. The latter can be caused to occur with greater than natural frequency by treating plants with the alkaloid drug **colchicine,** and plant breeders have produced many valuable polyploids in this way. Polyploids are often larger and more vigorous than diploids and may have larger flowers (Fig. 22.15) and heavier fruits, principally because of the increase in the number of genes involved in multiple gene heredity.

Many cultivated plants differ from their wild relatives in being polyploids rather than diploids. For example, wild strawberries are $2n$, whereas different cultivated varieties are $4n$, $6n$, $8n$, $10n$, and even $16n$. Three species of wheat have 14 ($2n$), 28 ($4n$), and 42 ($6n$) chromosomes; different chrysanthemums have 18, 36, 54, 72 and 90 chromosomes. Cultivated tetraploids include varieties of primrose, petunia, cosmos, zinnia, lily, marigold, apple, blackberry, watermelon, and cantaloupe. Some varieties of apple, pear, hyacinth, and tulip are among the cultivated triploids. Although all are sterile, they can be propagated vegetatively. In species that can be propagated only by seeds a triploid could not have offspring.

Fig. 22.15 Inflorescences of diploid (left) and tetraploid snapdragon plants. The tetraploid was produced by the use of colchicine.

PLANT BREEDING

The application of genetics to the breeding of better and more productive varieties of cultivated plants and domesticated animals has been of incalculable value to man and has contributed greatly to the ability of agriculture to provide adequate supplies of food for the ever-increasing human population of the Earth. Applied agricultural research really has only two basic goals: the production of improved varieties and the provision of the best possible environment for them.

Plant and animal breeding were practiced long before Mendel discovered the basic principles of inheritance, but the procedures were time consuming, rather haphazard, and frequently not too effective. Since most of our species of crop plants were already cultivated by man during the earliest days of recorded history, it seems likely that even prehistoric man engaged in the selection of desirable variations, if not in plant breeding. With the development of genetics during the present century, plant and animal breeding has been placed on a firm scientific basis, and breeders no longer have to work in the dark. Luther Burbank was perhaps the last of the productive plant breeders who worked without the benefit of basic genetic information about his plants, and he probably would have been even more productive if he had used the available genetic information.

Plant and animal breeding is essentially an exercise in accelerated and controlled evolution. First, plant breeders deliberately increase hereditary variation by hybridization (Fig. 22.16), the production of polyploidy through the use of colchicine (Fig. 22.15), the stimulation of mutation by radiation treatments, and the use of other techniques. Second, the breeders select the most desirable phenotypes and genotypes obtained in these ways for propagation. Occasionally, they may be fortunate enough to find a desirable natural hybrid or mutant. When plant breeders are dealing with plants that can be propagated vegetatively on a practical basis, they have a distinct advantage over animal breeders, since even sterile hybrids or mutants can be propagated and since it is not necessary to develop a pure (homozygous) strain that will breed true to form.

The improved varieties produced by plant breeders are too numerous to mention in detail, but every important farm crop, garden vegetable, orchard fruit, and ornamental species includes numerous varieties developed by scientific plant breeding. In general, the hereditary traits plant breeders strive to incorporate in their creations include vigor, high yields, desirable quality, and disease resistance. In recent years considerable interest has developed in the breeding of improved varieties of forest trees, particularly rapidly growing varieties, for use in making paper pulp. The breeding of rust resistant varieties of wheat may be cited as just one example of the immense importance of plant breeding to man. If no rust resistant varieties had been developed it is quite likely that today there would be a serious shortage of wheat. But the wheat rust fungus, like other plants, mutates and hybridizes and new types of the rust fungus that attack wheat resistant to previous types continue to appear. Consequently, new varieties of wheat resistant to the mutant rust must be bred. Similar genetic contests between plant breeders and mutating parasites are in progress with other species, too.

Fig. 22.16 ***Burpee's Red and Gold hybrid marigold (lower center) was produced by crossing an African marigold (left) with a dwarf French marigold (right). The African marigold is a diploid (2n = 24), the French marigold a tetraploid (4n = 48), and the hybrid a sterile triploid (3n = 36). In contrast with the usual situation, the diploid has larger inflorescences (heads) than the polyploids.***

One example of the recognition of the contributions of plant breeders to modern society was the awarding of a 1970 Nobel prize to Ernest Borlaug, who developed high-yield, dwarf wheats that have been a boon to developing nations with an inadequate food supply. Similar marked genetic improvements of rice and other important foods plants have been made by other plant breeders. The great increases in productivity brought about by plant breeding and improved cultural practices have been referred to as the "Green Revolution."

A word of caution needs to be urged regarding plant and animal breeding programs: genetic uniformity may court disaster. If advantageous genes are widely incorporated into a certain crop plant, a pathogen particularly adapted to the characters determined by these genes or linked to them may cause widespread destruction of the crop. The southern corn leaf blight epidemics of 1970 were promoted by such genetic uniformity bred into male-sterile corn. Plant breeders have been able to produce new strains of corn resistant to the blight, but the message is still clear: extensive genetic uniformity may be detrimental in the long run.

Another problem arises from the fact that with the breeding of superior new strains of crop plants, some of the older varieties are no longer cultivated and as a result often disappear. Although these older strains are not as

Fig. 22.17 Two fields of hybrid corn at different stages of development. Note the uniformity of the plants in both fields.

desirable in general as the newer ones, they do contain many useful genes that may be completely lost and so cannot be used in future plant breeding programs. Geneticists have urged that at least some plants of these currently unused strains be propagated year after year as a sort of gene bank, and progress is being made in this direction.

Table 22.4 **The Yield of Corn in the United States**

Year	Millions of Acres of Corn	Billions of Bushels of Corn	Average Yield, Bushels per Acre
1932	113	2.9	26
1946	90	3.3	37
1957	75	3.4	45
1963	68	4.1	60
1969	65	4.6	61

Hybrid Corn

In 1932 the average yield of corn in the United States was 26 bushels per acre and in 1963 the average yield was 60 bushels per acre. Although improved cultural practices contributed to this great increase in productivity, the increase was largely the result of one of the more important products of scientific plant breeding: hybrid corn (Fig. 22.17). The first substantial planting of hybrid corn by farmers was in 1933 and its use has increased until now almost 95% of the corn crop is hybrid corn. The average yields are far below those possible under the best environmental conditions. Many farmers have yields of over 80 bushels per acre and yields of almost 200 bushels per acre have been reported. Despite the great increase in productivity resulting from the use of hybrid corn, the total production of corn is only slightly more than in 1932 (Table 22.4).

The important point is that hybrid corn has made possible an increase in total production and also an increase in quality with the expenditure of much less labor and the release of some 38 million acres of farm land for other uses. At present there is a surplus of corn in this country, but it is quite likely that before the end of this century the increasing population will need all the food farmers can raise.

Unlike plants such as peas that are commonly self-pollinated, the wind-pollinated corn cannot easily be inbred and so pure breeding lines are difficult to establish and maintain. Furthermore, pure (homozygous) strains of corn are generally low in vigor and yield and small in size, at least partly as the result of numerous undesirable homozygous recessive traits. When, however, two pure lines are crossed, the hybrids are large vigorous plants with long, uniform ears and high yields (Fig. 22.18). Since they are an F_1 generation, the plants are also uniform both phenotypically and genotypically. The **hybrid vigor** found in corn has also been observed in many other species of plants and animals, particularly in those that are naturally cross fertilized and so are extensively heterozygous. Hybrid vigor is probably the result of a variety of different gene interactions and is rather complicated and difficult to explain satisfactorily, but it may result in part from the fact that most of the less desirable traits seem to be homozygous recessive genes. For example, pure line A may synthesize only small quantities of growth substance X because of genotype $xxYY$, whereas pure line B may synthesize only small amounts of growth substance Y because of genotype $XXyy$, so growth is restricted in the two pure lines for different reasons. The hybrid ($XxYy$) will be able to synthesize adequate quantities of both growth substances and so will be larger than either parent.

The first hybrid corn was produced in 1907 by the American botanist G. H. Shull, who developed several essentially homozygous parent (P_1) lines by self-pollination of corn plants and then crossed two different P_1 lines to obtain an F_1 hybrid generation. The hybrid plants were uniform phenotypically (as would be expected in an F_1 generation), exhibited

Fig. 22.18 (Left) A double-cross hybrid corn plant (center) with its two single-cross parents. (Right) An ear of corn from a double-cross hybrid plant (center) and ears from its two single cross parents.

great hybrid vigor, and were large and high yielding. However, the F_1 population was always small because the P_1 plants produced only a few small ears and consequently a limited supply of seeds containing F_1 embryos. Although the F_1 plants were high yielding, the grain they produced could not be sold to farmers for use as seed corn since it represented an extremely heterogeneous F_2 generation. For these reasons hybrid corn could not be made available to farmers for general use.

The problem of producing an adequate supply of hybrid seed corn for farmers was finally solved by the development of **double**

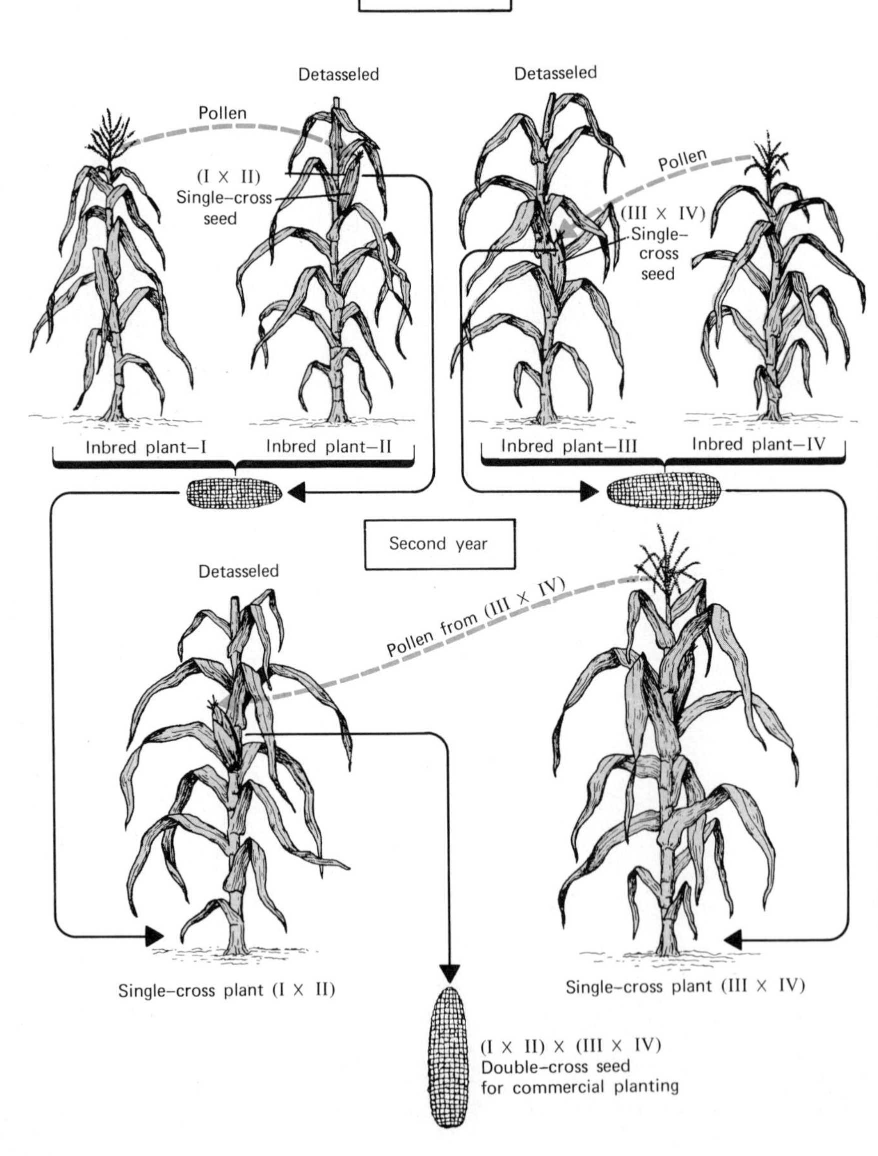

Fig. 22.19 The method of producing double-cross hybrid corn. The first year inbred parent strains I and II are crossed, as are inbred parent strains III and IV. In each case the grains from the detasseled (female) parent are planted the second year, the embryos developing into the single-cross hybrid plants (I ×II and III ×IV). The two single-cross hybrids are then crossed and the resulting double-cross grains are sold to farmers for planting.

cross hybrids, but it was 1933 before double cross hybrid seed corn became available to farmers. To produce double cross hybrids, four P_1 strains are required (Fig. 22.19). Strain I is then crossed with strain II, producing an F_1 hybrid that can be designated as (I × II). Similarly, strains III and IV are crossed to produce hybrid (III × IV). Then (I × II) plants are crossed with (III × IV) plants. The grains produced by this cross contain the double cross hybrid embryos, and it is these grains that are sold to farmers for planting.

Hybrid seed corn is expensive because of the extensive labor required to produce first the pure lines and then the hybrids, but farmers have found that the high yields of hybrid corn make it much more profitable to raise than the open pollinated varieties they previously used. Many different pure inbred P_1 lines have been produced, and from them many different kinds of hybrids adapted to different regions of the country have been developed.

The production of hybrid corn has been facilitated by the development of two different methods of controlling the formation of functional pollen, thus eliminating the task of detasseling the female parent plants. One method is genetic, the genes for male sterility being included in the genotype of one of the strains being crossed. The other method is biochemical: treatment of the plants with maleic hydrazide inhibits the development of mature stamens without inhibiting pistil development.

The great success of hybrid corn has resulted in the use of similar hybridizing techniques with many other crop plants and domestic animals, generally with marked increases in yield and quality. Hybrids of many species such as tobacco, tomato, cotton, chickens, turkeys, hogs, and cattle are now available or in the process of development.

RELATED READING

Anonymous. "Parasexual hybridization," *BioScience, 22,* 674, 1972.

Boyes, B. C. "The impact of Mendel," *BioScience, 16*(2), 85–92, 1966.

Curtis, B. C. and D.R. Johnston. "Hybrid wheat," *Scientific American, 220*(5), 21–29, May 1969.

Darrow, G. M. "Polyploidy in fruit improvement," *Scientific Monthly, 70,* 211–219, 1950.

Gabriel, M. L. and S. Fogel. *Great Experiments in Biology.* Englewood Cliffs, N. J.: Prentice-Hall, 1955, pp. 225–279.

Mangelsdorf, P. C. "Hybrid corn," *Scientific American, 185*(2), 39–47, August 1951.

Mangelsdorf, P. C. "Wheat," *Scientific American, 189*(1), 50–59, July 1953.

Mangelsdorf, P. C. "Ancestor of corn," *Science, 128,* 1313–1320, 1958.

Miller, J. "Genetic erosion: crop plants threatened by government neglect," *Science, 182,* 1231–1233, 1973.

Peters, J. A., ed. *Classic Papers in Genetics.* Englewood Cliffs, N. J.: Prentice-Hall, 1959.

Sigurbjörnsson, B. "Induced mutations, in plants," *Scientific American, 224*(1), 87–95, January 1971.

Zobel, B. J. "The genetic improvement of southern pines," *Scientific American, 225*(5), 94–103, November 1971.

23 THE MOLECULAR BASIS OF HEREDITY

As we have already noted, Mendel's unit factors were purely hypothetical and neither he nor any other biologist of his time had any knowledge of the physical or chemical basis of heredity or of the cellular structures carrying the unit factors. However, between 1875 and 1890 numerous investigations made evident the behavior of chromosomes during mitosis, meiosis, and fertilization, and following the rediscovery of Mendel's work in 1900, it became evident that Mendel's unit factors (or genes as they came to be called) were carried in the chromosomes. Cytoplasmic heredity was discovered later.

When it was found that the genes were in chromosomes, it became possible to pinpoint investigations of the chemical nature of genes. Even during Mendel's life, it was known that nuclei contained proteins. In 1869 the Swiss biochemist Friedrich Miescher found that when cells were treated with pepsin to hydrolyze the proteins the nuclei shrank but remained intact. He also found that the nuclear substance not hydrolyzed by pepsin, unlike proteins, had a substantial phosphorus content. Later the substance was found to be deoxyribonucleic acid (DNA). In 1914

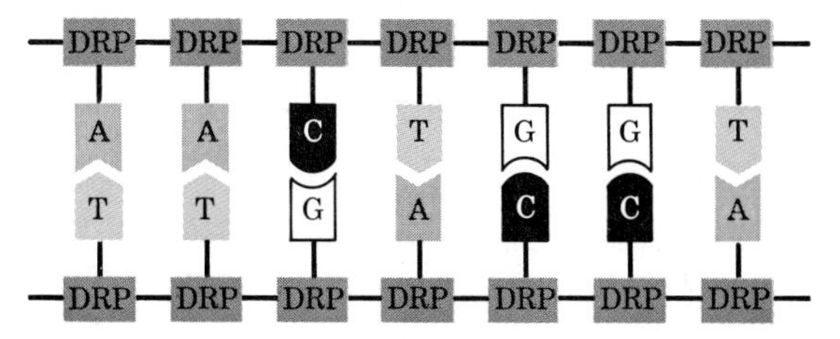

Fig. 23.1 Diagrammatic representation of a small portion of a DNA molecule. The two deoxyribose phosphate strands (DRP) provide the linear continuity of the molecule and the purines of one strand are linked to the pyrimidine, of the other strand by hydrogen bonding, thus holding together the two helical strands of the molecule. The paired nitrogenous bases (AT or TA and CG or GC) can occur in any sequence, providing the genetic code.

the German chemist Robert Feulgen devised a reagent that stains DNA (but no other cellular substance) red and later he found, by use of his reagent, that all the nuclear DNA was in the chromosomes. It then became evident that chromosomes were composed predominantly, if not entirely, of proteins and DNA and that the genes were composed of one of these substances, or perhaps of both.

When it was found that the amount of DNA in haploid cells was precisely half that in diploid cells of the species and that this was not necessarily true of the chromosomal proteins, it was suggested that the genetic material was DNA. However, most biologists continued to believe that the genetic material consisted of the chromosomal proteins because the diversity of the proteins could account for the large number of different genes of an organism. When the structure of DNA molecules was worked out they were found to consist of deoxyribose phosphate residues linked together in long chains with one of four nitrogenous bases (adenine, cytosine, guanine, or thymine, often shown as A, C, G, and T) bonded to each deoxyribose residue (Fig. 23.1). It was thought that these bases were repeated in this same order throughout the length of the DNA molecule (ACGT, ACGT, etc.) and that the order was the same in all species. Thus the DNA apparently did not have the molecular diversity essential for carrying genetic information.

DNA AS THE GENETIC MATERIAL

However, between 1928 and 1953 an increasing mass of experimental evidence was accumulated showing that DNA was indeed the substance that carried genetic information—that the genes consisted of DNA and not proteins. The evidence was provided largely by experiments with bacteria and viruses.

One of the later (1952) and more convincing of these experiments was conducted by Alfred Hershey and Martha Chase of the Carnegie Laboratories of Genetics. They used the bacterium *Escherichia coli* and a bacteriophage (bacterial virus) that attacks it (Fig. 7.21) and eventually kills it. The virus is composed of a protein coat and a DNA core (Fig. 7.22), and has a head and tail. When a bacteriophage attacks a bacterium the tail attaches to the bacterial wall and there is no evidence that the protein coat of the virus enters the bacterial cell, but within a few minutes the bacterium is synthesizing both new viral DNA and the protein coats (Fig. 23.2). Hundreds of new viral DNA and viral protein coats are thus produced, the bacterium is killed and ruptured, and the new viruses are released.

To determine with certainty whether the virus component injected into the bacteria was all DNA, all protein, or some of both, Hershey and Chase used radioactive tracers. Protein contains sulfur (while DNA does not) and DNA contains phosphorus (while protein does not). First, they cultured bacteria and bacteriophage in a medium containing both radioactive sulfur (^{35}S) and radioactive phosphorus (^{32}P). Then they infected bacteria in a culture without radioactive isotopes with radioactive bacteriophage. After a period of time long

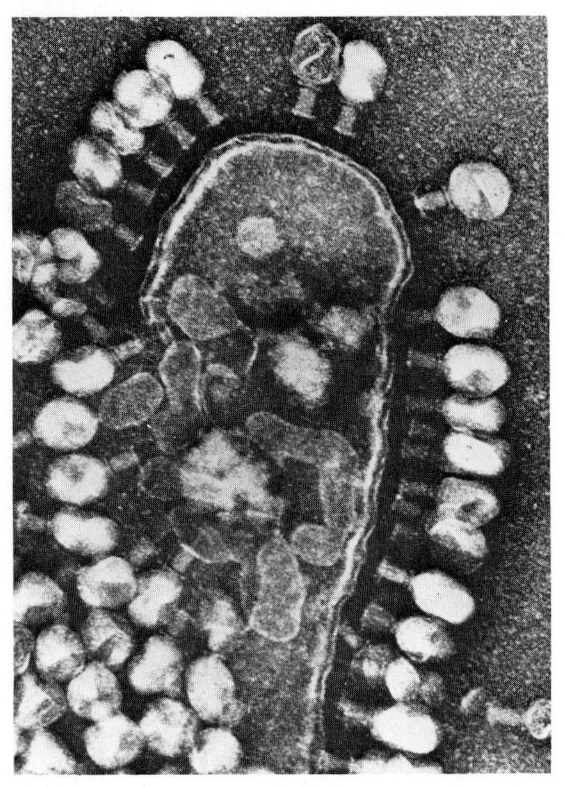

Fig. 23.2 Electron micrograph (about X86,000) of an Escherichia coli *with T-4 bacteriophage viruses on its wall. The cell has been disrupted by the virus and the viruses produced by the bacterium are pouring out.*

enough for the bacteria to produce new phage but before the disruption of the bacteria, they detached the viruses from the bacterial walls by use of a blender and analyzed the bacterial remains. These contained ^{32}P but no ^{35}S. The detached viral remains contained ^{35}S but no ^{32}P (Fig. 23.3). This made it clear that *only* DNA of the virus was injected into the bacteria and that the DNA provided the genetic information that caused the bacteria to produce copies of the bacteriophage virus.

When it was found that the four bases of DNA could occur in any sequence, not just a repeating A-C-G-T sequence, and that the sequence was different in different kinds of DNA, it became evident that DNA could have the variability essential for carrying genetic codes. It was also found that the four bases do not occur in equal amounts in DNA as had been thought. However, in 1950, E. Chargaff and his co-workers at Columbia University found that different kinds of DNA always contain the same amounts of adenine and thymine. They also contain the same amounts of guanine and cytosine, although the amounts of these are different from the amounts of adenine and thymine. The significance of this was not evident at the time.

STRUCTURE OF THE DNA MOLECULE

Before 1950, it had been established that DNA molecules are long chains of nucleotides, each nucleotide being composed of deoxyribose phosphate and one of four nitrogenous bases (adenine, thymine, cytosine, *or* guanine). However, essentially nothing was known about the structural organization of the DNA molecule nor about how it could replicate itself (as was essential if it were indeed the genetic material). In addition, there were no clues as to how DNA might function in controlling the expression of the hereditary potentialities.

The breakthrough in elucidating the structure of the DNA molecule began with X-ray diffraction analysis of DNA, particularly by M. H. F. Wilkins of King's College, London. He found that the diffraction patterns had three major periodicities (repeating units): 3.4 angstroms, 20 angstroms, and 34 angstroms (Fig. 23.4). On the basis of this information, Chargaff's finding about the equality of adenine and thymine and of guanine and cytosine, and other information about the chemistry of DNA, James D. Watson and Francis H. C. Crick of Cambridge University began efforts to elucidate the structural organization of the DNA molecule.

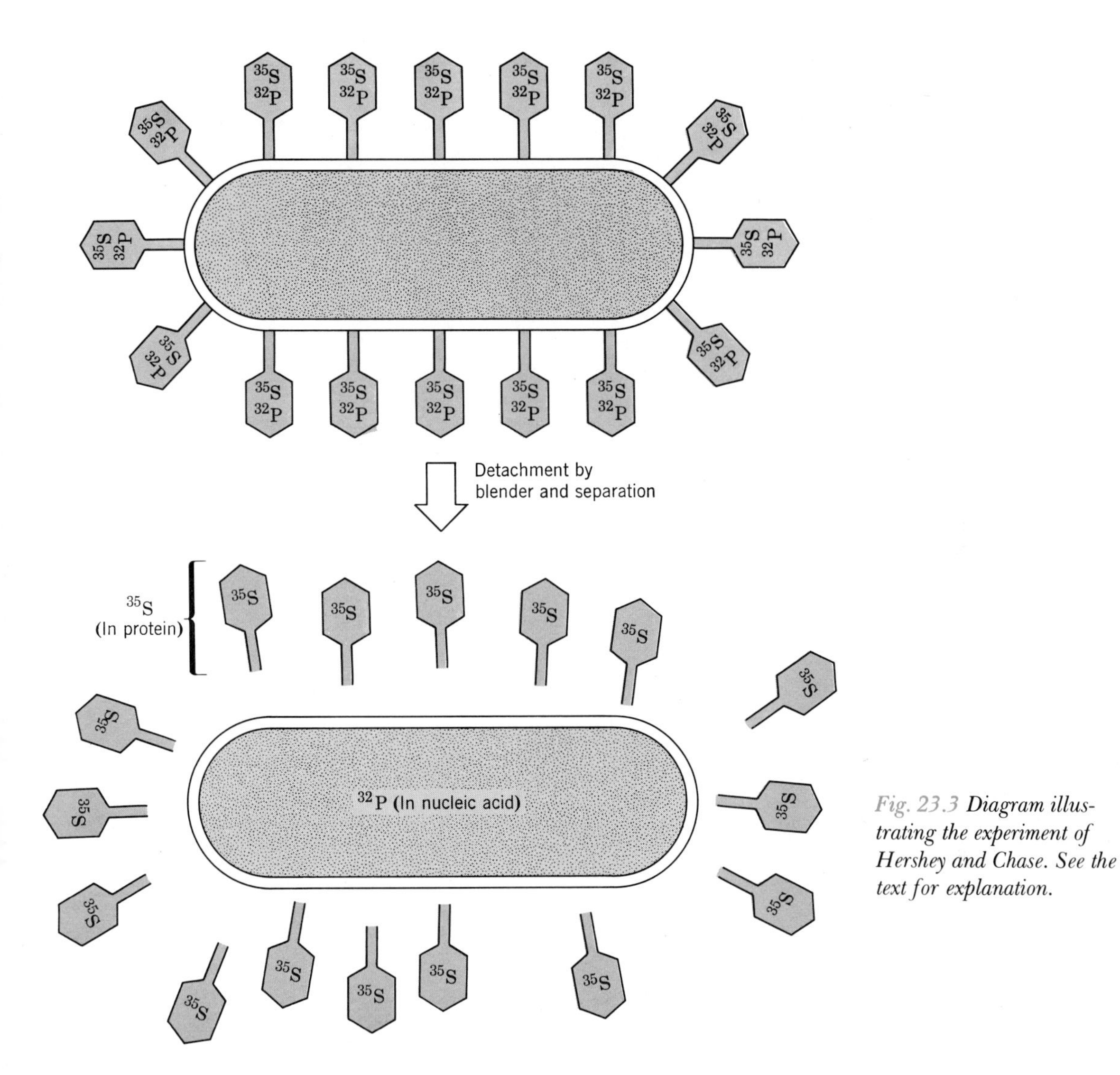

Fig. 23.3 Diagram illustrating the experiment of Hershey and Chase. See the text for explanation.

They built scale models of the DNA nucleotides and then tried to fit them together in a pattern that would be in accord with all the experimental data. They believed that the 3.4 angstrom periodicity corresponded with the distance between nucleotides in the chain and that the 20 angstrom periodicity corresponded with the width of the chain, but could not understand the 34 angstrom periodicity until it occurred to them that the chain might be coiled in a helix. Thus the 34 angstrom periodicity would correspond to successive turns of the helix and each turn would be ten nucleotides long. Since the known density of DNA was twice what it would be in a helix 20 angstroms wide and 34 angstroms long they concluded that the DNA molecule must be a double helix.

Watson and Crick then began making models of a double helical DNA molecule and found that the one that best fit the data had the two helical chains wound in opposite directions

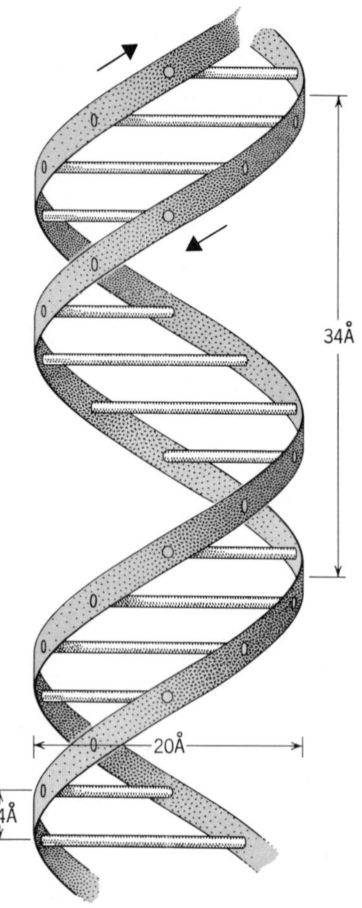

Fig. 23.4 Diagram of a small part of a DNA molecule giving its dimensions in Ångstrom units (Å). The horizontal bars represent the purine-pyrimidine pairs.

around a common center with the nitrogenous bases extending toward the center. Hydrogen bonds between the bases of the two chains could hold them together and maintain the double helix (Fig. 7.11). The scale model showed that a purine must be bonded to a pyrimidine. There was not enough space for two purines to bond together and too much to permit two pyrimidines to bond together. They also found that although adenine and cytosine were of the proper size, they could not be arranged so that there could be hydrogen bonding between them. The same was true of guanine and thymine. They concluded that adenine could bond only with thymine and that guanine could bond only with cytosine, thus explaining Chargaff's data. It evidently did not matter on which helix one member of a pair was—an A-T pair was just as feasible as a T-A pair, or a G-C as a C-G. Watson and Crick's report on their work was published in 1953 and has been substantiated by subsequent research. They, along with Wilkins, were awarded a Nobel prize in 1962 for their important contributions.

DNA REPLICATION

The Watson-Crick model of DNA structure suggested a means of precise replication of a DNA molecule, which is essential if it is to be a carrier of genetic information. Each cell derived from a parent cell by mitotic division has not only the same complete set of chromosomes as the parent cell but also the same genes and the same amount of DNA. Thus, prior to each mitotic division there must be a doubling of the amount of DNA and, even more important, the new DNA must have the same sequence of nucleotides in its chains as the parent DNA had.

Watson and Crick pointed out that if the two chains of a DNA molecule are separated by the breaking of the hydrogen bonds between the base pairs, each chain has the information needed for synthesizing a new chain that is identical with its old partner chain, since adenine can bond only with thymine and guanine only with cytosine. Each chain of the original pair could thus construct a new chain by bonding to itself a new complementary nucleotide from the available supply. A chain with the nucleotide sequence ACATGC would assemble against itself a new chain with the sequence TGTACG. At the same time, its old sister chain (with the sequence TGTACG) would be synthesizing a new chain with the sequence ACATGC. As this continued along the length of the two original DNA strands there would be two new DNA molecules produced, each with the same nucleotide sequence as the original parent DNA molecule (Fig. 23.5). One helix of each of the two new DNA molecules would be newly synthesized while the other helix would have come from the parent molecule. Of course, specific enzymes are required for this DNA synthesis.

Watson and Crick's proposal regarding DNA replication was purely hypothetical and was not supported by any experimental evidence, but within several years, support was provided by various experiments. Of these, we shall describe only the one conducted by the botanist, J. Herbert Taylor. He cultured germinating seedlings of broad beans (*Vicia faba*) in nutrient solutions containing radioactive thymidine (labeled with tritium (^3H), a radioactive isotope of hydrogen, which is selectively incorporated into DNA). The root tip cells incorporated the radioactive thymidine in their DNA, thus labeling it, and the chromosomes in which the DNA was located. After time for one cell division the seedlings were transferred to a solution containing nonradioactive thymidine and also colchicine, an alkaloid that prevents cell division but not chromosome duplication. The number of chromosome divisions could be determined by counting the number of chromosomes per cell under a microscope. The radioactivity of the chromosomes was determined by spreading a liquid photographic emulsion over cells on a microscope slide, allowing it to dry, and then developing it (all this in the dark). Since radioactivity (like light) causes darkening of the photographic emulsion, the chromosomes labeled with the radioactive thymidine then showed up as dark areas while the unlabeled chromosomes did not.

After one chromosome duplication in the nonradioactive medium both chromatids (daughter chromosomes) were labeled (dark), indicating that DNA replication does not result in the production of one intact old DNA molecule and one entirely new one (Fig. 23.6). However, when cells were allowed to proceed through two chromosome replications only one of the chromatids from each chromosome was labeled. Chromosomes resulting from a third replication were of two types: some had one of the two chromatids labeled but the remainder were not labeled at all. These results, like those of various other experiments, support the Watson-Crick hypothesis of DNA replication.

THE RIBONUCLEIC ACIDS

Although DNA carries the codes that determine the sequence in which the various amino acids are incorporated into proteins as they are being synthesized, it is now clear that DNA does not participate directly in protein synthesis. For one thing, DNA is restricted to the nucleus (except for the DNA in chloroplasts and mitochondria) whereas protein syntheses occurs primarily in the cytoplasm. The links between DNA and protein synthesis are provided by at least three kinds of ribonucleic acid (RNA). RNA differs from DNA in that its sugar component is ribose rather than deoxyribose, and it contains uracil rather than thymine, as well as in other ways that will be mentioned later. The three kinds of RNA involved in protein synthesis are messenger RNA (mRNA), ribosomal RNA (rRNA), and transfer RNA (tRNA).

Messenger RNA

The messenger RNA molecules consist of a single strand of nucleotides rather than the double helical strands characteristic of DNA. The mRNA is synthesized in the nucleus

Fig. 23.5 Diagram illustrating the duplication of DNA. Above: A portion of a DNA molecule before replication. Below: The molecule is progressively unzippering from the right end by breaking of the hydrogen bonds between A-T and C-G (adenine, thymine, cytosine, guanine) pairs. Each purine or pyrimidine is now free to bond with a nucleotide containing its complementary mate. Since the only possible pairings are A-T and C-G, each strand assembles an exact replica of the strand from which it has separated. The unzipping and replication continue, resulting in two new molecules identical with the parent molecule. Each new molecule has one strand from the parent molecule and one newly synthesized strand.

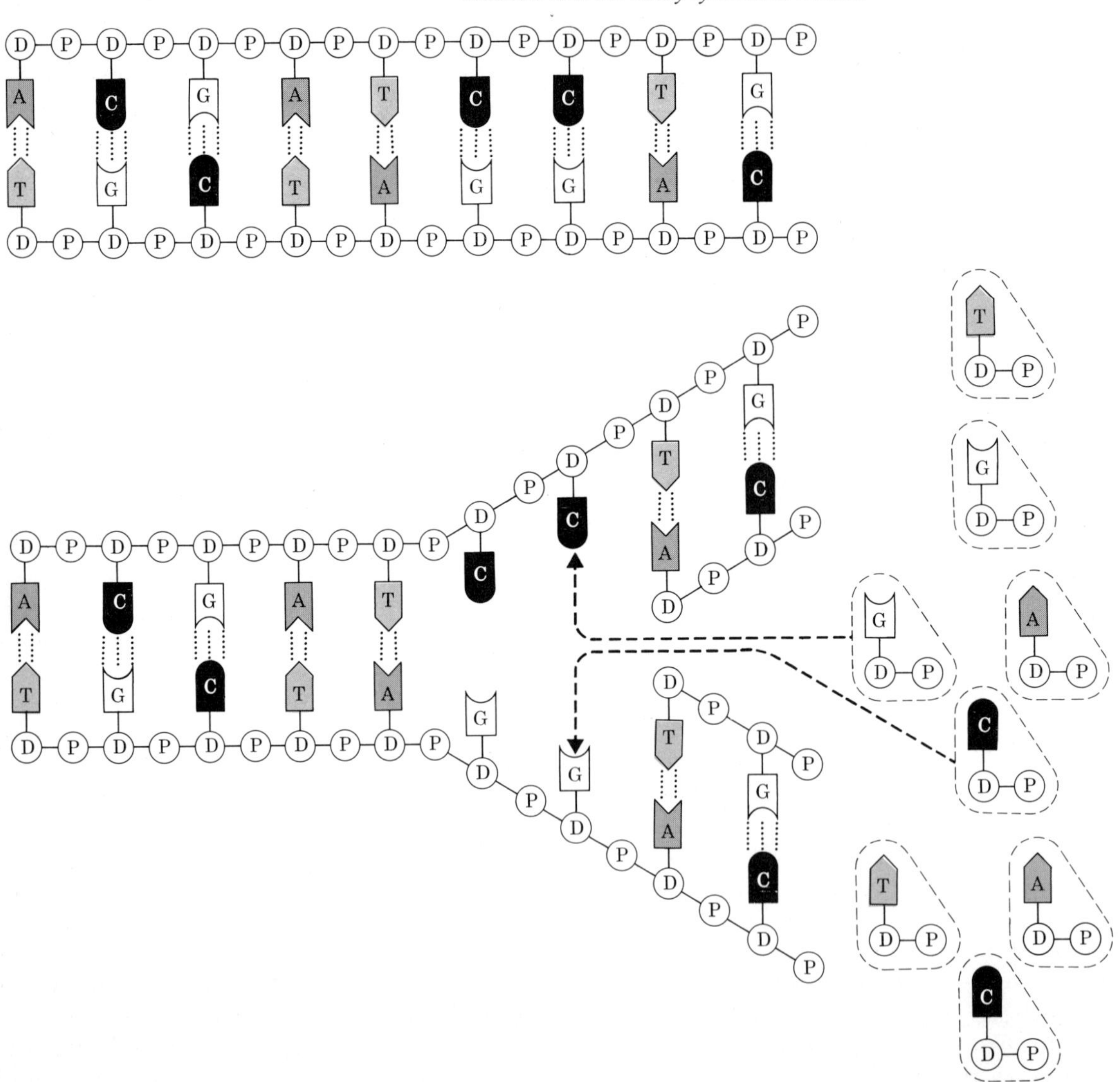

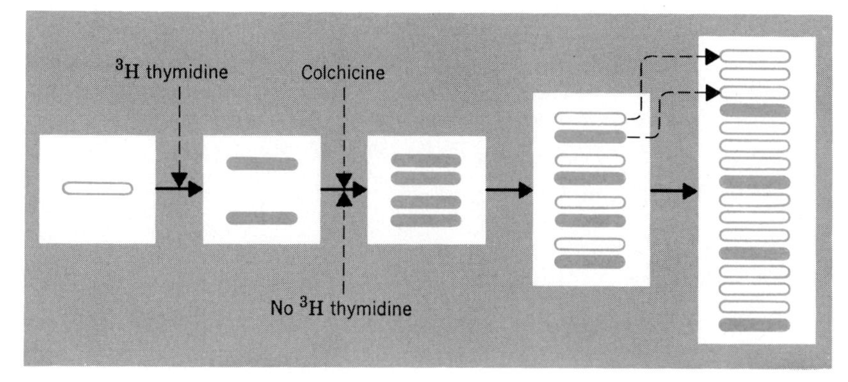

Fig. 23.6 Diagram illustrating Taylor's experiment. See the text for explanation. Only a single chromosome is shown in the diagram, and all the chromosomes resulting from four successive mitotic divisions are derivatives of the chromosome at the left. The dark chromosomes show radioactivity.

against a DNA template, using just one of the two strands of a DNA molecule, and so carries a code complementary to this particular DNA code. If the DNA is compared with a photographic negative, then the mRNA would be like a print made from the negative (Table 23.1). Like thymine, uracil can bond only with adenine. Thus, if a portion of the DNA nucleotide sequence were GTCTAGCGA the sequence in the mRNA molecule synthesized against it would be CAGAUCGCU. (Remember that in RNA uracil, rather than thymine, links with adenine.) DNA can serve as a template for the synthesis of mRNA only when the hydrogen bonds between the two strands of the DNA helix are broken. Each mRNA molecule thus carries the code from the DNA for the synthesis of a specific polypeptide. An enzyme or other protein may consist of only a single polypeptide, but generally is made of two or more polypeptides linked together.

Table 23.1 **Pairing of Nucleotides in the Replication of DNA and the Transcription of the DNA Code to mRNA and tRNA**

Replication		Transcription
DNA ←	—DNA—	→ RNA
T	A	U
A	T	A
G	C	G
C	G	C

Every cell of an organism contains one or more complete sets of DNA codes, that is, all the genes or hereditary potentialities possessed by the individual. However, any particular cell at a specific time is synthesizing only a fraction of the kinds of mRNA for which it has DNA codes. Consequently, the cell is producing only a fraction of the enzymes the individual is capable of producing. In the course of the life of a cell the production of a specific kind of mRNA may be terminated or the production of new kinds of mRNA may be initiated. The factors involved in turning on and off the synthesis of specific kinds of mRNA are not entirely clear, but it seems likely that in higher organisms the binding of histones (a type of protein) to DNA prevents the DNA from serving as a template for mRNA synthesis. The removal of histone from the portion of the DNA that constitutes a gene (or the addition of histone) can be brought about by hormones or other factors. The differential production of specific kinds of mRNA and their subsequently produced proteins is evidently an important factor in differentiation and development.

In some cases such as the beginning of seed germination, where there is a marked change in developmental patterns, a considerable number of enzymes (and, apparently, kinds of mRNA) not previously present are produced. In germinating seeds the enzymes include a variety of hydrolytic enzymes, and the production of these enzymes is promoted by gibberellin and inhibited by abscisic acid. It has been puzzling how a single hormone could either promote or inhibit the production of a whole complex of mRNAs and so enzymes, but discoveries beginning about 1971 may provide the answer. It has been found that portions of different DNA molecules have identical codes, even though they code for different genes (i.e., different kinds of mRNA) and so different enzymes. It seems possible that the repeated DNA codes may serve in the derepression of genes. Thus, a single hormone could promote the repression or depression of all the genes with the same repeated code, even though the rest of the code were different for each gene.

Ribosomal RNA

Messenger RNA cannot function in protein synthesis unless it is attached to a series of ribosomes (polysomes). Ribosomes (Chapter 8) are composed of RNA and proteins, each ribosome consisting of two subunits of unequal size. The mRNA is apparently bound to the smaller subunit while the larger subunit contains the enzymes that catalyze the peptide bonding of the amino acids. The ribosomal subunits are separate from one another when not involved in protein synthesis. Although ribosomes are essential for protein synthesis, it is the mRNA and not the rRNA that determines the kind of protein being synthesized. The ribosomes may be considered as somewhat analagous to a phonograph, which is essential for converting the coded information in the record groove into sound, but has nothing to do with the determination of the code. The record code is analagous to the mRNA code, while the complementary mold used in pressing the record corresponds to the DNA code.

Transfer RNA

Transfer RNA molecules are considerably smaller than mRNA molecules but have a more complex molecular structure. They are basically a single strand of RNA, but this is folded back on itself several times, resulting in a sort of cloverleaf pattern (Fig. 7.13). There are four short regions where there is hydrogen bonding between complementary bases, three loops with unpaired bases, and a free end to which a specific amino acid molecule can attach. Each kind of tRNA can bond with only one kind of amino acid. The loop opposite to the amino acid binding site has a triplet **anticodon** (three nucleotides complementary to a mRNA **codon**) that can attach only to the mRNA codon for the specific amino acid. For example, one of the mRNA codons for the amino acid asparagine is AAU. A tRNA molecule with the anticodon UUA can bond *only* with asparagine, thus assuring that the asparagine is introduced into the growing polypeptide chain at the precise point specified by the mRNA code. Another mRNA codon for asparagine is AAC. A tRNA molecule with the anticodon UUG is also capable of bonding only to asparagine. Thus tRNA plays a very important role in the incorporation of amino acids into the polypeptide chain at the exact place specified by the mRNA code.

THE GENETIC CODE

The DNA and RNA codes both consist of only four units: A, T, C, and G in DNA and A, U, C, G in mRNA, but they are capable of coding for 20 or more amino acids. A single base could code for only four amino acids (Fig. 23.7). Also, a two-base codon would be inadequate since there are only 16 possible combinations of four units in groups of two (4^2). In groups of three, there are 64 different combinations of the four units (4^3), more than enough to code for 20 amino acids. All possible combinations of four units in groups of four would provide 256 different combinations (4^4),

A	U	C	G	Unit Code 4 codes

AA	UA	CA	GA	Diad Code 16 codes (4^2)
AU	UU	CU	GU	
AC	UC	CC	GC	
AG	UG	CG	GG	

AAA	UAA	CAA	GAA	
AAU	UAU	CAU	GAU	
AAC	UAC	CAC	GAC	
AAG	UAG	CAG	GAG	
AUA	UUA	CUA	GUA	
AUU	UUU	CUU	GUU	
AUC	UUC	CUC	GUC	
UAG	UUG	CUG	GUG	Triplet Code 64 codes (4^3)
ACA	UCA	CCA	GCA	
ACU	UCU	CCU	GCU	
ACC	UCC	CCC	GCC	
ACG	UCG	CCG	GCG	
AGA	UGA	CGA	GGA	
AGU	UGU	CGU	GGU	
AGC	UGC	CGC	GGC	
AGG	UGG	CGG	GGG	

Fig. 23.7 Diagram showing the number of possible codes if the code consists of one, two, or three units that can combine in every possible sequence.

far too many and an unlikely situation. Thus, on purely theoretical grounds, it seemed probable that the DNA and RNA codons were triplets. A considerable amount of experimental evidence has substantiated this.

One problem that disturbed early investigators was how 64 different triplet codons could code precisely for only 20 or so amino acids. At one time it was thought that most of the triplets provided nonsense codes that did not specify any amino acid. However, it was then found that more than one triplet codon could code for a certain amino acid (Table 23.2). For example, there are six different triplets that code for leucine. There are a few (one to three) triplets that are called nonsense codes. However, these play a very important role since they signal the termination of a polypeptide chain. The nonsense triplets can be compared to a period at the end of a sentence or the word STOP in a telegram and are sometimes referred to appropriately as terminating triplets (or codons).

The deciphering of the genetic code (Table 23.2) was an important contribution that resulted from a series of relatively simple but elegant experiments. Ribosomes, tRNA, all the enzymes essential for protein synthesis, but no mRNA, were extracted from cells and placed in a suspension along with all of the amino acids used in protein synthesis. To this was added synthetic mRNA made in a test tube by supplying only one of the four nucleotides and an enzyme that links nucleotides together in the absence of a DNA template. When only uracil nucleotides were supplied the mRNA was UUUUUUUUU . . . and the only amino acid incorporated in the polypeptide chain was phenylalanine (UUU). When a mRNA consisting only of AAAAAA . . . was used, only lysine (AAA) was incorporated in the polypeptide. It was also found that CCC coded for proline and GGG for glycine. This was, of course, only a beginning in deciphering the code. It was much more difficult to synthesize mRNA from combinations of two or more different nucleotides and secure unequivocal results since these could join in several sequences and so code for several amino acids. The code was finally broken in 1964 and 1965 by Philip Leder and Marshall Nirenberg of the National Institutes of Health. They developed a technique for complexing RNA trinucleotides (triplets) with ribosomes, and were also able to synthesize each of the 64 possible triplets at will. Thus, they were able to determine the triplet codons for each amino acid. Their results, which have been slightly modified by subsequent investigations, are given in Table 23.2.

Despite the fact that only methionine and tryptophan have single codons and that there are as many as six codons for each of two

Table 23.2 **The mRNA Genetic Code** [a]

1	2: U	2: C	2: A	2: G	3
U	Phenylalanine	Serine	Tyrosine	Cystine	U
					C
	Leucine		*Nonsense*	*Nonsense*	A
				Tryptophan	G
C	Leucine	Proline	Histidine	Arginine	U
					C
			Glutamine		A
					G
A	Isoleucine	Threonine	Asparagine	Serine	U
					C
	Methionine		Lysine	Arginine	A
					G
G	Valine	Alanine	Aspartic acid	Glycine	U
					C
			Glutamic acid		A
					G

[a] To find the mRNA codes for any amino acid, locate the amino acid in the table, noting that some occupy more than one block. Read the first nucleotide of the triplet from column 1, the second from section 2 (reading down), and the third from column 3. For example, phenylalanine is coded for by UUU and UUC, proline by CC followed by any of the four nucleotides, and methionine by AUA and AUG. Questionable: UGA may code for cystine; UAG may not be nonsense.

amino acids (leucine and serine), it should be noted that the genetic code is not haphazard. In general, the synonymous codons differ only in their third nucleotide, and also when in the third position U and C are equivalent and A and G generally are equivalent. Of course, deciphering of the mRNA code also broke the DNA code since the two are complementary. For example, since the mRNA code for methionine is AUG the DNA code for it is TAC.

PROTEIN SYNTHESIS

We are now in a position to consider protein synthesis in somewhat greater detail than we did in Chapter 14. A prerequisite for protein synthesis is, of course, the presence of all the amino acids used in protein synthesis. Plants are capable of synthesizing all these amino acids. Before amino acids can partici-

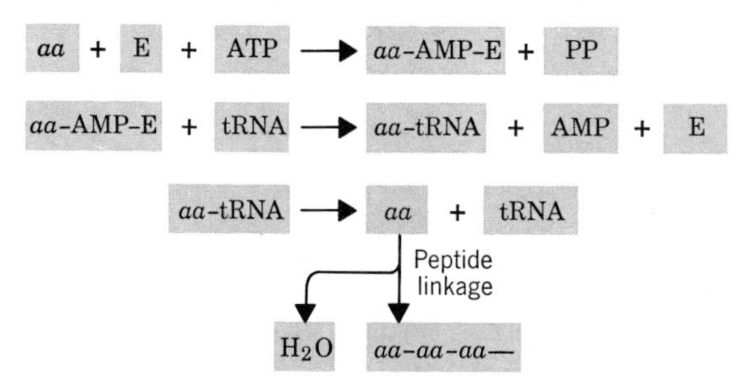

Fig. 23.8 The sequence of reactions as an amino acid (aa) is activated by combining with the activating enzyme (E) and ATP, is then attached to its specific t-RNA, and is finally incorporated in the growing polypeptide chain.

pate in protein synthesis, they must react with an activating enzyme and ATP. Two of the phosphate groups of the ATP are freed in this reaction, so the activated amino acid complex consists of the amino acid, the enzyme, and AMP (adenosine monophosphate). This complex then reacts with the tRNA specific for the amino acid, forming an amino acid-tRNA complex and releasing the AMP and the activating enzyme (Fig. 23.8).

All the different amino acids attached to their specific tRNA molecules are present near a mRNA molecule, but the tRNA can attach to the mRNA only at a point where the mRNA codon is complementary to a tRNA anticodon and where the mRNA codon is attached to a ribosome (Fig. 23.9). For example, a tRNA carrying leucine and having the anticodon GAA can attach to the mRNA only where there is a CUU codon in contact with a ribosome. The amino acid is then attached to the last amino acid incorporated in the growing polypeptide chain by a peptide linkage (Chapter 14) and its tRNA is released. As the ribosomes move along the mRNA molecule, they "read" its code and each of the ribosomes assembles the polypeptide specified by the code. This continues until a nonsense codon such as UAA is encountered, thus terminating the polypeptide, which is then freed from the ribosome. As noted before, the polypeptides are apparently assembled on the larger subunit of the ribosome while the mRNA is attached to the smaller one.

The polypeptide itself may constitute a protein molecule, but more generally two or more polypeptides (commonly different ones) are linked together, forming the complete molecule of an enzyme or other protein. Although little mention has been made of the enzymes involved in protein synthesis, it should be stressed that each of the steps in protein synthesis, as well as each of the steps in amino acid synthesis and activation and in the synthesis of the nucleic acids requires a specific enzyme to catalyze the reaction. It is interesting to note that each of these numerous enzymes was produced previously by the process of protein synthesis we have described.

GENES AND ENZYMES

It is now clear that all the hereditary potentialities of genes as prescribed by their various DNA codes are expressed by way of the specific enzymes for which they code. This is true whether the hereditary potentiality is for a physiological process such as starch synthesis or a structural characteristic such as plant height. In at least some cases, tall plants of a species differ from dwarf ones in that their genes code for all the enzymes needed for synthesis of adequate amounts of gibberellic acid, whereas this is not true of the dwarf plants. Every structural characteristic of a plant, such as the capacity for producing vascular tissue or of developing flowers of a specific type, is dependent upon the presence of a suitable complex of enzymes coded for by their genes.

A problem that once concerned geneticists was whether or not there was a one-to-one relationship between genes and enzymes, that is, whether each enzyme was coded for by a single gene. A variety of investigations has helped resolve the question. Among these

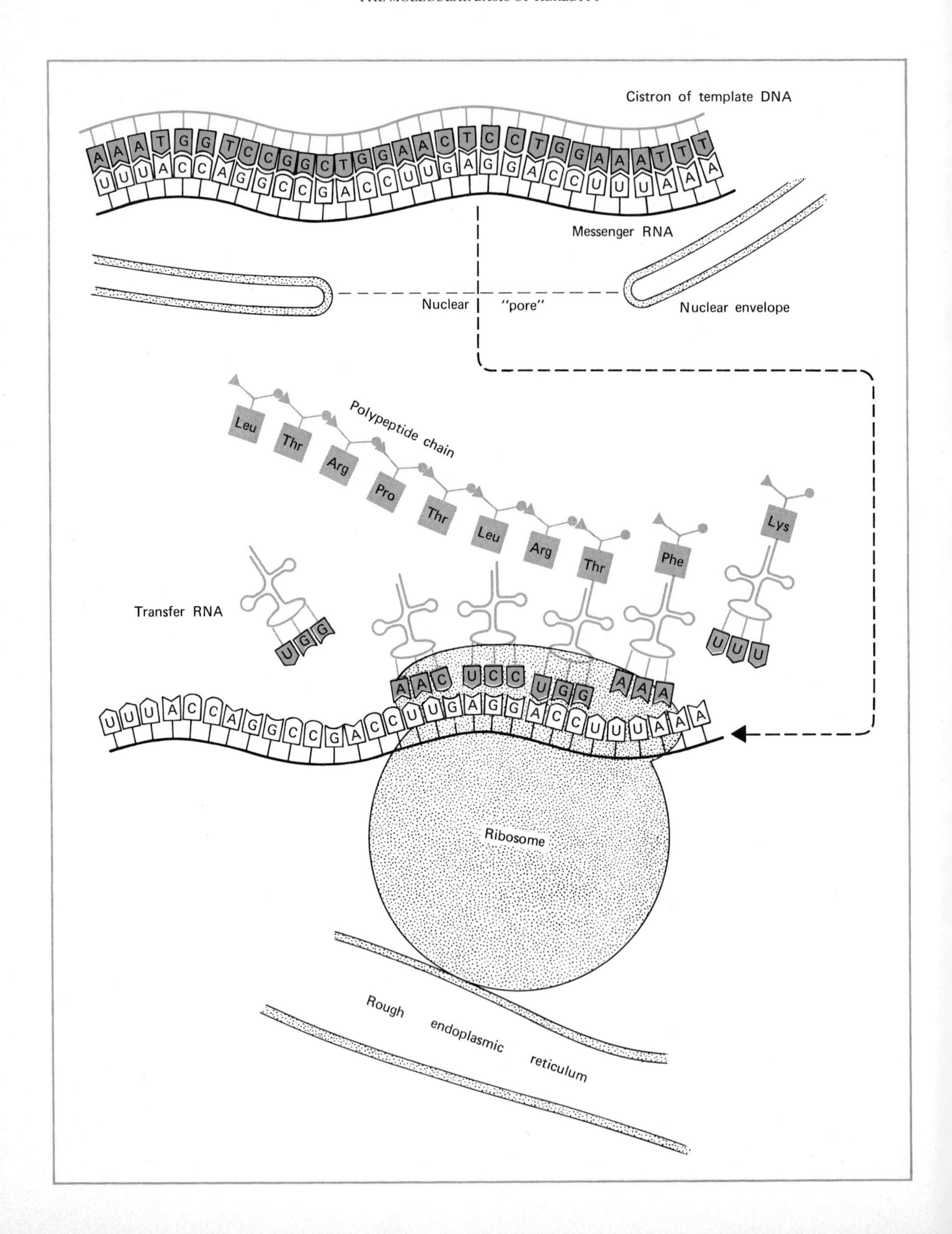
Cistron of template DNA
Messenger RNA
Nuclear
"pore"
Nuclear envelope
Polypeptide chain
Leu
Thr
Arg
Pro
Thr
Leu
Arg
Thr
Phe
Lys
Transfer RNA
UGG
AAC
UCC
UGG
AAA
UUU
Ribosome
Rough endoplasmic reticulum

Fig. 23.9 Diagram illustrating how the genetic code of DNA is transcribed to the complementary mRNA codons, which then (in cooperation with tRNA and ribosomes) determine the sequence of specific amino acids in protein molecules as prescribed by the DNA code. Thus, an organism can synthesize only those enzymes specified by its genes, and the enzymes in turn determine which metabolic processes the organism can carry on.

were experiments of Beadle and Tatum on *Neurospora,* an ascomycete mold that is commonly called pink bread mold. They received a 1958 Nobel prize for their important work, which was begun about 1940.

To give only one example, *Neurospora* generally can synthesize the essential amino acid **tryptophan** from **indole** and **serine,** the latter being another amino acid. The reaction is catalyzed by the enzyme **tryptophan synthetase.** Since most *Neurospora* plants can synthesize tryptophan, it is not necessary to supply this amino acid in the culture medium. Beadle and Tatum produced a mutant *Neurospora,* however, that would not grow well without an external supply of tryptophan, and found that the mutation had resulted in the inability of the fungus to produce the enzyme tryptophan synthetase. Crosses between the mutants and plants able to synthesize tryptophane resulted in the expected ratios of the two phenotypes among the offspring. Both indole and serine are, in the final analysis, synthesized from sugars and ammonia, and Beadle and Tatum found at least six other mutants unable to make tryptophan, not because they lacked tryptophan synthetase, but because they could not produce one of the enzymes essential for the various steps in the synthesis of either indole or serine. Many other examples of specific gene-enzyme relations have now been found in *Neurospora* and other plants and animals.

From all their numerous investigations Beadle and Tatum concluded that the genes of the normal plants controlled the production of the enzymes lacking in the mutants and also that there was a one gene to one enzyme relation. Various other invesitgators working with a variety of plants and animals reached similar conclusions. However, there is now evidence that some modification of the one-to-one relationship is necessary. If an enzyme or other protein is composed of two or more different polypeptides, a different gene would be necessary to code for each of the polypeptides, which is only logical. Thus, more than one gene may be required for the production of some enzymes, but the one-to-one relationship still holds as applied to genes and polypeptides.

WHAT IS A GENE?

Although there is now no doubt that a gene is essentially a DNA code, geneticists have used and are still using the term gene in various ways. Without getting involved in the complications, it seems most reasonable to consider a gene as a strand of DNA that codes for one specific polypeptide. It is probable that a gene is not a whole DNA molecule. A DNA molecule is usually composed of thousands of nucleotide pairs. Polypeptide molecules are usually composed of 300 to 500 amino acids and rarely more than 1000. This means that a gene coding for a polypeptide would generally be composed of 900 to 1500 nucleotide pairs and rarely more than 3000. Thus, each DNA molecule would contain numerous genes. While each DNA molecule is a continuous double helix, it seems probable that the genes in it are separated from one another by nonsense codons, which are complementary to the mRNA nonsense codons. Thus, a UAA nonsense codon in mRNA would appear as ATT in DNA. Although most of the genes in a chromosome code for polypeptides, there are some that serve as templates for tRNA and rRNA production and others that control gene activity, as described elsewhere.

It is still not known definitely how the molecules of DNA are oriented in the chromosomes. Since genes are located in chromosomes in linear order, it might be assumed that the DNA molecules are arranged end to end in single file the length of a chromosome. However, this would require that the chromosomes visible during mitosis and meiosis would have to be much longer and thinner than they are. Various theories have been proposed, including models that suggest the location of the DNA molecules crosswise of a chromosome, either parallel with one another or in zigzag fashion. However, the most plausible theory is that of DuPraw (Fig. 23.10). He proposes that the DNA strands in the chromosomes are highly folded fibers consisting of a single DNA double helix (that is highly coiled) per chromosome or chromatid. During chromosome replication in the interphase the fibers uncoil and partially unfold, making them much longer and thinner (and so not visible as the shorter and thicker chromosomes seen during mitosis). This theory explains how one long DNA strand could exist in the relatively short and thick mitotic chromosomes.

NUCLEIC ACIDS IN CYTOPLASMIC INHERITANCE

As we pointed out in Chapter 22, some genes are located in the cytoplasm rather than

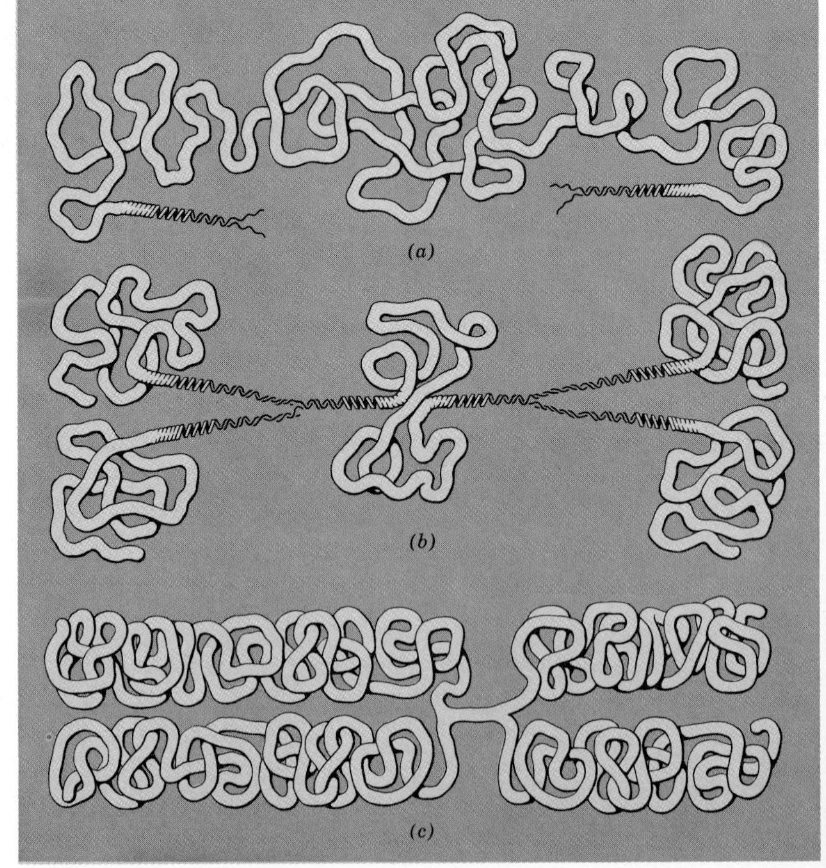

Fig. 23.10 Diagram of a folded fiber model of chromosome structure. (a) *Each chromosome has a single folded fiber 200 to 500 Å long, consisting of supercoiled DNA.* (b) *Replication of the chromosome occurs at several points where the DNA helix uncoils and separates.* (c) *Late replication at the centromere holds together the sister chromatids until they separate during mitosis. The highly folded fibers in* (c) *occur in the chromosomes visible during mitosis. (Redrawn from E. J. DuPraw,* Nature, 206, *338, 1965, with permission.)*

in the chromosomes of the nucleus. These cytoplasmic genes are located primarily in the chloroplasts and mitochondria but are also found in other plastids, the basal bodies of flagella, and centrioles. All these organelles are self-replicating and carry some of their own hereditary potentialities.

Chloroplasts and mitochondria contain everything needed to translate their genetic codes into polypeptides: DNA, mRNA, tRNA, ribosomes, and all the various enzymes required for nucleic acid and protein synthesis. However, their DNA is a naked strand not complexed with protein as with the chromosomal genes and it has a nucleotide ratio (CG:AT) quite different from that of the chromosomal genes. Also, the ribosomes are smaller than those in the surrounding cytoplasm. In both these respects the genetic apparatus of chloroplasts and mitochondria is strikingly similar to that of procaryotic cells (bacteria and blue-green algae). This, along with the fact that chloroplasts and mitochondria have a general structure and size range comparable with procaryotic cells, provides evidence in support of the theory held by some biologists that early in the course of evolution blue-green algae and bacteria entered eucaryotic cells and became symbiotic with them.

Although chloroplasts and mitochondria carry many of their own genes, and so are capable of producing many of their enzymes, they are by no means free from control by chromosomal genes. For example, the recessive albino trait of corn plants is definitely determined by chromosomal genes and follows the 3:1 Mendelian ratio in the F_2 generation. However, if all the chloroplasts and proplastids are removed from a *Euglena* cell, its chromosomal genes are incapable of restoring them even though the *Euglena* can survive heterotrophically and produce many new generations by asexual reproduction. If chloroplasts are introduced into one of these *Euglena* cells the chloroplasts become established, divide, and are present in all the offspring of the cell.

THE MOLECULAR BASIS OF GENE MUTATIONS

In Chapter 22, we pointed out that a gene mutation involved a chemical alteration of the gene. It should now be evident that this involves an alteration in the sequence of nucleotides in the DNA constituting the gene. The alteration may result from the deletion of one or more nucleotides, the addition of one or more nucleotides, or a substitution of one base for another in a nucleotide.

If a deletion occurs near the beginning of a gene, it can result in a drastic change in the code and so in the polypeptide produced. For example, suppose the first few triplets of a gene are GAC TAC CGA TTG. The polypeptide would then begin with leucine-methionine-alanine-asparagine. If the first T were deleted the code would then become GAC ~~T~~ACC GAT TG$^-$, and the polypeptide sequence would now be leucine-tryptophan-leucine-serine. (Any of the bases in the third position of the last codon would code for serine.) Thus, an entirely different polypeptide would be synthesized. It could possibly have enzyme activity different from that of the one originally coded for, but the chances are that it would lack enzyme activity. An enzyme can lose catalytic activity if there is a change in only one amino acid at one of its active sites, although changes in other parts of the enzyme molecule might not be critical.

If the fifth and sixth triplets of the above gene before the deletion were AAT and TTA, they would code for the addition of leucine and asparagine to the polypeptide chain. However, after the deletion the triplets would be ATT AA$^-$. Since ATT is a nonsense codon the polypeptide would be terminated prematurely.

Such a short polypeptide would be almost certain to lack enzyme activity.

A deletion near the end of a gene would have a much less drastic effect on the polypeptide produced, but might well result in a loss or change of enzyme activity. If, for example, the last five triplet codons of a gene were . . . . . . ACC ACA TCG GGA ATT they would code for the addition of tryptophan-cysteine-serine-proline and termination of the polypeptide. If the first A were deleted the code would become ~~A~~CCA CAT CGG GAA TT^{-} and would add an entirely different set of amino acids to the new polypeptide: glycine-valine-alanine-leucine. Note that the nonsense codon has been lost and so the polypeptide could be longer than the one originally coded for. The next amino acid would be asparagine if the next base in the DNA happened to be A or G or lysine if it happened to be T or C.

The addition of one or more nucleotides to a portion of DNA constituting a gene would alter the code as drastically as a deletion and, like a deletion would have the greatest effect near the beginning of the gene. For example, an original triplet sequence of AAA ACA TAG CCC GTA would code for phenylalanine-cysteine-isoleucine-glycine-histidine. If a C nucleotide were added at the end of the first triplet the codons would become AAA CAC ATA GCC CGT A^{--} and would incorporate phenylalanine-valine-tyrosine-argenine-alanine. If an entire triplet were either added or deleted between existing triplets, there would be a change in only one amino acid in the polypeptide chain, but even this could result in the loss or change in the enzymatic capability of the enzyme. Addition or deletion of any three nearby nucleotides would also result in a less drastic change in the code than if there were one, two or four deletions or additions. Suppose that the original codons were AAA ACA TAG CCC GTA AAA and that deletions occurred as follows: AAA ~~A~~C~~A~~ ~~T~~AG CCC GTA AAA. The code would now be AAA CAG CCC GTA AAA. The result would be the loss of one amino acid from the polypeptide (cystine) and substitution of valine for isolucine. Such a change might or might not result in an alteration of enzyme activity. Note that the rest of the code remains unaltered.

In general, the substitution of one base for another in a nucleotide is likely to have a much less drastic effect on the polypeptide coded for than an addition or deletion, especially if the substitution occurs in the third nucleotide of a triplet. Thus, if GAA were changed to GAT, GAC, or GAG it would still specify leucine. The fact that in a number of cases the third nucleotide of the triplet can be changed without affecting the amino acid coded for has undoubtedly resulted in a considerable reduction in the number of altered enzymes that might have occured otherwise. At the most, a substitution results only in a change in a single amino acid, but even this could result in the production of an enzymatically ineffective polypeptide. A change of ACA to ACC would substitute tryptophan for cystine. A change of ATA to ATT would be even more drastic since the codon for tyrosine would be changed to a nonsense codon that would terminate the polypeptide prematurely.

The fact that most mutations are recessives is in accord with the fact that most mutations result in the loss of enzyme activity. Many recessives do not code for an enzymatically active protein, as do their dominant alleles. Most mutant genes that are expressed phenotypically either result in traits that have reduced survival value (and may be lethal) or in traits that are neutral as regards survival. However, in the long course of evolution many mutations have occurred that have coded for new and different enzymes, and so new hereditary potentialities. An important example early in the history of evolution was the appearance of the large complex of enzymes essential for chlorophyll synthesis and photosynthesis. The evolution of vascular plants involved the appearance of numerous new genes that coded for the enzymes essential for the development of such structures as vascular tissues, roots, stems and leaves.

RELATED READING

Beadle, G. W. "The genes of men and molds," *Scientific American 179*(3) 30–39, September 1948.

Beadle, G. W. "Genes and chemical reactions in *Neurospora*," *Science, 129,* 1715–1719, 1959.

Goodenough, V. W. and R. P. Levine. "The genetic activity of mitochondria and chloroplasts," *Scientific American, 223*(5), 22–29, November 1970.

Jukes, T. H. "The genetic code," *American Scientist, 51,* 227–245, 1963.

Jukes, T. H. "The genetic code, II," *American Scientist, 53,* 477–487, 1965.

Ptashne, M. and W. Gilbert. "Genetic repressors," *Scientific American, 222*(6), 36–44, June 1970.

Taylor, J. H. "The duplication of chromosomes," *Scientific American, 198*(6), 36–42, June 1958.

Yanofsky, C. "Gene structure and protein structure," *Scientific American, 216*(5), 80–94, May 1967.

24 HEREDITARY VARIATION IN PLANT EVOLUTION

Biologists generally accept Charles Darwin's theory of natural selection (as modified by investigators during the present century) as the mechanism by which evolution occurs. Darwin first proposed his theory to the scientific community in a paper presented jointly with Alfred R. Wallace (who had independently reached the same conclusions about natural selection) to the Linnaean Society in 1858. The following year, Darwin published his important book, *On the Origin of Species by Natural Selection.* Actually, Darwin had first developed the concept of natural selection some 20 years earlier.

Stated briefly, the theory involved the following: (1) the reproductive potential of any species is greater than the actual rate of its population increase or the limited habitats available to the species; (2) this leads to a pressure of the population on the environment and to a struggle for existence (a concept derived from Thomas Malthus); (3) there is considerable variation within the population of a species, and so some of the members of the population are better adapted (fitted) to their particular environment than others; (4) those best fitted to the environment survive, while the others are eliminated by natural selection. Thus, over the years, some characteristics of the species survive and increase in frequency whereas others are eliminated, resulting in a change in the species.

The weakest part of Darwin's theory was his lack of understanding of the nature of variation in a species. Although he was not satisfied with the old idea that hereditary variations resulted from a blending of parental traits, he evidently had no knowledge of Gregor Mendel's work in genetics (which was being done at about the same time), and reached the conclusion that much of the variation resulted from the inheritance of acquired characteristics, which we now know does not occur. We also know that the variations subject to natural selection must be hereditary, not environmental, and that these hereditary variations can be any of the types considered in Chapter 22: gene mutations, chromosomal changes, recombinations, and polyploidy. However, recombinations of genes can occur only during sexual reproduction, so hereditary variation is less extensive in species that reproduce largely or entirely asexually. Also, we now know that hereditary variations resulting from mutations, recombinations, or the other genetic mechanisms can be much greater and more marked than the slight variations assumed by early biologists or even Darwin.

MUTATIONS

Gene mutations have undoubtedly been the major source of the hereditary variations and changes that in the long course of history have resulted in the evolution of the vast and diverse array of higher organisms from a relatively few simpler and smaller ones. The mutations have resulted in new genes that simply did not exist before. For example, the first photosynthetic plants had a whole complex of genes not present in their primitive heterotrophic ancestors. Also, vascular plants have numerous genes that provide the hereditary potentialities for many structures such as tracheids or vessel elements, sieve cells or sieve tube elements, leaves, stems, roots, and eventually flowers. Evidently the nonvascular plants from which the vascular plants evolved had no genes for such characteristics. Such mutant genes for fundamental traits of vascular plants were of great importance in the adaptation of the plants to land environments and have been preserved by natural selection over millions of years and throughout the large number of species of vascular plants of diverse groups.

Of course, there have been many mutations in the course of evolution that resulted in traits that were lethal or that made the plants less fitted for survival and so were eliminated by natural selection. However, some students of evolution have believed that random mutations could not account for the apparently directed course that evolution has taken in the various major groups of plants and animals over millions of years, despite the role of natural selection. They proposed that evolution was **orthogenetic,** that is, that in some way many of the mutations were channeled in certain directions that resulted in the progressive development (or reduction) of certain structures or characteristics. Present-day students of evolution generally ignore the theory of orthogenesis, primarily because there has never been an explanation as to how mutations could be directed. Nevertheless, there are various evolutionary trends such as the shift from woody stems in the more primitive members of various orders or families of angiosperms to herbaceous stems in species that are more advanced. In any event, evolution has not been random or haphazard even though mutations may have been.

Mutations also play an important role in hereditary variations subject to natural selection at the species and genus level, but at this level other kinds of hereditary variation such as chromosomal changes, recombinations, and polyploidy are perhaps more important. Mutations are generally recessive, and so in diploid plants may not result in a phenotypic expression for several generations, until zygotes receive the homozygous recessive condition. Recessive genes often result in the loss of a metabolic capability (e.g., anthocyanin synthesis or chlorophyll synthesis), which may or may not make the individual subject to elimination by natural selection. Natural selection has usually resulted in a genome that fits a plant to its environment, and so the chances are that a random mutation will make a plant less fit and subject to elimination by natural selection. However, recessives that are undesirable or lethal in the homozygous state (such as the albino gene in corn) are transmitted from generation to generation by heterozygous individuals. Of course, a mutation cannot be transmitted to offspring produced by sexual reproduction unless it occurs in a gamete, or a cell that gives rise to gametes.

CHROMOSOMAL CHANGES

All of the chromosomal changes discussed in Chapter 22 (inversions, translocations, and deletions) result in genetic changes that can result in phenotypic variations (sometimes marked) that may be subject to natural selection. In addition, phenotypic variation can result from the gain of one or more additional chromosomes, so that there are 3 of some homologous chromosomes, rather than the usual 2 in a diploid ($2n+1$, $2n+2$, etc.). Some cultivated varieties of cherries differ from one another in having extra chromosomes of different pairs. Also, some plants lack one chromosome of a homologous pair ($2n-1$). In some genera including *Oenothera* (evening primrose) and *Datura* (Jimsonweed), inversions, translocations, deletions, and extra chromosomes are unusually numerous and result in extensive phenotypic variation. A. F. Blakeslee found many such variations in the size and shape of the capsules (fruits) of Jimsonweed. Plants with altered chromosome structures are often sterile because of the complications of chromosome pairing and segregation at meiosis. Although the hereditary variations resulting from chromosomal changes can be subject to natural selection, they probably play their major role by making fertile crosses with some of the other members of the population difficult or impossible (because of difficulties during synapsis and segregation) and so isolating the population into several distinct species.

RECOMBINATIONS

At the species level, much of the hereditary variation within a population is a result of the recombination of genes in the course of sexual reproduction. Most species in nature have a large number of heterozygous alleles, and the chances are that there will be at least one pair of heterozygous genes on each chromosome pair, so variation by recombination of genes is limited only by the number of pairs of chromosomes. The degree of phenotypic variation depends on how the genes interact, that is, whether there is dominance, incomplete dominance, multiple gene heredity, or some other type of gene interaction. However, even when only dominance is considered the number of different phenotypes in a population that results from recombinations can be very large (Chapter 22).

The Hardy-Weinberg Law

No matter how extensive the hereditary variation resulting from recombinations may be, it is of no significance in evolution unless subjected to selective pressures. In 1908, G. H. Hardy of England and W. Weinberg of Germany demonstrated mathematically that in a large population, in a stable environment where there is free and random interbreeding among individuals, the frequency of each gene in the population will remain the same generation after generation (Fig. 24.1). The frequency of any gene will not change, regardless of whether the two alleles are of the same frequency initially (as in Fig. 24.1) or one has a higher frequency in the population than the other initially. Recessive genes will not decline in frequency, even though they are expressed phenotypically less frequently than the dominant gene. Since the gene frequency remains stable, so does the frequency of each phenotype.

Fig. 24.1 Genotypes and gene frequencies through four generations, illustrating the Hardy-Weinberg law. Note that the gene frequencies are the same in all four generations and that in the F_2 *and* F_3 *generations (and subsequent generations) the genotypic and phenotypic ratios are also the same. Conditions in which this genetic stability does not operate are described in the text.*

Generation	Genotypes	Gene frequencies
P_1	*AA*, *aa*	*A*:*a* = 1:1
F_1	*AaAa*	*A*:*a* = 1:1
F_2		*A*:*a* = 1:1

	A	*a*
A	*AA*	*Aa*
a	*Aa*	*aa*

(Genotypic ratio = 1:1:2)
(Phenotypic ratio = 3:1)

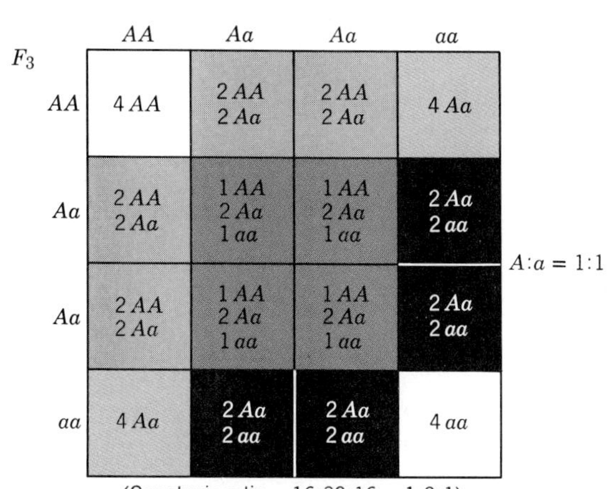

(Genotypic ratio = 16:32:16 = 1:2:1)
(Phenotypic ratio 48:16 = 3:1)

The interbreeding population must be large enough so that the statistical probability of the random recombinations is attained. In a small population, there may be a gene drift resulting in the increase or decrease of the frequency of a gene, although continued drift in one direction is likely to occur only if selection is operating. The stable gene frequency will be altered if there are mutations, and also if crossing is not random because of isolating mechanisms (which will be considered later). If a recessive gene is lethal when homozygous its frequency will be less than shown in Fig. 24.1, but the gene will be transmitted to the next generation by heterozygous parents and its lower frequency will be maintained. The stable gene frequency may also be altered if other members of the species migrate into or out of the interbreeding population.

However, the principal factor that will alter the gene frequency is natural selection. As long as the environment of the population is uniform and stable there is little natural selection or evolution, but if the environment changes or environmental gradients develop within the area occupied by the population, natural selection will begin and evolution will occur. The alterations may be in either the physical environment or the biological environment, as when a new parasite enters the community. Evolution is most rapid when there are marked changes in the environment. The ways in which natural selection operates will be considered later.

Hybridization

Hybridization between related species or the subspecies, races, and varieties of a species can occur in nature and may give rise to new species or races if the hybrids are fertile, if there are environmental niches in which the hybrids are better adapted than the parents, and if the new combination of genes becomes isolated and stabilized so that the hybrid population can survive as a distinct entity. Hybrids are frequently sterile because of meiotic failures or other reproductive factors, but some can reproduce sexually or by some means of vegetative propagation in nature. Of course, the F_2 generation will be extremely varied, providing many phenotypes subject to natural selection, and some of these may survive and become new subspecies or species. Hybrid vigor is one factor that may be involved in survival.

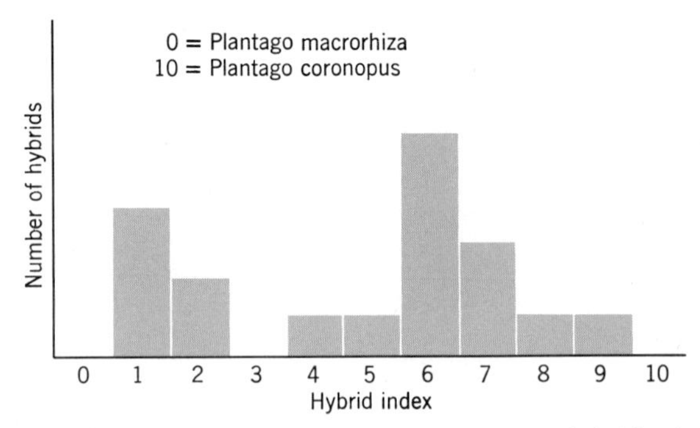

Fig. 24.2 The hybrid index of 40 introgressive hybrids of two species of Plantago *from Cap Matifou, Algeria. (Data of R. Gorenflot, Addisonia (n. s.), 4, 393–417, 1964.)*

In 1949 Edgar Anderson of the Missouri Botanical Garden identified a phenomenon he called **introgressive hybridization,** which involved crosses between two species. Since the fertile hybrids are much less numerous than the members of the two parent species, they are most likely to backcross with one or both of the parent species. With repeated backcrosses, there is a gradual infiltration of genes from one species into the other. The degree of infiltration can be estimated by determination of a **hybrid index.** As many contrasting characteristics of the two parents as possible are selected for comparison. These include such things as the size, shape, and hairiness of the leaves, the size, shape, and color of the petals, and the length of the stamens and pistils. For each character of one species an arbitrary value of zero is assigned, while a value of ten is assigned to each contrasting character of the other parent species, so one has an average of zero and the other ten for all characters. The average of the various hybrids will range from one to nine, depending on the degree to which they are similar to one parent or the other (Fig. 24.2). This indicates the extent to which each kind of hybrid has introgressed with one parent species or the other. There are also other means, such as scatter diagrams, of illustrating introgression, but these analyses of natural populations may still not make it certain whether the hybridization has been introgressive or not. In doubtful cases, experimental crosses and backcrosses are used.

Introgressive hybridization is very common among plants and has evidently played an important role in their evolution. The great diversity of hybrid phenotypes provides variations subject to natural selection. Some of the hybrid populations may be able to thrive in habitats not favorable to either parent species and so become essentially a new species, but others may persist in the same habitats as the parent species. Introgressive hybridization is an important factor contributing to the difficulty of drawing sharp dividing lines between the various species of a genus.

An interesting example of introgressive hybridization in a plant of great economic importance involves corn (*Zea mays*). The corn cultivated by prehistoric man was quite different from present-day corn, the plants, ears, and grains being smaller. Among other things the grains were in some varieties covered individually by husks. Growing as weeds in the cornfields were plants of teosinte (*Zea mexicana*). As a result of continuing introgressive hybridization between the two species, there was a much greater range of variation than there would have been otherwise, providing many new phenotypes subject to selection. However, in this case, selection was by man rather than nature. Phenotypes most desirable from an economic standpoint because of their larger grains and ears and other desirable characters eventually resulted in the modern types of corn.

POLYPLOIDY

Polyploidy (Chapter 22) has played an extremely important role in plant evolution and the origin of plant species. In contrast, polyploidy is rare in animals and of essentially no importance in their speciation. Up to 86% of the species of vascular plants in the arctic are polyploid, and the lowest percentage found in various floras is about 25%, as in West Africa. There is a lower incidence of polyploidy in trees and annuals than other plants, but almost every genus of any size has many polyploid species. In the genus *Rumex* (docks) with a basic n chromosome number of 10, for example, there are diploid species like *R. sanguineus* ($2n = 20$), tetraploids like *R. obtusifolius* ($2n = 40$), on through higher ploidys to *R. hydrolapathum* with an amazing $2n$ chromosome number of 200.

There are two types of polyploidy. One is **autopolyploidy,** in which there is self-pollination or cross-pollination with other members of the species that have essentially similar chromosome and gene complements, followed by chromosome doubling. The two sets of chromosomes in the diploid can be represented as *AA,* so the tetraploid would be *AAAA.* Autopolyploids can be induced by the use of colchicine and they also occur in nature but they play a minor role in species formation because they are usually sterile. The reason is that there are four homologous chromosomes of each kind (rather than the two in the diploid), and the pairing of three or four homologous chromosomes during meiosis results in haphazard distribution of the chromosomes to the meiospores.

The other type of polyploidy is **allopolyploidy,** which results from crosses between two species (or sometimes between two distantly related members of a species), followed by chromosome doubling. Here the two sets of chromosomes are so different (*AA* in one species and *BB* in another) that they do not synapse with each other during meiosis. Diploid hybrids of this type are sterile because their two sets of chromosomes (*AB*) can not synapse, and so meiosis fails. However, if the parents are tetrapolids (*AAAA* and *BBBB*) or if the diploid hybrid becomes tetraploid the tetraploid hybrids will be *AABB*. Thus, each chromosome has only one chromosome with which it can synapse, meiosis is successful, and the hybrid is fertile.

Fertile allopolyploids can result from crosses not only between related species but also between related genera within a family. Intergeneric allopolyploids have been produced experimentally, and have also occurred in nature. For example, cultivated wheat originated in Iraq from a cross between wild wheat (einkorn) and a related genus of wild grass. Both were diploids with an n chromosome number of 7. The two sets of chromosomes can be represented as *AA* and *BB,* respectively, and each had a $2n = 14$ chromosome number. The fertile tetraploid resulting from the cross between the genera was *AABB* with a $2n$ chromosome number of 28. Persian wheat is such a tetraploid. It later crossed with another related genus of wild grass (*CC*), resulting in a fertile hexaploid (*AABBCC*) with a $2n$ chromosome number of 42. Most of the cultivated varieties of wheat are hexaploids. Polyploidy is a very important factor in the evolution of plants at the species and genus levels, partly because it makes possible fertile hybrids between different species and different genera. Diploids resulting from such wide crosses are sterile because of the failure of meiosis, as are most of the autopolyploids.

However, sterile hybrids (or plants that are sterile for other reasons such as chromosomal changes) can survive in nature and establish new populations by another means: **apomixis.** This includes all kinds of vegetative propagation and also the production of seeds containing viable embryos that have arisen asexually rather than from a fertilized egg. Such asexual embryos may result from parthenogenesis, from another cell of the embryo sac, or from a cell of the nucellus that surrounds the embryo sac. Since species that can reproduce only by apomixis lack sexual reproduction, they lack the variability associated with sexual reproduction and are very uniform, as well as being isolated reproductively from related plants. Dandelions are apomictic, and many genera such as hawthorns (*Crataegus*) and blackberries (*Rubus*) have numerous apomictic species or races.

Polyploidy has a number of interesting features. For one thing, it generally results in sudden and marked phenotypic changes, in contrast to the gradual and less marked variation within an interbreeding species. Polyploids are more subject to wide ecological variation than diploids, probably because of the greater diversity of their genes. However, mutations that affect the phenotype are rarer in polyploids because of the larger number of alleles of each gene, which also tends to reduce the variation resulting from gene recombinations. Thus, although polyploidy results in many new species and races, it tends to stabilize established evolutionary lines and reduce the possibility of further extensive evolutionary changes.

ISOLATING MECHANISMS

It is difficult to define a species, but it can perhaps best be done by stating that a species is an actually or potentially interbreeding natural population that is reproductively isolated from other such populations. As long as a population freely interbreeds, it is a single species, regardless of how variable the members of the species may be as the result of recombinations. However, it should be noted that in some species of plants, such as apple (*Pyrus malus*), many individuals are self-sterile or cross-sterile with other individuals, even though the species as a whole can reproduce sexually.

If evolution is to result in new species, they must become reproductively isolated from related species. Apomixis is a perfect isolating mechanism, and any species that propagates itself asexually is a distinct entity with little hereditary variation. However, we are concerned here with isolating mechanisms that operate on sexual reproduction. For example, related species and genera are often isolated because their hybrids are sterile, but it should be noted that allopolyploidy overcomes this isolation. There are two kinds of reproductive isolation. The **prezygotic** mechanisms prevent pollination, and so fertilization and zygote formation. The **postzygotic** mechanisms involve zygote formation, but the zygotes are not viable or they develop into weak or sterile plants (generally hybrids).

Prezygotic Mechanisms

These are of three principal types. (1) Geographic or spatial isolation separates potentially interbreeding populations by major barriers such as oceans or mountains or simply by distance from one another. However, two populations in the same region may be isolated because they occupy different habitats. For example, scarlet oak lives in swamps or wet bottomlands while black oak occupies well-drained uplands. (2) Seasonal isolation results when two potentially interbreeding populations bloom and reproduce at different times of the year. (3) Mechanical isolation results from differences in flower structure that prevent cross-pollination, sometimes because they result in favoring different pollinating agents, or because the pollen of one species will not germinate on the stigma of the other or the pollen tube will not grow all the way to the ovules. There are many differences in flower structure that prevent cross-pollination, such as long stamens and short pistils in one species and the opposite in another.

Postzygotic Mechanisms

(1) Hybrid zygotes may fail to develop into viable embryos, or if they do the hybrids may be weak and may die before they reproduce. (2) The hybrids may thrive, but are unable to reproduce because of the failure of meiosis. Normal synapsis of the homologous chromosomes during meiosis and consequent formation of viable meiospores and gametes do not generally occur in diploid hybrids, triploid hybrids, or autopolyploids. Chromosomal changes may also result in failure of meiosis, and thus in a lack of viable gametes. (3) The F_1 hybrids may be fertile, but the F_2 generation may include many weak, nonviable, or sterile individuals.

TYPES OF NATURAL SELECTION

So far, we have centered our attention on the hereditary variations subject to natural selection, but now some mention should be made of the ways in which natural selection operates. There are three main types of natural selection

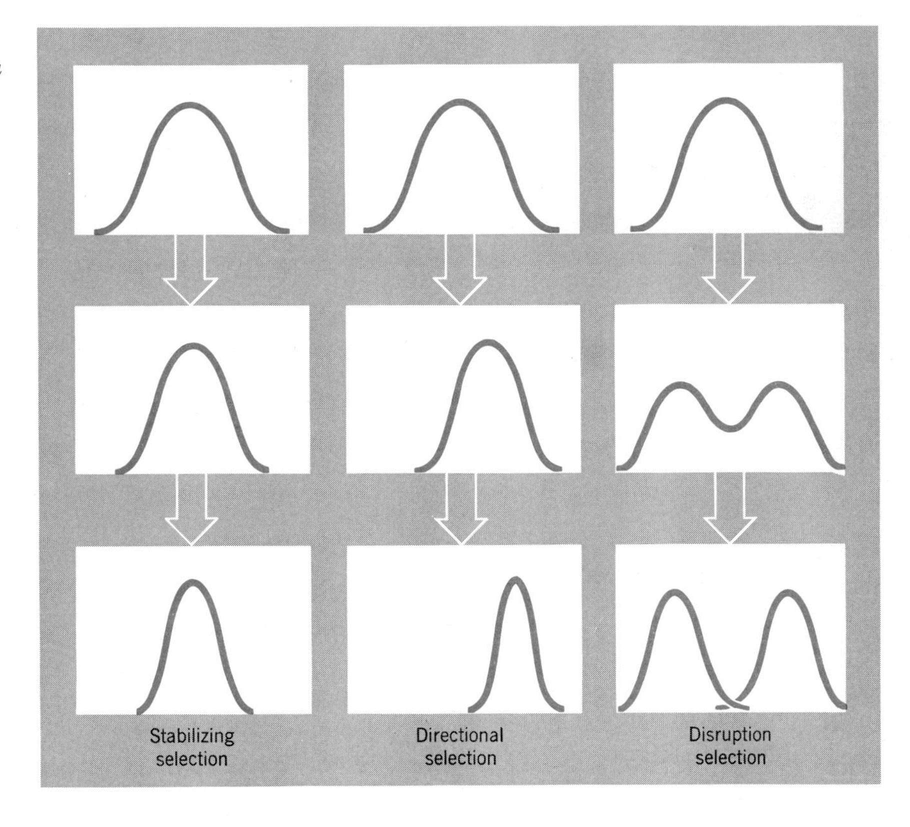

Fig. 24.3 Evolutionary changes in populations as a result of three kinds of natural selection.

(Fig. 24.3), which operate on the varied phenotypes and so in turn reduce the frequency of some of the genes in the population and increase the frequency of others. However, because of the various ways in which genes interact in producing phenotypic characters it is difficult to relate selection to a specific gene or set of genes.

Stabilizing selection occurs when the environment of a population remains quite constant over a period of time and throughout the range of the population. The bulk of the population is well adapted to the environment, and individuals at either extreme, whether resulting from mutations, recombinations, or other variations, are likely to be selected against. The result is a more homogenous and less variable population.

Directional selection occurs when there is a progressive change in the environment in one direction, such as from a humid climate toward a more xeric one. Such a change favors the phenotypic characteristics that make certain individuals better adapted to the changing environment, and thus a change in the gene frequencies and the complex of characteristics of the population as a whole. Over a period of time the changes can be extensive enough to result in essentially a new species.

Disruptive selection results when a previously stable environment throughout the range of the population becomes diverse, either because of changes in the physical environment or in the biological environment, as by the introduction of a new parasite. The result is likely to be the selection of the extreme phenotypes over the intermediate ones and so the breaking up of the previously homogenous population into distinct ones, each subject to different selective pressures.

Although natural selection operates to a great extent on the recombinations in a freely interbreeding population by disturbing the Hardy-Weinberg equilibrium, it should be recalled that it also operates on mutations, chromosomal changes, and the marked genetic changes resulting from introgressive hybridization and polyploidy.

RELATED READING

Becker, H. F. "Flowers, insects, and evolution," *Natural History, 72*(2), 38–45, February 1965.

Briggs, D. and S. M. Walters. *Plant Variation and Evolution.* New York: McGraw-Hill, 1969.

Hulse, J. and D. Spurgeon. "Triticale," *Scientific American, 231*(2), 72–80, 1974.

Stebbins, G. L. "From gene to character in higher plants," *American Scientist, 53,* 104–126, 1965.

Wills, C. "Genetic load," *Scientific American, 222*(3), 98–107, March 1970.

APPENDIX

SOME BASIC CHEMISTRY FOR BOTANY STUDENTS

ATOMS AND ELEMENTS

Although the word *atom* has been in everyday use since the advent of the atomic bomb, it is perhaps not generally recognized that all matter is composed of atoms. Atoms are extremely small particles, far too minute to be made visible even by an electron microscope. One gram of pure iron contains about 10.9×10^{22} (10,900,000,000,000,000,000,000 or almost 11 billion billion) atoms of iron. An **atom** may be defined as the smallest part of a chemical element that can participate in a chemical reaction. There are about 100 different chemical elements, among them such common substances as hydrogen, oxygen, nitrogen, carbon, iron, copper, sulfur, and lead. Elements cannot be decomposed chemically into other substances, and each element has atoms that differ from those of all other elements. Atoms have a rather wide range of sizes and weights, the lightest being the hydrogen atom and the heaviest naturally occurring atom being plutonium, which is 239 times as heavy as the hydrogen atom. The chemical elements of greatest biological interest and importance are listed in Table A.1.

As a matter of convenience, each chemical element has been assigned a symbol that generally consists of the first letter or two of its English name, for example, H for hydrogen, C for carbon, O for oxygen, P for phosphorus, and Ca for calcium. Some symbols, however, are derived from the Latin names of the elements as in Na for sodium, Fe for iron, K for potassium, Au for gold, and Cu for copper. These symbols are used in writing the equations for chemical reactions and are also used at times in text material in place of the names of elements. You will find it useful to know the symbols of the elements listed in Table A.1.

Atomic Structure

At one time atoms were considered to be indivisible particles, the term *atom* having been derived from a Greek word meaning "uncut"

Table A.1 **Chemical Elements of Botanical Importance, with Chemical Symbols and Some Common Ions of Each Element**

Element	Atomic Weight	Symbol	Bonds	Ions
Boron	10.8	B	3	BO_3^{---}
Calcium	40.1	Ca	2	Ca^{++}
Carbon	12.0	C	2, 4	CO_3^{--}
Copper	63.5	Cu	1, 2	Cu^{++}
Chlorine	35.5	Cl	1	Cl^-
Hydrogen	1.0	H	1	H^+, H_3O^+, OH^-
Iron	55.9	Fe	2, 3	Fe^{++}, Fe^{+++}
Magnesium	24.3	Mg	2	Mg^{++}
Manganese	54.9	Mn	2–7	Mn^{++}
Molybdenum	96.0	Mo	3, 4, 6	MoO_4^{--}
Nitrogen	14.0	N	3, 5	NO_3^-, NH_4^+
Oxygen	16.0	O	2	OH^-
Phosphorus	31.0	P	3, 5	$H_2PO_4^-$
Potassium	39.0	K	1	K^+
Sodium	23.0	Na	1	Na^+
Sulfur	32.1	S	2	SO_4^{--}
Zinc	65.4	Zn	2	Zn^{++}

Notes: All cations (positively charged ions) listed except H^+ or ions containing hydrogen (H_3O^+, NH_4^+) are those of metallic elements. Oxygen is a constituent of all the anions (negatively charged ions) listed except Cl^-, as well as of the OH^- ion listed after oxygen. Any anion in the table can combine with any cation, the number of charges corresponding with the combining proportions. Thus, potassium chloride would be KCl, potassium sulfate K_2SO_4, and potassium borate K_3BO_3; calcium chloride would be $CaCl_2$. The numbers of bonds each atom can form with other atoms is also given. The atomic weight of each element is given to the nearest tenth.

or "indivisible." However, chemists have known since the turn of the century that atoms are composed of still smaller particles. About a dozen different kinds of particles make up atoms, but of these only three need be considered here: **electrons, protons,** and **neutrons.** Electrons carry a unit negative electrical charge, protons a unit positive electrical charge, and neutrons are electrically neutral. A proton weighs only 17×10^{-25} gram; a neutron is just slightly heavier. Electrons are much lighter, 1837 electrons weighing as much as one proton.

The protons and neutrons of an atom are arranged rather compactly in the center of the atom and make up the **atomic nucleus** (Fig. A.1). The nucleus of an ordinary hydrogen atom consists only of a single proton, but the nuclei of all other kinds of atoms include both protons and neutrons.

The electrons of an atom are at relatively great distances from the nucleus, and are arranged in zones or shells. If an atomic nucleus were the size of a pea the nearest electron would be about 20 meters from it. Each shell can accommodate only a limited number of electrons. The shell nearest to the nucleus can accommodate only two electrons, but the second shell can accommodate 8, the third shell 18, and the fourth shell 32. The number of electrons in a shell of a particular kind of atom may range from 1 to the shell capacity of 2 or 8 (Table A.2). However, from the third shell outward there are subshells in each shell, the

Table A.2 **The Occupancy of Shells by Electrons in Some Common Element**

Atom	Atomic Number	1st Shell	2nd Shell	3rd Shell	4th Shell
H	1	1			
C	6	2	4		
N	7	2	5		
O	8	2	6		
F	9	2	7		
Ne	10	2	8		
Na	11	2	8	1	
P	15	2	8	5	
S	16	2	8	6	
Cl	17	2	8	7	
K	19	2	8	8	1

subshell capacities being 2, 8, 18, 32, and so forth. If a subshell capacity of 8 is reached in the outermost shell of an atom, and if the structure of this atom calls for still another electron, this electron will enter the next shell (see K in Table A.2).

As the distance of a shell from the nucleus of an atom increases, so does the energy of the electrons in the shell. However, the electrons of any one shell have different energies, depending on the subshell to which they belong.

All atoms of a certain element have the same number of electrons and protons. Hydrogen atoms have 1 electron and 1 proton,

Fig. A.1 Above: schematic representations of an atom of hydrogen (H) and an atom of oxygen (O). The nucleus of the H atom is a single proton, while the nucleus of the O atom consists of eight protons and eight neutrons (white). If the diagrams were to scale, the electrons of the first shell would be about 9 m from the nucleus and the protons and neutrons would be about 1837 times as large as the electrons. Below: an even more diagrammatic representation of a H atom, an O atom, and a molecule of water (HOH). In the water molecule the H electrons occupy the two unfilled electron sites in the outer shell of the O molecule. The two electrons are shared by the H and O atoms in covalent bonds.

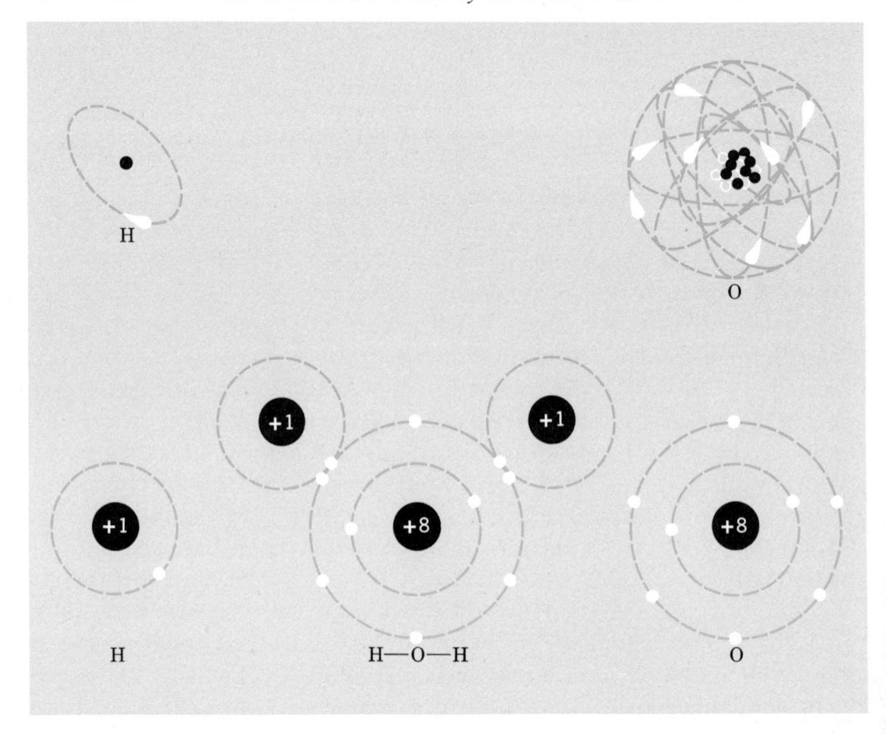

oxygen atoms have 8 electrons and 8 protons (Fig. A.1), carbon atoms have 6 electrons and 6 protons, and calcium atoms have 20 electrons and 20 protons. The distinct and characteristic properties of each element depend on the number of electrons and protons in the atoms of the element. Since the number of electrons in an atom is always equal to the number of protons, atoms are electrically neutral. The atomic number of an element is the number of protons (or the number of electrons) in its atoms (Table A.2).

Isotopes

Although all the atoms of a certain element have the same number of protons and electrons, some atoms of an element may have more neutrons in their nuclei than do others. For example, most hydrogen atoms have only a single proton in their nucleus, but some have 1 proton and 1 neutron and still others have 1 proton and 2 neutrons. Most carbon atoms have 6 protons and 6 neutrons in their nuclei, but some have 6 protons and 8 neutrons. The usual number of particles in oxygen nuclei is 8 protons and 8 neutrons, but a few oxygen atoms have 8 protons and 10 neutrons. The different varieties of atoms of any element are called **isotopes,** a particular isotope being composed of atoms that are uniform in neutron content and weight as well as in number of protons and electrons. The isotopes of an element differ from one another in the number of neutrons they contain (Table A.3). The isotopes of an element are designated by superscript numbers that give the total number of protons plus neutrons in each isotope. Only in the case of hydrogen are the isotopes of an element given different names. ^{1}H is hydrogen, ^{2}H is **deuterium,** and ^{3}H is **tritium.**

The atoms of most isotopes, including those that are most abundant in nature, are stable, but the atoms of some isotopes are unstable and disintegrate into stable isotopes of other elements by the emission of particles, radiation, or both. These are the **radioactive isotopes (radioisotopes).** Radioactive atoms may emit **alpha particles** (2 protons + 2 neutrons, that is a helium nucleus), or **beta particles** (rapidly moving electrons). Some radioactive atoms emit **gamma rays** (shortwave radiation similar to X-rays). Each kind of radioisotope emits only certain kinds of particles or radiation. For example, radioactive ^{14}C emits only beta particles from its atomic nuclei. Since a neutron may be considered as being composed of a proton plus an electron, loss of the beta particle (electron) reduces the number of neutrons in the nucleus from 8 to 7 and increases the number of protons from 6 to 7. Thus, ^{14}C is transformed into ^{14}N, the common stable isotope of nitrogen. Such a transformation of one element into another is referred to as **radioactive decay.**

Table A.3 **Some of the Isotopes of Selected Elements That Are Important Biologically**

Isotopes		Number of Protons	Number of Neutrons	Total: Protons + Neutrons
^{1}H	a	1	0	1
^{2}H		1	1	2
^{3}H	r	1	2	3
^{11}C	r	6	5	11
^{12}C	a	6	6	12
^{14}C	r	6	8	14
^{14}N	a	7	7	14
^{15}N		7	8	15
^{16}O	a	8	8	16
^{18}O		8	10	18
^{31}P	a	15	16	31
^{32}P	r	15	17	32
^{31}S	a	16	16	32
^{35}S	r	16	19	35
^{40}Ca	a	20	20	40
^{45}Ca	r	20	25	45

a = The isotope of the element that is abundant in nature.
r = Radioactive isotopes.

Each radioisotope has a characteristic rate of decay, measured in terms of **half life.** The half life of an isotope is the length of time required for half of its atoms to decay. ^{14}C decays very slowly, having a half life of about 5000 years, whereas radioactive ^{11}C has a half life of only 20 minutes. ^{32}P has a half life of about 14 days and ^{35}S a half life of about 87 days. Radioisotopes also differ from one another in the kinds of particles and waves emitted and the energy with which they are emitted, and thus in the penetrating power of the emissions.

Radioisotopes are of biological interest from two main points of view. (1) Their emissions may have harmful effects on organisms, causing disruption of metabolic processes, mutations, and abnormal growth and development. (2) Radioisotopes are used as tracers in various kinds of biological research and have made possible many important discoveries. For example, the intermediate steps in photosynthesis have been worked out by the identification of the radioactive substances present in plants supplied with $^{14}CO_2$. Also, much has been learned about the absorption and translocation of mineral elements in plants by the use of such radioactive isotopes as ^{32}P, ^{35}S, ^{45}Ca, and ^{59}Fe. To be useful as a tracer, an isotope must have a half life of at least several days. Unfortunately, the radioisotopes of nitrogen and oxygen have such short half lives that they are not suitable for tracer use, but stable isotopes of these elements can be used as tracers if an expensive instrument called a mass spectrometer is available for detecting and measuring them.

Bonding of Atoms

The atoms of most elements are capable of combining with other atoms of the same element or with atoms of other elements to form compounds. A fundamental and important feature of the process is that the atoms do not combine in just any proportion, but in very definite and fixed ratios. Ultimately, these combining ratios are determined by the capacities of the electron shells to accommodate electrons. Since each neutral atom possesses a different number of electrons, each element should have its own characteristic bonding capacity. The "driving force" for bonding is the completion of the occupancy of an electron shell (or subshell). For example, the hydrogen atom has a single electron in the first shell, which has a *capacity* for two electrons. Hydrogen forms chemical bonds in such a way that a second electron is added to complete that shell. Carbon, nitrogen, oxygen, and fluorine have four, five, six, and seven electrons, respectively, in the second shell. Bonds are formed by these elements in ways that the number of electrons in that shell becomes the filled number of eight. Neon, an inert gas, already possesses as a neutral atom eight electrons in the second shell and therefore has no tendency to form chemical bonds.

It is important to note that beyond hydrogen, the general rule is that the atoms do not achieve completely filled shells, but rather have the "magic number" of eight electrons in the outermost shell. This is a consequence of the energy levels of the subshells that will not be discussed here. Suffice it to say that all the rare gases from neon through radon, which do not normally form compounds, have this "closed shell of eight" electrons (octet) in their outermost shell.

There are two ways in which atoms can achieve completed electron shells, and this gives rise to two basic types of bonds. On the one hand an atom might gain or lose electrons completely to form **ions** with eight electrons in the outermost shell, or atoms can combine to share electrons in a way that both atoms have the completed octet. The first process gives rise to **ionic** or **electrovalent bonds;** the second, sharing process, gives rise to **covalent bonds.**

To illustrate ionic bonding, consider the

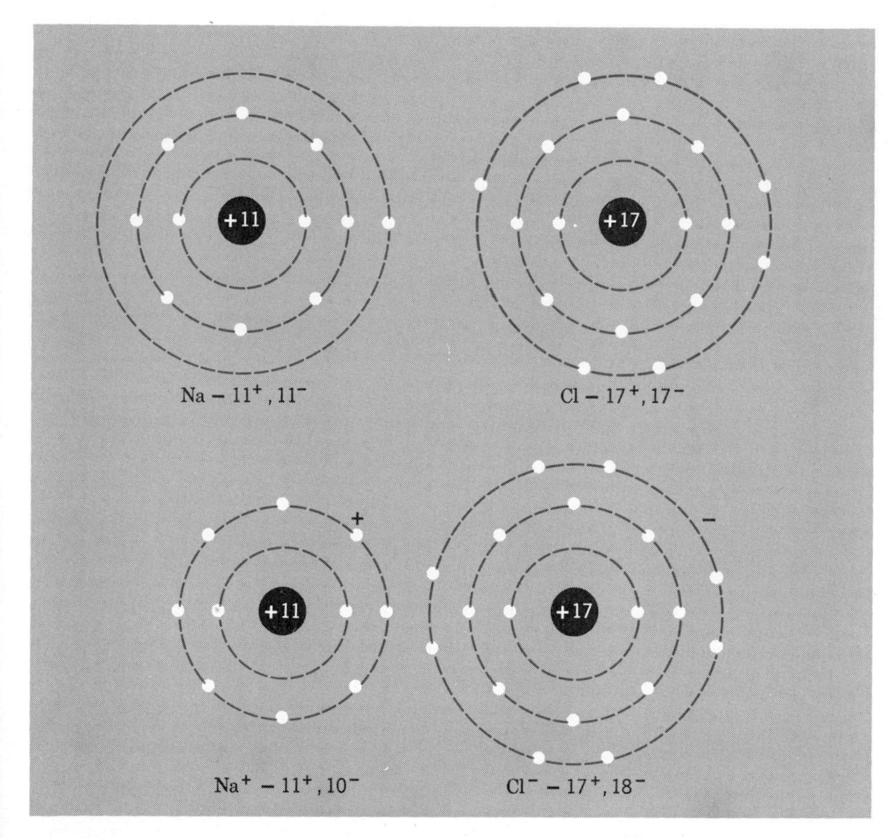

Fig. A.2 A sodium atom has 11 electrons and 11 protons and a chlorine atom has 17 of each. When they come in contact the single electron in the outer shell of the sodium atom occupies the vacant electron site of the chlorine molecule, forming an ion of sodium (Na^+) and an ion of chlorine (Cl^-). The oppositely charged ions attract each other, constituting a molecule of sodium chloride (*Na^+Cl^-, or less accurately NaCl*). *Bonds between ions such as these are electrovalent. In solution almost all of the Na^+ and Cl^- ions are separate, rather than in NaCl molecules.*

formation of sodium chloride from a sodium atom and a chlorine atom. The neutral chlorine atom has seven electrons in its third shell; it needs one more electron to complete its octet. Sodium has a single electron in its outermost shell; the simplest way that a completed outermost octet of electrons can be achieved for both atoms would be for the sodium atom to lose completely one electron and for the chlorine atom to acquire that electron completely. That is what happens and the product, sodium chloride, is an ionic compound (Fig. A.2). This material in the solid state is composed of ions arranged in a regular cubic pattern (Fig. A.3). When this material (common table salt) is dissolved in water, the ions break out of the lattice and move independently throughout the solvent water. Sodium chloride has, of course, quite different properties from either the soft and very reactive sodium metal or the greenish, poisonous chlorine gas.

Covalent compounds, on the other hand involve atoms rigidly bound together in units called **molecules.** The covalent bond is formed by a direct sharing of pairs of electrons by two atoms (Fig. A.4). Each atom so bound can count both electrons of the bond among its own for the purpose of achieving the eight electron arrangement. Water, for example, is composed of molecules containing two hydrogen atoms and one oxygen atom (Fig. A.1). Thus, there are two covalent bonds in the molecule, each bond consisting of two electrons and in each case one electron is provided for the bond from each atom. This sharing scheme allows each hydrogen to count two electrons in its electron shell (now filled) and the oxygen to count eight in its outer shell. Sometimes, in

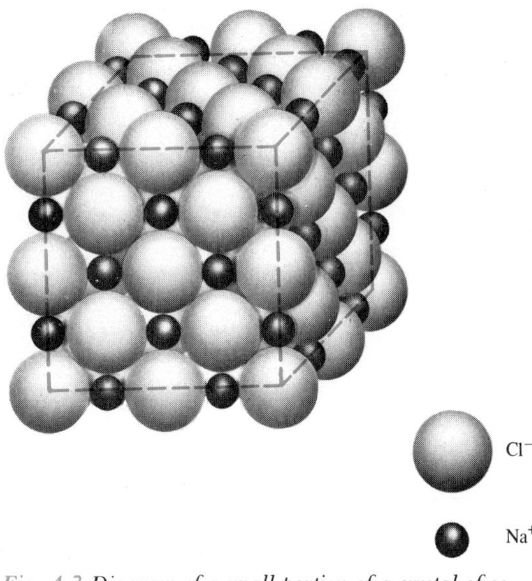

Fig. A.3 Diagram of a small portion of a crystal of sodium chloride (common table salt). The regular alternation of Na^+ and Cl^- ions is responsible for the cubical shape of the sodium chloride crystals.

order to achieve the octet, more than one pair of electrons may be shared by two atoms. In such a case, each electron pair constitutes a covalent bond and the molecule may possess double bonds or even triple bonds. Carbon dioxide is a good example.

MOLECULES AND COMPOUNDS

The particle formed by the bonding or union of two or more atoms is called a **molecule.** The atoms of many elements may unite with one another forming molecules. Thus, hydrogen gas consists of molecules of two hydrogen atoms each (H_2) while oxygen molecules consist of two atoms of oxygen held together by a covalent bond (O_2). When three atoms of oxygen are bonded into a molecule, the result is ozone (O_3), a gas with properties different from those of O_2.

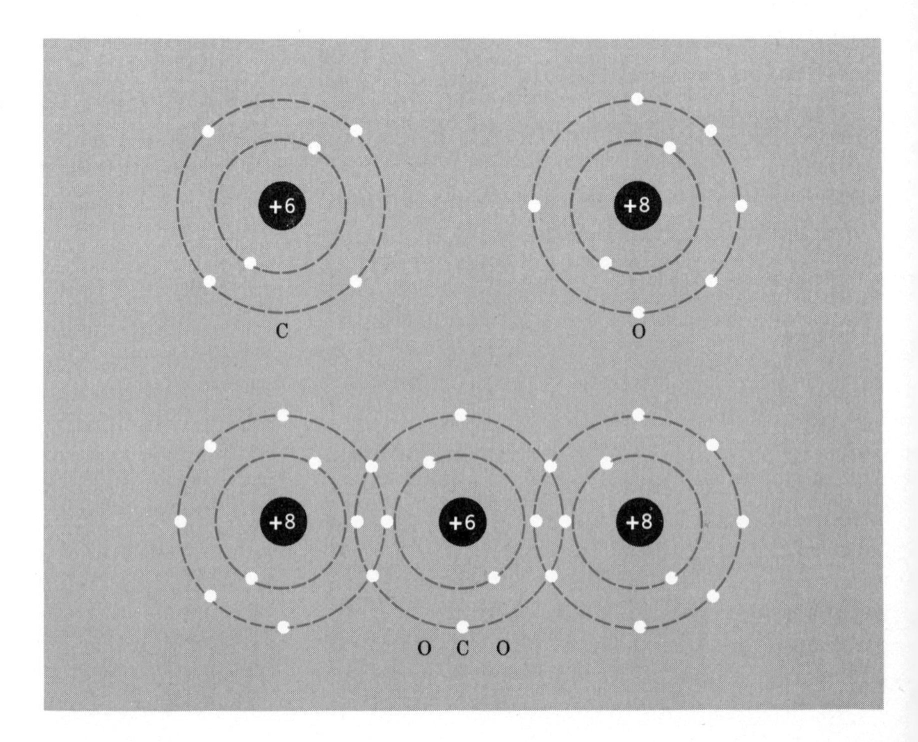

Fig. A.4 Atoms of carbon (C) and oxygen (O) and a molecule of carbon dioxide (CO_2) showing the shared electrons of the covalent bonds.

When atoms of two or more different elements are bonded together into a molecule the result is a **chemical compound.** A molecule is the smallest particle of a chemical compound, just as an atom is the smallest particle of a chemical element, although as has been noted a molecule may be dissociated into ions. It must be stressed that the properties of any chemical compound are very different from those of its constituent elements. The difference between the properties of sodium, chlorine, and sodium chloride has already been mentioned. As just one more example, water (H_2O) has very different properties from those of hydrogen or oxygen, the elements which compose it. Although there are only a hundred or so chemical elements, these may combine into many thousands of different chemical compounds.

The **formula** of a chemical compound consists of the symbols of the chemical elements composing it along with an indication of the number of each kind of atom present in the molecule. The simplest type of chemical formula **(empirical formula)** merely lists the numbers of each atom or ion present in the molecule as subscript figures, no subscript being used if only one atom is present. Thus, the formula of H_2O for water shows that a molecule of water is composed of two atoms of hydrogen and one atom of oxygen. The formula for table sugar (sucrose) is $C_{12}H_{22}O_{11}$ and tells us that each molecule of sugar contains 12 atoms of carbon, 22 of hydrogen, and 11 of oxygen. The formula of a substance with electrovalent bonds such as sodium chloride should really be written Na^+Cl^- to indicate that it is composed of ions, but for convenience the charges are frequently omitted and so we would write NaCl.

Another kind of formula, used particularly for organic compounds, is the **structural formula.** This is essentially a two-dimensional sketch of the molecule, showing the way the atoms are bonded to each other. Frequently the bonds are shown as lines connecting the symbols, or as dots representing the shared electrons of covalent bonds. For example, the structural formula of carbon dioxide might be written O=C=O or O :: C :: O and water would be H—O—H or H : O : H. The empirical formula of acetic acid is $C_2H_4O_2$, but this provides little information about its chemical nature. Its structural formula is much more illuminating:

```
      H
      |
   H—C—C=O
      |  |
      H  O—H
```

At times it is convenient and desirable to use **semistructural formulae.** The semistructural formula for acetic acid is CH_3COOH. This gives us the essential information about the structure of the compound, including the fact that it is an organic acid as indicated by the presence of a carboxyl (—COOH) group, and also it is simpler to write than the complete structural formula.

Sometimes, in biology in particular, compounds with extremely large and complex molecules are represented by a kind of shorthand using key letters in the name of the compound rather than chemical symbols. Thus, adenosine tri-phosphate is commonly written ATP and nicotinamide adenine dinucleotide is written NAD^+. Such chemical shorthand is convenient, but remember that the letters used do not represent chemical elements, but rather abbreviations of the names of the compounds, except that the H in a compound such as NADH *does* represent hydrogen. Another shorthand device often used is the representation of the phosphate radical ($—H_2PO_3$) merely as —P, or as Ⓟ. Other acronyms commonly used in botany are IAA (indoleacetic acid), ABA (abscisic acid), GA (gibberellic acid).

ELECTROLYTES AND NONELECTROLYTES

Substances with electrovalent bonds or partially electrovalent bonds are classed as **electrolytes** because their ions carry an electric current through the water in which they are dissolved. The positively charged ions **(cations)** moved to the negative electrode **(cathode)** where they pick up electrons and are converted into uncharged atoms; the negatively charged ions **(anions)** migrate to the positively charged electrode **(anode)** where they give up electrons and also become uncharged atoms. Thus, when an electric current is passed through melted sodium chloride, metallic sodium is deposited on the cathode as the Na^+ ions migrate to it and are discharged; the Cl^- ions are converted to chlorine gas at the anode. Electrolytes are mostly salts, acids, or bases.

Nonelectrolytes are substances that are not ionized when dissolved in water. A solution of a nonelectrolyte does not conduct an electric current. Among the nonelectrolytes are gases such as oxygen and hydrogen and most organic compounds, although some organic compounds such as the organic acids are electrolytes.

ACIDS, BASES, AND SALTS

Acids may be defined as substances that contribute hydrogen ions (H^+) when in solution. Since a H^+ ion is nothing but a proton (the H atom consisting of a single electron and a single proton), acids are sometimes defined as proton donors. However, H^+ ions almost immediately combine with water molecules forming **hydronium** (H_3O^+) ions, and it is these hydronium ions that give acids their characteristic properties. It is proper to speak of hydronium ions when dealing with acids, but for simplicity we shall follow the usual biological custom of speaking in terms of hydrogen ions. A **base** is a substance that contributes **hydroxyl** ions (OH^-) to a solution. When an acid and a base are present in the same solution they react with each other, forming a salt and water:

$$\underset{\text{hydrochloric acid}}{HCl} + \underset{\text{potassium hydroxide}}{KOH} \longrightarrow \underset{\text{potassium chloride}}{KCl} + \underset{\text{water}}{HOH}$$

$$\underset{\text{sulfuric acid}}{H_2SO_4} + \underset{\text{sodium hydroxide}}{2NaOH} \longrightarrow \underset{\text{sodium sulfate}}{Na_2SO_4} + \underset{\text{water}}{2HOH}$$

In such reactions the acid and base **neutralize** each other, and most of the H^+ and OH^- ions are removed by joining with one another and forming water molecules. Only one out of every 10 million water molecules is ionized, so few H^+ or OH^- ions remain in the solution.

ORGANIC AND INORGANIC COMPOUNDS

Chemical compounds may be classed as **organic** and **inorganic.** Frequently organic compounds are defined as those that contain carbon, all others being designated as inorganic. However, we shall use a somewhat more restrictive definition that limits organic compounds to those that contain both carbon and hydrogen (frequently along with oxygen and other elements). This definition excludes such carbon compounds as carbon dioxide (CO_2), carbon monoxide (CO), calcium carbonate ($CaCO_3$) and sodium cyanide (NaCN) that are really not organic compounds in the stricter sense of the term. Organic compounds were so named because it was thought that they could be synthesized only by living organisms. Although chemists are now able to synthesize some organic compounds from inorganic compounds in the laboratory, it is still true that organic compounds (in the more limited sense) found in nature at the present time have all been synthesized by plants or animals.

Because carbon atoms can bond with one another into rings and long chains, organic compounds may have larger and more complex molecules than inorganic compounds do. There are many more different kinds of organic compounds than there are inorganic compounds. For example, there could be many million different kinds of proteins alone. Organic compounds include the hydrocarbons (compounds composed only of carbon and hydrogen) of petroleum and natural gas, the carbohydrates, fats, and proteins that constitute the bulk of the dry weight of plant and animal tissues, and thousands of other compounds that are made by plants and animals and which play important roles in their structure and processes (see Chapter 7). The principal inorganic compounds of biological interest are water, acids, bases, salts, and several gases such as carbon dioxide and oxygen.

ISOMERS

Two or more compounds may contain exactly the same numbers and kinds of atoms and yet be different substances with different properties because of a difference in the arrangement of the atoms in the molecules. Such substances are known as **isomers.** Isomers are more common among organic than inorganic compounds. For example, there are many different sugars with the formula $C_6H_{12}O_6$. As a group they are referred to as hexoses or 6-carbon sugars. To show the differences in the various hexoses it is necessary to give their structural formulae. In the following example the structural formulae of two isomores of glucose, one of galactose, and one of fructose are shown. There are other isotopes of these hexoses, and also other hexoses, so many isomers of $C_6H_{12}O_6$ exist.

```
                                           H
 HC=O      O=CH        HC=O              HCOH
  |          |           |                 |
 HCOH      HOCH        HCOH               C=O
  |          |           |                 |
HOCH        HCOH      HOCH              HOCH
  |          |           |                 |
 HCOH      HOCH        HOCH              HCOH
  |          |           |                 |
 HCOH      HOCH        HCOH              HCOH
  |          |           |                 |
 HCOH      HOCH        HCOH              HCOH
  H          H           H                 H
D-glucose  L-glucose  D-galactose      D-fructose
```

The bonds between carbon atoms are indicated by lines, and the double bonds between carbon and oxygen atoms by double lines, but bond lines are not shown between the carbon atoms and the H and OH groups attached to them (although they exist, of course). Note that each C atom has four bonds, each O atom two (as C=O or C—O—H), and each H atom one. The mere difference in arrangement of the —H and —OH groups on the carbons results in compounds with different properties, as in the first three sugars given. Fructose further differs from the other three in that its C=O group is not on the end carbon. The two isomers of glucose are mirror images of each other. Such isomers are known as **stereoisomers.**

Although the structures for sugar molecules given above clearly show the differences between one sugar and another, the usual real structure of sugar molecules is more complicated. An —OH group in the molecule can react with the Keto (=C=O) group in the same molecule and so form a molecule with a ring (cyclic) structure, shown two ways:

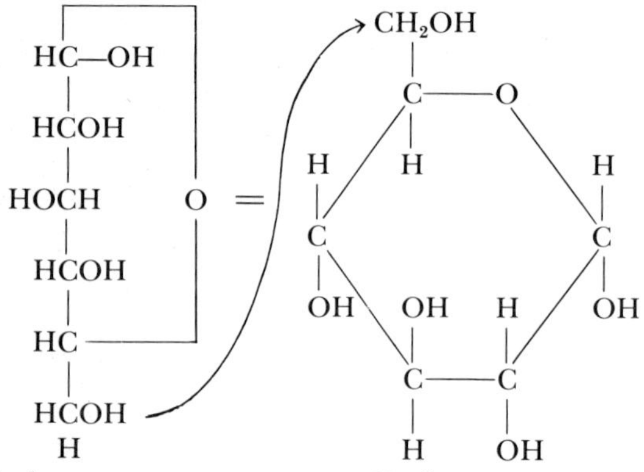

HC—OH
HCOH
HOCH   O =
HCOH
HC
HCOH
H

D-glucose

CH₂OH →
C—O
H H H
C C
OH OH H OH
C—C
H OH

D-glucose

In the crystalline (solid) state glucose exists entirely as the cyclic structure, and when dissolved in water it is more than 99% cyclic. In polymers of glucose such as starch and cellulose glucose always has the cyclic structure (see Figs. 7.6 and 7.7).

ATOMIC AND MOLECULAR WEIGHTS

The atomic weight of an element is the average weight of the atoms of that element as they occur in nature. The atomic weight is, of course, influenced by the percentage that each isotope of the element constitutes of the total. The scale of atomic weights is based on the assignment of atomic weight 12.000 to ^{12}C, the most abundant isotope of carbon. In proportion, the atomic weight of the ordinary and most abundant isotope of hydrogen is 1 (i.e., the carbon atom is 12 times as heavy). However, since there is always a very small amount of ^{2}H (deuterium) and ^{3}H (tritium) in any natural sample of hydrogen, the atomic weight of hydrogen is 1.008. The atomic weights of elements of biological interest are given in Table A.1, rounded off to the nearest tenth.

The **molecular weight** of a compound is the sum of the atomic weights of the elements composing the compound. For example, the molecular weight of water (H_2O) is $2 \times 1 + 16 = 18$. The molecular weight of a hexose sugar would be calculated as follows:

Element	Number of Atoms		Atomic Weight	Total
Carbon	6	×	12.0	= 72.0
Hydrogen	12	×	1.0	= 12.0
Oxygen	6	×	16.0	= 96.0
Molecular weight				=180.0

The **gram molecular weight** (or **mole**) of any compound is its molecular weight in grams. Thus a gram molecular weight of a hexose sugar is 180.162 grams, whereas a gram molecular weight of water is 18.016 grams. A mole of one substance contains the same number of molecules as a mole of any other substance. *MW* is a commonly used abbreviation for *molecular weight.*

CHEMICAL REACTIONS

Chemical elements and compounds can react with one another in various ways, the products of the reactions being other substances. The following main types of chemical reactions can be identified: exchange, condensation, decomposition, and rearrangement.

In **exchange reactions** one or more atoms of a molecule trade places with one or more atoms of another, forming two other compounds. The reactions between acids and bases are exchange reactions. Here is another:

$$\underset{\text{water}}{HOH} + \underset{\text{ammonium chloride}}{NH_4Cl} \longrightarrow \underset{\text{ammonium hydroxide}}{NH_4OH} + \underset{\text{hydrochloric acid}}{HCl}$$

In **condensation reactions** two or more molecules may combine and form a single larger molecule, with or without other products. The first reaction has a single product, while the second has two.

$$\underset{\text{carbon dioxide}}{CO_2} + \underset{\text{water}}{H_2O} \longrightarrow \underset{\text{carbonic acid}}{H_2CO_3}$$

$$\underset{\text{glucose phosphate}}{C_6H_{11}O_6 \cdot H_2PO_3} + \underset{\text{fructose}}{C_6H_{12}O_6} \longrightarrow \underset{\text{sucrose}}{C_{12}H_{22}O_{11}} + \underset{\text{phosphoric acid}}{H_3PO_4}$$

Decomposition reactions involve the breaking down of a molecule into two or more smaller molecules, and may be the reverse of the comparable condensation reactions:

$$\underset{\text{carbonic acid}}{H_2CO_3} \longrightarrow \underset{\text{carbon dioxide}}{CO_2} + \underset{\text{water}}{H_2O}$$

$$\underset{\text{sucrose}}{C_{12}H_{22}O_{11}} + \underset{\text{water}}{H_2O} \longrightarrow \underset{\text{glucose}}{C_6H_{12}O_6} + \underset{\text{fructose}}{C_6H_{12}O_6}$$

A decomposition in which water reacts with the substance being decomposed is known as **hydrolysis.** All digestive reactions are hydrolytic, for example, the digestion of sucrose as outlined in the last reaction above.

In **rearrangement reactions,** one isomer is converted to another by an internal rearrangement of atoms:

$$\underset{\text{3-phosphoglyceric acid}}{\begin{array}{c} COOH \\ | \\ HCOH \\ | \\ H_2CO{-}H_2PO_3 \end{array}} \longrightarrow \underset{\text{2-phosphoglyceric acid}}{\begin{array}{c} COOH \\ | \\ HCO{-}H_2PO_3 \\ | \\ H_2COH \end{array}}$$

The numbers in the names of these two compounds indicate which carbon holds the phosphate group ($—H_2PO_3$).

Balancing Equations

Equations must always be **balanced,** that is, there must be the same number of each kind of atom on one side of the arrow as on the other. In the sample reactions given, all the equations are balanced, as can be seen by counting the number of each kind of atom on each side of the equations. It so happened that in all the reactions given only one molecule of each substance is used or produced per molecule of the other substances in the reaction. However, in some reactions two or more molecules of one substance may react with only one molecule of another:

$$\underset{\text{sulfuric acid}}{H_2SO_4} + \underset{\text{potassium hydroxide}}{2KOH} \longrightarrow \underset{\text{water}}{2HOH} + \underset{\text{potassium sulfate}}{K_2SO_4}$$

This equation would not be balanced if only one molecule of KOH and one of HOH were included. The reason is that the sulfate ion (SO_4^{--}) has a negative charge of 2, or an excess of 2 electrons, and so combines with two cations such as H^+ or K^+ having a single charge or with one cation such as Ca^{++} with a double charge. The reaction of sulfuric acid with calcium hydroxide would be balanced as follows:

$$\underset{\text{sulfuric acid}}{H_2SO_4} + \underset{\text{calcium hydroxide}}{Ca(OH)_2} \longrightarrow \underset{\text{water}}{2HOH} + \underset{\text{calcium sulfate}}{CaSO_4}$$

The common ions present in plants are listed in Table A.1, and by observing the charges on each ion we should be able to determine its proper combining proportions with other ions of opposite charge and so balance equations in which the ions appear.

If an equation is balanced, the sum of the molecular weights on the left will equal the sum of the molecular weights on the right. For example:

$$\begin{array}{ccccc} H_2CO_3 & \longrightarrow & H_2O & + & CO_2 \\ 62 & = & 18 & + & 44 \end{array}$$

This is, of course, just a way of saying that matter is neither created nor destroyed during a chemical reaction, but is just converted into other substances. (This applies to *all* reactions, but not to radioactivity.)

Energy Transfers in Chemical Reactions

Every chemical reaction involves energy transfer. Energy is required to break chemical bonds and when new chemical bonds are formed, energy is released. If, in any chemical reaction, more energy is released than is used, there is a *net release* of energy from the reaction that may appear as heat, light, electricity or some other type of energy and may be used in doing work. If a reaction requires more energy to break bonds than is released when the new bonds are formed, then energy from some outside source such as heat, light, or electricity must be introduced to make the reaction proceed. In some reactions the energy released is almost or exactly the same as the energy used, and so there is little or no *net* energy release or consumption in the reaction. In any event, energy is neither created nor destroyed during a chemical reaction.

The energy unit usually used by chemists and biologists in quantitative considerations of the energy used and released in chemical reactions is the kilogram-calorie (abbreviated kcal or C). A kcal is the amount of heat energy required to raise the temperature of 1000 grams of water one degree centigrade (celsius). Physicists usually use the gram calorie (cal or c), which is the amount of heat required to raise the temperature of one gram of water one degree centigrade.

Oxidation-reduction reactions generally involve major net energy input or output and are very important in the processes of living organisms. Basically, a substance is *oxidized* when it loses electrons and this occurs (among other things) whenever a compound either gains oxygen or loses hydrogen. Oxidation results in a release of of energy, as in the release of heat during respiration or the release of heat and light when wood is burning. A substance is *reduced* when it gains electrons. This happens whenever the substance either gains hydrogen or loses oxygen. Reduction requires an input of energy. Photosynthesis is an important biological process in which the ultimate source of energy used in reducing CO_2 is light. Since electrons lost by one molecule are gained by others, all oxidation-reduction reactions involve both oxidation and reduction—some substances are reduced while others involved in the reaction are oxidized. However, it is customary among biologists to refer to oxidation-reduction reactions in which there is a net release of energy as oxidations and those in which there is a net input of energy from outside sources as reductions. Thus, respiration would be classed as an oxidation and photosynthesis as a reduction.

CATALYSTS AND ENZYMES

Many chemical reactions, particularly exchange reactions between substances with electrovalent bonds, take place spontaneously at ordinary room temperatures. Many other reactions, however, particularly between substances with covalent bonds (that are stronger and not so easily broken), proceed very slowly if at all under ordinary conditions. Even if the reaction is energy-releasing, the reacting molecules must be raised to a certain energy level by heating, application of high pressures, or other means before the reaction will proceed. Thus, wood must be raised to a certain temperature (its kindling point) before it will burn. Once the activation energy level is reached, the process will continue. The situation may be compared with an automobile going up a steep mountain. Much energy is required to reach the crest (the activation energy level), but once the downgrade is reached the automobile does not require further energy expenditure by its motor.

A **catalyst** is a substance that changes the rate of a chemical reaction without undergoing any permanent chemical change, i.e., the cata-

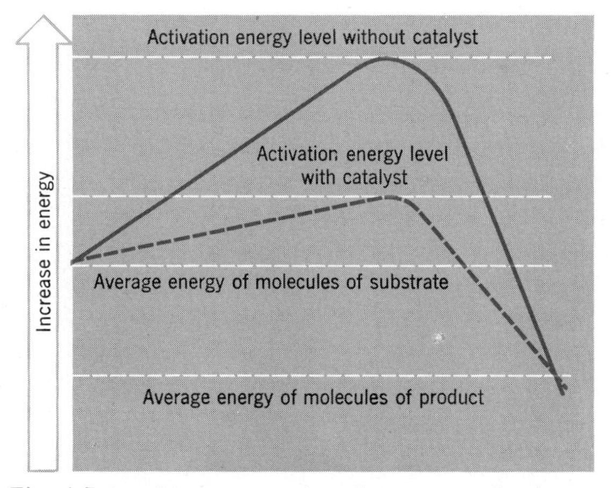

Fig. A.5 Graphic representation of an energy-releasing chemical reaction, showing the activation energy levels with and without a catalyst. Only a few of the reacting substrate molecules have a high enough energy to reach the high activating energy level without a catalyst and so the reaction proceeds very slowly. However, by supplying heat or another outside energy source, many more molecules have enough energy to clear the high energy hump, even without a catalyst. If a catalyst is present, its temporary combination with the substrate molecules results in a substance with a lower activation energy level, so the reaction proceeds more rapidly and at a much lower temperature than when no catalyst is present. Enzymes are the principal catalysts in organisms.

lyst is still present when the reaction is completed. Catalysts speed up reactions by reducing the activation energy level (Fig. A.5). In the automobile analogy, a comparable situation would be a deep cut through the mountain, thus permitting a car with little power to pass through the mountain, whereas it might not be able to climb the steep grade over the mountain. The presence of a suitable catalyst may permit a reaction to proceed at room temperature, whereas without the catalyst a high temperature might be required. Probably the commonest method of catalyst action is that in which the catalyst forms a temporary compound with one of the reacting substances (Fig. A.6), this compound having a lower activation energy level than the substance itself. Thus, the reaction $A + B \rightarrow AB$ may be very slow because of a high activation energy level. However, A may react readily with a catalyst (C), forming a compound AC with a lower activation energy level. Then the following reaction proceeds rapidly at ordinary temperatures:

$$AC + B \rightarrow AB + C$$

Enzymes are organic catalysts produced by plants and animals. Most of the biochemical reactions of organisms can proceed only if catalyzed by enzymes, so the enzymes produced by any particular organism determine just what reactions can go on in the organism. Enzymes are highly specific catalysts, each enzyme acting on only one type of chemical bond. Enzymes are proteins, although some enzymes are functional only when associated with a nonprotein **coenzyme.**

POLAR AND NONPOLAR COMPOUNDS

Although molecules as a whole are electrically neutral, the molecules of some nonelectrolytes as well as the molecules of all electrolytes have a negative charge on one part of the molecule and a positive charge on the other. Such molecules are said to be **polar.** It should be noted that despite the polarity of such molecules they are not ionized and remain intact. The polarity results from the fact that some of the atomic nuclei of the molecule attract more of the electrons close to them than do other atoms of the same molecule. The hydrogen side of a water molecule is positive and the oxygen side negative because the electrons are more strongly attracted by the oxygen nucleus. In addition to water, acids, bases, and salts there are many polar organic compounds such as the sugars and alcohols. **Nonpolar** molecules do not have differential charges. Among the nonpolar substances are hydrocarbons and

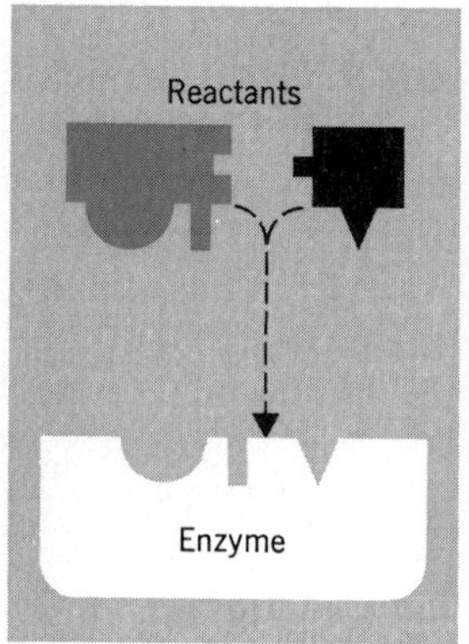

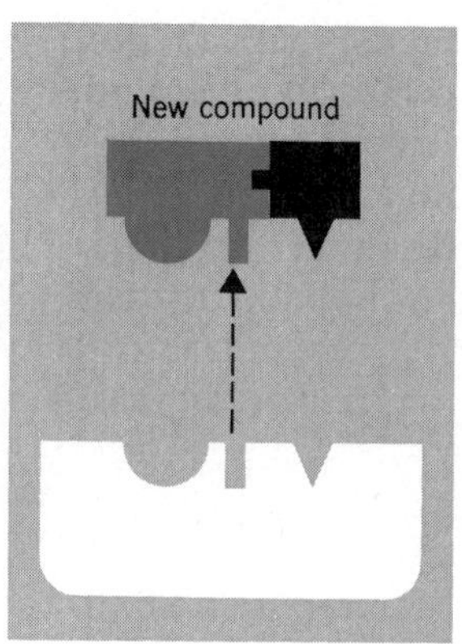

Fig. A.6 Diagram showing how the reactants in an enzymatic process fit the active sites in the enzyme protein molecule that catalyzes the reaction, forming a temporary enzyme-substrate complex. The shapes of the substrate and enzyme molecules are highly schematic, and are not intended to resemble those of any actual molecules.

fats. Some molecules such as those of the fatty acids are mostly nonpolar (the long hydrocarbon chain portion) but have polarity at one end (the —COOH group). In general, polar substances are soluble in water while nonpolar substances are soluble in fats and fat solvents. In the case of partially polar molecules, the nonpolar end is fat soluble and the polar end water soluble. This property has a number of biologically important consequences, which are discussed in the text.

HYDROGEN BONDS

Polar molecules tend to become oriented so that the positively charged part of one molecule is near the negatively charged part of the next one, because different charges attract and similar charges repel. In addition, the positively charged hydrogen portion of one molecule may be bound by this electromagnetic attraction to the negatively charged oxygen or nitrogen portion of an adjacent molecule, thus linking together two or more molecules. This is known as a **hydrogen bond.** Hydrogen bonds are much weaker than covalent bonds and can be broken quite easily. The bonding together of three water molecules is illustrated below, with the hydrogen bonds shown as dotted lines between the O and H atoms:

```
     O----H         H----O
    /      \       /      \
   H        H----O---- H    H
```

Hydrogen bonds play an important role in many compounds of biological importance. For example, the two helical strands of DNA molecules are held together by hydrogen bonds between the purines of one strand and the pyrimidines of the other (Fig. 7.11). Hydrogen bonding may also occur within a single protein molecule, resulting in the characteristic twisting and folding of the molecule. It is evident, then, that hydrogen bonds may occur within a molecule as well as between two or more molecules.

SOLUTIONS

A solution is a homogenous mixture of the molecules, ions, or atoms of two or more different substances. The dissolving medium is called the **solvent** and the substance dispersed through it is the **solute.** Here, we are concerned principally with liquid solvents, and of these, water is the most important biologically. However, some biological substances are insoluble in water but soluble in oils (liquid fats). The substances dissolved in water may be gases such as oxygen or carbon dioxide, solids such as sugar or salts, or liquids such as alcohol. Most chemical reactions that occur in plants or animals will proceed only when the substances are in solution, and indeed most substances present in cells are in solution (or are dispersed in water as clusters of molecules rather than as individual molecules as in a true solution).

When sugar is dissolved in water, the molecules making up a crystal separate from one another and become evenly dispersed through the water as individual molecules. This, of course, explains why the sugar is no longer visible after it is dissolved. When salt is dissolved in water, the Na^+ and Cl^- ions of the crystals separate from one another and become evenly dispersed through the water. Thus, in a sugar solution the dispersed particles are molecules whereas in a salt solution they are ions. Some solutes, acetic acid for example, are present in solution as a mixture of molecules and ions: CH_3COOH molecules and CH_3COO^- and H^+ ions. Some substances have molecules or ions that absorb some wavelengths of light more than others and so impart a characteristic color to the solution.

Expressing the Concentration of Solutions

The concentration of a solution may be expressed in a number of different units. A **percentage solution** is based on the percentage that the solvent constitutes of the solution. Thus, a 5% solution of sugar by **weight** could be made by dissolving 5 grams of sugar in 95 grams (or ml) of water, the total weight of the solution being 100 grams. A 15% solution of alcohol by **volume** could be made by dissolving 15 ml alcohol in 85 ml of water. Weight percentage solutions are used for dry solutes; liquid solutes are usually made up as percentage by volume.

Molar solutions are extensively used in scientific work. A 1-molar solution can be made by dissolving a gram molecular weight of the solute in enough water to make a liter of solution. The advantage of molar solutions is that a 1-molar solution of any substance contains just as many molecules as a 1-molar solution of any other substance, though of course there are more dispersed particles if the solute is ionized (like salt) than if the molecules are intact (like sugar).

Normal solutions are used principally for acids and bases. A 1-normal solution of an acid contains a gram equivalent weight (1.008) of replaceable or ionizable hydrogen per liter of solution. Thus, a 1-normal (1*N*) solution of an acid such as HCl that has only one replaceable H per molecule would also be a 1 molar (1*M*) solution. Acetic acid (CH_3COOH) also contains only 1 replaceable or ionizable H atom per molecule, even though there are 3 other H atoms in each molecule. However, a 1*N* solution of an acid such as H_2SO_4 that has 2 replaceable H atoms per molecule would be 0.5*M*, whereas a 1*N* solution of H_3PO_4 would be 0.33*M*. A 1*N* solution of a base contains a gram equivalent weight of OH (17.008 grams) per liter of solution. Thus, a 1*N* solution of NaOH or KOH would also be 1*M*, whereas a 1*N* solution of $Ca(OH)_2$ would be 0.5*M*. The advantage of using normality is that solutions of the same normality all have the same degree of total acidity or total alkalinity. Thus, 10 ml of 1*N*HCl would just neutralize 10 ml of 1*N* $Ca(OH)_2$ or any other 1*N* base.

pH or Hydrogen Ion Concentration

However, total acidity or alkalinity of a solution is generally of less biological interest than the concentration of hydrogen (H^+) or hydroxyl (OH^-) ions in the solution. The unit commonly used for expressing the **hydrogen ion concentration** of a solution is **pH,** which is defined as the negative of the logarithm of the hydrogen ion concentration in terms of normality (Fig. *A*.7). Thus, if a solution has a H^+ ion concentration of 0.00001*N* or $10^{-5}N$, its

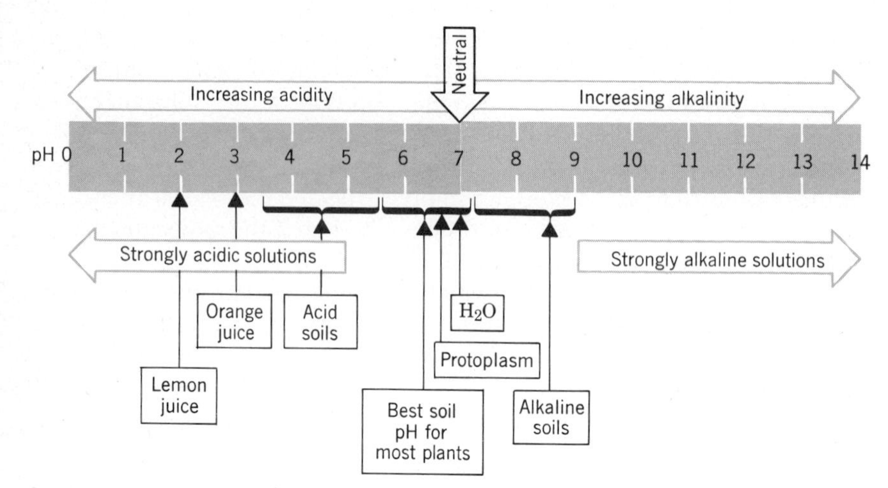

Fig. A.7 The logarithmic pH scale, with typical pH values for plant components and soils. The pH units are the negative logs of hydronium ion activity (e.g., pH 4 is 10^{-4}, so each pH unit is 10 times less acid than the next lower unit (e.g., pH 5 is 10 times less acid than pH 4 and 100 times less acid than pH3).

pH is 5, and a solution with a hydrogen ion concentration of 10^{-7} has a pH of 7. Since pH values are logarithms, each whole pH value is ten times more acid than the next higher value. Thus, pH 3 is ten times as acid as pH 4, one hundred times as acid as pH 5 and one thousand times as acid as pH 6. In a pH value such as 4.2, the 2 is the mantissa of the logarithm and should not be read as two-tenths. It is impossible to average pH values by adding them together and dividing by the number of values added, since when logarithms are added they are actually being multiplied.

It would not be necessary to use pH values if all acids were ionized to the same degree, but they are not. For example, the pH of a 0.1*N* solution of each of the following acids at 25°C is as follows: hydrochloric 1.1, sulfuric 1.2, citric 2.2, acetic 2.9, and boric 5.2. The strength of an acid depends not on its total acidity (concentration of molecules) but on the concentration of H^- ions. Although all the acids listed have the same total acidity, it is apparent that hydrochloric is a strong acid and boric a very weak one as shown by its high pH (i.e., a low H^+ ion concentration). A strong base in solution is highly ionized and thus has a high pH, a high OH^- ion concentration, and a low H^+ ion concentration.

As the H^+ ion concentration of a solution increases, the OH^- ion concentration decreases in proportion, and the product of the two always equals a constant value, 10^{-14}; thus pH is a measure of OH^- ion concentration as well as H^+ ion concentration. At pH 6 the H^+ ion concentration is $10^{-6}N$ and the OH^- ion concentration $10^{-8}N$, whereas at pH 7 both ions have a concentration of $10^{-7}N$. Thus, a solution of pH 7 is neutral, any solution below pH 7 is acid (having more H^+ ions than OH^- ions), and any solution over pH 7 is alkaline or basic (having more OH^- ions than H^+ ions). However, even a very strong acid of say pH 1 contains some OH^- ions ($10^{-13}N$) and a strong base of pH 14 has a few H^+ ions ($10^{-14}N$). Pure water has a pH of 7, which is equivalent to saying that one of every 10 million water molecules is dissociated into H^+ and OH^- ions.

Buffer Systems

If an acid is added to water, the pH decreases rapidly; if a base is added to water, the pH continues to rise as more is added. However, a solution may be buffered against marked and rapid changes in pH resulting from the addition of acids or bases. The most common type of **buffer system** consists of a weak acid plus one of its salts, for example, acetic, phosphoric, or carbonic acids. If a strong acid such as hydrochloric is added to a solution buffered by acetic acid and sodium acetate, the following reaction keeps the pH from dropping markedly:

$$H^+Cl^- + CH_3CO^-Na^+ \rightarrow CH_3COOH + Na^+Cl^-$$

Many of the hydrogen ions are thus incorporated into acetic acid molecules, relatively few of which are ionized, rather than remaining as highly ionized HCl molecules. If a strong base such as KOH is added to the buffered solution, the following reaction prevents a marked rise in pH by limiting the number of OH^- ions:

$$K^+OH^- + CH_3COOH \rightarrow CH_3COO^-K^+ + HOH$$

Many of the added OH^- ions thus become incorporated in very slightly ionized water molecules. Of course, if too many H^+ or OH^- ions are added to a buffered solution one of the components of the system (either the acetic acid or the sodium acetate in the example used) will be entirely consumed and then the pH will begin changing rapidly.

RELATED READING

Baker, J. J. W. and G. E. Allen. *Matter, Energy, and Life,* 2nd. ed. Reading, Mass.: Addison-Wesley, 1970.

MacInnes, D. A. *"pH," Scientific American, 184*(1), 40–43, January 1951.

White, E. H. *Chemical Background for the Biological Sciences,* 2nd. ed. Englewood Cliffs, N.J.: Prentice-Hall, 1970.

GLOSSARY

In writing this textbook, the authors have been at pains to define scientific terms at the place of their initial use in the text, often through the use of graphic illustrations. However, the meanings of unfamiliar terms are often difficult to remember when they are encountered at other points in the text, and a means for ready reference, as a sort of refresher, is helpful to the beginning student. Therefore, this glossary provides definitions having as broad applicability as possible. Most interested students, during their course of study (and afterwards, we hope), have occasion to write or speak about botanical matters; to assist in effective communication, the plural and adjectival forms of many terms used in botanical practice are provided, where those forms of the terms are unfamiliar.

abscisic acid: a plant hormone that inhibits growth and promotes abscission and dormancy. *Syn.* abscisin II, β-inhibitor.

abscission layer: layer of thin-walled cells extending across the base of petiole or stipe; breakdown of layer separates leaf or fruit from stem.

absorption spectrum: the amount of absorption of each wavelength of light by a pigment.

achene: a simple, dry, one-seeded, indehiscent fruit with the seed attached to ovulary wall at only one point.

acid: a substance that releases hydrogen ions (H^+) but not hydroxyl ions (OH^-) when it dissociates. Acids have a pH of less than 7.

action spectrum: the effectiveness of various wavelengths of light in bringing about light reactions, for example, photosynthesis, photoperiodism, and so forth.

active transport: movement of a substance across a cell membrane by the expenditure of energy by the cell, often against a diffusion gradient.

adaptation: adjustment of an organism to its environment; a structural or physiological modification supposedly accomplishing such adjustment.

adenine: a purine base that is a component of nucleic acids, ATP, some coenzymes, and other important metabolic substances.

adenosine triphosphate (ATP): a compound produced during photosynthesis and respiration from adenosine diphosphate (ADP) and phosphate. The energy in this bond is released readily when ATP is hydrolized to ADP and can be used in metabolic work. Adenosine monophosphate (AMP) has only one phosphate group and can be converted to ADP or ATP.

ADP, AMP, ATP: see adenosine triphosphate.

adsorption: the adhesion of molecules of a liquid, gas, or solute on the surfaces of a solid.

adventitious: referring to the origin of a structure in an unusual position on the plant, for example, roots from stem or leaf cuttings.

aeciospore: a binucleate, asexual spore of the rust fungi.

aerobic: referring to cells or organisms that require free oxygen (O_2) for respiration.

agar: a gelatinous substance extracted from certain species of algae; used as an emulsifier and stabilizer in commercial food manufacture and as substrate in culture media in bacteriology laboratories.

alga (*pl.*, **algae;** *adj.*, **algal**): a member of the large group of thallophytes containing chlorophyll.

algin: a gelatinous extract from the cell walls of brown algae; used as a stabilizer in commercial food processing.
alkali: a strong base, which see.
alkaloid (*adj.*, **alkaloidal**): a nitrogenous compound synthesized by plants and producing characteristic physiological effects on animals, for example, strychnine, caffeine, atropine, cocaine, morphine.
allele: one of the two or more forms of a gene found in the same location on homologous chromosomes.
allopolyploid: a polyploid in which one or more sets of chromosomes have come from each of two different species or varieties.
amino acid: an organic acid containing an amino ($-NH_2$) group in its molecule.
ammonification: the decomposition of amino acids resulting in production of ammonia.
amoeboid: resembling amoeba.
amylase: an enzyme that hydrolyzes starch.
amylopectin: a kind of starch with branched molecules.
amylose: a kind of starch with unbranched molecules.
anaphase: the stage in nuclear division in which sister chromatids move to opposite poles of the cell.
anaerobe (*adj.*, **anaerobic**): an organism capable of respiration in absence of free oxygen.
androecium: the aggregate of stamens in the flower of a seed plant.
anemophily (*adj.*, **anemophilous**): in seed plants, the characteristic in which the pollen is transferred by wind.
angiosperm: a seed plant in which the seeds are produced in a closed ovulary.
anion: an ion with a negative electrical charge, resulting from an excess of electrons over protons.
annual: a plant that completes its life cycle within one growing season and then dies.
annulus: a row of specialized cells in the sporangial wall of ferns, important in opening the sporangium and dispersal of spores.
anther: the pollen-bearing sacs (sporangia) borne on the stamen.
antheridium (*pl.*, **antheridia;** *adj.*, **antheridial**): the sperm-bearing organ of plants other than spermatophytes.
antheridogen: a plant hormone that induces development of antheridia.
Anthocerotae: a class of Division Bryophyta.
anthocyanin: a water-soluble pigment of a blue, purple, red, or pink color present in the vacuoles of plant cells; the color is influenced by pH.
antibiotic: a substance produced by an organism which, in low concentration, inhibits the growth of other organisms.
antipodal: referring to nuclei or cells at base of the embryo sac, opposite the egg and synergidae.

apical dominance: the suppression of growth of lateral buds by a terminal bud.

apomixis (*adj.*, **apomictic**): a pattern of reproduction superficially resembling sexual reproduction but without meiosis or fertilization; includes parthenogenesis, apospory, and apogamy, and sometimes extended to include vegetative propagation.

apogamy (*adj.*, **apogamous**): reproduction of a sporophyte from a cell of a gametophyte without fertilization.

apospory (*adj.*, **aposporous**): reproduction by a sporophyte from a spore-mother cell without meiosis, resulting in a diploid gametophyte.

archegonium (*pl.*, **archegonia**): the egg-bearing organ in which the egg is surrounded by sterile tissue.

ascocarp: the characteristic fruiting body of certain large representatives of ascomycetes, such as the cup-fungi and morels.

ascomycete: a member of a class of Eumycophyta in which meiospores are produced in a saclike sporangium, the ascus; some species reproduce by vegetative budding, as in yeast, or by conidiospores produced at the tips of the septate hyphae, as in *Aspergillus* and *Penicillium.*

ascospore: a meiospore produced within an ascus.

ascus (*pl.*, **asci**): the saclike reproductive cell characteristic of ascomycetes, in which after the fusion of two haploid nuclei followed by a reductional and two equational divisions, eight ascospores are formed.

asexual reproduction: reproduction that does not involve meiosis and fusion of gametes; includes fission, asexual sporulation, apomixis, and natural vegetative propagation.

association: a subdivision of a plant formation, for example, oak-hickory association of the Eastern Deciduous Forest Formation.

ATP: see adenosine triphosphate.

autopolyploid: a polyploid in which all the chromosomes come from closely related members of the same species.

autotrophism (*adj.*, **autotrophic**): a condition wherein an organism is capable of synthesizing its own foodstuffs by photo- or chemosynthesis.

auxin: a plant hormone essential for cell elongation and cell division, and having varied influences on plant growth and development.

axil (*adj.*, **axillary**): the upper angle between the petiole of a leaf and the stem, the position in which a branch bud develops.

B

bacillus (*pl.*, **bacilli**): a rod-shaped bacterium.

bacteria (*sing.*, **bacterium**): common name applied to members of Schizomycophyta (schizomycetes).

bacteriophage: a virus that attacks bacteria, causing disruption of the cells.

bar: a unit of pressure. One bar equals 75 cm Hg, 0.987 atmosphere, and 14.5 lb/in^2.

bark: the complex of tissues exterior to the cambium in a woody stem or root.

base: a substance that releases hydroxyl ions but not hydrogen ions and has a pH of more than 7.

basidiocarp: the spore-producing body of basidiomycetes.

basidiomycete: a member of a class of fungi in which meiospores are produced on a club-shaped hyphal tip, the basidium.

basidiospore: a meiospore borne upon a basidium.

basidium (*pl.,* **basidia**): characteristic reproductive cell of basidiomycetes, often club-shaped as in mushrooms or a short filament of four cells as in rust fungi, and bearing haploid basidiospores.

benthon: the aggregate of plants and animals growing attached to the bottom of bodies of water. Compare plankton.

berry: a fleshy-walled fruit derived from a single pistil, the seeds commonly embedded in a fleshy or pulpy mesocarp, endocarp, or placenta, for example, tomato.

biennial: a plant that completes its life cycle within two growing seasons and then dies.

binomial: the scientific name applied to an organism, consisting of the name of the genus and the name of the species to which it belongs.

biome: a community complex covering a large area with a characteristic climate and composed of distinctive vegetation, for example, grassland, forest.

biosphere: the portion of the Earth occupied by living organisms, including all communities on Earth.

bract: a usually small or specially modified leaf in close association with a flower or an inflorescence, for example, the white so-called "petals" of the dogwood and the red leaves of poinsettia.

blue-green alga (*pl.,* **blue-green algae**): a member of a class of algae characterized by the presence of the pigment phycocyanin.

bryophyte: a member of Division Bryophyta, embracing the mosses and liverworts.

bud: an undeveloped shoot usually protected by modified scale-leaves; may give rise to only a vegetative shoot, only a floral shoot, or a combination of the two (mixed bud). **Apical bud,** a bud arising at the apex of a stem or branch. **Lateral bud,** a bud produced in the axil of a leaf.

budding: a method of asexual reproduction in yeasts. Also used to describe a method of vegetative propagation in higher plants, involving the grafting of a bud upon a stock.

bud-scar: the scar left on a stem after the fall of a bud.

buffer solution: a solution that resists a change of pH on the addition of acids or bases.

bulb: a short, underground stem bearing many fleshy scalelike leaves, for example, onion.

bundle-scar: a scar remaining on a stem, formed by the separation of vascular tissue at leaf- or bud-fall.

C

callose: a carbohydrate constituent of cells walls, especially after wounding.

callus: a massive development of large, thin-walled cells resulting from wounding, as in the preparation of leaf- or stem-cuttings in vegetative propagation.

calorie: the quantity of heat energy required to raise the temperature of 1 gram of water 1°C. The unit used in metabolic measurements is the kilocalorie (kcal), equal to 1000 calories.

calyptra: a veillike covering of the developing sporophyte in bryophytes.

calyx: the lowermost whorl of the flower, composed of sepals.

cambium: a layer of meristematic cells producing secondary tissues. **Cork cambium** (phellogen), producing cork on its exterior and phelloderm on its interior. **Fascicular cambium,** an arc of the cambium ring included within a vascular bundle. **Interfascicular cambium,** an arc of the cambium ring extending across a pith ray and producing secondary ray tissue on both sides. **Vascular cambium** produces secondary phloem on the exterior and secondary xylem on the interior.

capsule: a simple, dry, dehiscent fruit usually composed of two or more carpels. The term is sometimes applied to the sporangium of the moss.

carbohydrate: an organic compound composed of C, H, O, generally in the ratio of 1:2:1. Carbohydrates include sugars, dextrins, and polysaccharides such as starch and cellulose.

carotene (carotin): a yellow, reddish, or orange pigment occurring in plastids.

carotenoids: fat-soluble plant pigments, including carotenes and xanthophylls.

carpel (*adj.,* **carpellary** the structural unit of a pistil, bearing ovules along its margin.

carrageenin: a mucilaginous extract derived from some species of algae, used as a stabilizer in some food manufacture.

catkin: a spikelike inflorescence usually bearing either staminate or pistillate flowers, for example, willows and oaks.

catalyst (*adj.,* **catalytic**): a substance that in low concentration speeds the rate of a chemical reaction but is not used up in the reaction.

cation: an ion with an excess of protons over electrons, thus having a positive electrical charge.

cell (*adj.,* **cellular**): the structural and physiological unit of living organisms.

cell-inclusions: particles of nonprotoplasmic substances, the products of cell metabolism, embedded in the cytoplasm of a cell, for example, starch, crystals, oil, water.

cellobiose: a disaccharide sugar used in the synthesis of cellulose.
cell-organelles: specialized protoplasmic structures of the protoplast of the cell, performing special functions in cell activity, for example, plastids, Golgi bodies, mitochondria, ribosomes.
cell-plate: a nonprotoplasmic layer laid down during cell division and separating the two new protoplasts.
cellulase: an enzyme that hydrolyzes cellulose.
cellulose (*adj.,* **cellulosic**): an insoluble polysaccharide carbohydrate that is the main constituent of the walls of most plant cells; formed from several hundred β-D glucose molecules.
cell wall: the cellulosic layer (often modified by the addition of lignin, chitin, cutin, suberin) secreted by and enveloping the protoplast of a plant cell. **Primary cell wall,** the first formed wall of most plant cells, usually cellulosic. **Secondary cell wall,** a later deposition of wall substance interior to the primary wall, present in cells of specialized function, such as support or protection.
centromere (kinetochore): the apparent point of attachment of spindle fibers to the chromosome in cell division.
chalaza: the region of attachment of the funiculus to the base of the ovule.
chemosynthesis: the synthesis of foods from carbon dioxide and water by certain species of bacteria, the required energy being provided by the oxidation of inorganic compounds rather than light.
chitin (*adj.,* **chitinous**): a cell wall component characteristic of many fungi; also composes the outer shell of crustaceans and insects.
chlorenchyma (*adj.,* **chlorenchymatous**): plant tissues containing chlorophyll.
chlorophyll: a green pigment of plant cells that absorbs light energy used in photosynthesis; exists in several forms, as chlorophyll *a, b, c, d, e.*
chlorophyte: a member of Division Chlorophyta, a green alga.
chloroplast: a specialized plastid containing chlorophyll.
chlorosis (*adj.,* **chlorotic**): loss or reduced development of chlorophyll, resulting in yellowish leaves. Chlorosis is a symptom of several mineral deficiencies and plant diseases.
chromatid: the half-chromosome visible during middle and late stages of mitosis.
chromatin: the bearer of hereditary characters in the chromosome; readily takes artificial dyes employed in microscopic study of cells.
chromocenter: a region of a chromosome that takes dye more heavily than other parts.
chromonema (*pl.,* **chromonemata**): a thread of chromatin embedded in the chromosomal matrix.
chromoplast: a specialized plastid containing yellow, orange, or red pigment.

chromosome: the heavily staining, specialized structure of the nucleus bearing genes and responsible largely for cell activity and hereditary potential.

chrysophyte: a yellow-green alga whose chlorophyll is more or less masked by yellow or brownish plastid pigments.

cilia (*sing.,* **cilium**): short protoplasmic hairs whose wavelike motion propel some unicellular organisms through water.

circadian rhythms: regular rhythms of activity and growth that occur at about 24-hour intervals.

class: a major group of organisms within a division, representing a small circle of affinity. A division may consist of several classes.

clone: a population of individuals derived from a single parent by asexual reproduction or vegetative propagation, and so having uniform hereditary potentialities (except for possible mutations).

club fungus: a basidiomycete. Reference is to the club-shaped spore-bearing basidium, for example, mushroom.

club moss: a common name applied to some species of lycophytes, for example, *Lycopodium* species.

coccus (*pl.,* **cocci**): a bacterium of spherical form.

codon: the sequence of three adjacent nucleotides in DNA that code for a specific amino acid.

coenocyte (*adj.,* **coenocytic**): a multinucleate filament not divided into separate protoplasts by cross walls.

coenzyme: a nonprotein organic compound essential for enzymatic action along with an enzyme protein, for example, NAD, CoA, ATP.

cohesion-tension mechanism: a mechanism of water-rise in plants involving the pulling up of the water columns in the xylem, resulting in a negative pressure (tension) and dependent on the cohesiveness of the water.

colchicine: an alkaloid extracted from the corm of the autumn crocus, *Colchicum autumnale;* used in experimental alteration of chromosome number in cytological studies.

coleoptile: a sheath enveloping the first emergent leaves in the germination of monocotyledons.

coleorrhiza: a sheath enveloping the embryonic root in the seed of grasses.

collenchyma (*adj.,* **collenchymatous**): a tissue composed of cells whose walls are unevenly thickened with cellulose; a primary supportive tissue of young or herbaceous plant stems.

colloid (*adj.,* **colloidal**): a finely subdivided material evenly dispersed in some medium.

community: a subdivision of an association characteristic of a certain habitat and consisting of a single or few species.

companion cell: a sister-cell of a sieve-tube element in the phloem of angiosperms.

compression: a type of fossil formed under great pressure, consisting of only the carbon residue and without cellular structure.

cone: an elongate structure consisting of small, closely set, overlapping, modified leaves or branches bearing sporangia, for example, a pine cone, the strobile of a club moss.

conidium (*pl.*, **conidia**): an asexual reproductive cell of certain species of fungi, produced by specialized modes of division of terminal cells of hyphae.

conifer (*adj.*, **coniferous**): cone-bearing; applied to members of Gymnospermae, such as pine, spruce, cycad.

coniferophyte: a gymnosperm such as pine, hemlock, fir, redwood, and others whose reproductive structures are cones (strobiles).

cork: a secondary covering tissue produced by the phellogen; the cell walls of cork cells are characteristically suberized.

corm: a short, fleshy underground stem similar to a bulb but with small scalelike leaves instead of fleshy scales, for example, crocus, gladiolus.

corolla: collectively, the petals of an angiospermous flower.

cortex (*adj.*, **cortical**): a primary parenchymatous tissue of the stem or root extending from epidermis to phloem in vascular plants.

cotyledon: a first seed-leaf in the embryo of spermatophytes, sometimes serving as a food-storage or absorbing organ in the embryo.

covalent bond: a chemical bond between two atoms, resulting from the sharing of two electrons.

crossing over: the exchange of corresponding gene segments between the chromatids of homologous chromosomes during meiosis.

cup fungus: an ascomycete whose spores are produced in a cup-shaped sporangiophore.

cuticle (*adj.*, **cuticular**): a waxy layer secreted on the exterior surface of epidermal cells, a protective layer against excessive water loss.

cutin: a waxy substance slightly permeable to water and gases, commonly modifying exterior cellulose walls of plant cells.

cyanophyte: a blue-green alga, member of Division Cyanophyta; characterized by the presence of the pigment phycocyanin in addition to chlorophyll.

cycadophyte: a primitive cone-bearing gymnosperm with foliage of a pronounced fernlike aspect.

cyclosis: the circulation of cytoplasm within a cell; also called cytoplasmic streaming.

cytokinesis: the division of the cytoplasm of a plant cell following division of the nucleus.

cytokinin: a plant hormone that promotes cell division and delays senescence.

cytology: the science dealing with all aspects of cell structure and behavior.

cytoplasm: the protoplast of the cell exclusive of the nucleus.

D

deciduous: referring to a plant that seasonally loses all of its leaves.
decomposition: the breaking down of plant and animal remains into inorganic substances by bacteria and fungi.
dehiscent: referring to sporangia or fruits that spontaneously rupture when mature, facilitating dispersal of spores or seeds.
denitrification: the conversion of nitrates into nitrogen gas (N_2) by certain species of bacteria.
deoxyribonucleic acid (DNA): the nucleic acid found in chromosomes, chloroplasts and mitochondria that carries the genetic code; DNA is self-replicating and controls the sequence of nucleotides in RNA.
deoxyribose: a five-carbon simple sugar with one less oxygen atom than ribose; ($C_5H_{10}O_4$).
desert: a type of habitat characterized by low water availability.
deuterium: the isotope of hydrogen with a neutron as well as a proton in its nucleus, and so an atomic weight of 2; heavy hydrogen; ^2H.
diastase: any enzyme that hydrolyzes starch; amylase.
diatom: a siliceous-walled unicellular plant, a member of Chrysophyta.
dicotyledon (*abbrev.*, **dicot**): an angiosperm whose embryo possesses two cotyledons. **Dicotyledoneae,** a subclass of angiosperms.
differentially permeable membrane: a membrane that is permeable to water but not to some solute particles; sometimes called a semipermeable membrane.
differentiation: the specialization of cells, tissues, and organs in the course of development of an organism.
diffusion: the net movement of the molecules of a substance from a region of higher to lower molecular activity of that substance as a result of the random movement of the molecules.
digestion: the hydrolysis of complex foods into simple, soluble foods through the action of hydrolytic enzymes.
dinoflagellate: unicellular flagellate member of a class of pyrrophytes, the golden-brown algae; important constituents of marine plankton.
dioecious: referring to flowering plants which bear their male and female reproductive parts on separate plants.
diploid: referring to cells or organisms whose nuclei contain a double set of chromosomes.
disaccharide: a sugar synthesized from two molecules of simple (monosaccharide) sugars.
division: a portion of the Plant Kingdom whose members reflect a major evolutionary trend.
DNA: see deoxyribonucleic acid.
dominant: the principal and generally most abundant species of plants in a community; a dominant gene.
dominant gene: a gene that has full phenotypic expression even though its allelic gene provides for a different phenotype.

dormancy: the cessation of growth of seeds and buds, even though the current environment may be favorable for growth, until the dormancy has been broken by a suitable complex of factors.
drupe: a simple, fleshy fruit with a stony endocarp, for example, peach, cherry.

E

ecology: the science dealing with interrelationships of organisms and environment.
ecosystem: the interacting system of a community of organisms and its physical environment.
ecotype: a population within a species differing from other members of the species because of environmental rather than hereditary variation.
edaphic: referring to conditions of the soil as they relate to plant growth and development.
egg: a female (mega-) gamete.
electrolyte: a substance that in solution dissociates into anions and cations, for example, acids, bases, salts.
electrovalent bond: a chemical bond between the atoms of a molecule resulting from the transfer of one or more electrons from one atom to another, as in acids, bases, and salts.
embryo: an immature sporophyte resulting from fertilization of the egg, such as contained in the seed of higher plants before germination.
embryo sac: the female gametophyte of angiosperms.
emulsion (*adj.*, **emulsoid**): a finely divided liquid dispersed in a liquid medium.
endocarp: innermost layer of the fruit wall.
endodermis: the innermost layer of the cortex, the cells often highly modified.
endoplasmic reticulum: the complex system of internal membranes within the cytoplasm.
endosperm: a nutritive tissue formed within the female gametophyte of seed plants, may be persistent in the seed as in grasses or consumed in the development of the embryo, as in beans.
entomophily (*adj.*, **entomophilous**): transfer of pollen by insect agent.
enzyme (*adj.*, **enzymatic**): a complex protein produced by living cells that influences the rate of metabolic processes.
ephemeral: referring to small, herbaceous plant species with a short life span.
epicotyl: the apical portion of the embryonic axis above the cotyledons.
epidermis: the surface layer of cells on all parts of the primary plant body, such as stems, roots, leaves, flowers, and fruits.
epigyny (*adj.*, **epigynous** arrangement of floral parts in which the ovulary is embedded in the torus, other floral parts thus appearing to arise from the top of the ovulary, for example, apple and daffodil.

epiphyte (*adj.*, **epiphytic**): a plant that grows on another plant but obtains no nourishment therefrom, for example, Spanish moss on oaks or lichens on tree trunks.

ester: an organic compound formed by reaction of an alcohol with an acid (often an organic acid) with the release of a molecule of water, corresponding with the formation of a salt by the reaction of an acid and a base.

ethylene: a simple hydrocarbon gas that acts as a plant growth substance, inhibiting growth and promoting the ripening of fleshy fruits.

etiolation: a condition of exaggerated stem growth, small leaf size, and absence of chlorophyll, resulting from very low light intensity.

eucaryote: an organism whose cells possess an organized nucleus.

euglenophyte: a member of the Division Euglenophyta, characterized by flagellate, unicellular organisms with an organized nucleus but lacking a rigid cell wall; important members of freshwater plankton.

eumycophyte: a member of the Division Mycophyta, known as the true fungi.

exocarp: the outermost layer of the fruit wall.

exine: the outer layer of the pollen grain wall, often intricately sculptured.

F

F_1: the first filial generation following a hybrid cross. The second and third generations are designated as F_2 and F_3.

family: a subdivision of an order, consisting of a group of closely related genera.

fat: a lipid that is synthesized from a molecule of glycerol and three molecules of fatty acids. *Syn.*, triglyceride.

fatty acid: a long chain organic acid with a terminal carboxyl ($-COOH$) group.

fermentation: an anaerobic process in which energy is released from foodstuffs by various microorganisms, for example, fermentation of sugar by yeasts with the production of ethyl alcohol.

fertilization: in the sexual reproductive process, the fusion of an egg and sperm, resulting in a diploid zygote.

fertilizer: a substance or mixture that contains one or more of the elements essential for plant growth (commonly N, P, and K) and is supplied to plants to prevent mineral deficiencies; erroneously called plant food.

fiber: an elongated, thick-walled supportive cell occurring in various parts of plant bodies: **cortical fibers** in the cortex, **phloem fibers** in the phloem, and **wood fibers** in the xylem.

fiber-tracheid: an elongated supportive cell intermediate in structure between a fiber and a tracheid, that is, a fiber with tracheidlike pits.

field capacity: the percentage of water that a soil contains after a capillary equilibrium has been established in the soil.

filament: the stalk of a stamen, usually threadlike, bearing an anther at its tip; also, used to describe a slender row of cells as in the algae.
Filicinae: ferns, a class of pterophytes.
fission: asexual reproduction of unicellular organisms resulting from the division of the cell into two daughter cells of equal size.
flagellum (*pl.,* **flagella**): a slender protoplasmic projection from a cell, as in motile spores, gametes, or unicellular organisms.
flora: the complex of species comprising the vegetation of a region.
flower: the characteristic reproductive structure of angiosperms consisting, as in a complete flower, of calyx, corolla, stamens, and pistils.
incomplete flower: one in which the calyx and/or corolla is lacking.
perfect flower: one having both stamens and pistils. **Imperfect flower,** one lacking stamens or pistils. **Regular flower** (actinomorphic), one that is radially symmetrical. **Irregular flower** (zygomorphic), one that is bilaterally symmetrical.
food: an organic compound that can be used by an organism as an energy source (respiration) or as a source of the organic constituents of cells (assimilation).
food chain: a series of organisms that secure food from one another, beginning with a photosynthetic plant and continuing through an herbivore and one or more levels of primary and secondary carnivores.
food web: the complex of food chains in a community, involving at each level all the organisms that secure food from each of the organisms at the next level.
florigen: a hypothetical plant hormone that brings about the initiation of flower buds.
follicle: a simple, dry fruit dehiscent along one suture.
forest: a plant formation in which the dominant plants are large or small trees.
formations: a regional aggregation of plant species in which one plant form (such as, trees, grass, shrubs, etc.) is dominant.
fossil: the remains, or evidence of remains, of organisms of past geologic ages variously occurring as impressions, compressions, mineralizations, in the Earth's crust.
frond: typically, the leaf of a fern, often variously subdivided.
fructose: a six-carbon monosaccharide used in synthesis of sucrose and inulin.
fruit: the matured ovulary of angiosperms, often associated with other floral parts; also applied to the reproductive structure of other groups of plants.
fucoxanthin: a brown pigment characteristic of the brown algae.
fungicide: a chemical substance employed to control the growth of fungi.
fungus, (*pl.,* **fungi;** *adj.,* **fungal**): exclusive of the bacteria, a thallophyte without chlorophyll.
funiculus: the stalk of the ovule in seed plants.

G

galactose: a six-carbon monosaccharide used in synthesis of lactose (milk sugar) and pectic compounds.

gall: an abnormal and often complex and highly sculptured structure developed by plants in response to the effects of parasites, such as insects, fungi, bacteria, or other organisms.

gametangium (*pl.*, **gametangia**): the reproductive organ producing gametes.

gamete (*adj.*, **gametic**): in sexual reproduction, a protoplast that fuses with another protoplast producing a zygote.

gametic number: the number of chromosomes of the gametes of a species.

gametophyte (*adj.*, **gametophytic**): a plant or phase of the life cycle of a sexually reproducing plant that produces gametes.

gel: a semisolid colloid in which both the dispersing medium (usually water) and the dispersed substance are continuous, for example, gelatin, jelly.

gemma (*pl.*, **gemmae**): a budlike mass of tissue produced upon the thallus of liverworts, as an agent in vegetative reproduction.

gene: a hereditary unit character that provides a specific hereditary potentiality that can be transmitted from generation to generation; a portion of a DNA molecule.

gene frequency: the relative abundance of any allele in a particular population of a species.

gene pool: all of the alleles in the members of a species.

genetics: the science of heredity.

genotype: the sum total of all the genes in an individual whether or not expressed phenotypically. The genes of an individual being considered in a specific genetic cross.

genus (*pl.*, **genera**; *adj.*, **generic**): a group of species structurally and genetically closely related; a group of related genera constitute a family.

geotropism (*adj.*, **geotropic**): the bending growth of roots, stems, or other plant parts, induced by gravity.

germination: the resumption of growth of a seed or spore after a characteristic period of dormancy.

gibberellin: a plant hormone that promotes elongation of intact plants and initiation of flowering of biennials and some long-day plants.

girdling: the removal of a ring of bark from a woody stem, resulting in prevention of translocation by the phloem in the girdled region. *Syn.*, ringing.

glucose: a monosaccharide sugar used in respiration and in the synthesis of disaccharides, starch, cellulose and other compounds.

glycolysis: a series of anaerobic reactions in which glucose is converted to pyruvic acid and some energy is transferred to ATP and NAD.

Golgi apparatus (Golgi body): a cell organelle consisting of a series of flat plates, believed to be associated with secretion of certain cell products. *Syn.,* dictyosomes.

grafting: a process in which the stem of one plant (the scion) is united with a root, or more commonly a stem with roots, of a related plant (the stock). Grafting is used for the propagation of various trees and shrubs and for experimental purposes.

grassland: a vegetational formation consisting predominantly of species of grass.

growth: an irreversible increase in size or weight of an organism; the term is sometimes used to include the differentiation and development of an organism.

guard cell: one of a pair lens-shaped cells of the epidermis of higher plants which together form a stoma.

guttation: the exudation of droplets of water from leaves.

gymnosperm: a member of Class Gymnospermae of Division Pterophyta in which the ovules and seeds are borne in an exposed manner, that is, not enclosed in a carpellary structure.

gynoecium: the carpel or aggregate of carpels comprising the pistil of an angiospermous flower.

H

habitat: the natural environment of a plant or animal, including both biological and physical factors.

haploid: referring to the presence in a cell nucleus of a single complete set of chromosomes (typically characteristic of the gametophytic phase of the life cycle); also applied to an organism or phase of its life cycle possessing a haploid set of chromosomes.

haustorium (*pl.,* **haustoria**): in fungi, an extension of the hyphae that penetrates a living host or a substrate and absorbs nutrients therefrom.

heartwood: in woody plants, the inactive, innermost portion of the xylem; the cells are usually modified chemically by deposition of resins, gums, or tannins, often accompanied by change of color.

helix: a coil with the same diameter throughout its length, as contrasted with the increasing diameter of a spiral from one end to the other.

hemicelluloses: cell wall polysaccharides generally synthesized from both hexoses and pentoses and lacking the fibrous structure of cellulose.

hepatic: a liverwort, member of Class Hepaticae of Division Bryophyta.

herb (*adj.,* **herbaceous**): a seed plant of essentially soft structure which develops no, or very little, secondary woody tissue.

herbal: a book containing names and descriptions of plants of known or suspected medicinal value.

herbivore (*adj.,* **herbivorous**): a plant-eating animal.

heredity: the transfer of structural and functional characteristics from parent to offspring.

heterogametes (*adj.*, **heterogametic**): gametes that differ from each other in size and behavior, for example, egg and sperm.

heterogamy (*adj.*, **heterogamous**): the condition of possessing gametes of two types (heterogametes).

heterospory (*adj.*, **heterosporous**): the condition of producing spores of two sizes and behaviors, for example, micro- and megaspores.

heterothallic: referring to heterogamous plants in which the micro- and megagametes are produced by different, individual gametophytes.

heterotroph (*adj.*, **heterotrophic**): an organism that cannot carry on photosynthesis or chemosynthesis and so must secure its food from other organisms.

heterozygous: referring to the presence of two different genes at a locus on the homologous chromosomes, for example, one recessive and one dominant gene.

hexose: a six-carbon sugar (monosaccharide).

hilum: the scar remaining on the seed after separation from the funiculus.

holdfast: an anchoring device as in several types of attached algae.

homologous chromosomes: members of a chromosome pair.

homospory (*adj.*, **homosporous**): the production of spores all of one type.

homothallic: referring to species in which male and female gametes are produced by the same plant body.

homozygous: referring to the presence of the same allelic genes at a locus on the homologous chromosomes, for example, two dominant genes or two recessive genes.

hormone: an organic compound synthesized in small amounts in one part of an organism and transported to other parts where it has marked influence on growth, development, and metabolic processes.

horsetail: a member of Division Sphenophyta.

host: an organism that furnishes support and nutrient to a parasite.

humus: organic matter in the soil, in a state of progressing decomposition.

hybrid: an offspring of a cross between members of two different strains, varieties, or species. An offspring of a cross between any two parents that differ in one or more hereditary characters, for example, tall x dwarf.

hydrogen bond: a weak chemical bond between a hydrogen atom already bonded to an oxygen or nitrogen atom of a compound and an oxygen or nitrogen atom of another molecule.

hydrogen ion concentration: see pH.

hydrophyte: an aquatic plant.

hydroponics: a horticultural practice of raising plants without soil, the plants being supplied with complete mineral nutrient solutions and some means of support such as vermiculite, sand, or gravel in which the roots can anchor the plant.

hymenium: a spore-bearing layer in the fruiting bodies of various fungi.

hypha (*pl.*, **hyphae;** *adj.*, **hyphal**): a filament of a fungus.

hypocotyl: the basal portion of the embryonic axis, below the cotyledons.

hypogyny (*adj.*, **hypogynous**): the condition in which stamens, petals, and sepals are situated on a conical or elongated torus, in that order, below the pistil.

I

IAA: see indoleacetic acid.

imbibition: diffusion of water into a substance with an affinity for water and the adsorption of water molecules on the molecules of the substance, causing the substance to swell.

imperfect fungi: fungi without a known method of sexual reproduction.

impression: a type of fossil in which sediments, consolidated into rock, preserve surface features of the buried plant or animal part.

indehiscent: referring to a condition in which fruits or sporangia do not open along definite lines or by special devices.

indoleacetic acid: the most abundant naturally occurring auxin.

indusium (*pl.*, **indusia**): in ferns, a membranous outgrowth of the leaf surface which covers or surrounds a sporangial cluster (sorus).

inferior ovulary: the apparent position of the pistil below the other floral parts. See epigyny.

inflorescence: the characteristic disposition of flowers on the stem.

insecticide: a chemical substance used for killing insects.

integument: a layer of tissue enveloping the nucellus of the ovule; different species may characteristically possess one integument, or two.

intercalary: referring to meristematic growth at positions on the stem other than the apex, for example, elongation of the stem just above the nodes, as in grasses and other monocoltyledons.

internode (*adj.*, **internodal**): a portion of the stem between successive nodes.

intine: the inner layer of the wall of a pollen grain.

involucre (*adj.*, **involudcral**): a whorl of bracts (modified leaves) surrounding or subtending an inflorescence.

ion (*adj.*, **ionic**): an electrically charged atom with either more protons than electrons (cation) or more electrons than protons (anion).

isogametes: gametes essentially alike in size and behavior.

isogamy (*adj.*, **isogamous**): the condition in which gametes appear to be alike in size and behavior.

isomer (*adj.*, **isomeric** one of a group of compounds with the same kinds and numbers of atoms but with different molecular structures, for example, glucose, fructose, galactose (all $C_6H_{12}O_6$).

isotope: one of several forms of an element, all having the same number of protons and electrons in their atoms, and the same properties, but having different numbers of neutrons in their atoms. Some isotopes are radioactive.

K

karyolymph: the undifferentiated protoplasmic content of the nucleus.

kelp: a member of the group of larger brown algae common along the coast.

kilogram calorie (kcal): see calorie.

kinetin: a purine derivative with cytokinin activity, but which is probably not synthesized by plants.

kinetochore: the specialized structure of a chromosome to which spindle fibers appear to be attached in nuclear division.

L

latex: a milky secretion produced by specialized cells of stem, root, leaf in plants belonging to several plant families, for example, milkweed, fig, dandelion, rubber tree.

leaf-axil: the upper angle formed by the petiole of the leaf and the stem to which it is attached.

leaf-gap: an opening in the vascular cylinder of the stem formed by the departure of a leaf-trace.

leaf primordium: a mass of meristematic tissue at the stem apex which will expand to become a leaf.

leaf-trace: the vascular supply to a leaf.

leaflet: a single subdivision of the blade of a compound leaf.

legume (*adj.*, **leguminous**): a simple, dry, dehiscent, monocarpellary fruit, dehiscing along two sutures; commonly applied to plants bearing legumes, such as clover, peas, beans.

lenticel: a commonly lens-shaped cluster of parenchymatous cells in the bark of woody stems that permits gaseous interchanges.

leucoplast: a colorless plastid.

liana: a plant that climbs on other plants for mechanical support, usually by means of twining stems.

lichen: a symbiotic association of an alga and a fungus.

life cycle: the entire history of a plant from its beginning as a discrete individual to its production of the next generation.

lignification: the deposition of lignin in the cellulose wall of certain plant cells.

lignin: an organic substance of complex composition secreted by the protoplast, impregnating the cellulose walls of certain types of plant cells and serving chiefly a supportive function, for example, as in sclereids, wood and phloem fibers.

lipid: an organic compound synthesized from fatty acids and alcohols by organisms, for example, fats, phospholipids, waxes.
liverwort: a member of Class Hepaticae of Division Bryophyta.
locule (*adj.,* **locular**): a cavity or compartment of an ovulary in which the ovules are situated.
lumen: the cavity inside the cell wall after the disappearance of the protoplast, as in vessel elements and tracheids.
lycophyte: a member of Division Lycophyta.

M

maltase: an enzyme that hydrolyzes maltose.
maltose: a disaccharide sugar synthesized from two molecules of α-D glucose and used in the synthesis of starch and glycogen.
materia medica: a list of drug substances used in medical practice.
megagamete: a female gamete (egg).
megasporangium: a sporangium that bears only megaspores.
megaspore: a spore which, upon germination, gives rise to a female (mega-) gametophyte.
megaspore mother cell: a diploid cell that produces four megaspores by meiosis.
megasporophyll: a leaf bearing megasporangia.
meiosis (*adj.,* **meiotic**): process by which a diploid spore mother cell gives rise to four haploid spores.
meiospore: a spore resulting from the meiotic divisions of a spore mother cell.
meristem (*adj.,* **meristematic**): an undifferentiated tissue whose cells are capable of cell division. The cells produced may mature into specialized types. **Apical meristem,** occurs at the apex of stem and branches and is responsible for elongation of those parts. **Lateral meristem,** a meristem occurring in the outer regions of a stem, and responsible for growth in diameter in woody plants. **Primary meristem,** a meristem responsible for developing the initial, basic structures of the plant body. **Secondary meristem,** a meristem arising in formerly mature tissue, responsible for the production, for example, of secondary xylem and cork in woody plants.
mesophyll: the interior parenchymatous tissue of a leaf.
mesophyte: a plant inhabiting regions of intermediate moisture availability. *Contrast:* hydrophyte, xerophyte, q.v.
messenger RNA (mRNA): the nucleic acid that carries the genetic code from DNA and on ribosomes determines the sequence of amino acids in polypetides being synthesized.
metabolism: all of the chemical reactions that occur within an organism.
metaphase: a stage of mitosis during which chromosomes are arranged at the equator of the cell.
micron: a unit of measurement equivalent to .001 millimeter.
microgamete: a male gamete.

microgametophyte: a gametophyte producing only male (micro-) gametes.
micropyle: a channel extending from the apex of the ovule (between the edges of the integuments) to the nucellus.
microsporangium: a sporangium that bears microspores.
microspore: a spore which, upon germination, produces a microgametophyte.
microspore mother cell: the diploid cell which, by meiosis, gives rise to four haploid microspores.
microsporophyll: a leaf (sometimes highly modified) bearing microsporangia.
middle lamella: the initial layer separating the two new protoplasts formed in cell division.
millimeter: a unit of measurement equivalent to .001 of a meter.
mildew: a superficial fungal growth developing on organic materials such as, leather and clothing in humid environments; commonly applied to certain fungal diseases of plants, such as mildew of grape.
mitochondrion (*pl.,* **mitochondria**): a specialized cell organelle associated with respiration.
mitosis (*adj.,* **mitotic**): nuclear division resulting in equal distribution of chromosomes to the daughter nuclei.
mold: a general term applied to fungal growth on organic substances.
monocaryotic: referring to a cell possessing a single nucleus.
monocotyledon (*abbrev.,* **monocot**): a plant whose embryo has a single cotyledon. **Monocotyledoneae,** a subclass of Angiospermae.
monoecious: referring to flowering plants that bear their male and female reproductive parts in separate flowers on the same plant.
monosaccharide: a simple sugar, that is, a sugar that cannot be hydrolyzed into sugars of lower molecular weight.
morphogenesis: the development of an organism, with special reference to structural differentiation.
morphology (*adj.,* **morphological**): the study of form and development.
moss: a member of Class Musci of Division Bryophyta.
mu, (μ): the Greek letter used to indicate a micron, .001 of a millimeter.
Musci: a class of Bryophyta, the mosses.
mutation: a chemical change in a DNA code resulting in a gene with altered hereditary potentialities.
mutualism: a relationship between two organisms that is advantageous to both of them; synonymous with symbiosis in strict sense of the term.
mycelium (*adj.,* **mycelial**): the mass of hyphae forming the vegetative portion of the body of a fungus.
mycology: a branch of botany dealing with the study of fungi.
mycorrhiza (*pl.,* **mycorrhizae**): the symbiotic complex of a fungus and the roots of some species of plants.

myxomycophyte: a slime mold, member of the Division Myxomycophyta.

N

NAD$^+$: nicotinamide adenine dinucleotide, a coenzyme that transfers hydrogen (protons and electrons) in respiration and other metabolic processes.

NADP$^+$: nicotinamide adenine dinucleotide phosphate, a coenzyme that transfers hydrogen (protons and electrons) in photosynthesis and other metabolic processes.

nanometer (nm): a metric unit of length; 1 nm equals one millionth of a millimeter.

necrosis: black spots or areas on leaves, stems or buds resulting from the death of cells, generally from certain mineral deficiencies.

nectar: a sugar-rich fluid secreted by special glands (nectaries), usually in or closely associated with flowers.

nitrate reduction: the conversion of nitrates to ammonium compounds.

nitrification: the conversion of ammonium compounds to nitrites and nitrates by certain species of bacteria and fungi.

nitrogen fixation: the conversion of nitrogen gas (N_2) into nitrogen compounds by some species of blue-green algae and bacteria, some of which are symbiotic with vascular plants.

nitrogenous base: an organic base containing nitrogen, for example, a purine or pyrimidine.

node (*adj.,* **nodal**): enlarged portion of a stem at which leaves, buds, and branches originate.

non-polar compound: a compound having molecules that lack differential electrical charges at their ends; nonpolar compounds are fat-soluble and include hydrocarbons.

non-septate: referring to fungal hyphae or algal filaments that lack cross-walls.

nucellus: the megasporangium in the ovule.

nuclear membrane: the double membrane surrounding the nucleus in eucaryotic plants.

nucleic acid: a large molecule organic acid composed of nucleotides linked together; DNA and RNA.

nucleolus: a specialized protoplasmic body in the nucleus.

nucleotide: a compound composed of a sugar (ribose or deoxyribose), phosphate, and a purine or pyrimidine.

nucleus: the specialized protoplasmic body bounded by a nuclear membrane, characteristic of all eucaryotic cells.

nut: a dry indehiscent one-seeded fruit.

O

oil: a fat that is liquid at room temperatures; or a viscous petroleum derivative, or any other oily liquid soluble in fat solvents.

oogamy: the condition in which the male and female gametes are different in form and behavior.

oogonium: in the thallophytes, the female gametangium.

operculum: in mosses, the cap at the apex of the sporangium.

order: a taxonomic category consisting of a group of closely related families.

organ: a part of a plant body performing a particular function, commonly composed of more than one tissue.

organism (*adj.*, **organismal**): an individual plant or animal that is capable of passing through its characteristic life cycle.

osmosis (*adj.*, **osmotic**): the movement of molecules of water (or other solvent) through a differentially permeable membrane.

ovulary: the basal portion of the pistil that contains the ovules and becomes the fruit. See hypogyny and epigyny.

ovulate: referring to a structure bearing ovules.

ovule: a rudimentary seed containing before fertilization the female gametophyte.

ovuliferous: see ovulate.

oxidation: a chemical reaction that releases energy from the substrate by loss of electrons, which may involve addition of oxygen or loss of hydrogen.

oxidative phosphorylation: the formation of ATP from ADP and phosphate at the expense of energy released during respiration.

P

P_1: the two parents in a hybrid cross.

parasite (*adj.*, **parasitic**): an organism that lives in or on another organism and derives its food from it, but does not benefit the host organism.

pathogen: a parasite that causes a disease.

palisade layer: elongated, ranked, chlorophyll-bearing cells occurring beneath the epidermis in leaves.

papain: a protein-hydrolyzing enzyme derived from the tropical pawpaw plant.

parenchyma (*adj.*, **parenchymatous**): a tissue composed of usually undifferentiated, spherical, thin-walled cells.

pectic compounds: polymers of 100 or more galacturonic acid (a galactose derivative) residues, which may be methylated (in pectin and protopectin) or form salts (as in calcium pectate); constituents of the primary cell wall and middle lamella; source of jellies.

pedicel: the stalk of the individual flowers in an inflorescence of few or many flowers.

peduncle: the stalk of an inflorescence consisting of a single flower, the basal stalk of a many flowered inflorescence.

penicillin: an antibiotic synthesized by the mold *Penicillium.*

pentose: a five-carbon simple sugar (monosaccharide).

peptide: a compound composed of two to several amino acids linked by peptide bonds between the -COOH group of one and the -NH_2 group of the next one.

perennial: a plant capable of living for several years, such as trees and shrubs, and also including bulbous and rhizomatous plants whose above ground stems and leaves are annually renewed.

perianth (floral envelop): the characteristic whorls of petals and/or sepals.

pericycle: the tissue of the root, usually parenchymatous, between the endodermis and the phloem.

perigyny (*adj.*, **perigynous**): a condition in which the gynoecium is centered in an open cup-shaped or flat extension of the torus, at the margins of which other floral whorls arise.

petal (*adj.*, **petaloid**): one of the constituent parts of the corolla.

petiole (*adj.*, **petiolate**): the stalk of a leaf.

petrifaction: a process occurring in formation of fossils in which the tissues of the organism become mineralized and turned to stone.

pH: the negative of the log of hydrogen ion concentration, used as a measure of the acidity or alkalinity of solutions; pH 0 is most acid, pH 14 is most basic, and pH 7 is neutral.

phaeophyte: a brown alga, member of the Division Phaeophyta.

phelloderm: a tissue composed of the cells cut off on the inside by division of the cells of the cork cambium or phellogen.

phellogen: the meristematic tissue giving rise externally to cork and internally to phelloderm in some woody plants.

phenotype: the observable characteristics of an organism resulting from the interaction between its hereditary potentialities (genotype) and the environmental factors.

phloem: a vascular tissue consisting of sieve tubes, companion cells (except gymnosperms), phloem parenchyma, and sometimes fibers; the chief pathway for downward passage of foodstuff.

phospholipid: a lipid synthesized from glycerol, two molecules of fatty acids, and a molecule of phosphoric acid or a phosphate-containing organic compound.

photoperiod: the optimal period of daily illumination required for normal growth and maturity of a plant.

photosynthesis: the absorption of light energy by chlorophyll and its conversion into chemical bond energy incorporated in molecules of sugars (and other organic compounds) synthesized from carbon dioxide and water.

phototropism: the bending or twisting of plant structures (including stems, petioles, and coleoptiles) toward more intense light because of the inhibition of growth by light.

phycobilins: types of water soluble pigments, associated with proteins, of red or blue color and serving as accessory photosynthetic pigments in blue-green or red algae.

phycocyanin: a blue phycobilin pigment synthesized by blue-green algae.

phycoerythrin: a red phycobilin pigment synthesized by red algae.

phycomycetes (algalike fungi): a class of the Eumycophyta (the true fungi).
phylogeny: the evolutionary history of a group of related organisms.
phylum: a primary division of the plant or animal kingdom reflecting a possible, or probable, evolutionary history; synonymous with Division.
phytobenthon: the members of an aquatic community of plants, growing attached.
phytochrome: a pale blue pigment present in plant cells that is converted from one form to another when it absorbs light and is involved in photomorphogenic reactions such as photoperiodism, seed germination, and leaf development.
phytohormone: a plant hormone.
phytoplankton: the free-floating plants in an aquatic habitat.
pileus: the expanded cap of fleshy fungi, as in mushrooms.
pistil: the female organ of the flower composed of one or more carpels and consisting of ovulary, style, and stigma.
pistillate: referring to a flower having one or more pistils but no stamens.
pit: a small channel or passage through a secondary wall of supportive or conductive cells formed by the nondeposition of secondary wall over limited areas of the primary wall. **Simple pit,** one whose lateral margins are essentially parallel. **Bordered pit,** one in which the sides are not parallel, the thickening secondary wall arching over the unthickened area of primary wall (the pit membrane). **Pit-pair,** opposite pits of adjacent cells, separated by the pit membrane consisting of the primary walls of adjacent cells.
pith: parenchymatous tissue lying interior to the vascular tissue of stems.
placenta: the tissue within the ovulary, which bears the ovules.
placentation: the characteristic manner in which placentae are disposed in the ovulary. **Axile placentation,** a placenta extending from base to apex of ovulary on the central axis. **Free-central placentation,** one in which the axile placenta is not attached to the side wall by septa. **Parietal placentation,** one in which the placentae are distributed upon the side walls of the ovulary.
plankton: the complex of free-floating plants and animals in an aquatic habitat.
plant acids: see polycarboxylic acids.
plant growth regulator: plant hormones and synthetic substances with hormonelike activity.
plasma membrane: a two-layered membrane defining the outer limits of the protoplast, an integral portion of the cell's membrane system.
plasmodesma (*pl.,* **plasmodesmata**): fine interconnecting protoplasmic threads between two adjacent protoplasts.
plasmodium (*pl.,* **plasmodia**): in the slime molds, an undifferentiated mass of protoplasm, multinucleate and without a cell wall.

plasmolysis: the separation of the cytoplasm from the wall of a cell as a result of osmotic movement of water out of the vacuole.

plastid: a cell organelle sometimes variously pigmented and concerned with metabolic processes.

plumule: the primordial tissues (leaf and stem) at the apex of the epicotyl.

polarity: a structural, chemical, or electrical difference between the two ends or sides of a cell, tissue, organ or organism.

polar compound: a compound that has differential electrical charges of its molecules; includes water-soluble compounds such as sugars as well as electrolytes.

pollen: the young microgametophytes, resulting from germination of microspores, in seed plants.

pollen mother cell: a diploid cell produced in the microsporangium (anther) which by meiosis gives rise to four microspores each of which becomes a pollen grain.

pollination: process of transfer of pollen from stamen to stigma, as in angiosperms, or as in gymnosperms, from male cone to female cone. **Cross pollination,** transfer of pollen to stigma of another individual. **Self-pollination,** transfer of pollen to stigma of the same individual.

polycarboxylic acids: organic acids with more than one carboxyl (-COOH) group; sometimes called plant acids; includes many acids important in metabolism, for example, citric, succinic, and oxalacetic acid.

polymer: a large organic molecule composed of many residues that are similar, for example, starch composed of glucose units.

polypeptide: a molecule composed of numerous amino acids linked by peptide bonds; proteins are composed of one or more polypeptides.

polyploidy: a condition where there are more than two complete sets of chromosomes per cell.

polysaccharide: a carbohydrate polymer composed of many monosaccharide molecules linked together, for example, starch, cellulose.

pome: simple, fleshy, indehiscent fruit in which ovulary is embedded in other floral parts, for example, apple, pear.

prairie: essentially treeless grassland vegetation.

primordium (*pl.,* **primordia**): an undifferentiated tissue that develops into an organ of the plant.

protease: a protein-digesting enzyme.

procaryote (*adj.,* **procaryotic**): a primitive organism such as a bacterium or blue-green alga in which an organized nucleus does not occur.

protein: a high molecular weight organic compound synthesized from one or more polypeptide molecules, that in turn are chains of amino acids linked by peptide bonds.

prophase: early stage in nuclear division in which chromatin becomes organized into definite chromosomes.

protein: a complex composed of combinations of various amino acids.

prothallium: in ferns, the gametophytic phase of the reproductive cycle.

protonema (*pl.*, **protonemata**): in mosses, a filamentous stage in development of the gametophyte.

protoplasm (*adj.*, **protoplasmic**): the living substance of plants or animals.

protoplast: the protoplasmic unit of a single cell.

provascular strand: a strand of primordial tissue in the growing point of a vascular plant which gives rise to primary vascular tissue.

pseudopodium (*pl.*, **pseudopodia**): in Myxomycetes, an armlike extension from a plasmodium by which the organism moves over a surface.

psilophyte: a primitive vascular plant, member of Division Psilophyta.

pteridosperms: primitive seed plants (seed ferns), members of Division Pterophyta, Class Gymnospermae; probably ancestors of the cycads.

pterophyte: a member of Division Pterophyta, including ferns, gymnosperms, and angiosperms.

puffball: a basidiomycete in which the basidiospores are produced in a spherical sporocarp.

purine: a nitrogenous base with a double ring structure, for example, adenine, guanine.

pyrenoid: a specialized body in the chloroplasts in some species of algae and liverworts, associated with starch formation.

pyrimidine: a nitrogenous base with a single ring structure, for example, cytosine, thymine, uracil.

pyrrophyte: a golden-brown alga, member of Division Pyrrophyta.

R

radicle: the portion of a plant embryo that develops into the primary root.

radioisotope (radioactive isotope): an unstable isotope of an element that disintegrates, emitting radiation.

raphides: needlelike crystals occurring in some plant cells.

ray: a vertical sheet of parenchymatous cells extending along a radius of a stem or root. **Pith ray,** a ray extending from pith to cortex and separating adjacent vascular bundles of the stem; may be lengthened by cambial action in secondary growth. **Vascular ray,** a narrow ray within a vascular bundle, extending through both phloem and xylem, or limited to either tissue (phloem ray and wood ray, respectively); may be initiated by or lengthened by cambial action.

recessive gene: a gene that is not expressed phenotypically when its dominant allele is present.

reduction: addition of energy to a substrate compound by the addition of electrons, which may involve addition of hydrogen or removal of oxygen.

reductional division: nuclear division in which homologous chromosomes are separated, resulting in formation of haploid nuclei.

regeneration: the replacement of missing tissues or organs by an organism.

reproduction: the process by which organisms give rise to the next generation.

respiration: the enzyme-controlled process in living organisms by which energy is released from food. Aerobic respiration, requires the availability of free oxygen. Anaerobic respiration, proceeds in the absence of free oxygen.

ribonucleic acids (RNA): nucleic acids containing ribose and uracil, synthesized against a DNA template, and carrying a code complementary to the DNA code. RNA of three kinds (messenger, transfer, and ribosomal) are involved in protein synthesis.

ribose: a five-carbon sugar used in the synthesis of RNA and other important compounds.

ribosomes: small cell organelles containing RNA.

ribulose: a five-carbon sugar that is an intermediate reactant in photosynthesis. Ribulose 1,5-diphosphate is a photosynthetic carbon dioxide acceptor.

rhizoid (*adj.*, **rhizoidal**): a cellular filament that performs the functions of a root, that is, fixation and absorption.

rhizome: an underground stem, usually elongated and often fleshy, as in iris.

rhodophyte: a red alga, member of Division Rhodophyta.

RNA: see ribonucleic acids.

root: the base of a plant axis, usually subterranean, serving the functions of anchorage and absorption. **Adventitious root,** a root arising from an unusual position on the plant, as in leaf- or stem-cuttings, or from a node as in many creeping plants. **Fibrous root,** a root system consisting of many branches arising from the base of the stem. **Tap root,** a root system consisting of a single, dominant, usually deeply penetrating root.

root cap: a protective mass of parenchymatous cells covering the apical meristem of a root.

root pressure: the pressure developed in roots as a result of osmosis and that sometimes causes water to rise in the xylem; also results in guttation and exudation (bleeding) from cut branches or stumps.

rust: a species of basidiomycetes requiring alternate hosts for completion of the life cycle; parasitic upon higher plants.

S

sac fungus: a member of Class Ascomycetes of Division Eumycophyta, in which meiospores are produced in a saclike cell.

salt: a compound formed by the reaction of an acid with a base.

samara: a simple, dry, indehiscent fruit with the pericarp extended to form a wing, for example, ash, maple.
sap: the solution of mineral salts, sugars, and other compounds in the water present in the xylem.
saprophyte (*adj.,* **saprophytic**): a plant that grows upon and obtains its nourishment from nonliving organic bodies.
sapwood: the outer portion of the secondary xylem actively engaged in conduction of sap.
savannah: a forest type of parklike aspect, consisting of scattered trees in grassland.
schizocarp: a dry fruit composed of two or more united carpels which separate at maturity.
schizomycophyte: a member of Division Schizomycophyta (bacteria).
sclereid: a cell of usually isodiametric form with heavily lignified secondary walls.
sclerenchyma (*adj.,* **sclerenchymatous**): a tissue consisting of supportive cells with heavily lignified secondary walls, for example, sclereids, fibers.
sclerotium: a hard mycelial mass replacing the grain in cereals, resulting from the infection of the developing grain by certain species of ascomycetes, for example, ergot.
seed: a matured and ripened ovule.
semi-permeable membrane: see differentially permeable membrane.
sepal: a member of the calyx, the lowermost whorls of a complete flower, usually enclosing other floral whorls in the bud.
septate: referring to a filament composed of a linear series of cells, as in certain algae and fungi, or to certain fibrous cells of woody plants, which are secondarily divided by cross-walls into compartments.
sessile: referring to a flower lacking a pedicel, or a leaf lacking a petiole.
shoot tension: see cohesion-tension mechanism.
shrub: a perennial woody plant of relatively small stature, with several stems arising from the base.
sieve cell: characteristic phloem element of primitive vascular plants, as in gymnosperms.
sieve plate: the perforated endwall of a sieve-tube element.
sieve tube: a linear series of sieve-tube elements characteristic of the phloem of angiosperms.
sieve-tube element: the cellular unit of a sieve tube; at maturity the protoplast lacks a nucleus.
slime mold: a member of Division Myxomycophyta.
smut: a species of basidiomycetes parasitic upon higher plants and requiring a single host. Compare rust.
sol: a liquid colloid that in some cases can be changed into a semisolid colloidal gel, or vice versa.
solute: a dispersed particle in a true solution; a dissolved substance.

solution: a liquid in which molecules of the dissolved substance (the solute) are dispersed among the molecules of the water or other substance.

solvent: a substance in which molecules of a dissolved substance are dispersed; water is the principal solvent in biological systems, but some biological substances are soluble in oils or other fat solvents.

soredium: the characteristic reproductive structure of lichens, consisting of a combination of algal cells and fungal hyphae.

sorus: in ferns, a cluster of sporangia borne upon a leaf.

species (*adj.,* **specific**): a group of individuals having many characteristics in common and usually freely interbreeding; a group of species constitutes a genus.

sperm: a male (micro-) gamete.

spermatophyte: a plant that bears seeds.

sphenophyte: a horsetail, member of Division Sphenophyta.

spindle: in nuclear division, a spindlelike figure composed of fiberlike lines on which the chromosomes appear to move from the equator of the cell to the poles.

spirillum (*pl.,* **spirilla**): a motile bacterium of spiral form, regularly flagellate.

sporangium: a saclike structure producing spores.

spore: a reproductive cell that gives rise directly to another plant as in the asexual reproduction of certain fungi and algae (mitospore), or to the gametophytic phase of the reproductive cycle as in sexually reproducing plants (meiospore).

sporocarp: the characteristic fruiting body of certain higher fungi, as in mushrooms, puffballs, and cup fungi.

sporophore: essentially synonymous with sporocarp.

sporophyll: a spore-bearing leaf.

sporophyte: a spore-producing plant; the phase in sexual reproduction in which meiosis occurs, giving rise to meiospores.

stamen (*adj.,* **staminate**): a member of the androecial whorl of a flower, consisting of filament and anther, and producing pollen.

staminate flower: an imperfect flower possessing stamens but no pistil.

starch: a complex insoluble polysaccharide composed of several hundred α-D glucose residues.

stele (*adj.,* **stelar**): the central vascular cylinder of roots and stems of higher plants.

steppe: grassland vegetation of low water availability occurring in cold climates.

stigma (*pl.,* **stigmata:** *adj.,* **stigmatic**): the pollen-receptive portion of a pistil; may be sessile upon the ovulary or borne upon a style.

stipule (*adj.,* **stipular**): a leaflike outgrowth arising from the base of a petiole; characteristic of certain families of angiosperms, in which the stipule or pair of them may be persistent or ephemeral.

stolon: a specialized branch arising at the base of the stem, growing along the soil surface and commonly rooting at the nodes where new aerial shoots arise, for example, strawberry.

stoma (*pl.*, **stomata;** *adj.*, **stomatal**): a small opening between specialized epidermal cells (guard cells) in the epidermis of leaves and herbaceous stems, through which gases are exchanged.

strobilus (strobile) (*pl.*, **strobili;** *adj.*, **strobilar**): an elongated axis bearing small, overlapping leaves or scales, forming a cone-shaped structure.

style: a slender upward extension of the pistil, bearing the pollen-receptive stigma.

suberin: a waxlike substance occurring in the walls of cork cells, rendering them impermeable to water and gases.

succession: the slow change in the species composition of a developing plant community, from the first colonization to the establishment of a climax community.

sucrase: an enzyme that hydrolyzes sucrose; invertase.

sucrose: a disaccharide sugar formed by the union of a molecule of glucose and one of fructose, for example, cane sugar, beet sugar, maple sugar.

suspension: a disperse system of solid particles large enough to settle out of the water (or other liquid) under the influence of gravity, for example, muddy water.

suspensor: a cell or small group of cells in the developing seed which, through growth, position the embryo within the embryo sac.

symbiont: a member of a symbiotic pair of organisms.

symbiosis (*adj.*, **symbiotic**): the living together of two different organisms. See parasite, mutualism.

synapsis (*adj.*, **synaptic**): the pairing of homologous chromosomes in meiosis.

synergid (*pl.*, **synergidae**): one of the pair of nuclei of the embryo sac which, with the egg, constitute the egg apparatus.

synthesis (*adj.*, **synthetic**): a chemical reaction in which a more complex compound is formed from two or more simpler ones.

T

taiga: a vegetational type of northern latitudes dominated by conifers.

tannin: a mixture of astringent substances extractable from bark.

tapetum: a spore-nourishing tissue lining the sporangial wall.

taxonomy: the science dealing with the identification and classification of plants.

tegmen: the inner seed coat.

teleology: purposeful behavior directed toward an identified end or goal.

teliospore: a black, two-celled, resistant spore characteristic of the rust fungi; a basidium may develop from each cell, the basidiospores infecting the alternate host.

telophase: the final stage of mitosis in which the daughter nuclei are reconstituted.

tendril: a slender, coiling stem outgrowth or modified leaf which assists in vertical support of climbing plants, as in grape and pea.

testa: the outer seed coat.

tetrad: a group of four meiospores resulting from meiosis.

tetraploid: a cell with four sets of chromosomes (4n) rather than the usual two sets (2n).

tetrose: a four-carbon monosaccharide (simple sugar).

thallophyte: a plant whose body is not differentiated into roots, stems, and leaves (algae and fungi).

thallus: a plant body undifferentiated into roots, stems, and leaves.

thermoperiodicity: the influence of alternating day and night temperatures on plant growth and development.

thylakoid: a saclike membranous structure in a chloroplast; a stack of thylakoids constituting a granum.

thymine: a pyrimidine in DNA but not RNA, where uracil is found instead.

tissue: a group of cells similar in structure and usually performing a specialized function.

torus: the basal structure which bears the organs of a single flower and in which vascular tissue is distributed to the floral whorls.

trace element: a mineral element essential for organisms in very small amount.

tracheid: an elongated cell of the xylem with tapering ends and lignified, pitted walls; at maturity the protoplast is lacking.

tracheophyte: a vascular plant.

translocation: the transport of solutes through the vascular tissues (phloem and xylem) of a plant; or the exchange of segments between nonhomologous chromosomes.

transpiration: the loss of water vapor from plants, generally through the stomata.

trichome: an outgrowth of the epidermis consisting of a short filament of cells.

triose: a three-carbon monosaccharide (simple sugar).

triploid: a cell with three complete sets of chromosomes (3n).

tritium: the radioactive isotope of hydrogen; ^{3}H.

trophic level: a step in the flow of food and energy through an ecosystem, that is, autotrophs, herbivores, and various levels of carnivores.

tropism: a growth movement in response to a directional environmental stimulus, for example, light, gravity.

tuber: an enlarged, fleshy, underground stem, for example, potato.

tundra: a vegetation type characteristic of colder regions, consisting of low-growing meadowlike vegetation.

turgor pressure: the pressure in a cell against its wall, resulting from the osmotic movement of water into the cell.

tylosis (*pl.*, **tyloses**): the growth of a parenchyma cell into the lumen of an adjacent vessel element through a pit in the vessel element wall, thus blocking the vessel.

U

uredospore: a red one-celled spore characteristic of the rust fungi, produced in summer on an infected host, and reinfecting other individuals of the host species.

unit factor: a gene.

uracil: a pyrimidine in RNA but not in DNA.

V

vacuole (*adj.*, **vacuolar**): a cell inclusion, especially in mature cells, consisting of water and dissolved substances and limited by a one-layered vacuolar membrane.

vascular bundle: a discrete strand of conductive tissue consisting of phloem and xylem. **Closed vascular bundle,** a bundle without cambium. **Open vascular bundle,** a bundle with cambium, and therefore capable of secondary growth.

vegetation: the characteristic plant cover of a region.

vein: a slender, threadlike vascular strand consisting of xylem and phloem and serving plant structures, such as leaves and flowers.

venation: the characteristic disposition of veins in a leaf, flower, or fruit.

vessel: a linear series of conductive cells characteristic of the xylem, functioning chiefly in the conduction of water and various solutes.

vessel element: the cellular unit of a vessel, the cell without a protoplast, with the side walls lignified and variously pitted and the endwalls variously perforated.

virus: a noncellular submicroscopic structure composed of nucleic acid and protein, active when in a suitable host cell, which transcribes its genetic code and replicates the virus.

VP (vapor pressure): the partial pressure of the water vapor in an atmosphere, generally expressed in millimeters of mercury (mmHg); the diffusion pressure or molecular activity of water vapor.

VPG (vapor pressure gradient): the difference in vapor pressure between two regions, for example, the intercellular spaces of a leaf and the atmosphere.

W

wall pressure: the back pressure of a cell wall against the turgor pressure within the cell, equal to the turgor pressure but opposite in direction.

wilting: the drooping of leaves, stems and other parts of a plant shoot as a result of loss of turgor by the cells. Permanent wilting results from deficient soil water. Temporary wilting results from a rate of transpiration in excess of the rate of water absorption when the soil water is abundant.

winter killing: the death of evergreens resulting from desiccation during warmer periods in the winter, when transpiration is rapid but when absorption from cold or frozen soil is low.

wood: the xylem portion of a stem. See heartwood and sapwood. **Primary wood,** xylem arising directly from the cells of the provascular strand. **Secondary wood,** xylem arising from activity of the cambium.

X

xanthophyll: a yellow carotenoid pigment found in chloroplasts and chromoplasts.

xerophyte (*adj.*, **xerophytic**): a plant capable of living in a very dry habitat.

xylem: the wood of a stem or a root, consisting of tracheids, vessels, fibers, and parenchyma.

Y

yeasts: members of a class of ascomycetes, usually single-celled; some species or varieties are important in fermentation processes, others are virulent pathogens in both plants and animals.

Z

zeatin: a cytokinin synthesized by corn plants.

zoospore: a motile spore.

zygospore: a thick-walled resistant zygote resulting from the fusion of isogametes, for example, as in *Rhizopus* and *Spirogyra.*

zygote: a protoplast resulting from the fusion of gametes.

PHOTO CREDITS

All Part Openers courtesy John R. Troughton, DSIR, Physics and Engineering Laboratory, New Zealand. From J. R. Troughton & F. B. Sampson PLANTS: A Scanning Electron Microscope Survey. Copyright © 1973 John Wiley & Sons, Inc.

CHAPTER TWO

Figure 2.1: FAO. **Figure 2.2:** Grant Heilman. **Figures 2.3** and **2.4:** Charles Perry Weimer. **Figures 2.5** and **2.6:** United Fruit Company. **Figure 2.7:** USDA. **Figure 2.8:** FAO. **Figure 2.9:** Grant Heilman. **Figure 2.10:** Charles Pfizer & Company. **Figure 2.11:** Drug Enforcement Administration, U.S. Department of Justice. **Figure 2.12:** Robert H. Wright/National Audubon Society. **Figures 2.13, 2.14** and **2.15:** Grant Heilman. **Figure 2.16:** USDA. **Figure 2.17:** Grant Heilman. **Figure 2.18:** USDA. **Figure 2.19:** Dr. Carl Hartley, USDA Forest Products Laboratory. **Figure 2.20:** U.S. Forest Service. **Figure 2.21:** FAO. **Figure 2.22:** U.S. Forest Service. **Figure 2.23:** Mimi Forsyth/Monkmeyer. **Figure 2.24:** USDA. **Figure 2.25:** FAO. **Figure 2.26:** David Van de Mark. **Figures 2.27** and **2.28:** U.S. Forest Service. **Figures 2.29, 2.30** and **2.31:** Weyerhauser. **Figure 2.32:** U.S. Forest Service. **Figure 2.33:** FAO.

CHAPTER FOUR

Figure 4.5: Grant Heilman. **Figure 4.7:** Courtesy Johns-Manville Products Corporation, Celite Division. **Figure 4.8:** Dr. F. E. Round, University of Bristol. **Figure 4.9:** Courtesy Johns-Manville Products Corporation, Celite Division. **Figure 4.10:** Courtesy Louise Keppler. **Figure 4.11** (bottom): R. H. Noailles. **Figure 4.12:** Dr. John D. Dodge, University of London. **Figure 4.16** (top): Dr. I. K. Ross. **Figure 4.19:** Courtesy The Fleischman Laboratories, Standard Brands Inc. **Figure 4.20** (top): Chicago Natural History Museum. **Figure 4.23:** Hal Harrison/Grant Heilman. **Figure 4.24:** (Top) Grant Heilman; (bottom) M. C. Noailles. **Figure 4.25:** (a) Runk/Schoenberger from Grant Heilman; (b) and (c): Hal Harrison/Grant Heilman. **Figure 4.26:** Louise Keppler. **Figure 4.27:** Jack Dermid. **Figure 4.30:** Photo courtesy Wilma Kane. **Figure 4.31:** (Top) R. H. Noailles: (bottom) Grant Heilman.

CHAPTER FIVE

Figure 5.2: Grant Heilman. **Figure 5.6:** Alan Pitcairn/Grant Heilman. **Figure 5.7:** Jack Dermid. **Figure 5.8:** Courtesy Australian News and Information Bureau. **Figure 5.10:** Chicago Natural History Museum. **Figure 5.11:** Grant Heilman. **Figure 5.12:** Chicago Natural History Museum. **Figure 5.13:** (Top left) R. H. Noailles; (top right and bottom left) Grant Heilman; (bottom right)

Hal Harrison/Grant Heilman. **Figure 5.14:** Chuck Abbott, Tucson, Arizona. **Figure 5.15:** USDA. **Figure 5.16:** Standard Oil Company of New Jersey.

CHAPTER SIX

Figures 6.1 and **6.2:** Chicago Natural History Museum. **Figures 6.4** and **6.5:** Courtesy E. S. Barghoorn, Harvard University. **Figures 6.6** and **6.7:** Chicago Natural History Museum.

CHAPTER SEVEN

Figure 7.12: M. Kunitz. **Figure 7.18:** C. A. Knight. **Figure 7.19:** Ralph W. G. Wyckoff, University of Arizona, Tucson. **Figure 7.20:** C. A. Knight. **Figure 7.21:** Carl Zeiss, Inc. **Figure 7.22** (left)**:** Courtesy Lee D. Simon, The Institute for Cancer Research, Philadelphia.

CHAPTER EIGHT

Figures 8.2 and **8.3:** Courtesy W. Gordon Whaley, University of Texas, Austin. **Figure 8.4:** Courtesy Dr. S. Granick, Rockefeller University and Dr. Keith Porter, University of Colorado, Boulder. **Figure 8.5:** Courtesy of University of Texas, Electron Microscope Laboratory. **Figure 8.7:** Courtesy E. H. Newcomer, University of Connecticut. **Figure 8.10:** Courtesy John H. Troughton. **Figure 8.12:** Courtesy Ralph W. G. Wyckoff, University of Arizona, Tucson. **Figure 8.13** (b)**:** Runk/Schoenberger from Grant Heilman. **Figure 8.22:** Courtesy Andrew Bajer, University of Oregon. From *Chromosoma* (1968), 25:249. Copyright Springer-Verlag, New York.

CHAPTER NINE

Figures 9.6, 9.7, 9.8 and **9.9:** Courtesy U.S. Forest Products Laboratory. **Figure 9.12:** Chicago Natural History Museum. **Figure 9.17:** Courtesy A. M. Winchester. **Figure 9.19:** Grant Heilman. **Figure 9.21:** Runk/Schoenberger from Grant Heilman. **Figure 9.32:** Courtesy T. Elliot Weier. From Weier, Stocking, Barbour BOTANY 5th edition. Copyright © 1974 John Wiley & Sons. **Figure 9.37:** Chicago Natural History Museum. **Figure 9.39:** Brookhaven National Laboratory. **Figures 9.41** and **9.42:** Courtesy Louise Keppler. **Figure 9.43:** Courtesy J. Arthur Herrick, Kent State University. **Figure 9.48:** Chicago Natural History Museum.

CHAPTER TEN

Figure 10.1: Courtesy John T. Bonner.

CHAPTER ELEVEN

Figure 11.5: Daniel Branton, *Proceedings of National Academy of Science U.S.* (1966), 552, p. 1048. **Figure 11.9:** Ross E. Hutchins.

CHAPTER TWELVE

Figure 12.2: Courtesy J. Arthur Herrick, Kent State University. **Figure 12.10:** Courtesy Henry Spitz. **Figure 12.12:** Courtesy L. C. Erickson, University of California, Riverside. **Figure 12.15:** Courtesy Richard Böhning, Ohio State University. **Figure 12.16:** Courtesy British Columbia Forest Service. **Figure 12.18:** V. A. Greulach.

CHAPTER THIRTEEN

Figure 13.1: Grant Heilman. **Figure 13.2:** Courtesy Fisher Scientific Company. **Figures 13.3** and **13.4:** Courtesy W. Rei Robbins. **Figure 13.5:** Courtesy R. C. Burrell. **Figures 13.6** and **13.7:** Courtesy Gerald Gerloff, University of Wisconsin. **Figure 13.8:** Courtesy O. Biddulph, Washington State University. **Figure 13.11:** Courtesy O. N. Allen, University of Wisconsin. **Figure 13.12:** Courtesy R. R. Herbert, R. D. Holsten & R. W. F. Hardy, Central Research Department, E. I. duPont de Nemours & Company. **Figure 13.13:** Courtesy O. N. Allen, University of Wisconsin. **Figure 13.15:** Kenneth Murray/Nancy Palmer.

CHAPTER FOURTEEN

Figure 14.1: Courtesy J. Arthur Herrick, Kent State University. **Figure 14.8** (bottom): Radioautograph by J. A. Bassham of Melvin Calvin's laboratory. **Figure 14.9** (bottom): Courtesy Herbert W. Israel, Cornell University. **Figure 14.10:** Courtesy Charles J. Arntzen. From Arntzen, Dilley, Crane, *Journal of Cell Biology* 43:16–31, 1969. **Figure 14.17:** Grant Heilman. **Figure 14.23:** Hugh Spencer. **Figure 14.24:** Grant Heilman. **Figure 14.30:** From BOTANY, 5th edition, by Carl L. Wilson, Walter E. Loomis and Taylor A. Steeves. Copyright 1952 © 1957, 1962, 1967, 1971 by Holt, Rinehart and Winston, Publishers. Reproduced by permission of Holt, Rinehart and Winston, Publishers.

CHAPTER FIFTEEN

Figure 15.10: From COLLEGE BOTANY, Revised Edition, by Harry J. Fuller and Oswald Tippo. Copyright, 1949, 1954, by Holt, Rinehart and Winston, Publishers. Reproduced by permission of Holt, Rinehart and Winston, Publishers. **Figure 15.11:** Courtesy of John G. Haesloop, Pfeiffer College. **Figure 15.12:** Courtesy K. V. Thimann, Crown College, University of California, Santa Cruz. **Figure 15.13:** Environmental Biology Program, Boyce Thompson Institute for Research, Yonkers, N.Y. **Figure 15.14:** Courtesy J. P. Nitsch. **Figure 15.15:** Environmental Biology Program, Boyce Thompson Institute for Plant Research, Yonkers, N.Y. **Figures 15.16** and **15.17:** Courtesy S. W. Wittwer. **Figure 15.21:** Courtesy Wayne McIlrath & D. R. Ergle, Northern Illinois University. **Figure 15.28:** Courtesy V. T. Stoutemyer. **Figure 15.31:** Courtesy L. C. Erickson, University of California, Riverside. **Figure 15.32:** Courtesy F. C. Steward, *Science* 143:23, 1964, with the permission of *Science.* Copyright 1964 by the American Association for the Advancement of Science. **Figures 15.33** and **15.34:** Courtesy A. C. Hildebrand & I. K. Vasil. From *Science* 150:889,

1965 with the permission of Science. Copyright 1965 by the American Association for the Advancement of Science. **Figure 15.35:** Courtesy A. H. Sparrow, Brookhaven National Laboratory. **Figure 15.36** and **15.37:** Environmental Biology Program, Boyce Thompson Institute for Plant Research, Yonkers, N.Y. **Figures 15:39, 15:42** and **15:43:** Agricultural Research Service, Plant Industry Station, USDA **Figure 15.45:** Courtesy Professor A. Takimoto, Kyoto University, Japan. **Figure 15.46:** Environmental Biology Program, Boyce Thompson Institute for Plant Research, Yonkers, N.Y. **Figure 15.47:** Courtesy H. C. Thompson, Cornell University.

CHAPTER SIXTEEN

Figure 16.1: Alan Pitcairn/Grant Heilman. **Figure 16.2:** U.S. Geological Survey. **Figure 16.7:** Courtesy Brookhaven National Laboratory. **Figure 16.8:** (Left) Grant Heilman; (right) Courtesy Phoenix Chamber of Commerce. **Figure 16.10:** Environmental Biology Program, Boyce Thompson Institute for Plant Research, Yonkers, N.Y. **Figure 16.11:** (Top) Grant Heilman; (bottom) H. L. Shantz, U.S. Forest Service. **Figure 16.13:** B. W. Muir, U.S. Forest Service. **Figure 16.14:** Daniel Todd, U.S. Forest Service. **Figure 16.15:** Grant Heilman. **Figure 16.16:** USDA Soil Conservation Service.

CHAPTER SEVENTEEN

Figure 17.1: Jack Dermid. **Figure 17.2:** Karl Weidmann/National Audubon Society. **Figure 17.3:** Ross E. Hutchins. **Figure 17.4:** (Left) Hal Harrison/Grant Heilman; (right) Jack Dermid. **Figure 17.5:** Courtesy C. H. Muller, University of California, Santa Barbara. **Figure 17.6:** U.S. Forest Service. **Figure 17.7:** Grant Heilman. **Figures 17.8** and **17.10:** USDA. **Figure 17.11:** Ross E. Hutchins. **Figure 17.12:** Courtesy B. C. Stone, University of Malaya. **Figure 17.14:** USDA. **Figure 17.15:** Ross E. Hutchins. **Figure 17.16:** Courtesy W. J. Robbins, New York Botanical Garden. **Figure 17.17:** J. A. Carlyle. Courtesy A. C. Braun, Rockefeller University. **Figures 17.19** and **17.20:** Crops Research Division, USDA. **Figure 17.24:** Courtesy Edward Hackskaylo, USDA.

CHAPTER EIGHTEEN

Figure 18.8: Courtesy Biological Section, Department of Public Lands, Brisbane, Australia. **Figure 18.10:** (Top left) Chicago Natural History Museum; (top and bottom right) Courtesy Harold Humm and Andrew E. Rehm, University of South Florida. **Figure 18.11:** (a), (b), (c), and (d) Courtesy Harold Humm; (e) and (f) Jack Dermid. **Figure 18.12:** Alan Pitcairn/Grant Heilman.

CHAPTER NINETEEN

Figure 19.2: Sheila Turner/Monkmeyer. **Figure 19.3:** (Top) U.S. Forest Service; (bottom) Grant Heilman. **Figure 19.4:** Chicago Natural History Museum. **Figure 19.5:** (Top) Courtesy J. Arthur Herrick; (bottom) Courtesy Northern

Pacific Railroad. **Figure 19.6:** (Top right) Chuch Abbott, Tucson, Arizona; (bottom right) Esther Henderson/Rapho-Photo Researchers; (top and bottom left) Alan Pitcairn/Grant Heilman. **Figure 19.7:** Pro Pix/Monkmeyer. **Figure 19.8:** Gordon S. Smith/National Audubon Society. **Figure 19.9:** Grant Heilman. **Figure 19.10:** Helen Barsky/Photo Researchers. **Figures 19.11, 19.12** and **19.13:** Grant Heilman.

CHAPTER TWENTY

Figure 20.1: Courtesy E. R. Squibb & Company. **Figure 20.11:** Courtesy J. Arthur Herrick, Kent State University.

CHAPTER TWENTY-ONE

Figure 21.14: (Top left) Courtesy J. H. Troughton: (top right) Courtesy Charles Neidorf. **Figure 21.20** (bottom): Rob Carr.

CHAPTER TWENTY-TWO

Figure 22.1: From the Photographic Archives of the Austrial National Library, by permission, all rights retained by the Library. **Figure 22.11:** Courtesy A. H. Sparrow, Brookhaven National Laboratory. **Figure 22.12:** Courtesy Merle T. Jenkins, USDA **Figure 22.14:** USDA **Figures 22.15** and **22.16:** Courtesy W. Atlee Burpee Seed Company. **Figures 22.17:** Grant Heilman. **Figure 22.18:** From BREEDING FIELD CROPS by John Poehlman. Copyright © 1959 by Holt, Rinehart and Winston, Inc. Reproduced by permission of Holt, Rinehart and Winston, Inc.

CHAPTER TWENTY-THREE

Figure 23.2: Courtesy Lee D. Simon, The Institute for Cancer Research, Philadelphia.

INDEX